MICROELECTRONICS
Digital and Analog
Circuits and Systems

McGraw-Hill Series in Electrical Engineering

Electronics and Electronic Circuits

Stephen W. Director, *Consulting Editor*

MICROELECTRONICS
Digital and Analog
Circuits and Systems

Jacob Millman, Ph.D.

Charles Batchelor Professor, Emeritus
Columbia University

INTERNATIONAL STUDENT EDITION

McGRAW-HILL INTERNATIONAL BOOK COMPANY

Auckland Bogotá Guatemala Hamburg Johannesburg Lisbon
London Madrid Mexico New Delhi Panama Paris San Juan
São Paulo Singapore Sydney Tokyo

TO MY WIFE
Sally

MICROELECTRONICS: Digital and Analog Circuits and Systems

INTERNATIONAL STUDENT EDITION

Copyright © 1979
Exclusive rights by McGraw-Hill Book Company Japan,
Ltd. for manufacture and export. This book cannot be
re-exported from the country to which it is consigned by
McGraw-Hill.

5th printing 1983

This book was set in Times Roman by Science Typographers, Inc.
The editors were Peter D. Nalle, Julienne V. Brown, and Madelaine Eichberg;
the production supervisor was Charles Hess.
New drawings were done by J & R Services, Inc.

Library of Congress Cataloging in Publication Data

Millman, Jacob, date
    Microelectronics.

    Includes bibliographies and index.
      1. Integrated circuits. 2. Microelectronics.
3. Electronic circuit design. 4. Digital electronics.
I. Title.
TK7874.M527   621.381′73   78-8552
ISBN 0-07-042327-X

When ordering this title use ISBN 0-07-066410-2

TOSHO PRINTING CO., LTD. TOKYO JAPAN

CONTENTS

12   Feedback Amplifier Characteristics   409

13   Frequency Response of Amplifiers   447

PREFACE

This book was written primarily as a text in modern electronics for electrical engineering students. Its comprehensive scope, in both breadth and depth, should also make it valuable to physics majors and to practicing engineers and scientists who wish to update their knowledge of the fast-changing field of microelectronics (integrated circuits).

The text is divided into three parts, thereby allowing it to be used in a number of different courses to suit the purpose and interest of the professor. Part 1 (Chapters 1–4) discusses *semiconductor device characteristics*, and is intended for students who have had no previous introduction to electronics. The physics and mathematics normally taught in the first year or two of an undergraduate curriculum are the only prerequisites required for an understanding of Part 1. These four chapters outline the properties of a semiconductor, explain fabrication of a *p-n* diode and a bipolar junction transistor (BJT) in monolithic integrated form, and include a discussion of their characteristics. Discrete devices no longer play such an important part in the design of today's electronic products, and hence the reader is introduced to the integrated circuit (IC) chip early in the book (Chapter 4).

Part 2 (Chapters 5 through 9) explores *digital circuits and systems*. There are a number of important reasons for introducing digital material before analog material:

1. Digital techniques, which involve only simple Boolean algebra, are easily learned by the student. The devices are either ON or OFF, resulting in very simple operation. Essentially the only characteristics that need be specified are switching speed and the loading on a gate. Analog considerations, on the other hand, are much more difficult to grasp since they involve frequency- and time-domain concepts, frequency compensation, and more detailed and complicated circuit analysis. Many small-signal-device parameters must be taken into consideration.

Part 2 requires no electrical engineering prerequisites since the very simple circuit analysis needed for digital networks is explained in Appendix C, Summary of Network Theory. Hence, a course covering *digital electronics* could be made available to sophomores.

2. Most students have learned how to program the digital computer (some learned in high school). Hence a very strong motivation for them to study electronics is to find out how digital hardware works.

3. In many universities (Columbia is one of them), only a one-semester electronics course is required for the computer science option in electrical engineering. Clearly, the scope should concentrate on digital electronics, and such a course can use Parts 1 and 2 for its text. These nine chapters contain somewhat more material than can be covered in one semester, which gives the instructor the freedom to omit those sections he or she feels are of less interest or importance.

4. Most electronics curriculum requirements include laboratory work. It is much simpler to design and perform digital, rather than analog, experiments. Such a laboratory course may be a corequisite for the digital course. This arrangement is not too successful with an analog laboratory because network theory courses are required as prerequisites.

5. Most new electronic systems are predominantly digital in nature.

Part 2 introduces small-scale integration (SSI) logic gates (AND, OR, NOT, NAND, . . .), and these are implemented into the various standard families (DTL, TTL, ECL, . . .). Then follows combinational systems, which are medium-scale integration (MSI), such as a binary adder, a digital comparator, a parity checker, a decoder/demultiplexer, a data selector/multiplexer, an encoder, and a read-only memory (ROM). As examples of sequential digital systems we consider FLIP-FLOPS (*S-R, J-K, T*, and *D* types) and use them as the building blocks for shift registers and counters.

Now that the student has gained some facility in the use of the bipolar transistor and has seen its application to digital systems, a new semiconductor device, the field-effect transistor, is introduced and exploited in logic gates. Finally, large-scale integration (LSI) systems with both MOSFETs and BJTs are studied in Chapter 9. These are principally memories and include dynamic MOS shift registers, MOS ROMs, erasable programmable read-only memories (EPROMs), programmable logic arrays (PLAs), random-access memories (RAMs), charge-coupled devices (CCDs), the microprocessor and microcomputer, and integrated-injection logic (I^2L).

Part 3 (Chapters 10 through 18) concentrates on *analog circuits and systems*. Methods of biasing a discrete BJT or FET are given, and stability of the operating point is discussed. The small-signal model for each device is obtained and used to calculate the performance of low-frequency single-stage and cascaded amplifiers.

Feedback concepts are introduced. The four standard feedback amplifier configurations are indicated and their characteristics are delineated. High-frequency transistor models are used to obtain the frequency response of amplifiers (with or without feedback).

The basic linear (analog) building block is the operational amplifier (OP AMP), and its characteristics and applications are described in the last four chapters. Monolithic analog design techniques for the OP AMP are presented in detail, including methods of frequency compensation to assure stability. The broad applications of OP AMPs which are discussed include: instrumentation amplifiers, analog computers, active filters, precision AC/DC converters, sample-and-hold systems, analog multiplexers and demultiplexers, logarithmic amplifiers, D/A and A/D converters, comparators, waveform generators, voltage time-base generators, sinusoidal oscillators, power amplifiers, and monolithic voltage regulators.

For a two-term electronics sequence, the second course would be based on Part 3. This material is too extensive for one course, which allows the professor the option of selecting those topics he or she wishes to emphasize.

Many curricula require only one *core* course in electronics, but then offer a number of optional electronics courses. This book contains enough material for a total of 3 one-semester courses.

If a professor wishes to consider analog material before digital material in a first electronics course, then Part 1 and selected portions of Part 3 could be used. However, Sections 8-1 through 8-6 of Part 2 on field-effect transistors should be included.

From the foregoing discussion it should be clear that the various topics in this book have been written in such a manner as to allow great latitude in the structuring of one or more courses, the content of which matches that desired by the instructor.

The number of components on a chip has doubled every year since 1959 when the planar transistor was introduced. With this ever-increasing component density on an IC chip the difference between an electronic circuit and a system becomes quite blurred. As a matter of fact an entire monolithic package, such as an OP AMP, is often referred to simply as a "device." In this book no attempt is made to distinguish unambiguously between *device*, *circuit*, or *system*; although, of course, a single transistor is clearly a device and a large-scale microelectronic chip merits the designation "system" or at least "subsystem."

Modern electronics engineers design a new product (for example, an instrument, a control, computer, or communication system, etc.) by interconnecting standard microelectronic chips so that the overall assembly will achieve the desired external objectives. They try to minimize the number of packages (and, hence, the cost) by using LSI and MSI wherever possible, and they only resort to SSI chips and discrete components (such as very large capacitors or resistors, inductors, transformers, transducers, etc.) whenever absolutely necessary. Clearly, engineers must know what IC chips are commercially available, what functions they perform, and their limitations.

In view of the foregoing facts the goal of this book is to take the reader step by step from a qualitative knowledge of the properties of a semiconductor to an understanding of the operation of devices (particularly the p-n diode, the BJT, the MOSFET, the CCD, and the I^2L gate), and finally to an appreciation for how these are combined monolithically to form microelectronic chips with distinct and useful input-output properties. A very broad variety of IC chips are studied in this book, including not only a description of what is fabricated within the silicon but also a deep understanding of the digital and/or analog functions which this chip can perform. After each circuit or system is studied, reference is made to a specific commercially available package which can give the desired operation (for example, digital multiplexing, analog comparison, digital-to-analog conversion, etc.). The practical limitations (due to temperature, voltage, power, loading, etc.) of real, rather than idealized, devices are explained. To appreciate these nonideal characteristics, manufacturer's specifications for representative discrete devices and IC chips are supplied (Appendix B). The depth of discussion, the broad choice of topics, and the practical emphasis is such as to prepare the student to do useful engineering immediately after he or she joins an electronics company.

This book is a thorough reorganization, rewriting, and updating of the material in Millman and Halkias "Integrated Electronics: Analog and Digital Circuits and Systems" (McGraw-Hill Book Company, New York, 1972). Many new topics have been added, including but not limited to the following: three-state output stage of a logic gate, higher-order demultiplexer and multiplexer, priority encoder, two-dimensional addressing of a ROM, word and address expansions of ROMs, a universal shift register, technological improvements in MOSFETs, inverters with nonsaturated or depletion loads, CMOS transmission gates, erasable programmable ROM, programmable logic array (PLA), dynamic RAM cells, charge-coupled devices (CCD), microprocessor, integrated-injection logic (I^2L), analog design techniques (current sources and repeaters, active load, level-shifting, and output stage of an OP AMP), sample-and-hold systems, analog multiplexer and demultiplexer, several A/D converter systems, voltage-controlled oscillator, positive-negative controlled-gain amplifier, retriggerable monostable multivibrator, voltage time-base generators, modulation of a square wave, power amplifiers (including thermal considerations), switching-regulated power supplies, and power FETs (VMOS). An attempt was made to present the state of the art of microelectronics as of early 1978 and to indicate some probable future developments.

To make room for the new material, some of the topics in "Integrated Electronics" were compressed or omitted completely. For example, the discussions of semiconductor device physics were drastically reduced, the biasing of discrete devices was deemphasized, the photoelectric effect in a semiconductor was deleted, the four-parameter low-frequency hybrid model was mentioned only briefly, and the discussions of amplifier noise, of tuned amplifiers, and of CRT character generators were omitted.

A brief historical survey of electronics and electronic industries is included in the Prologue (following this preface). It is hoped that the instructor and the student will read this fascinating history before beginning the study of the text.

Considerable thought was given to the pedagogy of presentation, to the explanation of device-circuit-system behavior, to the use of a consistent system of notation, to the care with which diagrams were drawn, to the illustrative examples worked out in detail in the text, and to the review questions at the end of each chapter. These review questions should be assigned as homework since they afford the students the opportunity to test themselves to see if they understand what they have read in the sections under consideration. The author has used these questions very successfully for about 30 percent of a quiz or an exam (the remaining 70 percent being quantitative problems).

Included are 717 homework problems, which will test the students' grasp of the fundamental concepts enunciated in the book and will give them experience in the analysis and design of electronic circuits and systems. In almost all numerical problems realistic parameter values and specifications have been chosen. Only a small percentage of the homework exercises are taken intact from "Integrated Electronics." Most of the problems are new or are modifications of those previously used. The answers to selected problems are found in Appendix E.

A solutions manual is available to an instructor who has adopted this text. Write to: College Division, McGraw-Hill Book Company, 1221 Avenue of the Americas, New York, NY 10020. Attention: Electrical Engineering Editor, 27th floor. As an added pedagogical aid, a set of 124 involved figures in the book are also available to the instructor. These may be used with a view graph during lectures on this subject matter.

The publishers sent a questionnaire to many of the professors who had adopted "Integrated Electronics," asking for desirable deletions, additions, revisions, etc., in this book. The present text reflects the replies to this questionnaire. I am especially grateful to Professor J. E. Steelman for his many helpful suggestions. Let me state specifically that the approach used in Section 13-3 follows notes which he sent to me. It is with great pleasure that I acknowledge the technical consultations and assistance of my son, Dr. J. T. Millman. In particular, he is responsible for the revision of Chapter 18. I also wish to express my great appreciation to Professor D. A. Hodges for his detailed review and very constructive criticism of Chapter 9. My thanks goes to Dr. T. V. Papathomas who was responsible for the preparation of the "Solutions Manual," and to Mrs. B. Lim for her skillful typing of the manuscript and the problem solutions.

Jacob Millman

PROLOGUE

A BRIEF HISTORY OF ELECTRONICS

This prologue to the text gives a historical perspective of the development of electronic devices which operate at frequencies from dc to hundreds of megahertz. It also includes a discussion of the growth of industries resulting from the exploitation of these devices into practical circuits and systems.

BACKGROUND

"Electronics" has different meanings to different people and in different countries. Hence, let me define the term in the sense that it is used here. "*Electronics* is the science and the technology of the passage of charged particles in a gas, in a vacuum, or in a semiconductor." Please note that particle motion confined within a metal only is not considered electronics.

Before *electronic engineering* came into existence, *electrical engineering* already flourished. Electrical engineering is the field which deals with devices that depend solely on the motion of electrons in metals; for example, a generator, a motor, a light bulb, or a telephone. The principal benefactors of these devices are the wire telephone or telegraph companies and the power industries.

Both electronic and electrical engineering owe their existence to the pioneering work in electricity and magnetism of scientific giants such as Coulomb, Ampere, Ohm, Gauss, Faraday, Henry, and Maxwell. Maxwell, in about 1865, put together the researches of the others into a consistent theory of electromagnetism, now called *Maxwell's equations*. Here is a historical example of theory being ahead of experiment, for although Maxwell's theory predicted that electromagnetic waves could be propagated in space and that light was such an electromagnetic wave, it was not until 23 years later (in 1888) that Hertz produced such radiation, using a spark-gap oscillator. In 1896 Marconi succeeded in transmitting Hertzian waves and detecting them at a distance of about two miles. Wireless telegraphy had its feeble origin in this experiment.

This history is divided into two periods of time, referred to simply as the *past* and *present*. By *past* is meant the era of the tube—the vacuum tube or the gas tube. The *present* starts with the invention of the transistors in 1948. Also included is a section speculating briefly on the *future*.

THE PAST

The beginning of electronics came in 1895 when H. A. Lorentz postulated the existence of discrete charges called *electrons*. Two years later J. J. Thompson found these electrons experimentally. In the same year (1897) Braun built what was probably the first electron tube, essentially a primitive cathode-ray tube.

Discovery of Vacuum Tubes

It was not until the start of the 20th century that electronics began to take technological shape. In 1904 Fleming invented the diode which he called a *valve*. It consisted of a heated wire which emitted electrons separated a short distance from a plate in a vacuum. For a positive voltage applied to the plate electrons were collected, whereas for a negative potential the current was reduced to zero. This valve was used as a detector of wireless signals. Two years later, Pickard tried a silicon crystal with a cat's whisker (a pointed wire pressed into silicon) as a detector. This was the first semiconductor diode. This device was very unreliable, was soon abandoned, and semiconductor electronics died a premature death in 1906.

The most important milestone in this early history of electronics came in the same year (1906) when De Forest put a third electrode (a grid) into the Fleming valve, and thus invented the triode tube which he called an *audion*. A small change in grid voltage resulted in a large plate-voltage change. Hence, the audion was the first amplifier. It took about five years to improve the vacuum in the audion and to add an efficient oxide-coated cathode in order to obtain a reliable electronic device. Thus the age of practical electronics began in about 1911. (By coincidence, I was born in the same year.)

Radio and Television

The first application of electronics was to radio and, simultaneously with the birth of electronics, the IRE (the Institute of Radio Engineers) was founded in the United States in 1912. It is a great tribute to the imagination of the early engineers that they realized immediately the importance of radio and formed this organization at the very beginning of radio communications. The American Institute of Electrical Engineers, which took care of the interest of conventional electrical engineers, had already been founded in 1884. Both societies combined in 1963 to become the IEEE (the Institute of Electrical and Electronic Engineers).

The first radio broadcasting station, KDKA, was built in 1920 by Westinghouse Electric Corporation in Pittsburgh, Pennsylvania. By 1924, just four short years later, there were 500 radio stations in the United States. The history of broadcasting (both radio and TV communications) can be divided into three main periods.

1907 to 1927 The devices available were simply diodes and triodes with filamentary-type cathodes. The circuits that were invented by the ingenuity of the engineers were cascaded amplifiers, regenerative amplifiers (Armstrong† in 1912), oscillators, heterodyning (Armstrong in 1917), and neutralization to prevent undesired oscillation in amplifiers.

1927 to 1936 The indirectly heated cathode was invented for the diode and triode. Two additional electrodes—a fourth and then a fifth—were introduced into the triode to form the screen-grid tube and the pentode, respectively. Also beam-power tubes and metal tubes were introduced during this period. With these new devices, engineers were able to invent the superheterodyne receiver, automatic gain control (AGC), single-knob tuning, and multiband operation. Radio was a flourishing business.

1936 to 1960 In this last period the new devices were closely spaced electrodes (for a high gain-bandwidth product), miniature glass tubes, and, toward the end of the period, color television tubes. Major Armstrong in 1933 invented frequency modulation. About five years later the first FM receiver was available. Electronic black and white television began in about 1930, and the most important name here is Zworykin of RCA. Ten years later, television, at least in the United States, was in fair use.

Commercial color television began around 1950, and many new functions had to be performed. Hence, the following circuits were invented: FM limiter, FM discriminator, automatic frequency control (AFC), saw-tooth waveform generator (linear deflection for a TV tube), synchronization, multiplexing, and inverse-feedback circuits (including operational amplifiers).

Electronic Industries

These can fit into one or more of the following four principal groups, which I shall call the four C's: C for Components, C for Communications, C for Control (or automation), and C for Computation. The components companies up to this time were those which came into existence to supply the various types

---

† Armstrong was an undergraduate at Columbia University at this time.

of tubes just described and others referred to later as well as the passive-circuit elements, such as resistors, capacitors, coils, transformers, etc.

The second C (communications) refers to the industry built up around AM and FM radio, hi-fi systems, and black and white as well as color TV receivers and transmitters.

The third C introduced (the C for control) was making itself evident in what was then referred to as "industrial electronics." Industrial electronics may be defined as "the use of electronic devices in the control and operation of machines in industry (other than in communication and computation)." The devices for industrial electronics were gaseous diodes and triodes (thyratrons), pool-cathode devices, such as the Mercury arc rectifier, and high-voltage and high-power tubes. The circuits for this period were power rectifiers, high-voltage rectifiers, power amplifiers, high-voltage transmitting circuits, induction and dielectric heating, power inverters (from dc to ac), measurements, motor control, and the control of industrial processes.

The computer (the fourth C) had barely made its appearance at this time and, hence, this industry is discussed in detail in the following.

THE PRESENT

This era begins with the invention of the transistor about 30 years ago.

Discovery of the Bipolar Junction Transistor

The history of this invention is interesting. M. J. Kelly, director of research (and later president of Bell Laboratories), had the foresight to realize that the telephone system needed electronic switching and better amplifiers. Vacuum tubes were not very reliable, principally because they generated a great deal of heat even when they were not being used, and, particularly, because filaments burned out and the tubes had to be replaced. In 1945 a solid-state physics group was formed. The following quote is from the authorization for work for this group: "The research carried out in this case has as its purpose the obtaining of new knowledge that can be used in the development of completely new and improved components and apparatus elements of communication systems." One of the most important specific goals was to try to develop a solid-state amplifier. The group consisted of theoretical and experimental physicists, a physical chemist, and an electronics engineer, and they collaborated with the metallurgists in the laboratory. These scientists were well aware of the theoretical research on metals and semiconductors already carried out by Block, Mott, Schottky, Slater, Sommerfeld, Van Vleck, Wigner, Wilson, and other physicists throughout the world. (I had the good fortune of doing graduate work under two of these professors, Sommerfeld and Slater.)

During an experiment in December 1947, two closely spaced gold-wire probes were pressed into the surface of a germanium crystal. It was found that the voltage output (with respect to the germanium base) at the "collector" probe was greater than the input to the "emitter" probe. Brattain and Bardeen immediately recognized that this was the effect they were looking for, and the solid-state amplifier (in the form of the point-contact transistor) was born. These first transistors were very bad. They had low gain and a low bandwidth and were noisy. Also, the parameters varied widely from device to device.

Shockley recognized that the difficulties were with the metallic point contacts. He proposed the junction transistor (Chapter 3) almost immediately and worked out the theory of its operation. Here was a device that had no pointed wire contacts and the transistor operation depended on diffusion instead of on the conduction current which was present in a tube. The new devices had charge carriers of both polarities operating simultaneously; they were bipolar devices. The carriers were electrons, which were well known, and other "strange particles" which were not well understood. Measurement showed their polarity to be opposite to that of the electrons, and hence they were equivalent to positive charges. These particles could only be explained with quantum-mechanical theory. They were called "holes" because they represented places in the crystal where electrons should have been but were missing. The current in a vacuum tube is limited by an electronic space charge built up in the vicinity of the thermionic emitter. This space charge, this cloud of electrons, repels any further electrons from being emitted. This phenomenon could not be present in a transistor because the theory predicted that the new device would be essentially neutral except for a thin, immobile space-charge layer close to a junction. Hence very large current densities could be expected from these new devices at low applied potentials. The possibility of obtaining important practical devices (*without heated filaments*) was immediately recognized.

From theoretical considerations it was known that the transistor could not be built reliably unless ultrapure single crystals were available. About two years later, Teal of Bell Labs grew single crystals of germanium and later silicon with much less than one part in a billion of impurity atoms. It was then possible to intentionally introduce controlled impurities called *donor* or *acceptor* atoms the extent of only one part in 100 millions. In this way they formed the junctions (Section 2-1) of the bipolar transistor. The first grown junction transistors appeared in 1950. The alloy junction process appeared the next year. Three short years after the discovery of amplification in a solid, the transistor was being produced commercially in 1951.

The Bell System made a most important corporate decision—not to keep these discoveries secret. It actually held symposia to share this knowledge with professors (who then could pass it on to students), and even with other companies. It offered to license its patents to any company that was interested in fabricating transistors. The tube companies such as Western Electric (which does the manufacturing for the Bell System), RCA, Westinghouse, and General Electric were the first to fabricate transistors, to be followed by many new components companies that saw the tremendous possibilities of this device. A current list of semiconductor device manufacturers is given in Appendix B-1.

By 1952, United States military funds were allotted to transistor research. The Armed Services was interested in using these devices mainly in missiles where small size and weight, low power, improved performance, and reliability (because of the absence of a filament) were of primary importance. This investment paid off very well. Solid-state components have virtually replaced tubes in almost all military and also commercial applications, except those involving exceedingly high voltage and power. Most universities in the United States no longer even mention the vacuum tube in their curricula.

Transistor characteristics vary greatly with changes in temperature. For germanium these variations become excessive for temperatures higher than about 75°C, whereas silicon can be used up to approximately 200°C. In 1954 production of the silicon transistor was announced by Texas Instruments. Today the vast majority of semiconductor transistors and other devices are fabricated from silicon.

In 1956 Bardeen, Brattain, and Shockley received the Nobel prize in physics. This was the first Nobel award ever given for the invention of an engineering device.

The Integrated Circuit

Shortly after joining Texas Instruments in 1958 Kilby conceived the monolithic idea, that is, the concept of building an entire circuit out of germanium or silicon. He used the bulk semiconductor to form a resistor, and he also fabricated a diffused-layer resistor (Section 4-9). He built a capacitor by using an oxide layer (for the dielectric) on silicon, and he also thought of the p-n junction capacitor (Section 4-10). To demonstrate the feasibility of his concept, he built a phase-shift oscillator and then a multivibrator using these resistors, capacitors, and a transistor, all made from germanium with thermally bonded gold connecting wires. However, in the patent application, he indicated that the components could be interconnected by laying down conducting material. In 1959 the *solid circuit* (later called the *integrated circuit*) was announced by Kilby at an IRE convention.

About this same time Noyce, then director of research and development at Fairchild Semiconductor (and now chairman of the board of Intel), also had the monolithic-circuit idea for making "multiple devices on a single piece of silicon in order to be able to make interconnections between devices as part of the manufacturing process, and thus reduce size, weight, etc., as well as cost per active element." He explained how devices could be isolated from one another with back-biased p-n diodes, how resistors could be fabricated, and how connections could be made by evaporating metal through holes in the oxide to interconnect the circuit components. (The idea of using back-to-back diodes for isolation of components was conceived independently and patented in 1959 by Lehovec, the research director of Sprague Electric Company). The first modern diffused transistors (Chapter 4) were developed by Hoerni at Fairchild in 1958. He was responsible for the planar process of passivating the junctions with an oxide layer on the surface. He used production photolithographic techniques and diffusion processes previously developed by Noyce and Moore. The real key to integrated-circuit manufacturing was the planar transistor and batch processing. By 1961, both Fairchild and Texas Instruments were producing integrated circuits commercially, and other companies soon joined them in IC fabrication. Today millions of transistors, passive components, and their interconnections are manufactured simultaneously (Chapter 4) in one production batch.

Field-Effect Transistors

Before the invention of the transistor a number of people studied the "field effect," that is, the change in conductivity of a solid caused by an applied transverse electric field. In fact, the bipolar transistor was discovered, as noted in the foregoing, during these field-effect investigations. The junction field-effect transistor was proposed by Shockley in 1951. However, early attempts to fabricate these devices failed because a stable surface could not be obtained. This difficulty was overcome with the discovery of the planar process and the surface passivation with silicon dioxide (glass, an excellent insulator). A metallic electrode (the *gate*) was placed over this thin (1,000 Å) SiO_2 layer. A voltage applied between the gate and the bulk silicon induced conductive charges near the surface. The gate extended laterally a few micrometers (referred to as the *channel*) between two electrodes (called the *source S* and *drain D*) and the current between S and D was controlled by the gate voltage. The first such metal-oxide semiconductor field-effect transistor (MOSFET, Chapter 8) was announced by Kahng and Atalla of Bell Laboratories in 1960. The reproducibility of these devices was poor. It took about five years to trace the difficulty to contaminants (principally sodium ions) in the SiO_2 and to learn how to eliminate them. Many technological improvements (Section 8-6) have been made in the basic MOSFET and this device now rivals the bipolar transistor in importance.

Charge-Coupled Device

It is possible to fabricate a long chain of closely spaced gate electrodes between S and D. Charge introduced from S can then be trapped under the first electrode and, by applying the proper voltage waveforms, this charge can be moved from the first to the second, and then from the second to the third electrode, etc. This so-called *charge-coupled device* (CCD) was invented by Boyle and Smith at Bell Laboratories in 1969 (Section 9-8).

Microelectronics

Reliability, speed of operation, and production yields have steadily improved, while cost, power consumption, and size have been reduced drastically for both bipolar junction transistor and MOSFET integrated circuits. The following approximate dates give some idea of the increase in the component (transistor, diode, resistor, or capacitor) count per IC silicon chip:

1951—*discrete transistors*

1960—*small-scale integration* (SSI), less than 100 components per chip

1966—*medium-scale integration* (MSI), more than 100 but less than 1,000 components per chip

1969—*large-scale integration* (LSI), more than 1,000 but less than 10,000 components per chip

1975—*very-large-scale integration* (VLSI), more than 10,000 components per chip

Moore, president of Intel, noted in 1964 (when he was director of research at Fairchild) that the number of components on a chip had doubled every year since 1959, when the planar transistor was introduced. He predicted correctly that this trend would continue. The size of a large silicon IC chip is only about 3 by 5 mm in area and only about 0.1 mm thick (about the thickness of one hair on your head). This chip might contain (in 1978) some 30,000 components, corresponding to 2,000 components/mm^2 or about 1 component/mil^2. These numbers are difficult to believe when one first learns of them, particularly since the IC's are manufactured in an industrial plant and not under research conditions in a laboratory. The term *microelectronics* is used to describe such high-density IC chips. There is now commercially available an entire computer on a single chip the area of which is about 6 by 6 mm. This microcomputer (Section 9-11) is a complete general-purpose digital-processing and control system and indicates what has been attained in (VLSI) microelectronics by 1977. In that year most of the world's $90 billion electronics industry depended on microelectronics.

The first IC's were digital logic circuits (Chapter 5), and these gates were interconnected to form combinational systems (Chapter 6), and sequential systems (Chapter 7). Starting in 1964, linear integrated circuits (Chapter 15) became available, and analog IC systems (Chapters 16 and 17) have flourished.

Semiconductor Memories

A number of transistor configurations have been devised for storing digital data. These are called *random access memories* (RAMs, Section 9-6), and the first LSI RAMs were sold commercially in 1970 by Intel and Fairchild. These early RAMs stored (approximately) 1,000 binary bits of information. By 1973 16,000-bit memories were introduced by Intel and Mostek, and 65,000-bit RAMs are expected to be commercially available in 1979. The CCD can be used as a circulating memory, and in 1977 65,000-bit CCDs were produced.

Integrated Injector Logic (I^{2}L)

Until recently the MOSFET occupied a very much smaller area than the BJT, but the latter was much faster than the former. In 1972 this situation changed when Berger and Wiedman at IBM (Germany) and Hart and Slob at Philips (Netherlands) invented a new bipolar transistor logic gate, called *integrated-injection logic* I^2L, Section 9-12. This is not a new device, but rather a new circuit configuration which uses standard BJT fabrication technology. The density of I^2L gates has increased to that of MOSFET gates, the speed is high, and the power is low. In 1977 Texas Instruments and Fairchild Semiconductor introduced I^2L chips commercially, and this logic is appearing in digital wrist watches, memories, and microprocessors. Technological improvements are now taking place in MOSFETs, which are increasing their density on a chip and increasing their speed. Hence, it is not evident that there will be a clear-cut victory of the I^2L over the MOSFET, or vice versa.

Communications and Control Industries

These industries adopted solid-state electronics, slowly at first, but now almost all equipment is transistorized, except those involving extremely high voltage or power. For medium voltage or power, discrete transistors are used. Such application of discrete transistors includes power switches (for paper punches or tape drives), motor controls, automobile ignition systems, TV deflection circuits, inverters, power supplies, and audio and radio-frequency output stages. For most other applications, the IC has taken over.

The communication industry has changed drastically because of microelectronics. In 1970 the transmission of data constituted a very small fraction of the volume of messages. In 1980 digital transmission is expected to equal voice (analog) transmission. Switching and memory in telephone systems are now performed by digital microelectronics. Active voice-frequency filters, which are analog, are also implemented with IC's. Obviously, communication satellites became feasible and economically viable because of microelectronics.

Similarly, the control industry has been affected drastically by microelectronics. Instrumentation, testing, automation of production processes, numerical control of machine tools, and energy management are predominantly digital- and computer-controlled, made possible because of integrated circuits.

The Computer Industry

The most dramatic outgrowth of the microelectronic revolution, however, was the creation of an entirely new industry—the computer industry. There has been a great deal of interest in computing machines for over 300 years. For example, in 1633 Schickhard in Germany described (in correspondence with his friend Kepler, the astronomer) a mechanical computer to do addition, subtraction, multiplication, and division. He designed a wheel with ten spokes on it, one spoke of which was longer than the others, and this wheel was placed mechanically next to another similar wheel. After the first wheel made ten angular increments, which corresponded to the ten digits, the large spoke engaged the next wheel, and it would turn one increment. In other words, he invented the *carry* in arithmetic. About the same time Pascal, 1642, and Leibnitz, 1671, also had similar ideas. But the first really serious effort to build a mechanical calculator was made about 200 years later (1833) by Babbage, a mathematics professor in England. However, the technology simply was not available to convert his ideas into a practicable machine.

The first working calculator was electromechanical, not electronic, and it was built by IBM engineers under the direction of Professor Aiken of Harvard about 100 years later in 1930. It was called the "IBM Automatic Sequence Controlled Calculator, Mark I." It was 17 m long, 3 m high, and was very clumsy. Yet it was used to make calculations for over 15 years. The first electronic calculator was built in 1946 by Eckert and Mauchly at the Moore School of Electrical Engineering of the University of Pennsylvania. It was called the ENIAC, an acronym for "Electronic Numerical Integrator and Computer." It was used for computation of ballistic tables for the armed forces, and it was not a general-purpose calculator. It contained 18,000 vacuum tubes. It occupied 40 racks of equipment, and it filled a room that was at least 10 by 13 m.

Also in 1946, IBM built the type 603, which was the first small electronic computer. Two years later IBM brought out the 604 which was the first large general-purpose electronic computer. It built and sold 4,000 of these machines in a period of 12 years. So 1948 was indeed the beginning of the computer industry; remember that in this same year the transistor was discovered.

At this period in time a number of universities, including Harvard, University of Pennsylvania, M.I.T., and Princeton were doing research in computer science, and were developing special-purpose computers needed by some governmental agencies. This work was funded by various branches of the Armed Services (the Signal Corps, The Ordnance Corps, the Office of Naval Research), the Bureau of Standards, and the Atomic Energy Commission. Many of the circuits, systems, and general ideas which originated at these universities were used in the commercial general-purpose digital computers which followed.

In 1954 the IBM 650 was built, and was considered the work horse of the industry at that time. About 1,000 of such machines were sold. This was called the first generation of computers, known as the tube version.

In 1959 there appeared the second-generation computer, the first transistorized machine. This was the IBM 7090/7094 series, and this dominated the computing market for many years. In 1965 the third-generation machine was built with integrated circuits, the now famous IBM system 360. In 1970 IBM introduced the 370 system, which included the first semiconductor memories.

In addition to these very expensive ($\sim$ $1 million) and versatile large machines, the advances in microelectronics have made possible a flourishing business in small, medium-priced ($\sim$ $10,000) minicomputers as well as much smaller, inexpensive ($\sim$ $300) microcomputers. Of course, the reader is familiar with the hand-held calculator, which does multiplication, division, addition, and subtraction and sells for only a few dollars. The impact of microelectronics was dramatically expressed by Noyce in the following 1977 quote:

Today's microcomputer at a cost of perhaps $300, has more computing capacity than the first large electronic computer, ENIAC. It is 20 times faster, has a larger memory, is thousands of times more reliable, consumes the power of a light bulb rather than that of a locomotive, occupies 1/30,000 the volume and costs 1/10,000 as much. It is available by mail order or from your local hobby shop.

Remember that this microcomputer is built on a silicon chip whose area is less than $\frac{1}{4}$ by $\frac{1}{4}$ in! Cochran and Boone of Texas Instruments applied for a patent for the microcomputer in 1971 and it was granted in 1978.

THE FUTURE

The foregoing completes the history of electronics and electronic industries up to 1978. There is already a start toward a merging of the computer and the communication industries which might be called "information manipulation." This includes storage of information, sorting, computation, information retrieval, and transmission of data. This combination of the computer and the communication fields will penetrate many disciplines. Applications will be made in the fields of law, medicine, biological sciences, engineering, library services, publishing, banking, reservation systems, management control, education, and defense.

Microelectronic applications to the automobile (electronic ignition systems, emission control, safety systems, and improvement of efficiency), to video games, to the home (systems for energy conservation, control of appliances, smart burglar alarms, the personal computer, and so forth), to medical equipment, and to federal electronics will be expanded tremendously.

I believe that silicon technology and the devices which result therefrom will dominate electronics for at least the next decade. The distinction between a device, a circuit, and a system will become more and more blurred. Furthermore, the principal electronic industries will continue to be the four C's: components, communications, computation, and control. But these will become exceedingly difficult to recognize as separate industries because they will overlap more and more.

The size of the electronics market in the United States (in *billions* of dollars) is given by the following statistics for 1977 and projected for 1981:

	1977	1981
Semiconductors (total)	3.25	5.14
IC's	2.22	3.99
Discretes	0.90	0.95
Optoelectronics	0.13	0.20
Tubes (receiving, power, and gaseous)	0.196	0.140

Source: Electronics, January 5, 1978, pp. 25–37.

Total electronic-equipment consumption in 1978 is expected to be $66 billion in the United States and $107 billion world wide.

GENERAL REFERENCES

1. Fiftieth Anniversary Issue: *Proceeding of the IRE*, vol. 50, no. 5, May, 1962.
2. Weiner, C.: How the Transistor Emerged, *IEEE Spectrum*, pp. 24–33, January, 1973.
3. Special Issue: Historical Notes on Important Tubes and Semiconductor Devices, *IEEE Trans. Electron Devices*, vol. ED-23, no. 7, July, 1976.
4. Wolff, M. F.: The Genesis of the Integrated Circuit, *IEEE Spectrum*, pp 45–53, August, 1976.
5. Special Issue: Microelectronics, *Scientific American*, September, 1977.

PART

ONE

SEMICONDUCTOR DEVICE
CHARACTERISTICS

SEMICONDUCTORS

The physical characteristics which allow us to distinguish among an insulator, a semiconductor, and a metal are discussed. The current in a metal is due to the flow of negative charges (*electrons*), whereas the current in a semiconductor results from the movement of both electrons and positive charges (*holes*). A semiconductor may be doped with impurity atoms so that the current is due predominantly either to electrons or to holes. The transport of the charges in a crystal under the influence of an electric field (a *drift* current), and also as a result of a nonuniform concentration gradient (a *diffusion* current), is investigated.

1-1 CHARGED PARTICLES

The charge, or quantity, of negative electricity and the mass of the electron have been found to be 1.60×10^{-19} C (coulomb) and 9.11×10^{-31} kg, respectively. The values of many important physical constants are given in Appendix A-1 and a list of conversion factors and prefixes is given in Appendix A-2. Some idea of the number of electrons per second that represents current of the usual order of magnitude is readily possible. For example, since the charge per electron is 1.60×10^{-19} C, the number of electrons per coulomb is the reciprocal of this number, or approximately, 6×10^{18}. Further, since a current of 1 A (ampere) is the flow of 1 C/s, then a current of only 1 pA (1 pico-ampere, or 10^{-12} A) represents the motion of approximately 6 million electrons per second. Yet a current of 1 pA is so small that considerable difficulty is experienced in attempting to measure it.

The charge of a positive ion is an integral multiple of the charge of the electron, although it is of opposite sign. For the case of singly ionized particles, the charge is equal to that of the electron. For the case of doubly ionized particles, the ionic charge is twice that of the electron.

The mass of an atom is expressed as a number that is based on the choice of the atomic weight of oxygen equal to 16. The mass of a hypothetical atom of atomic weight unity is, by this definition, one-sixteenth that of the mass of monatomic oxygen and has been calculated to be 1.66×10^{-27} kg. Hence, *to calculate the mass in kilograms of any atom, it is necessary only to multiply the atomic weight of the atom by* 1.66×10^{-27} *kg*.

In a semiconductor crystal such as silicon, two electrons are shared by each pair of ionic neighbors. Such a configuration is called a *covalent bond*. Under certain circumstances an electron may be missing from this structure, leaving a "hole" in the bond. These vacancies in the covalent bonds may move from ion to ion in the crystal and constitute a current equivalent to that resulting from the motion of free positive charges. The magnitude of the charge associated with the hole is that of a free electron. This very brief introduction to the concept of a hole as an effective charge carrier is elaborated upon in Sec. 1-5.

1-2 FIELD INTENSITY, POTENTIAL, ENERGY

By definition, *the force f* (*newtons*) *on a unit positive charge in an electric field is the electric field intensity* $\mathcal{E}$ *at that point*. Newton's second law determines the motion of a particle of charge q (coulombs), mass m (kilograms), moving with a velocity v (meters per second) in a field $\mathcal{E}$ (volts per meter).

$$f = q\mathcal{E} = m \frac{dv}{dt} \tag{1-1}$$

The mks (meter-kilogram-second) rationalized system of units is found to be most convenient for subsequent studies. Unless otherwise stated, this system of units is employed throughout this book.

Potential

By definition, *the potential V* (*volts*) *of point B with respect to point A is the work done* against *the field* in taking a unit positive charge from A to B. This definition is valid for a three-dimensional field. For a one-dimensional problem with A at x_0 and B at an arbitrary distance x, it follows that†

$$V \equiv - \int_{x_0}^{x} \mathcal{E} \, dx \tag{1-2}$$

where $\mathcal{E}$ now represents the X component of the field. Differentiating Eq. (1-2) gives

$$\mathcal{E} = - \frac{dV}{dx} \tag{1-3}$$

The minus sign shows that the electric field is directed from the region of higher

---

† The symbol $\equiv$ is used to designate "equal to by definition."

potential to the region of lower potential. In three dimensions, the electric field equals the negative gradient of the potential.

By definition, *the potential energy U (joules) equals the potential multiplied by the charge q under consideration*, or

$$U \equiv qV \qquad (1\text{-}4)$$

If an electron is being considered, q is replaced by $-q$ (where q is the *magnitude* of the electronic charge) and U has the same shape as V but is inverted.

The law of conservation of energy states that the total energy W, which equals the sum of the potential energy U and the kinetic energy $\frac{1}{2}mv^2$, remains constant. Thus, at any point in space,

$$W = U + \tfrac{1}{2}mv^2 = \text{constant} \qquad (1\text{-}5)$$

As an illustration of this law, consider two parallel electrodes (A and B of Fig. 1-1a) separated a distance d, with B at a negative potential V_d with respect to A. An electron leaves the surface of A with a velocity v_o in the direction toward B. How much speed v will it have if it reaches B?

From the definition, Eq. (1-2), it is clear that only differences of potential have meaning, and hence let us arbitrarily ground A, that is, consider it to be at zero potential. Then the potential at B is $V = -V_d$, and the potential energy is $U = -qV = qV_d$. Equating the total energy at A to that at B gives

$$W = \tfrac{1}{2}mv_o^2 = \tfrac{1}{2}mv^2 + qV_d \qquad (1\text{-}6)$$

This equation indicates that v must be less than v_o, which is obviously correct since the electron is moving in a repelling field. Note that the final speed v attained by the electron in this conservative system is independent of the form of the variation of the field distribution between the plates and depends only upon the magnitude of the potential difference V_d. Also, if the electron is to reach electrode B, its initial speed must be large enough so that $\frac{1}{2}mv_o^2 > qV_d$. Otherwise, Eq. (1-6) leads to the impossible result that v is imaginary. We wish to elaborate on these considerations now.

The Concept of a Potential-Energy Barrier

For the configuration of Fig. 1-1a with electrodes which are large compared with the separation d, we can draw (Fig. 1-1b) a linear plot of potential V versus distance x (in the interelectrode space). The corresponding potential energy U versus x is indicated in Fig. 1-1c. Since potential is the potential energy per unit charge, curve c is obtained from curve b by multiplying each ordinate by the charge on the electron (a negative number). Since the total energy W of the electron remains constant, it is represented as a horizontal line. The kinetic energy at any distance x equals the difference between the total energy W and the potential energy U at this point. This difference is greatest at O, indicating that the kinetic energy is a maximum when the electron leaves the electrode A. At the point P this difference is zero, which means that no kinetic energy exists,

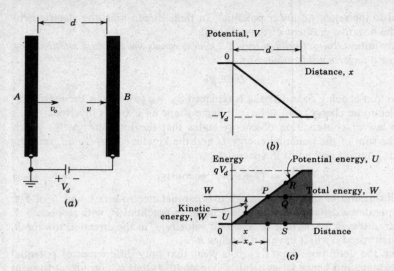

Figure 1-1 (a) An electron leaves electrode A with an initial speed v_o and moves in a retarding field toward plate B. (b) The potential. (c) The potential-energy barrier between electrodes.

so that the particle is at rest at this point. This distance x_o is the maximum that the electron can travel from A. At point P (where $x = x_o$) it comes momentarily to rest, and then reverses its motion and returns to A.

Consider a point such as S which is at a greater distance than x_o from electrode A. Here the total energy QS is less than the potential energy RS, so that the difference, which represents the kinetic energy, is negative. This is an impossible physical condition, however, since negative kinetic energy $(\frac{1}{2}mv^2 < 0)$ implies an imaginary velocity. We must conclude that the particle can never advance a distance greater than x_o from electrode A.

The foregoing analysis leads to the very important conclusion that the shaded portion of Fig. 1-1c can never be penetrated by the electron. Thus, at point P, the particle acts *as if* it had collided with a solid wall, hill, or barrier and the direction of its flight had been altered. *Potential-energy barriers* of this sort play an important role in the analyses of semiconductor devices.

It must be emphasized that the words "collides with" or "rebounds from" a potential "hill" are convenient descriptive phrases and that an actual encounter between two material bodies is not implied.

1-3 THE eV UNIT OF ENERGY

The joule (J) is the unit of energy in the mks system. In some engineering power problems this unit is very small, and a factor of 10^3 or 10^6 is introduced to convert from watts (1 W = 1 J/s) to kilowatts or megawatts, respectively.

However, in other problems, the joule is too large a unit, and a factor of 10^{-7} is introduced to convert from joules to ergs. For a discussion of the energies involved in electronic devices, even the erg is much too large a unit. This statement is not to be construed to mean that only minute amounts of energy can be obtained from electron devices. It is true that each electron possesses a tiny amount of energy, but as previously pointed out (Sec. 1-1), an enormous number of electrons are involved even in a small current, so that considerable power may be represented.

A unit of work or energy, called the *electron volt* (eV), is defined as follows:

$$1 \text{ eV} \equiv 1.60 \times 10^{-19} \text{ J}$$

Of course, any type of energy, whether it be electric, mechanical, thermal, etc., may be expressed in electron volts.

The name *electron volt* arises from the fact that, if an electron falls through a potential of 1 V, its kinetic energy will increase by the decrease in potential energy, or by

$$qV = (1.60 \times 10^{-19} \text{ C})(1 \text{ V}) = 1.60 \times 10^{-19} \text{ J} = 1 \text{ eV}$$

However, as mentioned above, the electron-volt unit may be used for any type of energy, and is not restricted to problems involving electrons.

A potential-energy barrier of E (electron volts) is equivalent to a potential hill of V (volts) if these quantities are related by

$$qV = 1.60 \times 10^{-19} E \tag{1-7}$$

Note that V and E are *numerically* identical but dimensionally different.

1-4 MOBILITY AND CONDUCTIVITY

In a metal the outer, or valence, electrons of an atom are as much associated with one ion as with another, so that the electron attachment to any individual atom is almost zero. Depending upon the metal, at least one, and sometimes two or three, electrons per atom are free to move throughout the interior of the metal under the action of applied fields.

Figure 1-2 is a two-dimensional schematic picture of the charge distribution within a metal. The shaded regions represent the net positive charge of the nucleus and the tightly bound inner electrons. The black dots represent the outer, or valence, electrons in the atom. It is these electrons that cannot be said to belong to any particular atom; instead, they have completely lost their individuality and can wander freely about from atom to atom in the metal. Thus a metal is visualized as a region containing a periodic three-dimensional array of heavy, tightly bound ions permeated with a swarm of electrons that may move about quite freely. This picture is known as the *electron-gas* description of a metal.

According to the electron-gas theory of a metal, the electrons are in continuous motion, the direction of flight being changed at each collision with

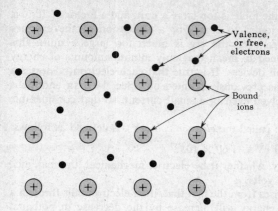

Figure 1-2 Schematic arrangement of the atoms in one plane in a metal, drawn for monovalent atoms. The black dots represent the electron gas, each atom having contributed one electron to this gas.

the heavy (almost stationary) ions. The average distance between collisions is called the *mean free path*. Since the motion is random, then, on an average, there will be as many electrons passing through a unit area in the metal in any direction as in the opposite direction in a given time. Hence the average current is zero.

Let us now see how the situation is changed if a constant electric field $\mathcal{E}$ (volts per meter) is applied to the metal. As a result of this electrostatic force, the electrons would be accelerated and the velocity would increase indefinitely with time, were it not for the collisions with the ions. However, at each inelastic collision with an ion, an electron loses energy and changes direction. The probability that an electron moves in a particular direction after a collision is equal to the probability that it travels in the opposite direction after colliding with an ion. Hence, as indicated in Fig. 1-3, the velocity of an electron increases linearly with time between collisions, and (on the average) its velocity is reduced to zero at each collision. A steady-state condition is reached where an average value of *drift speed v* is attained. This drift velocity is in the direction opposite to that of the electric field. The speed at a time t between collisions is at, where $a = q\mathcal{E}/m$ is the acceleration. Hence the average speed v is proportional to $\mathcal{E}$.

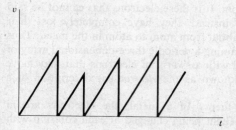

Figure 1-3 The speed of an electron in a crystal, subjected to an external electric field, increases linearly with time between collisions. Its velocity is effectively reduced to zero at each collision.

Thus

$$v = \mu \mathcal{E} \tag{1-8}$$

where μ (square meters per volt-second) is called the *mobility* of the electrons.

According to the foregoing theory, a steady-state drift speed has been superimposed upon the random thermal motion of the electrons. Such a directed flow of electrons constitutes a current. We now calculate the magnitude of the current.

Current Density

If N electrons are contained in a length L of conductor (Fig. 1-4), and if it takes an electron a time T sec to travel a distance of L m in the conductor, the total number of electrons passing through any cross section of wire in unit time is N/T. Thus the total charge per second passing any area, which, by definition, is the current in amperes, is

$$I \equiv \frac{Nq}{T} = \frac{Nqv}{L} \tag{1-9}$$

because L/T is the average, or *drift*, speed v m/s of the electrons. By definition, the current density, denoted by the symbol J, is the current per unit area of the conducting medium. That is, assuming a uniform current distribution,

$$J \equiv \frac{I}{A} \tag{1-10}$$

where J is in amperes per square meter, and A is the cross-sectional area (in meters) of the conductor. This becomes, by Eq. (1-9)

$$J = \frac{Nqv}{LA} \tag{1-11}$$

From Fig. 1-4 it is evident that LA is simply the volume containing the N electrons, and so N/LA is the electron concentration n (in electrons per cubic meter). Thus

$$n = \frac{N}{LA} \tag{1-12}$$

and Eq. (1-11) reduces to

$$J = nqv = \rho v \tag{1-13}$$

where $\rho \equiv nq$ is the charge density, in coulombs per cubic meter, and v is in meters per second.

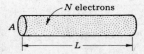

Figure 1-4 Pertaining to the calculation of current density.

This derivation is independent of the form of the conducting medium. Consequently, Fig. 1-4 does not necessarily represent a wire conductor. It may represent equally well a portion of a gaseous-discharge tube or a volume element of a semiconductor. Furthermore, neither ρ nor v need be constant, but may vary from point to point in space or may vary with time.

Conductivity

From Eqs. (1-13) and (1-8)

$$J = nqv = nq\mu\mathcal{E} = \sigma\mathcal{E} \tag{1-14}$$

where

$$\sigma = nq\mu \tag{1-15}$$

is the *conductivity* of the metal in (ohm-meter)$^{-1}$. From Eq. (1-14),

$$I = JA = \frac{\sigma AL\mathcal{E}}{L} = \frac{\sigma AV}{L} = \frac{V}{R} \tag{1-16}$$

where $V = L\mathcal{E}$ is the applied voltage across the length L, and R, the resistance of the conductor, is given by

$$R = \frac{L}{\sigma A} \tag{1-17}$$

Equation (1-16) is recognized as Ohm's law, namely, the conduction current is proportional to the applied voltage. As already mentioned, the energy which the electrons acquire from the applied field is, as a result of collisions, given to the lattice ions. Hence power is dissipated within the metal by the electrons, and the power density (Joule heat) is given by $J\mathcal{E} = \sigma\mathcal{E}^2$ (watts per cubic meter).

Example 1-1 A specimen of silicon is 2 cm long and has a square cross section 2×2 mm. The current is due to electrons whose mobility is 1,300 cm^2/V $\cdot$ s (Table 1-1). One volt impressed across the bar results in a current of 8 mA. (*a*) Calculate the concentration n of free electrons and (*b*) the drift velocity.

SOLUTION (*a*) n is obtained from Eq. (1-15), but we must first find σ. From Eq. (1-16)

$$\sigma = \frac{LI}{AV} = \frac{(2 \times 10^{-2} \text{ m})(8 \times 10^{-3} \text{ A})}{(2 \times 10^{-3} \text{ m})^2(1 \text{ V})} = 40 \ (\Omega \cdot \text{m})^{-1}$$

From Eq. (1-15),

$$n = \frac{\sigma}{q\mu} = \frac{40 \ (\Omega \cdot \text{m})^{-1}}{(1.60 \times 10^{-19} \text{ C})\left(1,300 \ \frac{\text{cm}^2}{\text{V} \cdot \text{s}} \times \frac{\text{m}^2}{10^4 \text{ cm}^2}\right)}$$

Table 1-1 Properties of silicon and germanium†

Property	Si	Ge
Atomic number	14	32
Atomic weight	28.1	72.6
Density, g/cm^3	2.33	5.32
Dielectric constant (relative)	12	16
Atoms/cm^3	5.0×10^{22}	4.4×10^{22}
E_{GO}, eV, at 0 K	1.21	0.785
E_G, eV, at 300 K	1.1	0.72
n_i at 300 K, cm^{-3}	1.5×10^{10}	2.5×10^{13}
Intrinsic resistivity at 300 K, $\Omega \cdot cm$	230,000	45
μ_n, $cm^2/V \cdot s$ at 300 K	1,300	3,800
μ_p, $cm^2/V \cdot s$ at 300 K	500	1,800
D_n, $cm^2/s = \mu_n V_T$	34	99
D_p, $cm^2/s = \mu_p V_T$	13	47

† G. L. Pearson and W. H. Brattain, History of Semiconductor Research, *Proc. IRE*, vol. 43, pp. 1794–1806, December, 1955. E. M. Conwell, Properties of Silicon and Germanium, Part II, *Proc. IRE*, vol. 46, no. 6, pp. 1281–1299, June, 1958.

Since 1 C/s is 1 A and 1 V/A is 1 Ω, then $[C/V \cdot s] = \Omega$ and $[n] = m^{-3}$. Hence,

$$n = 1.92 \times 10^{21}/m^3 = 1.92 \times 10^{15}/cm^3$$

(*b*) From Eq. (1-14),

$$v = \frac{J}{nq} = \frac{I}{Anq} = \frac{8 \times 10^{-3} \text{ A}}{(2 \times 10^{-3} \text{ m})^2 (1.92 \times 10^{21}/m^3)(1.60 \times 10^{-19} \text{ C})}$$

$$= 6.51 \text{ m/s}$$

1-5 ELECTRONS AND HOLES IN AN INTRINSIC SEMICONDUCTOR[1]

From Eq. (1-15) we see that the conductivity is proportional to the concentration n of free electrons. For a good conductor, n is very large ($\sim 10^{28}$ electrons/m^3); for an insulator, n is very small ($\sim 10^7$); and for a semiconductor, n lies between these two values. The valence electrons in a semiconductor are not free to wander about as they are in a metal, but rather are trapped in a bond between two adjacent ions, as explained below.

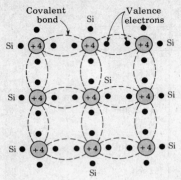

Figure 1-5 Crystal structure of silicon illustrated symbolically in two dimensions.

The Covalent Bond

Silicon and germanium are the two most important semiconductors used in electronic devices. The crystal structure of these materials consists of a regular repetition in three dimensions of a unit cell having the form of a tetrahedron with an atom at each vertex. This structure is illustrated symbolically in two dimensions in Fig. 1-5. Silicon has a total of 14 electrons in its atomic structure. Each atom in a silicon crystal contributes four valence electrons, so that the atom is tetravalent. The inert ionic core of the silicon atom carries a positive charge of $+4$ measured in units of the electronic charge. The binding forces between neighboring atoms result from the fact that each of the valence electrons of a silicon atom is shared by one of its four nearest neighbors. This *electron-pair*, or *covalent*, *bond* is represented in Fig. 1-5 by the two dashed lines which join each atom to each of its neighbors. The fact that the valence electrons serve to bind one atom to the next also results in the valence electrons being tightly bound to the nucleus. Hence, in spite of the availability of four valence electrons, the crystal has a low conductivity.

The Hole

At a very low temperature (say 0 K) the ideal structure of Fig. 1-5 is approached, and the crystal behaves as an insulator, since no free carriers of electricity are available. However, at room temperature, some of the covalent bonds will be broken because of the thermal energy supplied to the crystal, and conduction is made possible. This situation is illustrated in Fig. 1-6. Here an electron, which for the far greater period of time forms part of a covalent bond, is pictured as being dislodged, and therefore free to wander in a random fashion throughout the crystal. The energy E_G required to break such a covalent bond is about 1.1 eV for silicon and 0.72 eV for germanium at room temperature. The absence of the electron in the covalent bond is represented by the small circle in Fig. 1-6, and such an incomplete covalent bond is called a *hole*. The importance

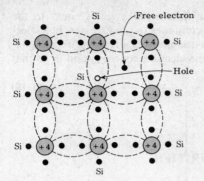

Figure 1-6 Silicon crystal with a broken covalent bond.

of the hole is that it may serve as a carrier of electricity comparable in effectiveness with the free electron.

The mechanism by which a hole contributes to the conductivity is qualitatively as follows: When a bond is incomplete so that a hole exists, it is relatively easy for a valence electron in a neighboring atom to leave its covalent bond to fill this hole. An electron moving from a bond to fill a hole leaves a hole in its initial position. Hence the hole effectively moves in the direction opposite to that of the electron. This hole, in its new position, may now be filled by an electron from another covalent bond, and the hole will correspondingly move one more step in the direction opposite to the motion of the electron. Here we have a mechanism for the conduction of electricity which does not involve *free* electrons. This phenomenon is illustrated schematically in Fig. 1-7, where a circle with a dot in it represents a completed bond, and an empty circle designates a hole. Figure 1-7a shows a row of 10 ions, with a broken bond, or hole, at ion 6. Now imagine that an electron from ion 7 moves into the hole at ion 6, so that the configuration of Fig. 1-7b results. If we compare this figure with Fig. 1-7a, it looks as if the hole in (a) has moved toward the right in (b) (from ion 6 to ion 7). This discussion indicates that the motion of the hole in one direction actually means the transport of a negative charge an equal distance in the opposite direction. So far as the flow of electric current is concerned, the hole behaves like a positive charge equal in magnitude to the electronic charge. We can consider that the holes are physical entities whose movement constitutes a flow of current. The heuristic argument that a hole behaves as a *free* positive charge carrier may be justified by quantum mechanics.[1] An experimental verification of this concept is given in Sec. 1-9.

Figure 1-7 The mechanism by which a hole contributes to the conductivity.

In a pure (*intrinsic*) semiconductor the number of holes is equal to the number of free electrons. Thermal agitation continues to produce new hole-electron pairs, whereas other hole-electron pairs disappear as a result of recombination. The hole concentration p must equal the electron concentration n, so that

$$n = p = n_i \qquad (1\text{-}18)$$

where n_i is called the *intrinsic concentration*.

1-6 DONOR AND ACCEPTOR IMPURITIES

If, to intrinsic silicon or germanium, there is added a small percentage of trivalent or pentavalent atoms, a *doped*, *impure*, or *extrinsic*, semiconductor is formed.

Donors

If the dopant has five valence electrons, the crystal structure of Fig. 1-8 is obtained. The impurity atoms will displace some of the silicon atoms in the crystal lattice. Four of the five valence electrons will occupy covalent bonds, and the fifth will be nominally unbound and will be available as a carrier of current. The energy required to detach this fifth electron from the atom is of the order of only 0.05 eV for Si or 0.01 eV for Ge. Suitable pentavalent impurities are antimony, phosphorus, and arsenic. Such impurities donate excess (negative) electron carriers, and are therefore referred to as *donor*, or *n*-type, impurities.

If intrinsic semiconductor material is "doped" with *n*-type impurities, not only does the number of electrons increase, but the number of holes decreases below that which would be available in the intrinsic semiconductor. The reason for the decrease in the number of holes is that the larger number of electrons present increases the rate of recombination of electrons with holes.

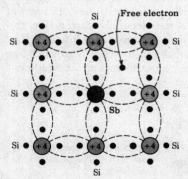

Figure 1-8 Crystal lattice with a silicon atom displaced by a pentavalent (antimony) impurity atom.

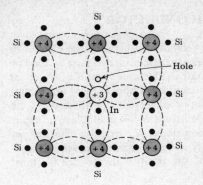

Figure 1-9 Crystal lattice with a silicon atom displaced by an atom of a trivalent (indium) impurity.

Acceptors

If a trivalent impurity (boron, gallium, or indium) is added to an intrinsic semiconductor, only three of the covalent bonds can be filled, and the vacancy that exists in the fourth bond constitutes a hole. This situation is illustrated in Fig. 1-9. Such impurities make available positive carriers because they create holes which can accept electrons. These impurities are consequently known as *acceptor*, or *p*-type, impurities. The amount of impurity which must be added to have an appreciable effect on the conductivity is very small. For example, if a donor-type impurity is added to the extent of 1 part in 10^8, the conductivity of silicon at 30°C is multiplied by a factor of 24,100.

The Mass-Action Law

We noted above that adding *n*-type impurities decreases the number of holes. Similarly, doping with *p*-type impurities decreases the concentration of free electrons below that in the intrinsic semiconductor. A theoretical analysis (Sec. 1-12) leads to the result that, under thermal equilibrium, the product of the free negative and positive concentrations is a constant independent of the amount of donor and acceptor impurity doping. This relationship is called the *mass-action law* and is given by

$$np = n_i^2 \tag{1-19}$$

The intrinsic concentration n_i is a function of temperature (Sec. 1-8).

We have the important result that the doping of an intrinsic semiconductor not only increases the conductivity, but also serves to produce a conductor in which the electric carriers are either predominantly holes or predominantly electrons. In an *n*-type semiconductor, the electrons are called the *majority carriers*, and the holes are called the *minority carriers*. In a *p*-type material, the holes are the majority carriers, and the electrons are the minority carriers.

1-7 CHARGE DENSITIES IN A SEMICONDUCTOR

Equation (1-19), namely, $np = n_i^2$, gives one relationship between the electron n and the hole p concentrations. These densities are further interrelated by the law of electrical neutrality, which we shall now state in algebraic form: Let N_D equal the concentration of donor atoms. Since, as mentioned above, these are practically all ionized, N_D positive charges per cubic meter are contributed by the donor ions. Hence the total positive-charge density is $N_D + p$. Similarly, if N_A is the concentration of acceptor ions, these contribute N_A negative charges per cubic meter. The total negative-charge density is $N_A + n$. Since the semiconductor is electrically neutral, the magnitude of the positive-charge density must equal that of the negative concentration, or

$$N_D + p = N_A + n \tag{1-20}$$

Consider an n-type material having $N_A = 0$. Since the number of electrons is much greater than the number of holes in an n-type semiconductor ($n \gg p$), then Eq. (1-20) reduces to

$$n \approx N_D \tag{1-21}$$

In an n-type material the free-electron concentration is approximately equal to the density of donor atoms.

The concentration p of holes *in the n-type semiconductor* is obtained from Eq. (1-19). Thus,

$$p = \frac{n_i^2}{N_D} \tag{1-22}$$

Similarly, *in a p-type semiconductor*,

$$p \approx N_A \tag{1-23}$$

and

$$n = \frac{n_i^2}{N_A} \tag{1-24}$$

It is possible to add donors to a p-type crystal or, conversely, to add acceptors to n-type material. If equal concentrations of donors and acceptors permeate the semiconductor, it remains intrinsic. The hole of the acceptor combines with the conduction electron of the donor to give no additional free carriers. Thus, from Eq. (1-20) with $N_D = N_A$, we observe that $p = n$, and from Eq. (1-19), $n^2 = n_i^2$, or $n = n_i =$ the intrinsic concentration.

An extension of the above argument indicates that if the concentration of donor atoms added to a p-type semiconductor exceeds the acceptor concentration ($N_D > N_A$), the specimen is changed from a p-type to an n-type semiconductor. [In Eqs. (1-21) and (1-22) N_D should be replaced by $N_D - N_A$.]

Generation and Recombination of Charges

In a pure (intrinsic) semiconductor the number of holes is equal to the number of free electrons. Thermal agitation, however, continues to generate g new hole-electron pairs per unit volume per second, while other hole-electron pairs disappear as a result of recombination; in other words, free electrons fall into empty covalent bonds, resulting in the loss of a pair of mobile carriers. On an average, a hole (an electron) will exist for $\tau_p(\tau_n)$ seconds before recombination. This time is called the *mean lifetime* of the hole and electron, respectively. These parameters are very important in semiconductor devices because they indicate the time required for electron and hole concentrations which have been caused to change to return to their equilibrium concentrations.

1-8 ELECTRICAL PROPERTIES OF Si AND Ge

A fundamental difference between a metal and a semiconductor is that the former is *unipolar* [conducts current by means of charges (electrons) of one sign only], whereas a semiconductor is *bipolar* (contains two charge-carrying "particles" of opposite sign).

Conductivity

One carrier is negative (the free electron), of mobility μ_n, and the other is positive (the hole), of mobility μ_p. These particles move in opposite directions in an electric field $\mathcal{E}$, but since they are of opposite sign, the current of each is in the same direction. Hence the current density J is given by (Sec. 1-4)

$$J = (n\mu_n + p\mu_p)q\mathcal{E} = \sigma\mathcal{E} \tag{1-25}$$

where $n=$ magnitude of free-electron (negative) concentration
   $p=$ magnitude of hole (positive) concentration
   $\sigma=$ conductivity

Hence
$$\sigma = (n\mu_n + p\mu_p)q \tag{1-26}$$

For the pure semiconductor, $n = p = n_i$, where n_i is the intrinsic concentration.

Intrinsic Concentration

With increasing temperature, the density of hole-electron pairs increases and, correspondingly, the conductivity increases. Theoretically[1,2] it is found that the intrinsic concentration n_i varies with T as

$$n_i^2 = A_o T^3 \epsilon^{-E_{GO}/kT} \tag{1-27}$$

where E_{GO} is the energy gap (the energy required to break a covalent bond) at 0 K in electron volts, k is the Boltzmann constant in eV/K (Appendix A-1), and A_o is a constant independent of T. The constants E_{GO}, μ_n, μ_p, and many other important physical quantities for silicon and germanium are given in Table 1-1. Note that silicon has of the order of 10^{22} atoms/cm^3, whereas at room temperature (300 K), $n_i \approx 10^{10}$/cm^3. Hence only 1 atom in about 10^{12} contributes a free electron (and also a hole) to the crystal because of broken covalent bonds. For germanium this ratio is about 1 atom in 10^9.

The Energy Gap

The energy E_G in a semiconductor depends upon temperature. Experimentally it is found that E_G decreases linearly with T. For silicon,[3]

$$E_G(T) = 1.21 - 3.60 \times 10^{-4}T \qquad (1\text{-}28)$$

and at room temperature (300 K), $E_G = 1.1$ eV. Similarly, for germanium,[3]

$$E_g(T) = 0.785 - 2.23 \times 10^{-4}T \qquad (1\text{-}29)$$

and at room temperature, $E_G = 0.72$ eV.

The Mobility

This parameter μ varies[3] as T^{-m} over a temperature range of 100 to 400 K. For silicon, $m = 2.5$ (2.7) for electrons (holes), and for germanium, $m = 1.66$ (2.33) for electrons (holes). The mobility is also found[4] to be a function of electric field intensity and remains constant only if $\mathscr{E} < 10^3$ V/cm in n-type silicon. For $10^3 < \mathscr{E} < 10^4$ V/cm, μ_n varies approximately as $\mathscr{E}^{-1/2}$. For higher fields, μ_n is inversely proportional to $\mathscr{E}$ and the carrier speed approaches the constant value of 10^7 cm/s.

Example 1-2 (*a*) Using Avogadro's number, verify the numerical value given in Table 1-1 for the concentration of atoms in silicon. (*b*) Find the resistivity of intrinsic silicon at 300 K. (*c*) If a donor-type impurity is added to the extent of 1 part in 10^8 silicon atoms, find the resistivity.

SOLUTION (*a*) A quantity of any substance equal to its molecular weight in grams is a *mole* of that substance. Further, a mole of any substance contains the same number of molecules as a mole of any other material. This number is called *Avogadro's number* and equals 6.02×10^{23} molecules per mole (Appendix A-1). Thus, for monatomic silicon (using Table 1-1),

$$\text{Concentration} = 6.02 \times 10^{23} \frac{\text{atoms}}{\text{mole}} \times \frac{1 \text{ mole}}{28.1 \text{ g}} \times \frac{2.33 \text{ g}}{\text{cm}^3}$$

$$= 4.99 \times 10^{22} \frac{\text{atoms}}{\text{cm}^3}$$

(b) From Eq. (1-26), with $n = p = n_i$,

$$\sigma = n_i q(\mu_n + \mu_p) = (1.5 \times 10^{10} \text{ cm}^{-3})(1.60 \times 10^{-19} \text{ C})(1{,}300 + 500) \frac{\text{cm}^2}{\text{V} \cdot \text{s}}$$

$$= 4.32 \times 10^{-6} \ (\Omega \cdot \text{cm})^{-1}$$

Resistivity $= \dfrac{1}{\sigma} = \dfrac{1}{4.32 \times 10^{-6}} = 2.31 \times 10^5 \ \Omega \cdot \text{cm}$

in agreement with the value in Table 1-1.

(c) If there is 1 donor atom per 10^8 silicon atoms, then $N_D = 4.99 \times 10^{14}$ atoms/cm^3. From Eq. (1-21) $n \approx N_D$ and from Eq. (1-22)

$$p = \frac{n_i^2}{N_D} = \frac{(1.5 \times 10^{10})^2}{4.99 \times 10^{14}} = 4.51 \times 10^5 \text{ holes/cm}^3$$

Since $n \gg p$, we can neglect p in calculating the conductivity. From Eq. (1-26)

$$\sigma = nq\mu_n = 4.99 \times 10^{14} \times 1.60 \times 10^{-19} \times 1{,}300 = 0.104 \ (\Omega \cdot \text{cm})^{-1}$$

The resistivity $= 1/\sigma = 1/0.104 = 9.62 \ \Omega \cdot \text{cm}$.

Note: The addition of 1 donor atom in 10^8 silicon atoms has multiplied the conductivity by a factor of $0.104/4.32 \times 10^{-6} = 24{,}100$.

1-9 THE HALL EFFECT

If a specimen (metal or semiconductor) carrying a current I is placed in a transverse magnetic field B, an electric field $\mathcal{E}$ is induced in the direction perpendicular to both I and B. This phenomenon, known as the *Hall effect*, is used to determine whether a semiconductor is n- or p-type and to find the carrier concentration. Also, by simultaneously measuring the conductivity σ, the mobility μ can be calculated.

The physical origin of the Hall effect is not difficult to find. If in Fig. 1-10 I is in the positive X direction and B is in the positive Z direction, a force will be exerted in the negative Y direction on the current carriers. The current I may be due to holes moving from left to right or to free electrons traveling from right to left in the semiconductor specimen. Hence, independently of whether the carriers are holes or electrons, they will be forced downward toward side 1 in Fig. 1-10. If the semiconductor is n-type material, so that the current is carried by electrons, these electrons will accumulate on side 1, and this surface becomes negatively charged with respect to side 2. Hence a potential, called the *Hall voltage*, appears between surfaces 1 and 2.

If the polarity of V_H is positive at terminal 2, then, as explained above, the carriers must be electrons. If, on the other hand, terminal 1 becomes charged

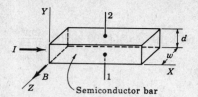

Figure 1-10 Pertaining to the Hall effect. The carriers (whether electrons or holes) are subjected to a magnetic force in the negative Y direction.

positively with respect to terminal 2, the semiconductor must be p-type. These results have been verified experimentally, thus justifying the bipolar (two-carrier) nature of the current in a semiconductor.

If I is the current in a p-type semiconductor, the carriers might be considered to be the *bound* electrons jumping from right to left. Then side 1 would become negatively charged. However, experimentally, side 1 is found to become positive with respect to side 2 for a p-type specimen. This experiment confirms the quantum-mechanical fact noted in Sec. 1-5 that the hole acts like a classical *free* positive-charge carrier.

Experimental Determination of Mobility

In the equilibrium state the electric field intensity $\mathcal{E}$ due to the Hall effect must exert a force on the carrier which just balances the magnetic force, or

$$q\mathcal{E} = Bqv \tag{1-30}$$

where q is the magnitude of the charge on the carrier, and v is the drift speed. From Eq. (1-3), $\mathcal{E} = V_H/d$, where d is the distance between surfaces 1 and 2. From Eq. (1-13), $J = \rho v = I/wd$, where J is the current density, ρ is the charge density, and w is the width of the specimen in the direction of the magnetic field. Combining these relationships, we find

$$V_H = \mathcal{E}d = Bvd = \frac{BJd}{\rho} = \frac{BI}{\rho w} \tag{1-31}$$

If V_H, B, I, and w are measured, the charge density ρ can be determined from Eq. (1-31).

If conduction is due primarily to charges of one sign, the conductivity σ is related to the mobility μ by Eq. (1-15), or

$$\sigma = \rho\mu \tag{1-32}$$

If the conductivity is measured together with the Hall voltage, the mobility can be determined from Eqs. (1-31) and (1-32).

Applications

Since V_H is proportional to B (for a given current I), then the Hall effect has been incorporated into a magnetic field meter. Conversely, if a known B field is imposed, the Hall effect can be utilized to implement an ammeter; for example, the Tektronix P6042, dc to 50 MHz, 1 mA to 10 A, clip-on current probe. Another instrument, called a *Hall-effect multiplier*, is available to give an output proportional to the product of two signals. If I is made proportional to one of the inputs and if B is linearly related to the second signal, then, from Eq. (1-31), V_H is proportional to the product of the two inputs.

1-10 THERMISTORS AND SENSISTORS

The conductivity of germanium (silicon) is found from Eq. (1-27) to increase approximately 6 (8) percent per degree increase in temperature. Such a large change in conductivity with temperature places a limitation upon the use of semiconductor devices in some circuits. On the other hand, for some applications it is exactly this property of semiconductors that is used to advantage. A semiconductor used in this manner is called a *thermistor*. Such a device finds extensive application in thermometry, in the measurement of microwave-frequency power, as a thermal relay, and in control devices actuated by changes in temperature. Silicon and germanium are not used as thermistors because their properties are too sensitive to impurities. Commercial thermistors consist of sintered mixtures of such oxides as NiO, Mn_2O_3, and Co_2O_3.

The exponential decrease in resistivity (reciprocal of conductivity) of a semiconductor should be contrasted with the small and almost linear increase in resistivity of a metal. An increase in the temperature of a metal results in greater thermal motion of the ions, and hence decreases slightly the mean free path of the free electrons. The result is a decrease in the mobility, and hence in conductivity. For most metals the resistance increases about 0.4 percent/°C increase in temperature. It should be noted that a thermistor has a negative coefficient of resistance, whereas that of a metal is positive and of much smaller magnitude. By including a thermistor in a circuit it is possible to compensate for temperature changes over a range as wide as 100°C.

A heavily doped semiconductor can exhibit a positive temperature coefficient of resistance, for under these circumstances the material acquires metallic properties and the resistance increases because of the decrease in carrier mobility with temperature. Such a device, called a *sensistor* (TG1/8 of Texas Instruments), has a temperature coefficient of resistance of +0.7 percent/°C (over the range from −50 to +125°C).

1-11 DIFFUSION

In addition to a conduction current, the transport of charges in a semiconductor may be accounted for by a mechanism called *diffusion*, not ordinarily encountered in metals. The essential features of diffusion are now discussed.

It is possible to have a nonuniform concentration of particles in a semiconductor. As indicated in Fig. 1-11, the concentration p of holes varies with

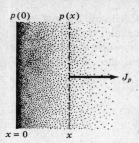

$p(0)$ $p(x)$

J_p

$x = 0$ x

Figure 1-11 A nonuniform concentration $p(x)$ results in a diffusion current density J_p.

distance x in the semiconductor, and there exists a concentration gradient, dp/dx, in the density of carriers. The existence of a gradient implies that if an imaginary surface (shown dashed) is drawn in the semiconductor, the density of holes immediately on one side of the surface is larger than the density on the other side. The holes are in a random motion as a result of their thermal energy. Accordingly, holes will continue to move back and forth across this surface. We may then expect that, in a given time interval, more holes will cross the surface from the side of greater concentration to the side of smaller concentration than in the reverse direction. This net transport of holes across the surface constitutes a current in the positive X direction. It should be noted that this net transport of charge is not the result of mutual repulsion among charges of like sign, but is simply the result of a statistical phenomenon. This diffusion is exactly analogous to that which occurs in a neutral gas if a concentration gradient exists in the gaseous container. The diffusion hole-current density J_p (amperes per square meter) is proportional to the concentration gradient, and is given by

$$J_p = -qD_p \frac{dp}{dx} \tag{1-33}$$

where D_p (square meters per second) is called the *diffusion constant* for holes. Since p in Fig. 1-11 decreases with increasing x, then dp/dx is negative and the minus sign in Eq. (1-33) is needed, so that J_p will be positive in the positive X direction. A similar equation exists for diffusion electron-current density [p is replaced by n, and the minus sign is replaced by a plus sign in Eq. (1-33)].

Einstein Relationship

Since both diffusion and mobility are statistical thermodynamic phenomena, D and μ are not independent. The relationship between them is given by the Einstein equation

$$\frac{D_p}{\mu_p} = \frac{D_n}{\mu_n} = V_T \tag{1-34}$$

where V_T is the "volt-equivalent of temperature," defined by

$$V_T \equiv \frac{\bar{k}T}{q} = \frac{T}{11,600} \tag{1-35}$$

where $\bar{k}$ is the Boltzmann constant in joules per degree Kelvin. Note the distinction between $\bar{k}$ and k; the latter is the Boltzmann constant in electron volts per degree Kelvin. (Numerical values of $\bar{k}$ and k are given in Appendix A-1. From Sec. 1-3 it follows that $\bar{k} = 1.60 \times 10^{-19}k$.) At room temperature (300 K), $V_T = 0.0259$ V, and $\mu = 38.6D$. Measured values of μ and computed values of D for silicon and germanium are given in Table 1-1.

Total Current

It is possible for both a potential gradient and a concentration gradient to exist simultaneously within a semiconductor. In such a situation the total hole current is the sum of the drift current [Eq. (1-14), with n replaced by p] and the diffusion current [Eq. (1-33)], or

$$J_p = q\mu_p p \mathcal{E} - qD_p \frac{dp}{dx} \tag{1-36}$$

Similarly, the net electron current is

$$J_n = q\mu_n n \mathcal{E} + qD_n \frac{dn}{dx} \tag{1-37}$$

1-12 THE POTENTIAL VARIATION WITHIN A GRADED SEMICONDUCTOR

Consider a semiconductor (Fig. 1-12a) where the hole concentration p is a function of x; that is, the doping is *nonuniform*, or graded.[5] Assume a steady-state situation and zero excitation; that is, no carriers are injected into the specimen from any *external* source. With no excitation there can be no *steady* movement of charge in the bar, although the carriers possess random motion due to thermal agitation. Hence the total hole current must be zero. (Also, the total electron current must be zero.) Since p is not constant, we expect a nonzero hole diffusion current. In order for the total hole current to vanish there must exist a hole drift current which is equal and opposite to the diffusion current. However, a conduction current requires an electric field, and hence we conclude that, as a result of the nonuniform doping, an electric field is generated within the semiconductor. We shall now find this field and the corresponding potential variation throughout the bar.

Setting $J_p = 0$ in Eq. (1-36) and using the Einstein relationship $D_p = \mu_p V_T$ [Eq. (1-34)], we obtain

$$\mathcal{E} = \frac{V_T}{p} \frac{dp}{dx} \tag{1-38}$$

If the doping concentration $p(x)$ is known, this equation allows the built-in field $\mathcal{E}(x)$ to be calculated. From $\mathcal{E} = -dV/dx$ we can calculate the potential

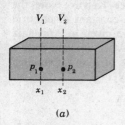

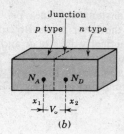

Junction

V_1 V_2

p type | n type

p_1 p_2

N_A N_D

x_1 x_2

x_1 x_2

V_o

(a)

(b)

Figure 1-12 (a) A graded semi-conductor: $p(x)$ is not constant. (b) One portion is doped uniformly with acceptor ions and the other section is doped uniformly with donor ions so that a metallurgical junction is formed. A contact potential V_o appears across this step-graded p-n junction.

variation. Thus

$$dV = -V_T \frac{dp}{p} \tag{1-39}$$

If this equation is integrated between x_1, where the concentration is p_1 and the potential is V_1 (Fig. 1-12a), and x_2, where $p = p_2$ and $V = V_2$, the result is

$$V_{21} \equiv V_2 - V_1 = V_T \ln \frac{p_1}{p_2} \tag{1-40}$$

Note that the potential difference between two points depends only upon the concentrations at these two points and is independent of their separation $x_2 - x_1$. Equation (1-40) may be put in the form

$$p_1 = p_2 \epsilon^{V_{21}/V_T} \tag{1-41}$$

This is the Boltzmann relationship of kinetic gas theory.

Mass-Action Law

Starting with $J_n = 0$ and proceeding as above, the Boltzmann equation for electrons is obtained.

$$n_1 = n_2 \epsilon^{-V_{21}/V_T} \tag{1-42}$$

Multiplying Eqs. (1-41) and (1-42) gives

$$n_1 p_1 = n_2 p_2 \tag{1-43}$$

This equation states that the product np is a constant independent of x, and hence of the amount of doping, under thermal equilibrium. For an intrinsic semiconductor, $n = p = n_i$, and hence $np = n_i^2$ which is the law of mass action introduced in Eq. (1-19).

An Open-circuited Step-graded Junction

Consider the special case indicated in Fig. 1-12b. The left half of the bar is p-type with a constant concentration N_A, whereas the right-half is n-type with a uniform density N_D. The dashed plane is a metallurgical (p-n) junction separating the two sections with different concentration. This type of doping, where the density changes abruptly from p- to n-type, is called *step grading*. The step-graded junction is located at the plane where the concentration is zero. The above theory indicates that there is a built-in potential between these two sections (called the *contact difference of potential V_o*). Equation (1-40) allows us to calculate V_o. Thus

$$V_o = V_{21} = V_T \ln \frac{p_{po}}{p_{no}} \tag{1-44}$$

because $p_1 = p_{po}$ = thermal-equilibrium hole concentration in p side and $p_2 = p_{no}$ = thermal-equilibrium hole concentration in n side. From Eq. (1-23) $p_{po} = N_A$, and from Eq. (1-22) $p_{no} = n_i^2/N_D$, so that

$$V_o = V_T \ln \frac{N_A N_D}{n_i^2} \tag{1-45}$$

The same expression for V_o is obtained from an analysis corresponding to that given above and based upon equating the total electron current I_n to zero (Prob. 1-29). The p-n junction, both open-circuited and with applied voltage, is studied in detail in Chap. 2.

1-13 RECAPITULATION

The fundamental principles governing the electrical behavior of semiconductors, discussed in this chapter, are summarized as follows:

1. Two types of mobile charge carriers (positive holes and negative electrons) are available. This bipolar nature of a semiconductor is to be contrasted with the unipolar property of a metal, which possesses only free electrons.
2. A semiconductor may be fabricated with donor (acceptor) impurities; so it contains mobile charges which are primarily electrons (holes).
3. The intrinsic concentration of carriers is a function of temperature. At room temperature, essentially all donors or acceptors are ionized.
4. Current is due to two distinct phenomena:
   (*a*) Carriers drift in an electric field (this conduction current is also available in a metal).
   (*b*) Carriers diffuse if a concentration gradient exists (a phenomenon which does not take place in a metal).
5. Carriers are continuously being generated (due to thermal creation of hole-electron pairs) and are simultaneously disappearing (due to recombination).
6. Across an open-circuited p-n junction there exists a contact difference of potential.

These basic concepts are applied in the next chapter to the study of the p-n junction diode.

REFERENCES

1. Shockley, W.: "Electrons and Holes in Semiconductors," D. Van Nostrand Company, Inc., Princeton, N. J., reprinted February, 1963. Yang, E. S.: "Fundamentals of Semiconductor Devices," Chap. 1, McGraw-Hill Book Company, New York, 1978.
2. Millman, J., and C. C. Halkias: "Integrated Electronics," sec. 19-5, McGraw-Hill Book Company, New York, 1972.
3. Morin, F. J., and J. P. Maita: Conductivity and Hall Effect in the Intrinsic Range of Germanium, *Phys. Rev.*, vol. 94, pp. 1525–1529, June, 1954.
   Morin, F. J., and J. P. Maita: Electrical Properties of Silicon Containing Arsenic and Boron, *Phys. Rev.*, vol. 96, pp. 28–35, October, 1954.
4. Sze, S. M.: "Physics of Semiconductor Devices," Fig. 29, p. 59, John Wiley & Sons, Inc., New York, 1969.
5. Gray, P. E., and C. L. Searle: "Electronic Principles: Physics, Models, and Circuits," John Wiley & Sons, Inc., New York, 1969.

REVIEW QUESTIONS

1-1 Define *potential energy* in words and as an equation.

1-2 Define an *electron volt*.

1-3 Give the electron-gas description of a metal.

1-4 (*a*) Define *mobility*.
(*b*) Give its dimensions.

1-5 (*a*) Define *conductivity*.
(*b*) Give its dimensions.

1-6 Is the temperature coefficient of resistance of a semiconductor positive or negative? Explain briefly.

1-7 Answer Rev. 1-6 for a metal.

1-8 Explain why a semiconductor acts as an insulator at 0 K and why its conductivity increases with increasing temperature.

1-9 What is the distinction between an intrinsic and an extrinsic semiconductor?

1-10 Define a *hole* (in a semiconductor).

1-11 Indicate pictorially how a hole contributes to conduction.

1-12 (*a*) Define *intrinsic concentration* of holes.
(*b*) What is the relationship between this density and the intrinsic concentration for electrons?
(*c*) What do these equal at 0 K?

1-13 Show (in two dimensions) the crystal structure of silicon containing a donor impurity atom.

1-14 Repeat Rev. 1-13 for an acceptor impurity atom.

1-15 Define (*a*) *donor*; (*b*) *acceptor* impurities.

1-16 A semiconductor is doped with both donors and acceptors of concentrations N_D and N_A, respectively. Write the equation or equations from which to determine the electron and hole concentrations (n and p).

1-17 Define *mean lifetime* of a carrier.

1-18 Explain physically the meaning of the following statement: An electron and a hole recombine and disappear.

1-19 Describe the *Hall effect*.

1-20 What properties of a semiconductor are determined from a Hall effect experiment?

1-21 Is the temperature coefficient of resistance of a semiconductor positive or negative? Explain briefly.

1-22 Answer Rev. 1-21 for a metal.

1-23 Define the *volt-equivalent of temperature*.

1-24 (*a*) Define *diffusion constant* for holes.
(*b*) Give its dimensions.

1-25 Repeat Rev. 1-24 for electrons.

1-26 (*a*) Write the equation for the net electron current in a semiconductor. What is the physical significance of each term?
(*b*) How is this equation modified for a metal?

1-27 (*a*) Define a *graded semiconductor*.
(*b*) Explain why an electric field must exist in a graded semiconductor.

1-28 Consider a step-graded junction under open-circuited conditions. Upon what four parameters does the contact difference of potential depend?

1-29 State the *mass-action law* as an equation and in words.

1-30 Explain why a contact difference of potential must develop across an open-circuited *p-n* junction.

JUNCTION-DIODE CHARACTERISTICS

In this chapter we demonstrate that if a junction is formed between a sample of p-type and one of n-type semiconductor, this combination possesses the properties of a rectifier. The volt-ampere characteristics of such a two-terminal device (called a *junction diode*) is studied. The capacitance across the junction is discussed.

Although the transistor is a triode (three-terminal) semiconductor, it may be considered as one diode biased by the current from a second diode. Hence most of the theory developed here is utilized in Chap. 3 in connection with the study of the transistor.

2-1 THE OPEN-CIRCUITED p-n JUNCTION

If donor impurities are introduced into one side and acceptors into the other side of a single crystal of a semiconductor, a p-n junction is formed, as in Fig. 1-12. Such a system is illustrated in more schematic detail in Fig. 2-1a. The donor ion is represented by a plus sign because, after this impurity atom "donates" an electron, it becomes a positive ion. The acceptor ion is indicated by a minus sign because, after this atom "accepts" an electron, it becomes a negative ion. Initially, there are nominally only p-type carriers to the left of the junction and only n-type carriers to the right.

Space-Charge Region

Because there is a density gradient across the junction, holes will initially diffuse to the right across the junction, and electrons to the left. We see that the positive

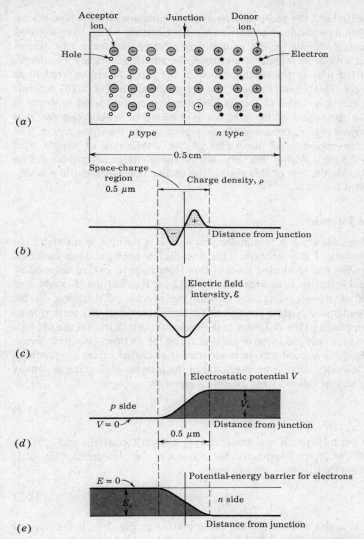

Figure 2-1 A schematic diagram of a *p-n* junction, including the charge density, electric field intensity, and potential-energy barriers at the junction. Since potential energy = potential × charge, the curve in (*d*) is proportional to the potential energy for a hole (a positive charge) and the curve in (*e*) is proportional to the negative of that in (*d*) (an electron is a negative charge). (It is assumed that the diode dimensions are large compared with the space charge region. Not drawn to scale.)

holes which neutralized the acceptor ions near the junction in the p-type silicon have disappeared as a result of combination with electrons which have diffused across the junction. Similarly, the neutralizing electrons in the n-type silicon have combined with holes which have crossed the junction from the p material. The unneutralized ions in the neighborhood of the junction are referred to as *uncovered charges*. The general shape of the charge density ρ (Fig. 2-1b) depends upon how the diode is doped (a step-graded junction is considered in detail in Sec. 2-6). Since the region of the junction is depleted of mobile charges, it is called the *depletion region*, the *space-charge region*, or the *transition region*. The thickness of this region is of the order of the wavelength of visible light (0.5 micron = 0.5 μm). Within this very narrow space-charge layer there are no mobile carriers. To the left of this region the carrier concentration is $p \approx N_A$, and to its right it is $n \approx N_D$.

Electric Field Intensity

The space-charge density ρ is zero at the junction. It is positive to the right and negative to the left of the junction. This distribution constitutes an electrical dipole layer, giving rise to electric lines of flux from right to left, corresponding to negative field intensity $\mathcal{E}$ as depicted in Fig. 2-1c. Equilibrium is established when the field is strong enough to restrain the process of diffusion. Stated alternatively, under steady-state conditions the drift hole (electron) current must be equal and opposite to the diffusion hole (electron) current so that the net hole (electron) current is reduced to zero—as it must be for an open-circuited device. In other words, there is no steady-state movement of charge across the junction.

The field intensity curve is proportional to the integral of the charge density curve. This statement follows from Poisson's equation

$$\frac{d^2V}{dx^2} = -\frac{\rho}{\epsilon} \tag{2-1}$$

where ϵ is the permittivity. If ϵ_r is the (relative) dielectric constant and ϵ_o is the permittivity of free space (Appendix A-1), then $\epsilon = \epsilon_r \epsilon_o$. Integrating Eq. (2-1) and remembering that $\mathcal{E} = -dV/dx$ gives

$$\mathcal{E} = \int_{x_o}^{x} \frac{\rho}{\epsilon} \, dx \tag{2-2}$$

where $\mathcal{E} = 0$ at $x = x_o$. Therefore the curve plotted in Fig. 2-1c is the integral of the function drawn in Fig. 2-1b (divided by ϵ).

Potential

The electrostatic-potential variation in the depletion region is shown in Fig. 2-1d, and is the negative integral of the function $\mathcal{E}$ of Fig. 2-1c. This variation constitutes a potential-energy barrier (Sec. 1-2) against the further diffusion of holes across the barrier. The form of the potential-energy barrier against the flow of electrons from the n side across the junction is shown in Fig. 2-1e. It is similar to that shown in Fig. 2-1d, except that it is inverted, since the charge on an electron is negative. Note the existence, across the depletion layer, of the *contact potential* V_o, discussed in Sec. 1-12.

Summary

Under open-circuited conditions the net hole current must be zero. If this statement were not true, the hole density at one end of the semiconductor would continue to increase indefinitely with time, a situation which is obviously physically impossible. Since the concentration of holes in the p side is much greater than that in the n side, a very large hole diffusion current tends to flow across the junction from the p to the n material. Hence an electric field must build up across the junction in such a direction that a hole drift current will tend to flow across the junction from the n to the p side in order to counterbalance the diffusion current. This equilibrium condition of zero resultant hole current allows us to calculate the height of the potential barrier V_o [Eq. 1-45)] in terms of the donor and acceptor concentrations. The numerical value for V_o is of the order of magnitude of a few tenths of a volt.

2-2 THE p-n JUNCTION AS A RECTIFIER[1]

The essential electrical characteristic of a p-n junction is that it constitutes a rectifier which permits the easy flow of charge in one direction but restrains the flow in the opposite direction. We consider now, qualitatively, how this diode rectifier action comes about.

Reverse Bias

In Fig. 2-2, a battery is shown connected across the terminals of a p-n junction. The negative terminal of the battery is connected to the p side of the junction, and the positive terminal to the n side. The polarity of connection is such as to cause both the holes in the p type and the electrons in the n type to move away from the junction. Consequently, the region of negative-charge density is spread to the left of the junction (Fig. 2-1b), and the positive-charge-density region is spread to the right. However, this process cannot continue indefinitely, because in order to have a steady flow of holes to the left, these holes must be supplied across the junction from the n-type silicon. And there are very few holes in the n-type side. Hence, nominally, zero current results. Actually, a small current does flow because a small number of hole-electron pairs are generated throughout the crystal as a result of thermal energy. The holes so formed in the n-type silicon will wander over to the junction where they are "pulled" across by the electric field. A similar remark applies to the electrons thermally generated in the p-type silicon. This small current is the diode *reverse*

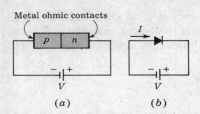

Figure 2-2 (*a*) A *p-n* junction biased in the reverse direction. (*b*) The rectifier symbol is used for the *p-n* diode.

saturation current, and its magnitude is designated by I_o. This reverse current will increase with increasing temperature [Eq. (2-5)], and hence the back resistance of a crystal diode decreases with increasing temperature. From the argument presented here, I_o should be independent of the magnitude of the reverse bias.

The mechanism of conduction in the reverse direction may be described alternatively in the following way: When no voltage is applied to the *p-n* diode, the potential barrier across the junction is as shown in Fig. 2-1*d*. When a voltage V is applied to the diode in the direction shown in Fig. 2-2, the height of the potential-energy barrier is increased by the amount qV. This increase in the barrier height serves to reduce the flow of majority carriers (i.e., holes in *p* type and electrons in *n* type). However, the minority carriers (i.e., electrons in *p* type and holes in *n* type), since they fall down the potential-energy hill, are uninfluenced by the increased height of the barrier. The applied voltage in the direction indicated in Fig. 2-2 is called the *reverse*, or *blocking*, *bias*.

Forward Bias

An external voltage applied with the polarity shown in Fig. 2-3 (opposite to that indicated in Fig. 2-2) is called a *forward* bias. An ideal *p-n* diode has zero ohmic voltage drop across the body of the crystal. For such a diode the height of the potential barrier at the junction will be lowered by the applied forward voltage V. The equilibrium initially established between the forces tending to produce diffusion of majority carriers and the restraining influence of the potential-energy barrier at the junction will be disturbed. Hence, for a forward bias, the holes cross the junction from the *p*-type into the *n*-type region, where they constitute an injected minority current. Similarly, the electrons cross the junction in the reverse direction and become a minority current injected into the *p* side. Holes traveling from left to right constitute a current in the *same* direction

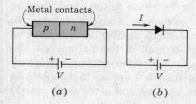

Figure 2-3 (*a*) A *p-n* junction biased in the forward direction. (*b*) The rectifier symbol is used for the *p-n* diode.

as electrons moving from right to left. Hence the resultant current crossing the junction is the *sum* of the hole and electron minority currents.

Ohmic Contacts

In Fig. 2-2 (2-3) we show an external reverse (forward) bias applied to a *p-n* diode. We have assumed that the external bias voltage appears directly across the junction and has the effect of raising (lowering) the electrostatic potential across the junction. To justify this assumption we must specify how electric contact is made to the semiconductor from the external bias circuit. In Figs. 2-2 and 2-3 we indicate metal contacts with which the homogeneous *p*-type and *n*-type materials are provided. We thus see that we have introduced two metal-semiconductor junctions, one at each end of the diode. We naturally expect a contact potential to develop across these additional junctions. However, we shall assume that the metal-semiconductor contacts shown in Figs. 2-2 and 2-3 have been manufactured in such a way that they are nonrectifying. In other words, the contact potential across these junctions is constant, independent of the direction and magnitude of the current. A contact of this type is referred to as an *ohmic contact*.

We are now in a position to justify our assumption that the entire applied voltage appears as a *change* in the height of the potential barrier. Inasmuch as the voltage across the metal-semiconductor ohmic contacts remains constant and the voltage drop across the bulk of the crystal is neglected, approximately the entire applied voltage will indeed appear as a change in the height of the potential barrier at the *p-n* junction.

The Short-circuited and Open-circuited *p-n* Junction

If the voltage V in Fig. 2-2 or 2-3 were set equal to zero, the *p-n* junction would be short-circuited. Under these conditions, as we show below, no current can flow ($I = 0$) and the electrostatic potential V_o remains unchanged and equal to the value under open-circuit conditions. If there were a current ($I \neq 0$), the metal would become heated. Since there is no external source of energy available, the energy required to heat the metal wire would have to be supplied by the *p-n* bar. The semiconductor bar, therefore, would have to cool off. Clearly, under thermal equilibrium the simultaneous heating of the metal and cooling of the bar is impossible, and we conclude that $I = 0$. Since under short-circuit conditions the sum of the voltages around the closed loop must be zero, the junction potential V_o must be exactly compensated by the metal-to-semiconductor contact potentials at the ohmic contacts. Since the current is zero, the wire can be cut without changing the situation, and the voltage drop across the cut must remain zero. If in an attempt to measure V_o we connected a voltmeter across the cut, the voltmeter would read zero voltage. In other words, it is not possible to measure contact difference of potential directly with a voltmeter.

Large Forward Voltages

Suppose that the forward voltage V in Fig. 2-3 is increased until V approaches V_o. If V were equal to V_o, the barrier would disappear and the current could be arbitrarily large, exceeding the rating of the diode. As a practical matter we can never reduce the barrier to zero because, as the current increases without limit, the bulk resistance of the crystal, as well as the resistance of the ohmic contacts, will limit the current. Therefore it is no longer possible to assume that all the voltage V appears as a change across the p-n junction. We conclude that, as the forward voltage V becomes comparable with V_o, the current through a real p-n diode will be governed by the ohmic-contact resistances and the crystal bulk resistance. Thus the volt-ampere characteristic becomes approximately a straight line.

2-3 THE VOLT-AMPERE CHARACTERISTIC

Theory [1-3] indicates that, for a p-n junction, the current I is related to the voltage V by the equation

$$I = I_o(\epsilon^{V/\eta V_T} - 1) \qquad (2-3)$$

A positive value of I means that current flows from the p to the n side. The diode is forward-biased if V is positive, indicating that the p side of the junction is positive with respect to the n side. The symbol η is unity for germanium and is approximately 2 for silicon at rated current.

The symbol V_T stands for the volt equivalent of temperature, and is given by Eq. (1-35), repeated here for convenience:

$$V_T \equiv \frac{T}{11,600} \qquad (2-4)$$

At room temperature ($T = 300$ K), $V_T = 0.026$ V = 26 mV.

The form of the volt-ampere characteristic described by Eq. (2-3) is shown in Fig. 2-4a. When the voltage V is positive and several times V_T, the unity in the parentheses of Eq. (2-3) may be neglected. Accordingly, except for a small range in the neighborhood of the origin, the current increases exponentially with voltage. When the diode is reverse-biased and $|V|$ is several times V_T, $I \approx -I_o$. The reverse current is therefore constant, independent of the applied reverse bias. Consequently, I_o is referred to as the *reverse saturation current*.

For the sake of clarity, the current I_o in Fig. 2-4 has been greatly exaggerated in magnitude. Ordinarily, the range of forward currents over which a diode is operated is many orders of magnitude larger than the reverse saturation current. To display forward and reverse characteristics conveniently, it is necessary, as in Fig. 2-4b, to use two different current scales. The volt-ampere characteristic shown in that figure has a forward-current scale in milliamperes and a reverse scale in microamperes.

The dashed portion of the curve of Fig. 2-4b indicates that, at a reverse-biasing voltage V_Z, the diode characteristic exhibits an abrupt and marked

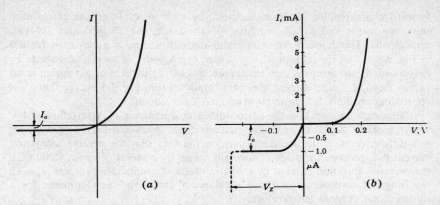

Figure 2-4 (*a*) The volt-ampere characteristic of an ideal *p-n* diode. (*b*) The volt-ampere characteristic for a germanium diode redrawn to show the order of magnitude of currents. Note the expanded scale for reverse currents. The dashed portion indicates breakdown at V_Z.

departure from Eq. (2-3). At this critical voltage a large reverse current flows, and the diode is said to be in the *breakdown* region, discussed in Sec. 2-9.

The Cutin Voltage V_γ

Both silicon and germanium diodes are commercially available. A number of differences between these two types are relevant in circuit design. The difference in volt-ampere characteristics is brought out in Fig. 2-5. Here are plotted the

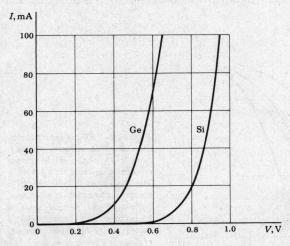

Figure 2-5 The forward volt-ampere characteristics of a germanium (1N3666) and a silicon (1N4153) diode at 25°C.

forward characteristics at room temperature of a general-purpose germanium switching diode and a fast-switching silicon diode, the 1N3666 and 1N4153, respectively. The diodes have comparable current ratings. A noteworthy feature in Fig. 2-5 is that there exists a *cutin*, *offset*, *break-point*, or *threshold*, voltage V_γ below which the current is very small (say, less than 1 percent of maximum rated value). Beyond V_γ the current rises very rapidly. From Fig. 2-5 we see that V_γ is approximately 0.2 V for germanium and 0.6 V for silicon.

Note that the break in the silicon-diode characteristic is offset about 0.4 V with respect to the break in the germanium-diode characteristic. The reason for this difference is to be found, in part, in the fact that the reverse saturation current in a germanium diode is normally larger by a factor of about 1,000 than the reverse saturation current in a silicon diode of comparable ratings. I_o is in the range of microamperes for a germanium diode and nanoamperes for a silicon diode at room temperature.

Since $\eta = 2$ for small currents in silicon, the current increases as $\epsilon^{V/2V_T}$ for the first several tenths of a volt and increases as ϵ^{V/V_T} only at higher voltages. This initial smaller dependence of the current on voltage accounts for the further delay in the rise of the silicon characteristic.

Logarithmic Characteristic

It is instructive to examine the family of curves for the silicon diode shown in Fig. 2-6. From Eq. (2-3), assuming that V is several times V_T, so that we may drop the unity, we have $\log I = \log I_o + 0.434V/\eta V_T$. We therefore expect in Fig. 2-6, where $\log I$ is plotted against V, that the plots will be straight lines. We do indeed find that at low currents the plots are linear and correspond to $\eta \approx 2$. At large currents an increment of voltage does not yield as large an increase of current as at low currents. The reason for this behavior is to be found in the ohmic resistance of the diode. At low currents the ohmic drop is negligible and

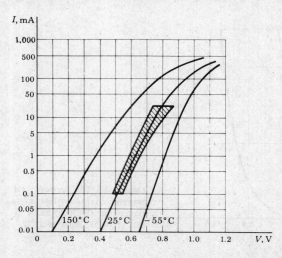

Figure 2-6 Volt-ampere characteristics at three different temperatures for a silicon diode (planar epitaxial passivated type 1N4153). The shaded area indicates 25°C limits of controlled conductance. Note that the vertical scale is logarithmic and encompasses a current range of 50,000 *(Courtesy of General Electric Company.)*

the externally impressed voltage simply decreases the potential barrier at the p-n junction. At high currents the externally impressed voltage is called upon principally to establish an electric field to overcome the ohmic resistance of the semiconductor material. Therefore, at high currents, the diode behaves more like a resistor than a diode, and the current increases linearly rather than exponentially with applied voltage.

The absolute maximum ratings and electrical characteristics of the silicon 1N4153 high-speed switching diode is given in the manufacturer's specification sheet (Appendix B-2).

Reverse Saturation Current

Many commercially available diodes exhibit an essentially constant value of I_o for negative values of V, as indicated in Fig. 2-4. On the other hand, some diodes show a very pronounced increase in reverse current with increasing reverse voltage. This variation in I_o results from leakage across the surface of the diode, and also from the additional fact that new charge carriers may be generated by collision in the transition region at the junction.

Example 2-1 Two p-n silicon diodes are connected in series opposing. A 5-V battery is impressed upon this series arrangement. Find the voltage across each junction at room temperature (300 K).

SOLUTION The current I is the same throughout a series circuit. Clearly, this current is in the forward direction through one diode and in the reverse direction through the second diode. Hence $I = I_o$ and the voltage V_1 across the forward-biased junction is, from Eq. (2-3),

$$I = I_o(\epsilon^{V_1/\eta V_T} - 1) = I_o$$

Independent of I_o we find (for $\eta = 2$ and $V_T = 0.026$ V) that

$$\epsilon^{V_1/0.052} = 2 \qquad V_1 = 0.693 \times 0.052 = 0.036 \text{ V}$$

The voltage V_2 across the reverse-biased diode is $V_2 = 5\text{-}0.036 = 4.964$ V. Note that only a small voltage is needed across the forward-biased junction to produce the small current I_o and that almost all of the applied voltage appears across the reverse-biased diode. Also note that V_1 is proportional to V_T and hence to the temperature.

2-4 THE TEMPERATURE DEPENDENCE OF THE V/I CHARACTERISTIC

The volt-ampere relationship (2-3) contains the temperature implicitly in the two symbols V_T and I_o. The theoretical[4] variation of I_o with T is 8 percent/°C for silicon and 11 percent/°C for germanium. The performance of commercial diodes is only approximately consistent with these results. The reason for the discrepancy is that, in a physical diode, there is a component of the reverse saturation current due to leakage over the surface. From experimental data we observe that the reverse saturation current increases approximately 7 percent/°C for both silicon and germanium. Since $(1.07)^{10} \approx 2.0$, we conclude that *the reverse saturation current approximately doubles for every 10°C rise in temperature*. If $I_o = I_{o1}$ at $T = T_1$, then at a temperature T, I_o is given by

$$I_o(T) = I_{o1} \times 2^{(T-T_1)/10} \qquad (2\text{-}5)$$

If the temperature is increased at a fixed voltage, the current increases. However, if we now reduce V, then I may be brought back to its previous value. It is found[4] that for either silicon or germanium (*at room temperature*)

$$\frac{dV}{dT} \approx -2.5 \text{ mV/°C} \qquad (2\text{-}6)$$

in order to maintain a constant value of I. It should also be noted that $|dV/dT|$ decreases with increasing T.

2-5 DIODE RESISTANCE

The static resistance R of a diode is defined as the ratio V/I of the voltage to the current. At any point on the volt-ampere characteristic of the diode (Fig. 2-5), the resistance R is equal to the reciprocal of the slope of a line joining the operating point to the origin. The static resistance varies widely with V and I and is not a useful parameter. The rectification property of a diode is indicated on the manufacturer's specification sheet (Appendix B-2) by giving the forward voltage V_F required to attain a given forward current I_F and also the reverse current I_R at a given reverse voltage V_R. Typical values for the silicon 1N4153 diode are $V_F = 0.8$ V at $I_F = 10$ mA (corresponding to $R_F = 80\ \Omega$) and $I_R = 0.05\ \mu A$ at $V_R = 50$ V (corresponding to $R_R = 1,000$ M).

For small-signal operation the *dynamic*, or *incremental, resistance r* is an important parameter, and is defined as the reciprocal of the slope of the volt-ampere characteristic, $r \equiv dV/dI$. The dynamic resistance is not a constant, but depends upon the operating voltage. For example, for a semiconductor diode, we find from Eq. (2-3) that the dynamic conductance $g \equiv 1/r$ is

$$g \equiv \frac{dI}{dV} = \frac{I_o \epsilon^{V/\eta V_T}}{\eta V_T} = \frac{I + I_o}{\eta V_T} \qquad (2\text{-}7)$$

For a reverse bias greater than a few tenths of a volt (so that $|V/\eta V_T| \gg 1$), g is extremely small and r is very large. On the other hand, for a forward bias greater than a few tenths of a volt, $I \gg I_o$, and r is given approximately by

$$r \approx \frac{\eta V_T}{I} \qquad (2\text{-}8)$$

The dynamic resistance varies inversely with current; at room temperature and for $\eta = 2$, $r = 52/I$, where I is in milliamperes and r in ohms. For a forward

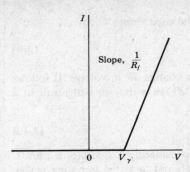

Figure 2-7 The piecewise linear characterization of a semiconductor diode.

current of 52 mA, the dynamic resistance is 1 Ω. The ohmic body resistance of the semiconductor may be of the same order of magnitude or even much higher than this value. Although r varies with current, in a small-signal model, it is reasonable to use the parameter r as a constant.

A Piecewise Linear Diode Characteristic

A large-signal approximation which often leads to a sufficiently accurate engineering solution is the *piecewise linear* representation. For example, the piecewise linear approximation for a semiconductor diode characteristic is indicated in Fig. 2-7. The break point is not at the origin, and hence V_γ is also called the *offset*, or *threshold*, *voltage*. The diode behaves like an open circuit if $V < V_\gamma$, and has a constant incremental resistance $r = dV/dI$ if $V > V_\gamma$. Note that the resistance r (also designated as R_f and called the *forward resistance*) takes on added physical significance even for this large-signal model, whereas the static resistance $R_F = V/I$ is not constant and is not useful.

The numerical values V_γ and R_f to be used depend upon the type of diode and the contemplated voltage and current swings. For example, from Fig. 2-5 we find that, for a current swing from cutoff to 10 mA with a germanium diode, reasonable values are $V_\gamma = 0.2$ V and $R_f = 20$ Ω, and for a silicon diode, $V_\gamma = 0.6$ V and $R_f = 15$ Ω. On the other hand, a better approximation for current swings up to 50 mA leads to the following values; germanium, $V_\gamma = 0.3$ V, $R_f = 6$ Ω; silicon, $V_\gamma = 0.65$ V, $R_f = 5.5$ Ω. For an avalanche diode, discussed in Sec. 2-9, $V_\gamma = V_Z$ and R_f is the dynamic resistance in the breakdown region.

2-6 SPACE-CHARGE, OR TRANSITION, CAPACITANCE[3] C_T

As mentioned in Sec. 2-1, a reverse bias causes majority carriers to move away from the junction, thereby uncovering more immobile charges. Hence the thickness of the space-charge layer at the junction increases with reverse voltage. This increase in uncovered charge with applied voltage may be considered a capacitive effect. We may define an incremental capacitance C_T by

$$C_T = \left| \frac{dQ}{dV} \right| \qquad (2\text{-}9)$$

where dQ is the increase in charge caused by a change dV in voltage. It follows from this definition that a change in voltage dV in a time dt will result in a current $i = dQ/dt$, given by

$$i = C_T \frac{dV}{dt} \qquad (2\text{-}10)$$

Therefore a knowledge of C_T is important in considering a diode (or a transistor) as a circuit element. The quantity C_T is referred to as the *transition-region*, *space-charge*, *barrier*, or *depletion-region*, *capacitance*. We now consider C_T quantitatively. As it turns out, this capacitance is not a constant, but depends upon the magnitude of the reverse voltage. It is for this reason that C_T is defined by Eq. (2-9) rather than as the ratio Q/V

A Step-graded Junction

Consider a junction in which there is an abrupt change from acceptor ions on one side to donor ions on the other side. Such a junction is formed experimentally, for example, by placing indium, which is trivalent, against n-type germanium and heating the combination to a high temperature for a short time. Some of the indium dissolves into the germanium to change the germanium from n to p type at the junction. Such a step-graded junction is called an *alloy*, or *fusion*, *junction*. A step-graded junction is also formed between emitter and base of an integrated transistor (Fig. 4-7). It is not necessary that the concentration N_A of acceptor ions equal the concentration N_D of donor impurities. As a matter of fact, it is often advantageous to have an unsymmetrical junction. Figure 2-8 shows the charge density as a function of distance from an alloy junction in which the acceptor impurity density is assumed to be much larger than the donor concentration. Since the net charge must be zero, then

$$N_A W_p = N_D W_n \qquad (2\text{-}11)$$

If $N_A \gg N_D$, then $W_p \ll W_n \approx W$. The relationship between potential and charge density is given by Eq. (2-1):

$$\frac{d^2V}{dx^2} = \frac{-qN_D}{\epsilon} \qquad (2\text{-}12)$$

The electric lines of flux start on the positive donor ions and terminate on the negative acceptor ions. Hence there are no flux lines to the right of the boundary $x = W_n$ in Fig. 2-8 and $\mathcal{E} = -dV/dx = 0$ at $x = W_n \approx W$. Integrating Eq. (2-12) subject to this boundary condition yields

$$\frac{dV}{dx} = \frac{-qN_D}{\epsilon}(x - W) = -\mathcal{E} \qquad (2\text{-}13)$$

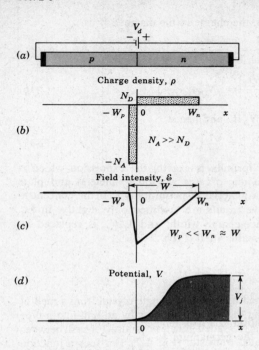

Figure 2-8 (a) A reverse-biased p-n step-graded junction. (b) The charge density. (c) The field intensity. (d) The potential variation with distance x.

Neglecting the small potential drop across W_p, we may arbitrarily choose $V = 0$ at $x = 0$. Integrating Eq. (2-13) subject to this condition gives

$$V = \frac{-qN_D}{2\epsilon}(x^2 - 2Wx) \qquad (2\text{-}14)$$

The linear variation in field intensity and the quadratic dependence of potential upon distance are plotted in Fig. 2-8c and d. These graphs should be compared with the corresponding curves of Fig. 2-1.

At $x = W$, $V = V_j =$ junction, or barrier, potential. Thus

$$V_j = \frac{qN_D W^2}{2\epsilon} \qquad (2\text{-}15)$$

In this section we have used the symbol V to represent the potential at any distance x from the junction. Hence, let us introduce V_d as the externally applied diode voltage. Since the barrier potential represents a reverse voltage, it is lowered by an applied forward voltage. Thus

$$V_j = V_o - V_d$$

where V_d is a negative number for an applied *reverse* bias and V_o is the contact potential (Fig. 2-1d). This equation and Eq. (2-15) confirms our qualitative conclusion that the thickness of the depletion layer increases with applied voltage. We now see that W varies as $V_j^{1/2} = (V_o - V_d)^{1/2}$.

If A is the area of the junction, the charge in the distance W is

$$Q = qN_DWA$$

The transition capacitance C_T, given by Eq. (2-9), is

$$C_T = \left| \frac{dQ}{dV_d} \right| = qN_DA \left| \frac{dW}{dV_j} \right| \tag{2-16}$$

From Eq. (2-15), $|dW/dV_j| = \epsilon/qN_DW$, and hence

$$C_T = \frac{\epsilon A}{W} \tag{2-17}$$

It is interesting to note that this formula is exactly the expression which is obtained for a parallel-plate capacitor of area A (square meters) and plate separation W (meters) containing a material of permittivity ϵ. If the concentration N_A is not neglected, the above results are modified only slightly. In Eq. (2-15) W represents the total space-charge width, and $1/N_D$ is replaced by $1/N_A + 1/N_D$. Equation (2-17) remains valid (Prob. 2-20).

A Linearly Graded Junction

A second form of junction is obtained by drawing a single crystal from a melt of silicon whose type is changed during the drawing process by adding first p-type and then n-type impurities. A linearly graded junction is also formed between the collector and base of an integrated transistor (Fig. 4-7). For such a junction the charge density varies gradually (almost linearly), as indicated in Fig. 2-9. If an analysis similar to that given above is carried out for such a junction, Eq. (2-17) is found to be valid where W equals the total width of the space-charge layer. However, it now turns out that W varies as $V_j^{1/3}$ instead of $V_j^{1/2}$ (Prob. 2-25).

Varactor Diodes

We observe from the above equations that the barrier capacitance is not a constant but varies with applied voltage. The larger the reverse voltage, the larger is the space-charge width W, and hence the smaller the capacitance C_T.

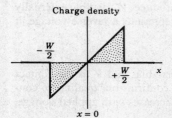

Figure 2-9 The charge-density variation versus distance at a linearly graded p-n junction.

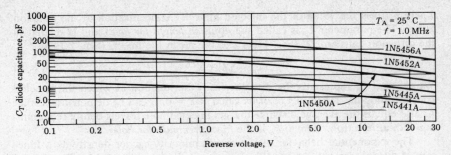

Figure 2-10 Diode transition capacitance C_T versus reverse voltage for diodes, types 1N5441A through 1N5456A. *(Courtesy of Motorola Semiconductor Products, Inc.)*

The variation is illustrated for typical diodes in Fig. 2-10. Such diodes exhibit a capacitance ratio of greater than 3 to 1 as the reverse voltage varies from 2 to 30 V. Values of C_T in the range from approximately 7 to 100 pF (with a series ohmic resistance of 1Ω) are available at a reverse bias of 5 V for the diodes of Fig. 2-10. Diodes made especially for the above applications which are based on the voltage-variable capacitance are called *varactors*, *varicaps*, or *voltacaps*.

The voltage-variable capacitance of a *p-n* junction biased in the reverse direction is useful in a number of circuits. One of these applications is voltage tuning of an LC resonant circuit. Other applications include self-balancing bridge circuits and special types of amplifiers, called *parametric amplifiers*.

In circuits intended for use with fast waveforms or at high frequencies, it is required that the transition capacitance be as small as possible, for the following reason: a diode is driven to the reverse-biased condition when it is desired to prevent the transmission of a signal. However, if the barrier capacitance C_T is large enough, the current which is to be restrained by the low conductance of the reverse-biased diode will flow through the capacitor (Sec. 2-13).

2-7 MINORITY-CARRIER STORAGE IN A DIODE

If the voltage across a diode is applied in the forward direction, the potential barrier at the junction is lowered and holes from the *p* side enter the *n* region. Similarly, electrons from the *n* type move into the *p* side. Define p_n as *the hole concentration in the n-type semiconductor*. If the small value of the thermally generated hole concentration is designated by p_{no}, then the *injected*, or *excess*, hole concentration p'_n is defined by $p'_n \equiv p_n - p_{no}$. As the holes diffuse into the *n* side, they encounter a plentiful supply of electrons and recombine with them. Hence, $p_n(x)$ decreases with the distance *x* into the *n* material. It is found that the excess hole density falls off exponentially with *x*:

$$p'_n(x) = p'_n(0)\epsilon^{-x/L_p} = p_n(x) - p_{no} \tag{2-18}$$

where $p_n'(0)$ is the value of the injected minority concentration at the junction $x = 0$. The parameter L_p is called the *diffusion length for holes* and is related to the diffusion constant D_p (Sec. 1-11) and the mean lifetime τ_p (Sec. 1-7) by

$$L_p = (D_p \tau_p)^{1/2} \tag{2-19}$$

We see that L_p represents the distance from the junction at which the injected concentration has fallen to $1/\epsilon$ of its value at $x = 0$. It can be demonstrated that L_p also equals the average distance that an injected hole travels before recombining with an electron. Hence, L_p is the *mean free path for holes*.

The exponential behavior of the excess minority-carrier density as a function of distance on either side of the junction is shown in Fig. 2-11a. The shaded area under the curve in the n-type (p-type) is proportional to the injected hole (electron) charge. Note that n_p denotes the electron concentration in the p-type material at a distance x from the junction and $n_p(0)$ is the value of this density at $x = 0$.

The Law of the Junction

In Sec. 2-2 it is pointed out that a forward bias V lowers the barrier height and allows more carriers to cross the junction. Hence $p_n(0)$ must be a function of V. From the Boltzmann relationship, Eq. (1-41), it seems reasonable that $p_n(0)$ should depend exponentially, upon V. It is found that

$$p_n(0) = p_{no} \epsilon^{V/V_T} \tag{2-20}$$

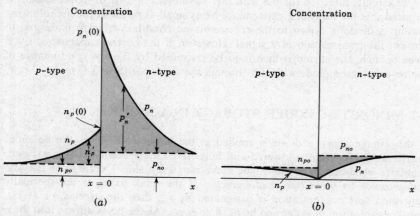

Figure 2-11 Minority-carrier density distribution as a function of the distance x from a junction. (a) A forward-biased junction; (b) a reverse-biased junction. The excess hole (electron) density $p_n' = p_n - p_{no}$ ($n_p' = n_p - n_{po}$) is positive in (a) and negative in (b). (The transition region is assumed to be so small relative to the diffusion length that it is not indicated in this figure.) Also note that these sketches are not drawn to scale since $p_n(0) \gg p_{no}$.

This relationship, which gives the hole concentration at the edge of the n region (at $x = 0$, just outside of the transition region) in terms of the thermal-equilibrium minority-carrier concentration p_{no} (far away from the junction) and the applied potential V, is called the *law of the junction*. A similar equation with p and n interchanged gives the electron concentration at the edge of the p region in terms of V.

Charge Storage under Reverse Bias

When an external voltage reverse-biases the junction, the steady-state density of minority carriers is as shown in Fig. 2-11b. Far from the junction the minority carriers are equal to their thermal-equilibrium values p_{no} and n_{po}, as is also the situation in Fig. 2-11a. As the minority carriers approach the junction they are rapidly swept across, and the density of minority carriers diminishes to zero at this junction. This result follows from the law of the junction, Eq. (2-20), since the concentration $p_n(0)$ reduces to zero for a negative junction potential V.

The injected charge under reverse bias is given by the shaded area in Fig. 2-11b. This charge is negative since it represents less charge than is available under conditions of thermal equilibrium with no applied voltage.

Diffusion Currents

From Eq. (1-33) it follows that the hole diffusion current $I_p(0)$ crossing the junction under forward bias is proportional to the slope at the origin of the p_n curve in Fig. 2-11a. The corresponding electron diffusion current $I_n(0)$ is proportional to the slope at the origin of the n_p curve in Fig. 2-11a. Theoretically[5] it can be shown that the minority-carrier drift current crossing the junction is negligible compared with the minority-carrier diffusion current. Hence, $I_p(0)$ represents the total current of holes moving from left to right across the junction, and $I_n(0)$ is the total current of electrons traveling from right to left across the junction. Therefore, the total diode current I is the sum of these two currents, or

$$I = I_p(0) + I_n(0) \tag{2-21}$$

The reverse saturation hole (electron) current is proportional to the slope at $x = 0$ of the $p_n(n_p)$ curves in Fig. 2-11b. The total reverse saturation current is the sum of these two currents and is negative.

Charge Control Description of a Diode

For simplicity of discussion we assume that one side of the diode, say, the p material, is so heavily doped in comparison with the n side that the current I crossing the junction is due entirely to holes moving from the p to the n side, or

$I = I_p(0)$. From Eq. (1-33)

$$I_p(x) = -AqD_p \frac{dp_n}{dx} = \frac{AqD_p p_n'(0)}{L_p} \epsilon^{-x/L_p} \qquad (2\text{-}22)$$

where Eq. (2-18) is used for $p_n(x)$. The hole current I is given by $I_p(x)$ in Eq. (2-22), with $x = 0$, or

$$I = \frac{AqD_p p'(0)}{L_p} \qquad (2\text{-}23)$$

The excess minority charge Q exists only on the n side, and is given by the shaded area in the n region of Fig. 2-11a multiplied by the diode cross section A and the electronic charge q. Hence, from Eq. (2-18),

$$Q = \int_0^\infty Aqp'(0)\epsilon^{-x/L_p} \, dx = AqL_p p'(0) \qquad (2\text{-}24)$$

Eliminating $p'(0)$ from Eqs. (2-23) and (2-24) yields

$$I = \frac{Q}{\tau} \qquad (2\text{-}25)$$

where $\tau \equiv L_p^2/D_p \equiv \tau_p$ = mean lifetime for holes [Eq. (2-19)].

Equation (2-25) is an important relationship, referred to as the *charge-control description of a diode*. It states that the diode current (which consists of holes crossing the junction from the p to the n side) is proportional to the stored charge Q of excess minority carriers. The factor of proportionality is the reciprocal of the decay time constant (the mean lifetime τ) of the minority carriers. Thus, in the steady state, *the current I supplies minority carriers at the rate at which these carriers are disappearing because of the process of recombination.*

The charge-control characterization of a diode describes the device in terms of the current I and the stored charge Q, whereas the equivalent-circuit characterization uses the current I and the junction voltage V. One immediately apparent advantage of this charge-control description is that the exponential relationship between I and V is replaced by the linear dependence of I on Q. The charge Q also makes a simple parameter, the sign of which determines whether the diode is forward- or reverse-biased. The diode is forward-biased if Q is positive and reverse-biased if Q is negative.

2-8 DIFFUSION CAPACITANCE

For a forward bias a capacitance which is much larger than the transition capacitance C_T considered in Sec. 2-6 comes into play. The origin of this larger capacitance lies in the injected charge stored near the junction outside the transition region (Fig. 2-11a). It is convenient to introduce an incremental capacitance, defined as the rate of change of injected charge with voltage, called the *diffusion*, or *storage*, *capacitance* C_D.

Static Derivation of C_D

We now make a quantitative study of C_D. From Eqs. (2-25) and (2-7)

$$C_D \equiv \frac{dQ}{dV} = \tau \frac{dI}{dV} = \tau g = \frac{\tau}{r} \qquad (2\text{-}26)$$

where the diode incremental conductance is $g \equiv dI/dV$. Substituting the expression for the diode incremental resistance $r = 1/g$ given in Eq. (2-8) into Eq. (2-26) yields

$$C_D = \frac{\tau I}{\eta V_T} \qquad (2\text{-}27)$$

We see that *the diffusion capacitance is proportional to the current I*. In the derivation above we have assumed that the diode current I is due to holes only. If this assumption is not satisfied, Eq. (2-26) gives the diffusion capacitance C_{D_p} due to holes only, and a similar expression can be obtained for the diffusion capacitance C_{D_n} due to electrons. The total diffusion capacitance can then be obtained as the sum of C_{D_p} and C_{D_n} (Prob. 2-27).

For a reverse bias, g is very small and C_D may be neglected compared with C_T. For a forward current, on the other hand, C_D is usually much larger than C_T. For example, for silicon ($\eta = 2$) at $I = 26$ mA, $C_D = \tau/2$. If, say, $\tau = 20$ μs, then $C_D = 10$ μF, a value which is about a million times larger than the transition capacitance.

Despite the large value of C_D, the time constant rC_D (which is of importance in circuit applications) may not be excessive because the dynamic forward resistance $r = 1/g$ is small. From Eq. (2-26),

$$rC_D = \tau \qquad (2\text{-}28)$$

Hence the diode time constant equals the mean lifetime of minority carriers, which lies in range of nanoseconds to hundreds of microseconds.

2-9 BREAKDOWN DIODES[3]

The reverse-voltage characteristic of a semiconductor diode, including the breakdown region, is redrawn in Fig. 2-12a. Diodes which are designed with adequate power-dissipation capabilities to operate in the breakdown region may be employed as voltage-reference or constant-voltage devices. Such diodes are known as *avalanche*, *breakdown*, or *Zener diodes*. They are used as voltage regulators (Sec. 10-4) to keep the load voltage essentially constant at the value V_Z, independent of variations in load current or supply voltage. The diode will continue to regulate until the circuit operation requires the diode current to fall to I_{ZK}, in the neighborhood of the knee of the diode volt-ampere curve. The upper limit on diode current is determined by the power-dissipation rating of the diode.

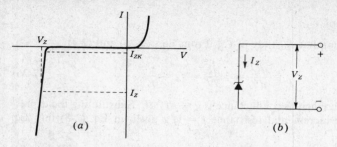

Figure 2-12 (*a*) The volt-ampere characteristic of an avalanche, or Zener, diode. (*b*) The symbol used for a breakdown diode.

Avalanche Multiplication

Two mechanisms of diode breakdown for increasing reverse voltage are recognized. Consider the following situation: A thermally generated carrier (part of the reverse saturation current) falls down the junction barrier and acquires energy from the applied potential. This carrier collides with a crystal ion and imparts sufficient energy to disrupt a covalent bond. In addition to the original carrier, a new electron-hole pair has now been generated. These carriers may also pick up sufficient energy from the applied field, collide with another crystal ion, and create still another electron-hole pair. Thus each new carrier may, in turn, produce additional carriers through collision and the action of disrupting bonds. This cumulative process is referred to as *avalanche multiplication*. It results in large reverse currents, and the diode is said to be in the region of *avalanche breakdown*.

Zener Breakdown

Even if the initially available carriers do not acquire sufficient energy to disrupt bonds, it is possible to initiate breakdown through a direct rupture of the bonds. Because of the existence of the electric field at the junction, a sufficiently strong force may be exerted on a bound electron by the field to tear it out of its covalent bond. The new hole-electron pair which is created increases the reverse current. Note that this process, called *Zener breakdown*, does not involve collisions of carriers with the crystal ions (as does avalanche multiplication).

The field intensity $\mathcal{E}$ increases as the impurity concentration increases, for a fixed applied voltage (Prob. 2-28). It is found that Zener breakdown occurs at a field of approximately 2×10^7 V/m. This value is reached at voltages below about 6 V for heavily doped diodes. For lightly doped diodes the breakdown voltage is higher, and avalanche multiplication is the predominant effect. Nevertheless, the term *Zener* is commonly used for the *avalanche*, or *breakdown*, *diode* even at higher voltages. Silicon diodes operated in avalanche breakdown are available with maintaining voltages from several volts to several hundred volts and with power ratings up to 50 W.

Temperature Characteristics

A matter of interest in connection with Zener diodes, as with semiconductor devices generally, is their temperature sensitivity. The temperature coefficient is given as the percentage change in reference voltage per centigrade degree change in diode temperature. These data are supplied by the manufacturer. The coefficient may be either positive or negative and will normally be in the range $\pm$ 0.1 percent/°C. If the reference voltage is above 6 V, where the physical mechanism involved is avalanche multiplication, the temperature coefficient is positive. However, below 6 V, where true Zener breakdown is involved, the temperature coefficient is negative.

Dynamic Resistance and Capacitance

A matter of importance in connection with Zener diodes is the slope of the diode volt-ampere curve in the operating range. If the reciprocal slope $\Delta V_Z/\Delta I_Z$, called the *dynamic resistance*, is r, then a change ΔI_Z in the operating current of the diode produces a change $\Delta V_Z = r\Delta I_Z$ in the operating voltage. Ideally, $r = 0$, corresponding to a volt-ampere curve which, in the breakdown region, is precisely vertical. The variation of r at various currents for a series of avalanche diodes of fixed power-dissipation rating and various voltages show a rather broad minimum in the range 6 to 10 V. This minimum value of r is of the order of magnitude of a few ohms. However, for values of V_Z below 6 V or above 10 V, and particularly for small currents ($\sim$ 1 mA), r may be of the order of hundreds of ohms.

Some manufacturers specify the minimum current I_{ZK} (Fig. 2-12a) below which the diode should not be used. Since this current is on the knee of the above curve, where the dynamic resistance is large, then for currents lower than I_{ZK} the regulation will be poor. Some diodes exhibit a very sharp knee even down into the microampere region.

The capacitance across a breakdown diode is the transition capacitance, and hence varies inversely as some power of the voltage. Since C_T is proportional to the cross-sectional area of the diode, high-power avalanche diodes have very large capacitances. Values of C_T from 10 to 10,000 pF are common.

Temperature-compensated Reference Diodes

These devices consist of a reverse-biased Zener diode with a positive temperature coefficient, combined in a single package with a forward-biased diode whose temperature coefficient is negative. As an example, the Motorola 1N8241 silicon 6.2-V reference diode has a temperature coefficient of ± 0.0005 percent/°C at 7.5 mA over the range -55 to $+100$°C. The dynamic resistance is only 10 Ω. The voltage stability with time of some of these reference diodes is comparable with that of conventional standard cells.

When a high-voltage reference is required, it is usually advantageous (except of course with respect to economy) to use two or more diodes in series rather

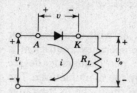

Figure 2-13 The basic diode circuit. The anode (the p side) of the diode is marked A, and the cathode (the n side) is labeled K.

than a single diode. This combination will allow higher voltage, higher dissipation, lower temperature coefficient, and lower dynamic resistance.

2-10 THE DIODE AS A CIRCUIT ELEMENT

The basic diode circuit, indicated in Fig. 2-13, consists of the device in series with a load resistance R_L and an input-signal source v_i. This circuit is now analyzed to find the instantaneous current i and the instantaneous diode voltage v, when the instantaneous input voltage is v_i.

The Load Line

From Kirchhoff's voltage law (KVL),†

$$v = v_i - iR_L \tag{2-29}$$

where R_L is the magnitude of the load resistance. This one equation is not sufficient to determine the two unknowns v and i in this expression. However, a second relation between these two variables is given by the static characteristic of the diode (Fig. 2-5). In Fig. 2-14 is indicated the simultaneous solution of Eq. (2-29) and the diode characteristic. The straight line, which is represented by Eq. (2-29), is called the *load line*. The load line passes through the points $i = 0$, $v = v_i$, and $i = v_i/R_L$, $v = 0$. That is, the intercept with the voltage axis is v_i, and with the current axis is v_i/R_L. The slope of this line is determined, therefore, by R_L; the negative value of the slope is equal to $1/R_L$. The point of intersection A of the load line and the static curve gives the current i_A that will flow under these conditions. This construction determines the current in the circuit and the diode voltage v_A when the instantaneous input potential is v_i.

The Transfer Characteristic

The curve which relates the output voltage v_o to the input v_i of any circuit is called the *transfer*, or *transmission*, *characteristic*. Since in Fig. 2-14 $v_o = iR_L$,

---

† Summary of the elementary circuit theory which is used in the analysis of the electronic circuits discussed in this book is given in Appendix C.

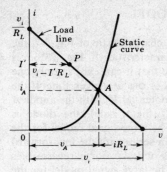

Figure 2-14 The intersection A of the load line with the diode static characteristic gives the current i_A corresponding to an instantaneous input voltage v_i.

then for this particular circuit the transfer curve is a plot of $i_A R_L$ versus v_i, where i_A is the current corresponding to v_i (Fig. 2-14).

It must be emphasized that, regardless of the shape of the static volt ampere characteristic or the waveform of the input signal, the resultant output wave-shape can always be found graphically (at low frequencies) from the transfer curve.

2-11 THE LOAD-LINE CONCEPT

We now show that the use of the load-line construction allows the graphical analysis of many circuits involving devices which are much more complicated than the p-n diode. The external circuit at the output of almost all devices consists of a dc (constant) supply voltage V in series with a load resistance R_L, as indicated in Fig. 2-15. Since KVL applied to this output circuit yields

$$v = V - iR_L \qquad (2\text{-}30)$$

we once again have a straight-line relationship between output current i and output (device) voltage v. The load line passes through the point $i = 0$, $v = V$ and has a slope equal to $-1/R_L$ *independently of the device characteristics.* A p-n junction diode or an avalanche diode possesses a single volt-ampere characteristic at a given temperature. However, most other devices must be described by a family of curves.

The volt-ampere characteristics of a transistor are discussed in the following chapter. The output circuit is identical with that in Fig. 2-15, and the graphical analysis begins with the construction of the load line.

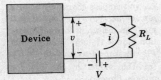

Figure 2-15 The output circuit of most devices consists of a supply voltage V in series with a load resistance R_L.

2-12 THE PIECEWISE LINEAR DIODE MODEL

If the reverse resistance R_r is included in the diode characteristic of Fig. 2-7, the piecewise linear and continuous volt-ampere characteristic of Fig. 2-16a is obtained. The diode is a *binary* device, in the sense that it can exist in only one of two possible states; that is, the diode is either ON or OFF at a given time. If the voltage applied across the diode exceeds the cutin potential V_γ with the anode A (the p side) more positive than the cathode K (the n side), the diode is forward-biased and is said to be in the ON state. The large-signal model for the ON state is indicated in Fig. 2-16b as a battery V_γ in series with the low forward resistance R_f (of the order of a few tens of ohms or less). For a reverse bias ($v < V_\gamma$) the diode is said to be in its OFF state. The large-signal model for the OFF state is indicated in Fig. 2-16c as a large reverse resistance R_r (of the order of several hundred kilohms or more).

Analysis of Diode Circuits Using the Piecewise Linear Model

Consider a circuit containing several diodes, resistors, supply voltages, and sources of excitation. A general method of analysis of such a circuit consists in assuming (guessing) the state of each diode. For the ON state, replace the diode by a battery V_γ in series with a forward resistance R_f, and for the OFF state replace the diode by the reverse resistance R_r (which can usually be taken as infinite), as indicated in Fig. 2-16b and c. After the diodes have been replaced by these piecewise linear models, the entire circuit is linear and the currents and voltages everywhere can be calculated using Kirchhoff's voltage and current laws. The assumption that a diode is ON can then be verified by observing the sign of the current through it. If the current is in the forward direction (from anode to cathode), the diode is indeed ON and the initial guess is justified. However, if the current is in the reverse direction (from cathode to anode), the assumption that the diode is ON has been proved incorrect. Under this circumstance the analysis must begin again with the diode assumed to be OFF.

Analogous to the above trial-and-error method, we test the assumption that a diode is OFF by finding the voltage across it. If this voltage is either in the

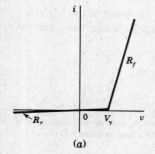

(a)

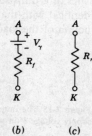

(b)　(c)

Figure 2-16 (a) The piecewise linear volt-ampere characteristic of a p-n diode. (b) The large-signal model in the ON, or forward, direction (anode A more positive than V_γ with respect to the cathode K). (c) The model in the OFF, or reverse, direction ($v < V_\gamma$).

reverse direction or in the forward direction but with a voltage less than V_γ, the diode is indeed OFF. However, if the diode voltage is in the forward direction and exceeds V_γ, the diode must be ON and the original assumption is incorrect. In this case the analysis must begin again by assuming the ON state for this diode.

The above method of analysis will be employed in the study of the diode circuits of the following example and throughout this text.

Example 2-2 Calculate the output voltage v_o in the circuit of Fig. 2-17a for the following values of input voltages. (a) $v_1 = v_2 = 0$; (b) $v_1 = V$, $v_2 = 0$; and (c) $v_1 = v_2 = V$.

SOLUTION (a) If both inputs are 0, the voltage across each diode is 0. Since in order for a diode to conduct, it must be forward-biased by at least the cutin voltage V_γ (Fig. 2-16), neither diode conducts. Hence, all currents are 0 and $v_o = 0$.

(b) Assume $D1$ is ON and $D2$ is OFF. Following the rules given above (with $R_r = \infty$), the network reduces to that given in Fig. 2-17b. Applying KVL to the circuit, we obtain

$$- V + IR_s + V_\gamma + IR_f + IR = 0 \qquad (2\text{-}31)$$

or
$$I = \frac{V - V_\gamma}{R_s + R_f + R}$$

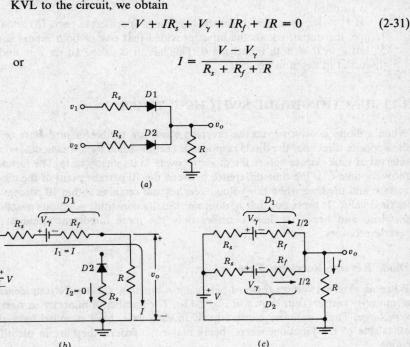

(a)

(b)   (c)

Figure 2-17 (a) An illustrative diode example. The circuit model assuming (b) that $D1$ is ON and $D2$ is OFF and (c) that both diodes are conducting.

and $v_o = IR$. If $R \gg R_s + R_f$, then $v_o \approx V - V_\gamma$. The output voltage is approximately equal to the input voltage minus the cutin diode voltage.

Note that the current is in the forward direction in $D1$ and, hence, the assumption that $D1$ is ON is justified. Also note that the voltage across $D2$ is v_o and is in the reverse direction, which verifies that $D2$ is indeed OFF.

(c) Since v_1 and v_2 are both excited by a voltage V, we assume that each diode conducts so that the equivalent circuit in Fig. 2-17c results. From symmetry the current in each diode has the same value (labeled $I/2$ in the circuit). From KCL (Kirchhoff's current law) the load current is I. From KVL we have

$$-V + \frac{IR_s}{2} + V_\gamma + \frac{IR_f}{2} + IR = 0 \tag{2-32}$$

or

$$I = \frac{V - V_\gamma}{(R_s + R_f)/2 + R}$$

and $v_o = IR \cong V - V_\gamma$ if $R \gg R_s + R_f$. Since the current $I/2$ is in the forward direction through each diode, the assumption that both diodes are ON is justified.

If $V \gg V_\gamma$, then $v_o = V$ for the conditions specified in parts (b) and (c). Hence, the output equals the input provided that one *or* both inputs equal V, but $v_o = 0$, if both inputs are 0. This circuit is called an OR *gate* and is discussed in detail in Sec. 5-2.

2-13 JUNCTION-DIODE SWITCHING TIMES

When a diode is driven from the reversed condition to the forward state or in the opposite direction, the diode response is accompanied by a transient, and an interval of time elapses before the diode recovers to its steady state. The forward recovery time t_{fr} is the time difference between the 10 percent point of the diode voltage and the time when this voltage reaches and remains within 10 percent of its final value. It turns out that t_{fr} does not usually constitute a serious practical problem, and hence we here consider only the more important situation of reverse recovery.

Diode Reverse Recovery Time

When an external voltage forward-biases a *p-n* junction, the steady-state density of minority carriers is as shown in Fig. 2-11a. The number of minority carriers is very large. These minority carriers have, in each case, been supplied from the other side of the junction, where, being majority carriers, they are in plentiful supply.

If the external voltage is suddenly reversed in a diode circuit which has been carrying current in the forward direction, the diode current will not immediately fall to its steady-state reverse-voltage value. The current cannot attain its steady-state value until the minority-carrier distribution, which at the moment of voltage reversal had the form in Fig. 2-11a, reduces to the distribution in Fig. 2-11b. Until such time as the *injected*, or *excess*, *minority-carrier density* $p_n - p_{no}$ (or $n_p - n_{po}$) has dropped nominally to zero, the diode will continue to conduct easily, and the current will be determined by the external resistance in the diode circuit.

Storage and Transition Times

The sequence of events which accompanies the reverse biasing of a conducting diode is indicated in Fig. 2-18. We consider that the voltage in Fig. 2-18b is

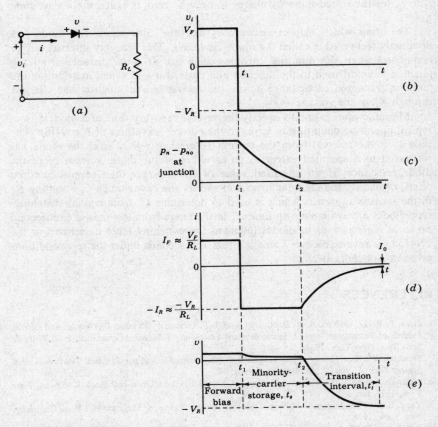

Figure 2-18 The waveform in (b) is applied to the diode circuit in (a); (c) the excess carrier density at the junction; (d) the diode current; (e) the diode voltage.

applied to the diode-resistor circuit in Fig. 2-18a. For a long time, and up to the time t_1, the voltage $v_i = V_F$ has been in the direction to forward-bias the diode. The resistance R_L is assumed large enough so that the drop across R_L is large in comparison with the drop across the diode. Then the current is $i \approx V_F/R_L \equiv I_F$. At the time $t = t_1$ the input voltage reverses abruptly to the value $v_i = -V_R$. For the reasons described above, the current does not drop to zero, but instead reverses and remains at the value $i \approx -V_R/R_L \equiv -I_R$ until the time $t = t_2$. At $t = t_2$, as is seen in Fig. 2-18c, the minority-carrier density p_n at $x = 0$ has reached its equilibrium state p_{no}. If the diode ohmic resistance is R_d, then at the time t_1 the diode voltage falls slightly [by $(I_F + I_R)R_d$] but does not reverse. At $t = t_2$, when the excess minority carriers in the immediate neighborhood of the junction have been swept back across the junction, the diode voltage begins to reverse and the magnitude of the diode current begins to decrease. The interval t_1 to t_2, for the stored-minority charge to become zero, is called the *storage time* t_s.

The time which elapses between t_2 and the time when the diode has nominally recovered is called the *transition time* t_t. This recovery interval will be completed when the minority carriers which are at some distance from the junction have diffused to the junction and crossed it and when, in addition, the junction transition capacitance across the reverse-biased junction has charged through R_L to the voltage $-V_R$.

Manufacturers normally specify the reverse recovery time of a diode t_{rr} in a typical operating condition in terms of the current waveform of Fig. 2-18d. The time t_{rr} is the interval from the current reversal at $t = t_1$ until the diode has recovered to a specified extent in terms either of the diode current or of the diode resistance. If the specified value of R_L is larger than several hundred ohms, ordinarily the manufacturers will specify the capacitance C_L shunting R_L in the measuring circuit which is used to determine t_{rr}. Commercial switching-type diodes are available with times t_{rr} in the range from less than a nanosecond up to as high as 1 μs in diodes intended for switching large currents. For the 1N4153 the reverse recovery time is a few nanoseconds under the test conditions given in Appendix B-2.

REFERENCES

1. Gray, P. E., D. DeWitt, A. R. Boothroyd, and J. F. Gibbons: "Physical Electronics and Circuit Models of Transistors," vol. 2, Semiconductor Electronics Education Committee, John Wiley & Sons, Inc., New York, 1964.
   Shockley, W.: The Theory of p-n Junctions in Semiconductor and p-n Junction Transistors, *Bell System Tech. J.*, vol. 28, pp. 435-489, July, 1949.
2. Millman, J. and C. C. Halkias: "Integrated Electronics," McGraw-Hill Book Company, New York, 1972, sec. 3-3.
3. Yang, E. S.: "Fundamentals of Semiconductor Devices," chap. 4, McGraw-Hill Book Company, New York, 1978.
4. Ref. 2, sec. 19-11.
5. Ref. 2, sec 2-11.

REVIEW QUESTIONS

2-1 Consider an open-circuited p-n junction. Sketch curves as a function of distance across the junction of space charge, electric field, and potential.

2-2 (a) What is the order of magnitude of the space-charge width at a p-n junction?

(b) What does this space charge consist of—electrons, holes, neutral donors, neutral acceptors, ionized donors, ionized acceptors, etc.?

2-3 (a) For a reverse-biased diode, does the transition region increase or decrease in width?

(b) What happens to the junction potential?

2-4 Explain why the p-n junction contact potential *cannot* be measured by placing a voltmeter across the diode terminals.

2-5 Explain physically why a p-n diode acts as a rectifier.

2-6 (a) Write the volt-ampere equation for a p-n diode.

(b) Explain the meaning of each symbol.

2-7 Plot the volt-ampere curves for germanium and silicon to the same scale, showing the cutin value for each.

2-8 (a) How does the reverse saturation current of a p-n diode vary with temperature?

(b) How does the diode voltage (at constant current) vary with temperature?

2-9 How does the dynamic resistance r of a diode vary with (a) current; (b) temperature?

(c) What is the order of magnitude of r for silicon at room temperature and for a dc current of 1 mA?

2-10 (a) Sketch the piecewise linear characteristic of a diode.

(b) What are the approximate cutin voltages for silicon and germanium?

2-11 Consider a step-graded p-n junction with equal doping on both sides of the junction ($N_A = N_D$). Sketch the charge density, field intensity, and potential as a function of distance from the junction for a reverse bias.

2-12 (a) How does the transition capacitance C_T vary with the depletion-layer width?

(b) With the applied reverse voltage?

(c) What is the order of magnitude of C_T?

2-13 What is a *varactor diode*?

2-14 Plot the minority-carrier concentration as a function of distance from a p-n junction in the n side only for (a) a forward-biased junction; (b) a negatively biased junction. Indicate the excess concentration and note where it is positive and where negative.

2-15 Under steady-state conditions the diode current is proportional to a charge Q.

(a) What is the physical meaning of the factor of proportionality?

(b) What charge does Q represent—transition layer charge, injected minority-carrier charge, majority-carrier charge, etc.?

2-16 (a) How does the diffusion capacitance C_D vary with dc diode current?

(b) What does the product of C_D and the dynamic resistance of a diode equal?

2-17 (a) Draw the volt-ampere characteristic of an avalanche diode.

(b) What is meant by the *knee* of the curve?

(c) By the dynamic resistance?

(d) By the temperature coefficient?

2-18 Describe the physical mechanism for avalanche breakdown.

2-19 Describe the physical mechanism for Zener breakdown.

2-20 Describe a temperature-compensated reference diode.

2-21 You are given the V-I output characteristic in graphical form of a new device.

(*a*) Sketch the circuit using this device which will require a load-line construction to determine i and v.

(*b*) Is the load line vertical, horizontal, at 135° or 45° for infinite load resistance;

(*c*) for zero load resistance?

2-22 (*a*) Draw the piecewise linear volt-ampere characteristic of a *p-n* diode.

(*b*) What is the circuit model for the ON state; (*c*) the OFF state?

2-23 Consider a circuit consisting of a diode D, a resistance R, and a signal source v_i in series. Define (*a*) static characteristic; (*b*) *transfer*, or *transmission*, characteristic.

2-24 What is meant by the *minority-carrier storage time* of a diode?

2-25 A diode in series with a resistor R_L is forward-biased by a voltage V_F. After a steady state is reached, the input changes to $-V_R$. Sketch the current as a function of time. Explain qualitatively the shape of this curve.

BIPOLAR TRANSISTOR CHARACTERISTICS

The physical behavior of a semiconductor triode, called a *bipolar junction transistor* (BJT), is given. The volt-ampere characteristics of this device are studied. It is demonstrated that the transistor is capable of producing amplification. For the transistor operating in either the active region or in saturation, the method of analysis for obtaining the currents and voltages is explained. Typical voltage values are given for the several possible modes of operation.

3-1 THE JUNCTION TRANSISTOR[1]

A junction transistor consists of a silicon (or germanium) crystal in which a layer of *n*-type silicon is sandwiched between two layers of *p*-type silicon. Alternatively, a transistor may consist of a layer of *p*-type between two layers of *n*-type material. In the former case the transistor is referred to as a *p-n-p* transistor, and in the latter case, as an *n-p-n* transistor. The semiconductor sandwich is extremely small, and is hermetically sealed against moisture inside a metal or plastic case. Manufacturing techniques and constructional details for several transistor types are described in Sec. 3-4.

The two types of transistor are represented in Fig. 3-1a. The representations employed when transistors are used as circuit elements are shown in Fig. 3-1b. The three portions of a transistor are known as *emitter*, *base*, and *collector*. The arrow on the emitter lead specifies the direction of current flow when the emitter-base junction is biased in the forward direction. In *both* cases, however, the emitter, base, and collector currents, I_E, I_B, and I_C, respectively, are assumed

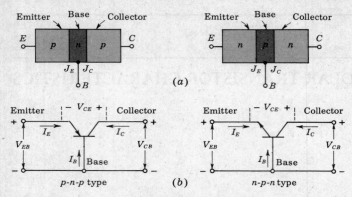

Emitter Base Collector

Emitter Base Collector

J_E J_C

J_E J_C

p-n-p type

n-p-n type

(a)

(b)

Figure 3-1 (a) A p-n-p and an n-p-n transistor. The emitter (collector) junction is $J_E(J_C)$. (b) Circuit representation of the two transistor types.

positive when the currents flow *into* the transistor. The symbols V_{EB}, V_{CB}, and V_{CE} are the emitter-base, collector-base, and collector-emitter voltages, respectively. (More specifically, V_{EB} represents the voltage *drop* from emitter to base.)

Open-circuited Transistor

If no external biasing voltages are applied, all transistor currents must be zero. The potential barriers at the junctions adjust to the contact difference of potential V_o—given in Fig. 2-1d (a few tenths of a volt)—required so that no free carriers cross each junction. If we assume a completely symmetrical junction (emitter and collector regions having identical physical dimensions and doping concentrations), the barrier height is identical at the emitter junction J_E and at the collector junction J_C, as indicated in Fig. 3-2a. The narrow space-charge regions at the junctions have been neglected.

Under open-circuited conditions, the minority concentration is constant within each section and is equal to its thermal-equilibrium value, n_{po} in the p-type emitter and collector regions and p_{no} in the n-type base, as shown in Fig. 3-2b. Since the transistor may be looked upon as a p-n diode followed by an n-p diode, much of the theory developed in Chap. 2 for the junction diode will be

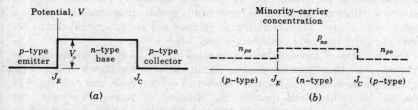

(a)

(b)

Figure 3-2 (a) The potential and (b) the minority-carrier density in each section of an open-circuited symmetrical p-n-p transistor.

used to explain the physical behavior of the transistor, when voltages are applied so as to disturb it from the equilibrium situation pictured in Fig. 3-2.

The Transistor Biased in the Active Region

We may now begin to appreciate the essential features of a transistor as an active circuit element by considering the situation depicted in Fig. 3-3a. Here a p-n-p transistor is shown with voltage sources which serve to bias the emitter-base junction in the forward direction and the collector-base junction in the reverse direction. The potential variation through the biased transistor is indicated in Fig. 3-3b. The dashed curve applies to the case before the application of external biasing voltages (Fig. 3-2a), and the solid curve to the case after the biasing voltages are applied. The externally applied voltages appear, essentially, across the junctions. Hence, as shown in Fig. 3-3b, the forward biasing of the emitter junction lowers the emitter-base potential barrier by $|V_{EB}|$, whereas the reverse biasing of the collector junction increases the collector-base potential barrier by $|V_{CB}|$. The lowering of the emitter-base barrier permits minority-carrier injection; that is, holes are injected into the base, and electrons are injected into the emitter region. The excess holes diffuse across the n-type base, where the electric field intensity $\mathcal{E}$ is zero, to the collector junction. At J_C the

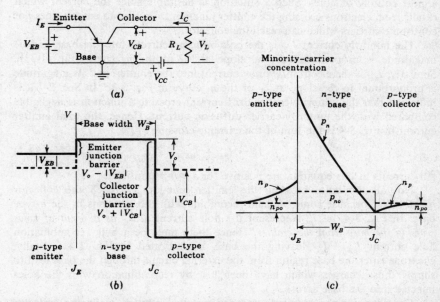

Figure 3-3 (a) A p-n-p transistor biased in the active region (the emitter is forward-biased and the collector is reverse-biased). (b) The potential variation through the transistor. The narrow depletion regions at the junctions are negligibly small. (c) The minority-carrier concentration in each section of the transistor. It is assumed that the emitter is much more heavily doped than the base.

field is positive and large ($\mathcal{E} = - dV/dx \gg 0$), and hence holes are accelerated across this junction. In other words, the holes which reach J_C fall down the potential barrier, and are therefore *collected* by the collector. Since the applied potential across J_C is negative, then from the law of the junction, Eq. (2-20), p_n is reduced to zero at the collector as shown in Fig. 3-3c. Similarly, the reverse collector-junction bias reduces the collector electron density n_p to zero at J_C. The minority-carrier-density curves pictured in Fig. 3-3c should be compared with the corresponding concentration plots for the forward- and reverse-biased p-n junction given in Fig. 2-11.

3-2 TRANSISTOR CURRENT COMPONENTS

In Fig. 3-4 we show the various current components which flow across the forward-biased emitter junction and the reverse-biased collector junction. The emitter current I_E consists of hole current I_{pE} (holes crossing from emitter into base, designated *forward injection*) and electron current I_{nE} (electrons crossing from base into the emitter, designated *reverse injection*). In a commercial transistor the doping of the emitter is made much larger than the doping of the base. This feature ensures (in a p-n-p transistor) that the emitter current consists almost entirely of holes. Such a situation is desirable since the current which results from electrons crossing the emitter junction from base to emitter does not contribute carriers which can reach the collector.

The minority current I_{pE} is the hole *diffusion* current into the base and its magnitude is proportional to the slope at J_E of the p_n curve [Eq. (1-33)]. Similarly, I_{nE} is the electron *diffusion* current into the emitter, and its magnitude is proportional to the slope at J_E of the n_p curve in Fig. 3-3c. In Sec. 2-7 it is indicated that the minority-carrier drift current crossing a junction is negligible compared with the minority-carrier diffusion current. Hence, the total emitter current in Fig. 3-4 is the sum of the currents crossing J_E,

$$I_E = I_{pE} + I_{nE} \tag{3-1}$$

All currents in this equation are positive for a p-n-p transistor.

Not all the holes crossing the emitter junction J_E reach the collector junction J_C, because some of them recombine with the electrons in the n-type base. In Fig. 3-4, let I_{pC} represent the hole current at J_C *as a result of holes crossing the base from the emitter*. Hence there must be a bulk recombination hole current $I_{pE} - I_{pC}$ leaving the base, as indicated in Fig. 3-4 (actually, electrons enter the base region from the external circuit through the base lead to supply those charges which have been lost by recombination with the holes injected into the base across J_E).

Consider, for the moment, an open-circuited emitter, while the collector junction remains reverse-biased. Then I_C must equal the reverse saturation current I_{CO} of the back-biased diode at J_C. This *reverse* current consists of two components, as shown in Fig. 3-4, I_{nCO} consisting of electrons moving from the

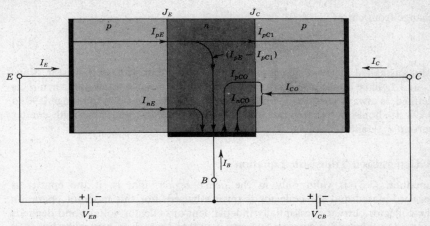

Figure 3-4 Transistor current components for a forward-biased emitter junction and a reversed-biased collector junction. If a current has a subscript $p(n)$, it consists of holes (electrons) moving in the same (opposite) direction as the arrow indicating the current direction.

p to the n region across J_C and a term, I_{pCO}, resulting from holes crossing from n to p across J_C.

$$- I_{CO} = I_{nCO} + I_{pCO} \tag{3-2}$$

(The minus sign is chosen arbitrarily so that I_C and I_{CO} will have the same sign.) The magnitude of I_{nCO} is proportional to the slope at J_C of the n_p distribution in Fig. 3-3c.

Since $I_E = 0$ under open-circuit conditions, no holes are injected across J_E, and, hence, none can reach J_C from the emitter. Clearly, I_{pCO} results from the small concentration of holes generated thermally within the base.

Now let us return to the situation depicted in Fig. 3-4, where the emitter is forward-biased. Now the total hole current crossing J_C is the sum of I_{pC} and I_{pCO}, and its magnitude is proportional to the slope at J_C of the p_n distribution in Fig. 3-3c. The complete collector current is given by

$$I_C = I_{CO} - I_{pC} = I_{CO} - \alpha I_E \tag{3-3}$$

where α is defined as the fraction of the total emitter current [given in Eq. (3-1)] which represents holes which have traveled from the emitter across the base to the collector. For a p-n-p transistor, I_E is positive and both I_C and I_{CO} are negative, which means that the current in the collector lead is in the direction opposite to that indicated by the arrow of I_C in Fig. 3-4. For an n-p-n transistor these currents are reversed.

Large-Signal Current Gain α

From Eq. (3-3) it follows that α may be defined as the ratio of the negative of the collector-current increment from cutoff ($I_C = I_{CO}$) to the emitter-current change from cutoff ($I_E = 0$), or

$$\alpha \equiv -\frac{I_C - I_{CO}}{I_E - 0} \tag{3-4}$$

Alpha is called the *large-signal current gain* of a common-base transistor. Since I_C and I_E have opposite signs (for either a *p-n-p* or a *n-p-n* transistor), then α, as defined, is always positive. Typical numerical values of α lie in the range 0.90 to 0.998. It should be pointed out that α is not a constant, but varies with emitter current I_E, collector voltage V_{CB}, and temperature.

A Generalized Transistor Equation

Equation (3-3) is valid only in the *active region*, that is, if the emitter is forward-biased and the collector is reverse-biased. For this mode of operation the collector current is essentially independent of collector voltage and depends only upon the emitter current. Suppose now that we seek to generalize Eq. (3-3) so that it may apply not only when the collector junction is substantially reverse-biased, but also for any voltage across J_C. To achieve this generalization we need only replace I_{CO} by the current in a *p-n* diode (that consisting of the base and collector regions). This current is given by the volt-ampere relationship of Eq. (2-3), with I_o replaced by $-I_{CO}$ and V by V_C, where the symbol V_C represents the drop across J_C from the *p* to the *n* side. The complete expression for I_C for any V_C and I_E is

$$I_C = -\alpha I_E + I_{CO}(1 - \epsilon^{V_C/V_T}) \tag{3-5}$$

Note that if V_C is negative and has a magnitude large compared with V_T, Eq. (3-5) reduces to Eq. (3-3). The physical interpretation of Eq. (3-5) is that the *p-n* junction diode current crossing the collector junction is augmented by the fraction α of the current I_E flowing in the emitter.

3-3 THE TRANSISTOR AS AN AMPLIFIER

A load resistor R_L is in series with the collector supply voltage V_{CC} of Fig. 3-3a. A small voltage change ΔV_i between emitter and base causes a relatively large emitter-current change ΔI_E. We define by the symbol α' that fraction of this current change which is collected and passes through R_L, or $\Delta I_C = \alpha' \Delta I_E$. The change in output voltage across the load resistor

$$\Delta V_L = -R_L \Delta I_C = -\alpha' R_L \Delta I_E \tag{3-6}$$

may be many times the change in input voltage ΔV_i. Under these circumstances, the voltage amplification $A \equiv \Delta V_L / \Delta V_i$ will be greater than unity, and the transistor acts as an amplifier. If the dynamic resistance of the emitter junction is r_e, then $\Delta V_i = r_e \Delta I_E$, and

$$A \equiv -\frac{\alpha' R_L \Delta I_E}{r_e \Delta I_E} = -\frac{\alpha' R_L}{r_e} \tag{3-7}$$

From Eq. (2-7), $r_e = 52/I_E$, where I_E is the quiescent emitter current in milliamperes. For example, if $r_e = 40$ Ω (for $I_E = 1.3$ mA), $\alpha' = -1$, and $R_L = 3,000$ Ω, $A = +75$. This calculation is oversimplified, but in essence it is correct and gives a physical explanation of why the transistor acts as an amplifier. The transistor provides power gain as well as voltage or current amplification. From the foregoing explanation it is clear that current in the low-resistance input circuit is transferred to the high-resistance output circuit. The word "transistor," which originated as a contraction of "transfer resistor," is based upon the above physical picture of the device.

The Parameter α'

The parameter α' introduced above is defined as the ratio of the change in the collector current to the change in the emitter current at constant collector-to-base voltage and is called the *negative of the small-signal short-circuit current transfer ratio, or gain*. More specifically,

$$\alpha' \equiv \frac{\Delta I_C}{\Delta I_E}\bigg|_{V_{CB}} \tag{3-8}$$

On the assumption that α is independent of I_E, then from Eq. (3-3) it follows that $\alpha' = -\alpha$.

3-4 TRANSISTOR CONSTRUCTION[2, 3]

Four basic techniques have been developed for the manufacture of diodes, transistors, and other semiconductor devices. Consequently, such devices may be classified into one of the following types: grown, alloy, diffusion, or epitaxial.

Grown Type

The *n-p-n* grown-junction transistor is illustrated in Fig. 3-5a. It is made by drawing a single crystal from a melt of silicon or germanium whose impurity concentration is changed during the crystal-drawing operation by adding *n*- or *p*-type atoms as required.

Alloy Type

This technique, also called the *fused* construction, is illustrated in Fig. 3-5b for a *p-n-p* transistor. The center (base) section is a thin wafer of *n*-type material. Two small dots of indium are attached to opposite sides of the wafer, and the whole structure is raised for a short time to a high temperature, above the melting point of indium but below that of germanium. The indium dissolves the germanium beneath it and forms a saturation solution. On cooling, the germanium contact with the base material recrystallizes, with enough indium concentration to change it from *n* to *p* type. The collector is made larger than

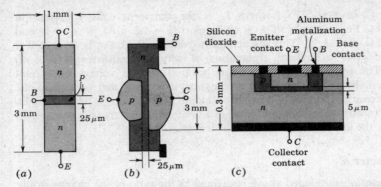

Figure 3-5 Construction of transistors. (*a*) Grown (*n-p-n*), (*b*) alloy (*p-n-p*), and (*c*) diffused planar (*n-p-n*) types. (The dimensions are approximate, and the figures are not drawn to scale.)

the emitter, so that the collector subtends a large angle as viewed from the emitter. Because of this geometrical arrangement, almost all of the emitter current follows a diffusion path toward the collector rather than to the base.

Diffusion Type

This technique consists of subjecting a semiconductor wafer to gaseous diffusions of both *n*- and *p*-type impurities to form both the emitter and the collector junctions. A *planar* silicon transistor of the diffusion type is illustrated in Fig. 3-5*c*. In this process (described in greater detail in Chap. 4 on integrated-circuit fabrication), the base-collector junction area is determined by a diffusion mask. The emitter is then diffused on the base through a different mask. A thin layer of silicon dioxide is grown over the entire surface and photoetched, so that aluminum contacts can be made for the emitter and base leads (Fig. 3-5*c*). Because of the passivating action of this oxide layer, most surface problems are avoided and very low leakage currents result. There is also an improvement in the current gain at low currents and in the noise figure.

Epitaxial Type

The epitaxial technique (Sec. 4-3) consists of growing a very thin, high-purity, single-crystal layer of silicon on a heavily doped substrate of the same material. This augmented crystal forms the collector on which the base and emitter may be diffused (Fig. 4-2). All microelectronic bipolar transistors are fabricated in this manner (Chap. 4).

The foregoing techniques may be combined to form a large number of methods for constructing transistors. For example, there are *diffused-alloy* types, *grown-diffused* devices, *alloy-emitter–epitaxial-base* transistors, etc. The special features of transistors of importance at high frequencies are discussed in Chap. 13. The volt-ampere characteristics at low frequencies of all types of junction transistors are essentially the same, and the discussion to follow applies to them all.

Finally, because of its historical significance, let us mention the first type of transistor to be invented. This device consists of two sharply pointed tungsten wires pressed against a semiconductor wafer. However, the reliability and reproducibility of such point-contact transistors are very poor, and as a result these transistors are no longer of practical importance.

3-5 THE COMMON-BASE (CB) CONFIGURATION

In Fig. 3-3*a*, a *p-n-p* transistor is shown in a *grounded-base* configuration. This circuit is also referred to as a *common-base*, or CB, configuration, since the base is common to the input and output circuits. For a *p-n-p* transistor the largest current components are due to holes. Since holes flow from the emitter to the collector and down toward ground out of the base terminal, then, referring to the polarity conventions of Fig. 3-1, we see that I_E is positive, I_C is negative, and I_B is negative. For a forward-biased emitter junction, V_{EB} is positive, and for a reverse-biased collector junction, V_{CB} is negative. For an *n-p-n* transistor all current and voltage polarities are the negative of those for a *p-n-p* transistor. In summary, *in the active region, for either an n-p-n or a p-n-p transistor, the current in the emitter is positive in the direction of the arrow on the emitter lead. Also, the sign of I_B is the same as that of I_C and opposite to that of I_E.*

The Output Characteristics

From Eq. (3-5) we see that the output (collector) current I_C is completely determined by the input (emitter) current I_E and the output (collector-to-base) voltage $V_{CB} = V_C$. This output relationship is given in Fig. 3-6 for a typical *p-n-p* silicon transistor and is a plot of collector current I_C versus collector-to-base voltage drop V_{CB}, with emitter current I_E as a parameter. The curves of Fig. 3-6 are known as the *output*, or *collector*, *static characteristics*.

A qualitative understanding of the form of the output characteristics is not difficult if we consider the fact that the transistor consists of two diodes placed in series "back to back" (with the two cathodes connected together). In the active region the input diode (emitter-to-base) is biased in the forward direction so that I_E is positive and hence I_C is negative. Note, as in Fig. 3-6, that it is customary to plot along the abscissa and to the right that polarity of V_{CB} which reverse-biases the collector junction even if this polarity is negative. If $I_E = 0$, the collector current is $I_C = I_{CO}$. For other values of I_E, the output-diode reverse current is augmented by the fraction of the input-diode forward current which reaches the collector.

The curves indicate that when the collector-to-base voltage exceeds the output-junction *cutin* or *threshold* voltage ($V_{CB} > +0.6$ V), the collector diode conducts and I_C increases rapidly (I_C becomes less negative). In this region the

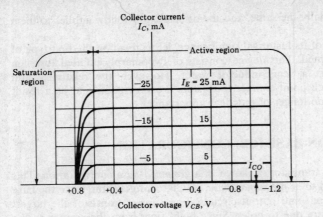

Figure 3-6 The CB output characteristics of the 2N2907A p-n-p silicon transistor around zero collector voltage. The characteristics are almost horizontal for $V_{CB} < 0.6\ V$.

transistor is said to be *in saturation*. Note that I_C and I_{CO} are negative for a p-n-p transistor and positive for an n-p-n transistor.

The Input Characteristics

These curves are plots of emitter-to-base voltage V_{EB} versus emitter current I_E (or vice-versa), with collector-to-base voltage V_{CB} as a parameter. This set of curves is referred to as the *input*, or *emitter*, static characteristics. The input characteristics of Fig. 3-7 represent simply the forward characteristic of the

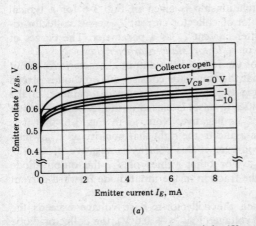

(a)

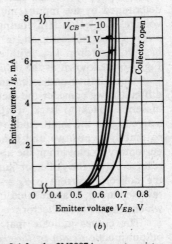

(b)

Figure 3-7 (a) Common-base input characteristics (V_{EB} versus I_E) for the 2N2907A p-n-p transistor. (b) The same characteristics plotted with V_{EB} horizontal and I_E vertical. Note the similarity to a diode curve.

emitter-to-base diode for various collector voltages. A noteworthy feature of the input characteristics is that there exists a *cutin*, *offset*, or *threshold*, voltage V_γ below which the emitter current is very small. In general, V_γ is approximately 0.5 V for silicon transistors (Fig. 3-7) and 0.1 V for germanium. We digress now to discuss a phenomenon which is used to account for the shapes of the transistor characteristics.

The Early Effect, or Base-Width Modulation[4]

In Fig. 3-3 the narrow space-charge regions in the neighborhood of the junctions are neglected. This restriction is now to be removed. From Eq. (2-15) we note that the width W of the depletion region of a diode increases with the magnitude of the reverse voltage. Since the emitter junction is forward-biased but the collector junction is reverse-biased in the active region, then in Fig. 3-3 the barrier width at J_E is negligible compared with the space-charge width W in the base at J_C.

The transition region at a junction is the region of uncovered charges on both sides of the junction at the positions occupied by the impurity atoms. As the reverse voltage applied across the junction increases, the transition region penetrates deeper into the collector and base. If the metallurgical base width is W_B, then the *effective electrical base width is* $W'_B = W_B - W$, *where W is the depletion width at* J_C. This modulation of the effective base width by the collector voltage is known as the *Early effect*. The decrease in W'_B with increasing reverse collector voltage has three consequences: First, there is less chance for recombination within the base region. Hence α increases with increasing $|V_{CB}|$. Second, the concentration gradient of minority carriers p_n is increased within the base. Since the hole current injected across the emitter is proportional to the gradient of p_n at J_E, then I_E increases with increasing reverse collector voltage. Third, for extremely large voltages, W'_B may be reduced to zero, causing voltage breakdown in the transistor. This phenomenon of *punch-through* is discussed further in Sec. 3-12.

The shape of the input characteristics can be understood if we consider the fact that an increase in magnitude of collector voltage will, by the Early effect, cause the emitter current to increase, with V_{EB} held constant. Thus the curves shift as $|V_{CB}|$ increases, as noted in Fig. 3-7. The curve with the collector open represents the characteristic of the forward-biased emitter diode. Compare with the silicon curve in Fig. 2-5.

Active Region

In this region *the collector junction is biased in the reverse direction and the emitter junction in the forward direction*. Consider first that the emitter current is zero. Then the collector current is small and equals the reverse saturation current I_{CO} (nanoamperes for silicon and microamperes for germanium) of the collector junction considered as a diode. Suppose now that a forward emitter current I_E is caused to flow in the emitter circuit. Then a fraction $-\alpha I_E$ of this current will reach the collector, and I_C is therefore given by Eq. (3-3). In the active region, the collector current is essentially independent of collector voltage and depends only upon the emitter current. However, because of the Early effect, we note in Fig. 3-6 that there actually is a small (perhaps 0.5 percent) increase in $|I_C|$ with $|V_{CB}|$. Because α is less than, but almost equal to, unity, the magnitude of the collector current is (slightly) less than that of the emitter current.

Saturation Region

The region to the left of the ordinate, $V_{CB} \approx 0.6$ V, and above the $I_E = 0$ characteristic, in which *both emitter and collector junctions are forward-biased*, is called the *saturation* region. The forward biasing of the collector accounts for the large change in collector current with small changes in collector voltage. For a forward bias, I_C increases exponentially with voltage according to the diode relationship [Eq. (2-3)]. A forward bias means that the collector p material is made positive with respect to the base n side, and hence that hole current flows from the p side across the collector junction to the n material. This hole flow corresponds to a positive change in collector current. Hence the collector current increases rapidly, as indicated in Fig. 3-6.

Cutoff Region

The characteristic for $I_E = 0$ passes through the origin, but is otherwise similar to the other characteristics. This characteristic is not coincident with the voltage axis, though the separation is difficult to show because I_{CO} is only a few nanoamperes for silicon. The region below the $I_E = 0$ characteristic, for which the *emitter and collector junctions are both reverse-biased*, is referred to as the *cutoff* region. The temperature characteristics of I_{CO} are discussed in Sec. 3-7.

3-6 THE COMMON-EMITTER (CE) CONFIGURATION

Most transistor circuits have the emitter, rather than the base, as the terminal common to both input and output. Such a *common-emitter* (CE), or *grounded-emitter*, configuration is indicated in Fig. 3-8. In the common-emitter (as in the common-base) configuration, the input current and the output voltage are taken as the independent variables, whereas the input voltage and output current are the dependent variables.

The physical operation of a transistor is somewhat easier to understand if reference is made to a *p-n-p* rather than to an *n-p-n* device. Hence, the preceding CB discussion centered around the *p-n-p* transistor. However, since most designs using bipolar transistors are of the *n-p-n* silicon type, we now concentrate on the CE characteristics of the *industry standard n-p-n-type* 2N2222A discrete transistor. [Integrated-circuit (IC) transistors, discussed in the next chapter, are also predominantly of the *n-p-n* variety.]

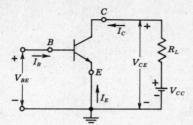

Figure 3-8 An n-p-n transistor common-emitter configuration.

The p-n-p 2N2907A and the n-p-n 2N2222A are complementary transistors. They have (almost) identical characteristics except that the signs of the currents and voltages of one are the negative of the other. For example, if in Fig. 3-6 I_C is made positive, I_E negative, and V_{CB} positive to the right of the origin, these curves will represent the CB output characteristics of the 2N2222A.

The Output Characteristics†

In Fig. 3-9 the abscissa is the collector-to-emitter voltage V_{CE}, the ordinate is the collector current I_C, and the curves are given for various values of base current I_B. For a fixed value of I_B, the collector current is not a very sensitive value of V_{CE}. However, the slopes of the curves of Fig. 3-9 are larger than in the common-base characteristics of Fig. 3-6. Observe also that the base current is much smaller than the emitter current. In Fig. 3-9 we have selected $R_L = 500\ \Omega$ and a supply $V_{CC} = 10$ V and have superimposed the corresponding load line on the output characteristics. The method of constructing a load line is identical with that explained in Sec. 2-10 in connection with a diode.

The output characteristic curves may be divided into three regions, just as was done for the CB configuration. The first of these, the *active region*, is discussed here, and the *cutoff* and *saturation regions* are considered in the next two sections.

In the active region *the collector junction is reverse-biased and the emitter junction is forward-biased*. In Fig. 3-9 the active region is the area to the right of the ordinate $V_{CE} =$ a few tenths of a volt and above $I_B = 0$. In this region the transistor output current responds most sensitively to an input signal. If the transistor is to be used as an amplifying device without appreciable distortion, it must be restricted to operate in this region.

The common-emitter characteristics in the active region are readily understood qualitatively on the basis of our earlier discussion of the common-base configuration. From Kirchhoff's current law (KCL) applied to Fig. 3-8, the base current is

$$I_B = -(I_C + I_E) \tag{3-9}$$

---

† Transistor output and input characteristics are no longer supplied by the transistor manufacturer since they are seldom used in either digital or analog design. However, these characteristics are necessary for an understanding of the transistor. The device characteristics shown in this chapter were obtained experimentally.

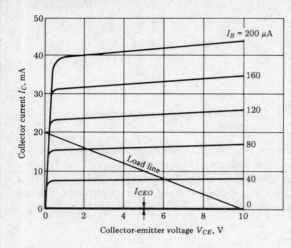

Figure 3-9 Common-emitter output characteristics of a 2N2222A *n-p-n* silicon transistor. A load line corresponding to $V_{CC} = 10$ V and $R_L = 500$ is superimposed.

Combining this equation with Eq. (3-3), we find

$$I_C = \frac{I_{CO}}{1 - \alpha} + \frac{\alpha I_B}{1 - \alpha} \tag{3-10}$$

If we define β by

$$\beta \equiv \frac{\alpha}{1 - \alpha} \tag{3-11}$$

then Eq. (3-10) becomes

$$I_C = (1 + \beta)I_{CO} + \beta I_B \tag{3-12}$$

Note that usually $I_B \gg I_{CO}$, and hence $I_C \approx \beta I_B$ in the active region.

If α were truly constant, then, according to Eq. (3-10), I_C would be independent of V_{CE} and the curves of Fig. 3-9 would be horizontal. Assume that, because of the Early effect, α increases by only one-tenth of 1 percent, from 0.995 to 0.996, as $|V_{CE}|$ increases from a few volts to 10 V. Then the value of β increases from $0.995/(1 - 0.995) = 200$ to $0.996/(1 - 0.996) = 250$, or about 25 percent. This numerical example illustrates that a very small change (0.1 percent) in α is reflected in a very large change (25 percent) in the value of β. It should also be clear that a slight change in α has a large effect on β, and hence upon the common-emitter curves. Therefore the common-emitter characteristics are normally subject to a wide variation even among transistors of a given type.

The Input Characteristics

In Fig. 3-10*a* the abscissa is the base current I_B, the ordinate is the base-to-emitter voltage V_{BE}, and the curves are given for various values of collector-to-emitter voltage V_{CE}. In Fig. 3-10*b* the axes are interchanged. We observe that,

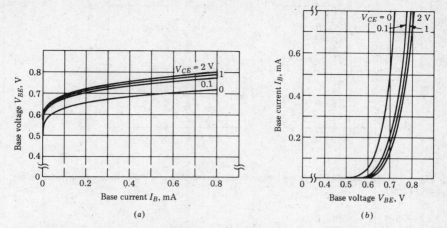

Figure 3-10 (a) Common-emitter input characteristics (V_{BE} versus I_B) for the 2N2222A n-p-n transistor. (b) The same characteristics plotted with V_{BE} horizontal and I_B vertical. Note the similarity to a diode curve.

with the collector shorted to the emitter and the emitter forward-biased, the input characteristic is essentially that of a forward-biased diode. If V_{BE} becomes zero, then I_B will be zero, since under these conditions both emitter and collector junctions will be short-circuited. In general, increasing $|V_{CE}|$ with constant V_{BE} causes a decrease in base width W'_B due to the Early effect, and results in a decreasing recombination base current. These considerations account for the shape of input characteristics shown in Fig. 3-10.

The input characteristics for germanium transistors are similar in form to those in Fig. 3-10. The only notable difference in the case of germanium is that the curves break away from zero current in the range 0.1 to 0.2 V, rather than in the range 0.5 or 0.6 V as for silicon. A reasonable value for V_{BE} in the active region is 0.7 V for silicon and 0.2 V for germanium.

Example 3-1 (a) Find the transistor currents in the circuit of Fig. 3-11a. A silicon transistor with $\beta = 100$ and $I_{CO} = 20$ nA $= 2 \times 10^{-5}$ mA is under consideration. (b) Repeat part a if a 2-kΩ emitter resistor is added to the circuit, as in Fig. 3-11b.

SOLUTION (a) Since the base is forward-biased, the transistor is not cut off. Hence it must be either in its active region or in saturation. Assume that the transistor operates in the active region. From KVL applied to the base circuit of Fig. 3-11a (with I_B expressed in milliamperes), we have

$$- 5 + 200\,I_B + V_{BE} = 0$$

As noted above, a reasonable value for V_{BE} is 0.7 V in the active region, and

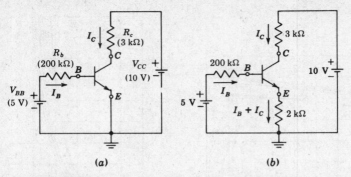

Figure 3-11 An example illustrating how to determine whether or not a transistor is operating in the active region.

hence

$$I_B = \frac{5 - 0.7}{200} = 0.0215 \text{ mA}$$

Since $I_{CO} \ll I_B$, then $I_C \approx \beta I_B = 2.15$ mA.

We must now justify our assumption that the transistor is in the active region, by verifying that the collector junction is reverse-biased. From KVL applied to the collector circuit we obtain

$$- 10 + 3\,I_C + V_{CB} + V_{BE} = 0$$

or

$$V_{CB} = 10 - (3)(2.15) - 0.7 = +2.85 \text{ V}$$

For an n-p-n device a positive value of V_{CB} represents a reverse-biased collector junction, and hence the transistor is indeed in its active region.

Note that I_B and I_C in the active region are independent of the collector circuit resistance R_c. Hence, if R_c is increased sufficiently above 3 kΩ, then V_{CB} changes from a positive to a negative value, indicating that the transistor is no longer in its active region. The method of calculating I_B and I_C when the transistor is in saturation is given in Sec. 3-9.

(b) The current in the emitter resistor of Fig. 3-11b is

$$I_B + I_C \approx I_B + \beta I_B = 101\,I_B$$

assuming $I_{CO} \ll I_B$. Applying KVL to the base circuit yields

$$- 5 + 200 I_B + 0.7 + (2)(101\,I_B) = 0$$

or

$$I_B = 0.0107 \text{ mA} \qquad I_C = 100\,I_B = 1.07 \text{ mA}$$

Note that $I_{CO} = 2 \times 10^{-5}$ mA $\ll I_B$, as assumed.

To check for active circuit operation, we calculate V_{CB}. Thus

$$V_{CB} = -3\,I_C + 10 - (2)(101\,I_B) - 0.7$$

$$= - (3)(1.07) + 10 - (2)(101)(0.0107) - 0.7 = +3.93 \text{ V}$$

Since V_{CB} is positive, this (n-p-n) transistor is in its active region.

3-7 THE CE CUTOFF CURRENTS

Cutoff in a transistor is given by the condition $I_E = 0$ and $I_C = I_{CO}$. It is important to note that a transistor is *not* at cutoff if the base is open-circuited. From Eqs. (3-9) and (3-10), if $I_B = 0$, then $I_E = - I_C$ and

$$I_C = - I_E = \frac{I_{CO}}{1 - \alpha} \equiv I_{CEO} \qquad (3\text{-}13)$$

The actual collector current with collector junction reverse-biased and base open-circuited is designated by the symbol I_{CEO}. Since, even in the neighborhood of cutoff, α may be as large as 0.9 for germanium, then $I_C \approx 10\, I_{CO}$ at zero base current. Accordingly, in order to cut off the transistor, it is not enough to reduce I_B to zero. Instead, it is necessary to reverse-bias the emitter junction slightly. It is found[5] that a reverse-biasing voltage of the order of 0.1 V established across the emitter junction will ordinarily be adequate to cut off a germanium transistor. In silicon, at collector currents of the order of I_{CO}, α is very nearly zero because of recombination[6, 7] in the emitter-junction transition region. Hence, even with $I_B = 0$, we find, from Eq. (3-13), that $I_C = I_{CO} = - I_E$, so that the transistor is very close to cutoff. In silicon, cutoff occurs[5] at $V_{BE} \approx 0$ V corresponding to a base short-circuited to the emitter. *In summary, cutoff means that* $I_E = 0$, $I_C = I_{CO}$, $I_B = - I_C = - I_{CO}$, *and* V_{BE} *is a reverse voltage whose magnitude is of the order of* 0 V *for a silicon and* 0.1 V *for a germanium transistor.*

The Reverse Collector Saturation Current I_{CBO}

The collector current in a physical transistor (a real, nonidealized, or commercial device) when the emitter current is zero is designated by the symbol I_{CBO}. Two factors cooperate to make $|I_{CBO}|$ larger than $|I_{CO}|$. First, there exists a leakage current which flows, not through the junction, but around it and across the surfaces. The leakage current is proportional to the voltage across the junction. The second reason why $|I_{CBO}|$ exceeds $|I_{CO}|$ is that new carriers may be generated by collision in the collector-junction transition region, leading to avalanche multiplication of current and eventual breakdown. But even before breakdown is approached, this *multiplication* component of current may attain considerable proportions (Fig. 3-17).

At 25°C, I_{CBO} for a silicon transistor whose power dissipation is in the range of some hundreds of milliwatts is of the order of nanoamperes. Under similar conditions a germanium transistor has an I_{CBO} in the range of microamperes. The temperature sensitivity of I_{CBO} is the same as that of the reverse saturation current I_O of a *p-n* diode (Sec. 2-4). Specifically, it is found that I_{CBO} approximately doubles for every 10°C increase in temperature for both Si and Ge. However, because of the lower absolute value of I_{CBO} in silicon, these transistors may be used up to a junction temperature of about 200°C, whereas germanium transistors are limited to about 100°C.

In addition to the change of reverse saturation current with temperature, there may also be a wide variability (by a factor of 100 or more) of I_{CBO} among samples of a given transistor type. Accordingly, the manufacturers specification sheets (Appendix B-3) lists the maximum value of I_{CBO}. A low-power silicon transistor is considered "leaky" if I_{CBO} exceeds 10 nA at 25°C.

Emitter Junction Reverse-biased Currents

If V_{EB} is reverse-biased so as to reduce the collector current to zero, the *emitter cutoff current* is designated by I_{EBO}. For a specified value of V_{CE} and $V_{EB\text{(OFF)}}$ (reverse-biased) the *collector cutoff current* and the *base cutoff current* are designated by I_{CEX} and I_{BL}, respectively. The maximum values of these currents are also listed in the specification sheets, and are of the same order of magnitude as I_{CBO}.

3-8 THE CE SATURATION REGION

In the saturation region *the collector junction (as well as the emitter junction) is forward-biased by at least the cutin voltage*. Since the voltage V_{BE} (or V_{BC}) across a forward-biased junction has a magnitude of only a few tenths of a volt, then $V_{CE} = V_{BE} - V_{BC}$ is also only a few tenths of a volt at saturation. Hence, in Fig. 3-9, the saturation region is very close to the zero-voltage axis, where all the curves merge and fall rapidly toward the origin. A load line has been superimposed on the characteristics of Fig. 3-9 corresponding to a resistance $R_L = 500\ \Omega$ and a supply voltage of 10 V. We note that in the saturation region the collector current is approximately independent of base current, for given values of V_{CC} and R_L. Hence we may consider that the onset of saturation takes place at the knee of the transistor curves in Fig. 3-9.

Saturation Voltages

We are not able to read the collector-to-emitter saturation voltage, $V_{CE\text{(sat)}}$, with any precision from the plots of Fig. 3-9. We refer instead to the characteristics shown in Fig. 3-12. In these characteristics the 0- to 0.5-V region of Fig. 3-9 has been expanded, and we have superimposed the same load line as before, corresponding to $R_L = 500\ \Omega$. We observe from Figs. 3-9 and 3-12 that V_{CE} and I_C no longer respond appreciably to base current I_B, after the base current has attained the value 0.12 mA. At this current the transistor enters saturation. For $I_B = 0.12$ mA, $|V_{CE}| \approx 0.19$ V. At $I_B = 0.20$ mA, $|V_{CE}|$ has dropped to $|V_{CE}| \approx 0.12$ V. Larger magnitudes of I_B will, of course, decrease $|V_{CE}|$ slightly further.

It is clear from Fig. 3-12 that $V_{CE\text{(sat)}}$ depends somewhat on the values of I_B and I_C. The manufacturer's specification sheets give this dependence in graphical form. For the 2N2222A the values of $V_{CE\text{(sat)}}$ and $V_{BE\text{(sat)}}$ versus I_C for $I_C/I_B = 10$ is indicated in Fig. 3-13.

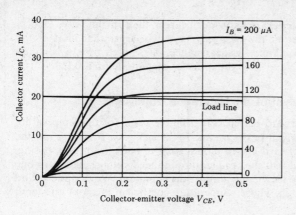

Figure 3-12 Saturation-region common-emitter characteristics of a 2N2222A transistor. A load line corresponding to $V_{CC} = 10$ V and $R_L = 500\ \Omega$ is superimposed.

Saturation Resistance

For a transistor operating in the saturation region, a quantity of interest is the ratio $V_{CE(sat)}/I_C$. This parameter is called the *common-emitter saturation resistance*, variously abbreviated R_{CS}, R_{CES}, or $R_{CE(sat)}$. To specify R_{CS} properly, we must indicate the operating point at which it was determined. For example, from Fig. 3-12, we find that, at $I_C = 20$ mA and $I_B = 0.20$ mA, $R_{CS} \approx 0.12/(20 \times 10^{-3}) = 6\ \Omega$. The usefulness of R_{CS} stems from the fact, as appears in Fig. 3-12, that to the left of the knee each of the plots, for fixed I_B, may be approximated, at least roughly, by a straight line.

The Base-spreading Resistance $r_{bb'}$

Recalling that the base region is very thin (Fig. 3-5), we see that the current which enters the base region across the emitter junction must flow through a long narrow path to reach the base terminal. The cross-sectional area for current

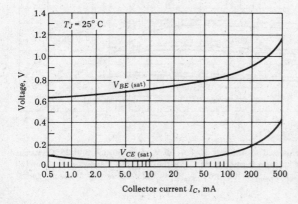

Figure 3-13 Saturation voltages for the 2N2222A silicon transistor versus collector current for $I_C/I_B = 10$. *(Courtesy of Motorola, Inc.)*

flow in the collector (or emitter) is very much larger than in the base. Hence, usually the ohmic resistance of the base is very much larger than that of the collector or emitter. The dc ohmic base resistance, designated by $r_{bb'}$, is called the *base-spreading resistance* and is of the order of magnitude of 100 Ω.

The Temperature Coefficient of the Saturation Voltages

Since both junctions are forward-biased, a reasonable value for the temperature coefficient of $V_{BE(active)}$, $V_{BE(sat)}$ or $V_{BC(sat)}$ is -2.5 mV/°C. In saturation the transistor consists of two forward-biased diodes back to back in series opposing. Hence it is to be anticipated that the temperature-induced voltage change in one junction will be canceled by the change in the other junction. We do indeed find such to be the case for $V_{CE(sat)}$ whose temperature coefficient is about one-tenth that of $V_{BE(sat)}$.

The DC Current Gain h_{FE}

A transistor parameter of interest is the ratio I_C/I_B, where I_C is the collector current and I_B is the base current. This quantity is designated by β_{dc} or h_{FE}, and is known as the (negative of the) *dc beta*, the *dc forward current transfer ratio*, or the *dc current gain*.

In the saturation region, the parameter h_{FE} is a useful number and one which is usually supplied by the manufacturer when a switching transistor is involved. We know $|I_C|$, which is given approximately by V_{CC}/R_L, and a knowledge of h_{FE} tells us the minimum base current (I_C/h_{FE}) which will be needed to saturate the transistor. The parameter $\beta \approx h_{FE}$ is also important for the proper biasing of a discrete transistor in the active region (Sec. 11-3).

The variation of h_{FE} with collector current for the 2N2222A is given in Fig. 3-14 over values of I_C from 0.5 to 500 mA. Although h_{FE} decreases both at small

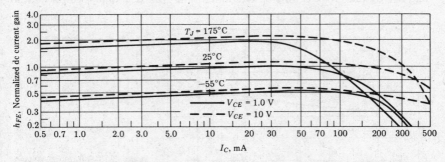

Figure 3-14 Plots of dc gain (normalized to unity at $V_{CE} = 1.0$ V, $I_C = 30$ mA and 25°C) versus collector current at three different junction temperatures T_J for the 2N2222A transistor. At $I_C = 150$ mA, $V_{CE} = 10$ V, and $T_J = 25$°C, $h_{FE(min)} = 100$ and $h_{FE(max)} = 300$. *(Courtesy of Motorola, Inc.)*

and large currents, this parameter is fairly constant over a wide range of values. Note the increase in h_{FE} as V_{CE} increases from 1.0 to 10 V because of the Early effect. In the figure h_{FE} is normalized to unity at $I_C = 30$ mA and $V_{CE} = 1.0$ V at 25°C. The absolute value of this parameter has a wide spread of values even for a transistor of a particular type. For example, the specification sheet for the 2N2222A indicates a minimum value of $h_{FE} = 100$ and a maximum of 300 at $I_C = 150$ mA. Commercially available transistors have values of h_{FE} as high as 1,200 at 0.1 mA for a low-level application (2N5089) and as low as 5 at 30 A for a high-power transistor (2N5301).

The parameter h_{FE} increases with an increase in temperature. From Fig. 3-14 it is found that h_{FE} increases ~ 0.6 percent/°C for the 2N2222A transistor.

3-9   TYPICAL TRANSISTOR—JUNCTION VOLTAGE VALUES

The characteristic plotted in Fig. 3-15 of output current I_C as a function of input voltage V_{BE} for an *n-p-n* silicon transistor is quite instructive and indicates the several regions of operation for a CE transistor circuit. The numerical values indicated are typical values obtained experimentally or from theoretical equations.[5] Let us examine the various portions of the transfer curve of Fig. 3-15.

The Cutoff Region

Cutoff is defined, as in Sec. 3-7, to mean $I_E = 0$ and $I_C = I_{CO}$, and it is found that a *reverse* bias $V_{BE(\text{cutoff})} = 0$ V (0.1 V) will cut off a silicon (germanium) transistor.

What happens if a larger reverse voltage than $V_{BE(\text{cutoff})}$ is applied? It turns out that if V_{BE} is reverse-biased and much larger than V_γ, that the collector current falls slightly below I_{CO} and that the emitter current *reverses* but remains small in magnitude (less than I_{CO}).

Short-circuited Base

Suppose that, instead of reverse-biasing the emitter junction, we connect the base to the emitter so that $V_E = V_{BE} = 0$. As indicated in Fig. 3-15, $I_C \equiv I_{CES}$ does not increase greatly over its cutoff value I_{CO}.

Open-circuited Base

If instead of a shorted base we allow the base to "float" so that $I_B = 0$, we obtain the $I_C \equiv I_{CEO}$ given in Eq. (3-13). At low currents $\alpha \approx 0$ (0.9) for Si (Ge), and hence $I_C \approx I_{CO}$ (10 I_{CO}) for Si (Ge). The values of V_{BE} calculated for this open-base condition ($I_C = -I_E$) are a few tens of millivolts of *forward* bias, as indicated in Fig. 3-15.

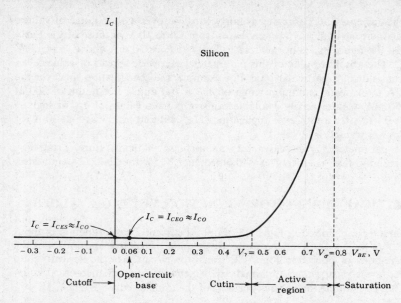

Figure 3-15 Plot of collector current against base-to-emitter voltage for a silicon *n-p-n* transistor. (I_C is not drawn to scale.) A similar plot (with different numerical values, indicated in Table 3-1) is valid for a germanium transistor.

The Cutin Voltage

The volt-ampere characteristic between base and emitter at constant collector-to-emitter voltage (Fig. 3-10) is not unlike the volt-ampere characteristic of a simple junction diode. When the emitter junction is reverse-biased, the base current is very small, being of the order of nanoamperes or microamperes, for silicon and germanium, respectively. When the emitter junction is forward-biased, again, as in the simple diode, no appreciable base current flows until the emitter junction has been forward-biased to the extent where $|V_{BE}| \geqslant |V_\gamma|$, where V_γ is called the *cutin voltage*. Since the collector current is nominally proportional to the base current, no appreciable collector current will flow until an appreciable base current flows. Therefore a plot of collector current against base-to-emitter voltage will exhibit a cutin voltage, just as does the simple diode.

In principle, a transistor is in its active region whenever the base-to-emitter voltage is on the forward-biasing side of the cutoff voltage, which occurs at a reverse voltage of 0 V for silicon and 0.1 V for germanium. In effect, however, a transistor enters its active region when $V_{BE} > V_\gamma$.

We may estimate the cutin voltage V_γ by assuming that $V_{BE} = V_\gamma$ when the collector current reaches, say, 1 percent of the maximum (saturation) current in the CE circuit of Fig. 3-8. Typical values of V_γ are 0.5 V for silicon and 0.1 V for germanium.

Table 3-1 Typical *n-p-n* transistor-junction voltages† at 25°C

Material	$V_{CE(\text{sat})}$	$V_{BE(\text{sat})} \equiv V_{\sigma}$	$V_{BE(\text{active})}$	$V_{BE(\text{cutin})} \equiv V_{\gamma}$	$V_{BE(\text{cutoff})}$
Si	0.2	0.8	0.7	0.5	0.0
Ge	0.1	0.3	0.2	0.1	−0.1

† The temperature variation of these voltages is discussed in Sec. 3-8.

Saturation Voltages

Manufacturers specify saturation values of input and output voltages in a number of different ways, such as in Fig. 3-13. The saturation voltages depend not only on the operating point but also on the semiconductor material and on the type of transistor construction.

Summary

The voltages referred to above and indicated in Fig. 3-15 are summarized in Table 3-1. The entries in the table are appropriate for an *n-p-n* transistor. For a *p-n-p* transistor the signs of all entries should be reversed. Observe that the total range of V_{BE} between cutin and saturation is rather small, being only 0.3 V. The voltage $V_{BE(\text{active})}$ has been located somewhat arbitrarily, but nonetheless reasonably, near the midpoint of the active region in Fig. 3-15.

> **Example 3-2** (*a*) The circuits of Fig. 3-11*a* and *b* are modified by changing the base-circuit resistance from 200 to 50 kΩ (as indicated in Fig. 3-16). If $h_{FE} = 100$, determine whether or not the silicon transistor is in saturation and find I_B and I_C. (*b*) Repeat with the 2 kΩ emitter resistance added.
>
> SOLUTION Assume that the transistor is in saturation. Using the values $V_{BE(\text{sat})}$ and $V_{CE(\text{sat})}$ in Table 3-1, the circuit of Fig. 3-16*a* is obtained.

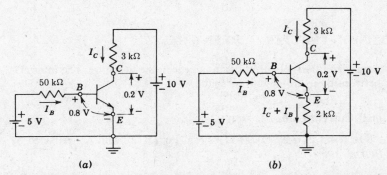

(*a*)   (*b*)

Figure 3-16 An example illustrating how to determine whether or not a transistor is operating in the saturation region.

Applying KVL to the base circuit gives

$$-5 + 50I_B + 0.8 = 0$$

or
$$I_B = \frac{4.2}{50} = 0.0840 \text{ mA}$$

Applying KVL to the collector circuit yields

$$-10 + 3I_C + 0.2 = 0$$

or
$$I_C = \frac{9.8}{3} = 3.267 \text{ mA}$$

The minimum value of base current required for saturation is

$$I_{B(\text{min})} = \frac{I_C}{h_{FE}} = \frac{3.267}{100} = 0.0327 \text{ mA}$$

Since $I_B = 0.0840 < I_{B(\text{min})} = 0.0327$ mA, we have verified that the transistor is in saturation.

(b) If the 2-kΩ emitter resistance is added, the circuit becomes that in Fig. 3-16b. Assume that the transistor is in saturation. Applying KVL to the base and collector circuits, we obtain

$$-5 + 50I_B + 0.8 + 2(I_C + I_B) = 0$$
$$-10 + 3I_C + 0.2 + 2(I_C + I_B) = 0$$

If these simultaneous equations are solved for I_C and I_B, we obtain

$$I_C = 1.96 \text{ mA} \qquad I_B = 0.00550 \text{ mA}$$

Since $I_{B(\text{min})} = I_C/h_{FE} = 0.0196$ mA $> I_B = 0.00550$, the transistor is *not* in saturation. Hence the device must be operating in the active region. Proceeding exactly as we did for the circuit of Fig. 3-11b (but with the 200 kΩ replaced by 50 kΩ), we obtain

$$I_C = 1.71 \text{ mA} \qquad I_B = 0.0171 \text{ mA} = 17.1 \text{ μA} \qquad V_{CB} = 0.735 \text{ V}$$

3-10 COMMON-EMITTER CURRENT GAIN

Three different definitions of current gain appear in the literature. The interrelationships between these are now to be found.

Large-Signal Current Gain β

We define β in terms of α by Eq. (3-11). From Eq. (3-12), with I_{CO} replaced by I_{CBO}, we find

$$\beta = \frac{I_C - I_{CBO}}{I_B - (-I_{CBO})} \tag{3-14}$$

In Sec. 3-7 we define *cutoff* to mean that $I_E = 0$, $I_C = I_{CBO}$, and $I_B = -I_{CBO}$. Consequently, Eq. (3-14) gives the ratio of the collector-current increment to the base-current change from cutoff to I_B, and hence β *represents the* (negative of the) *large-signal current gain of a common-emitter transistor*. This parameter is of primary importance in connection with the biasing and bias stability of discrete transistor circuits, as discussed in Chap. 11.

DC Current Gain h_{FE}

In Sec. 3-8 we define the dc current gain by

$$\beta_{dc} \equiv \frac{I_C}{I_B} \equiv h_{FE} \qquad (3\text{-}15)$$

In that section it is noted that h_{FE} is most useful in connection with determining whether or not a transistor is in saturation. In general, the base current (and hence the collector current) is large compared with I_{CBO}. Under these conditions the large-signal and the dc betas are approximately equal; then $h_{FE} \approx \beta$.

Small-Signal Current Gain h_{fe}

We define β' as the ratio of a collector-current increment ΔI_C for a small base-current change ΔI_B at a given quiescent operating point, (at a fixed collector-to-emitter voltage V_{CE}), or

$$\beta' \equiv \frac{\Delta I_C}{\Delta I_B}\bigg|_{V_{CE}} = h_{fe} \qquad (3\text{-}16)$$

Clearly, β' is (the negative of) the *small-signal* current gain. If β were independent of current, we see from Eq. (3-15) that $\beta' = \beta \approx h_{FE}$. Although Fig. 3-14 indicates that h_{FE} is a function of current, over a wide range of currents it is fairly constant, and h_{fe} differs from h_{FE} by about 20 percent in the active region. It should be emphasized that $h_{fe} \rightarrow 0$ in the saturation region because $\Delta I_C \rightarrow 0$ for a small increment ΔI_B. In the active region h_{fe} is the most important parameter used for calculating the low-frequency amplification of a transistor circuit (Chap. 11).

3-11 INVERTED MODE OF OPERATION

Three normal modes of operation of a transistor have been considered, depending on the polarities of the junction voltages J_E and J_C, as follows:

1. J_E forward-biased and J_C reverse-biased—normal, active.
2. J_E forward-biased and J_C forward-biased—saturation.
3. J_E reverse-biased and J_C reverse-biased—cutoff.

A fourth possibility exists, and is called the *inverted (or reverse) mode* of operation.

4. J_E reverse-biased and J_C forward-biased—inverted, active.

Note that if E and C are interchanged that mode 4 becomes identical with mode 1. In other words, if the collector of a transistor is treated as the emitter and vice-versa, and if this modified transistor is biased in its active region, then the inverted mode of operation results. If a transistor is fabricated symmetrically so that the emitter and collector have the same geometry and impurity doping, then the inverted-mode volt-ampere characteristics are identical with the normal active-mode volt-ampere curves. Such a transistor behaves as a symmetrical bidirectional switch because it can conduct current in either direction through the transistor, depending upon the polarity of the applied voltage.

Usually [Fig. 3-5(*b* or *c*)] the emitter is doped much more heavily than the collector, and the area of the collector is much greater than that of the emitter. This construction results in a high current gain for the normal-active connection and a very low current gain for the inverted operation. Therefore, the latter is used in practice only in special situations (Sec. 9-12), and by far the largest number of applications makes use of the transistor in the normal mode.

3-12 TRANSISTOR RATINGS

There are various precautions which must be taken when using a transistor. The ratings (listed in the manufacturer's specification sheets, Appendix B-3) which must not be exceeded are the maximum values of the collector current, the maximum collector power dissipation, and the maximum output and input junction voltages.

Maximum Collector Current

Even if power and voltage ratings are not exceeded, there is an absolute maximum value of current-handling capacity of the collector, associated with the junction area and the wire bonds that connect the transistor terminals to the external leads. This rating, which determines the maximum allowable saturation current, is 800 mA for the 2N2222A transistor.

Maximum Power Dissipation, P_D

Device destruction can occur if the collector junction is subjected to excessive power. For the 2N2222A device, $P_D = 0.5$ W at an ambient temperature of 25°C, whereas for a power transistor 2N5671 (Appendix B-8) $P_D = 140$ W.

Maximum Output Voltage Rating

Even if the rated collector current or the rated dissipation of a transistor is not exceeded, there is an upper limit to the maximum allowable collector-junction voltage since, at high voltages, there is the possibility of voltage breakdown in the transistor. Two types of breakdown are possible, *avalanche breakdown*, discussed in Sec. 2-9, and *reach-through*, discussed below.

Avalanche Multiplication

The maximum reverse-biasing voltage which may be applied before breakdown between the collector and emitter terminals of the transistor, under the condition that the base lead be open-circuited, is represented by the symbol BV_{CEO}. Breakdown may occur because of avalanche multiplication of the current I_{CO} that crosses the collector junction. For 2N2222A the CE characteristics extending into the breakdown region are shown in Fig. 3-17 and $BV_{CEO} \approx 50$ V. The specification sheets list the minimum value of BV_{CEO} as 40 V. For the CB configuration the output breakdown voltage BV_{CBO} is usually about twice BV_{CEO}. If the base is returned to the emitter through a resistor R, then the breakdown voltage, designated by BV_{CER}, will lie between BV_{CEO} and BV_{CBO}. In other words, the maximum allowable collector-to-emitter voltage depends not only on the transistor but also on the circuit in which it is used.[8]

Reach-through

The second mechanism by which a transistor's usefulness may be terminated as the collector voltage is increased is called *punch-through*, or *reach-through*, and

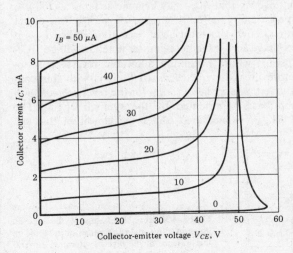

Figure 3-17 Common-emitter characteristics of the 2N2222A transistor extended into the breakdown region.

results from the increased width of the collector-junction transition region with increased collector-junction voltage (the Early effect, Sec. 3-5).

The transition region at a junction is the region of uncovered charges on both sides of the junction at the positions occupied by the impurity atoms. As the voltage applied across the junction increases, the transition region penetrates deeper into the base. Since the base is very thin, it is possible that, at moderate voltages, the transition region will have spread completely across the base to reach the emitter junction. The emitter barrier is now smaller than the normal value $V_o - |V_{EB}|$ because the collector voltage has "reached through" the base region. This lowering of the emitter-junction voltage may result in an excessively large emitter current, thus placing an upper limit on the magnitude of the collector voltage.

Punch-through differs from avalanche breakdown in that it takes place at a fixed voltage [given by V_j in Eq. (2-15), with $W = W_B$] between collector and base, and is not dependent on circuit configuration. In a particular transistor, the voltage limit is determined by punch-through or breakdown, whichever occurs at the lower voltage.

Maximum Input Voltage Rating

Consider the circuit configuration of Fig. 3-18, where V_{BB} represents a biasing voltage intended to keep the transistor cutoff. Assume that the transistor is just at the point of cutoff, with $I_E = 0$, so that $I_B = -I_{CBO}$. If we require that at cutoff $V_{BE} \approx 0$ V, then the condition of cutoff requires that

$$V_{BE} = -V_{BB} + R_b I_{CBO} \leqslant 0 \tag{3-17}$$

As an extreme example consider that R_b is, say, as large as 100 kΩ and that we want to allow for the contingency that I_{CBO} may become as large as 100 μA, as might occur with a large-geometry power transistor or with a low-power device at elevated temperatures. Then V_{BB} must be at least 10 V. When I_{CBO} is small, the magnitude of the voltage across the base-emitter junction will be 10 V. Hence we must use a transistor whose maximum allowable reverse base-to-emitter junction voltage before breakdown exceeds 10 V. It is with this contingency in mind that a manufacturer supplies a rating for the reverse *breakdown voltage* between emitter and base, represented by the symbol BV_{EBO}. The subscript O indicates that BV_{EBO} is measured under the condition that the

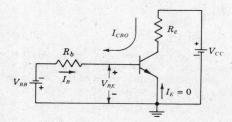

Figure 3-18 Reverse biasing of the emitter junction to maintain the transistor in cutoff in the presence of the reverse saturation current I_{CBO} through R_b.

collector current is zero. Breakdown voltages BV_{EBO} may be as high as some tens of volts or as low as 0.5 V. If $BV_{EBO} = 6$ V (as it is for the 2N2222A transistor), then V_{BB} must be chosen to have a maximum value of 6 V.

3-13 ADDITIONAL TRANSISTOR CHARACTERISTICS

A very important use for a bipolar transistor is as an amplifier of low-amplitude signals. For this application small-signal, low-frequency parameters are introduced in Chap. 11 and high-frequency parameters are defined in Chap. 13. Small-signal models based on these parameters are then developed and used to calculate the gain of amplifiers over a frequency range from zero to tens of megahertz.

. A large-amplitude pulse may be applied to the base of a transistor to drive it into saturation and then (at the end of the pulse) to turn it off again. Under such circumstances the output current does not follow the input waveform exactly, but there are nonzero rise and fall times and the pulse is stretched out in time. This phenomenon is discussed in detail in the next section. Both the small-signal and the switching parameters for the 2N2222A transistor are indicated in Appendix B-3.

3-14 TRANSISTOR SWITCHING TIMES[9]

We consider the transistor circuit shown in Fig. 3-19a, driven by the pulse waveform shown in Fig. 3-19b. This waveform makes transitions between the voltage levels V_2 and V_1. At V_2 the transistor is at cutoff, and at V_1 the transistor is in saturation. The input waveform v_i is applied between base and emitter through a resistor.

The response of the collector current i_c to the input waveform, together with its time relationship to that waveform, is shown in Fig. 3-19c. The current does not immediately respond to the input signal. Instead, there is a delay, and the time that elapses during this delay, together with the time required for the current to rise to 10 percent of its maximum (saturation) value $I_{CS} \approx V_{CC}/R_L$, is called the *delay time* t_d. The current waveform has a nonzero *rise time* t_r, which is the time required for the current to rise through the active region from 10 to 90 percent of I_{CS}. The total *turn-on* time t_{ON} is the sum of the delay and rise time, $t_{ON} \equiv t_d + t_r$. When the input signal returns to its initial state at $t = T$, the current again fails to respond immediately. The interval which elapses between the transition of the input waveform and the time when i_c has dropped to 90 percent of I_{CS} is called the *storage time* t_s. The storage interval is followed by the *fall time* t_f, which is the time required for i_c to fall from 90 to 10 percent of I_{CS}. The *turnoff time* t_{OFF} is defined as the sum of the storage and fall times, $t_{OFF} \equiv t_s + t_f$. We shall consider now the physical reasons for the existence of each of these times. The actual calculation of the time intervals (t_d, t_r, t_s, and t_f)

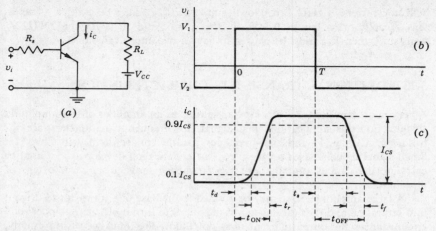

Figure 3-19 The pulse waveform in (*b*) drives the transistor in (*a*) from cutoff to saturation and back again. (*c*) The collector-current response to the driving input pulse.

is complex, and the reader is referred to Ref. 9. Numerical values (in nanoseconds) of delay time, rise time, storage time, and fall time for the silicon transistor 2N2222A under specified conditions are given in Appendix B-3.

The Delay Time

Three factors contribute to the delay time. First, when the driving signal is applied to the transistor input, a nonzero time is required to charge up the emitter-junction transition capacitance so that the transistor may be brought from cutoff to the active region. Second, even when the transistor has been brought to the point where minority carriers have begun to cross the emitter junction into the base, a time interval is required before these carriers can cross the base region to the collector junction and be recorded as collector current. Finally, some time is required for the collector current to rise to 10 percent of its maximum.

Rise Time and Fall Time

The rise time and the fall time are due to the fact that if a base-current step is used to saturate the transistor or return it from saturation to cutoff, the transistor collector current must traverse the active region. The collector current increases or decreases along an exponential curve whose time constant τ_r can be shown to be given by $\tau_r = h_{FE}(C_c R_c + 1/\omega_T)$, where C_c is the collector transition capacitance and ω_T is the radian frequency at which the current gain is unity (Sec. 13-7).

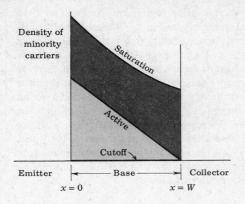

Figure 3-20 Minority-carrier concentration in the base for cutoff, active, and saturation conditions of operation.

Storage Time

The failure of the transistor to respond to the trailing edge of the driving pulse for the time interval t_s (indicated in Fig. 3-19c) results from the fact that a transistor in saturation has a saturation charge of excess minority carriers stored in the base. The transistor cannot respond until this saturation excess charge has been removed. The stored charge density in the base is indicated in Fig. 3-20 under various operating conditions.

The concentration of minority carriers [Eq. (2-20)] in the base region decreases linearly from $p_{no}\epsilon^{V_E/V_T}$ at $x = 0$ to $p_{no}\epsilon^{V_C/V_T}$ at $x = W$, as indicated in Fig. 3-21b. In the cutoff region, both V_E and V_C are negative, and p_n is almost zero everywhere. In the active region, V_E is positive and V_C negative, so that p_n is large at $x = 0$ and almost zero at $x = W$. Finally, in the saturation region, where V_E and V_C are both positive, p_n is large everywhere, and hence a large amount of minority-carrier charge is stored in the base. These densities are pictured in Fig. 3-20.

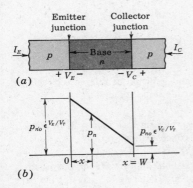

Figure 3-21 The minority-carrier density in the base region. The concentrations at $x = 0$ and $x = W$ are governed by the law of the junction [(Eq. 2-20)].

Consider that the transistor is in its saturation region and that at $t = T$ an input step is used to turn the transistor off, as in Fig. 3-19. Since the turnoff process cannot begin until the abnormal carrier density (the heavily shaded area of Fig. 3-20) has been removed, a relatively long storage delay time t_s may elapse before the transistor responds to the turnoff signal at the input. In an extreme case this storage-time delay may be many times the rise or fall time through the active region. It is clear that when transistor switches are to be used in an application where speed is at a premium, it is advantageous to reduce the storage time. By adding a capacitor C across the base resistor (Fig. 3-19), an impulsive current will flow out of the base at the time T at the end of the pulse. If C is properly chosen,[9] this impulsive current will instantaneously reduce t_s to zero. A method for preventing a transistor from saturating, and thus eliminating storage time, is given in Sec. 4-8 (the Schottky transistor).

REFERENCES

1. Shockley, W.: The Theory of p-n Junctions in Semiconductors and p-n Junction Transistors, *Bell System Tech. J.*, vol. 28, pp. 435–489, July, 1949.
   Moll, J. L.: Junction Transistor Electronics, *Proc. IRE*, vol. 43, pp. 1807–1819, December, 1955.
   Yang, E. S.: "Fundamentals of Semiconductor Devices," chap. 9, McGraw-Hill Book Company, New York, 1978.
2. Phillips, A. B.: "Transistor Engineering," chap. 1, McGraw-Hill Book Company, New York, 1962.
3. Texas Instruments, Inc.: J. Miller (ed.), "Transistor Circuit Design," chap. 1, McGraw-Hill Book Company, New York, 1963.
4. Early, J. M.: Effects of Space-charge Layer Widening in Junction Transistors, *Proc. IRE*, vol. 40, pp. 1401–1406, November, 1952.
5. Millman J., and C. C. Halkias: "Integrated Electronics: Analog and Digital Circuits and Systems," sec. 19-15, McGraw-Hill Book Company, New York, 1972.
6. Sah, C. T., R. N. Noyce, and W. Shockley: Carrier-generation and Recombination in p-n Junctions and p-n Junction Characteristics, *Proc. IRE*, vol. 45, pp. 1228–1243, September 1957.
   Pritchard, R. L.: Advances in the Understanding of the P-N Junction Triode, *Proc. IRE*, vol. 46, pp. 1130–1141, June, 1958.
7. Ref. 2, pp. 236–237.
8. Millman, J., and H. Taub: "Pulse, Digital, and Switching Waveforms," Sec. 6-9, McGraw-Hill Book Company, New York, 1965.
9. Ref. 8, chap. 20.

REVIEW QUESTIONS

3-1 Draw the circuit symbol for a p-n-p transistor and indicate the reference directions for the three currents and the reference polarities for the three voltages.

3-2 Repeat Rev. 3-1 for an n-p-n transistor.

3-3 For a p-n-p transistor biased in the active region, plot (in each region E, B, and C) (*a*) the potential variation; (*b*) the minority-carrier concentration.
   (*c*) Explain the shapes of the plots in (*a*) and (*b*).

3-4 (*a*) For a p-n-p transistor biased in the active region, indicate the various electron and hole current components crossing each junction and entering (or leaving) the base terminal.

(b) Which of the currents is proportional to the gradient of p_n at J_E and J_C, respectively?

(c) Repeat part b with p_n replaced by n_p.

(d) What is the physical origin of the several current components crossing the base terminal?

3-5 (a) From the currents indicated in Rev. 3-4 obtain an expression for the collector current I_C. Define each symbol in this equation.

(b) Generalize the equation for I_C in part a so that it is valid even if the transistor is not operating in its active region.

3-6 (a) Define the *current gain* α in words and as an equation.

(b) Repeat part a for the parameter α'.

3-7 Describe the fabrication of an alloy transistor.

3-8 For a p-n-p transistor in the active region, what is the sign (positive or negative) of I_E, I_C, I_B, V_{CB}, and V_{EB}?

3-9 Repeat Rev. 3-8 for an n-p-n transistor.

3-10 (a) Sketch a family of CB output characteristics for a transistor.

(b) Indicate the active, cutoff, and saturation regions.

(c) Explain the shapes of the curves qualitatively.

3-11 (a) Sketch a family of CB input characteristics for a transistor.

(b) Explain the shapes of the curves qualitatively.

3-12 Explain *base-width modulation* (the Early effect).

3-13 Explain qualitatively the three consequences of base-width modulation.

3-14 Define the following regions in a transistor: (a) active; (b) saturation; (c) cutoff.

3-15 (a) Draw the circuit of a transistor in the CE configuration.

(b) Sketch the output characteristics.

(c) Indicate the active, saturation, and cutoff regions.

3-16 (a) Sketch a family of CE input characteristics.

(b) Explain the shape of these curves qualitatively.

3-17 (a) Derive the expression for I_C versus I_B for a CE transistor configuration in the active region.

(b) For $I_B = 0$, what is I_C?

3-18 (a) What is the order of magnitude of the reverse collector saturation current I_{CBO} for a silicon transistor?

(b) How does I_{CBO} vary with temperature?

3-19 Repeat Rev. 3-18 for a germanium transistor.

3-20 Why does I_{CBO} differ from I_{CO}?

3-21 (a) Define *saturation resistance* for a CE transistor.

(b) Give its order of magnitude.

3-22 (a) Define *base-spreading resistance* for a transistor.

(b) Give its order of magnitude.

3-23 What is the order of magnitude of the temperature coefficients of $V_{BE(sat)}$, $V_{BC(sat)}$, and $V_{CE(sat)}$?

3-24 (a) Define h_{FE}.

(b) Plot h_{FE} versus I_C.

3-25 (a) Give the order of magnitude of V_{BE} at cutoff for a silicon transistor.

(b) Repeat part a for a germanium transistor.

(c) Repeat parts a and b for the cutin voltage.

3-26 Is $|V_{BE(\text{sat})}|$ greater or less than $|V_{CE(\text{sat})}|$? Explain.

3-27 (a) What is the range in volts for V_{BE} between cutin and saturation for a silicon transistor?

(b) Repeat part a for a germanium transistor.

3-28 What is the collector current relative to I_{CO} in a silicon transistor if (a) the base is short-circuited to the emitter? (b) the base floats?

(c) Repeat parts a and b for a germanium transistor.

3-29 Consider a transistor circuit with resistors R_b, R_c, and R_e in the base, collector, and emitter legs, respectively. The biasing voltages are V_{BB} and V_{CC} in base and collector circuits, respectively.

(a) Outline the method for finding the quiescent currents, assuming that the transistor operates in the active region.

(b) How do you test to see if your assumption is correct?

3-30 Repeat Rev. 3-29, assuming that the transistor is in saturation.

3-31 For a CE transistor define (in words and symbols) (a) β; (b) $\beta_{dc} = h_{FE}$; (c) $\beta' = h_{fe}$.

3-32 (a) For what condition is $\beta \approx h_{FE}$?

(b) For what condition is $h_{FE} \approx h_{fe}$?

3-33 (a) Define the three normal modes of operation.

(b) Define the inverted-active mode of operation.

3-34 Discuss the two possible sources of breakdown in a transistor as the collector-to-emitter voltage is increased.

3-35 What is the meaning of the symbol BV_{EBO}?

3-36 List the four maximum ratings specified by the manufacturer of a transistor.

3-37 A pulse waveform drives an n-p-n transistor from cutoff into saturation and then back to cutoff.

(a) Draw the output current waveshape, lined up in time with the input voltage.

(b) Indicate the following times on your sketch: *delay*, *rise*, ON, *storage*, *fall*, and OFF.

3-38 (a) What is the physical origin of *storage time*?

(b) Is it important in turning a transistor ON or OFF? Explain.

(c) Draw the minority-carrier concentration in the base; in the active region and in saturation.

INTEGRATED CIRCUITS: FABRICATION AND CHARACTERISTICS

An integrated circuit consists of a single-crystal chip of silicon, typically 50 by 50 mils in cross section,† containing both active and passive elements and their interconnections. Such circuits are produced by the same processes used to fabricate individual transistors and diodes. These processes include epitaxial growth, masked impurity diffusion, oxide growth, and oxide etching, using photolithography for pattern definition. A method of batch processing is employed which offers excellent repeatability and is adaptable to the production of large numbers of integrated circuits at low cost. In this chapter we describe the basic processes involved in fabricating an integrated circuit.

4-1 INTEGRATED-CIRCUIT (MICROELECTRONIC) TECHNOLOGY

The fabrication of integrated circuits is based on materials, processes, and design principles which constitute a highly developed semiconductor (planar-diffusion) technology. The basic structure of an integrated circuit is shown in Fig. 4-1b, and consists of four distinct layers of material. The bottom layer ① (6 mils thick) is p-type silicon and serves as a *substrate* or *body* upon which the integrated circuit is to be built. The second layer ② is thin (typically 5 to 25 μm) n-type material which is grown as a single-crystal extension of the substrate. All active and passive components are built within the thin n-type layer using a series of diffusion steps. These components are transistors, diodes, capacitors, and resistors, and they are made by diffusing p-type and n-type impurities. The most complicated component fabricated is the transistor, and all other elements are constructed with one or more of the processes required to make a transistor.

---

† 1 mil = 0.001 in = 25.4 μm = 0.0254 mm.

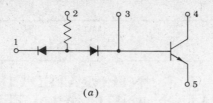

(a)

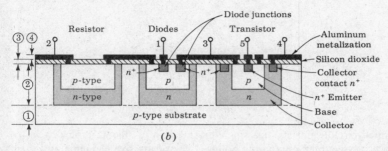

(b)

Figure 4-1 (a) A circuit containing a resistor, two diodes, and a transistor. (b) Cross-sectional view of the circuit in (a) when transformed into a monolithic form (not drawn to scale). The four layers are ① substrate, ② n-type crystal containing the integrated circuit, ③ silicon dioxide, and ④ aluminum metalization. *(After Phillips.[2])*

In the fabrication of all the above elements it is necessary to distribute impurities in certain precisely defined regions within the second (n-type) layer. The selective diffusion of impurities is accomplished by using SiO_2 as a barrier which protects portions of the wafer against impurity penetration. Thus the third layer of material ③ is silicon dioxide, and it also provides protection of the semiconductor surface against contamination. In the regions where diffusion is to take place, the SiO_2 layer is etched away, leaving the rest of the wafer protected against diffusion. To permit selective etching, the SiO_2 layer must be subjected to a photolithographic process, described in Sec. 4-4. Finally, a fourth metallic (aluminum) layer ④ is added to supply the necessary interconnections between components.

We are now in a position to appreciate some of the significant advantages of microelectronic technology. Let us consider a 2 by 2 in wafer divided into 1,600 chips of surface area 50 by 50 mils. We demonstrate in this chapter that a reasonable area under which a component (say, a transistor) is fabricated is 50 $mils^2$. Hence each chip (each integrated circuit) contains 50 separate components, and there are $50 \times 1,600 = 80,000$ components on each wafer.

If we process 20 wafers in a batch, we can manufacture 32,000 integrated circuits simultaneously, and these contain 1,600,000 components. Some of the chips will contain faults due to imperfections in the manufacturing process, but if the *yield* (the percentage of fault-free chips per wafer) is only 20 percent, then 6,400 good chips containing 320,000 circuit components are mass-produced in a single batch! †

The following advantages are offered by integrated-circuit technology as compared with discrete components interconnected by conventional techniques:

1. Low cost (due to the large quantities processed).
2. Small size.
3. High reliability. (All components are fabricated simultaneously, and there are no soldered joints.)
4. Improved performance. (Because of the low cost, more complex circuitry may be used to obtain better functional characteristics.)
5. Matched devices. Since all transistors are manufactured simultaneously by the same processes, the corresponding parameters of these devices as well as the temperature variation of their characteristics have essentially the same magnitudes (the parameters track well with temperature).

In the next sections we examine the processes required to fabricate an integrated circuit.

4-2 BASIC MONOLITHIC INTEGRATED CIRCUITS[1–4]

We now examine in some detail the various techniques and processes required to obtain the circuit of Fig. 4-1*a* in an integrated form, as shown in Fig. 4-1*b*. This configuration is called a monolithic integrated circuit because it is formed on a single silicon chip. The word "monolithic" is derived from the Greek *monos*, meaning "single," and *lithos*, meaning "stone." Thus a monolithic circuit is built into a single stone, or single crystal.

In this section we describe qualitatively a complete epitaxial-diffused fabrication process for integrated circuits. In subsequent sections we examine in more detail the epitaxial, photographic, and diffusion processes involved. The circuit of Fig. 4-1*a* is chosen for discussion because it contains typical components: a resistor, diodes, and a transistor. These elements (and also capacitors with small values of capacitances) are the components encountered in integrated circuits. The monolithic circuit is formed by the steps indicated in Fig. 4-2 and described below.

---

† The above numbers are actually quite conservative. Using the microelectronic techniques described in Chaps. 8 and 9, a component density about eight times that assumed above has been achieved on a very much larger chip. For example, an entire microprocessor (Intel 8085) containing 6,200 transistors is commercially available (in 1977) on a single chip whose dimensions are 164 by 222 mils. Hence the average area per transistor is only about 6 mils² compared with the 50 mils² used in the above calculations.

Step 1. Crystal Growth of the Substrate[4]

A tiny crystal of silicon is attached to a rod and lowered into a crucible of molten silicon to which acceptor impurities have been added. As the rod is very slowly pulled out of the melt under carefully controlled conditions, a single p-type crystal ingot of the order of 3 in (7.5 cm) in diameter and 20 in (50 cm) long is grown. The ingot is subsequently sliced into round wafers approximately 6 mils thick to form the substrate upon which all integrated components will be fabricated. One side of each wafer is lapped and polished to eliminate surface imperfections before proceeding with the next process.

Step 2. Epitaxial Growth

An n-type epitaxial layer, typically 5 to 25 μm thick, is grown into a p-type substrate which has a resistivity of approximately 10 $\Omega \cdot$ cm, corresponding to $N_A = 1.4 \times 10^{15}$ atoms/cm^3. The epitaxial process described in Sec. 4-3 indicates that the resistivity of the n-type epitaxial layer can be chosen independently of that of the substrate. Values of 0.1 to 0.5 $\Omega \cdot$ cm are chosen for the n-type layer. After polishing and cleaning, a thin layer (0.5 μm = 5,000 Å) of oxide, SiO$_2$, is formed over the entire wafer, as shown in Fig. 4-2a. The SiO$_2$ is grown by exposing the epitaxial layer to an oxygen or steam atmosphere while being heated to about 1000°C. Silicon dioxide has the fundamental property of preventing the diffusion of impurities through it. Use of this property is made in the following steps.

Step 3. Isolation Diffusion

In Fig. 4-2b the wafer is shown with the oxide removed in four different places on the surface. This removal is accomplished by means of a photolithographic etching process described in Sec. 4-4. The remaining SiO$_2$ serves as a mask for the diffusion of acceptor impurities (in this case, boron). The wafer is now subjected to the so-called *isolation diffusion*, which takes place at the temperature and for the time interval required for the p-type impurities to penetrate the n-type epitaxial layer and reach the p-type substrate. We thus leave the shaded n-type regions in Fig. 4-2b. These sections are called *isolation islands*, or *isolated regions*, because they are separated by two back-to-back p-n junctions. Their purpose is to allow electrical isolation between different circuit components. For example, it will become apparent later in this section that a different isolation region must be used for the collector of each separate transistor. The p-type substrate must always be held at a negative potential with respect to the isolation islands in order that the p-n junctions be reverse-biased. If these diodes were to become forward-biased in an operating circuit, then, of course, the isolation would be lost.

It should be noted that the concentration of acceptor atoms ($N_A \approx 5 \times 10^{20}$ cm^{-3}) in the region between isolation islands will generally be much higher (and hence indicated as p^+) then in the p-type substrate. The reason for this higher density is to prevent the depletion region of the reverse-biased isolation-to-substrate junction from extending into p^+-type material (Sec. 2-6) and possibly connecting two isolation islands.

Parasitic Capacitance It is now important to consider that these isolation regions, or junctions, are connected by a significant barrier, or transition capacitance C_{Ts}, to the p-type substrate, which capacitance can affect the operation of the circuit. Since C_{Ts} is an undesirable by-product of the isolation process, it is called the *parasitic capacitance*.

The parasitic capacitance is the sum of two components, the capacitance C_1 from the bottom of the n-type region to the substrate (Fig. 4-2b) and C_2 from the sidewalls of the isolation islands to the p^+ region. The bottom component, C_1, results from an essentially step junction due to the epitaxial growth (Sec. 4-3), and hence varies inversely as the square root of the voltage V between the isolation region and the substrate (Sec. 2-6). The sidewall capacitance C_2 is associated with a diffused graded junction, and it varies as $V^{-1/3}$. For this component the junction area is equal to the perimeter of the isolation region times the thickness y of the epitaxial n-type layer. The total capacitance is of the order of a few picofarads.

Step 4. Base Diffusion

During this process a new layer of oxide is formed over the wafer, and the photolithographic process is used again to create the pattern of openings shown in Fig. 4-2c. The p-type impurities (boron) are diffused through these openings. In this way are formed the transistor base regions as well as resistors, the anode of diodes, and junction capacitors (if any). It is important to control the depth of this diffusion so that it is shallow and does not penetrate to the substrate. The resistivity of the base layer will generally be much higher than that of the isolation regions.

Step 5. Emitter Diffusion

A layer of oxide is again formed over the entire surface, and the masking and etching processes are used again to open windows in the p-type regions, as shown in Fig. 4-2d. Through these openings are diffused n-type impurities (phosphorus) for the formation of transistor emitters, the cathode regions for diodes, and junction capacitors.

Additional windows (such as W_1 and W_2 in Fig. 4-2d) are often made into the n regions to which a lead is to be connected, using aluminum as the ohmic contact, or interconnecting metal. During the diffusion of phosphorus a heavy concentration (called n^+) is formed at the points where contact with aluminum is to be made. Aluminum is a p-type impurity in silicon, and a large concentration of phosphorus prevents the formation of a p-n junction when the aluminum is alloyed to form an ohmic contact.[3,5]

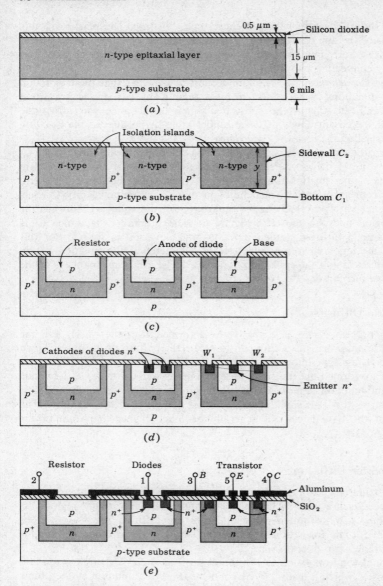

Figure 4-2 The steps involved in fabricating a monolithic circuit (not drawn to scale). (*a*) Epitaxial growth; (*b*) isolation diffusion; (*c*) base diffusion; (*d*) emitter diffusion; (*e*) aluminum metalization.

Step 6. Aluminum Metalization

All *p-n* junctions and resistors for the circuit of Fig. 4-1*a* have been formed in the preceding steps. It is now necessary to interconnect the various components of the integrated circuit as dictated by the desired circuit. To make these connections, a fourth set of windows is opened into a newly formed SiO_2 layer, as shown in Fig. 4-2*e*, at the points where contact is to be made. The interconnections are made first, using vacuum deposition of a thin even coating of aluminum over the entire wafer. The photoresist technique is now applied to etch away all undesired aluminum areas, leaving the desired pattern of interconnections shown in Fig. 4-2*e* between resistors, diodes, and transistors.

In production a large number (several hundred) of identical circuits are manufactured simultaneously on a single wafer (Fig. 4-3). After the metalization process has been completed, the wafer is scribed with a diamond-tipped tool and separated into individual chips. Each chip is then mounted on a ceramic wafer and is attached to a suitable header. The package leads are connected to the integrated circuit by stitch bonding[1] of a 1-mil aluminum or gold wire from the terminal pad on the circuit to the package lead. Most of the labor cost of an IC is in the packaging and testing (which cannot be done in a batch process).

Summary

In this section the epitaxial-diffused method of fabricating microcircuits is described. We have encountered the following processes:

1. Crystal growth of a substrate
2. Epitaxial layer growth
3. Silicon dioxide growth
4. Photoetching
5. Diffusion
6. Vacuum evaporation of aluminum

Figure 4-3 A semiconductor wafer about 2 in (5.1 cm) in diameter which includes almost 600 monolithic IC chips. *(Courtesy of IBM, Inc.)*

Using these techniques, it is possible to produce the following elements on the same chip: transistors, diodes, resistors, capacitors, and aluminum interconnections. Processes 3, 4, and 5 enumerated in the foregoing are repeated several times. For example, to fabricate a transistor, five masks are required: the first for the isolation diffusion, the second for the base diffusion, the third for the emitter diffusion, the fourth for location of the ohmic contacts through the SiO_2, and the fifth for removing the undesired aluminum areas, so as to leave only the necessary interconnections.

4-3 EPITAXIAL GROWTH[1, 3]

The epitaxial process grows a thin film of single-crystal silicon from the gas phase as a continuation of an existing crystal wafer of the same material. The basic chemical reaction used to describe the epitaxial growth of pure silicon is the hydrogen reduction of silicon tetrachloride:

$$SiCl_4 + 2H_2 \overset{1200°C}{\rightleftarrows} Si + 4HCl \tag{4-1}$$

Since it is required to produce epitaxial films of specific impurity concentrations, it is necessary to introduce impurities such as phosphine (PH_3) for n-type doping or biborane (B_2H_6) for p-type doping into the silicon tetrachloride-hydrogen gas stream. An apparatus (which allows simple and precise impurity control) for the production of an epitaxial layer consists of a long cylindrical quartz tube encircled by a radio-frequency induction coil. The silicon wafers are placed on a rectangular graphite rod called a *boat*. The boat is inserted in the reaction chamber, and the graphite is heated inductively to about 1200°C. A control console permits the introduction and removal of various gases required for the growth of appropriate epitaxial layers. Thus it is possible to form an almost abrupt step p-n junction similar to the junction shown in Fig. 2-8.

4-4 MASKING AND ETCHING[1, 3]

The monolithic technique described in Section 4-2 requires the selective removal of the SiO_2 to form openings through which impurities may be diffused. The photoetching method used for this removal is illustrated in Fig. 4-4. During the photolithographic process the wafer is coated with a uniform film of a photosensitive emulsion (such as the Kodak *photoresist* KPR). A large black-and-white layout of the desired pattern of openings is made and then reduced photographically. This negative, or stencil, of the required dimensions is placed as a mask over the photoresist, as shown in Fig. 4-4. By exposing the KPR to ultraviolet light through the mask, the photoresist becomes polymerized under the transparent regions of the stencil. The mask is now removed, and the wafer is "developed" by using a chemical (such as trichloroethylene) which dissolves the unexposed (unpolymerized) portions of the photoresist film and leaves the

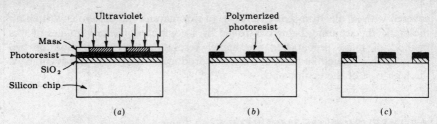

Figure 4-4 Photoetching technique. (*a*) Masking and exposure to ultraviolet radiation. (*b*) The photoresist after development. (*c*) After etching.

surface pattern as shown in Fig. 4-4*b*. The emulsion which was not removed in development is now *fixed*, or *cured*, so that it becomes resistant to the corrosive etches used next. The chip is immersed in an etching solution of hydrofluoric acid, which removes the oxide from the areas through which dopants are to be diffused. Those portions of the SiO_2 which are protected by the photoresist are unaffected by the acid (Fig. 4-4*c*). After diffusion of impurities, the resist mask is removed (stripped) with a chemical solvent (hot H_2SO_4) coupled with a mechanical abrasion process.

The making of a photographic mask[3] involves complicated and expensive processes. After the circuit layout has been determined, a large-scale drawing is made showing the locations of the openings to be etched in the SiO_2 for a particular process step (say, for the isolation diffusion). The width of the diffusion path is typically 1 mil and it is desired to control this width to a tolerance of ± 10 percent. Hence, the drawing is made to a magnified scale of about 500 : 1, resulting in line widths of 0.5 ± 0.05 in (~ 1 ± 0.1 cm), which are more manageable dimensions for a draftsman. For an IC with a surface area of 50 × 50 mils, this magnification results in a drawing 25 × 25 in(~ 60 × 60 cm). For complex circuits, the layout may be carried out by use of computer-aided graphics.

Artwork is often made of the drawing by using clear mylar coated with a red plastic (called *Rubylith*). Cuts are made in the red coating and this plastic is peeled off in the regions where diffusion of impurities are desired. This Rubylith pattern is photographed by a large camera and reduced by factors of 5 or 10 several times until a demagnification of 500 times has taken place, resulting in an exact size photographic master. This master is used to produce multiple images in two dimensions (Fig. 4-3) on a photographic plate by means of a precision step-and-repeat printer. This glass plate is the mask (Fig. 4-4) used for the photoresist operation.

The smallest features that can be formed by the photolithographic process described in the foregoing is limited by the wavelength of light, because of diffraction. Electron beams have much smaller wavelengths than radiation and are capable of defining much smaller areas. Hence, electron-beam lithography[4] is now used in the production of masks. A narrow electron beam scans a mask covered with an electron-sensitive resist. In this manner the pattern is written on the mask, the scanning being controlled by a computer. The advantages of this method of mask preparation are higher resolution, the elimination of two photographic reduction steps, and shorter production time. The disadvantage is the high cost of the equipment.

4-5 DIFFUSION OF IMPURITIES[3]

The most important process in the fabrication of integrated circuits is the diffusion of impurities into the silicon chip. Reasonable diffusion times (2 hours) require high diffusion temperatures ($\sim 1000°C$). Therefore a high-temperature diffusion furnace, having a closely controlled temperature over the length of the hot zone of the furnace, is standard equipment in a facility for the fabrication of integrated circuits. About 20 wafers are placed on a quartz carrier inside the quartz tube of the furnace. Impurity sources in connection with diffusion ovens can be gases, liquids, or solids. For example, the impurity gas for boron diffusion is B_2H_6, for phosphorous it is PH_3 and for arsenic it is AsH_3. An inert carrier gas (such as nitrogen) brings the impurity atoms to the surface of the wafers where they can diffuse into the silicon.

Lateral Diffusion[3]

For the sake of simplicity of drawing, the cross-sectional diagrams in this chapter are all shown with vertical diffusion edges. However, when a hole is opened in the SiO_2 and impurities are introduced, they will diffuse laterally the same distance that they do vertically. Hence, the impurity will spread out under the passivating oxide surface layer and the junction profiles should be drawn more realistically as shown in Fig. 4-5

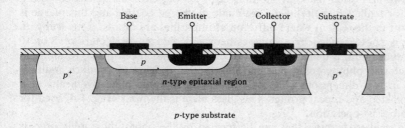

Figure 4-5 The cross section of an *n-p-n* transistor indicating curved junction profiles as a result of lateral diffusion.

4-6 TRANSISTORS FOR MONOLITHIC CIRCUITS[1, 6]

A planar transistor made for monolithic integrated circuits, using epitaxy and diffusion, is shown in Fig. 4-6a. Here the collector is electrically separated from the substrate by the reverse-biased isolation diodes. Since the anode of the isolation diode covers the back of the entire wafer, it is necessary to make the collector contact on the top, as shown in Fig. 4-6a. It is now clear that the isolation diode of the integrated transistor has two undesirable effects: it adds a parasitic shunt capacitance to the collector and a leakage current path. In addition, the necessity for a top connection for the collector increases the collector-current path and thus increases the collector resistance and $V_{CE(sat)}$. All these undesirable effects are absent from the discrete epitaxial transistor shown in Fig. 4-6b. What is then the advantage of the monolithic transistor? A significant improvement in performance arises from the fact that integrated transistors are located physically close together and their electrical characteristics are closely matched. For example, integrated transistors spaced within 30 mils (0.75 mm) have V_{BE} matching of better than 5 mV with less than 10 μV/°C drift and an h_{FE} match of ± 10 percent. These matched transistors make excellent difference amplifiers (Sec. 15-3).

The electrical characteristics of a transistor depend on the size and geometry of the transistor, doping levels, diffusion schedules, and the basic silicon material. Of all these factors the size and geometry offer the greatest flexibility for design. The doping levels and diffusion schedules are determined by the standard processing schedule used for the desired transistors in the integrated circuit.

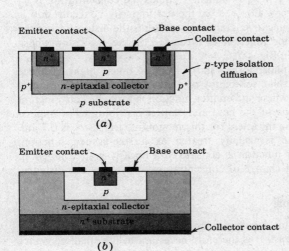

Figure 4-6 Comparison of cross sections of (a) a monolithic integrated circuit transistor with (b) a discrete planar epitaxial transistor. (For a top view of the transistor in a see Fig. 4-8.)

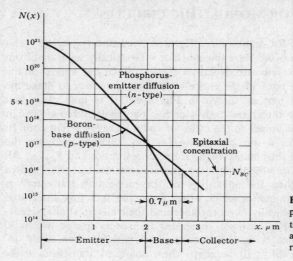

Figure 4-7 A typical impurity profile in a monolithic integrated transistor. [Note that $N(x)$, atoms/cm³, is plotted on a logarithmic scale.]

Impurity Profiles for Integrated Transistors

Figure 4-7 shows a typical impurity profile for a monolithic integrated circuit transistor. The background, or epitaxial-collector, concentration N_{BC} is shown as a dashed line in Fig. 4-7. The concentration N of boron is high (5×10^{18} atoms/cm³) at the surface and falls off with distance into the silicon as indicated in Fig. 4-7. At that distance $x = x_j$, at which N equals the concentration N_{BC}, the net impurity density is zero. For $x < x_j$, the net impurity concentration is positive, and for $x > x_j$, it is negative. Hence x_j represents the distance from the surface at which the collector junction is formed. For the transistor whose impurity profile is indicated in Fig. 4-7, $x_j = 2.7 \, \mu$m.

The emitter diffusion (phosphorus) starts from a much higher surface concentration (close to the solid solubility) of about 10^{21} atoms/cm³, and is diffused to a depth of 2 μm, where the emitter junction is formed. This junction corresponds to the intersection of the base and emitter distributions of impurities. We now see that the base thickness for this monolithic transistor is 0.7 μm. The emitter-to-base junction is usually treated as a step-graded junction, whereas the base-to-collector junction is considered a linearly graded junction, because of the slower rate of change of concentration with respect to distance.

Monolithic Transistor Layout

The physical size of a transistor determines the parasitic isolation capacitance as well as the junction capacitance. It is therefore necessary to use small-geometry transistors if the integrated circuit is designed to operate at high frequencies or high switching speeds. The geometry of a typical monolithic transistor is shown

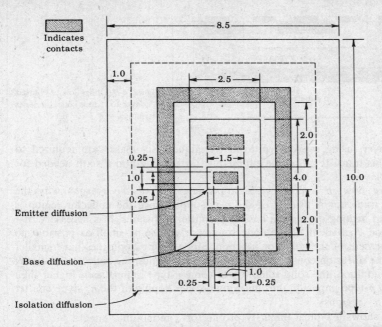

Figure 4-8 A typical double-base stripe geometry of an integrated-circuit transistor. Dimensions are in mils. (For a side view of the transistor see Fig. 4-6.) *(Courtesy of Motorola Monitor.)*

in Fig. 4-8. The emitter rectangle measures 1 by 1.5 mils, and is diffused into a 2.5- by 4.0-mil base region. Contact to the base is made through two metalized stripes on either side of the emitter. The rectangular metalized area forms the ohmic contact to the collector region. The rectangular collector contact of this transistor reduces the saturation resistance. The substrate in this structure is located about 1 mil below the surface. Since diffusion proceeds in three dimensions, it is clear that the *lateral-diffusion* distance (Fig. 4-5) will also be 1 mil. The dashed rectangle in Fig. 4-8 represents the substrate area and is 6.5 by 8 mils.

Buried Layer[1, 3]

We noted above that the integrated transistor, because of the top collector contact, has a higher collector series resistance than a similar discrete-type transistor. One common method of reducing the collector series resistance is by means of a heavily doped n^+ "buried" layer sandwiched between the p-type substrate and the n-type epitaxial collector, as shown in Fig. 4-9. The buried-layer structure can be obtained by diffusing the n^+ layer into the substrate before the n-type epitaxial collector is grown or by selectively growing the

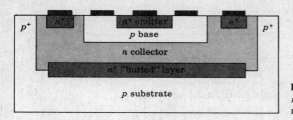

Figure 4-9 Utilization of "buried" n^+ layer to reduce collector series resistance.

n^+-type layer, using masked epitaxial techniques. Six masks are required to fabricate this transistor; the five enumerated in Sec. 4-2 and a sixth needed for the buried layer.

We are now in a position to appreciate one of the reasons why the integrated transistor is usually of the *n-p-n* type. Since the collector region is subjected to heating during the base and emitter diffusions, it is necessary that the diffusion coefficient of the collector impurities be as small as possible, to avoid movement of the collector junction. Since *n*-type impurities have smaller values of the diffusion constant D than *p*-type impurities, the collector is usually *n*-type. In addition, the solid solubility of some *n*-type impurities is higher than that of any *p*-type impurity, thus allowing heavier doping of the n^+-type emitter and other n^+ regions.

Lateral *p-n-p* Transistor[3, 5]

The standard integrated-circuit transistor is an *n-p-n* type, as we have already emphasized. In some applications it is required to have both *n-p-n* and *p-n-p* transistors on the same chip. The lateral *p-n-p* structure shown in Fig. 4-10 is the most common form of the integrated *p-n-p* transistor. This *p-n-p* uses the standard diffusion techniques as the *n-p-n*, but the last *n* diffusion (used for the *n-p-n* transistor) is eliminated. While the *p* base for the *n-p-n* transistor is made, the two adjacent *p* regions are diffused for the emitter and collector of the *p-n-p* transistor shown in Fig. 4-10. Note that the current flows *laterally* from emitter to collector.

The *p-n-p* transistor has inferior characteristics compared with those of the *n-p-n* device. The tolerance of the base thickness (width) in Fig. 4-10 is determined by the lateral diffusion of the *p*-type impurities as well as by

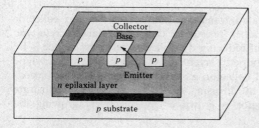

Figure 4-10 A *p-n-p* lateral transistor.

photographic limitations during mask making and alignment. Consequently, the base width is much larger than that of an *n-p-n* transistor, and the current gain of the *p-n-p* transistor is usually very low (0.5 to 5) instead of 50 to 300 for the *n-p-n* device. However, with improved processing control it is possible to obtain a gain as high as 100 for the lateral *p-n-p* transistor. Since the resistivity of the base region of the *n-p-n* resistor is relatively high, the collector and emitter resistances are large. Because of the long thickness of the base, these transistors have poor high-frequency response (Sec. 13-5).

Vertical *p-n-p* Transistor[3.5]

This transistor uses the substrate for the *p* collector; the *n* epitaxial layer for the base; and the *p* base of the standard *n-p-n* transistor as the emitter of this *p-n-p* device. We have already emphasized that the substrate must be connected to the most negative potential in the circuit. Hence a vertical *p-n-p* transistor can be used only if its collector is at a fixed negative voltage. Such a configuration is called an *emitter follower*, and is discussed in Sec. 11-10.

The vertical *p-n-p* transistor is a poor match for the *n-p-n* transistor. From Figs. 4-2 and 4-7 the base width is $15 - 2.7 = 12.3$ μm compared with 0.7 μm for the *n-p-n* transistor. However, if the epitaxial layer is made much thinner than the 15 μm indicated in Fig. 4-2, it is possible to obtain values of current gain as large as 100 for the vertical *p-n-p* device.

Triple-diffused *p-n-p* Transistor

If an extra *p*-type diffusion is added (after the *n*-type diffusion) to the processes described in Fig. 4-2, it is possible to obtain a *p-n-p* transistor. Besides the additional fabrication step required, there are serious design[3] considerations which limit the usefulness of the triple-diffused *p-n-p* transistor.

Supergain *n-p-n* Transistor[3, 5]

If the emitter is diffused into the base region so as to reduce the effective base width almost to the point of *punch-through* (Sec. 3-12), the current gain may be increased drastically (typically, 5,000). However, the breakdown voltage is reduced to a very low value (say, 5 V). If such a transistor in the CE configuration is operated in series with a standard integrated CB transistor (such a combination is called a *cascode* arrangement), the superhigh gain can be obtained at very low currents and with breakdown voltages in excess of 50 V.

4-7 MONOLITHIC DIODES[1]

The diodes utilized in integrated circuits are made by using transistor structures in one of five possible connections (Prob. 4-5). The three most popular diode structures are shown in Fig. 4-11. They are obtained from a transistor structure

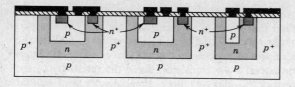

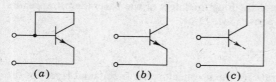

Figure 4-11 Cross section of various diode structures. (a) Emitter-base diode with collector shorted to base; (b) emitter-base diode with collector open; (c) collector-base diode (no emitter diffusion).

by using the emitter-base diode, with the collector short-circuited to the base (a); the emitter-base diode, with the collector open (b); and the collector-base diode, with the emitter open-circuited (or not fabricated at all) (c). The choice of the diode type used depends upon the application and circuit performance desired. Collector-base diodes have the higher collector-base voltage-breakdown rating of the collector junction (~ 12 V minimum), and they are suitable for common-cathode diode arrays diffused within a single isolation island, as shown in Fig. 4-12a. Common-anode arrays can also be made with the collector-base diffusion, as shown in Fig. 4-12b. A separate isolation is required for each diode, and the anodes are connected by metalization.

The emitter and base regions are very popular for the fabrication of diodes provided that the reverse-voltage requirement of the circuit does not exceed the lower base-emitter breakdown voltage (~ 7V). Common-anode arrays can easily be made with the emitter and base diffusions by using a multiple-emitter transistor within a single isolation area, as shown in Fig. 4-13. The collector may

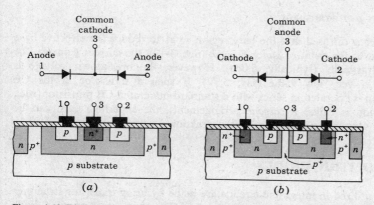

Figure 4-12 Diode pairs. (a) Common-cathode pair and (b) common-anode pair, using collector-base diodes.

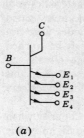
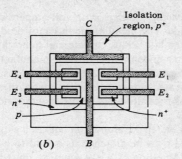

Figure 4-13 A multiple-emitter *n-p-n* transistor. (*a*) Schematic, (*b*) monolithic surface pattern. If the base is connected to the collector, the result is a multiple-cathode diode structure with a common anode.

be either open or shorted to the base. The diode pair in Fig. 4-11 is constructed in this manner, with the collector floating (open).

The multiple-emitter transistor is an important device in logic circuits (Secs. 5-11 and 6-8). As many as 64 emitters have been fabricated in such devices.

Diode Characteristics

The forward volt-ampere characteristics of the three diode types discussed above are shown in Fig. 4-14. It will be observed that the diode-connected transistor (emitter-base diode with collector short-circuited to the base) provides the highest conduction for a given forward voltage. The reverse recovery time for this diode is also smaller, one-third to one-fourth that of the collector-base diode.

4-8 THE METAL-SEMICONDUCTOR CONTACT[3.7]

Two types of metal-semiconductor junctions are possible, *ohmic* and *rectifying*. The former is the type of contact desired when a lead is to be attached to a semiconductor. On the other hand, the rectifying contact results in a metal-semi-

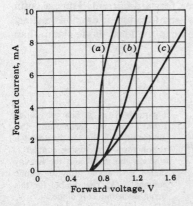

Figure 4-14 Typical diode volt-ampere characteristics for the three diode types of Fig. 4-11. (*a*) Base-emitter (collector shorted to base); (*b*) base-emitter (collector open); (*c*) collector-base (emitter open). (*Courtesy of Fairchild Semiconductor.*)

conductor diode (called a *Schottky barrier*), with volt-ampere characteristics very similar to those of a *p-n* diode. The metal-semiconductor diode was investigated many years ago, but until the late 1960s commercial Schottky diodes were not available because of problems encountered in their manufacture. It has turned out that most of the fabrication difficulties are due to surface effects; by employing the surface-passivated integrated-circuit techniques described in this chapter, it is possible to construct almost ideal metal-semiconductor diodes very economically.

As mentioned in Sec. 4-2 (step 5), aluminum acts as a *p*-type impurity when in contact with silicon. If Al is to be attached as a lead to *n*-type Si, an ohmic contact is desired and the formation of a *p-n* junction must be prevented. It is for this reason that n^+ diffusions are made in the *n* regions near the surface where the Al is deposited (Fig. 4-2d). On the other hand, if the n^+ diffusion is omitted and the Al is deposited directly upon the *n*-type Si, an equivalent *p-n* structure is formed, resulting in an excellent metal-semiconductor diode. In Fig. 4-15 contact 1 is a Schottky barrier, whereas contact 2 is an ohmic (nonrectifying) contact, and a metal-semiconductor diode exists between these two terminals, with the anode at contact 1. Note that the fabrication of a Schottky diode is actually simpler than that of a *p-n* diode, which requires an extra (*p*-type) diffusion.

The external volt-ampere characteristic of a metal-semiconductor diode is essentially the same as that of a *p-n* junction, but the physical mechanisms involved are more complicated. Note that in the forward direction electrons from the *n*-type Si cross the junction into the metal, where electrons are plentiful. In this sense, this is a majority-carrier device, whereas minority carriers account for a *p-n* diode characteristic. As explained in Sec. 2-13, there is a delay in switching a *p-n* diode from ON to OFF because the minority carriers stored at the junction must first be removed. Schottky diodes have a negligible storage time t_s because the current is carried predominantly by majority carriers. (Electrons from the *n* side enter the aluminum and become indistinguishable from the electrons in the metal, and hence are not "stored" near the junction.)

It should be mentioned that the voltage drop across a Schottky diode is much less than that of a *p-n* diode for the same forward current. Thus, a cutin voltage of about 0.3 V is reasonable for a metal-semiconductor diode as against 0.6 V for a *p-n* barrier. Hence the former is closer to the ideal diode clamp than the latter.

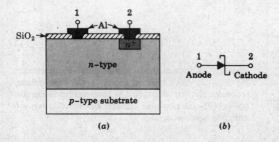

(a)

(b)

Figure 4-15 (*a*) A Schottky diode formed by IC techniques. The aluminum and the lightly doped *n* region form a rectifying contact 1, whereas the metal and the heavily doped n^+ region form an ohmic contact 2. (*b*) The symbol for this metal-semiconductor diode.

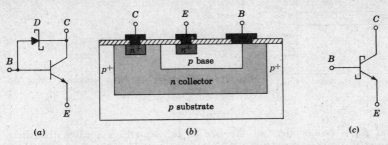

Figure 4-16 (*a*) A transistor with a Schottky-diode clamp between base and collector to prevent saturation. (*b*) The cross section of a monolithic IC equivalent to the diode-transistor combination in (*a*). (*c*) The Schottky transistor symbol, which is an abbreviation for that shown in (*a*).

The Schottky Transistor

To obtain the fastest circuit operation, a transistor must be prevented from entering saturation (Sec. 3-14). This condition can be achieved, as indicated in Fig. 4-16*a*, by using a Schottky diode as a clamp between the base and collector. If an attempt is made to saturate this transistor by increasing the base current, the collector voltage drops, *D* conducts, and the base-to-collector voltage is limited to about 0.4 V. Since the collector junction is forward-biased by less than the cutin voltage (≈ 0.5 V), the transistor does *not* enter saturation (Sec. 3-8).

As indicated in Fig. 4-16*b*, the aluminum metalization for the base lead is allowed to make contact also with the *n*-type collector region (but without an intervening n^+ section). This simple procedure forms a metal-semiconductor diode between base and collector. The device in Fig. 4-16*b* is equivalent to the circuit of Fig. 4-16*a*. This is referred to as a *Schottky transistor*, and is represented by the symbol in Fig. 4-16*c*.

4-9 INTEGRATED RESISTORS[1]

A resistor in a monolithic integrated circuit is very often obtained by utilizing the bulk resistivity of one of the diffused areas. The *p*-type base diffusion is most commonly used, although the *n*-type emitter diffusion is also employed. Since these diffusion layers are very thin, it is convenient to define a quantity known as the *sheet resistance* R_S.

Sheet Resistance

If, in Fig. 4-17, the width *w* equals the length *l*, we have a square *l* by *l* of material with resistivity ρ, thickness *y*, and cross-sectional area $A = ly$. The resistance of this conductor (in ohms per square) is

$$R_S = \frac{\rho l}{ly} = \frac{\rho}{y} \tag{4-2}$$

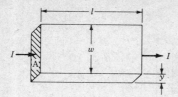

Figure 4-17 Pertaining to sheet resistance, ohms per square.

Note that R_S is independent of the size of the square. Typically, the sheet resistance of the base and emitter diffusions whose profiles are given in Fig. 4-7 is 200 Ω/square and 2.2 Ω/square, respectively.

The construction of a base-diffused resistor is shown in Fig. 4-1 and is repeated in Fig. 4-18a. A top view of this resistor is shown in Fig. 4-18b. The resistance value may be computed from

$$R = \frac{\rho l}{yw} = R_S \frac{l}{w} \tag{4-3}$$

where l and w are the length and width of the diffused area, as shown in the top view. For example, a base-diffused-resistor stripe 1 mil wide and 10 mils long contains 10 (1 by 1 mil) squares, and its value is $10 \times 200 = 2,000$ Ω. Empirical[1,2] corrections for the end contacts are usually included in calculations of R.

Resistance Values

Since the sheet resistance of the base and emitter diffusions is fixed, the only variables available for diffused-resistor design are stripe length and stripe width. Stripe widths of less than 1 mil (0.001 in) are not normally used because a line-width variation of 0.0001 in due to mask drawing error or mask misalignment or photographic-resolution error can result in 10 percent resistor-tolerance error.

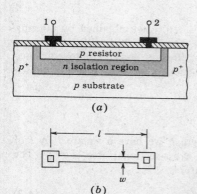

Figure 4-18 A monolithic resistor. (a) Cross-sectional view; (b) top view.

The range of values obtainable with diffused resistors is limited by the size of the area required by the resistor. Practical range of resistance is 20 Ω to 30 kΩ for a base-diffused resistor and 10 Ω to 1 kΩ for an emitter-diffused resistor. The tolerance which results from profile variations and surface geometry errors[1] is as high as ±10 percent of the nominal value at 25°C, with ratio tolerance of ±1 percent. For this reason the design of integrated circuits should, if possible, emphasize *resistance ratios rather than absolute values*. The temperature coefficient for these heavily doped resistors is positive (for the same reason that gives a positive coefficient to the silicon sensistor, discussed in Sec. 1-10) and is + 0.06 percent/°C from −55 to 0°C and +0.20 percent/°C from 0 to 125°C.

Equivalent Circuit

A model of the diffused resistor is shown in Fig. 4-19, where the parasitic capacitances of the base-isolation (C_1) and isolation-substrate (C_2) junctions are included. In addition, it can be seen that a parasitic *p-n-p* transistor exists, with the substrate as collector, the isolation *n*-type region as base, and the resistor *p*-type material as the emitter. The collector is reverse-biased because the *p*-type substrate is at the most negative potential. It is also necessary that the emitter be reverse-biased to keep the parasitic transistor at cutoff. This condition is maintained by placing all resistors in the same isolation region and *connecting the n-type isolation region surrounding the resistors to the most positive voltage present in the circuit*. Typical values of h_{fe} for this parasitic transistor range from 0.5 to 5.

Epitaxial Resistor[3]

The sheet resistance of the collector epitaxial region is about six times that of the base diffusion. Hence, it is possible to fabricate higher value resistances by using the epitaxial layer. Such a resistor is defined by the isolation diffusion which surrounds the resistor (Fig. 4-18). These sidewall effects become important and, to maintain accuracy of resistance values, the isolation diffusion must be carefully controlled.

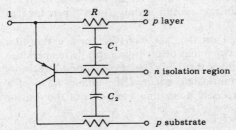

Figure 4-19 The equivalent circuit of a diffused resistor.

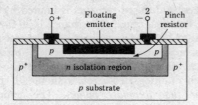

Figure 4-20 Cross-sectional view of a pinch resistor.

Pinch Resistor

Consider what happens to the resistance of Fig. 4-18 if an emitter diffusion is added as in Fig. 4-20. The n-type material does not contribute to the conduction because to do so the current from 1 to 2 would have to flow through the n-p diode at contact 2 in the reverse direction. In other words, only the very-small-diode reverse saturation current passes through the n-emitter material. With the reduction in the conduction path cross section of the p material (*pinching*), the resistance must increase. Resistances in excess of 50 kΩ may be obtained, although their exact values are not highly controllable. Pinch resistors are nonlinear since the resistance depends upon the impressed voltage (Sec. 8-3).

The same voltage limitation of reverse base-emitter breakdown BV_{EBO} (~ 6 V) must apply for pinch resistors, since they are identical in construction to the base-emitter junction with the emitter lead absent. This is not a serious problem since such a resistor is usually used in low-voltage biasing applications across a forward-biased base-emitter junction.

Thin-Film Resistors[1, 3]

A technique of vapor thin-film deposition can also be used to fabricate resistors for integrated circuits. The metal (usually nichrome NiCr) film is deposited (to a thickness of less than 1 μm) on the silicon dioxide layer, and masked etching is used to produce the desired geometry. The metal resistor is then covered by an insulating layer, and apertures for the ohmic contacts are opened through this insulating layer. Typical sheet-resistance values for nichrome thin-film resistors are 40 to 400 Ω/square, resulting in resistance values from about 20 Ω to 50 kΩ. The temperature coefficient of a nichrome resistor is only about 5 percent that of a diffused resistor.

Tantalum may also be deposited on the SiO₂ layer to form a thin-film resistor. Its sheet resistance is about twice that of nichrome and its temperature coefficient may be much smaller than the NiCr resistor. Diffused resistors cannot be adjusted after they are fabricated, whereas the ohmic value of a thin-film resistor may be trimmed precisely by cutting away part of the resistor with a laser beam.

4-10 INTEGRATED CAPACITORS[1, 2, 3]

Capacitors in integrated circuits may be obtained by utilizing the transition capacitance of a reverse-biased p-n junction or by a thin-film technique.

Junction Capacitors

A cross-sectional view of a junction capacitor is shown in Fig. 4-21a. The capacitor is formed by the reverse-biased junction J_2, which separates the epitaxial n-type layer from the upper p-type diffusion area. An additional junction J_1 appears between the n-type epitaxial plane and the substrate, and a parasitic capacitance C_1 is associated with this reverse-biased junction. The equivalent circuit of the junction capacitor is shown in Fig. 4-21b, where the desired capacitance C_2 should be as large as possible relative to C_1. The value of C_2 depends on the junction area and impurity concentration. This junction is essentially linearly graded. The series resistance R (10 to 50 Ω) represents the resistance of the n-type layer.

It is clear that the substrate must be at the most negative voltage so as to minimize C_1 and isolate the capacitor from other elements by keeping junction J_1 reverse-biased. It should also be pointed out that the junction capacitor C_2 is polarized since the p-n junction J_2 must always be reverse-biased.

MOS and Thin-Film Capacitors

A metal-oxide semiconductor (MOS) nonpolarized capacitor is indicated in Fig. 4-22a. This structure is a parallel-plate capacitor with SiO_2 (whose thickness is 500 Å) as the dielectric. A surface thin film of metal (aluminum) is the top plate. The bottom plate consists of the heavily doped n^+ region that is formed during the emitter diffusion. The equivalent circuit of the MOS capacitor is

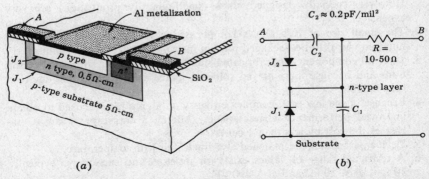

Figure 4-21 (a) Junction monolithic capacitor. (b) Equivalent circuit. *(Courtesy of Motorola, Inc.)*

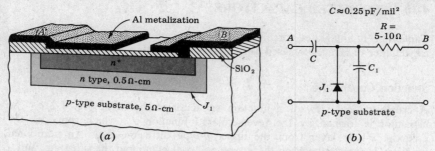

Figure 4-22 An MOS capacitor. (*a*) The structure; (*b*) the equivalent circuit.

shown in Fig. 4-22*b*, where C_1 denotes the parasitic capacitance of the collector-substrate junction, and R is the small series resistance of the n^+ region.

The capacitance of either the junction or the MOS capacitor is about 0.25 pF/mil^2. An increase in this value by a factor of about 10 is possible with the use of tantalum. A controlled growth of tantalum pentoxide is used for the dielectric, and metalic tantalum is deposited for the top plate (since aluminum is soluble in Ta_2O_5). The increased capacitance is obtained at the expense of additional processing steps.

4-11 CHARACTERISTICS OF INTEGRATED COMPONENTS

Based on our discussion of IC technology, we can summarize the significant characteristics of integrated circuits as follows:

1. Standard IC's (those stocked by manufacturers) are very low cost; For example, the National Semiconductor LM741 OP AMP containing 21 transistors, 1 diode, and 12 resistors sold for about 50 cents (in 1977). However, specially designed chips (small-quantity production) are very expensive.
2. The small size of IC's allows complicated systems (consisting of several hundred chips) to be packaged into an instrument of practical volume.
3. Since all components are fabricated simultaneously under controlled conditions and because there are no soldered joints, a microelectronic device is very reliable.
4. Because of the low cost, complex circuitry on a chip may be used to obtain improved performance characteristics. Adding a transistor to an IC increases the cost by less than 1 cent!
5. Device parameters are matched and track well with temperature.
6. A restricted range of values exists for resistors and capacitors. Typically, $10\ \Omega \leqslant R \leqslant 50\ k\Omega$ and $C \leqslant 200$ pF.
7. Poor tolerances are obtained in fabricating resistors and capacitors of specific magnitudes. For example, ± 20 percent of absolute values is typical.

Resistance ratio tolerance can be specified to ± 1 percent because all resistors are made at the same time using the same techniques.

8. Components have high-temperature coefficients and may also be voltage-sensitive.
9. High-frequency response is limited by parasitic capacitances.
10. No practical inductors or transformers can be integrated.
11. Because extra steps are required in the fabrication of thin-film resistors and capacitors, their cost increases and the yield decreases. Hence, these thin-film devices should be used only if their special characteristics are required.

In the next section we examine some of the design rules for the layout of monolithic circuits.

4-12 MONOLITHIC-CIRCUIT LAYOUT[1, 3, 8]

In this section we describe how to transform the discrete logic circuit of Fig. 4-23a into the layout of the monolithic circuit shown in Fig. 4-24.

Design Rules for Monolithic Layout

The following are reasonable rules to be used in the fabrication of an IC:

1. In the layout, allow an isolation border equal to twice the epitaxial thickness in order to take lateral diffusion into account.

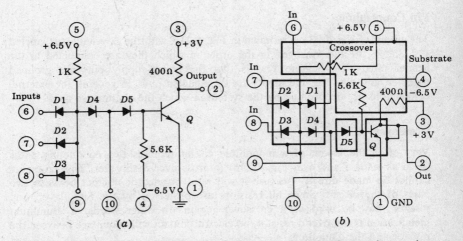

Figure 4-23 (*a*) A DTL gate. (*b*) The schematic redrawn to indicate the 10 external connections arranged in the sequence in which they will be brought out to the header pins. The isolation regions are shown in heavy outline.

2. Since the isolation diffusion occupies an appreciable area of the chip, then the number of isolation islands should be minimized.

3. Place all resistors in the same isolation island and return that isolation region to *the most positive potential* in the circuit.

4. Place all transistors *whose collectors are tied together* into the same isolation island. For most circuits each transistor will be in a separate island.

5. Connect the substrate (Fig. 4-5) to *the most negative potential* of the circuit.

6. Use 1-mil widths for diffused emitter regions and $\frac{1}{2}$-mil widths for base (and collector) contacts and spacings.

7. For resistors, use widest possible designs consistent with die-size limitations. Resistances which must have a close ratio must have the same width and be placed close to one another.

8. Determine component and metallization geometries from the performance requirements of the circuit. For example, the transistor in the output stage of an amplifier would have a larger area than the other transistors if the output stage is to supply the maximum current.

9. Use an aluminum pad of area at least $3 \times 3 = 9$ mils^2 to which to bond wires for connection to the header pins.

10. Keep all metalization runs as short and wide as possible, particularly at the emitter and collector connections of a saturating transistor.

11. Optimize the layout arrangement to maintain the smallest possible chip size.

12. Redraw the schematic to satisfy the required pin connections with the minimum number of crossovers.

13. Use an alignment pattern on the artwork so as to simplify the registration of successive masks.

Pin Connections

The circuit of Fig. 4-23a is redrawn in Fig. 4-23b, with the external leads labeled 1, 2, 3, . . . , 10 and arranged in the order in which they are connected to the header pins. The diagram reveals that the power-supply pins are grouped together, and also that the inputs are on adjacent pins. In general, the external connections are determined by the system in which the circuits are used.

Crossovers

Very often the layout of a monolithic circuit requires two conducting paths (such as leads 5 and 6 in Fig. 4-23b) to cross over each other. This crossover cannot be made directly because it will result in electric contact between two parts of the circuit. Since all resistors are protected by the SiO$_2$ layer, any resistor may be used as a crossover region. In other words, if aluminum metalization is run over a resistor, no electric contact will take place between the resistor and the aluminum.

Sometimes the layout is so complex that additional crossover regions may be required. A diffused structure which allows a crossover is obtained as follows:[3] During the emitter fabrication n^+ impurities are diffused along a line in the epitaxial region and contact windows are opened at each end of the line.

This process forms a "diffused wire." Aluminum is deposited on the insulating SiO_2 (between the two end contacts) in a line perpendicular to the diffused section so as to form a connecting wire for some other part of the circuit. Thus, the two wires (one of aluminum and the other of n^+ material) cross over each other without making electrical contact. The diffused wire is called a *buried crossover*.

Isolation Islands

The number of isolation islands is determined next. Since the transistor collector requires one isolation region, the heavy rectangle has been drawn in Fig. 4-23*b* around the transistor. It is shown connected to the output pin 2 because this isolation island also forms the transistor collector. Next, all resistors are placed in the same isolation island, and the island is then connected to the most positive voltage in the circuit, for reasons discussed in Sec. 4-9.

To determine the number of isolation regions required for the diodes, it is necessary first to establish which kind of diode will be fabricated. In this case, because of the low forward drop shown in Fig. 4-14, it was decided to make the

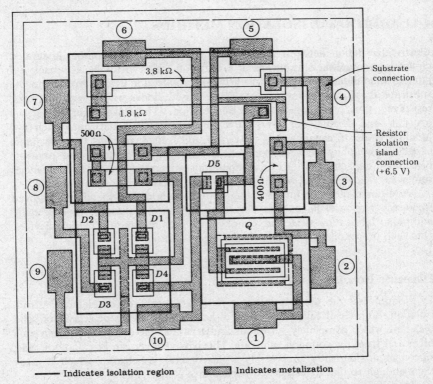

Figure 4-24 Monolithic design layout for the circuit of Fig. 4-23. *(Courtesy of Motorola Monitor, Phoenix, Ariz.)*

common-anode diodes of the emitter-base type with the collector shorted to the base. Since the "collector" is at the "base" potential, it is required to have a single isolation island for the four common-anode diodes. Finally, the remaining diode is fabricated as an emitter-base diode, with the collector open-circuited, and thus it requires a separate isolation island.

The Fabrication Sequence

The final monolithic layout is determined by a trial-and-error process, having as its objective the smallest possible die size. This layout is shown in Fig. 4-24. The reader should identify the four isolation islands, the three resistors, the five diodes, and the transistor. It is interesting to note that the 5.6-kΩ resistor has been achieved with a 2-mil-wide 1.8-kΩ resistor in series with a 1-mil-wide 3.8-kΩ resistor. To conserve space, the resistor was folded back on itself. In addition, two metalizing crossovers ran over this resistor. From a layout such as shown in Fig. 4-24, the manufacturer produces the masks required for the fabrication of the monolithic integrated circuit.

4-13 ADDITIONAL ISOLATION METHODS

Electrical isolation between the different elements of a monolithic integrated circuit is accomplished by means of a diffusion which yields back-to-back p-n junctions, as indicated in Sec. 4-2. With the application of bias voltage to the substrate, these junctions represent reverse-biased diodes with a very high back resistance, thus providing adequate dc isolation. (The leakage current is ~ 1 pA/mil^2 and doubles for every 10°C in temperature.) But since each p-n junction is also a capacitance (~ 0.06 pF/mil^2), there remains that inevitable capacitive coupling between components and the substrate. These parasitic distributed capacitances thus limit monolithic integrated circuits to frequencies somewhat below those at which corresponding discrete circuits can operate.

Additional methods for achieving better isolation, and therefore improved frequency response, have been developed.[9] However, these methods require additional fabrication steps and, hence, result in much more expensive devices. One such process is described in the following section.

Dielectric Isolation

In this process[9] the diode-isolation concept is discarded completely. Instead, isolation, both electrical and physical, is achieved by means of a layer of solid dielectric which completely surrounds and separates the components from each other and from the common substrate. This passive layer can be silicon dioxide, silicon monoxide, ruby, or possibly a glazed ceramic substrate which is made thick enough so that its associated capacitance is negligible.

In a dielectric isolated integrated circuit it is possible to fabricate readily p-n-p and n-p-n transistors within the same silicon substrate. It is also simple to have both fast and charge-storage diodes and also both high- and low-frequency transistors in the same chip through selective gold diffusion—a process prohibited by conventional techniques due to the rapid rate at which gold diffuses through silicon unless impeded by a physical barrier such as a dielectric layer.

An isolation method pioneered by RCA is referred to as SOS (silicon-on-sapphire). On a single-crystal sapphire substrate an n-type silicon layer is grown heteroepitaxially. By etching away selected portions of the silicon, isolated islands are formed (interconnected only by the high-resistance sapphire substrate).

REFERENCES

1. Motorola, Inc. (R. M. Warner, Jr., and J. N. Fordemwalt, eds.): "Integrated Circuits," McGraw-Hill Book Company, New York, 1965.
2. Phillips, A. B.: Monolithic Integrated Circuits, *IEEE Spectrum*, vol. 1, no. 6, pp. 83–101, June, 1964.
3. Hamilton, D. J., and W. G. Howard: "Basic Integrated Circuit Engineering," McGraw-Hill Book Company, New York
   Yang, E. S.: "Fundamentals of Semiconductor Devices," McGraw-Hill Book Company, New York, 1978.
4. Oldham, W. G.: The Fabrication of Microelectronic Circuits, *Scientific American*, vol. 237, no. 3, pp. 111–128, September, 1977.
5. Hunter, L. P.: "Handbook of Semiconductor Electronics," 2d ed., sec. 8, McGraw-Hill Book Company, New York, 1962.
6. King, D., and L. Stern: Designing Monolithic Integrated Circuits, *Semicond. Prod. Solid State Technol.*, March, 1965.
7. Yu, A. Y. C.: The Metal-semiconductor Contact: An Old Device with a New Future, *IEEE Spectrum*, vol. 7, no. 3, pp. 83-89, March, 1970.
8. Phillips, A. B.: Designing Digital Monolithic Integrated Circuits, *Motorola Monitor*, vol. 2, no. 2, pp. 18-27, 1964.
9. Hamilton, D. J., and W. G. Howard, Ref. 3, pp. 83–88.

REVIEW QUESTIONS

4-1 What are the five advantages of integrated circuits?

4-2 List the six steps involved in fabricating a monolithic integrated circuit (IC).

4-3 List the six basic processes involved in the fabrication of a microcircuit.

4-4 Describe *epitaxial growth*.

4-5 (*a*) Describe the *photoetching process*.

 (*b*) How many masks are required to complete an IC? List the function performed by each mask.

4-6 (*a*) Describe the *diffusion process*.

 (*b*) What is meant by an *impurity profile*?

4-7 (*a*) How is the surface layer of SiO_2 formed?

 (*b*) How thick is this layer?

 (*c*) What are the reasons for forming the SiO_2 layers?

4-8 Explain how isolation between components is obtained in an IC.

4-9 How are the components interconnected in an IC?

4-10 Explain what is meant by *parasitic capacitance* in an IC.

4-11 Give the order of magnitude of (*a*) the substrate thickness; (*b*) the epitaxial thickness; (*c*) the base width (thickness); (*d*) the diffusion time; (*e*) the diffusion temperature; (*f*) the surface area of a transistor; (*g*) the chip size.

4-12 Sketch the cross section of an IC transistor.

4-13 Sketch the cross section of a discrete planar epitaxial transistor.

4-14 List the advantages and disadvantages of an IC versus a discrete transistor.

4-15 Define *buried layer*. Why is it used?

4-16 Describe a *lateral p-n-p* transistor. Why is its current gain low?

4-17 Describe a *vertical p-n-p* transistor. Why is it of limited use?

4-18 Describe a *supergain* transistor.

4-19 (*a*) How are IC diodes fabricated?

(*b*) Sketch the cross sections of two types of emitter-base diodes.

4-20 Sketch the cross section of a diode pair using collector-base regions if (*a*) the cathode is common and (*b*) the anode is common.

4-21 Sketch the top view of a multiple-emitter transistor. Show the isolation, collector, base, and emitter regions.

4-22 How is an aluminum contact made with *n*-type silicon so that it is (*a*) ohmic; (*b*) rectifying?

4-23 Why is storage time eliminated in a metal-semiconductor diode?

4-24 What is a *Schottky transistor*? Why is storage time eliminated in such a transistor? Are there any extra fabrication steps required to produce such a transistor? Explain.

4-25 Sketch the cross section of an IC Schottky transistor.

4-26 (*a*) Define *sheet resistance* R_S.

(*b*) What is the order of magnitude of R_S for the base region and also for the emitter region?

(*c*) Sketch the cross section of an IC resistor.

(*d*) What are the order of magnitudes of the smallest and the largest values of an IC resistance?

4-27 (*a*) Sketch the equivalent circuit of a base-diffused resistor, showing all parasitic elements.

(*b*) What must be done (externally) to minimize the effect of the parasitic elements?

4-28 Describe a *thin-film resistor*.

4-29 (*a*) Sketch the cross section of a junction capacitor.

(*b*) Draw the equivalent circuit, showing all parasitic elements.

4-30 Repeat Rev. 4-29 for an MOS capacitor.

4-31 (*a*) What are the two basic distinctions between a junction and an MOS capacitor?

(*b*) What is the order of magnitude of the capacitance per square mil?

4-32 (*a*) To what voltage is the substrate connected? Why?

(*b*) To what voltage is the isolation island containing the resistors connected? Why?

(*c*) Can several transistors be placed in the same isolation island? Explain.

4-33 List six important characteristics of integrated components.

4-34 List six design rules for monolithic-circuit layout.

TWO

DIGITAL CIRCUITS AND SYSTEMS

DIGITAL CIRCUITS

Even in a large-scale digital system, such as a computer, or a data-processing, control, or digital-communication system, there are only a few basic operations which must be performed. These operations, to be sure, may be repeated very many times. The four circuits most commonly employed in such systems are known as the OR, AND, NOT, and FLIP-FLOP. These are called *logic* gates, or circuits, because they are used to implement Boolean algebraic equations (as we shall soon demonstrate). This algebra was invented by G. Boole in the middle of the nineteenth century as a system for the mathematical analysis of logic.

This chapter discusses in detail the first three basic logic circuits mentioned above. These basic gates are combined into FLIP-FLOPS and other digital-system building blocks in Chaps. 7, 8, and 9.

5-1 DIGITAL (BINARY) OPERATION OF A SYSTEM

A digital system functions in a binary manner. It employs devices which exist only in two possible states. A transistor is allowed to operate at cutoff or in saturation, but not in its active region. A node may be at a high voltage of, say 4 ± 1 V or at a low voltage of, say, 0.2 ± 0.2 V, but no other values are allowed (Fig. 5-1). Various designations are used for these two quantized states, and the most common are 1 or 0, high or low, and true or false. Binary arithmetic and mathematical manipulation of switching or logic functions are best carried out with the classification, which involves two symbols, 0 (zero) and 1 (one).

The binary system of representing numbers will now be explained by making reference to the familiar *decimal system*. In the latter the base is 10 (ten), and ten numerals, 0, 1, 2, 3, . . . , 9, are required to express an arbitrary number. To write numbers larger than 9, we assign a meaning to the *position* of a numeral in an array of numerals. For example, the number 1,264 (one thousand two hundred sixty four) has the meaning

$$1,264 \equiv 1 \times 10^3 + 2 \times 10^2 + 6 \times 10^1 + 4 \times 10^0$$

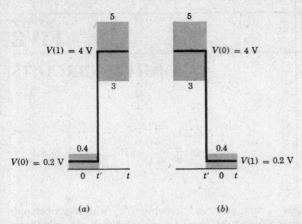

Figure 5-1 Illustrating the definitions of (a) positive and (b) negative logic. A transition from one state to the other occurs at $t = t'$.

Thus the individual digits in a number represent the coefficients in an expansion of the number in powers of 10. The digit which is farthest to the right is the coefficient of the zeroth power, the next is the coefficient of the first power, and so on.

In the *binary system* of representation the base is 2, and only the two numerals 0 and 1 are required to represent a number. The numerals 0 and 1 have the same meaning as in the decimal system, but a different interpretation is placed on the position occupied by a digit. In the binary system the individual digits represent the coefficients of powers of *two* rather than *ten* as in the

Table 5-1 Equivalent numbers in decimal and binary notation

Decimal notation	Binary notation	Decimal notation	Binary notation
0	00000	11	01011
1	00001	12	01100
2	00010	13	01101
3	00011	14	01110
4	00100	15	01111
5	00101	16	10000
6	00110	17	10001
7	00111	18	10010
8	01000	19	10011
9	01001	20	10100
10	01010	21	10101

decimal system. For example, the decimal number 19 is written in the binary representation as 10011 since

$$10011 \equiv 1 \times 2^4 + 0 \times 2^3 + 0 \times 2^2 + 1 \times 2^1 + 1 \times 2^0$$
$$= \quad 16 \quad + \quad 0 \quad + \quad 0 \quad + \quad 2 \quad + \quad 1 \quad = 19$$

A short list of equivalent numbers in decimal and binary notation is given in Table 5-1.

A binary digit (a 1 or a 0) is called a *bit*. A group of bits having a significance is a *byte*, *word*, or *code*. For example, to represent the 10 numerals $(0, 1, 2, \ldots, 9)$ and the 26 letters of the English alphabet would require 36 different combinations of 1's and 0's. Since $2^5 < 36 < 2^6$, then a minimum of 6 bits per bite are required in order to accommodate all the alphanumeric characters. In this sense a byte is sometimes referred to as a *character*.

Logic Systems

In a *dc*, or *level-logic*, system a bit is implemented as one of two voltage levels. If, as in Fig. 5-1*a*, the more positive voltage is the 1 level and the other is the 0 level, the system is said to employ dc *positive* logic. On the other hand, a dc *negative*-logic system, as in Fig. 5-1*b*, is one which designates the more negative voltage state of the bit as the 1 level and the more positive as the 0 level. It should be emphasized that the absolute values of the two voltages are of no significance in these definitions. In particular, the 0 state need not represent a zero voltage level (although in some systems it might).

The parameters of a physical device (for example, $V_{CE(\text{sat})}$ of a transistor) are not identical from sample to sample, and they also vary with temperature. Furthermore, ripple or voltage spikes may exist in the power supply or ground leads, and other sources of unwanted signals, called *noise*, may be present in the circuit. For these reasons the digital levels are not specified precisely, but as indicated by the shaded regions in Fig. 5-1, each state is defined by a voltage range about a designated level, such as 4 ± 1 V and 0.2 ± 0.2 V.

In a *dynamic*, or *pulse-logic*, system a bit is recognized by the presence or absence of a pulse. A 1 signifies the existence of a positive pulse in a dynamic positive-logic system; a negative pulse denotes a 1 in a dynamic negative-logic system. In either system a 0 at a particular input (or output) at a given instant of time designates that no pulse is present at that particular moment.

5-2 THE OR GATE

An OR gate has two or more inputs and a single output, and it operates in accordance with the following definition: *The output of an* OR *assumes the 1 state if one or more inputs assume the 1 state.* The *n* inputs to a logic circuit will be designated by $A, B, \ldots, N$ and the output by Y. It is to be understood that

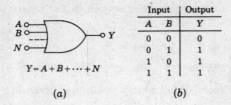

$$Y = A + B + \cdots + N$$

(a)

Input		Output
A	B	Y
0	0	0
0	1	1
1	0	1
1	1	1

(b)

Figure 5-2 (a) The standard symbol for an OR gate and its Boolean expression. (b) The truth table for a two-input OR gate.

each of these symbols may assume one of two possible values, either 0 or 1. A standard symbol for the OR circuit is given in Fig. 5-2a, together with the Boolean expression for this gate. The equation is to be read "Y equals A or B or $\cdots$ or N." Instead of defining a logical operation in words, an alternative method is to give a *truth table* which contains a tabulation of all possible input values and their corresponding outputs. It should be clear that the two-input truth table of Fig. 5-2b is equivalent to the above definition of the OR operation.

In a *diode-logic* (DL) system the logical gates are implemented by using diodes. A diode OR for negative logic is shown in Fig. 5-3. The generator source resistance is designated by R_s. We consider first the case where the supply voltage V_R has a value equal to the voltage $V(0)$ of the 0 state for dc logic.

If all inputs are in the 0 state, the voltage across each diode is $V(0) - V(0) = 0$. Since, in order for a diode to conduct, it must be forward-biased by at least the cutin voltage V_γ (Fig. 2-7), none of the diodes conducts. Hence the output voltage is $v_o = V(0)$, and Y is in the 0 state.

If now input A is changed to the 1 state, which for negative logic is at the potential $V(1)$, less positive than the 0 state, then $D1$ will conduct. The output becomes

$$v_o = V(0) - [V(0) - V(1) - V_\gamma] \frac{R}{R + R_s + R_f} \tag{5-1}$$

where R_f is the diode forward resistance. Usually R is chosen much larger than $R_s + R_f$. Under this restriction

$$v_o \approx V(1) + V_\gamma \tag{5-2}$$

Hence the output voltage exceeds the more negative level $V(1)$ by V_γ (approxi-

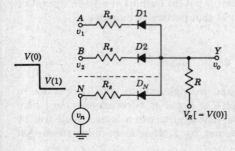

Figure 5-3 A diode OR circuit for negative logic. [It is also possible to choose the supply voltage such that $V_R > V(0)$, but that arrangement has the disadvantage of drawing standby current when all inputs are in the 0 state.]

mately 0.6 V for silicon or 0.2 V for germanium). Furthermore, the step in output voltage is *smaller* by V_γ than the change in input voltage.

From now on, unless explicitly stated otherwise, we shall assume $R \gg R_s$ and ideal diodes with $R_f = 0$ and $V_\gamma = 0$. The output, for input A excited, is then $v_o = V(1)$, and the circuit has performed the following logic: if $A = 1$, $B = 0, \ldots, N = 0$, then $Y = 1$, which is consistent with the OR operation.

For the above excitation, the output is at $V(1)$, and each diode, except $D1$, is back-biased. Hence the presence of signal sources at $B, C, \ldots, N$ does not result in an additional load on generator A. Since the OR configuration minimizes the interaction of the sources on one another, this gate is sometimes referred to as a *buffer* circuit. Since it allows several independent sources to be applied at a given node, it is also called a (nonlinear) *mixing* gate.

If two or more inputs are in the 1 state, the diodes connected to these inputs conduct and all other diodes remain reverse-biased. The output is $V(1)$, and again the OR function is satisfied. If for any reason the level $V(1)$ is not identical for all inputs, *the most negative value of* $V(1)$ *(for negative logic) appears at the output*, and all diodes except one are nonconducting.

A positive-logic OR gate uses the same configuration as that in Fig. 5-3, except that all diodes must be reversed. *The output now is equal to the most positive level* $V(1)$ [or more precisely is smaller than the most positive value of $V(1)$ by V_γ]. If a dynamic logic system is under consideration, *the output-pulse magnitude is* (approximately) *equal to the largest input pulse* (regardless of whether the system uses positive or negative logic).

A second mode of operation of the OR circuit of Fig. 5-3 is possible if V_R is set equal to a voltage more positive than $V(0)$ by at least V_γ. For this condition *all diodes conduct in the 0 state, and* $v_o \approx V(0)$ *if* $R \gg R_s + R_f$. If one or more inputs are excited, then the diode connected to the most negative $V(1)$ conducts, the output equals this value of $V(1)$, and all other diodes are back-biased. Clearly, the OR function has been satisfied.

Boolean Identities

If it is remembered that A, B, and C can take on only the value 0 or 1, the following equations from Boolean algebra pertaining to the OR ($+$) operation are easily verified:

$$A + B + C = (A + B) + C = A + (B + C) \tag{5-3}$$

$$A + B = B + A \tag{5-4}$$

$$A + A = A \tag{5-5}$$

$$A + 1 = 1 \tag{5-6}$$

$$A + 0 = A \tag{5-7}$$

These equations may be justified by referring to the definition of the OR operation, to a truth table, or to the action of the OR circuits discussed above.

5-3 THE AND GATE

An AND gate has two or more inputs and a single output, and it operates in accordance with the following definition: *The output of an AND assumes the 1 state if and only if all the inputs assume the 1 state*. A symbol for the AND circuit is given in Fig. 5-4a, together with the Boolean expression for this gate. The equation is to be read "*Y* equals *A* and *B* and ... and *N*." [Sometimes a dot (·) or a cross (×) is placed between symbols to indicate the AND operation.] It may be verified that the two-input truth table of Fig. 5-4b is consistent with the above definition of the AND operation.

A diode-logic (DL) configuration for a negative AND gate is given in Fig. 5-5a. To understand the operation of the circuit, assume initially that all source resistances R_s are zero and that the diodes are ideal. If *any* input is at the 0 level $V(0)$, the diode connected to this input conducts and the output is clamped at the voltage $V(0)$, or $Y = 0$. However, if *all* inputs are at the 1 level $V(1)$, then all diodes are reverse-biased and $v_o = V(1)$, or $Y = 1$. Clearly, the AND operation has been implemented. The AND gate is also called a *coincidence circuit*.

A positive-logic AND gate uses the same configuration as that in Fig. 5-5a, except that all diodes are reversed. This circuit is indicated in Fig. 5-5b and should be compared with Fig. 5-3. It is to be noted that the symbol $V(0)$ in Fig. 5-3 designates the same voltage as $V(1)$ in Fig. 5-5b because each represents the upper binary level. Similarly, $V(1)$ in Fig. 5-3 equals $V(0)$ in Fig. 5-5b, since both represent the lower binary level. Hence these two circuits are identical, and we conclude that a *negative* OR *gate is the same circuit as a positive* AND *gate*. This result is not restricted to diode logic, and by using Boolean algebra, we show in Sec. 5-7 that it is valid independently of the hardware used to implement the circuit.

In Fig. 5-5b it is possible to choose V_R to be more positive than $V(1)$. If this condition is met, all diodes will conduct upon a coincidence (all inputs in the 1 state) and the output will be clamped to $V(1)$. The output impedance is low in this mode of operation, being equal to $(R_s + R_f)/n$ in parallel with R. On the other hand, if $V_R = V(1)$, then all diodes are cut off at a coincidence, and the output impedance is high (equal to R). If for any reason not all inputs have the same upper level $V(1)$, then the output of the positive AND gate of Fig. 5-5b will equal $V(1)_{min}$, the *least* positive value of $V(1)$. Note that the diode connected to

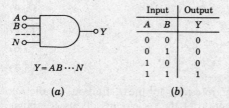

$$Y = AB \cdots N$$

(a)

Input		Output
A	*B*	*Y*
0	0	0
0	1	0
1	0	0
1	1	1

(b)

Figure 5-4 (a) The standard symbol for an AND gate and its Boolean expression; (b) The truth table for a two-input AND gate.

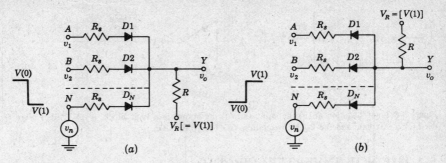

Figure 5-5 A diode-logic AND circuit for (a) negative logic and (b) positive logic.

$V(1)_{min}$ conducts, clamping the output to this minimum value of $V(1)$ and maintaining all other diodes in the reverse-biased condition. If, on the other hand, V_R is smaller than all inputs $V(1)$, then all diodes will be cut off upon coincidence and the output will rise to the voltage V_R. Similarly, if the inputs are pulses, then *the output pulse will have an amplitude equal to the smallest input amplitude* [provided that V_R is greater than $V(1)_{min}$].

Boolean Identities

Since A, B, and C can have only the value 0 or 1, the following expressions involving the AND operation may be verified:

$$ABC = (AB)C = A(BC) \tag{5-8}$$

$$AB = BA \tag{5-9}$$

$$AA = A \tag{5-10}$$

$$A1 = A \tag{5-11}$$

$$A0 = 0 \tag{5-12}$$

$$A(B + C) = AB + AC \tag{5-13}$$

These equations may be proved by reference to the definition of the AND operation, to a truth table, or to the behavior of the AND circuits discussed above. Also, by using Eqs. (5-11), (5-13), and (5-6), it can be shown that

$$A + AB = A \tag{5-14}$$

Similarly, it follows from Eqs. (5-13), (5-10), and (5-6) that

$$A + BC = (A + B)(A + C) \tag{5-15}$$

We shall have occasion to refer to the last two equations later.

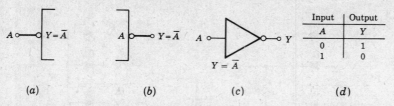

Figure 5-6 Logic negation at (a) the input and (b) the output of a logic block. (c) A symbol often used for a NOT gate and the Boolean equation. (d) The truth table.

5-4 THE NOT (INVERTER) CIRCUIT

The NOT circuit has a single input and a single output and performs the operation of *logic negation* in accordance with the following definition: *The output of a NOT circuit takes on the* 1 *state if and only if the input does* not *take on the* 1 *state.* The standard to indicate a *logic negation* is a small circle drawn at the point where a signal line joins a logic symbol. Negation at the input of a logic block is indicated in Fig. 5-6a and at the output in Fig. 5-6b. The symbol for a NOT gate and the Boolean expression for negation are given in Fig. 5-6c. The equation is to be read "Y equals NOT A" or "Y is the complement of A." [Sometimes a prime (') is used instead of the bar (–) to indicate the NOT operation.] The truth table is given in Fig. 5-6d.

A circuit which accomplishes a logic negation is called a NOT circuit, or, since it inverts the sense of the output with respect to the input, it is also known as an *inverter*. In a truly binary system only two levels $V(0)$ and $V(1)$ are recognized, and the output, as well as the input, of an inverter must operate between these two voltages. When the input is at $V(0)$, the output must be at $V(1)$, and vice versa. Ideally, then, a NOT circuit inverts a signal while preserving its shape and the binary levels between which the signal operates.

The transistor circuit of Fig. 5-7 implements an inverter for positive logic having a 0 state of $V(0) = V_{CE(\text{sat})} = 0.2$ V and a 1 state of $V(1) = 12$ V. If the input is low, $v_i = V(0)$, then the parameters are chosen so that the Q is OFF, and hence $v_o = V_{CC} = V(1)$. On the other hand, if the input is high, $v_i = V(1)$, then

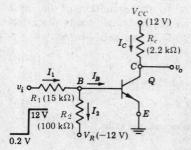

Figure 5-7 An INVERTER for positive logic. A similar circuit using a *p-n-p* transistor is used for a negative-logic NOT circuit.

the circuit parameters are picked so that Q is in saturation and then $v_o = V_{CE(sat)} = V(0)$. A detailed calculation of quiescent conditions is made in the following example.

Example 5-1 If the silicon transistor in Fig. 5-7 has a minimum value of h_{FE} of 30, find the output levels for input levels of 0.2 V and 12 V, obtained from a preceding gate.

SOLUTION For $v_i = V(0) = 0.2$ V the open-circuited base voltage V_B is, using superposition,

$$V_B = -12 \times \frac{15}{100 + 15} + 0.2 \times \frac{100}{100 + 15} = -1.391 \text{ V}$$

Since a bias of about 0 V is adequate to cut off a silicon emitter junction (Table 3-1, page 79), then Q is indeed cut off. Hence $v_o = 12$ V for $v_i = 0$.

For $v_i = V(1) = 12$ V let us verify the assumption that Q is in saturation. The minimum base current required for saturation is

$$I_{B(min)} = \frac{I_C}{h_{FE}}$$

It is usually sufficiently accurate to use the approximate values for the saturation junction voltages given in Table 3-1, which for silicon are $V_{BE(sat)} = 0.8$ V and $V_{CE(sat)} = 0.2$ V. With these values

$$I_C = \frac{12 - 0.2}{2.2} = 5.364 \text{ mA} \qquad I_{B(min)} = \frac{5.36}{30} = 0.179 \text{ mA}$$

$$I_1 = \frac{12 - 0.8}{15} = 0.747 \text{ mA} \qquad I_2 = \frac{0.8 - (-12)}{100} = 0.128 \text{ mA}$$

and

$$I_B = I_1 - I_2 = 0.747 - 0.128 = 0.619 \text{ mA}$$

Since this value exceeds $I_{B(min)}$, Q is indeed in saturation and the drop across the transistor is $V_{CE(sat)}$. Hence $v_o = 0.2$ V for $v_i = 12$ V, and the circuit has performed the NOT operation.

Transistor Limitations

There are certain transistor characteristics as well as certain circuit features which must particularly be taken into account in designing transistor inverters.

1. *The back-bias emitter-junction voltage V_{EB}.* This voltage must not exceed the emitter-to-base breakdown voltage BV_{EBO} specified by the manufacturer. For the type 2N2222A, $BV_{EBO} = 6$ V.
2. *The dc current gain h_{FE}.* Since h_{FE} decreases with decreasing temperature, the circuit must be designed so that at the lowest expected temperature the

transistor will remain in saturation. The maximum value of R_1 is determined principally by this condition.

3. *The reverse collector saturation current* I_{CBO}. Since $|I_{CBO}|$ increases about 7 percent/°C (doubles every 10°C), we cannot continue to neglect the effect of I_{CBO} at high temperatures. At cutoff the emitter current is zero and the base current is I_{CBO} (in a direction opposite to that indicated as I_B in Fig. 5-7). Let us calculate the value of I_{CBO} which just brings the transistor to the point of cutoff. If we assume, as in Table 3-1, that at cutoff, $V_{BE} = 0$ V, then

$$I_1 = 0.2/15 = 0.0133 \text{ mA}$$

The drop across the 100-kΩ resistor is $100\, I_2 = 12$ V and

$$I_{CBO} = I_2 - I_1 = 0.12 - 0.013 = 0.107 \text{ mA}$$

The ambient temperature at which $I_{CBO} = 0.107$ mA $= 107\ \mu$A is the maximum temperature at which the inverter will operate satisfactorily. A silicon transistor can be operated at temperatures in excess of 185°C.

Boolean Identities

From the basic definitions of the NOT, AND, and OR connectives we can verify the following Boolean identities:

$$\overline{\overline{A}} = A \tag{5-16}$$

$$\overline{A} + A = 1 \tag{5-17}$$

$$\overline{A} A = 0 \tag{5-18}$$

$$A + \overline{A} B = A + B \tag{5-19}$$

Example 5-2 Verify Eq. (5-19).

SOLUTION Since $B + 1 = 1$ and $A1 = A$, then

$$A + \overline{A} B = A(B + 1) + \overline{A} B = AB + A + \overline{A} B = (A + \overline{A})B + A = B + A$$

where use is made of Eq. (5-17).

5-5 THE INHIBIT (ENABLE) OPERATION

A NOT circuit preceding one terminal (S) of an AND gate acts as an *inhibitor*. This modified AND circuit implements the logical statement. *If $A = 1$, $B = 1, \ldots, M = 1$, then $Y = 1$ provided that $S = 0$. However, if $S = 1$, then the coincidence of $A, B, \ldots, M$ is inhibited (disabled), and $Y = 0$.* Such a configuration is also called an *anticoincidence* circuit. The logical block symbol is drawn in Fig. 5-8a, together with its Boolean equation. The equation is to be read "Y

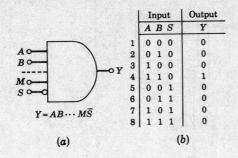

		Input			Output
		A	B	S	Y
1		0	0	0	0
2		0	1	0	0
3		1	0	0	0
4		1	1	0	1
5		0	0	1	0
6		0	1	1	0
7		1	0	1	0
8		1	1	1	0

$$Y = AB \cdots M\bar{S}$$

(a)

(b)

Figure 5-8 (a) The logic block and Boolean expression for an AND with an enable terminal S. (b) The truth table for $Y = AB\bar{S}$. The column on the left numbers the eight possible input combinations.

equals *A and B and . . . and M and not S.*" The truth table for a three-input AND gate with one inhibitor terminal (S) is given in Fig. 5-8b.

The terminal S is also called a *strobe* or an *enable input*. The enabling bit $S = 0$ allows the gate to perform its AND logic, whereas the inhibiting bit $S = 1$ causes the output to remain at $Y = 0$, independently of the values of the input bits.

It is possible to have a two-input AND, one terminal of which is inhibiting. This circuit satisfies the logic: "The output is true (1) if input A is true (1) provided that B is not true (0) [or equivalently, provided that B is false (0)]." Another possible configuration is an AND with more than one inhibit terminal.

5-6 THE EXCLUSIVE OR CIRCUIT

An EXCLUSIVE OR gate obeys the definition: *The output of a two-input* EXCLUSIVE OR *assumes the* 1 *state if one and only one input assumes the* 1 *state.* The standard symbol for an EXCLUSIVE OR is given in Fig. 5-9a and the truth table in Fig. 5-9b. The circuit of Sec. 5-2 is referred to as an INCLUSIVE OR if it is desired to distinguish it from the EXCLUSIVE OR.

The above definition is equivalent to the statement: "If $A = 1$ or $B = 1$ but not simultaneously, then $Y = 1$." In Boolean notation,

$$Y = (A + B)(\overline{AB}) \tag{5-20}$$

This function is implemented in logic diagram form in Fig. 5-10a.

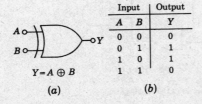

Input		Output
A	B	Y
0	0	0
0	1	1
1	0	1
1	1	0

$$Y = A \oplus B$$

(a)

(b)

Figure 5-9 (a) The standard symbol for an EXCLUSIVE OR gate and its Boolean expression. (b) The truth table.

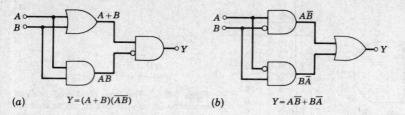

(a) $\quad Y=(A+B)(\overline{AB})$

(b) $\quad Y=A\overline{B}+B\overline{A}$

Figure 5-10 Two logic block diagrams for the EXCLUSIVE OR gate.

A second logic statement equivalent to the definition of the EXCLUSIVE OR is the following: "If $A = 1$ and $B = 0$, or if $B = 1$ and $A = 0$, then $Y = 1$." The Boolean expression is

$$Y = A\overline{B} + B\overline{A} \tag{5-21}$$

The block diagram which satisfies this logic is indicated in Fig. 5-10b.

An EXCLUSIVE OR is employed within the arithmetic section of a computer. Another application is as an *inequality comparator*, *matching circuit*, or *detector* because, as can be seen from the truth table, $Y = 1$ only if $A \neq B$. This property is used to check for the inequality of two bits. If bit A is not identical with bit B, then an output is obtained. Equivalently, "If A and B are both 1 or if A and B are both 0, then no output is obtained, and $Y = 0$." This latter statement may be put into Boolean form as

$$Y = \overline{AB + \overline{A}\,\overline{B}} \tag{5-22}$$

This equation leads to a third implementation for the EXCLUSIVE OR block, which is indicated by the logic diagram of Fig. 5-11a. An *equality detector* gives an output $Z = 1$ if A and B are both 1 or if A and B are both 0, and hence

$$Z = \overline{Y} = AB + \overline{A}\,\overline{B} \tag{5-23}$$

where use was made of Eq. (5-16). If the output Z is desired, the negation in Fig. 5-11a may be omitted or an additional inverter may be cascaded with the output of the EXCLUSIVE OR.

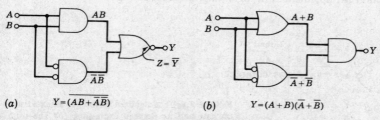

(a) $\quad Y=\overline{(AB+\overline{A}\overline{B})}$

(b) $\quad Y=(A+B)(\overline{A}+\overline{B})$

Figure 5-11 Two additional logic block diagrams for the EXCLUSIVE OR gate.

A fourth possibility for this gate is

$$Y = (A + B)(\overline{A} + \overline{B}) \qquad (5\text{-}24)$$

which may be verified from the definition or from the truth table. This logic is depicted in Fig. 5-11b.

It should be noted that a two-input EXCLUSIVE-OR behaves as a *controlled inverter* or an inverter with a strobe input. Thus, if A is the input and $B = S$ is the strobe, then from the truth table of Fig. 5-9 it follows that $Y = \overline{A}$ if $S = 1$, whereas $Y = A$ if $S = 0$.

We have demonstrated that there often are several ways to implement a logical circuit. In practice one of these may be realized more advantageously than the others. Boolean algebra is sometimes employed for manipulating a logic equation so as to transform it into a form which is better from the point of view of implementation in hardware. In the next section we shall verify through the use of Boolean algebra that the four expressions given above for the EXCLUSIVE OR are equivalent.

Two-Level Logic

Digital design often calls for several gates (AND, OR, or combinations of those) feeding into an OR (or AND) gate. Such a combination is known as *two-level* (*or two-wide*) *logic*. The EXCLUSIVE OR circuits of Figs. 5-10 and 5-11 are examples of two-level logic. In the discussion of combinational systems in Chap. 6 it is found that the most useful logic array consists of several ANDS which feed an OR which is followed by a NOT gate. This cascade of gates (for example, Fig. 5-11a) is called an AND-OR-INVERT (AOI) configuration. The detailed circuit topology for an AOI is given in Fig. 6-1.

5-7 DE MORGAN'S LAWS

The following two binary equations are known as De Morgan's theorems:

$$\overline{ABC \cdots} = \overline{A} + \overline{B} + \overline{C} + \cdots \qquad (5\text{-}25)$$

$$\overline{A + B + C + \cdots} = \overline{A}\,\overline{B}\,\overline{C} \cdots \qquad (5\text{-}26)$$

To verify Eq. (5-25) note that if all inputs are 1, then each side of the equation equals 0. On the other hand, if one (or more than one) input is 0, then each side of Eq. (5-25) equals 1. Hence, for all possible inputs the right-hand side of the equation equals the left-hand side. Equation (5-26) is verified in a similar manner. De Morgan's laws complete the list of basic Boolean identities. For each future reference, all these relationships are summarized in Table 5-2.

With the aid of Boolean algebra we shall now demonstrate the equivalence of the four EXCLUSIVE OR circuits of the preceding section. Using Eq. (5-25), it is immediately clear that Eq. (5-20) is equivalent to Eq. (5-24). Now the latter

Table 5-2 Summary of basic Boolean identities

Fundamental laws

OR	AND	NOT
$A + 0 = A$	$A0 = 0$	$A + \bar{A} = 1$
$A + 1 = 1$	$A1 = A$	$A\bar{A} = 0$
$A + A = A$	$AA = A$	$\bar{\bar{A}} = A$
$A + \bar{A} = 1$	$A\bar{A} = 0$	

Associative laws
$$(A + B) + C = A + (B + C) \qquad (AB)C = A(BC)$$

Commutative laws
$$A + B = B + A \qquad AB = BA$$

Distributive law
$$A(B + C) = AB + AC$$

De Morgan's laws
$$\overline{AB \cdots} = \bar{A} + \bar{B} + \cdots$$
$$\overline{A + B + \cdots} = \bar{A}\bar{B} \cdots$$

Auxiliary identities
$$A + AB = A \qquad A + \bar{A}B = A + B$$
$$(A + B)(A + C) = A + BC$$

equation can be expanded with the aid of Table 5-2 as follows:

$$(A + B)(\bar{A} + \bar{B}) = A\bar{A} + B\bar{A} + A\bar{B} + B\bar{B} = B\bar{A} + A\bar{B} \qquad (5\text{-}27)$$

This result shows that the EXCLUSIVE OR of Eq. (5-21) is equivalent to that of Eq. (5-24).

It follows from De Morgan's laws that *to find the complement of a Boolean function change all* OR *to* AND *operations, all* AND *to* OR *operations, and negate each binary symbol.* If this procedure is applied to Eq. (5-22), the result is Eq. (5-24), if use is made of the identity $\bar{\bar{A}} = A$.

With the aid of De Morgan's law we can show that *an* AND *circuit for positive logic also operates as an* OR *gate for negative logic.* Let Y be the output and $A, B, \ldots, N$ be the inputs to a positive AND so that

$$Y = AB \cdots N \qquad (5\text{-}28)$$

Then, by Eq. (5-25),

$$\bar{Y} = \bar{A} + \bar{B} + \cdots + \bar{N} \qquad (5\text{-}29)$$

If the output and all inputs of a circuit are complemented so that a 1 becomes a 0 and vice versa, then positive logic is changed to negative logic (refer to Fig. 5-1). Since Y and $\bar{Y}$ represent the *same* output terminal, A and $\bar{A}$ the *same* input terminal, etc., the circuit which performs the positive AND logic in Eq. (5-28) also operates as the negative OR gate of Eq. (5-29). Similar reasoning is used to verify that the same circuit is either a negative AND or a positive OR, depending upon

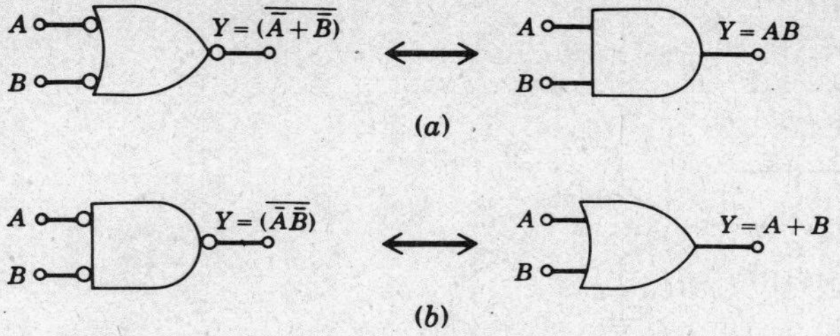

Figure 5-12 (*a*) An OR is converted into an AND by inverting all inputs and also the output. (*b*) An AND becomes an OR if all inputs and the output are complemented.

how the binary levels are defined. We verified this result for diode logic in Sec. 5-3, but the present proof is independent of how the circuit is implemented.

It should now be clear that it is really not necessary to use all three connectives OR, AND, and NOT. The OR and the NOT are sufficient because, from the De Morgan law of Eq. (5-25), the AND can be obtained from the OR and the NOT, as is indicated in Fig. 5-12*a*. Similarly, the AND and the NOT may be chosen as the basic logic circuits, and from the De Morgan law of Eq. (5-26), the OR may be constructed as shown in Fig. 5-12*b*. This figure makes clear once again that an OR (AND) circuit negated at input and output performs the AND (OR) logic.

5-8 THE NAND AND NOR DIODE-TRANSISTOR LOGIC (DTL) GATES

In Fig. 5-10*a* the negation before the second AND could equally well be put at the output of the first AND without changing the logic. Such an AND-NOT sequence is also present in Fig. 5-12*b* and in many other logic operations. This negated AND is called a NOT-AND, or a NAND, gate. The logic symbol, Boolean expression, and truth table for the NAND are given in Fig. 5-13. The NAND may be implemented by placing a transistor NOT circuit *after* a diode AND as in Fig. 5-14. Circuits involving diodes and transistors as in Fig. 5-14 are called *diode-transistor logic* (DTL) *gates*.

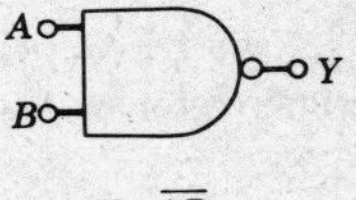

Input		Output
A	B	Y
0	0	1
0	1	1
1	0	1
1	1	0

$Y = \overline{AB}$

(*a*)   (*b*)

Figure 5-13 (*a*) The logic symbol and Boolean expression for a two-input NAND gate. (*b*) The truth table.

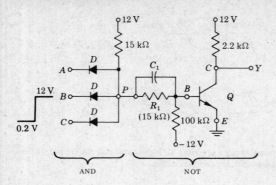

Figure 5-14 A three-input positive NAND (or negative NOR) gate.

The capacitor C_1 across R_1 in Fig. 5-14 is added to improve the transient response of the inverter. This capacitor aids in the removal of the minority-carrier saturation charge stored in the base when the signal changes abruptly between logic states. A transistor cannot come out of saturation until this charge is taken from the base region. The interval required to remove the saturation charge is called the *storage time* t_s. A discussion of this phenomenon is given in Sec. 3-14.

Example 5-3 Verify that the circuit of Fig. 5-14 is a positive NAND if the inputs are obtained from the outputs of similar NAND gates. Silicon transistors and diodes are used. Assume that the drop across a conducting diode is 0.7 V. Find the minimum value of h_{FE} for proper operation of the gate.

SOLUTION Consider that at least one input is low. We must verify that the output is high. The low input now comes from a transistor in saturation, and $V_{CE(sat)} \approx 0.2$ V. The open-circuit voltage at the base of Q is, from Fig. 5-15a, using superposition,

$$V_B = -12\frac{15}{100 + 15} + 0.9\frac{100}{100 + 15} = -0.782 \text{ V}$$

which cuts off Q and $Y = 1$, as it should.

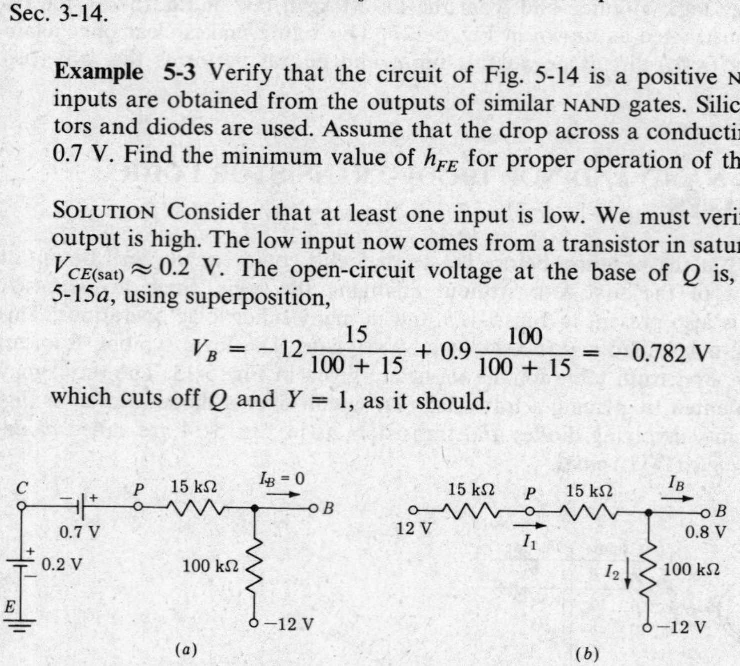

Figure 5-15 Relating to calculations in the circuit of Fig. 5-14. (a) At least one input is low. (b) All inputs are high.

If all inputs are at $V(1) = 12$ V, assume that all diodes are reverse-biased and that the transistor is in saturation. We shall now verify that these assumptions are indeed correct. If Q is in saturation, then with $V_{BE} = 0.8$ V, the voltage at P is obtained from Fig. 5-14 or Fig 5-15b. Using superposition,

$$V_P = 12 \frac{15}{15 + 15} + 0.8 \frac{15}{15 + 15} = 6.40 \text{ V}$$

Hence, with 12 V at each input, all diodes are reverse-biased by $12 - 6.40 = 5.60$ V. From Fig. 5-15b and Fig. 5-14,

$$I_1 = \frac{12 - 0.8}{15 + 15} = 0.373 \text{ mA} \qquad I_2 = \frac{0.8 + 12}{100} = 0.128 \text{ mA}$$

$$I_B = 0.373 - 0.128 = 0.245 \text{ mA} \qquad I_C = \frac{12 - 0.2}{2.2} = 5.364 \text{ mA}$$

$$h_{FE(\text{min})} = \frac{I_C}{I_B} = \frac{5.364}{0.245} = 21.9$$

If $h_{FE} \geqslant 21.9$, then Q will indeed be in saturation and the output is $V_{CE(\text{sat})} = 0.2$ V $= V(0)$, as it should be if all the inputs are high.

A NOR Gate

A negation following an OR is called a NOT-OR, or a NOR gate. The logic symbol, Boolean expression, and truth table for the NOR are given in Fig. 5-16. A positive NOR circuit is implemented by a cascade of a diode OR and a transistor INVERTER.

The circuit of Fig. 5-14 employs *diode-transistor logic* (DTL). The NAND and NOR may also be implemented in other configurations, as is indicated in Secs. 5-9 through 5-14. With the aid of De Morgan's laws, it can be shown that, regardless of the hardware involved, a positive NAND is also a negative NOR, whereas a negative NAND may equally well be considered a positive NOR.

It is clear that a single input NAND is a NOT. Also, a NAND followed by a NOT is an AND. In Sec. 5-7 it is pointed out that all logic can be performed by using only the two connectives AND and NOT. Therefore we now conclude that, by repeated use of the NAND circuit alone, any logical function can be carried out. A similar argument leads equally well to the result that all logic can be performed by using only the NOR circuit.

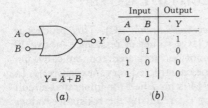

Input		Output
A	B	Y
0	0	1
0	1	0
1	0	0
1	1	0

$Y = \overline{A + B}$

(a)   (b)

Figure 5-16 (a) The logic symbol and Boolean expression for a two-input NOR gate. (b) The truth table.

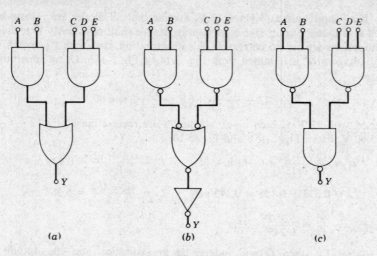

Figure 5-17 A two-level AND-OR is equivalent to a NAND-NAND configuration.

Example 5-4 Verify that two-level AND-OR topology is equivalent to a NAND-NAND system.

SOLUTION The AND-OR logic is indicated in Fig. 5-17a. Since $X = \overline{\overline{X}}$, then inverting the output of an AND and simultaneously negating the input to the following OR does not change the logic. These modifications are made in Fig. 5-17b. We have also negated the output of the OR gate and, at the same time, have added an INVERTER to Fig. 5-17b, so that once again the logic is unaffected. An OR gate negated at each terminal is an AND circuit (Fig. 5-12a). Since an AND followed by an INVERTER is a NAND then Fig. 5-17c is equivalent to Fig. 5-17b. Hence, the NAND-NAND of Fig. 5-17c is equivalent to the AND-OR of Fig. 5-17a.

If any of the inputs in Fig. 5-17 are obtained from the output of another gate then the resultant topology is referred to as *three-level logic*.

5-9 MODIFIED (INTEGRATED-CIRCUIT) DTL GATES[1, 2]

Most logic gates are fabricated as an *integrated circuit* (IC). This process is described in Chap. 4, where it is found that large values of resistance (above 30 kΩ) and of capacitance (above 100 pF) cannot be fabricated economically. On the other hand, transistors and diodes may be constructed very inexpensively. In view of these facts, the NAND gate of Fig. 5-14 is modified for integrated-circuit implementation by eliminating the capacitor C_1, reducing the resistance values drastically, and using diodes or transistors to replace resistors wherever possible. At the same time the power-supply requirements are simplified so that only a single 5-V supply is used. The resulting circuit is indicated in Fig. 5-18.

The operation of this positive NAND gate is easily understood qualitatively. If at least one of the inputs is low (the 0 state), the diode D connected to this input conducts and the voltage V_P at point P is low. Hence diodes D_1 and D_2 are nonconducting, $I_B = 0$, and the transistor is OFF. Therefore the output of Q is high and Y is in the 1 state. This logic satisfies the first three rows of the truth table in Fig. 5-13. Consider now the case where all inputs are high (1) so that all input diodes D are cut off. Then V_P tries to rise toward V_{CC}, and a base current I_B results. If I_B is sufficiently large, Q is driven into saturation and the output Y falls to its low (0) state, thus satisfying the fourth row of the truth table.

This NAND gate is considered quantitatively in the following illustrative example. The necessity for using two diodes D_1 and D_2 in series is explained. False logic can be caused by switching transients, power-supply noise spikes, coupling between leads, etc. The noise voltage at the input which will cause the circuit to malfunction when the output is in the 0(1) state is called the *noise-margin NM*(0) [*NM*(1)]. These noise margins are calculated below.

Example 5-5 (*a*) For the transistor in Fig. 5-18 assume (Table 3-1) that $V_{BE(\text{sat})} = 0.8$ V, $V_\gamma = 0.5$ V, and $V_{CE(\text{sat})} = 0.2$ V. The drop across a conducting diode is 0.7 V and V_γ (diode) $= 0.6$ V. The inputs of this switch are obtained from the outputs of similar gates. Verify that the circuit functions as a positive NAND and calculate $h_{FE(\text{min})}$. (*b*) Will the circuit operate properly if $D2$ is not used? (*c*) Calculate NM(0). (*d*) Calculate NM(1). Assume, for the moment, that Q is not loaded by a following stage.

SOLUTION (*a*) The logic levels are $V_{CE(\text{sat})} = 0.2$ V for the 0 state and $V_{CC} = 5$ V for the 1 state. If at least one input is in the 0 state, its diode conducts and $V_P = 0.2 + 0.7 = 0.9$ V. Since, in order for $D1$ and $D2$ to be

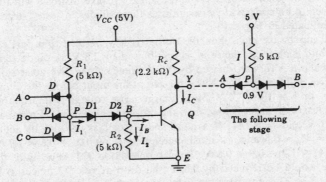

Figure 5-18 An integrated positive DTL NAND gate.

conducting, a voltage of $(2)(0.7) = 1.4$ V is required, these diodes are cut off, and $V_{BE} = 0$. Since the cutin voltage of Q is $V_\gamma = 0.5$ V, then Q is OFF, the output rises to 5 V, and $Y = 1$. This confirms the first three rows of the NAND truth table.

If all inputs are at $V(1) = 5$ V, then we shall assume that all input diodes are OFF, that D_1 and D_2 conduct, and that Q is in saturation. If these conditions are true, the voltage at P is the sum of two diode drops plus $V_{BE(sat)}$ or $V_P = 0.7 + 0.7 + 0.8 = 2.2$ V. The voltage across each input diode is $5 - 2.2 = 2.8$ V in the reverse direction, thus justifying the assumption that D is OFF. We now find $h_{FE(min)}$ to put Q into saturation.

$$I_1 = \frac{V_{CC} - V_P}{R_1} = \frac{5 - 2.2}{5} = 0.560 \text{ mA}$$

$$I_2 = \frac{V_{BE(sat)}}{R_2} = \frac{0.8}{5} = 0.160 \text{ mA.}$$

$$I_B = I_1 - I_2 = 0.560 - 0.160 = 0.400 \text{ mA}$$

$$I_C = \frac{V_{CC} - V_{CE(sat)}}{R_c} = \frac{5 - 0.2}{2.2} = 2.182 \text{ mA}$$

and

$$h_{FE(min)} = \frac{I_C}{I_B} = \frac{2.182}{0.400} = 5.46$$

If $h_{FE} > h_{FE(min)}$, then $Y = V(0)$ for all inputs at $V(1)$, thus verifying the last line in the truth table in Fig. 5-13.

(b) If at least one input is at $V(0)$, then $V_P = 0.2 + 0.7 = 0.9$ V. Hence, if only one diode $D1$ is used between P and B, then $V_{BE} = 0.9 - 0.6 = 0.3$ V, where 0.6 V represents the diode cutin voltage. Since the cutin base voltage is $V_\gamma = 0.5$ V, then theoretically Q is cut off. However, this is not a very conservative design because a small (> 0.2 V) spike of noise will turn Q ON. An even more conservative design uses three diodes in series, instead of the two indicated in Fig. 5-18.

(c) If all inputs are high, then the output is low, $V(0)$. From part (a), $V_P = 2.2$ V and each input diode is reverse-biased by 2.8 V. A diode starts to conduct when it is forward-biased by 0.6 V. Hence a negative noise spike in excess of $2.8 + 0.6 = 3.4$ V must be present at the input before the circuit malfunctions, or $NM(0) - 3.4$ V. Such a large noise voltage is improbable.

(d) If at least one input is low, then the output is high, $V(1)$. From part (a), $V_P = 0.9$ V and Q is OFF. If a noise spike just takes Q into its active region, $V_{BE} = V_\gamma = 0.5$ and V_P must increase to $0.5 + 0.6 + 0.6 = 1.7$ V. Hence $NM(1) = 1.7 - 0.9 = 0.8$ V. If only one diode $D1$ were used, the noise voltage would be reduced by 0.6 V (the drop across $D2$ at cutin) to $0.8 - 0.6 = 0.2$ V. This confirms the value obtained in part (b).

Fan-out

In the foregoing discussion we have unrealistically assumed that the NAND gate is unloaded. If it drives N similar gates, we say that the *fan-out* is N. The output transistor now acts as a *sink* for the current in the input to the gates it drives. In other words, when Q is in saturation ($Y = 0$), the input current I in Fig. 5-18 of a following stage adds to the collector current of Q. Assume that all the input diodes to a following stage (which is now considered to be a *current source*) are high except the one driven by Q. Then the current in this diode is $I = (5 - 0.9)/5 = 0.820$ mA. This current is called a *standard load*. The total collector current of Q is now $I_C = 0.820N + 2.182$ mA, where 2.182 mA is the unloaded collector current found in part a of the preceding example. Since the base current is almost independent of loading, I_B remains at its previous value of 0.400 mA. If we assume a reasonable value for $h_{FE(\min)}$ of 30, the fan-out is given by $I_C = h_{FE}I_B$, or

$$I_C = 0.820N + 2.182 = (30)(0.400) = 12.00 \text{ mA} \tag{5-30}$$

and $N = 11.97$. Since N must be an integer, a conservative choice is $N = 11$. Of course, the current rating of Q must not be exceeded.

The fan-out may be increased considerably by replacing $D1$ by a transistor $Q1$, as indicated in Fig. 5-19. When $Q1$ is conducting, it is in its active region and *not* in saturation. This statement follows from the fact that the current in the 2-kΩ resistance is in the direction to reverse-bias the collector junction of the *n-p-n* transistor $Q1$. Since the emitter current of $Q1$ supplies the base current of $Q2$, then $Q2$ is driven by a much higher base current than is Q in Fig. 5-18. For the same $h_{FE(\min)}$ of the output transistors in Figs. 5-18 and 5-19, it is clear that the latter circuit has the larger collector current, and hence the larger fan-out.

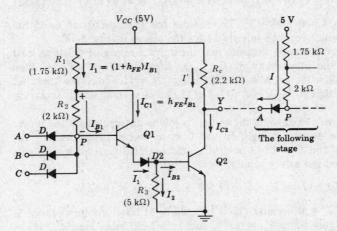

Figure 5-19 A modified integrated positive DTL NAND gate with increased fan-out.

Example 5-6 If $h_{FE(\text{min})} = 30$, calculate the fan-out N for the NAND gate of Fig. 5-19. From Table 3-1, $V_{\text{BE(active)}} = 0.7$ V. (b) Calculate the average power P dissipated by the gate.

SOLUTION (a) As with the circuit of Fig. 5-18, if any input is low, then $V_P = 0.9$ V, and both $Q1$ and $D2$ are OFF. Hence $V_{BE2} = 0$, $Q2$ is OFF, and $Y = 1$. If, however, all inputs are high, the input diodes are OFF, $Q2$ goes into saturation, and

$$V_P = V_{BE1(\text{active})} + V_{D2} + V_{BE2(\text{sat})} = 0.7 + 0.7 + 0.8 = 2.2 \text{ V}$$

Since $Q1$ is in its active region, $I_{C1} = h_{FE}I_{B1}$. As indicated in Fig. 5-19, the current in R_2 is I_{B1} (remember that each D is cut off), and the current in R_1 is $I_1 = I_{B1} + I_{C1} = (1 + h_{FE})I_{B1}$. Applying KVL between V_{CC} and V_P, we have, for $h_{FE} = 30$,

$$5 - 2.2 = (1.75)(31)I_{B1} + 2I_{B1} \tag{5-31}$$

or

$$I_{B1} = 0.0498 \text{ mA} \qquad I_1 = (31)(0.0498) = 1.543 \text{ mA}$$

$$I_2 = 0.8/5 = 0.160 \text{ mA} \qquad I_{B2} = 1.543 - 0.160 = 1.383 \text{ mA}$$

The unloaded collector current of $Q2$ is $I' = (5 - 0.2)/2.2 = 2.182$ mA. For *each* gate which it drives, $Q2$ must sink a standard load of

$$I = \frac{5 - 0.7 - 0.2}{1.75 + 2} = 1.093 \text{ mA}$$

Since the maximum collector current is $h_{FE}I_{B2}$, and $I_{C2} = IN + I'$, then

$$I_{C2} = (30)(1.383) = 1.093N + 2.182 = 41.49 \text{ mA} \tag{5-32}$$

and $N = 35.96$. Choose $N = 35$. The fan-out has been increased to 35 for the same h_{FE} which resulted in only 11 for the circuit of Fig. 5-18.

The above calculation assumes that the current rating of $Q2$ is at least 41.5 mA. On the other hand, if $I_{C2(\text{max})}$ is limited to, say, 15 mA, then $1.093N + 2.182 = 15$, or $N = 11.73$. Choose $N = 11$. To drive these 11 gates requires that $h_{FE(\text{min})} = I_{C2}/I_{B2} = 15/1.383 = 10.8$, which is a very small number.

(b) The power $P(0)$ when the output is low is different from the power $P(1)$ when the output is high. In the 0 state, we obtained in part (a) $I_1 = 1.543$ mA and $I' = 2.182$ mA. Since the power dissipated by the gate must come from the power supply,

$$P(0) = (I_1 + I')V_{CC} = (1.543 + 2.182)(5) = 18.62 \text{ W}$$

In the 1 state, $I' = 0$ because $Q2$ is OFF. Since at least one input diode is conducting in this state, $V_P = 0.9$ V and $I_1 = 1.093$ mA. Hence, $P(1) = (1.093)(5) = 5.47$ W.

If we assume that in a particular system this gate is equally likely to be in either state, then the average power is

$$P_{av} = \frac{P(0) + P(1)}{2} = \frac{18.62 + 5.47}{2} = 12.04 \text{ W}$$

Note that the output voltages are almost independent of the fan-out. The low-level $V(0) = V_{CE(\text{sat})}$ does not vary appreciably with I_{C2} (and hence N). The high-level $V(1) = V_{CC} - I''R_c$ where I'' is the reverse saturation current of the fan-out diodes (which are reverse-biased in the 1 state). At room temperature, $I''R_c \ll V_{CC}$ and $V(1) \approx V_{CC}$, independent of the fan-out. However, at highly elevated temperatures and large values of N, the high-level $V(0)$ may fall appreciably below V_{CC} (Prob. 5-42).

The fan-in M of a logic gate gives the number of inputs to the switch. For example, in Fig. 5-19, $M = 3$.

5-10 HIGH-THRESHOLD-LOGIC (HTL) GATE[2]

In an industrial environment the noise level is quite high because of the presence of motors, high-voltage switches, on-off control circuits, etc. By using a higher supply voltage (15 V instead of 5 V) and a 6.9-V Zener diode in place of $D2$ in the DTL gate of Fig. 5-19, this circuit is converted into the high-noise-immunity gate of Fig. 5-20. The resistances are increased in Fig. 5-20 with respect to those in Fig. 5-19, so that approximately the same currents are obtained in both circuits. The noise margin obtained with this circuit is typically 7 V (Prob. 5-46).

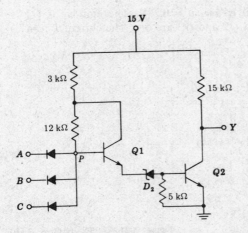

Figure 5-20 A high-threshold-logic NAND gate.

5-11 TRANSISTOR-TRANSISTOR-LOGIC (TTL) GATE[1,3]

The fastest-saturating logic circuit is the transistor-transistor-logic gate (TTL, or T^2L), shown in Fig. 5-21. This switch uses a multiple-emitter transistor which is easily and economically fabricated using integrated-circuit techniques (Sec. 4-7). The TTL circuit has the topology of the DTL circuit of Fig. 5-18, with the emitter junctions of $Q1$ acting as the input diodes D of the DTL gate and the collector junction of $Q1$ replacing the diode $D1$ of Fig. 5-18. The base-to-emitter diode of $Q2$ is used in place of the diode $D2$ of the DTL gate, and both circuits have an output transistor ($Q3$ or Q).

The explanation of the operation of the TTL gate parallels that of the DTL switch. Thus, if at least one input is at $V(0) = 0.2$ V, then

$$V_P = 0.2 + 0.7 = 0.9 \text{ V}$$

For the collector junction of $Q1$ to be forward-biased and for $Q2$ and $Q3$ to be ON requires about $0.7 + 0.7 + 0.7 = 2.1$ V. Hence these are OFF; the output rises to $V_{CC} = 5$ V, and $Y = V(1)$. On the other hand, if all inputs are high (at 5 V), the input diodes (the emitter junctions) are reverse-biased and V_P rises toward V_{CC} and drives $Q2$ and $Q3$ into saturation. Then the output is $V_{CE(sat)} = 0.2$ V, and $Y = V(0)$ (and V_P is clamped at about 2.3 V).

Input Transistor Action

The explanation given in the preceding paragraph assumes that $Q1$ acts like isolated back-to-back diodes and not as a transistor. The above conclusions are also reached if the transistor behavior of $Q1$ is taken into consideration.

Condition I. *At least one input is low,* $v_i = 0.2$ V. The emitter of $Q1$ is forward-biased and we assume that $Q2$ and $Q3$ are OFF. The current $I_{CI}(= I')$

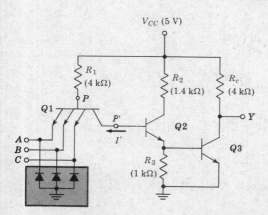

Figure 5-21 An IC positive TTL NAND gate. (Neglect the diodes in the shaded block.)

into the collector P' must be the current from emitter to base of $Q2$. Hence, I_{C1} equals the reverse saturation current of the emitter-junction diode of $Q2$. Since this current is very small (a few nanoamperes), $I_{BI} \gg I_{CI}/h_{FE}$ and $Q1$ is in saturation. The voltage at P' equals $V_{CE(\text{sat})} + v_i = 0.2 + 0.2 = 0.4$ V. This voltage is too small to put $Q2$ and $Q3$ ON. This argument justifies our assumptions that $Q2$ and $Q3$ are OFF and, therefore, $Y = V(1) = V_{CC}$.

Condition II: *All inputs are high.* The emitters of $Q1$ are reverse-biased, whereas the collector is forward-biased, because the p-type base is connected to the positive 5-V supply (through the 4-kΩ resistor). Hence, $Q1$ is operating in the inverted mode (Sec. 3-11). The inverted-current gain h_{FEI} for an IC transistor is very small (< 1). The input current (now the collector current of the inverted transistor) is $h_{FEI}I_{B1}$ The current I' (now the emitter current of the inverted transistor) is $-(1 + h_{FEI})I_{B1}$. This large current saturates $Q2$ and $Q3$ and $Y = V(0)$. This concludes the argument that Fig. 5-21 obeys NAND logic.

Low Storage Time

We now show that, because of the transistor behavior of $Q1$ during turnoff, the storage time t_s (Sec. 3-14) is reduced considerably. Note that the base voltage of $Q2$, which equals the collector voltage of $Q1$, is at $0.8 + 0.8 = 1.6$ V during saturation of $Q2$ and $Q3$. If now any input drops to 0.2 V, then $V_P = 0.9$ V, and hence the base of $Q1$ is at 0.9 V. At this time the collector junction is reverse-biased by $1.6 - 0.9 = 0.7$ V, the emitter junction is forward-biased, and $Q1$ *is in its active region.* The large collector current I' of $Q1$ now quickly removes the stored charge in $Q2$ and $Q3$. It is this transistor action which gives TTL the highest speed of any saturated logic. It is not until all the charge is removed from $Q3$ and $Q2$ (so that these transistors go OFF) that $Q1$ saturates, as discussed in Condition I.

Input Clamping Diodes

These diodes (shown in the shaded block in Fig. 5-21) are often included from each input to ground, with the anode grounded. These diodes are effectively out of the circuit for positive input signals, but they limit negative voltage excursions at the input to a safe value. These negative signals may arise from ringing caused by lead inductance resonating with shunt capacitance.

Wired Logic

It is possible to connect the outputs of several TTL or DTL gates together, as in Fig. 5-22, to perform additional logic (called *wired* or *collector logic*) without additional hardware. If positive NAND logic is under consideration, this connection is called a *wired*-AND, *phantom*-AND, *dotted*-AND, or *implied*-AND. Thus, if both $Y_1 = 1$ and $Y_2 = 1$, then $Y = 1$, whereas if $Y_1 = 0$ and/or $Y_2 = 0$, then $Y = 0$.

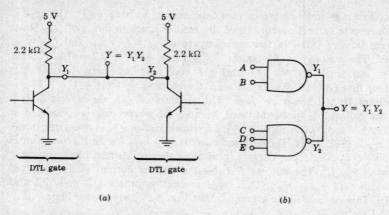

(a)

(b)

Figure 5-22 (a) Wired-AND logic is obtained by connecting the outputs of positive NAND gates together. (b) A two-input and a three-input NAND gate wired-AND together to perform the logic in Eq. (5-33).

The circuit of Fig. 5-21 also represents a negative NOR gate; and connecting two outputs together as in Fig. 5-22 now represents negative wired-OR logic. If Y_1 and/or Y_2 is in the low state (which is now the 1 state), then Y is also in the low state ($Y = 1$), whereas if Y_1 and Y_2 are both high (the 0 state), then Y is also high ($Y = 0$).

Consider two positive NAND gates wired-AND together as in Fig. 5-22b. Then $Y_1 = \overline{AB}$ and $Y_2 = \overline{CDE}$. Hence

$$Y = Y_1 Y_2 = (\overline{AB})(\overline{CDE}) = \overline{AB + CDE} \qquad (5\text{-}33)$$

where use is made of De Morgan's law. Note that the wired-AND has led to an implementation of the AOI two-level logic (Sec. 5-6). Because of the + sign in Eq. (5-33), the connection in Fig. 5-22 is often incorrectly referred to as a positive wired-OR.

Note that the wired-AND connection places the collector resistors in Fig. 5-22a in parallel. This reduction in resistance increases the power dissipation in the ON state. In order to avoid this condition *open-collector* gates (R_c omitted from Fig. 5-21) are available specifically for wired-AND applications. Of course, after the open-collector outputs of several gates are tied together, a passive pullup resistor R_c must be added externally.

5-12 OUTPUT STAGES

In the discussion of fan-out (Sec. 5-9), two dc (static) conditions are taken into account; namely, the output transistor must saturate when loaded by N gates and the current rating of this sink transistor must not be exceeded. Another (dynamic) condition is now considered.

At the output terminal of the DTL or TTL gate there is a capacitive load C_L, consisting of the capacitances of the reverse-biased diodes of the fan-out gates and any stray wiring capacitance. If the collector-circuit resistor of the inverter is R_c (called a *passive pull-up*), then, when the output changes from the low to the high state, the output transistor is cut off and the capacitance charges exponentially from $V_{CE(sat)}$ to V_{CC}. The time constant $R_c C_L$ of this waveform may introduce a prohibitively long delay time into the operation of these gates.

The output delay may be reduced by decreasing R_c, but this will increase the power dissipation when the output is in its low state and the voltage across R_c is $V_{CC} - V_{CE(sat)}$. A better solution to this problem is indicated in Fig. 5-23, where a transistor acts as an *active pull-up* circuit, replacing the passive pull-up resistance R_c. This output configuration is called a *totem-pole* amplifier because the transistor $Q4$ "sits" upon $Q3$. It is also referred to as a power-driver, or power-buffer, output stage.

The transistor $Q2$ acts as a *phase splitter*, since the emitter voltage is out of phase with the collector voltage (for an increase in base current, the emitter voltage increases and the collector voltage decreases). We now explain the operation of this driver circuit in detail, with reference to the TTL gate of Fig. 5-23.

The output is in the low-voltage state when $Q2$ and $Q3$ are driven into saturation. For this state we should like $Q4$ to be OFF. Is it? Note that the collector voltage V_{CN2} of $Q2$ with respect to ground N is given by

$$V_{CN2} = V_{CE2(sat)} + V_{BE3(sat)} = 0.2 + 0.8 = 1.0 \text{ V} \tag{5-34}$$

Since the base of $Q4$ is tied to the collector of $Q2$, then $V_{BN4} = V_{CN2} = 1.0$ V. *If the output diode DO were missing*, the base-to-emitter voltage of $Q4$

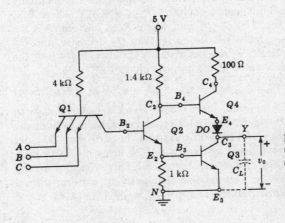

Figure 5-23 A TTL gate with a totem-pole output driver. This same active pullup circuit ($Q4$, DO, and 100 Ω) may be used in place of R_c in the DTL gate of Fig. 5-19, if $D2$ is replaced by transistor $Q2$.

would be

$$V_{BE4} = V_{BN4} - V_{CE3(sat)} = 1.0 - 0.2 = 0.8 \text{ V}$$

which would put $Q4$ into saturation. Under these circumstances the steady current through $Q4$ would be

$$\frac{V_{CC} - V_{CE4(sat)} - V_{CE3(sat)}}{100} = \frac{5 - 0.2 - 0.2}{100} \text{ A} = 46 \text{ mA} \qquad (5\text{-}35)$$

which is excessive and wasted current. The necessity for adding DO is now clear. With it in place, the sum of V_{BE4} and V_{DO} is 0.8 V. Hence both $Q4$ and DO are at cutoff. In summary, if C_L is at the high voltage $V(1)$ and the gate is excited, $Q4$ and DO go OFF, and $Q3$ conducts. Because of its large active-region current, $Q3$ quickly discharges C_L, and as v_o approaches $V(0)$, $Q3$ enters saturation. The bottom transistor $Q3$ of the totem pole is referred to as a *current sink*, which discharges C_L.

Assume now that with the output at $V(0)$, there is a change of state, because one of the inputs drops to its low state. Then $Q2$ is turned OFF, which causes $Q3$ to go to cutoff because V_{BE3} drops to zero. The output v_o remains momentarily at 0.2 V because the voltage across C_L cannot change instantaneously. Now $Q4$ goes into saturation and DO conducts, as we can verify:

$$V_{BN4} = V_{BE4(sat)} + V_{DO} + v_o = 0.8 + 0.7 + 0.2 = 1.7 \text{ V}$$

and the base and collector currents of $Q4$ are

$$I_{B4} = \frac{V_{CC} - V_{BN4}}{1.4} = \frac{5 - 1.7}{1.4} = 2.36 \text{ mA}$$

$$I_{C4} = \frac{V_{CC} - V_{CE4(sat)} - V_{DO} - v_o}{0.1} = \frac{5 - 0.2 - 0.7 - 0.2}{0.1} = 39.0 \text{ mA}$$

Hence, if h_{FE} exceeds $h_{FE(min)} = I_{C4}/I_{B4} = 39.0/2.36 = 16.5$, then $Q4$ is in saturation. The transistor $Q4$ is referred to as a *source*, supplying current to C_L. As long as $Q4$ remains in saturation, the output voltage rises exponentially toward V_{CC} with the very small time constant $(100 + R_{CS4} + R_f)C_L$, where R_{CS4} is the saturation resistance (Sec. 3-8) of $Q4$, and where R_f (a few ohms) is the diode forward resistance. As v_o increases, the currents in $Q4$ decrease, and $Q4$ comes out of saturation and finally v_o reaches a steady state when $Q4$ is at the cutin condition. Hence the final value of the output voltage is

$$v_o = V_{CC} - V_{BE4(cutin)} - V_{DO(cutin)} \approx 5 - 0.5 - 0.6 = 3.9 \text{ V} = V(1) \quad (5\text{-}36)$$

If the 100-Ω resistor were omitted, there would result a faster change in output from $V(0)$ to $V(1)$. However, the 100-Ω resistor is needed to limit the current spikes during the turn-on and turn-off transients. In particular, $Q3$ does not turn off (because of storage time) as quickly as $Q4$ turns on. With both totem-pole transistors conducting at the same time, the supply voltage would be short-circuited if the 100-Ω resistor were missing. The peak current drawn from the supply during the transient is limited to $I_{C4} + I_{B4} = 39 + 2.4 \approx 41$ mA if the 100-Ω resistor is used. These current spikes generate noise in the power-supply distribution system, and also result in increased power consumption at high frequencies.

Alternative Output Stages

From the foregoing discussion it should be clear that the diode DO can be moved from the emitter into the base lead of $Q4$. This configuration is used by some manufacturers.

The diode in the base of $Q4$ referred to in the preceding paragraph may be the base-to-emitter diode of an additional transistor, such as $Q5$ in Prob. 5-51.

The manufacturer's specification sheet for the TTL gate is given in Appendix B-6.

Wired Logic

It should be emphasized that the wired-AND connection must *not* be used with the totem-pole driver circuit. If the output from one gate is high while that from a second gate is low, and if these two outputs are tied together, we have exactly the situation just discussed in connection with transient current spikes. Hence, if the wired-AND were used, the power supply would deliver a *steady* current of 41 mA under these circumstances.

Three-State Output[4]

It is often necessary to expand the capability of a digital system by combining a number of identical packages (Fig. 6-27). Consider such a design where the nth output Y_n corresponds to Y_{n1} from chip 1, to Y_{n2} from chip 2, to Y_{n3} from chip 3, etc. Depending upon the specified logic, it is required that either Y_{n1}, Y_{n2}, or Y_{n3}, etc. (but only one of these) appear at an output Y_n. This result is obtained by connecting all leads Y_{n1}, Y_{n2}, Y_{n3}, etc., together (referred to as *wire* OR-*ing* or as the OR-*tied connection*) and by enabling only the ith chip during the interval when Y_{ni} is to appear at Y_n. The TTL totem-pole output stage (Fig. 5-23), modified to include such an enable, is indicated in Fig. 5-24a and the corresponding open-collector output circuit is shown in Fig. 5-24b.

In Fig. 5-24a if the *enable* or *chip select* (CS) signal is low, $D1$ and $D2$ are OFF, and the output is either in state 1 or state 0, depending upon whether the input data are 0 or 1. However, if CS is high, then $D1$ and $D2$ are ON, these diodes clamp $Q3$ and $Q4$ OFF, and the output Y is effectively an open-circuit. This condition, referred to as *the high-impedance third state*, allows OR-ing of the outputs from the several packages. The circuit of Fig. 5-24b operates in a similar 3-state fashion. However, the manufacturers designate the configuration in Fig. 5-24a as the tristate (TS) output and that in Fig. 5-24b as the open-collector (OC) output.

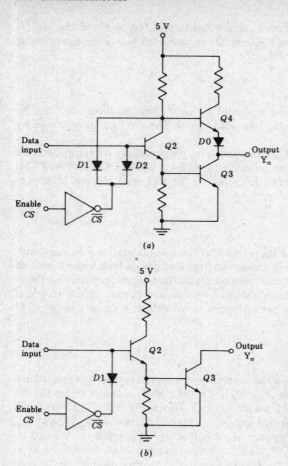

Figure 5-24 Three-state output stages controlled by an enable, strobe, or chip select CS input. (*a*) Totem-pole stage. (*b*) Open-collector stage. For these circuits the chip is selected if $CS = 0$. If an additional inverter is fabricated in cascade with the enable terminal, the chip is selected if $CS = 1$.

5-13 RESISTOR-TRANSISTOR LOGIC (RTL)[1, 2] AND DIRECT-COUPLED TRANSISTOR LOGIC (DCTL)[1, 5]

There are commercially available several other logic families, two of which are discussed in this section. The configuration for a *resistor-transistor logic* gate is indicated in Fig. 5-25*a*, which represents a three-input positive NOR gate with a fan-out of 5. If any input is high, the corresponding transistor is driven into saturation and the output is low, $v_o = V_{CE(\text{sat})} \approx 0.2 \text{ V} = V(0)$. However, if all inputs are low, then all input transistors are cut off by $V_\gamma - V(0) = 0.5 - 0.2 = 0.3 \text{ V}$ and the output v_o is high. (Note the low noise margin.) The preceding two statements confirm that the gate performs positive NOR (or negative NAND) logic.

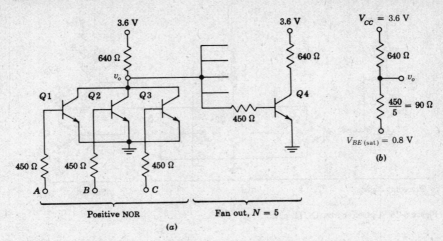

Figure 5-25 (a) An RTL positive NOR gate with a fan-in of 3 and a fan-out of 5. (b) The equivalent circuit from which to calculate v_o in the high state.

The value of v_o depends upon the fan-out. For example, if $N = 5$, then the output of the NOR gate is loaded by five 450-Ω resistors in parallel (or 90 Ω), which is tied to $V_{BE(\text{sat})} \approx 0.8$ V, as shown in Fig. 5-25b. Under these circumstances (using superposition),

$$v_o = \frac{3.6 \times 90}{90 + 640} + \frac{0.8 \times 640}{90 + 640} = 1.14 \text{ V} \tag{5-37}$$

This voltage must be large enough so that the base current can drive each of the five transistors into saturation. Since

$$I_B = \frac{1.14 - 0.8}{0.45} = 0.755 \text{ mA} \qquad I_C = \frac{3.6 - 0.2}{0.64} = 5.31 \text{ mA}$$

then the circuit will operate properly if $h_{FE} > h_{FE(\text{min})} = 5.31/0.76 = 7.0$.

Direct-coupled Transistor Logic (DCTL)

This configuration is the same as RTL, except that the base resistors are omitted. In Fig. 5-26 the fan-in is 3 and the fan-out is 2.

To verify that the circuit implements positive NOR logic, consider first that all inputs are in the 0 state. Because this low voltage to an input (say, to $Q1$) comes from a saturated transistor (Q') of a preceding state,

$$v_1 = V_{CE(\text{sat})} = V(0)$$

Since this voltage is 0.2 V for a saturated silicon transistor, and since the cutin voltage $V_\gamma \approx 0.5$ V, $Q1$ will conduct very little (although the noise margin is only $0.5 - 0.2 = 0.3$ V as it is with RTL logic). Since the current in $Q1$ is almost

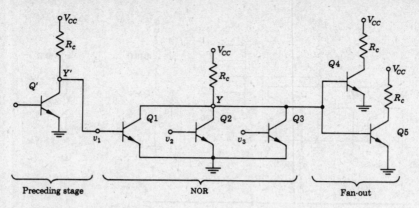

Figure 5-26 A positive NOR DCTL gate.

zero, the output Y tries to rise to V_{CC} and $Q4$ and $Q5$ go into saturation. Hence the output Y is clamped at

$$V_{BE(\text{sat})} = V(1) \approx 0.8 \text{ V}$$

for silicon. Thus, with all inputs in the low state, the output is in the high state. Note that the high state is only 0.8 V, independent of V_{CC}.

Consider now that at least one input v_1 is in the high state. Since $Q1$ is fed from Q', Q' is cut off and $Q1$ is driven into saturation. Under these circumstances the output Y is $V_{CE(\text{sat})} = V(0)$. If more than one input is excited, the output will certainly be low. Hence we have confirmed that the NOR function is satisfied.

There are a number of difficulties with DCTL: (1) The reverse saturation current for all fan-in transistors adds in the common collector-circuit resistor R_c. At high enough temperatures the total $I_{CBO}R_c$ drop may be large enough so that the output Y is too low to drive the fan-out transistors into saturation. (2) Because of the direct connection, the base current is almost equal to the collector current (for $V_{CC} \gg V_{CE(\text{sat})}$ and $V_{CC} \gg V_{BE(\text{sat})}$). With a transistor so heavily driven into saturation, very large stored base charge will result, with a corresponding detrimental effect on the switching speed. (3) Since the voltage levels are so low—the total output-voltage step is only of the order of 0.6 V for silicon—then spurious (noise) spikes can be troublesome. (4) The bases of the fan-out transistors are connected together. Since the input characteristics can never be identical, let us assume that $Q4$ has a much lower V_{BE} for a given I_B than does $Q5$. Under these circumstances, $Q4$ will "hog" most of the base current, and it is possible that $Q5$ may not even be driven into saturation. Hence transistors suitable for DCTL must have very close control on uniformity of input characteristics, very low values of I_{CBO}, as large a differential as possible between $V_{BE(\text{sat})}$ and $V_{CE(\text{sat})}$, a large h_{FE}, and a small storage time.

The advantages of DCTL are: (1) Only one low-voltage supply (operation with 1.5 V is possible) is needed; (2) transistors with low breakdown voltages may be used; and (3) the power dissipation is low.

In DCTL a NOR or NAND circuit is possible in which the transistors are in series (totem-pole fashion) rather than in parallel as in Fig. 5-26. Because of the difficulties mentioned above, DCTL is not often used with bipolar transistors, but it is the standard logic with field-effect transistors (Sec. 8-8). Also, a modification of DCTL, called *integrated injection logic* (I^2L), using bipolar technology makes an excellent gate for large-scale integration (LSI); see Sec. 9-11.

5-14 EMITTER-COUPLED LOGIC (ECL)[1, 6]

Consider the configuration of Fig. 5-27a which is called a *difference amplifier* (DIFF AMP) because the output is proportional to the difference between the two input voltages v_1 and v_2. This circuit is discussed in detail in Chap. 15 where its analog behavior is studied. This same configuration exhibits digital properties, and a logic family based upon this building block is called *emitter-coupled logic* (ECL) or *current-mode logic* (CML).

If $v_1 = v_2$ then, from symmetry, the transistor currents are equal. However if v_1 exceeds v_2 by about 0.1 V, then it can be shown (Sec. 15-4) that $Q1$ is ON and $Q2$ is OFF. Conversely, if v_1 is less than v_2 by approximately 0.1 V, then $Q1$ is OFF and $Q2$ is ON. The transfer characteristic is indicated in Fig. 5-27b. We find the emitter current remains essentially constant and that this current is switched from transistor $Q1$ to the other $Q2$ as v_1 varies from 0.1 V above

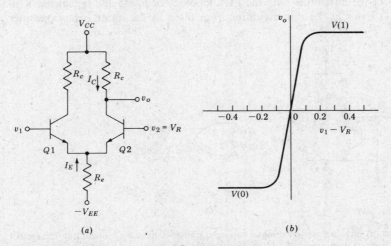

(a) (b)

Figure 5-27 (a) A difference amplifier. (b) The transfer characteristic.

$v_2 = V_R$ = reference voltage to 0.1 V below V_R. Except for a very narrow range of input voltage v_1, the output v_o takes on only one of two possible values and, hence, behaves as a digital circuit. The two logic levels are easily found. If $Q2$ is OFF, $v_o = V_{CC} = V(1)$. When $Q2$ is ON, then the parameter values are chosen so that $Q2$ remains in its active region. Then $v_o = V_{CC} - I_C R_c = V(0)$. If we neglect the base current then from Fig. 5-27a,

$$I_C = -I_E = \frac{V_R - V_{BE(\text{active})} + V_{EE}}{R_e} \qquad (5\text{-}38)$$

(Transistor $Q2$ will be in its active region if the collector junction V_{CB} is reverse-biased.) Since in the DIFF AMP neither transistor is allowed to go into saturation, then storage time is eliminated and, hence, the ECL is the fastest of all logic families (Table 5-3); a propagation delay (Sec. 5-15) as low as 0.5 ns per gate is possible.

A two-input OR (and also NOR) gate is drawn in Fig. 5-28a. This circuit is obtained from Fig. 5-27 by using two transistors in parallel at the input. Consider positive logic. If both A and B are low, then neither $Q1$ nor $Q2$ conducts whereas $Q3$ is in its active region. Under these circumstances Y is low and Y' is high. If either A or B is high, then the emitter current switches to the input transistor the base of which is high, and the collector current of $Q3$ drops approximately to zero. Hence Y goes high and Y' drops in voltage. Note that OR logic is performed at the output Y and NOR logic at Y', so that $Y' = \bar{Y}$. The logic symbol for such an OR gate with both true and false outputs is indicated in Fig. 5-28b. The availability of complementary outputs is clearly an advantage to the logic design engineer since it avoids the necessity of adding gates simply as inverters. The basic Motorola ECL (MECL II) three-input gate is shown in Fig. 5-29.

One of the difficulties with the ECL topology of Fig. 5-28a is that the $V(0)$ and $V(1)$ levels at the outputs differ from those at the inputs. Hence emitter

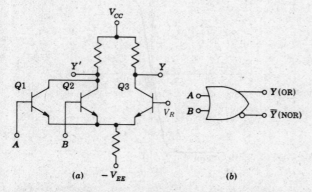

(a) $-V_{EE}$ (b)

Figure 5-28 (a) DIFF AMP converted into a two-input emitter-coupled logic circuit. (b) The symbol for a two-input OR/NOR gate.

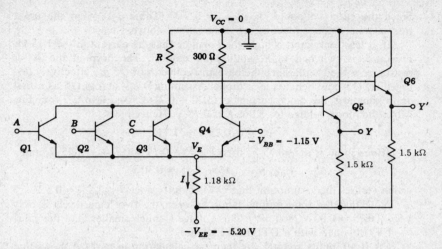

Figure 5-29 A three-input ECL OR/NOR gate, with no dc-level shift between input and output voltages.

followers $Q5$ and $Q6$ are used at the outputs to provide the proper dc-level shifts, as shown in Fig. 5-29. The reference voltage $V_R = -V_{BB}$ is obtained from a temperature-compensated network (not indicated). The quantitative operation of the gate is given in the following illustrative problem.

Example 5-7 (*a*) What are the logic levels at output Y of the ECL gate of Fig. 5-29? Assume a drop of 0.7 V between base and emitter of a conducting transistor. (*b*) Calculate the noise margins. (*c*) Verify that a conducting transistor is in its active region (*not* in saturation). (*d*) Calculate R so that the logic levels at Y' are the complements of those at Y. (*e*) Find the average power dissipated by the gate.

SOLUTION (*a*) If all inputs are low, then assume transistors $Q1$, $Q2$, and $Q3$ are cut off and $Q4$ is conducting. The voltage at the common emitter is

$$V_E = -1.15 - 0.7 = -1.85 \text{ V}$$

The current I in the 1.18-kΩ resistance is

$$I = \frac{-1.85 + 5.20}{1.18} = 2.84 \text{ mA}$$

Neglecting the base current compared with the emitter current, I is the current in the 300-Ω resistance and the output voltage at Y is

$$v_Y = -0.3I - V_{BE5} = -(0.3)(2.84) - 0.7 = -1.55 \text{ V} = V(0)$$

If all inputs are at $V(0) = -1.55$ V and $V_E = -1.85$ V, then the base-to-emitter voltage of an input transistor is

$$V_{BE} = -1.55 + 1.85 = 0.30 \text{ V}$$

Since the cutin voltage is $V_{BE(\text{cutin})} = 0.5$ V (Table 3-1), then the input transistors are nonconducting, as was assumed above.

If at least one input is high, then assume that the current in the 1.18-kΩ resistance is switched to R, and $Q4$ is cut off. The drop in the 300-Ω resistance is then zero. Since the base and collector of $Q5$ are effectively tied together, $Q5$ now behaves as a diode. Assuming 0.7 V across $Q5$ as a first approximation, the diode current is $(5.20 - 0.7)/1.5 = 3.0$ mA. From Fig. 4-14a the diode voltage for 3.0 mA is 0.75 V. Hence

$$v_Y = -0.75 \text{ V} = V(1)$$

If one input is at -0.75 V, then $V_E = -0.75 - 0.7 = -1.45$ V, and

$$V_{BE4} = -1.15 + 1.45 = 0.30 \text{ V}$$

which verifies the assumption that $Q4$ is cutoff; since $V_{BE(\text{cutin})} = 0.5$ V.

Note that the total output swing between the two logic levels is only $1.55 - 0.75 = 0.80$ V (800 mV). This voltage is much smaller than the value (~ 4 V) obtained with a DTL or TTL gate.

(*b*) If all inputs are at $V(0)$, then the calculation in part (*a*) shows that an input transistor is within $0.50 - 0.30 = 0.20$ V of cutin. Hence a positive noise spike of 0.20 V will cause the gate to malfunction.

If one input is at $V(1)$, then we find in part (*a*) that $V_{BE4} = 0.30$ V. Hence a negative noise spike at the input of 0.20 V drops V_E by the same amount and brings V_{BE4} to 0.5 V, or to the edge of conduction. Note that the noise margins are quite small (± 200 mV) and are equal in magnitude.

(*c*) From part (*a*) we have that, when $Q4$ is conducting, its collector voltage with respect to ground is the drop in the 300-Ω resistance, or $V_{C4} = -(0.3)(2.84) = -0.85$ V. Hence the collector junction voltage is

$$V_{CB4} = V_{C4} - V_{B4} = -0.85 + 1.15 = +0.30 \text{ V}$$

For an *n-p-n* transistor this represents a reverse bias, and $Q4$ must be in its active region.

If any input, say A, is at $V(1) = -0.75$ V $= V_{B1}$, then $Q1$ is conducting and the output $Y' = \overline{Y} = V(0) = -1.55$ V. The collector of $Q1$ is more positive than $V(0)$ by V_{BE6}, or

$$V_{C1} = -1.55 + 0.7 = -0.85 \text{ V}$$

and $\qquad V_{CB1} = V_{C1} - V_{B1} = -0.85 + 0.75 = -0.10 \text{ V}$

For an *n-p-n* transistor this represents a forward bias, but one whose magnitude is less than the cutin voltage of 0.5 V. Therefore $Q1$ is *not* in saturation; it is in its active region.

(*d*) If input A is at $V(1)$, then $Q1$ conducts and $Q4$ is OFF. Then

$$V_E = V(1) - V_{BE1} = -0.75 - 0.7 = -1.45 \text{ V}$$

$$I = \frac{V_E + V_{EE}}{1.18} = \frac{-1.45 + 5.20}{1.18} = 3.18 \text{ mA}$$

This value of I is about 10 percent larger than that found in part (a). In part (c) we find that, if $Y' = \overline{Y}$, then $V_{C1} = -0.85$ V. This value represents the drop across R if we neglect the base current of $Q1$. Hence

$$R = \frac{0.85}{3.18} = 0.267 \text{ k}\Omega = 267 \text{ }\Omega$$

This value of R ensures that, if an input is $V(1)$, then $Y' = V(0)$. If all inputs are at $V(0) = -1.55$ V, then the current through R is zero and the output is -0.75 V $= V(1)$, independent of R.

   Note that, if I had remained constant as the input changed state (true current-mode switching), then R would be identical to the collector resistance (300 Ω) of $Q4$. The above calculation shows that R is slightly smaller than this value.

   (e) If the input is low, $I = 2.84$ mA [part (a)], whereas if the input is high, $I = 3.18$ mA [part (d)]. The average I is $\frac{1}{2}(2.84 + 3.18) = 3.01$ mA. Since $V(0) = -0.75$ V and $V(1) = -1.55$ V, the currents in the two emitter followers are

$$\frac{5.20 - 0.75}{1.50} = 2.97 \text{ mA} \qquad \text{and} \qquad \frac{5.20 - 1.55}{1.50} = 2.43 \text{ mA}$$

The total power supply current drain is $3.01 + 2.97 + 2.43 = 8.41$ mA and the power dissipation is $(5.20)(8.41) = 43.7$ mW.

Since the current drain from the power supply varies very little as the input switches from one state to the other, power line spikes (of the type discussed in Sec. 5-11 for TTL gates) are virtually nonexistent.

Fan-out

The input resistance can be considered infinite if all inputs are low so that all input transistors are cut off. If an input is high, then $Q4$ is OFF, and the input resistance corresponds to a transistor with an emitter resistor $R_e = 1.18$ kΩ, and from Eq. (11-45) a reasonable estimate is $R_i \approx 100$ kΩ. The output resistance is that of an emitter follower (or a diode) and a reasonable value is $R_o \approx 15$ Ω. Since the input resistance is very high and the output resistance is very low, a large fan-out is possible at low frequencies.

   Because ECL gates are intended for high-speed operation, the dc fan-out is not important. Rather the fan-out is determined by the fact that capacitive loading slows down the gate operation. If C is the input capacitance per gate and N is the fan-out, then the total capacitance shunting the emitter-follower driver $Q5$ is NC. This capacitance is charged rapidly through the low output resistance when $Q5$ is ON. However, consider the situation where $v_o = V(1)$ and the input to the emitter follower falls. Since the voltage across a capacitor cannot change instantaneously, $Q5$ is cut off. Hence v_o falls toward $-V_{EE}$ with a time constant $1.5 \times 10^{-3} NC$ and N is determined by the maximum allowable transition time between states.

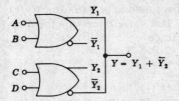

Figure 5-30 An implied-OR connection at the output of two ECL gates.

If the outputs of two or more ECL gates are tied together as in Fig. 5-30 then wired-OR logic (Sec. 5-11) is obtained (Prob. 5-57). Open-emitter gates are available for use in this application.

Summary

The principal characteristics of the ECL gate are summarized below:

Advantages

1. Since the transistors do not saturate, then the highest speed of any logic family is available.
2. Complementary outputs are available.
3. Current switching spikes are not present in the power supply leads.
4. Outputs can be tied together to give the implied-OR function.
5. There is little degradation of parameters with variations in temperature.
6. The number of functions available is high.
7. Data transmission over long distances by means of balanced twisted-pair 50-Ω lines is possible.[6]

Disadvantages

1. A small voltage difference (800 mV) exists between the two logic levels and the noise margins are only ±200 mV.
2. The power dissipation is high relative to the other logic families.
3. Level shifters are required for interfacing with other families.
4. Capacitive loading limits the fan-out.

5-15 COMPARISON OF LOGIC FAMILIES

An exhaustive comparison of each logic configuration is extremely difficult because we must take into account all the following characteristics: (1) Speed (propagation time delay); (2) noise immunity; (3) fan-in and fan-out capabilities; (4) power supply requirements; (5) power dissipation per gate; (6) operating temperature range; (7) number of functions available; and (8) cost. Items 1 and 7 require some explanation; all others have already been defined or are self-evident.

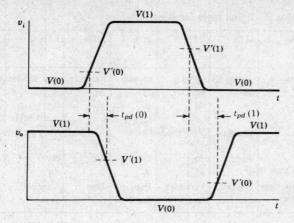

Figure 5-31 Pertaining to the definitions of the propagation delay times.

Propagation Delay

As the input voltage to a positive NAND gate rises from $V(0)$ toward $V(1)$, then at some *switching threshold voltage* $V'(0)$ (Fig. 5-31), conditions within the gate are modified, so that a change of state of the output from $V(1)$ to $V(0)$ is initiated. Similarly, as the input falls from $V(1)$ toward $V(0)$, then at some other *switching threshold voltage* $V'(1)$, the initiation of the change of state from $V(0)$ to $V(1)$ takes place. We now define (as in Fig. 5-31) the propagation delay ON time $t_{pd}(0)$ as the interval between the time when the input v_i reaches $V'(0)$ and the output falls to $V'(1)$. Also, the propagation delay OFF time $t_{pd}(1)$ is defined as the interval between the time when the input equals $V'(1)$ and the output rises to $V'(0)$. Because of minority-carrier storage time, $t_{pd}(1) > t_{pd}(0)$. Hence the *propagation delay time* t_{pd} is usually defined as the average of these two times, or

$$t_{pd} = \tfrac{1}{2}\left[\, t_{pd}(0) + t_{pd}(1)\,\right] \tag{5-39}$$

In passing, we note that some authors arbitrarily assume the two threshold voltages to be equal: $V'(0) = V'(1) = \tfrac{1}{2}[V(0) + V(1)]$.

A reduction in propagation delay time usually means the use of smaller circuit resistances and, hence, an increase in power dissipation. Hence, a figure of merit often used in the comparison of logic gate is the product $t_{pd}P_{av}$, which is called the *delay-power* or *speed-power product*.

Functions

The basic AND, OR, NAND, and NOR gates are combined in one integrated chip in various combinations to perform specific functions. The *building blocks* in Table 5-3 are used in the following chapters in connection with digital systems, which include binary adders, decoders, counters, shift registers, FLIP-FLOPS, etc.

Table 5-3 Comparison of the major IC digital logic families[1, 6]

Logic Parameter	RTL	DTL	HTL	TTL	ECL	MOS	CMOS
Basic gate*	NOR	NAND	NAND	NAND	OR-NOR	NAND	NOR or NAND
Fan-out†	5	8	10	10	25	20	> 50
Power dissipated‡ per gate, mW	12	8	55	10	40	1	0.01 static 1 at 1 MHz
Noise immunity	Nominal	Good	Excellent	Very good	Good	Nominal	Very good
Propagation delay§ per gate, ns	12	30	90	10	2	100	50
Clock rate,¶ MHz	8	12	4	15	60	2	10
Number of functions	High	Fairly high	Nominal	Very high	High	Low	Very high
Reference	Sec. 5-13	Sec. 5-8	Sec. 5-10	Sec. 5-11	Sec. 5-14	Sec. 8-8	Sec. 8-9

*   Positive logic.
†   Worst-case number of inputs that the gate can drive.
‡   Typical; affected by temperature and frequency of operation.
§   Typical for a nominal fan-out.
¶   Maximum frequency at which FLIP-FLOPS operate. The actual clock rate is from one-half to one-tenth the frequency listed.

Logic Family Characteristics

A comparison of standard IC digital logic families is made in Table 5-3. In addition to the five types already discussed, *metal-oxide-semiconductor* (MOS) *logic* is included. Since MOS logic uses the field-effect transistor (FET), the discussion of these gates is postponed until Chap. 8, where the MOSFET is introduced.

The most popular family (the industry standard) is TTL. The voltage levels of DTL and TTL are compatible and, hence, these two gates may be intermixed in a logic system. However, TTL is rapidly replacing DTL because the former has better noise immunity, a smaller propagation delay, and more functions (different IC chips) are available than for the latter.

The RTL gate (the first logic family manufactured) is no longer used in new systems because its fan-out is low, the output depends upon the fan-out, the noise margins are small, and the output swing is low (not compatible with TTL). The brief discussion of RTL in Sec. 5-13 is included for historical reasons and because it leads to a convenient introduction of DCTL, which is used in MOS and CMOS (complementary MOS) logic. The overwhelming choice for lowest power is the CMOS family.

The ECL gate (which is nonsaturating) is the fastest logic gate, with a propagation delay of only about a nanosecond. However, the power dissipation is highest of all the other families (except for HTL), and the voltage swing is low (less than 1 V).

Table 5-4 Typical TTL performance characteristics[7]

	Series	Gates			FLIP-FLOPS
		Propagation delay time, ns	Power per gate, mW	Delay-power product, pJ	Clock-input frequency, MHz
S	Schottky, high-speed	3	19	57	dc to 125
H	high-power, high-speed	6	22	132	dc to 50
LS	Schottky, low-power	9.5	2	19	dc to 45
–	standard	10	10	100	dc to 35
L	low-power, low-speed	33	1	33	dc to 3

The MOS gate has very low power dissipation and it occupies an extremely small chip area. Hence, a very high packing density is possible, and MOS is used principally for very long shift registers and for large memories (Chap. 9). However, these devices have long propagation delays and, hence, are limited to systems operating at frequencies under 1 MHz.

The TTL entries in Table 5-3 are for the standard gate. The TTL family is also available in four other series emphasizing either high speed, low power, or both, as indicated in Table 5-4. These gates differ from one another in the numerical values for the resistances in Fig. 5-23 and in that some use *Schottky transistors* to increase speed by preventing saturation (Sec. 3-14 and Sec. 4-8).

From Texas Instruments, Inc., there is available the 74 series which may be used at temperatures between 0 and 70°C, and the 54 series which is valid over the military range from −55 to 125°C. A designation used with a TTL package could be TI-74LS10, which means that the IC may be operated only in the range 0 to 70°C, and that it is a low-power Schottky chip. The number ten identifies the function (a triple three-input NAND gate). In this book we shall abbreviate this designation to TI-10; it being understood that it may be of the 74 or 54 type and may belong to any one of the five series listed in Table 5-4. It should also be noted that many IC's are available with either totem-pole (Fig. 5-23), tristate, or open-collector outputs (Fig. 5-24).

The LS family has become the industry's favorite TTL. The LS has virtually replaced the L, and the S has replaced the H family. The highest speed is obtained with ECL and the lowest power with CMOS.

REFERENCES

1. Taub, H., and D. Schilling: "Digital Integrated Electronics," McGraw-Hill Book Company, New York, 1977.
   Grinich, V. H., and H. G. Jackson: "Introduction to Integrated Circuits," McGraw-Hill Book Company, New York, 1975.

2. Garrett, L. S.: Integrated-circuit Digital Logic Families, Part I, RTL, DTL, and HTL Devices, *IEEE Spectrum*, vol. 7, pp. 46–56, October, 1970.
3. Garrett, L. S.: Integrated-Circuit Digital Logic Families, Part II, TTL Devices, *IEEE Spectrum*, vol. 7, pp. 63–72, November, 1970.
4. Monolithic Memories, Inc.: "Data Book," 1977.
5. Beter, R. H., W. E. Bradley, R. B. Brown, and M. Rubinoff: Directly Coupled Transistor Circuits, *Electronics*, vol. 28, no. 6, pp. 132–136, June, 1955.
6. Garrett, L. S.: Integrated-Circuit Digital Logic Families, Part III, ECL and MOS Devices, *IEEE Spectrum*, vol. 7, p. 41, December, 1970.
7. Texas Instruments, Inc.: "The TTL Data Book for Design Engineers," 1974.

REVIEW QUESTIONS

5-1 Express the following decimal numbers in binary form: (*a*) 26; (*b*) 100; (*c*) 1024.

5-2 Define (*a*) *positive logic*; (*b*) *negative logic*.

5-3 Define an OR gate and give its truth table.

5-4 Draw a positive-diode OR gate and explain its operation.

5-5 Evaluate the following expressions: (*a*) $A + 1$; (*b*) $A + A$; (*c*) $A + 0$.

5-6 Define an AND gate and give its truth table.

5-7 Draw a positive-diode AND gate and explain its operation.

5-8 Evaluate the following: (*a*) $A1$; (*b*) AA; (*c*) $A0$; (*d*) $A + AB$.

5-9 Define a NOT gate and give its truth table.

5-10 Draw a positive-logic NOT gate and explain its operation.

5-11 Evaluate the following expressions: (*a*) $\bar{\bar{A}}$; (*b*) $\bar{A}A$; (*c*) $\bar{A} + A$.

5-12 Define an INHIBITOR and give the truth table for $AB\bar{S}$.

5-13 Define an EXCLUSIVE OR and give its truth table.

5-14 Show two logic block diagrams for an EXCLUSIVE OR.

5-15 Verify that the following Boolean expressions represent an EXCLUSIVE OR: (*a*) $AB + \bar{A}\bar{B}$; (*b*) $(A + B)(\bar{A} + \bar{B})$.

5-16 State the two forms of De Morgan's laws.

5-17 Show how to implement an AND with OR and NOT gates.

5-18 Show how to implement an OR with AND and NOT gates.

5-19 Define a NAND gate and give its truth table.

5-20 Draw a positive NAND gate with diodes and a transistor (DTL) and explain its operation.

5-21 Define a NOR gate and give its truth table.

5-22 Repeat Rev. 5-20 for a positive NOR gate.

5-23 Draw the circuit of an IC DTL gate and explain its operation.

5-24 Define (*a*) *fan-out*; (*b*) *fan-in*; (*c*) *standard load*; (*d*) *current sink*; (*e*) *current source*.

5-25 What logic is performed if the outputs of two DTL gates are connected together? Explain.

5-26 How does high-threshold logic (HTL) differ from DTL?

5-27 Draw the circuit of a TTL gate and explain its operation.

5-28 Draw a totem-pole output buffer with a TTL gate. Explain its operation.

5-29 Explain the function of a 3-state TTL gate.

5-30 Draw the circuit of an RTL gate and explain its operation for positive logic.

5-31 Repeat Rev. 5-30 for negative logic.

5-32 Draw a DCTL circuit and explain its operation.

5-33 List three advantages and three disadvantages of DCTL gates.

5-34 (*a*) Sketch a two-input OR (and also NOR) ECL gate.
(*b*) What parameters determine the noise margin?
(*c*) Why are the two collector resistors unequal?
(*d*) Explain why power line spikes are virtually nonexistent.

5-35 List and discuss at least four advantages and four disadvantages of the ECL gate.

5-36 Define (*a*) two *threshold voltages*; (*b*) *propagation delay* ON *time*; (*c*) *propagation delay* OFF *time*; (*d*) *propagation delay time*.

COMBINATIONAL DIGITAL SYSTEMS

A digital system is constructed from very few types of basic network configurations, these elementary types being used over and over again in various topological combinations. As emphasized in Sec. 5-8, it is possible to perform all logic operations with a single type of circuit (for example, a NAND gate). A digital system must store binary numbers in addition to performing logic. To take care of this requirement, a memory cell, called a FLIP-FLOP, is introduced in the next chapter.

Theoretically, any digital system can be constructed entirely from NAND gates and FLIP-FLOPS. Some functions (such as binary addition) are present in many systems, and hence the combination of gates and/or FLIP-FLOPS required to perform this function is available on a single chip. These integrated circuits form the practical (commercially available) basic building blocks for a digital system. The number of such different IC's is not large, and these packages† perform the following functions: binary addition, decoding (demultiplexing), data selection (multiplexing), counting, storage of binary information (memories and registers), digital-to-analog (D/A) and analog-to-digital (A/D) conversion, and a number of other operations. Those building blocks which depend upon combinational logic are described in this chapter.[1]

6-1 STANDARD GATE ASSEMBLIES[2]

Since the fundamental gates are used in large numbers even in a relatively simple digital system, they are not packaged individually; rather, several gates

---

† The terms *package, chip,* and *IC* are used interchangeably.

are constructed within a single chip. The following list of standard digital IC gates is typical, but far from exhaustive:

Quad two-input NAND	Quad two-input NOR
Triple three-input NAND	Quad two-input AND
Dual four-input NAND	Dual two-wide, two-input AOI
Single eight-input NAND	Two-wide, four-input AOI
Hex inverter buffer	Four-wide, 4-2-3-2-input AOI

These combinations are available in most logic families (TTL, DTL, etc.) listed in Sec. 5-15. The limitation on the number of gates per chip is usually set by the number of pins available. The most common package is the *dual-in-line* (plastic or ceramic) package, which has 14 leads, 7 brought out to each side of the IC (Fig. 6-1c). The dimensions of the assembly, which is much larger than the chip size, are approximately 0.8 by 0.3 by 0.2 in. A schematic of the triple three-input NAND is shown in Fig. 6-1a. Note that there are $3 \times 3 = 9$ input leads, 3 output leads, a power-supply lead, and a ground lead; a total of 14 leads are used.

In Fig. 6-1b is indicated the dual two-wide, two-input AOI (Sec. 5-6). This combination needs 4 input leads and 1 output lead per AOI, or 10 for the dual array. If 1 power-supply lead and 1 ground lead are added, we see that 12 of the 14 available pins are used.

The circuit diagram for this AOI gate is given in Fig. 6-2, implemented in TTL logic. The operation of this network should be clear from the discussion in

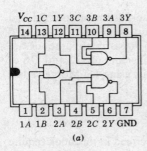

(a)

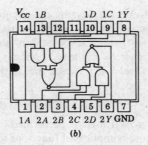

(b)

(c)

Figure 6-1 The lead connections (top view) of (a) the TI-10 triple three-input NAND. (b) The TI-51 dual two-wide, two-input AOI gate. (No connections are to be made to pins 11 and 12.) (c) A dual-in-line package, DIP.

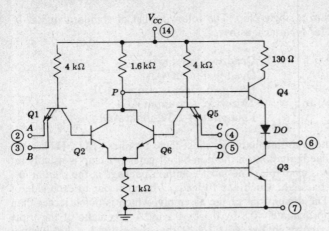

Figure 6-2 The circuit configuration for a TTL AND-OR-INVERT gate.

Chap. 5. Thus $Q1$ and the input to $Q2$ (corresponding to the similarly numbered transistors of Fig. 5-21) constitute an AND gate. The identical arrangement of $Q5$ and $Q6$ constitutes a second AND gate. Since the collectors of $Q2$ and $Q6$ are tied together at P, the output at this node corresponds to either the inputs 2 AND 3 OR 4 AND 5. Also, because of the inversion through a transistor, the NOT operation is performed at P. The result is AND-OR-INVERT (AOI) logic $(\overline{AB + CD})$. Finally, note that $Q3$, DO, and $Q4$ form the totem-pole output stage of Fig. 5-23.

An alternative way of analyzing the circuit of Fig. 6-2 is to consider $Q1$ and $Q2$ (with the output at P) to constitute a NAND circuit. Similarly, $Q5$ and $Q6$ form a second NAND gate. The outputs of these two NAND configurations are short-circuited together by the lead connecting the collectors of $Q2$ and $Q6$ to form a wired-AND (Sec. 5-11). Hence the output at P is, using De Morgan's law [Eq. (5-26)]

$$\overline{(AB)}\ \overline{(CD)} = \overline{AB + CD}$$

which confirms that AOI logic is performed.

Some of the more complicated functions to be described in this book require in excess of 14 pins, and these IC's are packaged with 16, 20, 24, and up to 64 leads.

The standard combinations considered in this section are examples of *small-scale integration* (SSI). Less than about 12 gates (~ 100 components) on a chip is considered SSI. The FLIP-FLOPS discussed in Sec. 7-3 are also SSI packages. Most other functions (using BJTs) discussed in this chapter are examples of *medium-scale integration* (MSI), defined to have more than 12, but less than 100, gates per chip. The BJT memories of Sec. 6-9 and the MOSFET arrays of Chap. 9 may contain in excess of 100 gates ($> 1,000$ components) and are defined as *large-scale integration* (LSI).

Design Philosophy

An electronics engineer should design a system so as to use standard IC's for as many subsystems as possible. He or she must attempt to minimize the required number of packages (and hence the total cost[3]). A single MSI is used in place of a number of SSI chips which could perform the same function. Similarly, an LSI package is used in the system wherever this IC can replace several MSI chips. In summary, in designing a digital system, it should be defined in terms of standard MSI and LSI packages. Discrete gates (SSI) should be used only for "interfaces" (also called the "glue") which may be required between the subsystem IC's.

A list of manufacturers of IC's is given in Appendix B-1. These companies have available data books, handbooks, and application notes which are invaluable to the system designer since they keep him up to date on new packages and applications as they become available. The most important functions performed by MSI chips are given in Chaps. 6 and 7. After introducing the MOSFET in Chap. 8, LSI packages are discussed in Chap. 9. The most versatile LSI system is the microcomputer, a programmable *computer on a chip*. As discussed in Sec. 9-11 the design of a complicated logic system is usually based upon the use of the microcomputer.

6-2 BINARY ADDERS[4]

A digital computer must obviously contain circuits which will perform arithmetic operations, i.e., addition, subtraction, multiplication, and division. The basic operations are addition and subtraction, since multiplication is essentially repeated addition, and division is essentially repeated subtraction.

Suppose we wish to sum two numbers in decimal arithmetic and obtain, say, the hundreds digit. We must add together not only the hundreds digit of each number but also a carry from the tens digit (if one exists). Similarly, in binary arithmetic we must add not only the digit of like significance of the two numbers to be summed, but also the carry bit (should one be present) of the next lower significant digit. This operation may be carried out in two steps: first, add the two bits corresponding to the 2^n digit, and then add the resultant to the carry from the 2^{n-1} digit. A two-input adder is called a *half adder*, because to complete an addition requires two such half adders.

We show how a *half adder* is constructed from the basic logic gates. A half adder has two inputs—A and B—representing the bits to be added, and two outputs—D (for the digit of the same significance as A and B represent) and C (for the carry bit).

Half Adder

The symbol for a half adder is given in Fig. 6-3a, and the truth table in Fig. 6-3b. Note that the D column gives the sum of A and B as long as the sum can be represented by a single digit. When, however, the sum is larger than can be

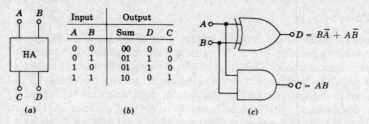

Figure 6-3 (*a*) The symbol for a half adder. (*b*) The truth table for the digit D and the carry C. (*c*) The implementation for D with an EXCLUSIVE OR gate and for C with an AND gate.

represented by a single digit, then D gives the digit in the result which is of the same significance as the individual digits being added. Thus, in the first three rows of the truth table, D gives the sum of A and B directly. Since the decimal equation "1 plus 1 equals 2" is written in binary form as "01 plus 0̅1̅ equals 10," then in the last row $D = 0$. Because a 1 must now be carried to the place of next higher significance, $C = 1$.

From Fig. 6-3*b* we see that D obeys the EXCLUSIVE OR function and C follows the logic of an AND gate. These functions are indicated in Fig. 6-3*c*, and may be implemented in many different ways with the circuitry discussed in Chap. 5. For example, the EXCLUSIVE OR gate can be constructed with any of the four topologies of Sec. 5-6 and in any of the logic families in Table 5-3. The configuration in Fig. 5-10*b* ($Y = A\bar{B} + B\bar{A}$) is implemented in TTL logic with the AOI circuit of Fig. 6-2. The inverter for B (or A) is a single-input NAND gate. Since Y has an AND-OR (rather than an AND-OR-INVERT) topology, a transistor inverter is placed between node P and the base of $Q4$ of Fig. 6-2.

Parallel Operation

Two multidigit numbers may be added serially (one column at a time) or in parallel (all columns simultaneously). Consider parallel operation first. For an N-digit binary number there are (in addition to a common ground) N signal leads in the computer for each number. The nth line for number A (or B) is excited by A_n (or B_n), the bit for the 2^n digit ($n = 0, 1, \ldots, N - 1$).

Full Adder

In integrated circuit implementation, addition is performed using a complete adder, which (for reasons of economy of components) is not constructed from two half adders. The symbol for the nth full adder (FA) is indicated in Fig. 6-4*a*. The circuit has three inputs: the addend A_n, the augend B_n, and the input carry C_{n-1} (from the next lower bit). The outputs are the sum S_n (sometimes designated Σ_n) and the output carry C_n. A parallel 4-bit adder is indicated in Fig. 6-4*b*. Since FA0 represents the least significant bit (LSB), it has no input carry; hence $C_{-1} = 0$.

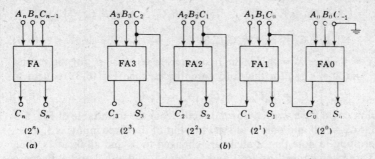

Figure 6-4 (*a*) The symbol for a full adder. (*b*) A 4-bit parallel binary adder constructed from cascaded full adders.

The circuitry within the block FA may be determined from Fig. 6-5, which is the truth table for adding 3 binary bits. From this table we can verify that the Boolean expressions for S_n and C_n are given by

$$S_n = \bar{A}_n \bar{B}_n C_{n-1} + \bar{A}_n B_n \bar{C}_{n-1} + A_n \bar{B}_n \bar{C}_{n-1} + A_n B_n C_{n-1} \qquad (6\text{-}1)$$

$$C_n = \bar{A}_n B_n C_{n-1} + A_n \bar{B}_n C_{n-1} + A_n B_n \bar{C}_{n-1} + A_n B_n C_{n-1} \qquad (6\text{-}2)$$

Note that the first term of S_n corresponds to line 1 of the table, the second term to line 2, the third term to line 4, and the last term to line 7. (These are the only rows where $S_n = 1$.) Similarly, the first term of C_n corresponds to the line 3 (where $C_n = 1$), the second term to the line 5, etc.

The AND operation ABC is sometimes called the *product* of A and B and C. Also, the OR operation $+$ is referred to as *summation*. Hence expressions such as those in Eqs. (6-1) and (6-2) represent a *Boolean sum of products*. Such an equation is said to be in a *standard*, or *canonical*, *form*, and each term in the equation is called a *minterm*. A minterm contains the product of all Boolean variables, or their complements.

The expression for C_n can be simplified considerably as follows: Since $Y + Y + Y = Y$, then Eq. (6-2), with $Y = A_n B_n C_{n-1}$, becomes

Line	Inputs			Outputs	
	A_n	B_n	C_{n-1}	S_n	C_n
0	0	0	0	0	0
1	0	0	1	1	0
2	0	1	0	1	0
3	0	1	1	0	1
4	1	0	0	1	0
5	1	0	1	0	1
6	1	1	0	0	1
7	1	1	1	1	1

Figure 6-5 Truth table for a three-input adder. The lines are numbered decimally as if $A_n B_n C_{n-1}$ represents a 3-bit binary number with C_{n-1} equal to the LSB and A_n equal to the MSB (most significant bit).

$$C_n = \left(\bar{A}_n B_n C_{n-1} + A_n B_n C_{n-1} \right) + \left(A_n \bar{B}_n C_{n-1} + A_n B_n C_{n-1} \right)$$
$$+ \left(A_n B_n \bar{C}_{n-1} + A_n B_n C_{n-1} \right) \tag{6-3}$$

Since $\bar{X} + X = 1$ where $X = A_n$ for the first parentheses, $X = B_n$ for the second parentheses, and $X = C_{n-1}$ for the third parentheses, then Eq. (6-3) reduces to

$$C_n = B_n C_{n-1} + C_{n-1} A_n + A_n B_n \tag{6-4}$$

This expression could have written down directly from the truth table of Fig. 6-5 by noting that $C_n = 1$ if and only if at least two out of the three inputs is 1.

It is interesting to note that if all 1s are changed to 0s and all 0s to 1s, then lines 0 and 7 are interchanged, as are 1 and 6, 2 and 5, and also 3 and 4. Because this switching of 1s and 0s leaves the truth table unchanged, whatever logic is represented by Fig. 6-5 is equally valid if all inputs and outputs are complemented. Therefore Eq. (6-3) is true if all variables are negated, or

$$\bar{C}_n = \bar{B}_n \bar{C}_{n-1} + \bar{C}_{n-1} \bar{A}_n + \bar{A}_n \bar{B}_n \tag{6-5}$$

This same result is obtained (Prob. 6-3) by Boolean manipulation of Eq. (6-4).

By evaluating $D_n \equiv (A_n + B_n + C_{n-1}) \bar{C}_n$ and comparing the result with Eq. (6-1), we find that $S_n \equiv D_n + A_n B_n C_{n-1}$, or

$$S_n = A_n \bar{C}_n + B_n \bar{C}_n + C_{n-1} \bar{C}_n + A_n B_n C_{n-1} \tag{6-6}$$

Equations (6-4) and (6-6) are implemented in Fig. 6-6 using AOI gates of the type shown in Fig. 6-2.

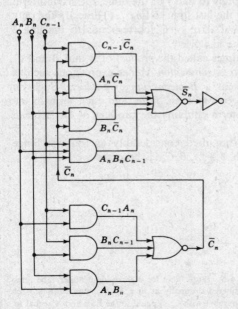

Figure 6-6 Block-diagram implementation of the nth stage of a full adder.

MSI Adders

There are commercially available 1-bit, 2-bit, and 4-bit full adders, each in one package. In Fig. 6-7 is indicated the logic topology for 2-bit addition. The inputs to the first stage are A_0 and B_0; the input marked C_{-1} is grounded. The output is the sum S_0. The carry C_0 is connected internally and is not brought to an output pin. This 2^0 stage (LSB) is identical with that in Fig. 6-6 with $n = 0$.

Since the carry from the first stage is C_0, it should be negated before it is fed to the 2^1 stage. However, the delay introduced by this inversion is undesirable, because the limitation upon the maximum speed of operation is the propagation delay (Sec. 5-15) of the carry through all the bits in the adder. The NOT-gate delay is eliminated completely in the carry by connecting $\overline{C}_0$ directly to the

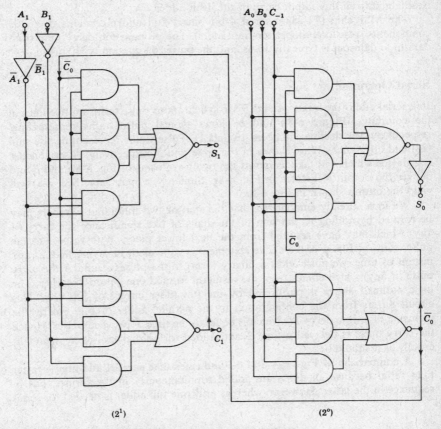

Figure 6-7 Logic diagram of an integrated 2-bit full adder (TI-82).

following stage and by complementing the inputs A_1 and B_1 before feeding these to this stage. This latter method is used in Fig. 6-7. Note that now the outputs S_1 and C_1 are obtained directly without requiring inverters. The logic followed by this second stage for the carry is given by Eq. (6-5), and for the sum by the modified form of Eq. (6-6), where each symbol is replaced by its complement.

In a 4-bit adder C_1 is not brought out but is internally connected to the third stage, which is identical with the first stage. Similarly, the fourth and second stages have identical logic topologies. A 4-bit adder requires a 16-pin package: 8 inputs, 4 sum outputs, a carry output, a carry input, the power-supply input, and ground. The carry input is needed only if two arithmetic units are cascaded; for example, cascading a 2-bit with a 4-bit adder gives the sum of two 6-bit numbers. If the 2-bit unit is used for the 2^4 and 2^5 digits, then 4 must be added to all the subscripts in Fig. 6-7. For example, C_{-1} is now called C_3 and is obtained from the output carry of the 4-bit adder.

The MSI chip (TI-283†) for a 4-bit binary full adder contains over 200 components (resistors, diodes, or transistors). The propagation delay time from data-in to data-out is typically 16 ns, and the power dissipation is 310 mW.

Serial Operation

In a serial adder the inputs A and B are synchronous pulse trains on two lines in the computer. Figure 6-8a and b shows typical pulse trains representing, respectively, the decimal numbers 13 and 11. Pulse trains representing the sum (24) and difference (2) are shown in Fig. 6-8c and d, respectively. A serial *adder* is a device which will take as inputs the two waveforms of Fig. 6-8a and b and deliver the output waveform in Fig. 6-8c. Similarly, a *subtractor* (Sec. 6-3) will yield the output shown in Fig. 6-8d.

We have already emphasized that the sum of two multidigit numbers may be formed by adding to the sum of the digits of like significance the carry (if any) which may have resulted from the next lower place. With respect to the pulse trains of Fig. 6-8, the above statement is equivalent to saying that, at any instant of time, we must add (in binary form) to the pulses A and B the carry pulse (if any) which comes from the resultant formed one period T earlier. The logic outlined above is performed by the full-adder circuit of Fig. 6-9. This circuit differs from the configuration in the parallel adder of Fig. 6-4 by the inclusion of a time delay TD which is equal to the time T between pulses. Hence the carry pulse is delayed a time T and added to the digit pulses in A and B, exactly as it should be.

A comparison of Figs. 6-4 and 6-9 indicates that parallel addition is faster than serial because all digits are added simultaneously in the former, but in sequence in the latter. However, whereas only one full adder is needed for serial

---

† The specific designations (Sec. 5-15) given in this chapter refer to Texas Instrument units.[2] However, equivalent units are available from other vendors (Appendix B-1).

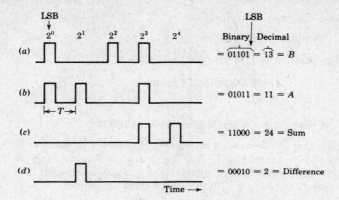

Figure 6-8 (*a, b*) Pulse waveforms representing numbers *B* and *A*. (*c, d*) Waveforms representing sum and difference. (LSB = least significant bit.)

arithmetic, we must use a full adder for each bit in parallel addition. Hence parallel addition is much more expensive than serial operation.

The time delay unit TD is a type D FLIP-FLOP, and the serial numbers A_n, B_n, and S_n are stored in *shift registers* (Secs. 7-3 and 7-4).

6-3 ARITHMETIC FUNCTIONS

In this and the next two sections other arithmetic units besides the adder are discussed. These include the *subtractor*, the *ALU*, the *multiplier*, the digital comparator, and the parity checker.

Binary Subtraction[4]

The process of subtraction (*B* minus *A*) is equivalent to addition if the complement $\overline{A}$ of the subtrahend is used. To justify this statement consider the following argument (applied specifically to a 4-bit number). The NOT function

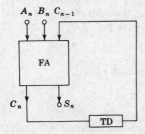

Figure 6-9 A serial binary full adder.

changes a 1 to a 0, and vice versa. Therefore†

$$A \text{ plus } \bar{A} = 1111$$

and

$$A \text{ plus } \bar{A} \text{ plus } 1 = 1111 \text{ plus } 0001 = 10000$$

so that

$$A = 10000 \text{ minus } \bar{A} \text{ minus } 1$$

Finally,

$$B \text{ minus } A = (B \text{ plus } \bar{A} \text{ plus } 1) \text{ minus } 10000 \qquad (6\text{-}7)$$

This equation indicates that to subtract a 4-bit number A from a 4-bit number B it is only required to add B, $\bar{A}$, and 1 (a 2^0 bit). The operation B minus A must yield a 4-bit answer. The term "minus 10000" in Eq. (6-7) infers that the addition (B plus $\bar{A}$ plus 1) results in a fifth bit, which must be ignored.

Example 6-1 Verify Eq. (6-7) for $B = 1100$ and $A = 1001$ (decimal 12 and 9).

SOLUTION

$$B \text{ plus } \bar{A} \text{ plus } 1 = 1100 \text{ plus } 0110 \text{ plus } 0001 = 10011$$

The four (less significant) bits 0011 represent decimal 3 and the fifth bit 1 is a generated carry. Since, in decimal notation, B minus $A = 12 - 9 = 3$, then the correct answer is obtained by evaluating the sum in the parentheses of Eq. (6-7), provided that the carry is ignored.

In Eq. (6-7) the 1 in 10000 is the output carry $C_3 = 1$ from the 4-bit adder, and may be used to supply the 1 which must be added to $\bar{A}$. This bit is called the *end-around carry* (EAC) because this carry out is fed back to the carry input C_{-1} (Fig. 6-7) of the least significant bit of A. This process of subtraction by means of a 4-bit parallel adder is indicated schematically in Fig. 6-10.

The 1s complement method of subtraction just described is valid only if B is greater than A, so that a positive difference results and a carry is generated from (B plus $\bar{A}$ plus 1). If B is less than A, then the most significant bit (MSB) of B (which differs from the corresponding bit of A) is 0 and that of A is 1. Since $\bar{A} = 0$, the MSB of (B plus $\bar{A}$) is 0. Hence no carry results from the sum (B plus $\bar{A}$ plus 0001), and the method indicated in Fig. 6-10 must be modified. We now demonstrate that if no carry results in the system of Fig. 6-10, the correct answer for B minus A is negative, and is obtained by forming the sum (B plus $\bar{A}$) and by complementing the sum digits S_0, S_1, S_2, and S_3. From Eq. (6-7)

$$B \text{ minus } A = (B \text{ plus } \bar{A}) \text{ minus } 1111$$

$$= \text{minus}\big[\, 1111 \text{ minus}(B \text{ plus } \bar{A})\,\big]$$

$$= \text{minus}\big(\, \overline{B \text{ plus } \bar{A}}\,\big) \qquad (6\text{-}8)$$

† To avoid confusion with the OR operation, the word *plus (minus)* is used in place of $+(-)$ in the following equations.

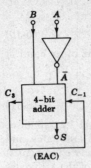

Figure 6-10 A simplified sketch of a 4-bit parallel adder used as a subtractor.

because 1111 minus a 4-bit binary number is the complement of the number. *In summary:* To subtract A from B form the sum $(B$ plus $\bar{A})$ and check to see if a carry exists. If it does, the difference $(B$ minus $A)$ is positive and is given by S in Fig. 6-10. However, if the carry is zero, then the difference is negative and is given by $\bar{S}$. The circuits for testing for a carry and for obtaining the complement of S when the EAC is missing are not shown in Fig. 6-10.

Arithmetic Logic Unit (ALU)/Function Generator

Subtraction may be accomplished by using an ALU such as the TI-181 (or TI-S381). Using four (or three) function select lines, the following operations can be performed on two 4-bit numbers: B minus A, A minus B, A plus B, $A \oplus B$, $A + B$, AB, $A = B$, $A > B$, and a number of other arithmetic and logic operations. These MSI packages are large (24 pins) and have the complexity of 85 equivalent gates (~ 800 components).

Binary Multipliers

The 16-pin TI-285 package is used to perform parallel multiplication of 4×4 bits and to produce a 4-bit output. Used in conjunction with the TI-284 chip an 8-bit product is obtained in about 40 ns. The same operation can be obtained with a single high-complexity 20-pin TI-S274 chip.

6-4 DIGITAL COMPARATOR

It is sometimes necessary to know whether a binary number A is greater than, equal to, or less than another number B. The system for making this determination is called a *magnitude digital* (or *binary*) *comparator*. Consider single bit numbers first. As mentioned in Sec. 5-6, the EXCLUSIVE-NOR gate is an *equality detector* because

$$E = \overline{A\bar{B} + \bar{A}B} = \begin{cases} 1 & A = B \\ 0 & A \neq B \end{cases} \qquad (6\text{-}9)$$

The condition $A > B$ is given by

$$C = A\bar{B} = 1 \qquad (6\text{-}10)$$

because if $A > B$, then $A = 1$ and $B = 0$, so that $C = 1$. On the other hand, if $A = B$ or $A < B(A = 0, B = 1)$, then $C = 0$.

Similarly, the restriction $A < B$ is determined from

$$D = \bar{A}B = 1 \qquad (6\text{-}11)$$

The logic block diagram for the nth bit drawn in Fig. 6-11 has all three desired outputs C_n, D_n, and E_n. It consists of two inverters, two AND gates, and the AOI circuit of Fig. 6-2. Alternatively, Fig. 6-11 may be considered to consist of an EXCLUSIVE-NOR and two AND gates. (Note that the outputs of the AND gates in the AOI block of Fig. 6-2 are not available, and hence additional AND gates must be fabricated to give C_n and D_n.)

Consider now a 4-bit comparator. $A = B$ requires that

$$A_3 = B_3 \quad \text{and} \quad A_2 = B_2 \quad \text{and} \quad A_1 = B_1 \quad \text{and} \quad A_0 = B_0$$

Hence the AND gate E in Fig. 6-12 described by

$$E = E_3 E_2 E_1 E_0 \qquad (6\text{-}12)$$

implies $A = B$ if $E = 1$ and $A \neq B$ if $E = 0$. (Assume that the input E' is held high; $E' = 1$.)

The inequality $A > B$ requires that

$$A_3 > B_3 \qquad \text{(MSB)}$$

or $\quad A_3 = B_3 \quad$ and $\quad A_2 > B_2$

or $\quad A_3 = B_3 \quad$ and $\quad A_2 = B_2 \quad$ and $\quad A_1 > B_1$

or $\quad A_3 = B_3 \quad$ and $\quad A_2 = B_2 \quad$ and $\quad A_1 = B_1 \quad$ and $\quad A_0 > B_0$

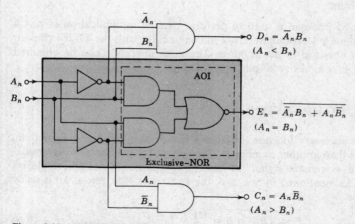

Figure 6-11 A 1-bit digital comparator.

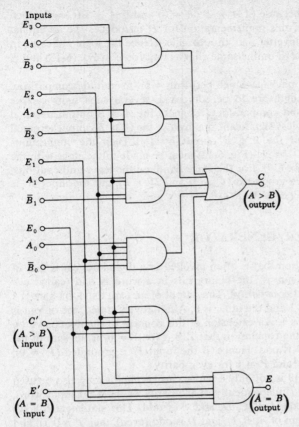

Figure 6-12 A 4-bit magnitude comparator. (Assume that $C' = 0$ and $E' = 1$.) If $E = 1$, then $A = B$, and if $C = 1$, then $A > B$. If $D = 1$, then $A < B$, where D has the same logic topology as C but with A and B interchanged. The inputs $\bar{A}_n$, B_n, and D' ($A < B$) are not indicated. The inputs E_n are obtained from Fig. 6-11.

The above conditions are satisfied by the Boolean expression

$$C = A_3\bar{B}_3 + E_3A_2\bar{B}_2 + E_3E_2A_1\bar{B}_1 + E_3E_2E_1A_0\bar{B}_0 \qquad (6\text{-}13)$$

if and only if $C = 1$. The AND-OR gate for C is indicated in Fig. 6-12. (Assume that $C' = 0$.)

The condition that $A < B$ is obtained from Eq. (6-13) by interchanging A and B. Thus

$$D = \bar{A}_3B_3 + E_3\bar{A}_2B_2 + E_3E_2\bar{A}_1B_1 + E_3E_2E_1\bar{A}_0B_0 \qquad (6\text{-}14)$$

implies that $A < B$ if and only if $D = 1$. This portion of the system is obtained from Fig. 6-12 by changing A to ·B, B to A, and C to D. Alternatively, D may be obtained from $D = \overline{EC}$ because, if $A \neq B$ ($E = 0$) and if $A \not> B(C = 0)$, then $A < B(D = 1)$. However, this implementation for D introduces the additional propagation delay of an inverter and an AND gate. Hence the logic indicated in Eq. (6-14) for D fabricated on the same chip as that for C in Eq. (6-13) and E in Eq. (6-12).

The TI-85 is an MSI package which performs 4-bit-magnitude comparison. If numbers of greater length are to be compared, several such units can be cascaded. Consider an 8-bit comparator. Designate the $A = B$ output terminal of the stage handling the less significant bits by E_L, the $A > B$ output terminal of this stage by C_L, and the $A < B$ output by D_L. Then the connections $E' = E_L$, $C' = C_L$, and $D' = D_L$ (Fig. 6-12) must be made to the stage with the more significant bits (Prob. 6-9). For the stage handling the less significant bits, the outputs C' and D' are grounded ($C' = 0$ and $D' = 0$) and the input E' is tied to the supply voltage ($E' = 1$). Why?

6-5 PARITY CHECKER/GENERATOR

Another arithmetic operation that is often invoked in a digital system is that of determining whether the sum of the binary bits in a word is odd (called *odd parity*) or even (designated *even parity*). The output of an EXCLUSIVE-OR gate is 1 if and only if one input is 1 and the other is 0. Alternatively stated, the output is 1 if the sum of the digits is 1. An extension of this concept to the EXCLUSIVE-OR tree of Fig. 6-13 leads to the conclusion that $Z = 1$ (or $Y = 0$) if the sum of the input bits A, B, C, and D is odd. Hence, if the input P' is grounded ($P' = 0$), then $P = 0$ for odd parity and $P = 1$ for even parity.

The system of Fig. 6-13 is not only a parity checker, but it may also be used to generate a parity bit P. Independently of the parity of the 4-bit input word, the parity of the 5-bit code A, B, C, D, and P is odd. This statement follows from the fact that if the sum of A, B, C, and D is odd (even), then P is 0(1), and therefore the sum of A, B, C, D, and P is always odd.

The use of a parity code is an effective way of increasing the reliability of transmission of binary information. As indicated in Fig. 6-14, a parity bit P_1 is

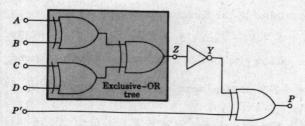

Figure 6-13 An odd-parity checker, or parity-bit generator system, for a 4-bit input word. Assume $P' = 0$ and then $P = 0(1)$ represents odd (even) parity.

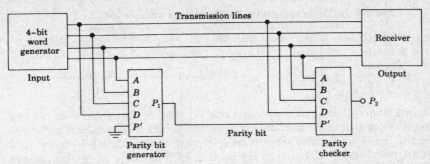

Figure 6-14 Binary data transmission is tested by generating a parity bit at the input to a line and checking the parity of the transmitted bits plus the generated bit at the receiving end of the system.

generated and transmitted along with the N-bit input word. At the receiver the parity of the augmented $(N + 1)$-bit signal is tested, and if the output P_2 of the checker is 0, it is assumed that no error has been made in transmitting the message, whereas $P_2 = 1$ is an indication that (say, due to noise) the received word is in error. Note that only errors in an odd number of digits can be detected with a single parity check.

An MSI 8-bit parity generator/checker is available (TI-180) with control inputs so that it may be used in either odd- or even-parity applications (Prob. 6-14). For words of length greater than 8 bits, several such units may be cascaded (Prob. 6-15).

The MSI unit TI-86 contains four two-input EXCLUSIVE-OR gates.

6-6 DECODER/DEMULTIPLEXER[4]

In a digital system, instructions as well as numbers are conveyed by means of binary levels or pulse trains. If, say, 4 bits of a character are set aside to convey instructions, then 16 different instructions are possible. This information is *coded* in binary form. Frequently a need arises for a multiposition switch which may be operated in accordance with this code. In other words, for each of the 16 codes, one and only one line is to be excited. This process of identifying a particular code is called *decoding*.

Binary-Coded-Decimal (BCD) System

This code translates decimal numbers by replacing each decimal digit with a combination of 4 binary digits. Since there are 16 distinct ways in which the 4 binary digits can be arranged in a row, any 10 combinations can be used to represent the decimal digits from 0 to 9. Thus we have a wide choice of BCD codes. One of these, called the "natural binary-coded-decimal," is the 8421 code

Table 6-1 BCD representation for the decimal number 264

Weighting factor	800	400	200	100	80	40	20	10	8	4	2	1
BCD code	0	0	1	0	0	1	1	0	0	1	0	0
Decimal digits		2				6				4		

illustrated by the first 10 entries in Table 5-1. This is a weighted code because the decimal digit in the 8421 code is equal to the sum of the products of the bits in the coded word times the successive powers of two starting from the right (LSB). We need N 4-bit sets to represent in BCD notation an N-digit decimal number. The first 4-bit set on the right represents units, the second represents tens, the third hundreds, and so on. For example, the decimal number 264 requires three 4-bit sets, as shown in Table 6-1. Note that this three-decade BCD code can represent any number between 0 and 999; hence it has a resolution of 1 part in 1,000, or 0.1 percent. It requires 12 bits, which in a straight binary code can resolve one part in $2^{12} = 4,096$, or 0.025 percent.

BCD-to-Decimal Decoder

Suppose we wish to decode a BCD instruction representing one decimal digit, say 5. This operation may be carried out with a four-input AND gate excited by the 4 BCD bits. For example, the output of the AND gate in Fig. 6-15 is 1 if and only if the BCD inputs are $A = 1$ (LSB), $B = 0$, $C = 1$, and $D = 0$. Since this code represents the decimal number 5, the output is labeled "line 5."

A BCD-to-decimal decoder is indicated in Fig. 6-16. This MSI unit (TI-42A) has four inputs, A, B, C, and D, and 10 output lines. (Ignore the dashed lines, for the moment.) In addition, there must be a ground and a power-supply connection, and hence a 16-pin package is required. The complementary inputs $\bar{A}$, $\bar{B}$, $\bar{C}$, and $\bar{D}$ are obtained from inverters on the chip. Since NAND gates are used, an output is 0 (low) for the correct BCD code and is 1 (high) for any other (invalid) code. The system in Fig. 6-15 is also referred to as a "4-to-10-line decoder" designating that a 4-bit input code selects 1 of 10 output lines. In other words, the decoder acts as a 10-position switch which responds to a BCD input instruction.

It is sometimes desired to decode only during certain intervals of time. In such applications an additional input, called a *strobe*, is added to each NAND gate. All strobe inputs are tied together and are excited by a binary signal S, as indicated by the dashed lines in Fig. 6-16. If $S = 1$, a gate is *enabled* and

Figure 6-15 The output is 1 if the BCD input is 0101 and is 0 for any other input instruction.

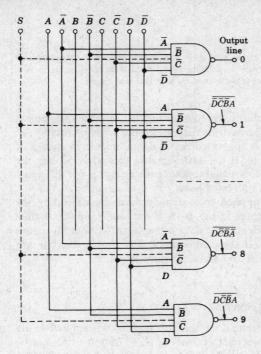

Figure 6-16 A BCD-to-decimal decoder; Assume that $S = 1$. (Lines 2 to 7 are not indicated.) The dashed lines convert the system into a demultiplexer if S represents the input signal.

decoding takes place, whereas if $S = 0$, no coincidence is possible and decoding is inhibited. The strobe input can be used with a decoder having any number of inputs or outputs.

Demultiplexer

A *demultiplexer* is a system for transmitting a binary signal (serial data) on one of N lines, the particular line being selected by means of an address. A

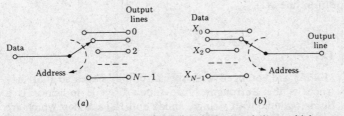

Figure 6-17 Mechanical analog of (a) a demultiplexer and (b) a multiplexer.

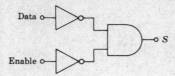

Data o—

Enable o—

o S

Figure 6-18 A decoder is converted into a demultiplexer (with an enabling input) if the S terminal in Fig. 6-16 is obtained from the above AND gate output.

single-pole N-position rotary switch connected as in Fig. 6-17a is the mechanical analog of such a demultiplexer. The address determines the angle of rotation of the arm of the switch. A decoder is converted into a demultiplexer by means of the dashed connections in Fig. 6-16. If the data signal is applied at S, then the output will be the complement of this signal (because the output is 0 if all inputs are 1) and will appear only on the addressed line.

An enabling signal may be applied to a demultiplexer by cascading the system of Fig. 6-16 with that indicated in Fig. 6-18. If the *enable* input is 0, then S is the complement of the data. Hence, the data will appear (without inversion) on the line with the desired code. If the enable input is 1, $S = 0$, the data are inhibited from appearing on any line and all inputs remain at 1.

4-to-16-Line Decoder/Demultiplexer

If an address corresponding to a decimal number in excess of 9 is applied to the inputs in Fig. 6-16, this instruction is rejected; that is, all 10 outputs remain at 1. If it is desired to select 1 of 16 output lines, the system is expanded by adding 6 more NAND gates and using all 16 codes possible with 4 binary bits.

The TI-154 is a 4-to-16-line decoder/demultiplexer. It has 4 address lines, 16 output lines, an enable input, a data input, a ground pin, and a power-supply lead, so that a 24-pin package is required.

A dual 2-to-4-line (TI-155) and a 3-to-8-line (TI-138) decoder/demulti plexer are also available in individual IC packages.

A 1-to-2-line demultiplexer is constructed from 2 two-input NAND gates. The zero-output line comes from the NAND whose inputs are S and $\bar{A}$, whereas the one-output line is connected to the NAND whose inputs are S and A. The "address" A is called the *control* input because, if $A = 0(1)$, the complement of the data ($\bar{S}$) appear on line 0(1).

Decoder/Lamp Driver

Some decoders (TI-46A) are equipped with special output stages so that they can drive lamps such as the Burroughs Nixie tube. A Nixie indicator is a cold-cathode gas-discharge tube with a single anode and 10 cathodes, which are wires shaped in the form of numerals 0 to 9. These cathodes are connected to output lines 0 to 9, respectively, and the anode is tied to a fixed supply voltage.

The decoder/lamp driver/Nixie indicator combination makes visible the decimal number corresponding to the BCD number applied. Thus, if the input is 0101, the numeral 5 will glow in the lamp.

A decoder for seven-segment numerals made visible by using light-emitting diodes is discussed in Sec. 6-11.

Higher-Order Demultiplexers[3]

If the number of output lines N exceeds 16, the demultiplexers discussed in the foregoing for $N = 16$, 8, 4, or 2 are arranged in a "tree" formation to yield the desired number of output lines. For example, for $N = 32$, we can use a demultiplexer with the "trunk" $N_1 = 4$ and four "branches" $N_2 = 8$ as indicated in Fig. 6-19. Note that the total number of output lines is $N = N_1 N_2 = 32$. Lines 0 through 7 are decoded by demultiplexer N_{20}, whereas N_{21} decodes the next eight lines, etc.

For $ED = 01$, lines 8 thru 15 are decoded in sequence as the address CBA changes from 000 to 001 to $\cdots$ to 111. For example, line 12 is decoded by the address $EDCBA = 01100$, which is the binary representation for decimal 12.

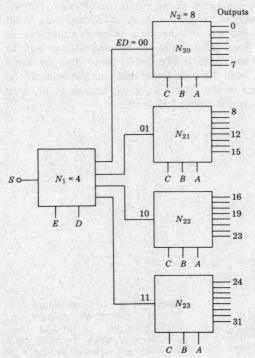

Figure 6-19 Thirty two-output demultiplexer tree, where N_1 is a four-output and N_2 is an eight-output demultiplexer.

Line 19 is decoded by $EDCBA = 10011$, etc. Since there are two 2-to-4-line decoders in one package, a total of $4\frac{1}{2}$ equivalent packages are needed for the system in Fig. 6-19. It is also possible to design this system with $N_1 = 8$, $N_2 = 4$ (Prob. 6-17) or with $N_1 = 2$, $N_2 = 16$, etc. The proper design choice is indicated by total cost.

A demultiplexer with 64 outputs can be designed with $N_1 = N_2 = 8$, for a total of 9 packages. Why? For very large values of N, higher-level branching is required (Prob. 6-18), where each output in Fig. 6-19 is an input to another demultiplexer.

6-7 DATA SELECTOR/MULTIPLEXER

The function performed by a *multiplexer* is to select 1 out of N input data sources and to transmit the selected data to a single information channel. The N-position switch connected as in Fig. 6-17b is the mechanical analog of a multiplexer. Compare Fig. 6-17a and b. Since in a demultiplexer there is only one input line and these data are caused to appear on 1 out of N output lines, a multiplexer performs the inverse process of a demultiplexer.

The demultiplexer of Fig. 6-16 is converted into a multiplexer by making the following two changes: (1) Add a NAND gate whose inputs include all N outputs of Fig. 6-16 and (2) augment each NAND gate with an individual data input $X_0, X_1 \cdots X_N$. The logic system for a 4-to-1-line data selector/multiplexer is drawn in Fig. 6-20. This AND-OR logic is equivalent to the NAND-NAND logic as described in the above steps 1 and 2. (See Fig. 5-17.) Note that the same decoding configuration is used in both the multiplexer and demultiplexer. If the select code is 01, then X_1 appears at the output Y, if the address is 11, then $Y = X_3$, etc., provided that the system is enabled ($S = 0$).

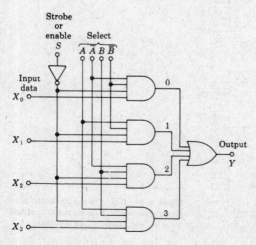

Figure 6-20 A 4-to-1-line multiplexer. Two such systems are packaged as TI-153. Note that A is the LSB. The complementary inputs $\overline{A}$ and $\overline{B}$ are obtained from inverters on the chip.

The following data selector/multiplexers are available: 16-to-1-line (TI-150), one per package; 8-to-1-line (TI-151A), one per package; 4-to-1-line (TI-153), two per package; and 2-to-1-line (TI-157), four per package. The 1-out-of-16 multiplexer is a 24-pin IC with 16 data inputs, a 4-bit select code, a strobe input, one output, a power-supply load, and a ground terminal. For this 16-to-1-line data selector, Fig. 6-20 is extended from 4 four-input AND gates to 16 six-input AND gates.

Parallel-to-Serial Conversion

Consider a 16-bit word available in parallel form so that X_0 represents the 2^0 bit, X_1 the 2^1 bit, etc. By means of a counter (Sec. 7-5), it is possible to change the select code so that it is 0000 for the first T seconds, 0001 for the next T seconds, 0010 for the third interval T, etc. With such excitation of the address, the output of the multiplexer will be X_0 for the first T seconds, X_1 for the next interval T, X_2 for the third period, etc. The output Y is a waveform which represents serially the binary data applied in parallel at the input. In other words, a parallel-to-serial conversion is accomplished of one 16-bit word. This process takes $16T$ seconds.

In a digital system, such as a computer, a data communication system, etc., a pulse train is often required for testing or control (gating) purposes. Such a *sequence generator* is obtained by means of the parallel-to-series converter. Any desired waveform may be obtained by properly choosing the input data X.

Sequential Data Selection

By changing the address with a counter in the manner indicated in the preceding paragraph, the operation of an electromechanical stepping switch is simulated. If the data inputs are pulse trains, this information will appear sequentially on the output channel: in other words, pulse train X_0 will appear for T seconds, followed by X_1 for the next T seconds, etc. If the number of data sources is M, then X_0 is again selected during the interval $MT < t < (M + 1)T$.

Higher-Order Multiplexers

If the number of input lines exceeds 16, then the logic block diagram assumes a topology which is the inverse to that shown in Fig. 6-19. For example, to select 1 out of 32 data inputs, the system in Fig. 6-21 may be used. Multiplexer N_{20} places the data inputs X_0 through X_7 in sequence onto line L_0 as the address CBA changes from 000 to 001 to $\cdots$ to 111. Similarly N_{21} transmits the data X_8 through X_{15} onto line L_1 as CBA sequences from 000 through 111. Specifically, if the address is $CBA = 100$, then X_4 appears on L_0, X_{12} on L_1, X_{20} on L_2, and X_{28} on L_3. If it is desired that X_{20} be transmitted to the output, then ED must equal 10 so that N_1 will select the data on line L_2. In summary, for the

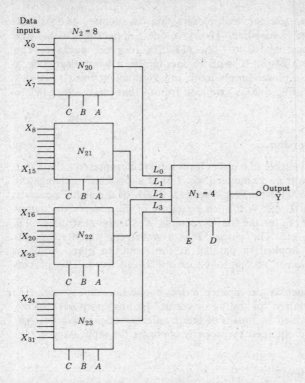

- **Figure 6-21** A 32-to-1-line data selector/multiplexer, where N_2 is an 8-to-1-line and N_1 is a 4-to-1-line multiplexer.

address $EDCBA = 10100$, the multiplexer transfers the input data X_{20} to the output line Y. An alternative solution using two 16-data-input multiplexers to obtain 1 out of 32 data inputs is given in Prob. 6-21.

Note that the total number of input lines $N = N_2 N_1$. For very large values of N, a third level of multiplexers N_3 may be necessary. The outputs from N_3 are connected to the inputs of N_2, and the outputs from N_2 are the inputs to N_1. The system selects 1 out of $N = N_3 N_2 N_1$ inputs (Prob. 6-24).

Combinational Logic

The Boolean expression for the output Y of the multiplexer in Fig. 6-20 is

$$Y = X_0 \overline{B}\overline{A} + X_1 \overline{B} A + X_2 B \overline{A} + X_3 B A \qquad (6\text{-}15)$$

As noted in Sec. 6-2, combinational logic of three variables is represented by a Boolean sum of products of A, B, and C. Each minterm is of the form CBA or the complements of these variables. Hence, a multiplexer can satisfy any combinational-logic equation if the proper choices are made for the X inputs.

Thus it may be required that $X = C$ or $X = \overline{C}$. If terms contain both C and $\overline{C}$, then $X = C + \overline{C} = 1$, and if a term is missing, then $X = 0$.

Example 6-2 Generate the following combinational-logic equation using a four-input multiplexer:

$$Y = C\overline{B}\,\overline{A} + \overline{C}\,\overline{B}\,\overline{A} + C\overline{B}A + \overline{C}BA \qquad (6\text{-}16)$$

SOLUTION Since $\overline{B}\,\overline{A}$ represents decimal 0, the coefficient of $\overline{B}\,\overline{A}$ is X_0. Hence $X_0 = C + \overline{C} = 1$. Since $\overline{B}A$ represents decimal 1, the factor multiplying $\overline{B}A$ is X_1. Hence $X_1 = C$. Since BA represents 3, $X_3 = \overline{C}$. Since $B\overline{A}$, which represents 2, is missing from the equation, $X_2 = 0$. In summary,

$$X_0 = 1 \qquad X_1 = C \qquad X_2 = 0 \qquad \text{and} \qquad X_3 = \overline{C}$$

If these values of X are used in the multiplexer of Fig. 6-20, the output Y will equal the combinational logic in Eq. (6-16).

In the foregoing example, a standard Boolean equation in three variables is generated with a 4-to-1-line multiplexer. In general, a combinational-logic equation in N variables is generated by using a 2^{N-1}-input data selector.

6-8 ENCODER

A decoder is a system which accepts an M-bit word and establishes the state 1 on one (and only one) of 2^M output lines (Sec. 6-6). In other words, a decoder identifies (recognizes) a particular code. The inverse process is called *encoding*. An encoder has a number of inputs, only one of which is in the 1 state, and an N-bit code is *generated*, depending upon which of the inputs is excited.

Consider, for example, that it is required that a binary code be transmitted with every stroke of an alphanumeric keyboard (a typewriter or teletype). There are 26 lowercase and 26 capital letters, 10 numerals, and about 22 special characters on such a keyboard so that the total number of codes necessary is approximately 84. This condition can be satisfied with a minimum of 7 bits ($2^7 = 128$, but $2^6 = 64$). Let us modify the keyboard so that, if a key is depressed, a switch is closed, thereby connecting a 5-V supply (corresponding to the 1 state) to an input line. A block diagram of such an encoder is indicated in Fig. 6-22. Inside the shaded block there is a rectangular array (or matrix) of wires, and we must determine how to interconnect these wires so as to generate the desired codes.

To illustrate the design procedure for constructing an encoder, let us simplify the above example by limiting the keyboard to only 10 keys, the numerals 0, 1, . . . , 9. A 4-bit output code is sufficient in this case, and let us choose BCD words for the output codes. The truth table defining this encoding is given in Table 6-2. Input W_n ($n = 0, 1, \ldots, 9$) represents the nth key. When $W_n = 1$, key n is depressed. Since it is assumed that no more than one key is activated simultaneously, then in any row every input except one is a 0. From

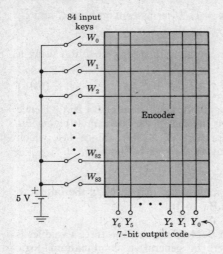

84 input
keys

5 V

Y_6 Y_5 Y_2 Y_1 Y_0
7–bit output code

Figure 6-22 A block diagram of an encoder for generating an output code (word) for every character on a keyboard.

this truth table we conclude that $Y_0 = 1$ if $W_1 = 1$ or if $W_3 = 1$ or if $W_5 = 1$ or if $W_7 = 1$ or if $W_9 = 1$. Hence, in Boolean notation,

$$Y_0 = W_1 + W_3 + W_5 + W_7 + W_9 \tag{6-17}$$

Similarly,

$$Y_1 = W_2 + W_3 + W_6 + W_7$$
$$Y_2 = W_4 + W_5 + W_6 + W_7 \tag{6-18}$$
$$Y_3 = W_8 + W_9$$

The OR gates in Eqs. (6-17) and (6-18) are implemented with diodes in Fig. 6-23. (Compare with Fig. 5-3, but with the diodes reversed, because we are now

Table 6-2 The truth table for encoding the decimal numbers 0 to 9

Inputs										Outputs			
W_9	W_8	W_7	W_6	W_5	W_4	W_3	W_2	W_1	W_0	Y_3	Y_2	Y_1	Y_0
0	0	0	0	0	0	0	0	0	1	0	0	0	0
0	0	0	0	0	0	0	0	1	0	0	0	0	1
0	0	0	0	0	0	0	1	0	0	0	0	1	0
0	0	0	0	0	0	1	0	0	0	0	0	1	1
0	0	0	0	0	1	0	0	0	0	0	1	0	0
0	0	0	0	1	0	0	0	0	0	0	1	0	1
0	0	0	1	0	0	0	0	0	0	0	1	1	0
0	0	1	0	0	0	0	0	0	0	0	1	1	1
0	1	0	0	0	0	0	0	0	0	1	0	0	0
1	0	0	0	0	0	0	0	0	0	1	0	0	1

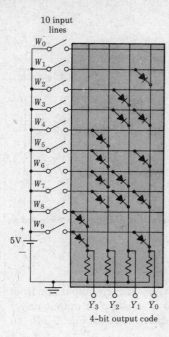

Figure 6-23 An encoding matrix to transform a decimal number into a binary code (BCD). The key W_0 may be omitted since it is implied that the output is 0000 unless one of the other nine keys is activated.

considering positive logic.) An encoder array such as that in Fig. 6-23 is called a *rectangular diode matrix*.

Incidentally, a decoder can also be constructed as a rectangular diode matrix (Prob. 6-30). This statement follows from the fact that a decoder consists of AND gates (Fig. 6-15), and it is possible to implement AND gates with diodes (Fig. 5-5b).

Each diode of the encoder of Fig. 6-23 may be replaced by the base-emitter diode of a transistor. If the collector is tied to the supply voltage V_{CC}, then an emitter-follower OR gate results. Such a configuration is indicated in Fig. 6-24a for the output Y_2. Note that if either W_4 or W_5 or W_6 or W_7 is high, then the emitter-follower output is high, thus verifying that $Y_2 = W_4 + W_5 + W_6 + W_7$, as required by Eq. (6-18).

Only one transistor (with multiple emitters) is required for each encoder input. The base is tied to the input line, and each emitter is connected to a different output line, as dictated by the encoder logic. For example, since in Fig. 6-23 line W_7 is tied to three diodes whose cathodes go to Y_0, Y_1, and Y_2, then this combination may be replaced by the three-emitter transistor $Q7$ connected as in Fig. 6-24b. The maximum number of emitters that may be required equals the number of bits in the output code. For the particular encoder sketched in Fig. 6-23, $Q1$, $Q2$, $Q4$, and $Q8$ each have one emitter, $Q3$, $Q5$, $Q6$, and $Q9$ have two emitters each, and $Q7$ has three emitters.

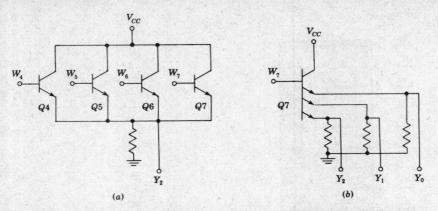

Figure 6-24 (*a*) An emitter-follower OR gate. (*b*) Line W_7 in the encoder of Fig. 6-23 is connected to the base of a three-emitter transistor.

Output Stages

A bipolar encoder uses the standard TTL output stages. If each output line from the encoder goes to the data input in Fig. 5-24*a*, a totem-pole output driver results. If an encoder output line goes to the data input in Fig. 5-24*b* an open-collector output is obtained.

Priority Encoder

We now remove the restriction that only one key is depressed at any given time. If several keys are simultaneously pushed accidentally, then let us give priority to the highest-order data line and encode it. For example, if W_5 and W_6 are simultaneously activated, then it is desired that the output correspond to W_6. The truth table for a 10-to-4-line priority encoder is given in Table 6-3. An X in the table means that this entry is *irrelevant*. It may be either a 1 or a 0 and, hence, X is referred to as a *don't care state*. However, now a 0 in the table must be taken into account, whereas in Table 6-2 the zeros could be ignored because Table 6-2 is uniquely determined by the 1s along the diagonal.

The Boolean expression for Y_1, obtained from Table 6-3, is

$$Y_1 = \overline{W}_9\overline{W}_8\overline{W}_7\overline{W}_6\overline{W}_5\overline{W}_4\overline{W}_3W_2 + \overline{W}_9\overline{W}_8\overline{W}_7\overline{W}_6\overline{W}_5\overline{W}_4W_3$$
$$+ \overline{W}_9\overline{W}_8\overline{W}_7W_6 + \overline{W}_9\overline{W}_8W_7 \tag{6-19}$$

This equation can be simplified considerably. Note that

$$Y_1 = \overline{W}_9\overline{W}_8(\overline{W}_7B + W_7) \tag{6-20}$$

where
$$B \equiv \overline{W}_6\overline{W}_5\overline{W}_4\overline{W}_3W_2 + \overline{W}_6\overline{W}_5\overline{W}_4W_3 + W_6 \tag{6-21}$$

Table 6-3 A priority encoder (10-line decimal to 4-line BCD)

Inputs										Output			
W_9	W_8	W_7	W_6	W_5	W_4	W_3	W_2	W_1	W_0	Y_3	Y_2	Y_1	Y_0
0	0	0	0	0	0	0	0	0	1	0	0	0	0
0	0	0	0	0	0	0	0	1	X	0	0	0	1
0	0	0	0	0	0	0	1	X	X	0	0	1	0
0	0	0	0	0	0	1	X	X	X	0	0	1	1
0	0	0	0	0	1	X	X	X	X	0	1	0	0
0	0	0	0	1	X	X	X	X	X	0	1	0	1
0	0	0	1	X	X	X	X	X	X	0	1	1	0
0	0	1	X	X	X	X	X	X	X	0	1	1	1
0	1	X	X	X	X	X	X	X	X	1	0	0	0
1	X	X	X	X	X	X	X	X	X	1	0	0	1

From Eq. (5-19) with $A = W_7$, we obtain

$$Y_1 = \overline{W}_9\overline{W}_8(W_7 + B) \tag{6-22}$$

From Eq. (6-21),

$$B = \overline{W}_6 C + W_6 = W_6 + C \tag{6-23}$$

where Eq. (5-19) is used again, and where

$$C \equiv \overline{W}_5\overline{W}_4\overline{W}_3 W_2 + \overline{W}_5\overline{W}_4 W_3 = \overline{W}_5\overline{W}_4\left(\overline{W}_3 W_2 + W_3\right)$$

$$= \overline{W}_5\overline{W}_4(W_3 + W_2) \tag{6-24}$$

From Eqs. (6-22), (6-23), and (6-24),

$$Y_1 = \overline{W}_9\overline{W}_8\left(W_7 + W_6 + \overline{W}_5\overline{W}_4 W_3 + \overline{W}_5\overline{W}_4 W_2\right) \tag{6-25}$$

A NOR gate is used to generate $\overline{W}_9\overline{W}_8 = \overline{W_9 + W_8}$ (De Morgan's law) and a four-wide (two-two-four-four-input) AND-OR gate is required to generate Y_1. Proceeding in a similar manner, the combinational logic for Y_0, Y_2, and Y_3 is found (Probs. 6-31 and 6-32).

The above logic is fabricated on an MSI chip (TI-147) which priority encodes 10-line decimal to 4-line BCD. Applications include encoding of small keyboards, analog-to-digital conversion (Sec. 16-13), and controlling computer priority interrupts. The TI-148 package encodes eight data lines to a three-line binary (octal code).

6-9 READ-ONLY MEMORY (ROM)[5, 6]

Consider the problem of converting one binary code into another. Such a code-conversion system (designated ROM and sketched in Fig. 6-25*a*) has M inputs ($X_0, X_1, \ldots, X_{M-1}$) and N outputs ($Y_0, Y_1, \ldots, Y_{N-1}$), where N may be

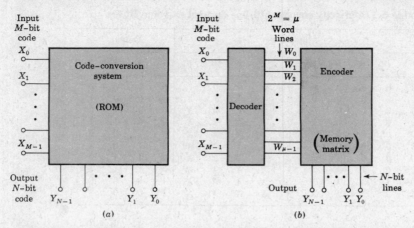

Figure 6-25 (*a*) A block diagram of a system for converting one code into another; a read-only memory (ROM). (*b*) A ROM may be considered to be a decoder for the input code followed by an encoder for the output code.

greater than, equal to, or less than M. A definite M-bit code is to result in a specific output code of N bits. This code translation is achieved, as indicated in Fig. 6-25b, by first decoding the M inputs onto $2^M \equiv \mu$ word lines ($W_0, W_1, \ldots, W_{\mu-1}$) and then encoding each line into the desired output word. If the inputs assume all possible combinations of 1s and 0s, then μ N-bit words are "read" at the output (not all these 2^M words need be unique, since it may be desirable to have the same output code for several different input words).

The functional relationship between output and input words is built into hardware in the encoder block of Fig. 6-25. Since this information is thus stored permanently, we say that the system has *nonvolatile memory*. The *memory elements* are the diodes in Fig. 6-23 or the emitters of transistors in Fig. 6-24. The output word for any input code may be read as often as desired. However, since the stored relationship between output and input codes cannot be modified without adding or subtracting memory elements (hardware), this system is called a *read-only memory*, abbreviated ROM.

Code Converters

The truth table for translating from a binary to a Gray code[7] is given in Table 6-4. In the progression from one line to the next of the Gray code, 1 and only 1 bit is changed from 0 to 1, or vice versa. (This property does not uniquely define a code, and hence a number of Gray codes may be constructed.) The input bits [(1) in Table 6-4] are decoded in a ROM into the word lines $W_0, W_1, \ldots, W_{15}$, as indicated in Fig. 6-25b, and then are encoded [(2) in Table 6-4] into the desired Gray code $Y_3 Y_2 Y_1 Y_0$. The W's are the minterm outputs of the decoder.

Table 6-4 Conversion from a binary to a Gray code [(1) to (2)] and from a Gray to a binary code [1) to (3)]

(1) Inputs				Decoded word	(2) Gray code outputs				(3) Binary code outputs			
X_3	X_2	X_1	X_0	W_n	Y_3	Y_2	Y_1	Y_0	Y_3	Y_2	Y_1	Y_0
0	0	0	0	W_0	0	0	0	0	0	0	0	0
0	0	0	1	W_1	0	0	0	1	0	0	0	1
0	0	1	0	W_2	0	0	1	1	0	0	1	1
0	0	1	1	W_3	0	0	1	0	0	0	1	0
0	1	0	0	W_4	0	1	1	0	0	1	1	1
0	1	0	1	W_5	0	1	1	1	0	1	1	0
0	1	1	0	W_6	0	1	0	1	0	1	0	0
0	1	1	1	W_7	0	1	0	0	0	1	0	1
1	0	0	0	W_8	1	1	0	0	1	1	1	1
1	0	0	1	W_9	1	1	0	1	1	1	1	0
1	0	1	0	W_{10}	1	1	1	1	1	1	0	0
1	0	1	1	W_{11}	1	1	1	0	1	1	0	1
1	1	0	0	W_{12}	1	0	1	0	1	0	0	0
1	1	0	1	W_{13}	1	0	1	1	1	0	0	1
1	1	1	0	W_{14}	1	0	0	1	1	0	1	1
1	1	1	1	W_{15}	1	0	0	0	1	0	1	0

For example,

$$W_0 = \bar{X}_3\bar{X}_2\bar{X}_1\bar{X}_0 \qquad W_5 = \bar{X}_3 X_2 \bar{X}_1 X_0 \qquad W_9 = X_3 \bar{X}_2 \bar{X}_1 X_0 \qquad (6\text{-}26)$$

From the truth table (Table 6-4), we conclude that

$$Y_0 = W_1 + W_2 + W_5 + W_6 + W_9 + W_{10} + W_{13} + W_{14} \qquad (6\text{-}27)$$

This equation is implemented by connecting eight diodes with their cathodes all tied to Y_0 and their anodes connected to the decoder lines W_1, W_2, W_5, W_6, W_9, W_{10}, W_{13}, and W_{14}, respectively (or the base-emitter diodes of transistors may be used in an analogous manner to form an emitter-follower OR gate, as in Fig. 6-24a). Similarly, from Table 6-4, we may write the Boolean expressions for the other output bits. For example,

$$Y_3 = W_8 + W_9 + W_{10} + W_{11} + W_{12} + W_{13} + W_{14} + W_{15} \qquad (6\text{-}28)$$

Consider the inverse code translation, from Gray to binary. The Gray code inputs [(1) in Table 6-4] are arranged in the order W_0, W_1, ..., W_{15} (corresponding to decimal numbers 0 to 15). The binary code corresponding to a given input word W_n is listed as the output code [(3) in Table 6-4] for that line. For example, from (1) and (2) of Table 6-4 at line W_{14}, we find that the Gray code 1001 corresponds to the binary code 1110, and this relationship is maintained in Table 6-4 [(1) and (3)] on line W_9. From this table we obtain the relationship between binary output (3) and Gray input (1) bits. For example,

$$Y_0 = W_1 + W_2 + W_4 + W_7 + W_8 + W_{11} + W_{13} + W_{14} \qquad (6\text{-}29)$$

This equation defines how the memory elements are to be arranged in the encoder. Note that the ROM for conversion from a binary to a Gray code uses the same decoding arrangement as that for conversion from a Gray to a binary code. However, the encoders are completely different. In other words, the IC chips for these two ROMs are quite distinct since individual masks must be used for the encoder matrix of memory elements.

Programming the ROM

Consider a 256-bit bipolar ROM TI-88A arranged in 32 words of 8 bits each. The decoder input is a 5-bit binary select code, and its outputs are the 32 word lines. The encoder consists of 32 transistors (each base is tied to a different line) and with 8 emitters in each transistor. The customer fills out the truth table he wishes the ROM to satisfy, and then the vendor makes a mask for the metallization so as to connect one emitter of each transistor to the proper output line, or alternatively, to leave it floating. For example, for the Gray-to-binary-code conversion, Eq. (6-29) indicates that one emitter from each of transistors $Q1$, $Q2$, $Q4$, $Q7$, $Q8$, $Q11$, $Q13$, and $Q14$ is connected to line Y_0, whereas the corresponding emitter on each of the other transistors $Q0$, $Q3$, $Q5$, $Q6 \ldots$ is left unconnected. The process just described is called *custom programming* or *mask programming*, of a ROM. Note that *hardware* (not *software*) programming is under consideration. If the sales demand for a particular code is sufficient, this ROM becomes available as an "off-the-shelf" item.

For small quantities of a ROM, the mask cost may be prohibitive ($\sim \$800$), and also the delivery time may be too long. Hence manufacturers (Appendix B-1) supply *field-programmable* ROMs,[5, 8] abbreviated PROMs. Such an IC chip has an encoder matrix made with all connections which may possibly be required. For example, the 256-bit memory discussed above is constructed as a PROM with 32 transistors, each having eight emitters (labeled E_0, E_1, $\ldots$, E_7) and with E_0 from each transistor tied to output Y_0, E_1 to Y_1, etc. In series with each emitter there is incorporated a narrow polycrystalline silicon[6] or nichrome strip which acts as a fuse and opens up when a current in excess of a maximum value is passed through this memory element. The user can easily "blow," "burn," or "zap" in the field those memory-element fuses which must be opened in order that the ROM perform the desired functional relationships between output and input.

An instrument called a *PROM programmer* is available specifically in order to make this task quite simple. Clearly, once a bipolar ROM has been programmed by opening the desired fusible links, the program cannot be changed. However, it is possible to erase the program of MOSFET ROMs and to write a new program electrically (Sec. 9-4).

Diode matrices are also available with fusible links, and these can be used for the encoder portion of a PROM, or also as a decoder.

6-10 TWO-DIMENSIONAL ADDRESSING OF A ROM

Many manufacturers (Appendix B-1) supply bipolar ROMs in sizes from 256 to 4,096 bits, with either four- or eight-output lines. (Since 4,096 is close to 4,000, a 4,096-bit ROM is referred to as a 4-kb memory.) Larger ROMs are also available and are examples of large-scale integration (LSI). For example, Monolithic Memories #6255, with $M = N = 10$, is a $1024 \times 10 = 10,240$-bit (10-kb) memory and #6275, with $M = 11$ and $N = 8$, is a $2,048 \times 8 = 16,384$-bit (16-kb) bipolar memory. The time required for a valid output to appear on the bit lines from the time an input address is applied to the memory is defined as the *access time*. For bipolar ROMs access times of 50 to 100 ns are obtained. LSI ROMs using MOSFETs (Chap. 8) are discussed in Sec. 9-3.

For a ROM with a large number of inputs, the decoding arrangement in Fig. 6-25 is impractical. Consider, for example, a $512 \times 4 = 2,048$-bit ROM ($M = 9$ and $N = 4$). A total of 512 NAND gates are required in the decoder, one for each word line. Considerable economy results if the topology of Fig. 6-26 is

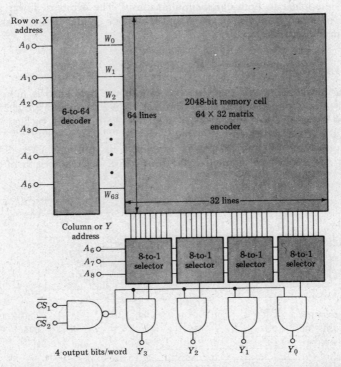

Figure 6-26 A 2-kb bipolar ROM (512×4 bits) with two-dimensional addressing (Intel 3602/3622). Note that the column address $A_8A_7A_6$ is applied to all four selectors/multiplexers. The *chip select* input CS is used for enabling purposes.

used. A 6-bit X (row) address generates 64 horizontal lines. If 32 Y (vertical) lines are used in the memory matrix, the total number of bits is $64 \times 32 = 2,048$ as required. However, since only four output lines are specified then four 8- to 1-line selectors are used. A 3-bit column address feeds each multiplexer. This arrangement is called X-Y or *two-dimensional addressing*. Note that now 64 NAND gates are needed for the decoder and $4 \times 9 = 36$ for the selectors from the NAND-NAND (AND-OR) configuration of Fig. 6-20 (it is clear that an 8-input selector requires nine gates). The total of $64 + 36 = 100$ NANDs for X-Y addressing is a tremendous saving over 512 NANDs required by the decoder arrangement of Fig. 6-25 for the same size ROM. From Fig. 6-26, 64 transistors with 32 emitters each are needed for the encoder whereas in Fig. 6-25 there are 512 transistors, each with 4 emitters.

Word Expansion

Increasing the number of bits per word is easily accomplished. For example, a 512-$\times$8-bit ROM is obtained by using two 512×4 ROMs. The addressing in Fig. 6-26 is applied to both chips simultaneously. The 4 bits of lower significance are obtained from one package and the 4 bits of higher significance are taken from the second chip.

Address Expansion

To obtain additional words (with no increase in the byte size) is more complicated. For example, to obtain 1,024 words of 4 bits each requires four 256×4 ROMs, an external 2-to-4-line decoder, and the use of 3-state OR-tied output stages (Fig. 5-24), as indicated in Fig. 6-27.

The operation of this system is explained as follows. The addresses $A_7 \cdots A_0$ are applied in parallel to the four 256×4 ROMs, each of whose

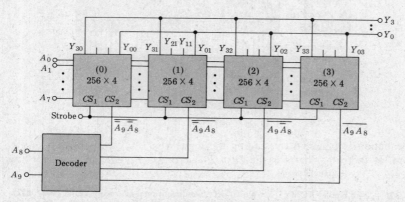

Figure 6-27 Address expansion allows four 256×4 ROMs to become a $1,024 \times 4$-bit memory.

outputs are OR-tied together. A 2-bit address A_9A_8 is applied to a decoder (Fig. 6-16), whose four outputs control the four chip selects CS_2. For example, if $A_9 = 0$ and $A_8 = 1$, then $CS_2 = 0$ for chip 1 and $CS_2 = 1$ for all other packages. Hence, chip 1 is enabled, whereas the other three packages present a high impedance at each output (Fig. 5-24). Hence, only the outputs $Y_0 = Y_{01}$, $Y_1 = Y_{11}$, $Y_2 = Y_{21}$, and $Y_3 = Y_{31}$ from package 1 appear at the output. Each combination of 1s and 0s in an address $A_9A_8 \cdots A_1A_0$ yields one 4-bit output code, for a total of 1,024 4-bit words in the memory.

6-11 ROM APPLICATIONS

As emphasized in the preceding section, a ROM is a code-conversion unit. However, many different practical systems represent a translation from one code to another. The most important of these ROM applications are discussed below.

Look-up Tables

Routine calculations such as trigonometric functions, logarithms, exponentials, square roots, etc., are sometimes required of a computer. If these are repeated often enough, it is more economical to include a ROM as a *look-up table*, rather than to use a subroutine or a software program to perform the calculation. A look-up table for $Y = \sin X$ is a code-conversion system between the input code representing the argument X in binary notation (to whatever accuracy is desired) and the output code giving the corresponding values of the sine function. Clearly, any calculation for which a truth table can be written may be implemented with a ROM—a different ROM for each truth table.

Sequence Generators

If in a digital system, P pulse trains are required for testing or control purposes, these may be obtained by using P multiplexers connected for parallel-to-serial conversion (Sec. 6-7). A more economical way to supply these binary sequences is to use a ROM with P outputs and to change the address by means of a counter. As mentioned in Sec. 6-7, the input to the encoder changes from W_0 to W_1 to W_2, etc., every T seconds. Under this excitation the output Y_1 of the ROM represented by Table 6-4 (for Gray code to binary code conversion) is

$$Y_1 = 1100001100111100 \quad \text{(LSB)} \quad (6\text{-}30)$$

This equation is obtained by reading the digits in the Y_1 column from top to bottom. It indicates that for the first $2T$ seconds, Y_1 remains low; for the following $4T$ seconds, Y_1 is high; for the next $2T$ seconds, Y_1 is low; for the next $2T$ seconds, Y_1 is high; for the following $4T$ seconds, Y_1 is low; for the last $2T$ seconds, Y_1 is high; and after these $16T$ seconds, this sequence is repeated (as long as pulses are fed to the counter).

Simultaneously with Y_1, three other synchronous pulse trains, Y_0, Y_2, and Y_3, are created. In general, the number of sequences obtained equals the number of outputs from the ROM. Any desired serial binary waveforms are generated if the truth table is properly specified, i.e., if the ROM is correctly programmed.

Waveform Generator[6]

If the output of the digital sequence generator is converted into an analog voltage, then a *waveform generator* is obtained. Consider a 256×8-bit ROM sequenced by means of an 8-bit counter. Each step of the counter represents $\frac{360}{256} = 1.406°$ of the waveform. The ROM is programmed so the outputs Y_0 to Y_8 give the digital number corresponding to the analog amplitude at each step. The ROM outputs feed into a D/A (digital-to-analog converter, Sec. 16-12), and the output gives the analog waveform desired. This output changes in small discrete steps (each less than $\frac{1}{2}$ percent of full scale) and hence some simple filtering may be desireable.

Seven-segment Visible Display

It is common practice to make visible the reading of a digital instrument (a frequency meter, digital voltmeter, etc.) by means of the seven-segment numeric indicator sketched in Fig. 6-28a. A wide variety[9] of readouts are commercially available. A solid-state indicator in which the segments obtain their luminosity from light-emitting gallium arsenide or phosphide diodes is operated at low voltage and low power and hence may be driven directly from IC logic gates.

The first 10 displays in Fig. 6-28b are the numerals 0 to 9, which, in the digital instrument, are represented in BCD form. Such a 4-bit code has 16 possible states, and the displays 10 to 15 of Fig. 6-28b are unique symbols used to identify a nonvalid BCD condition.

The problem of converting from a BCD input to the seven-segment outputs of Fig. 6-28 is easily solved using a ROM. If an excited (luminous) segment is identified as state 0 and a dark segment as the 1 state, then truth table 6-5 is obtained. This table is verified as follows: For word W_0 (corresponding to

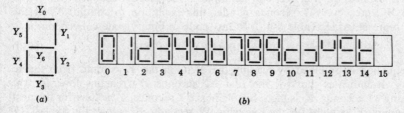

Figure 6-28 (a) Identification of the segments in a seven-segment LED visible indicator. (b) The display which results from each of the sixteen 4-bit input codes.

Table 6-5 Conversion from a BCD to a seven-segment-indicator code

Binary-coded-decimal inputs				Decoded word	Seven-segment-indicator code outputs						
$S_3 = D$	$X_2 = C$	$X_1 = B$	$X_0 = A$	W_n	Y_6	Y_5	Y_4	Y_3	Y_2	Y_1	Y_0
0	0	0	0	W_0	1	0	0	0	0	0	0
0	0	0	1	W_1	1	1	1	1	0	0	1
0	0	1	0	W_2	0	1	0	0	1	0	0
0	0	1	1	W_3	0	1	1	0	0	0	0
0	1	0	0	W_4	0	0	1	1	0	0	1
0	1	0	1	W_5	0	0	1	0	0	1	0
0	1	1	0	W_6	0	0	0	0	0	1	1
0	1	1	1	W_7	1	1	1	1	0	0	0
1	0	0	0	W_8	0	0	0	0	0	0	0
1	0	0	1	W_9	0	0	1	1	0	0	0
1	0	1	0	W_{10}	0	1	0	0	1	1	1
1	0	1	1	W_{11}	0	1	1	0	0	1	1
1	1	0	0	W_{12}	0	0	1	1	1	0	1
1	1	0	1	W_{13}	0	0	1	0	1	1	0
1	1	1	0	W_{14}	0	0	0	0	1	1	1
1	1	1	1	W_{15}	1	1	1	1	1	1	1

numeral 0) we see from Fig. 6-28 that $Y_6 = 1$ and all other Y's are 0. For word W_4 (corresponding to the numeral 4) $Y_0 = Y_3 = Y_4 = 1$ and $Y_1 = Y_2 = Y_5 = Y_6 = 0$, and so forth. The ROM is programmed as explained in Sec. 6-9 to satisfy this truth table. For example,

$$Y_0 = W_1 + W_4 + W_6 + W_{10} + W_{11} + W_{12} + W_{14} + W_{15} \qquad (6\text{-}31)$$

It should be pointed out that a ROM may not use the smallest number of gates to carry out a particular code conversion. Consider Eq. (6-31) written as a sum of products. Replacing the minterm W_1 by $\bar{X}_3\bar{X}_2\bar{X}_1X_0 \equiv \bar{D}\bar{C}\bar{B}A$ and using analogous expressions for the outputs of the other decoders, Eq. (6-31) becomes

$$Y_0 = \bar{D}\bar{C}\bar{B}A + \bar{D}C\bar{B}\bar{A} + \bar{D}CB\bar{A} + D\bar{C}B\bar{A} + D\bar{C}BA$$
$$+ DC\bar{B}\bar{A} + DCB\bar{A} + DCBA \qquad (6\text{-}32)$$

There are a number of algebraic and graphical techniques[10, 11] and computer programs for minimizing such Boolean expressions. Note, for example, that the second and third minterms can be simplified to

$$\bar{D}C\bar{B}\bar{A} + \bar{D}CB\bar{A} = \bar{D}C\bar{A}$$

because $\bar{B} + B = 1$. Proceeding in this manner (Prob. 6-42), the following minimized form of Y_0 is obtained:

$$Y_0 = \bar{D}\bar{C}\bar{B}A + C\bar{A} + DB \qquad (6\text{-}33)$$

Using the minimized expressions for $Y_0, Y_1, \ldots, Y_6$ results in some saving (about 20 percent) of components over those required in the ROM. A chip fabricated in this manner (for example, TI-64A) is designated a "BCD-to-seven-segment decoder/driver."

Minimization of Boolean equations (particularly if the number of variables in each product exceeds five) is tedious and time-consuming. The engineering man-hours cost for minimization and for designing the special IC chip to realize the savings in components must be compared with that of simply programming an existing ROM. Unless a tremendous number of units are to be manufactured (and particularly if the matrix size is large), the ROM is the more economical procedure.

Combinational Logic

If N logic equations of M variables are given in the sum-of-products canonical form, these equations may be implemented with an M-input, N-output ROM. As explained above, this is an economical solution if M and N are large (particularly if M is large). However, in the logic design of one stage of a full adder, where $M = 3$ and $N = 2$ (small numbers), and where this unit is sold in considerable quantities, using distinct gate combinations as in Fig. 6-6 is more economical than using a ROM.

Character Generator[12]

Alphanumeric characters may be "written" on the face of a cathode-ray tube (a television-type display) with the aid of a ROM.

Stored Programs

Control programs (for example, in a pocket calculator) are permanently stored in ROM.

REFERENCES

1. Grinich, V. H., and H. G. Jackson: "Introduction to Integrated Circuits," McGraw-Hill Book Company, New York, 1975.
2. Texas Instruments Inc.: "The TTL Data Book for Design Engineers," and Supplement, 1974.
3. Blakeslee, T. R.: "Digital Design with Standard MSI and LSI," John Wiley and Sons, Inc., New York, 1975.
4. Texas Instruments Staff: "Designing with TTL Integrated Circuits," McGraw-Hill Book Company, New York, 1971.
5. Rostky, G.: On Semiconductor Memories, *Electron. Design*, vol. 19, pp. 50–63, September 16, 1971.
6. Intel Corporation: "Memory Design Handbook," 1975.
7. Sifferlen, T. P., and V. Vartanian: "Digital Electronics with Engineering Applications," pp. 216–218, Prentice-Hall, Inc., Englewood Cliffs, N.J., 1970.

8. Vimari, D. C.: Field-programmable Read-only Memories and Applications, *Computer Design*, vol. 9, pp. 49–54, December, 1970.

9. Baasch, T. L.: Selecting Alphanumeric Readouts, *Electron. Prod.*, pp. 31–37, December 21, 1970.

10. McCluskey, E. J., Jr.: "Introduction to the Theory of Switching Circuits," McGraw-Hill Book Company, New York, 1965.
   Wickes, W. E.: "Logic Design with Integrated Circuits," chap. 3, John Wiley & Sons, Inc., New York, 1968. Ref. 7, chap. 1.

11. Peatman, J. B.: "The Design of Digital Systems," chap. 3, McGraw-Hill Book Company, New York, 1971.
   Taub, H., and D. Schilling: "Digital Integrated Electronics," chap. 3, McGraw-Hill Book Company, New York, 1977.

12. Millman, J., and C. C. Halkias: "Integrated Electronics: Analog and Digital Circuits and Systems," sec. 17-21, McGraw-Hill Book Company, New York, 1972.

REVIEW QUESTIONS

6-1 (*a*) How many input leads are needed for a chip containing quad two-input NOR gates? Explain.

(*b*) Repeat part *a* for dual two-wide, two-input AOI gates.

6-2 Define SSI, MSI, and LSI.

6-3 Draw the circuit configuration for an IC TTL AOI gate. Explain its operation.

6-4 (*a*) Find the truth table for the *half adder*.

(*b*) Show the implementation for the digit *D* and the carry *C*.

6-5 Show the system of a 4-bit parallel binary adder, constructed from single-bit full adders.

6-6 (*a*) Draw the truth table for a three-input adder. Explain clearly the meaning of the input and output symbols in the table.

(*b*) Write the Boolean expressions for the sum and the carry. (Do not simplify these.)

6-7 (*a*) Show the system for a *serial* binary full adder.

(*b*) Explain the operation.

6-8 (*a*) Consider two 4-bit numbers *A* and *B* with $B > A$. Verify that to subtract *A* from *B* it is only required to add B, $\bar{A}$, and 1.

(*b*) Indicate in simple form a 4-bit subtractor obtained from a full adder.

6-9 Consider two 1-bit numbers *A* and *B*. What are the logic gates required to test for (*a*) $A = B$; (*b*) $A > B$; and (*c*) $A < B$?

6-10 (*a*) Consider two 4-bit numbers *A* and *B*. If $E = 1$ represents the equality $A = B$, write the Boolean expression for *E*. Explain.

(*b*) If $C = 1$ represents the inequality $A > B$, write the Boolean expression for *C*. Explain.

6-11 Show the system for a 4-bit odd-parity checker.

6-12 (*a*) Show a system for increasing the reliability of transmission of binary information, using a parity checker and generator.

(*b*) Explain the operation of the system.

6-13 Write the decimal number 538 in the BCD system.

6-14 (*a*) Define a *decoder*.

(*b*) Show how to decode the 4-bit code 1011 (LSB).

6-15 (*a*) Define a *demultiplexer*.

(*b*) Show how to convert a decoder into a demultiplexer.

(*c*) Indicate how to add a strobe to this system.

6-16 Draw the logic block diagram for a 1-to-32-output demultiplexer tree using a "trunk" with four output lines. Indicate the correct addressing.

6-17 (*a*) Define a *multiplexer*.

(*b*) Draw a logic block diagram of a 4-to-1-line multiplexer.

6-18 Show how a multiplexer may be used as (*a*) a *parallel-to-serial converter*, and (*b*) a *sequential data selector*.

6-19 Draw the logic block diagram for a 32-to-1-line selector/multiplexer. Use selectors with a maximum of 8 input lines. Indicate the correct addressing.

6-20 (*a*) Define an *encoder*.

(*b*) Indicate a diode matrix encoder to transform a decimal number into a binary code.

6-21 (*a*) Indicate an encoder matrix using emitter followers. In particular, for an encoder to transform a decimal number into a binary code, show the connections (*b*) to the output Y_2 and (*c*) to the line W_4.

6-22 (*a*) Define a *priority* encoder.

(*b*) Show the truth table for a 4-to-2-line priority encoder.

6-23 (*a*) Define a *read-only memory*.

(*b*) Show a block diagram of a ROM.

(*c*) What is stored in the memory?

(*d*) What hardware constitutes the memory elements?

6-24 Indicate a block diagram of a 512×4-bit ROM, using two-dimensional addressing. Use a 64×32 matrix encoder.

6-25 (*a*) Write the truth table for converting from a binary to a Gray code.

(*b*) Write the first six lines of the truth table for converting a Gray into a binary code.

6-26 Explain what is meant by *mask-programming* a ROM.

6-27 (*a*) Explain what is meant by a PROM.

(*b*) How is the programming done in the field?

6-28 List three ROM applications and explain these very briefly.

6-29 (*a*) What is a *seven-segment visible display*?

(*b*) Show the following two lines in the conversion table from BCD to seven-segment-indicator code: 0001 and 0101.

SEQUENTIAL DIGITAL SYSTEMS

Systems which operate in synchronism with a train of pulses and which possess memory are discussed in this chapter. These include FLIP-FLOPS, shift registers, and counters.

7-1 A 1-BIT MEMORY[1-6]

All the systems discussed in Chap. 6 were based upon combinational logic; the outputs at a given instant of time depend only upon the values of the inputs at the same moment. Such a system is said to have no memory. Note that a ROM is a combinational circuit and, according to the above definition, it has no memory. *The memory of a ROM refers to the fact that it "memorizes" the functional relationship between the output variables and the input variables*. It does *not* store bits of information.

A 1-bit Storage Cell

The basic digital memory circuit is obtained by cross-coupling two NOT circuits $N1$ and $N2$ (single-input NAND gates) in the manner shown in Fig. 7-1a. The output of each gate is connected to the input of the other, and this feedback combination is called a *latch*. The most important property of the latch is that it can exist in one of two stable states, either $Q = 1$ ($\overline{Q} = 0$), called the 1 state, or $Q = 0$ ($\overline{Q} = 1$), referred to as the 0 state. The existence of these stable states is consistent with the interconnections shown in Fig. 7-1a. For example, if the output of $N1$ is $Q = 1$, so also is A_2, the input to $N2$. This inverter then has the state 0 at its output $\overline{Q}$. Since $\overline{Q}$ is tied to A_1, then the input of $N1$ is 0, and the

205

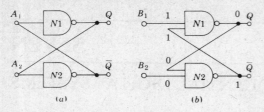

Figure 7-1 (*a*) A 1-bit memory or latch. (*b*) The latch provided with means for entering data into the cell.

corresponding output is $Q = 1$. This result confirms our original assumption that $Q = 1$. A similar argument leads to the conclusion that $Q = 0$; $\bar{Q} = 1$ is also a possible state. It is readily verified that the situation in which both outputs are in the same state (both 1 or both 0) is not consistent with the interconnection.

Since the configuration of Fig. 7-1*a* has two stable states, it is also called a binary, or bistable circuit. Since it may store one bit of information (either $Q = 1$ or $Q = 0$), it is a 1-*bit memory unit*, or a 1-*bit storage cell*. This information is locked, or latched, in place, which accounts for the name *latch*.

Suppose it is desired to store a specific state, say $Q = 1$, in the latch. Or conversely, we may wish to remember the state $Q = 0$. We may "write" a 1 or 0 into the memory cell by changing the NOT gates of Fig. 7-1*a* to two-input NAND gates, $N1$ and $N2$, and by feeding this latch through two inputs B_1 and B_2 as in Fig. 7-1*b*. If we assume $B_1 = 1$ and $B_2 = 0$, then the state of each gate input and output is indicated on the diagram. Since $Q = 0$ we have verified that to enter a 0 into the memory required inputs $B_1 = 1$ and $B_2 = 0$. In a similar manner it can be demonstrated that to store a 1, it is necessary that the inputs be $B_1 = 0$ and $B_2 = 1$. If $B_1 = 1$ and $B_2 = 1$ these two leads may be removed from the NAND gates without effecting the logic. In other words, *the state of the latch is unaffected by the input combination $B_1 = B_2 = 1$*; if $Q = 1$ (0) before this set of inputs is applied, the output will remain $Q = 1$ (0) after the inputs change to $B_1 = B_2 = 1$. Note also that $B_1 = B_2 = 0$ is not allowed (Prob. 7-1).

A Chatterless Switch[5]

In a digital system it is often necessary to push a key in order to introduce a 1 or a 0 at a particular node. Most switches *chatter* or *bounce* several times before they settle down into the closed position. The storage cell just described may be used to give a single change of state when the key first closes, regardless of the chatter which follows. Consider the situation in Fig. 7-2, where the single-pole, double-throw switch is initially in the position to ground B_2 so that $B_2 = 0$. Neglecting the gate input current, the voltage at B_1 is 5 V, which is considered to be the 1-state voltage, so that initially $B_1 = 1$. At $t = t_1$ the key is pushed and the pole moves from 2 toward 1. The waveform for B_2 is shown in Fig. 7-2*b*. It takes an interval of time $t' = t_2 - t_1$ for the pole to reach contact 1 so that B_1 changes from 1 to 0 at $t = t_2$. However, as indicated in Fig. 7-2*c* the contact chatters and the connection is broken in the intervals between t_3 and t_4 and between t_5 and t_6 (two bounces are assumed in drawing this figure). The output

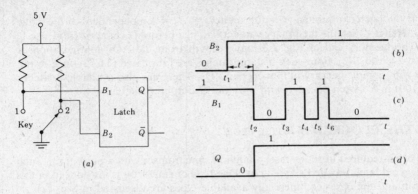

Figure 7-2 (*a*) The latch of Fig. 7-1*b* effectively causes the contact of a key to be without chatter. The states (waveforms) (*b*) B_2, (*c*) B_1, and (*d*) Q as a function of time.

Q of the latch is indicated in Fig. 7-2*d* and is consistent with the logic discussed above; namely, $Q = 0$ if $B_1 = 1$ and $B_2 = 0$: $Q = 1$ if $B_1 = 0$ and $B_2 = 1$: Q is unaffected (does not change) if $B_1 = B_2 = 1$. Note that the latch has rendered the switch chatterless, since the output Q shows a single change of state from 0 to 1 at the first instant when B_1 is grounded. In a similar manner it can be shown that Q will show a single step from 1 to 0 if the switch blade is thrown from B_1 to B_2, even if there is bouncing at the contact B_2 (Prob. 7-2). For applications like the chatterless switch an IC containing four latches is available (TI-279). Since B_1 is labeled $\overline{S}$ and B_2 is called $\overline{R}$ this package contains quadruple $\overline{S}$-$\overline{R}$ latches.

The Bistable Latch

The addition of two NAND gates preceding N_1 and N_2 in Fig. 7-1*b* together with an inverter results in a system (Fig. 7-3) for the storage of one bit of binary information. When the enable input is high ($G = 1$), then the input data D are transferred to the output Q. This statement is readily verified, based upon the logic which the latch $N1$-$N2$ satisfies. Thus, if $D = 0$, $S = 0$, $R = 1$, $B_1 = 1$, $B_2 = 0$, and $Q = 0$. Similarly, if $D = 1$, $S = 1$, $R = 0$, $B_1 = 0$, $B_2 = 1$, and $Q = 1$. As long as $G = 1$, any change in the data D appears at Q.

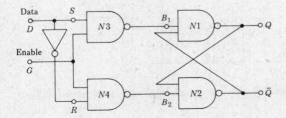

Figure 7-3 A bistable latch transfers the input data D to the output Q, if $G = 1$.

If the latch is inhibited ($G = 0$), then $B_1 = B_2 = 1$ independent of the value of D. Hence Q retains the binary value it had just before G changed from 1 to 0.

The memory cell of Fig. 7-3 may be built from the AOI configuration of Fig. 6-2 (Prob. 7-3). Four such bistable latches are fabricated (TI-75) on a 16-pin chip with complementary outputs (Q and $\overline{Q}$). There are also available 8 latches (TI-100) in a 24-pin package with only the Q output available.

7-2 THE CLOCKED S-R FLIP-FLOP[1-6]

It is often required to set or reset a latch in synchronism with a pulse train. Such a triggered latch is called a FLIP-FLOP. One type of FLIP-FLOP is introduced in this section and the other commercially available types are discussed in Sec. 7-3.

A Sequential System

Many digital systems are pulsed or clocked; i.e., they operate in synchronism with a pulse train of period T, called the system *clock* (abbreviated Ck), such as that indicated in Fig. 7-4. The pulse width t_p is assumed small compared with T. The binary values at each node in the system are assumed to remain constant in each interval between pulses. A transition from one state of the system to another may take place only with the application of a clock pulse. Let Q_n be the output (0 or 1) at a given node in the nth interval (bit time n) preceding the nth clock pulse (Fig. 7-4). Then Q_{n+1} is the corresponding output in the interval immediately after the nth pulse. Such a system where the values Q_1, Q_2, Q_3, . . . , of Q_n are obtained in time sequence at intervals T is called a *sequential* (to distinguish it from a *combinational*) logic system. The value of Q_{n+1} may depend upon the nodal values during the previous (nth) bit time. Under these circumstances a sequential circuit possesses memory.

The S-R FLIP-FLOP

If the enable terminal in Fig. 7-3 is used for the clock (Ck) input and if the inverter is omitted to allow two data inputs S (*set*) and R (*reset*), the S-R clocked FLIP-FLOP of Fig. 7-5 results. The gates $N1$ and $N2$ form a latch,

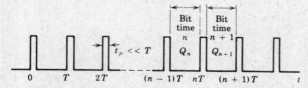

Figure 7-4 The output of a master oscillator used as a clock-pulse train to synchronize a digital sequential system.

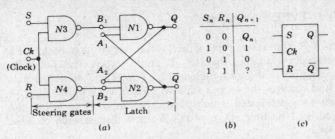

Figure 7-5 (*a*) An *S-R* clocked FLIP-FLOP. (*b*) The truth table (the question mark in the last row indicates that this state cannot be predicted). (*c*) The logic symbol.

whereas $N3$ and $N4$ are the *control*, or *steering*, gates which program the state of the FLIP-FLOP after the pulse appears.

Note that between clock pulses ($Ck = 0$), the outputs of $N3$ and $N4$ are 1 independently of the values of R or S. Hence the circuit is equivalent to the latch of Fig. 7-1*a*. If $Q = 1$, it remains 1, whereas if $Q = 0$, it remains 0. In other words, *the* FLIP-FLOP *does not change state between clock pulses*; it is invariant within a bit time.

Now consider the time $t = nT(+)$ when a clock pulse is present ($Ck = 1$). If $S = 0$ and $R = 0$, then the outputs of $N3$ and $N4$ are 1. By the argument given in the preceding paragraph, the state Q_n of the FLIP-FLOP does not change. Hence, after the pulse passes (in the bit time $n + 1$), the state Q_{n+1} is identical with Q_n. If we denote the values of R and S in the interval just before $t = nT$ by R_n and S_n, then $Q_{n+1} = Q_n$ if $S_n = 0$ and $R_n = 0$. This relationship is indicated in the first row of the truth table of Fig. 7-5*b*.

If $Ck = 1$, $S_n = 0$ and $R_n = 1$, then $B_1 = 1$ and $B_2 = 0$, so that the situation is that pictured in Fig. 7-1*b* and the output state is 0. Hence, after the clock pulse passes (at bit time $n + 1$), we find $Q_{n+1} = 0$, confirming the third row of the truth table. If R and S are interchanged and if simultaneously Q is interchanged with $\overline{Q}$, then the logic diagram of Fig. 7-5*a* is unaltered. Hence the second row of Fig. 7-5*b* follows from the third row.

If $Ck = 1$, $S_n = 1$, and $R_n = 1$, then the outputs of the NAND gates $N3$ and $N4$ are both 0. Hence the input B_1 of $N1$ as well as B_2 of $N2$ is 0, so that the outputs of *both* $N1$ and $N2$ must be 1. This condition is logically inconsistent with our labeling the two outputs Q and $\overline{Q}$. We must conclude that the output transistor ($Q3$ of Fig. 5-23) of each gate $N1$ and $N2$ is cut off, resulting in both outputs being high (1). At the end of the pulse the inputs at B_1 and B_2 rise from 0 toward 1. Depending upon which input increases faster and upon circuit parameter asymmetries, either the stable state $Q = 1$ ($\overline{Q} = 0$) or $Q = 0$ ($\overline{Q} = 1$) will result. Therefore we have indicated a question mark for Q_{n+1} in the fourth row of the truth table of Fig. 7-5*b*. This state is said to be *indeterminate*, *ambiguous*, or *undefined*, and the condition $S_n = 1$ and $R_n = 1$ is forbidden; it must be prevented from taking place.

7-3 J-K, T, AND D-TYPE FLIP-FLOPS[1-6]

In addition to the S-R FLIP-FLOP, three other variations of this basic 1-bit memory are commercially available: the J-K, T, and D types. The J-K FLIP-FLOP removes the ambiguity in the truth table of Fig. 7-5b. The T FLIP-FLOP acts as a toggle switch and changes the output state with each clock pulse; $Q_{n+1} = \overline{Q_n}$. The D type acts as a delay unit which causes the output Q to follow the input D, but delayed by 1 bit time; $Q_{n+1} = D_n$. We now discuss each of these three FLIP-FLOP types.

The J-K FLIP-FLOP

This building block is obtained by augmenting the S-R FLIP-FLOP with two AND gates $A1$ and $A2$ (Fig. 7-6a). Data input J and the output $\overline{Q}$ are applied to $A1$. Since its output feeds S, then $S = J\overline{Q}$. Similarly, data input K and the output Q are applied to $A2$, and hence $R = KQ$. The logic followed by this system is given in the truth table of Fig. 7-6b. This logic can be verified by referring to Table 7-1. There are four possible combinations for the two data inputs J and K. For each of these there are two possible states for Q, and hence Table 7-1 has eight rows. From the J_n, K_n, Q_n, and $\overline{Q_n}$ bits in each row, $S_n = J_n\overline{Q_n}$ and $R_n = K_nQ_n$ are calculated and are entered into the fifth and sixth columns of the table. Using these values of S_n and R_n and referring to the S-R FLIP-FLOP truth table of Fig. 7-5b, the seventh column is obtained. Finally, column 8 follows from column 7 because $Q_n = 1$ in row 4, $Q_n = 0$ in row 5, $\overline{Q_n} = 1$ in row 7, and $\overline{Q_n} = 0$ in row 8.

Columns 1, 2, and 8 of Table 7-1 form the J-K FLIP-FLOP truth table of Fig. 7-6b. Note that *the first three rows of a J-K are identical with the corresponding rows for an S-R truth table* (Fig. 7-5b). However, the ambiguity of the state $S_n = 1 = R_n$ is now replaced by $Q_{n+1} = \overline{Q_n}$ for $J_n = 1 = K_n$. *If the two data inputs in the J-K FLIP-FLOP are high, the output will be complemented by the clock pulse.*

It is really not necessary to use the AND gates $A1$ and $A2$ of Fig. 7-6a, since the same function can be performed by adding an extra input terminal to each NAND gate $N3$ and $N4$ of Fig. 7-5a. This simplification is indicated in Fig. 7-7. (Ignore the dashed inputs; i.e., assume that they are both 1.) Note that Q and $\overline{Q}$ at the inputs are obtained by the feedback connections (drawn heavy) from the outputs.

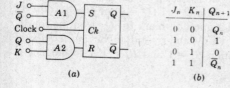

J_n	K_n	Q_{n+1}
0	0	Q_n
1	0	1
0	1	0
1	1	$\overline{Q_n}$

(a)

(b)

Figure 7-6 (a) An S-R FLIP-FLOP is converted into a J-K FLIP-FLOP. (b) The truth table.

Table 7-1 Truth table for Fig. 7-6a

Column	1	2	3	4	5	6	7	8
Row	J_n	K_n	Q_n	$\overline{Q}_n$	S_n	R_n	Q_{n+1}	
1	0	0	0	1	0	0	Q_n ⎫	Q_n
2	0	0	1	0	0	0	Q_n ⎬	
3	1	0	0	1	1	0	1 ⎫	1
4	1	0	1	0	0	0	Q_n ⎬	
5	0	1	0	1	0	0	Q_n ⎫	0
6	0	1	1	0	0	1	0 ⎬	
7	1	1	0	1	1	0	1 ⎫	$\overline{Q}_n$
8	1	1	1	0	0	1	0 ⎬	

Preset and Clear

The truth table of Fig. 7-6b tells us what happens to the output with the application of a clock pulse, as a function of the data inputs J and K. However, the value of the output before the pulse is applied is arbitrary. The addition of the dashed inputs in Fig. 7-7 allows the initial state of the FLIP-FLOP to be assigned. For example, it may be required to *clear* the latch, i.e., to specify that $Q = 0$ when $Ck = 0$.

The clear operation may be accomplished by programming the clear input to 0 and the *preset* input to 1; $Cr = 0$, $Pr = 1$, $Ck = 0$. Since $Cr = 0$, the output of $N2$ (Fig. 7-7) is $\overline{Q} = 1$. Since $Ck = 0$, the output of $N3$ is 1, and hence all inputs to $N1$ are 1 and $Q = 0$, as desired. Similarly, if it is required to preset the latch into the 1 state, it is necessary to choose $Pr = 0$, $Cr = 1$, $Ck = 0$. The preset and clear data are called *direct*, or *asynchronous*, inputs; i.e., they are not in synchronism with the clock, but may be applied at any time in between clock pulses. Once the state of the FLIP-FLOP is established asynchronously, the direct inputs must be maintained at $Pr = 1$, $Cr = 1$, before the next pulse arrives in

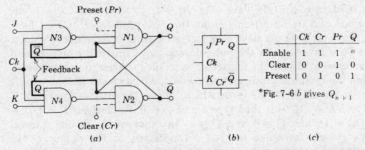

	Ck	Cr	Pr	Q
Enable	1	1	1	*
Clear	0	0	1	0
Preset	0	1	0	1

*Fig. 7-6 b gives Q_{n+1}

(a)   (b)   (c)

Figure 7-7 (a) A J-K FLIP-FLOP. (b) The logic symbol. (c) The necessary conditions for synchronous operation (row 1) or for asynchronous clearing (row 2) or presetting (row 3).

order to *enable* the FLIP-FLOP. The data $Pr = 0$, $Cr = 0$, must not be used since they lead to an ambiguous state. Why?

The logic symbol for the J-K FLIP-FLOP is indicated in Fig. 7-7b, and the inputs for proper operation are given in Fig. 7-7c.

The Race-Around Condition

There is a possible physical difficulty with the J-K FLIP-FLOP constructed as in Fig. 7-7. Truth table 7-1 is based upon combinational logic, which assumes that the inputs are independent of the outputs. However, because of the feedback connection Q ($\bar{Q}$) at the input to K (J), the input will change during the clock pulse ($Ck = 1$) if the output changes state. Consider, for example, that the inputs to Fig. 7-7 are $J = 1$, $K = 1$, and $Q = 0$. When the pulse is applied, the output becomes $Q = 1$ (according to row 7 of Table 7-1), this change taking place after a time interval Δt equal to the propagation delay (Sec. 5-15) through two NAND gates in series in Fig. 7-7. Now $J = 1$, $K = 1$, and $Q = 1$, and from row 8 of Table 7-1, we find that the input changes back to $Q = 0$. Hence we must conclude that for the duration t_p (Fig. 7-4) of the pulse (while $Ck = 1$), the output will oscillate back and forth between 0 and 1. At the end of the pulse ($Ck = 0$), the value of Q is ambiguous.

The situation just described is called a *race-around condition*. It can be avoided if $t_p < \Delta t < T$. However, with IC components the propagation delay is very small, usually much less than the pulse width t_p. Hence the above inequality is *not* satisfied, and the output is indeterminate. Lumped delay lines can be used in series with the feedback connections of Fig. 7-7 in order to increase the loop delay beyond t_p, and hence to prevent the race-around difficulty. A more practical IC solution is now to be described.

The Master-Slave J-K FLIP-FLOP

In Fig. 7-8 is shown a cascade of two S-R FLIP-FLOPS with feedback from the output of the second (called the *slave*) to the input of the first (called the *master*). Positive clock pulses are applied to the master, and these are inverted before being used to excite the slave. For $Pr = 1$, $Cr = 1$, and $Ck = 1$, the master is enabled and its operation follows the J-K truth table of Fig. 7-6b. Furthermore, since $\overline{Ck} = 0$, the slave S-R FLIP-FLOP is inhibited (cannot change state), so that Q_n is invariant for the pulse duration t_p. Clearly, the race-around difficulty is circumvented with the master-slave topology. After the pulse passes, $Ck = 0$, so that the master is inhibited and $\overline{Ck} = 1$, which causes the slave to be enabled. The slave is an S-R FLIP-FLOP, which follows the logic in Fig. 7-5b. If $S = Q_M = 1$ and $R = \bar{Q}_M = 0$, then $Q = 1$ and $\bar{Q} = 0$. Similarly, if $S = Q_M = 0$ and $R = \bar{Q}_M = 1$, then $Q = 0$ and $\bar{Q} = 1$. In other words, in the interval between clock pulses, the value of Q_M is transferred to the output Q. In summary, during a clock pulse the output Q does not change but Q_M follows J-K logic; at the end of the pulse, the value of Q_M is transferred to Q.

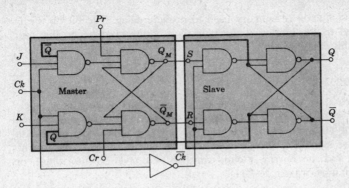

Figure 7-8 A J-K master-slave FLIP-FLOP.

It should be emphasized that the data J and K must remain constant for the pulse duration or an erroneous output may result (Prob. 7-9). The 16-pin MSI package (TI-76) contains two independent J-K master-slave FLIP-FLOPS. Some commercially available FLIP-FLOPS also have internal AND or AOI (TI-71) gates at the inputs to provide multiple J and K inputs, thereby avoiding the necessity of external gates in applications where these may be required.

The D-type FLIP-FLOP

If a J-K FLIP-FLOP is modified by the addition of an inverter as in Fig. 7-9a, so that K is the complement of J, the unit is called a D (*delay*) FLIP-FLOP. From the J-K truth table of Fig. 7-6b, $Q_{n+1} = 1$ for $D_n = J_n = \bar{K}_n = 1$ and $Q_{n+1} = 0$ for $D_n = J_n = \bar{K}_n = 0$. Hence $Q_{n+1} = D_n$. The output Q_{n+1} after the pulse (bit time $n + 1$) equals the input D_n before the pulse (bit time n), as indicated in the truth table of Fig. 7-9c. If the FLIP-FLOP in Fig. 7-9a is of the S-R type, the unit also functions as a D-type latch with the clock (Ck) replaced by the enable (G) of Fig. 7-3. There is no ambiguous state because $J = K = 1$ is not possible.

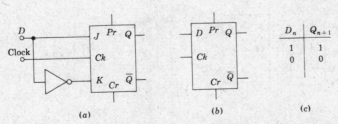

Figure 7-9 (*a*) A J-K FLIP-FLOP is converted into a D-type latch. (*b*) The logic symbol. (*c*) The truth table.

The D-type FLIP-FLOP is a binary used to provide delay. The bit on the D line is transferred to the output at the next clock pulse, and hence this unit functions as a 1-bit delay device.

The T-type FLIP-FLOP

This unit changes state with each clock pulse, and hence it acts as a toggle switch. If $J = K = 1$, then $Q_{n+1} = \bar{Q}_n$, so that the J-K FLIP-FLOP is converted into a T-type FLIP-FLOP. In Fig. 7-10a such a system is indicated with a data input T. The logic symbol is shown in Fig. 7-10b, and the truth table in Fig. 7-10c. The S-R- and the D-type latches can also be converted into toggle, or complementing, FLIP-FLOPS (Prob. 7-11).

Summary

Four FLIP-FLOP configurations S-R, J-K, D, and T are important. The logic satisfied by each type is repeated for easy reference in Table 7-2. An IC FLIP-FLOP is driven synchronously by a clock, and in addition it may (or may not) have direct inputs for asynchronous operation, preset (Pr) and clear (Cr). A direct input can be 0 only in the interval between clock pulses when $Ck = 0$. When $Ck = 1$, both asynchronous inputs must be high; $Pr = 1$ and $Cr = 1$. The inputs must remain constant during a pulse width, $Ck = 1$. For a master-slave FLIP-FLOP the output Q remains constant for the pulse duration and changes only after Ck changes from 1 to 0, at the *negative-going* (*trailing*) *edge* of the pulse. It is also possible[6] to design a J-K FLIP-FLOP so that the output changes at the *positive-going* (*leading*) *edge* of the pulse. The TI-70 chip is a positive-edge-triggered J-K FLIP-FLOP with AND gates at the J and K inputs. The TI-174 contains six (hex) positive-edge-triggered D-type FLIP-FLOPS.

The toggle, or complementing, FLIP-FLOP is not available commercially because a J-K can be used as a T type by connecting the J and K inputs together (Fig. 7-10).

The FLIP-FLOP is available in all the IC digital families, and the maximum frequencies of operation are given in Table 5-3.

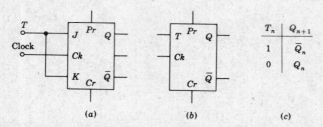

(a) (b) (c)

Figure 7-10 A J-K FLIP-FLOP is converted into a T-type FLIP-FLOP, with a data input T. (b) The logic symbol. (c) The truth table.

Table 7-2 FLIP-FLOP **truth tables**

S-R			J-K			D		T		Direct inputs			
S_n	R_n	Q_{n+1}	J_n	K_n	Q_{n+1}	D_n	Q_{n+1}	T_n	Q_{n+1}	Ck	Cr	Pr	Q
0	0	Q_n	0	0	Q_n	1	1	1	$\overline{Q}_n$	0	1	0	1
1	0	1	1	0	1	0	0	0	Q_n	0	0	1	0
0	1	0	0	1	0					1	1	1	†
1	1	?	1	1	$\overline{Q}_n$								
Fig. 7-5			Fig. 7-8			Fig. 7-9		Fig. 7-10					

† Refer to truth table S-R, J-K, D, or T for Q_{n+1} as a function of the inputs.

7-4 SHIFT REGISTERS[1-6]

Since a binary is a 1-bit memory, then n FLIP-FLOPS can store an n-bit word. This combination is referred to as a *register*. To allow the data in the word to be read into the register serially, the output of one FLIP-FLOP is connected to the input of the following binary. Such a configuration, called a *shift register*, is indicated in Fig. 7-11. Each FLIP-FLOP is of the S-R (or J-K) master-slave type. Note that the stage which is to store the most significant bit (MSB) is converted into a D-type latch (Fig. 7-9) by connecting S and R through an inverter. The 5-bit shift register indicated in Fig. 7-11 is available on a single chip in a 16-pin package (medium-scale integration). We shall now explain the operation of this system by assuming that the serial data 01011 (LSB) is to be registered. (The least significant bit is the right-most digit, which in this case is a 1.)

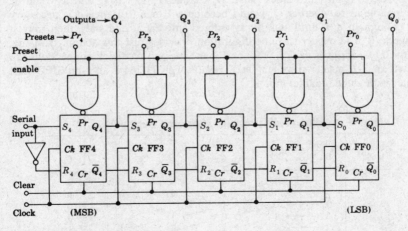

Figure 7-11 A 5-bit shift register (TI-96).

Series-in, Parallel-out (SIPO) Register

The FLIP-FLOPS are cleared by applying a 0 to the *clear* input (while the preset enable is low) so that every output $Q_0, Q_1, \ldots, Q_4$ is 0. Then Cr is set to 1 and Pr is held constant at 1 (by keeping the preset enable at 0). The serial data train and the synchronous clock are now applied. The least significant bit (LSB) is entered into the master latch of FF4 when Ck changes from a 0 to a 1 by the action of a D-type FLIP-FLOP. After the clock pulse, the 1 is transferred to the slave latch of FF4 and $Q_4 = 1$, while all other outputs remain at 0.

At the second clock pulse the state of Q_4 is transferred to the master latch of FF3 by the action of an S-R FLIP-FLOP. Simultaneously, the next bit (a 1 in the 01011 word) enters the master of FF4. After the second clock pulse the bit in each master transfers to its slave and $Q_4 = 1$, $Q_3 = 1$, and the other outputs remain 0. The readings of the register *after* each pulse are given in Table 7-3. For example, after the third pulse, Q_3 has shifted to Q_2, Q_4 to Q_3, and the third input bit (0) has entered FF4, so that $Q_4 = 0$. We may easily follow this procedure and see that by registering each bit in the MSB FLIP-FLOP and then shifting to the right to make room for the next digit, the input word becomes installed in the register after the nth clock pulse (for an n-bit code). Of course, the clock pulses must stop at the moment the word is registered. Each output is available on a separate line, and they may be read simultaneously. Since the data entered the system serially and came out in parallel, this shift register is a *serial-to-parallel* converter. It is also referred to as a *series-in, parallel-out* (*SIPO*) *register*. A *temporal code* (a time arrangement of bits) has been changed to a *spacial code* (information stored in a static memory).

Master-slave FLIP-FLOPS are required because of the race problem between stages (Sec. 7-3). If all FLIP-FLOPS were to change states simultaneously, there would be an ambiguity as to what data would transfer from the preceding stage. For example, at the third clock pulse, Q_4 changes from 1 to 0, and it would be questionable as to whether Q_3 would become a 1 or a 0. Hence it is necessary that Q_4 remain a 1 until this bit is entered into FF3, and only then may it change to 0. The master-slave configuration provides just this action. If in Fig. 7-8, the $J(K)$ input is called $S(R)$ and if the (heavy) feedback connections are omitted an S-R master-slave FLIP-FLOP results. The TI-164 is an 8-bit SIPO shift register with gated (enable) inputs.

Table 7-3 Reading of shift register after each clock pulse

Clock pulse	Word bit	Q_4	Q_3	Q_2	Q_1	Q_0
1	1 ⟶ 1		0	0	0	0
2	1 ⟶ 1	⟶ 1		0	0	0
3	0 ⟶ 0	⟶ 1	⟶ 1		0	0
4	1 ⟶ 1	⟶ 0	⟶ 1	⟶ 1		0
5	0 ⟶ 0	⟶ 1	⟶ 0	⟶ 1	⟶ 1	

Series-in, Series-out (SISO) Register

We may take the output at Q_0 and read the register serially if we apply n clock pulses, for an n-bit word. After the nth pulse each FLIP-FLOP reads 0. Note that the shift-out clock rate may be greater or smaller than the original pulse frequency. Hence here is a method for changing the spacing in time of a binary code, a process referred to as *buffering*.

The TI-91 MSI package is an 8-bit SISO register with gated inputs and complementary outputs. Since a SISO chip requires only one data input pin and one data output pin, independent of the number of bits to be stored, a very long (for example, a 1,024-bit) shift register is possible with LSI (Sec. 9-1).

Parallel-in, Series-out (PISO) Register

Consider the situation where the word bits are available in parallel, e.g., at the outputs from a ROM (Sec. 6-9). It is desired to present this code, say 01011, in serial form.

The LSB is applied to Pr_0, the 2^1 bit to $Pr_1, \ldots$ so that $Pr_0 = 1$, $Pr_1 = 1$, $Pr_2 = 0$, $Pr_3 = 1$, and $Pr_4 = 0$. The register is first cleared by $Cr = 0$, and then $Cr = 1$ is maintained. A 1 at the *preset enable* input activates all kth input NAND gates for which $Pr_k = 1$. The preset of the corresponding kth FLIP-FLOP is $Pr = 0$, and this stage is therefore preset to 1 (Table 7-2). In the present illustration FF0, FF1, and FF3 are preset and the input word 01011 is written into the register, all bits in parallel, by the preset enable pulse.

As explained above, the stored word may be read serially at Q_0 by applying five clock pulses. This is a *parallel-to-serial*, or a *spacial-to-temporal, converter*. The TI-165 package is a *parallel-in, series-out* (PISO), 8-bit register.

Parallel-in, Parallel-out (PIPO) Register

The data are entered as explained above by applying a 1 at the preset enable, or *write*, terminal. It is then available in parallel form at the outputs $Q_0, Q_1, \ldots$. If it is desired to *read* the register during a selected time, each output Q_k is applied to one input of a two-input AND gate N_k, and the second input of each AND is excited by a read pulse. The output of N_k is 0 except for the pulse duration, when it reads 1 if $Q_k = 1$. (The gates N_k are not shown in Fig. 7-11.)

Note that in this application the system is not operating as a shift register since there is no clock required (and no serial input). Each FLIP-FLOP is simply used as an isolated 1-bit read/write memory.

Right-Shift, Left-Shift (Bidirectional) Register

Some commercial shift registers are equipped with gates which allow shifting the data from right to left as well as in the reverse direction. One application for such a system is to perform multiplication or division by multiples of 2, as will

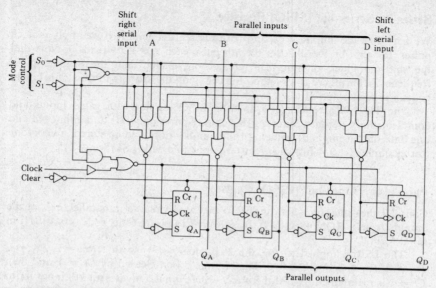

Figure 7-12 The logic diagram for the 4-bit (16-pin) TI-194 bidirectional shift register. *(Courtesy of Texas Instruments, Inc.).*

now be explained. Consider first a right-shift register as in Fig. 7-11 and that the serial input is held low.

Assume that a binary number is stored in a shift register, with the least-significant bit stored in FF0. Now apply one clock pulse. Each bit then moves to the next lower significant place, and hence is divided by 2. The number now held in the register is half the original number, provided that FF0 was originally 0. Since the 2^0 bit is lost in the shift to the right, then if FF0 was originally in the 1 state, corresponding to the decimal number 1, after the shift the register is in error by the decimal number 0.5. The next clock pulse causes another division by 2, etc.

Consider now that the system is wired so that each clock pulse causes a shift to the left. Each bit now moves to the next higher significant digit, and the number stored is multiplied by 2.

The logic diagram for the TI-194 4-bit bidirectional shift register is given in Fig. 7-12. This is a *universal* register because it can function in all the modes discussed in this section; SIPO, SISO, PISO, PIPO, and as a bidirectional register. It has two control inputs, S_0 and S_1, which allow the four operational modes listed in Table 7-4 to be realized. The verification of this table is considered in Prob. 7-14. The 8-bit TI-198 universal shift register has the same structure as that indicated in Fig. 7-12. It has the equivalent of 87 gates, and comes in a 24-pin package.

Table 7-4 Modes of operation of a universal register

$S_0^\dagger$	$S_1^\dagger$	Operational mode
0	0	Inhibit clock
1	1	Parallel input ‡
1	0	Shift right
0	1	Shift left

† S_0 and S_1 should be changed only while the clock input is high.

‡ Data are loaded into a FLIP-FLOP after a clock pulse. During loading, serial data flow is inhibited.

Digital Delay Line

A shift register may be used to introduce a time delay Δ into a system, where Δ is an integral multiple of the clock period T. Thus an input pulse train appears at the output of an n-stage register delayed by a time $(n - 1)T = \Delta$.

Sequence Generator[2]

An important application of a shift register is to generate a binary sequence. This system is also called a *word, code,* or *character, generator*. The shift register FLIP-FLOPS are preset to give the desired code. Then the clock applies shift pulses, and the output of the shift register gives the temporal pattern corresponding to the specified sequence. Clearly, we have just described a parallel-in, series-out register. For test purposes it is often necessary that the code be repeated continuously. This mode of operation is easily obtained by feeding the output Q_0 of the register back into the serial input to form a "reentrant shift register." Such a configuration is called a *dynamic,* or *circulating, memory,* or a *shift-register read-only memory*.

A sequence generator may also be obtained from a multiplexer (Sec. 6-7) and a number of simultaneous sequences may be generated using a ROM (Sec. 6-9).

Shift-Register Ring Counter[1]

Consider the 5-bit shift register (Fig. 7-11) with Q_0 connected to the serial input. Such a circulating memory forms a *ring counter*. Assume that all FLIP-FLOPS are cleared and then that FF0 is preset so that $Q_0 = 1$ and $Q_4 = Q_3 = Q_2 = Q_1 = 0$. The first clock pulse transfers the state of FF0 to FF4, so that after the pulse $Q_4 = 1$ and

$$Q_3 = Q_2 = Q_1 = Q_0 = 0$$

Succeeding pulses will transfer the state 1 progressively around the ring. The count is read by noting which FLIP-FLOP is in state 1; no decoding is necessary.

Consider a ring counter with N stages. If the interval between triggers is T, then the output from any binary stage is a pulse train of period NT, with each pulse of duration T. The output pulse of one stage is delayed by a time T from a pulse in the preceding stage. These pulses may be used where a set of sequential gating waveforms is required. Thus a ring counter is analogous to a stepping switch, where each triggering pulse causes an advance of the switch by one step.

Since there is one output pulse for each N clock pulses, the counter is also a *divide-by-N* unit, or an $N : 1$ *scaler*. Typically, TTL shift-register counters operate at frequencies as high as 25 MHz.

Twisted-Ring Counter[1]

The topology where $\bar{Q}_0$ (rather than Q_0) is fed back to the input of the shift register is called a *twisted-ring, switched tail, moebius*, or *Johnson, counter*. This system is a $2N : 1$ scaler. To verify this statement consider that initially all stages in Fig. 7-11 are in the 0 state. Since $S_4 = \bar{Q}_0 = 1$, the first pulse puts FF4 into the 1 state; $Q_4 = 1$, and all other FLIP-FLOPS remain in the 0 state. Since now $S_3 = Q_4 = 1$ and S_4 remains in the 1 state, then after the next pulse there results $Q_4 = 1$, $Q_3 = 1$, $Q_2 = 0$, $Q_1 = 0$, and $Q_0 = 0$. In other words, pulse 1 causes only Q_4 to change state, and pulse 2 causes only Q_3 to change from 0 to 1. Continuing the analysis, we see that pulses 3, 4, and 5 cause Q_2, Q_1, and Q_0, in turn, to switch from the 0 to the 1 state. At the end of five pulses all FLIP-FLOPS are in the 1 state.

After pulse 5, $S_4 = \bar{Q}_0$ changes from 1 to 0. Hence the sixth pulse causes Q_4 to change to 0. The seventh pulse resets Q_3 to 0, and so on, until, at the tenth pulse, all stages have been returned to the 0 state, and the counting cycle is complete. We have demonstrated that this five-stage twisted-ring configuration is a $10 : 1$ counter. To read the count requires a 5-to-10-line decoder, but because of the unique waveforms generated, only two-input AND gates are required (Prob. 7-16).

MOS shift registers are considered in Sec. 9-1.

7-5 RIPPLE (ASYNCHRONOUS) COUNTERS[1-6]

The ring counters discussed in the preceding section do not make efficient use of the FLIP-FLOPS. A $5 : 1$ counter (or $10 : 1$ with the Johnson ring) is obtained with five stages, whereas five FLIP-FLOPS define $2^5 = 32$ states. By modifying the interconnections between stages (*not* using the shift-register topology), we now demonstrate that n binaries can function as a $2^n : 1$ counter.

Ripple Counter

Consider a chain of four J-K master-slave FLIP-FLOPS with the output Q of each stage connected to the clock input of the following binary, as in Fig. 7-13. The

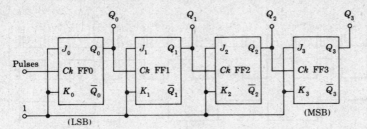

Figure 7-13 A chain of FLIP-FLOPs connected as a ripple counter (TI-93). The TI-393 is a dual 4-bit binary counter MSI package which operates up to a maximum frequency of 25 MHz.

pulses to be counted are applied to the clock input of FF0. For all stages J and K are tied to the supply voltage, so that $J = K = 1$. This connection converts each stage to a T-type FLIP-FLOP (Fig. 7-10), with $T = 1$.

It should be recalled that, for a T-type binary with $T = 1$, the master changes state every time the waveform at its clock input changes from 0 to 1 and that the new state of the master is transferred to the slave when the clock falls from 1 to 0. This operation requires that

1. Q_0 changes state at the *falling* edge of each pulse.
2. All other Q's make a transition when and only when the output of the preceding FLIP-FLOP changes from 1 to 0. This negative transition "ripples" through the counter from the LSB to the MSB.

Following these two rules, the waveforms in Fig. 7-14 are obtained. Table 7-5 lists the states of all the binaries of the chain as a function of the number of externally applied pulses. This table may be verified directly by comparison with the waveform chart of Fig. 7-14. Note that in Table 7-5 the FLIP-FLOPS have been ordered in the reverse direction from their order in Fig. 7-13. We observe that the ordered array of states 0 and 1 in any row in Table 7-5 is precisely the

Table 7-5 States of the FLIP-FLOPS in Fig. 7-13

Number of input pulses	FLIP-FLOP outputs				Number of input pulses	FLIP-FLOP outputs			
	Q_3	Q_2	Q_1	Q_0		Q_3	Q_2	Q_1	Q_0
0	0	0	0	0					
1	0	0	0	1	9	1	0	0	1
2	0	0	1	0	10	1	0	1	0
3	0	0	1	1	11	1	0	1	1
4	0	1	0	0	12	1	1	0	0
5	0	1	0	1	13	1	1	0	1
6	0	1	1	0	14	1	1	1	0
7	0	1	1	1	15	1	1	1	1
8	1	0	0	0	16	0	0	0	0

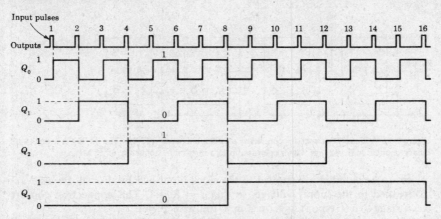

Figure 7-14 Waveform chart for the 4-stage ripple counter. Note that after pulse 5 $Q_0 = 1$, $Q_1 = 0$, $Q_2 = 1$, and $Q_3 = 0$. These binary outputs correspond to the decimal number 5.

binary representation of the number of input pulses as given in Table 5-1. Thus the chain of FLIP-FLOPS *counts* in the binary system.

A chain of n binaries will count up to the number 2^n before it resets itself into its original state. Such a chain is referred to as a counter *modulo* 2^n. To read the counter, the 4-bit words (numbers) in Table 7-5 are recognized with a decoder, which in turn drives visible numerical indicators (Sec. 6-11). Spikes are possible in any counter unless all FLIP-FLOPS change state simultaneously. To eliminate the spikes at the decoder output, a strobe pulse is used (S in Fig. 6-16) so that the counter is read only after the spikes have decayed and a steady state is reached.

Up-Down Counter

A counter which can be made to count in either the forward or reverse direction is called an *up-down*, a *reversible*, or a *forward-backward*, counter. Forward counting is accomplished, as we have seen, when the trigger input of a succeeding binary is coupled to the Q output of a preceding binary. The count will proceed in the reverse direction if the coupling is made instead to the $\overline{Q}$ output, as we shall now verify.

If a binary makes a transition from state 0 to 1, the output $\overline{Q}$ will make a transition from state 1 to 0. This negative-going transition in $\overline{Q}$ will induce a change in state in the succeeding binary. Hence, for the reversing connection, the following rules apply:

1. FLIP-FLOP FF0 makes a transition at each externally applied pulse.
2. Each of the other binaries makes a transition when and only when the preceding FLIP-FLOP goes from state 0 to state 1.

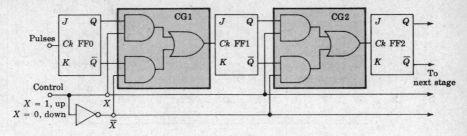

Figure 7-15 An up-down ripple counter. (It is understood that $J = K = 1$.)

If these rules are applied to any of the numbers in Table 7-5, the next smaller number in the table results. For example, consider the number 12, which is 1100 in binary form. At the next pulse, the rightmost 0 (corresponding to Q_0) becomes 1. This change of state from 0 to 1 causes Q_1 to change state from 0 to 1, which in turn causes Q_2 to change state from 1 to 0. This last transition is in the direction not to affect the following binary, and hence Q_3 remains in state 1. The net result is that the counter reads 1011, which represents the number 11. Since we started with 12 and ended with 11, a reverse count has taken place.

The logic block diagram for an up-down counter is indicated in Fig. 7-15. For simplicity in drawing, no connections to J and K are indicated. For a ripple counter it is always to be understood that $J = K = 1$ as in Fig. 7-13. The two-level AND-OR gates CG1 and CG2 between stages constitutes a two-input multiplexer which controls the direction of the counter. Note that this logic combination is equivalent to a NAND-NAND configuration (Fig. 5-17). If the input X is a 1 (0), then Q ($\overline{Q}$) is effectively connected to the following FLIP-FLOP and pulses are added (subtracted). In other words, $X = 1$ converts the system to an *up* counter and $X = 0$ to a *down* counter. The control X may not be changed from 1 to 0 (or 0 to 1) between input pulses, because a spurious count may be introduced by this transition. (The synchronous counter of Fig. 7-17 does not have this difficulty and hence up-down counters are operated synchronously, Sec. 7-6.)

Divide-by-N Counter

It may be desired to count to a base N which is not a power of 2. We may prefer, for example, to count to the base 10, since the decimal system is the one with which we are most familiar. To construct such a counter, start with a ripple chain of n FLIP-FLOPS such that n is the smallest number for which $2^n > N$. Add a feedback gate so that at count N all binaries are reset to zero. This feedback circuit is simply a NAND gate whose output feeds all *clear* inputs in parallel. Each input to the NAND gate is a FLIP-FLOP output Q which becomes 1 at the count N.

Let us illustrate the above procedure for a decade counter. Since the smallest value of n for which $2^n > 10$ is $n = 4$, then four FLIP-FLOPS are required. The decimal number 10 is the binary number 1010 (LSB), and hence

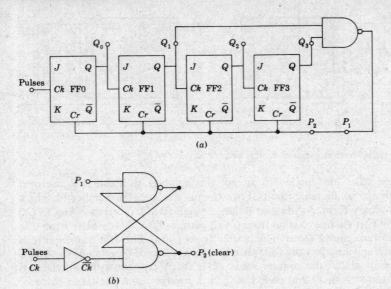

(a)

(b)

Figure 7-16 (a) A decade counter ($J = K = 1$). (b) A latch to eliminate resetting difficulties (due to unequal internal delays).

$Q_0 = 0$, $Q_1 = 1$, $Q_2 = 0$, and $Q_3 = 1$. The inputs to the feedback NAND gate are therefore Q_1 and Q_3, and the complete circuit is shown in Fig. 7-16a. Note that after the tenth pulse Q_1 and Q_3 both go to 1, the output of the NAND gate becomes 0, and all FLIP-FLOPS are cleared (reset to 0). (Note that Q_1 and Q_3 first become 1 and then return to 0 after pulse 10, thereby generating a narrow spike.)

If the propagation delay from the clear input to the FLIP-FLOP output varies from stage to stage, the clear operation may not be reliable. In the above example, if FF3 takes appreciably longer time to reset than FF1, then when Q_1 returns to 0, the output of the NAND gate goes to 1, so that $Cr = 1$ and Q_3 will not reset. Wide variations in reset propagation time may occur if the counter outputs are unevenly loaded. A method of eliminating the difficulty with resetting is to use a latch to memorize the output of the NAND gate at the Nth pulse. The lead in Fig. 7-16a between the NAND output P_1 and the clear input P_2 is opened, and the circuit drawn in Fig. 7-16b is inserted between these two points. The operation of the latch is considered in detail in Prob. 7-20. The TI-90 decade counter, which does not require a latch, is indicated in Prob. 7-22. Two such counters are available in a single package (TI-390). A 12 : 1 counter (TI-92) is considered in Prob. 7-24.

A divide-by-6 counter is obtained using a 3-bit ripple counter, and since for $N = 6$, $Q_1 = 1 = Q_2$, then Q_1 and Q_2 are the inputs to the feedback NAND gate.

Similarly, a divide-by-7 counter requires a three-input NAND gate with inputs Q_0, Q_1, and Q_2.

In some applications it is important to be able to program the count (the value of N) of a divide-by-N counter, either by means of switches or through control data inputs at the preset terminals. Such a *programmable* or *presettable* counter is indicated in the figure of Prob. 7-25.

Consider that it is required to count up to 10,000 and to indicate the count visually in the decimal system. Since $10,000 = 10^4$, then four decade-counter units, such as in Fig. 7-16, are cascaded. A BCD-to-decimal decoder/lamp driver (Sec. 6-6) or a BCD-to-seven-segment display decoder (Sec. 6-11) is used with each unit to make visible the four decimal digits giving the count.

7-6 SYNCHRONOUS COUNTERS[1, 5, 6]

The *carry propagation delay* is the time required for a counter to complete its response to an input pulse. The carry time of a ripple counter is longest when each stage is in the 1 state. For in this situation, the next pulse must cause all previous FLIP-FLOPS to change state. Any particular binary will not respond until the preceding stage has nominally completed its transition. The clock pulse effectively "ripples" through the chain. Hence the carry time will be of the order of magnitude of the sum of the propagation delay times (Sec. 5-15) of all the binaries. If the chain is long, the carry time may well be longer than the interval between input pulses. In such a case, it will not be possible to read the counter between pulses.

If the asynchronous operation of a counter is changed so that all FLIP-FLOPS are clocked simultaneously (synchronously) by the input pulses, the propagation delay time may be reduced considerably. Repetition rate is limited by the delay of any one FLIP-FLOP plus the propagation times of any control gates required. Typically, the maximum frequency of operation of a 4-bit synchronous counter (TI-S168) is 55 MHz, which is about twice that of a ripple counter. Another advantage of the synchronous counter is that no decoding spikes appear at the output since all FLIP-FLOPS change state at the same time. Hence no strobe pulse is required when decoding a synchronous counter.

Series Carry

A 5-bit synchronous counter is indicated in Fig. 7-17. Each FLIP-FLOP is a T type, obtained by tying the J terminal to the K terminal of a J-K FLIP-FLOP (Fig. 7-10). If $T = 0$, there is no change of state when the binary is clocked, and if $T = 1$, the FLIP-FLOP output is complemented with each pulse.

The connections to be made to the T inputs are deduced from the waveform chart of Fig. 7-14.

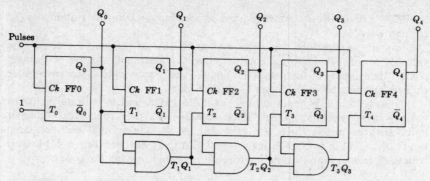

Figure 7-17 A 5-bit synchronous counter with series carry ($J = K = T$).

Q_0 toggles with each pulse: $\qquad\qquad T_0 = 1$

Q_1 complements only if $Q_0 = 1$: $\qquad\qquad T_1 = Q_0$

Q_2 becomes $\overline{Q_2}$ only if $Q_0 = Q_1 = 1$: $\qquad T_2 = Q_0 Q_1$

Q_3 toggles only if $Q_0 = Q_1 = Q_2 = 1$: $\qquad T_3 = Q_0 Q_1 Q_2$

Extending this logic to Q_4, we conclude that $T_4 = Q_0 Q_1 Q_2 Q_3$. Therefore the T logic is given by

$$T_0 = 1 \qquad T_1 = Q_0 \qquad T_2 = T_1 Q_1 \qquad T_3 = T_2 Q_2 \qquad T_4 = T_3 Q_3 \quad (7\text{-}1)$$

Clearly, the two-input AND gates of Fig. 7-17 perform this logic.

The minimum time T_{min} between pulses is the interval required for each J and K node to reach its steady-state value and is given by

$$T_{min} = T_F + (n - 2)T_G \qquad\qquad (7\text{-}2)$$

where T_F is the propagation delay of one FLIP-FLOP, and T_G is the propagation delay of one AND gate (actually, a NAND gate plus an INVERTER). The maximum pulse frequency for series carry is the reciprocal of T_{min}.

Parallel Carry

Since the carry passes through all the control gates in series in Fig. 7-17, this is a synchronous counter with *series*, or *ripple*, *carry*. The maximum frequency of operation can be improved by using parallel, or *look-ahead*, *carry*, where the toggle input to each binary comes from a multi-input AND gate excited by the outputs from every preceding FLIP-FLOP. From Eq. (7-1) it follows that

$$T_1 = Q_0 \qquad T_2 = Q_0 Q_1 \qquad T_3 = Q_0 Q_1 Q_2 \qquad T_4 = Q_0 Q_1 Q_2 Q_3 \quad (7\text{-}3)$$

Hence T_4 is obtained from a four-input AND gate fed by Q_0, Q_1, Q_2, and Q_3. Clearly, for parallel carry,

$$T_{min} = T_F + T_G \qquad\qquad (7\text{-}4)$$

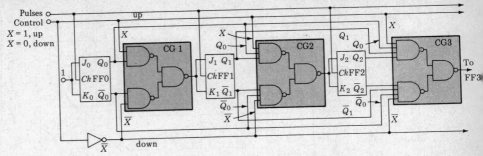

Figure 7-18 An UP-DOWN synchronous counter with parallel carry. The control X may be changed from UP to DOWN (or vice versa) between input pulses without introducing spurious counts, because the counter responds only upon the application of a clock pulse.

which may be considerably smaller than the corresponding time for series carry given by Eq. (7-2), particularly if n is large (high division ratios).

The disadvantages of a parallel-carry counter are: (1) The large fan-in of the gates; the gate feeding T_k requires k inputs. (2) The heavy loading of the FLIP-FLOPS at the beginning of the chain; the fan-out of Q_0 is $n - 1$, since it must feed the carry gates of every succeeding stage.

Up-Down Synchronous Counter with Parallel Carry

As explained in the preceding section, a counter is reversed if $\overline{Q}$ is used in place of Q in the coupling from stage to stage. Hence a synchronous up-down counter is obtained if the control gates CG of Fig. 7-15 are interposed between the FLIP-FLOPS of Fig. 7-17. This change to an UP-DOWN synchronous counter is made in Fig. 7-18, where CG is now indicated as a NAND-NAND gate (equivalent to the AND-OR logic of Fig. 7-15). Note that CG1 is identical in Figs. 7-15 and 7-18. All control gates in the ripple counter are two-input gates, whereas in the synchronous counter the fan-in for CG2 is 3, for CG3 is 4, etc. The extra input leads to the gates, as required by Eq. (7-3), are used for the parallel carry. In other words, the CG blocks in Fig. 7-18 perform both the up-down and the parallel-carry logic.

Synchronous Decade Counter

Design of a system which is to divide by a number that is not a multiple of 2 is much more difficult for a synchronous than for a ripple counter. Control matrices (Karnaugh maps) are used[1, 3, 6, 7] to simplify the procedure.

With a great deal of patience and intuition, the design may be carried out from direct observation of the waveform chart. Consider, for example, the synthesis of a synchronous decade counter with parallel carry. The waveform chart is that given in Fig. 7-14 except that *after the tenth pulse all waveforms return to* 0. Since $Q_0 = 0$ and $Q_2 = 0$ after the tenth pulse, FF0 and FF2 are

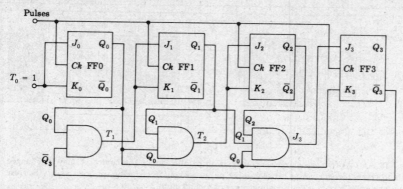

Figure 7-19 A synchronous decade counter with parallel carry.

excited as in the 16 : 1 synchronous counter. Hence, from Eq. (7-1),

$$T_0 = J_0 = K_0 = 1 \qquad T_2 = J_2 = K_2 = Q_0 Q_1 \qquad (7-5)$$

We note from Fig. 7-14 that FF1 toggles if $Q_0 = 1$. However, to prevent Q_1 from going to 1 after the tenth pulse, it is inhibited by Q_3. These statements are equivalent to the statement

$$T_1 = J_1 = K_1 = Q_0 \bar{Q}_3 \qquad (7-6)$$

Finally, we wish FF3 to change state from 0 to 1 after the eighth pulse and to return to 0 after the tenth pulse. If

$$J_3 = Q_0 Q_1 Q_2 \qquad K_3 = Q_0 \qquad (7-7)$$

then the desired logic is followed because $Q_0 = Q_1 = Q_2 = 1$, so that $J_3 = 1$, $K_3 = 1$, before pulse 8, whereas $Q_0 = 1$, $Q_1 = 0$, and $Q_2 = 0$, so that $J_3 = 0$, $K_3 = 1$, before pulse 10. The implementation of Eqs. (7-5) to (7-7) is given in the logic block diagrams of Fig. 7-19.

Synchronous up-down decade counters are available commercially (for example, TI-192 or TI-168) on a single MSI chip, as are 4-bit binary counters such as TI-193 or TI-169. The FLIP-FLOPS are provided with *preset* (so that they are programmable) and *clear* inputs, not indicated in Fig. 7-18. Division by a number other than 2, 5, 6, 10, 12 or a power of 2 is not commercially available and must be designed as explained in the foregoing.

7-7 APPLICATIONS OF COUNTERS

Many systems, including digital computers, data handling, and industrial control systems, use counters. We describe briefly some of the fundamental applications.

Direct Counting

Direct counting finds application in many industrial processes. Counters will operate with reliability where human counters fail because of fatigue or limitations of speed. It is required, of course, that the event which is to be counted first be converted into an electrical signal, but this requirement usually imposes no important limitation. For example, objects may be counted by passing them single-file on a conveyor belt between a photoelectric cell and a light source.

The *preset* input allows control of industrial processes. The counter may be preset so that it will deliver an output pulse when the count reaches a predetermined number. Such a counter may be used, for example, to count the number of pills dropped into a bottle. When the preset count is attained, the output pulse is used to divert the pills to the next container and at the same time to reset the counter for counting the next batch.

Divide-by-N

There are many applications where it is desired to change the frequency of a square wave from f to f/N, where N is some multiple of 2. From the waveforms of Fig. 7-14 it is seen that a counter performs this function.

If instead of square waves it is required to use narrow pulses or spikes for system synchronization, these may be obtained from the waveforms of Fig. 7-14. A small RC coupling combination at the counter output, as in Fig. 7-20a, causes a positive pulse to appear at each transition from 0 to 1 and a negative pulse at each transition from 1 to 0, as in Fig. 7-20c. If now we count only the positive pulses, as in Fig. 7-20d (the negative pulses are eliminated, by using a diode, as in Fig. 7-20a), it appears that each binary divides by 2 the number of positive pulses applied to it. The four FLIP-FLOPS together accomplish a division by a factor $N = 2^4 = 16$. A single positive pulse will appear at the output for each 16 pulses applied at the input. A chain of n binaries used for this purpose of

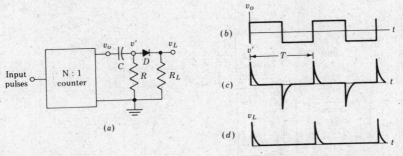

Figure 7-20 (a) An $N : 1$ counter loaded by a network which converts the square wave output (b) to pulses (c) or (d). If the input frequency is f, the spacing between the positive pulses is $T = N/f$.

dividing or scaling down the number of pulses is referred to as a *scaler*. Thus a chain of four FLIP-FLOPS constitutes a scale-of-16 circuit, etc.

Measurement of Frequency

The basic principle by which counters are used for the precise determination of frequency is illustrated in Fig. 7-21. The input signal whose frequency is to be measured is converted into pulses by means of the zero crossing detector (Sec. 17-2) and applied through an AND gate to a counter. To determine the frequency, it is now only required to keep the gate open for transmission for a known time interval. If, say, the gating time is 1 s, the counter will yield the frequency directly in cycles per second (hertz). The *clock* for timing the gate interval is an accurate crystal oscillator whose frequency is, say, 1 MHz. The crystal oscillator drives a scale-of-10^6 circuit which divides the crystal frequency by a factor of 1 million. The divider output consists of a 1-Hz signal whose period is as accurately maintained as the crystal frequency. This divider output signal controls the gating time by setting a toggle FLIP-FLOP to the 1 state for 1 s. The system is susceptible to only slight errors. One source of error results from the fact that a variation of ± 1 count may be obtained, depending on the instant when the first and last pulses occur in relation to the sampling time. Beyond these, of course, the accuracy depends on the accuracy of the crystal oscillator.

Measurement of Time

The time interval between two pulses may also be measured with the circuit of Fig. 7-21. The FLIP-FLOP is now converted into set-reset type, with the first input pulse applied to the S terminal, the second pulse to the R terminal, and no connection made to Ck. With this configuration the first pulse opens the AND gate for transmission and the second pulse closes it. The crystal-oscillator signal (or some lower frequency from the divider chain) is converted into pulses, and these are passed through the gate into the counter. The number of counts

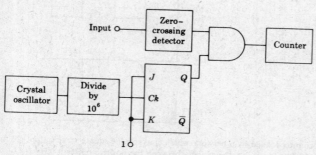

Figure 7-21 A system for measuring frequency by means of a counter.

recorded is proportional to the length of time the gate is open and hence gives the desired time interval.

Measurement of Distance

In radar or sonar systems a pulse is transmitted and a reflected pulse is received delayed by a time T. Since the velocity of light (or sound) is known, a measurement of the interval T, as outlined above, gives the distance from the transmitter to the object from which the reflection was received.

Measurement of Speed

A speed determination may be converted into a time measurement. For example, if two photocell-light-source combinations are set a fixed distance apart, the average speed of an object passing between these points is inversely proportional to the time interval between the generated pulses. Projectile velocities have been measured in this manner.

Digital Computer

In a digital computer a problem is solved by subjecting data to a sequence of operations in accordance with the program of instructions introduced into the computer. Counters may be used to count the operations as they are performed and to call forth the next operation from the memory when the preceding one has been completed.

REFERENCES

1. Texas Instruments Staff: "Designing with TTL Integrated Circuits," McGraw-Hill Book Company, New York, 1971.
2. Sifferlen, T. P., and V. Vartanian: "Digital Electronics with Engineering Applications," Prentice-Hall, Inc., Englewood Cliffs, N. J., 1970.
3. Peatman, J. B.: "Design of Digital Systems," McGraw-Hill Book Company, New York, 1971.
4. Millman, J., and H. Taub: "Pulse, Digital, and Switching Waveforms," chap. 10, McGraw-Hill Book Company, New York, 1965.
5. Grinich, V. H., and H. G. Jackson: "Introduction to Integrated Circuits," McGraw-Hill Book Company, New York, 1975.
6. Taub, H., and D. Schilling: "Digital Integrated Electronics," McGraw-Hill Book Company, New York, 1977.

REVIEW QUESTIONS

7-1 (a) Define a latch.
    (b) Show how to construct this unit from NOT gates.
    (c) Verify that the circuit of part (b) has two stable states.

7-2 Modify the latch of Rev. 7-1 so that data may be entered into the latch, by means of an enable input.

7-3 (a) Define a sequential system.
  (b) How does it differ from a combinational system?

7-4 (a) Sketch the logic system for a clocked *S-R* FLIP-FLOP.
  (b) Verify that the state of the system does not change in between clock pulses.
  (c) Give the truth table.
  (d) Justify the entries in the truth table.

7-5 (a) Augment an *S-R* FLIP-FLOP with two AND gates to form a *J-K* FLIP-FLOP.
  (b) Give the truth table.
  (c) Verify part (b) by making a table of J_n, K_n, Q_n, $\overline{Q}_n$, S_n, R_n, and Q_{n+1}

7-6 Explain what is meant by a *race-around* condition in connection with the *J-K* FLIP-FLOP of Rev. 7-5.

7-7 (a) Draw a clocked *J-K* FLIP-FLOP system and include *preset* (*Pr*) and *clear* (*Cr*) inputs.
  (b) Explain the clear operation.

7-8 (a) Draw a *master-slave J-K* FLIP-FLOP system.
  (b) Explain its operation and show that the race-around condition is eliminated.

7-9 (a) Show how to convert a *J-K* FLIP-FLOP into a *delay* (*D*-type) unit
  (b) Give the truth table.
  (c) Verify this table.

7-10 Repeat Rev. 7-9 for a *toggle* (*T*-type) FLIP-FLOP.

7-11 Give the truth tables for each FLIP-FLOP type: (a) *S-R*; (b) *J-K*; (c) *D*; and (d) *T*. What are the direct inputs *Pr* and *Cr* and the clock *Ck* for (e) presetting; (f) clearing; and (g) normal clocked operation?

7-12 (a) Define a *register*.
  (b) Construct a shift register from *S-R* FLIP-FLOPS.
  (c) Explain its operation.

7-13 (a) Explain why there may be a race condition in a shift register.
  (b) How is this difficulty bypassed?

7-14 Explain how a shift register is used as a converter from (a) *serial-to-parallel* data and (b) *parallel-to-serial* data.

7-15 Explain how a shift register is used as a *sequence generator*.

7-16 Explain how a shift register is used as a circulating *read-only memory*.

7-17 (a) Explain how a shift register is used as a *ring counter*.
  (b) Draw the output waveform from each FLIP-FLOP of a three-stage unit.

7-18 (a) Sketch the block diagram for a *Johnson* (*twisted ring*) *counter*.
  (b) Draw the output waveform from each FLIP-FLOP of a three-stage unit.
  (c) By what number *N* does this system divide?

7-19 (a) Draw the block diagram of a *ripple counter*.
  (b) Sketch the waveform at the output of each FLIP-FLOP for a three-stage counter.
  (c) Explain how this waveform chart is obtained.
  (d) By what number *N* does this system divide?

7-20 (a) Draw the block diagram for an *up-down counter*.
  (b) Explain its operation.

7-21 Explain how to modify a ripple counter so that it divides by N, where N is *not* a power of 2.

7-22 (*a*) Draw the block diagram of a decade ripple counter.

(*b*) Explain its operation.

7-23 Repeat Rev. 7-22 for a divide-by-6 ripple counter.

7-24 What is the advantage of a *synchronous counter* over a ripple counter?

7-25 (*a*) Draw the block diagram of a four-stage synchronous counter with *series carry*.

(*b*) Explain its operation.

(*c*) What is the maximum frequency of operation? Define the symbols in your equation.

7-26 (*a*) Repeat Rev. 7-25 if the counter uses *parallel carry*.

(*b*) What are the advantages and disadvantages of a parallel-carry counter?

7-27 Explain how to measure frequency by means of a counter.

7-28 List six applications of counters. Give no explanations.

FIELD-EFFECT TRANSISTORS

The field-effect transistor[1-5] is a semiconductor device which depends for its operation on the control of current by an electric field. There are two types of field-effect transistors, the *junction field-effect transistor* (abbreviated JFET, or simply FET) and the *metal-oxide-semiconductor field-effect transistor* (MOSFET).

The principles on which these devices operate, as well as the differences in their characteristics, are examined in this chapter. Digital circuits making use of field-effect transistors are also presented.

The field-effect transistor differs from the bipolar junction transistor in the following important characteristics:

1. Its operation depends upon the flow of majority carriers only. It is therefore a *unipolar* (one type of carrier) device.
2. It is simpler to fabricate and occupies less space in integrated form. Hence, the packing density may be extremely high (tens of thousands of MOSFETs per chip).
3. It can be connected as a resistor load so that a digital system consisting only of MOS devices (and no other components) is possible.
4. It has a high input resistance, typically in excess of 10^{10} Ω, allowing very high fanout.
5. It can be used as a symmetrical bilateral switch.
6. By means of the charge stored on small internal capacitances, it functions as a memory device.
7. It is less noisy than a bipolar transistor.

8. It exhibits no offset voltage at zero drain current, and hence makes an excellent signal chopper.[6]

The main disadvantage of the FET is its relatively small gain-bandwidth product in comparison with that which can be obtained with a bipolar transistor. The BJT operates at higher speeds than are possible with the FET. The principal applications of MOSFETs are as LSI (large-scale integration) arrays such as semiconductor memories, long shift registers, and microprocessors (Chap. 9). MOSFETs have made possible such complicated and yet relatively inexpensive instruments as the hand-held calculator and the digital wrist watch.

8-1 THE JUNCTION FIELD-EFFECT TRANSISTOR

The structure of an *n-channel* field-effect transistor is shown in Fig. 8-1. Ohmic contacts are made to the two ends of a semiconductor bar of *n*-type material (if *p*-type silicon is used, the device is referred to as a *p-channel* FET). Current is caused to flow along the length of the bar because of the voltage supply connected between the ends. This current consists of majority carriers, which in this case are electrons. A simple side view of a JFET is indicated in Fig. 8-1a and a more detailed sketch is shown in Fig. 8-1b. The circuit symbol with current and voltage polarities marked is given in Fig. 8-2. The following FET notation is standard.

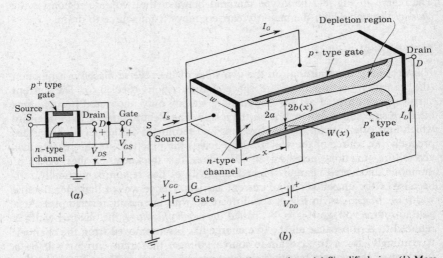

Figure 8-1 The basic structure of an *n*-channel field-effect transistor. (*a*) Simplified view. (*b*) More detailed drawing. The normal polarities of the drain-to-source (V_{DD}) and gate-to-source (V_{GG}) supply voltages are shown. In a *p*-channel FET the voltages would be reversed.

Source

The *source* S is the terminal through which the majority carriers enter the bar. Conventional current entering the bar at S is designated by I_S.

Drain

The *drain* D is the terminal through which the majority carriers leave the bar. Conventional current entering the bar at D is designated by I_D. The drain-to-source voltage is called V_{DS}, and is positive if D is more positive than S. In Fig. 8-1, $V_{DS} = V_{DD}$ = drain supply voltage. The symbols V_{DD}, V_{SS}, and V_{GG} are always positive and represent the *magnitude* of the drain, source, and gate, respectively,

Gate

On both sides of the *n*-type bar of Fig. 8-1, heavily doped (p^+) regions of acceptor impurities have been formed by alloying, by diffusion, or by any other procedure available for creating *p-n* junctions. These impurity regions are called the *gate G*. Between the gate and source a voltage $V_{GS} = -V_{GG}$ is applied in the direction to reverse-bias the *p-n* junction. Conventional current entering the bar at G is designated I_G.

Channel

The region in Fig. 8-1 of *n*-type material between the two gate regions is the *channel* through which the majority carriers move from source to drain.

FET Operation

It is necessary to recall that on the two sides of the reverse-biased *p-n* junction (the transition region) there are space-charge regions (Sec. 2-6). The current carriers have diffused across the junction, leaving only uncovered positive ions on the *n* side and negative ions on the *p* side. The electric lines of field intensity which now originate on the positive ions and terminate on the negative ions are precisely the source of the voltage drop across the junction. As the reverse bias across the junction increases, so also does the thickness of the region of immobile uncovered charges. The conductivity of this region is nominally zero because of the unavailability of current carriers. Hence we see that the effective width of the *channel* in Fig. 8-1 will decrease with increasing reverse bias. At a gate-to-source voltage $V_{GS} = V_P$, called the *pinchoff voltage*, the channel width is reduced to zero because all the free charge has been removed from the channel. Accordingly, for a fixed drain-to-source voltage, the drain current will be a function of the reverse-biasing voltage across the gate junction. The term *field effect* is used to describe this device because the mechanism of current control is the *effect* of the extension, with increasing reverse bias, of the *field* associated with the region of uncovered charges.

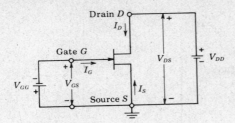

Figure 8-2 Circuit symbol for an n-channel FET. (For a p-channel FET the arrow at the gate junction points in the opposite direction.) For an n-channel FET, I_D and V_{DS} are positive and V_{GS} is negative. For a p-channel FET, I_D and V_{DS} are negative and V_{GS} is positive.

The FET Circuit Symbol

The common-source circuit, symbol, and polarity conventions for a FET are indicated in Fig. 8-2. The direction of the arrow at the gate of the junction FET in Fig. 8-2 indicates the direction in which gate current would flow if the gate junction were forward-biased. We note that the n-channel FET requires zero or negative gate bias and positive drain voltage. The p-channel FET requires opposite voltage polarities. We can remember supply polarities by using the channel type, p or n, to designate the polarity of the source side of the drain supply.

Since the FET is a symmetrical device, either end of the channel may be used as the source. In other words, the *field-effect transistor is a bidirectional device*; the direction of the current in the channel is determined by the polarity of the voltage across the channel.

8-2 THE JFET VOLT-AMPERE CHARACTERISTICS

The drain characteristics for a typical n-channel FET shown in Fig. 8-3 give I_D against V_{DS}, with V_{GS} as a parameter. To see qualitatively why the characteristics have the form shown, consider, say, the case for which $V_{GS} = 0$. For $I_D = 0$, the channel between the gate junctions is entirely open. In response to a small applied voltage V_{DS}, the n-type bar acts as a simple semiconductor resistor, and the current I_D increases linearly with V_{DS}. With increasing current, the ohmic voltage drop along the n-type channel region reverse-biases the gate junction, and the conducting portion of the channel begins to constrict. Because of the ohmic drop along the length of the channel itself, the constriction is not uniform, but is more pronounced at distances farther from the source, as indicated in Fig. 8-1. Eventually, a voltage V_{DS} is reached at which the channel is "pinched off." This is the voltage, not too sharply defined in Fig. 8-3, where the current I_D begins to level off and approach a constant value. It is, of course, in principle not possible for the channel to close completely and thereby reduce the current I_D to zero. For if such, indeed, could be the case, the ohmic drop required to provide the necessary back bias would itself be lacking. Note that each characteristic curve has an *ohmic* or *nonsaturation region* for small values of V_{DS}, where I_D is proportional to V_{DS}. Each also has a *constant-current* or *current saturation region* for large values of V_{DS}, where I_D responds very slightly to V_{DS}.

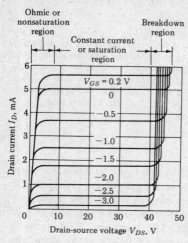

Figure 8-3 Output characteristics of a 2N4869 n-channel depletion-mode JFET. *(Courtesy of Siliconix, Inc.)*

If now a gate voltage V_{GS} is applied in the direction to provide additional reverse bias, pinchoff will occur for smaller values of $|V_{DS}|$, and the maximum drain current will be smaller. This feature is brought out in Fig. 8-3. Note that a plot for a silicon FET is given even for $V_{GS} = +0.2$ V, which is in the direction of forward bias. We note from Table 3-1 that, the gate current will be very small, because at this gate voltage the Si junction voltage is less than the cutin voltage $V_\gamma = 0.5$ V. The saturation current for $V_{GS} = 0$ is designated by the symbol I_{DSS}, and this current for a commercial JFET may range from 0.1 mA to several hundred milliamperes.

We now examine the *ohmic, saturation, breakdown,* and *cutoff* regions in more detail.

The ON Resistance $r_{DS(ON)}$

Assume, first, that a small voltage V_{DS} is applied between drain and source. The resulting small drain current I_D will then have no appreciable effect on the channel profile. Under these conditions we may consider the effective channel cross section A to be constant throughout its length. Hence $A = 2bw$, where $2b$ is the channel width corresponding to zero drain current for a specified V_{GS}, and w is the channel dimension perpendicular to the b direction, as indicated in Fig. 8-1.

Since no current flows in the depletion region, then, using Ohm's law [Eq. (1-14)], we obtain for the drain current

$$I_D = Aq N_D \mu_n \mathcal{E}_x = 2b(x)wq N_D \mu_n \mathcal{E}_x = 2bwq N_D \mu_n \frac{V_{DS}}{L} \qquad (8-1)$$

where L is the length of the channel. Equation (8-1) describes the volt-ampere characteristics of Fig. 8-3 for very small V_{DS}, and it suggests that under these conditions the FET behaves like an ohmic resistance whose value is determined by V_{GS}. The ratio V_{DS}/I_D at the origin is called the ON *drain resistance* $r_{DS(ON)}$. For a JFET we obtain from Eq. (8-1), with $V_{GS} = 0$ and hence $b = a$,

$$r_{DS(ON)} = \frac{L}{2awqN_D\mu_n} \qquad (8-2)$$

For commercially available n-channel FETs and MOSFETs values of $r_{DS(ON)}$ ranging from a few ohms to several hundred ohms are listed in the manufacturer's specification sheets. Since the mobility for holes is less than that for electrons, $r_{DS(ON)}$ is much higher for p-channel than for n-channel FETs. This parameter is important in switching applications where the FET is driven heavily ON. A bipolar transistor has an R_{CS} which is only a few ohms and, hence, is of the same order of magnitude as $r_{DS(ON)}$ for an n-channel FET. However, a bipolar transistor has the disadvantage for chopper applications[6] of possessing an offset voltage, whereas the FET characteristics pass through the origin, $I_D = 0$ and $V_{DS} = 0$.

The Pinchoff Region

We now consider the situation where an electric field $\mathcal{E}_x$ appears along the x axis. If a substantial drain current I_D flows, the drain end of the gate is more reverse-biased than the source end, and hence the boundaries of the depletion region are not parallel to the longitudinal axis of the channel, but converge as shown in Fig. 8-1. A qualitative explanation is given in the following paragraph of what takes place within the channel as the applied drain voltage is increased and pinchoff is approached.

As V_{DS} increases, $\mathcal{E}_x$ and I_D increase, whereas $b(x)$ decreases because the channel narrows, and hence the current density $J = I_D/2b(x)w$ increases. We now see that complete pinchoff ($b = 0$) cannot take place because, if it did, J would become infinite, which is a physically impossible condition. If J were to increase without limit, then, from Eq. (8-1), so also would $\mathcal{E}_x$, provided that μ_n remains constant. It is found experimentally,[7,8] however, that the mobility is a function of electric field intensity and remains constant only for $\mathcal{E}_x < 10^3$ V/cm in n-type silicon. For moderate fields, 10^3 to 10^4 V/cm, the mobility is approximately inversely proportional to the square root of the applied field. For still higher fields, such as are encountered at pinchoff, μ_n is inversely proportional to $\mathcal{E}_x$. In this region the drift velocity of the electrons ($v_x = \mu_n\mathcal{E}_x$) remains constant, and Ohm's law is no longer valid. From Eq. (8-1) we now see that both I_D and b remain constant, thus explaining the constant-current portion of the V-I characteristic of Fig. 8-3.

What happens[8] if V_{DS} is increased beyond pinchoff, with V_{GS} held constant? As explained above, the minimum channel width $b_{min} = \delta$ has a small nonzero constant value. This minimum width occurs at the drain end of the bar. As V_{DS} is increased, this increment in potential causes an increase in $\mathcal{E}_x$ in an adjacent channel section toward the source. Referring to Fig. 8-4, the velocity-limited region L' increases with V_{DS}, whereas δ remains at a fixed value.

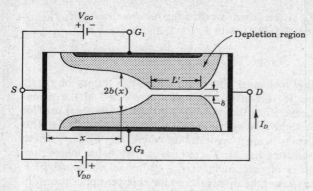

Figure 8-4 After pinchoff, as V_{DS} is increased, then L' increases but δ and I_D remain essentially constant. (G_1 and G_2 are tied together.)

Breakdown

The maximum voltage that can be applied between any two terminals of the FET is the lowest voltage that will cause avalanche breakdown (Sec. 2-9) across the gate junction. From Fig. 8-3 it is seen that avalanche occurs at a lower value of $|V_{DS}|$ when the gate is reverse-biased than for $V_{GS} = 0$. This is caused by the fact that the reverse-bias gate voltage adds to the drain voltage, and hence increases the effective voltage across the gate junction. The breakdown voltage between drain and source with the gate short-circuited to the source, designated by BV_{DSS}, is given in the manufacturer's specification sheets (Appendix B-5). Its voltage range is 20-50 V.

Cutoff

With a physical FET device the same leakage drain current $I_{D\text{(OFF)}}$ still exists even under the cutoff condition $|V_{GS}| > |V_P|$. The *gate reverse current*, also called the *gate cutoff current*, designated I_{GSS}, gives the gate-to-source current, with the drain short-circuited to the source for $|V_{GS}| > |V_P|$. A manufacturer specifies maximum values of $I_{D\text{(OFF)}}$ and I_{GSS}. Each of these is in the range from 0.1 to a few nanoamperes at 25°C and increases by a factor of about 1,000 (to a few microamperes) at a temperature of 150°C.

Applications of JFETs

These devices are packaged either singly or two per chip (a matched pair). They make excellent digital or analog switches, the input stages to a differential amplifier, mixers, and oscillators. They are also used as special-purpose amplifiers; for example, an electrometer (very high input resistance) amplifier, a low-noise amplifier, a buffer voltage follower (high input and low output resistance), and a high-frequency (VHF-UHF) amplifier. A FET may be connected to function as a voltage-controlled resistance (Sec. 11-20).

8-3 FABRICATION OF JFETs[9]

The structure shown in Fig. 8-1 is not practical because of the difficulties involved in diffusing impurities into both sides of a semiconductor wafer.

An *n*-Channel JFET

Figure 8-5 shows a single-ended-geometry JFET where diffusion is from one side only. Figure 8-5*a* is a cross-sectional view in the plane *AA'* of the top view of Fig. 8-5*b*. The substrate is of *p*-type material onto which an *n*-type channel is epitaxially grown (Sec. 4-3). A *p*-type gate is then diffused into the *n*-type channel. The diffused gate is of very low resistivity material, allowing the depletion region to spread mostly into the *n*-type channel.

A *p*-Channel FET

This type of device is obtained from the pinch-resistor structure of Fig. 4-20 by adding a metallic connection to the emitter. This contact is the gate lead, whereas terminals 1 and 2 serve as source and drain. The channel is the thin *p* region which is the pinch resistor in Fig. 4-20. Note that this *p*-channel JFET requires two *p*-diffusion operations in contrast to the *n*-channel FET of Fig. 8-5.

It should be clear now why the pinch resistor is highly nonlinear. Its volt-ampere characteristic is given by the FET curve of Fig. 8-3, corresponding to a gate voltage equal to the floating emitter potential of Fig. 4-20. The

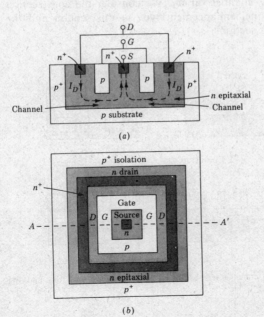

(a)

(b)

Figure 8-5 Single-ended geometry of an *n*-channel junction JFET. (*a*) Cross-sectional view; (*b*) top view.

resistance is equal to the reciprocal slope (dV_{DS}/dI_D) of the volt-ampere curve and, hence, its value depends on the magnitude of V_{DS}. As V_{DS} increases so that pinching causes the channel cross section to decrease and current saturation to occur, the resistance increases to very high values.

8-4 THE ENHANCEMENT METAL-OXIDE-SEMICONDUCTOR FIELD-EFFECT TRANSISTOR (MOSFET)

In a junction field-effect transistor an electric field is applied to the channel through a p-n diode. A basically different field-effect device is obtained by using a metal gate electrode separated by an oxide layer from the semiconductor channel. This metal-oxide-semiconductor (MOS) arrangement allows an electric field to affect the channel, if an external potential is applied between the gate and substrate. Such a device is called a MOSFET or MOS transistor and is of much greater importance than the JFET. As a matter of fact, the largest portion of the IC components industry consists of MOS devices even though the MOSFET achieved volume production only a few years ago (in 1967).

There are two types of MOS transistors. The *depletion* MOSFET has a behavior similar to that of the JFET; at zero gate voltage and a fixed drain voltage, the current is a maximum and then it decreases with applied gate potential (of the proper polarity) as in Fig. 8-3. The second kind of device, called the *enhancement* MOSFET exhibits no current at zero gate voltage and the output current increases with an increase in gate potential. Both types can exist in either the p-channel or the n-channel variety. We consider the construction and characteristics of a p-channel enhancement type in this section and the depletion MOS in the following section.

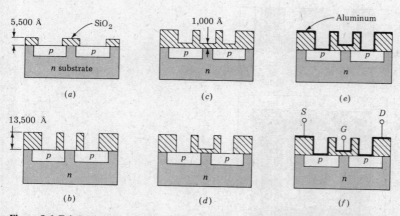

Figure 8-6 Fabrication of a p-channel enhancement MOSFET. One diffusion and four masking steps are required. (Not drawn to scale.)

The Enhancement Structure

The fabrication[10] of this device is indicated in simplified form in Fig. 8-6. The processes involved are those discussed in detail in Chap. 4. Starting with an n-type substrate, a 5,500-Å oxide layer is grown on the surface. Using the standard masking and etching processes, two openings are made into the SiO_2 through which p-type impurities are diffused. These fabrication steps result in the structure of Fig. 8-6a. The two p regions are the source and drain. A thick layer of oxide ($\sim$ 13,500 Å) is grown over the surface and a second masking and etching result in three openings in the oxide (Fig. 8-6b). A thin layer ($\sim$ 1,000 Å) of SiO_2 is added for the gate oxide (Fig. 8-6c). A third mask allows the oxide covering the source and drain regions to be etched away (Fig. 8-6d). Aluminum is then evaporated over the entire surface (Fig. 8-6e). Final masking and etching remove the undesired aluminum so as to delete the interconnections between source, gate, and drain (Fig. 8-6f).

A somewhat more realistic cross section of the p-channel MOSFET (designated PMOS) of Fig. 8-6f is drawn in Fig. 8-7, where lateral diffusion and etching are indicated, and approximate dimensions in micrometers are given. The channel length is L and the width of the channel w (not shown) is the dimension into the paper. The chip area of a MOSFET is 3 mils^2 or less, which is only about 5 percent of that required by a bipolar junction transistor.

The metal area of the gate, in conjunction with the insulating dielectric oxide layer and the semiconductor channel, form a parallel-plate capacitor. The insulating layer of silicon dioxide (0.1 μm thick) is the reason why this device is also called the *insulated-gate* field-effect transistor (IGFET). This layer results in an extremely high input resistance (10^{10} to 10^{15} Ω) for the MOSFET.

Volt-Ampere Enhancement Characteristics

If we ground the substrate and source and set $V_{DS} = 0$ in the structure of Fig. 8-7 and apply a negative voltage at the gate, an electric field will be directed

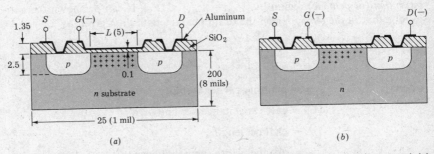

Figure 8-7 Enhancement in a p-channel MOSFET (PMOS). (Dimensions are in micrometers.) (*a*) $V_{DS} = 0$ and V_{GS} is negative. The induced mobile positive charges (holes) are located in a very thin layer at the surface of the n-type semiconductor. (Not drawn to scale.) (*b*) V_{DS} negative and V_{GS} at the same value as in (*a*).

perpendicularly through the oxide. This field will end on "induced" positive charges near the semiconductor surface, as shown in Fig. 8-7a. Since the n-type substrate contains very few holes, the positive surface charges are primarily holes obtained from the p-type source and drain. These mobile positive charges, which are minority carriers in the n-type substrate, form an "inversion layer." Such an inversion layer is formed only if V_{GS} exceeds a threshold level $V_{GS(th)}$ (abbreviated V_T).† As the magnitude of the negative voltage on the gate increases beyond V_T, the induced positive charges in the semiconductor increase. The region beneath the oxide now is a p-type channel, the conductivity increases, and current flows from source to drain through the induced channel when a negative potential is applied between drain and source. Thus the drain current is "enhanced" by the negative gate voltage, and such a device is called an *enhancement-type* MOS.

Consider now the situation where V_{GS} is held at a constant negative value and V_{DS} is increased in magnitude. As indicated in Fig. 8-8a, the drain current increases linearly with drain voltage for small values of V_{DS}. As $|V_{DS}|$ increases, the drop across the channel increases in magnitude and, hence, the voltage across the gate oxide at the drain side of the channel $V_{DG} = (V_{DS} - V_{GS})$ decreases. This reduced potential difference lowers the field across the drain end of the dielectric, which results in fewer inversion charges in this portion of the induced channel, as indicated in Fig. 8-7b. The channel is being "pinched off," $|I_D|$ increases much more slowly with respect to increases in $|V_{DS}|$ than in the ohmic region near the origin, and current saturation sets in.

Analytic Expressions for the Volt-Ampere Characteristics

For $V_{DS} = 0$ an inversion channel exists between S and D if $|V_{GS}| > |V_T|$ and pinchoff occurs (zero free charge at the drain end of the channel) if $|V_{GS}| \leqslant |V_T|$. If $V_{DS} \neq 0$, then an inversion layer is present if $|V_{GS} - V_{DS}| > |V_T|$, and no free charges are present if $|V_{GS} - V_{DS}| \leqslant |V_T|$. Hence the *ohmic* (also called the *triode* or *nonsaturation*) *region* is defined by $|V_{GS} - V_{DS}| > |V_T|$ and the *saturation* (or *pentode*) region by $|V_{GS} - V_{DS}| \leqslant |V_T|$. It can be shown theoretically[1-5] that *in the ohmic region* the drain characteristics are given by

$$I_D = \frac{\mu C_o w}{2L} [2(V_{GS} - V_T)V_{DS} - V_{DS}^2] \tag{8-3}$$

where

μ = majority carrier mobility

C_o = gate capacitance per unit area

L = channel length

w = channel width (perpendicular to L)

---

† In this chapter the threshold voltage should not be confused with the volt-equivalent of temperature V_T of Sec. 1-11.

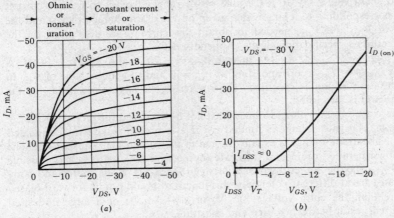

Figure 8-8 (*a*) The drain characteristics and (*b*) the transfer curve (drain current versus gate voltage, for $V_{DS} = -30$ V) of the 3N163 *p*-channel enhancement-type MOSFET. (*Courtesy of Siliconix, Inc.*) The characteristics for an *n*-channel MOSFET are similar except that all voltages and currents are positive (see Fig. 8-18).

The dividing line between the two regions (ohmic and saturation) is given by substituting $V_{GS} - V_{DS} = V_T$ into Eq. (8-3), which reduces to

$$I_D = \frac{\mu C_o w}{2L} V_{DS}^2 \tag{8-4}$$

A plot of I_D versus V_{DS} is a quadratic curve which separates the ohmic from the saturation region.

Ideally I_D remains constant at the value given in Eq. (8-4) for all values of V_{DS} in the saturation region. If V_{DS} is replaced by $V_{GS} - V_T$ in Eq. (8-4), then

$$I_D \equiv I_{DS} = \frac{\mu C_o w}{2L} (V_{GS} - V_T)^2 \tag{8-5}$$

This equation represents the transfer characteristic *in the saturation* region where $|V_{DS}| > |V_{GS} - V_T|$. Theoretically, in this region I_{DS} is independent of V_{DS} and increases as the square of $|V_{GS} - V_T|$. For $|V_{GS} - V_{DS}| < |V_T|$ the inversion layer is pinched off at the drain end of the channel, as indicated in Fig. 8-7*b*.

Specifications

The volt-ampere drain characteristics of a *p*-channel enhancement-mode MOSFET are given in Fig. 8-8*a*, and its transfer curve in Fig. 8-8*b*. The current I_{DSS} at $V_{GS} \geqslant 0$ (called the *drain cutoff current* $I_{D(OFF)}$) is very small, of the order of 1 nA. As V_{GS} is made negative, the current $|I_D|$ increases slowly at first, and then much more rapidly with an increase in $|V_{GS}|$. The manufacturer indicates the gate-source *threshold voltage* $V_{GS(th)} = V_T$, at which $|I_D|$ reaches some defined small value, say 10 μA. Value of V_T ranges from 1 to 6 V. A current $I_{D(ON)}$, corresponding to a large value given on the drain characteristics, and the values of V_{GS} and V_{DS}, needed to obtain this current, are also given on the manufacturer's specification sheets (Appendix B-5).

In Fig. 8-8b, $I_{D(ON)} = -44$ mA at $V_{GS} = -20$ V and $V_{DS} = -30$ V. A current $I_{D(OFF)} = I_{DSS}$, corresponding to $V_{GS} = 0$ and to a large value of V_{DS} in the saturation region is specified. The ON resistance $r_{DS(ON)} = V_{DS}/I_D$ in the linear region is also listed.

There are several voltages which must not be exceeded for proper MOSFET operation. The gate voltage is limited by gate-to-body breakdown (designated BV_{GBS}). The drain voltage is usually limited to the value which causes avalanche breakdown between drain and substrate (designated BV_{DSS}). It is also possible that BV_{DSS} may be limited by *punch-through* (Sec. 3-12). If the source and body (substrate) are at different potentials, this junction should not be forward-biased by more than the cutin voltage (0.5 V) of a *p-n* diode or a large undesired current will result between source and substrate.

MOSFET Gate Protection

Since the SiO_2 layer of the gate is extremely thin, it may easily be damaged by excessive voltage. An accumulation of charge on an open-circuited gate may result in a large enough field to punch through the dielectric. To prevent this damage some MOS devices are fabricated with a Zener diode between gate and substrate. In normal operation this diode is open and has no effect upon the circuit. However, if the voltage at the gate becomes excessive, then the diode breaks down and the gate potential is limited to a maximum value equal to the Zener voltage.

Self-Isolation

Usually the source and body are tied together and therefore the source-substrate diode is cut off. The drain potential polarity is such that the drain-to-substrate diode is also cut off (Fig. 8-7b). Clearly, *no isolation island is required for a MOS transistor* and the current is confined to the channel between *D* and *S*. For a BJT the isolation diffusion (Sec. 4-2) occupies an extremely large percentage of the transistor area. It is the lack of an isolation border, the completely different geometry, and the fact that diffusion resistors are not required (Sec. 8-7) that accounts for the fact that the packing density of MOSFETs is about 20 times that of bipolar transistor IC's.

n-Channel Enhancement MOSFET

If the body of a MOS transistor is *p*-type silicon and if two *n* regions, separated by the channel length, are diffused into the substrate to form the source and drain, an *n*-channel enhancement device (designated NMOS) is obtained. The structure in Fig. 8-7 would represent an NMOS device if the *p* and *n* regions were interchanged and if positive voltages were applied to the gate and drain (with respect to the source). The induced mobile channel surface charges would then be electrons. The NMOS volt-ampere characteristics are similar to those in Fig. 8-8, but the current and voltage polarities are opposite to those in that figure. (The output curves for an *n*-channel enhancement MOSFET are given in Fig. 8-16.)

Comparison of *p*- with *n*-Channel FETs

The *p*-channel enhancement FET, shown in Fig. 8-7 was first used in MOS systems because it was much easier to produce than the *n*-channel device. Most of the contaminants in MOS fabrication are mobile ions which are positively charged and are trapped in the oxide layer between gate and substrate. In an *n*-channel enhancement device the gate is normally positive with respect to the substrate and, hence, the positively charged contaminants collect along the interface between the SiO_2 and the silicon substrate. The positive charge from this layer of ions behaves in the same way as the positive gate bias in the formation of an *n*-type channel, which tends to make the transistor turn on prematurely. In *p*-channel devices the positive contaminant ions are pulled to the opposite side of the oxide layer (to the aluminum-SiO_2 interface) by the negative gate voltage and there they cannot affect the channel.

The hole mobility in silicon and at normal field intensities is approximately $500 \text{ cm}^2/\text{V} \cdot \text{s}$. On the other hand, electron mobility is about $1,300 \text{ cm}^2/\text{V} \cdot \text{s}$. Thus the *p*-channel device will have more than twice the ON resistance of an equivalent *n*-channel unit of the same geometry and at the same operating conditions. In other words, the *p*-channel device must have more than twice the area of the *n*-channel device to achieve the same resistance. Therefore *n*-channel MOS circuits can be smaller for the same complexity than *p*-channel circuits. The higher packing density of the *n*-channel MOS also makes it faster in switching applications due to the smaller junction areas. The operating speed is limited primarily by the internal *RC* time constants, and the capacitance is directly proportional to the junction cross sections. Since the applied gate voltage and the drain supply are positive for an *n*-channel enhancement MOS, this device is TTL compatible. For all of the above reasons it is clear that *n*-channel MOS IC's are more desirable than the *p*-channel circuits. The technological difficulties in the fabrication of *n*-channel devices have been overcome and volume production began in 1974.[11] The cost of these chips has dropped drastically with demand and *p*-channel MOSFETs have become almost obsolete, except for CMOS (Sec. 8-9).

8-5 THE DEPLETION MOSFET

A second type of MOSFET can be made if, to the basic structure of Fig. 8-7, a channel is diffused between the source and the drain, with the same type of impurity as used for the source and drain diffusion. Let us now consider such an

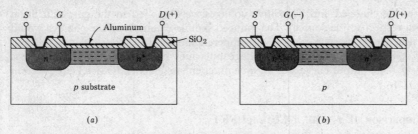

Figure 8-9 The structure of an n-channel depletion MOSFET (NMOS.) (*a*) $V_{GS} = 0$ V. (*b*) V_{GS} is negative which induces positive charges into the channel and these recombine with the free electrons to reduce the majority carriers.

n-channel structure, shown in Fig. 8-9*a*. The minus signs are intended to indicate the free electrons in the channel near the surface. For V_{DS} positive an appreciable drain current I_{DSS} flows for zero gate-to-source voltage $V_{GS} = 0$. If the gate voltage is made negative, positive charges are induced in the channel through the SiO$_2$ of the gate capacitor. Since the current in an FET is due to majority carriers (electrons for an n-type material), the induced positive charges make the channel less conductive, and the drain current drops as V_{GS} is made more negative. The recombination of charge in the channel causes an effective depletion of majority carriers, which accounts for the designation *depletion* MOSFET. Note in Fig. 8-9*b* that, because of the voltage drop due to the drain current, the channel region nearest the drain is more depleted than is the volume near the source. This phenomenon is analogous to that of pinchoff occurring in a JFET at the drain end of the channel (Fig. 8-1). As a matter of fact, the volt-ampere characteristics of the depletion-mode MOS and the JFET are quite similar.

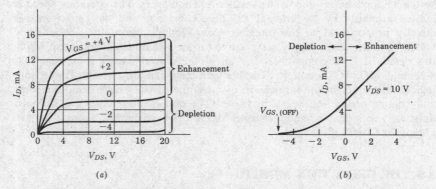

Figure 8-10 (*a*) The drain characteristics and (*b*) the transfer curve (for $V_{DS} = 10$ V) for the 2N3631 n-channel MOSFET which may be used in either the enhancement or the depletion mode. (*Courtesy of Siliconix, Inc.*)

A MOSFET of the depletion type just described may also be operated in an enhancement mode. It is only necessary to apply a positive gate voltage so that negative charges are induced into the n-type channel. In this manner the conductivity of the channel increases and the current rises above I_{DSS}. The volt-ampere characteristics of this device are indicated in Fig. 8-10a, and the transfer curve is given in Fig. 8-10b. The depletion and enhancement regions, corresponding to V_{GS} negative and positive, respectively, should be noted. The manufacturer sometimes indicates the *gate-source cutoff voltage* $V_{GS(OFF)}$ at which I_D is reduced to some specified negligible value at a recommended V_{DS}. This gate voltage corresponds to the pinch-off voltage V_P of a JFET.

The foregoing discussion is applicable in principle also to the p-channel MOSFET. For such a device the signs of all currents and voltages in the volt-ampere characteristics of Fig. 8-10 must be reversed.

8-6 TECHNOLOGICAL IMPROVEMENTS[10, 11]

Present-day MOSFET fabrication differs from the structure of Fig. 8-6 in several processing steps, materials, and geometry. These techniques, which lead to improved device performance, are described briefly in this section.

The value of V_T for the n-channel standard MOSFET is typically 3 to 6 V, and it is common to use a power-supply voltage of 12 V for the drain supply. This large voltage is incompatible with the 5-V supply used in bipolar integrated circuits. In general, a low threshold voltage allows (1) the use of a small power-supply voltage and hence less power, (2) compatible operation with bipolar devices, and (3) higher packing densities. Four methods are used to lower the magnitude of V_T and to yield more ideal MOSFET characteristics.

The ⟨100⟩ Orientation

The standard p-channel enhancement MOS (Fig. 8-7) uses silicon which has been sliced in the ⟨111⟩ crystal direction—the same orientation as that employed to fabricate bipolar transistors. If a crystal is utilized in the ⟨100⟩ direction, it is found that a value of V_T results which is about -1.5 to -2.5 V. However, the hole mobility is somewhat lower for the ⟨100⟩ direction than for the ⟨111⟩ orientation.

The Silicon Nitride Process

This technique uses ⟨111⟩ silicon for higher mobility, but adds a layer of Si_3N_4 to the SiO_2 dielectric. In Fig. 8-6c the oxidation is terminated when the gate thickness is 200 Å. Then an extra processing step is added which consists of the introduction of nitrogen and allows a layer of silicon nitride to form to a thickness of 800 Å. The dielectric constant of Si_3N_4 is 7.5 whereas that of SiO_2 is 4. The increased overall dielectric constant reduces $|V_T|$ to about 2 to 3 V. Such a device is designated by MNOS, and may be either a p- or an n-channel MOSFET.

The Silicon Gate

Polycrystalline silicon doped with phosphorus is conductive and is used as the gate electrode instead of aluminum. This change of material reduces V_T to about 1 to 2 V. The processing sequence of Fig. 8-6 is rearranged and modified in fabricating an n-channel silicon gate MOS, as follows:

A thick oxide layer ($\sim 10,000$ Å) is grown over the p-type silicon substrate. Masking and etching remove the oxide over an area large enough to include the source, gate, and drain. A thin-oxide ($\sim 1,000$ Å) gate dielectric is grown on the surface (Fig. 8-11a).

Polycrystaline silicon is deposited over the entire area in an epitaxial-type oven. A second masking and etching operation defines the gate area and the openings required for source and drain (Fig. 8-11b).

In a diffusion oven phosphorus is allowed to enter the polysilicon and to convert it into a conductor (the *silicon gate*). At the same time the phosphorus diffuses into the substrate to form the drain and source (Fig. 8-11c).

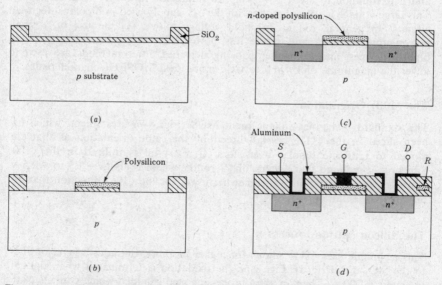

Figure 8-11 Fabrication of an n-channel silicon gate enhancement MOSFET. One diffusion and four masking steps are required. Compare part (d) with Fig. 8-9. The buried resistor R in part (d) is described in the text. (Not drawn to scale.)

Oxide is grown over the top surface, completely covering it. Masking and etching open up contact holes for S, G, and D. Aluminum is evaporated over the entire surface. A final masking and etching process removes the undesired aluminum so as to form the electrode contacts (Fig. 8-11d).

The silicon-gate IGFET has other advantages over the metal-gate device, in addition to that of a lower threshold voltage. Note the *self-alignment* of the gate with the source and drain in Fig. 8-11c. In the conventional fabrication of a metal-gate MOS a photomasking step (Fig. 8-6b) is necessary to align the gate with the source and drain which have already been formed. An overlap of 0.2 mil is necessary to ensure that the gate extends from the S to the D regions. This overlap increases the capacitance C_{gs} between G and S and also C_{gd} between G and D. These capacitances lower the speed of operation and increase the power consumption. Therefore the self-aligned silicon gate improves the performance of the device. The sketches in Fig. 8-11 are idealized in that lateral diffusion is ignored. However, the overlap from lateral diffusion is less than one-half of that necessary to ensure mask alignment, and hence appreciable capacitance reduction is obtained with the silicon gate.

The self-aligning property is not possible with a metal gate. If, for example, the polysilicon in Fig. 8-11b were replaced by aluminum and this chip were placed in a diffusion oven at a temperature of about 1000°C, the metal would melt.

While the polysilicon gate is being fabricated, other regions (such as R in Fig. 8-11d) can simultaneously be transformed into n-doped silicon of low resistance ($\sim 100\ \Omega$ per square). These conductors serve as *buried crossovers* (Sec. 4-12). In addition to aluminum stripes on the SiO_2 surface they can be used to interconnect different portions of a complex IC. A considerable reduction in chip size results. (Note that R should also be indicated in Fig. 8-11b and c but is omitted for simplicity in the discussion of the silicon-gate fabrication processes.)

Ion Implantation

This technique may also be used to lower the threshold voltage V_T. Ions of proper dopant, such as boron for a p-channel or phosphorous for an n-channel, are accelerated to an energy of up to 300 keV and used to bombard the silicon-wafer target before aluminization (Fig. 8-6d). The energy of the ions determines the depth of penetration, and this energy is chosen so that the boron ions penetrate only the thin gate oxide, but not the thick oxide of Fig. 8-6d. The p-type ions injected into the substrate reduces the n-type carrier concentration and, hence, reduces V_T (because it becomes easier to form a p-type channel with an applied gate voltage).

If the ion implantation process described in the foregoing is continued for a long enough time, the channel under the gate may be converted into p-type silicon. Under these circumstances the device becomes a depletion-mode MOSFET rather than an enhancement device. By making use of the proper

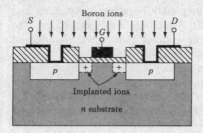

Figure 8-12 Self-aligned gate by means of ion implantation.

masking it is possible to use ion implantation to fabricate *on the same chip* some devices which are depletion and others which are enhancement IGFETs.

It is possible to use ion implantation to obtain self-alignment between the gate and the source and drain regions. In the standard *p*-channel enhancement MOSFET (Fig. 8-6) the mask opening for the gate aluminization is made larger than the channel length so as to avoid alignment inaccuracies. This device, but with a metal gate whose length is smaller than the distance from *S* to *D*, is shown in Fig. 8-12, bombarded by boron ions. The boron cannot penetrate the metal and, hence, ions enter the channel only in the end sections, marked by the small dashed rectangles with a + sign. As a result of this ion implantation the source and drain regions have been extended laterally so that the *S* and *D* edges line up with the edges of the metallic gate. Since there is no overlap of these electrodes, there is a drastic reduction in C_{gd} and C_{gs}.

If in Fig. 8-12 the *n*-type material is changed to *p*-type (and vice versa), and if phosphorus or arsenic ions are used in place of boron, then an *n*-channel enhancement MOSFET with a self-aligned gate can be fabricated.

State of the Art

As a result of the foregoing technical improvements and manufacturing experience the *n*-channel MOSFET is evolving into a much smaller device with improved operating characteristics. A high-resistivity substrate of 50 Ω · cm, *p*-type material is used. A polysilicon gate gives self-alignment with source and drain. By using arsenic ion implantation the threshold voltage V_T is reduced and maintained stable. Photolithographic and etch-control requirements are severe,[12] and x-ray lithography and electron-beam techniques are coming into use.

By scaling the MOSFET dimensions downward properly,[13] it is possible to obtain not only higher component density but also higher speed and lower power dissipation. The result is a great reduction in the speed-power product, thus challenging the performance of bipolar gates (Sec. 5-15). Table 8-1 shows what had been achieved by Intel Corporation by 1977 and what is expected by 1980. These newer transistors are called HMOS (*high-performance* MOSFETs). Note that, to optimize the characteristics of an HMOS, the supply voltage must be reduced as the device size diminishes.

Table 8-1 Evolution of MOS device scaling[13]

Device/circuit parameter	Enhancement-mode NMOS 1972	Depletion-mode NMOS 1976	HMOS 1977	MOS 1980
Channel length, L (μm)	6	6	3.5	2
Lateral diffusion, L_D (μm)	1.4	1.4	0.6	0.4
Junction depth, X_j (μm)	2.0	2.0	0.8	0.8
Gate-oxide thickness, T_{ox} (Å)	1,200	1,200	700	400
Power supply voltage, V_{CC} (V)	4–15	4–8	3–7	2–4
Shortest gate delay, τ (ns)	12–15	4	1	0.5
Gate power, P_D (mW)	1.5	1	1	0.4
Speed-power product (pJ)	18	4	1	0.2

An alternative short-channel technique uses V-shaped notches or grooves etched into a vertical *n-p-n* silicon structure (similar to that shown in Fig. 18-22). This device called VMOS, manufactured by American Microsystems and Texas Instruments, achieved approximately the same speed-power product (1 pJ) in 1977 as an HMOS. The VMOS fabrication is more complex than that for HMOS. Also, the former is an asymmetrical device, whereas the latter has symmetrical bidirectional characteristics.

8-7 MOSFET INVERTERS[14]

The most common applications of MOS devices are digital, such as logic gates (discussed in this chapter) and registers, or memory arrays (Chap. 9). Because of the gate-to-drain and gate-to-source and substrate parasitic capacitances, MOSFET circuits are slower than corresponding bipolar circuits. However, the lower power dissipation and higher density of fabrication make MOS devices attractive and economical for many applications.

Circuit Symbols

It is possible to bring out the connection to the substrate externally so as to have a tetrode device. The circuit symbols used by several manufacturers are indicated in Fig. 8-13. Often the substrate or body lead *B* is omitted from the symbol as in Fig. 8-13a, and it is then understood to be connected to the source internally. The enhancement-type MOSFET symbol of Fig. 8-13c uses a dashed line to indicate the channel.

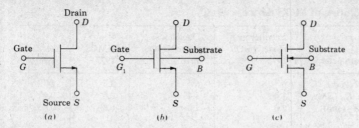

Figure 8-13 Three circuit symbols for an *n*-channel MOSFET (NMOS). (*a*) and (*b*) can be either depletion or enhancement types, whereas (*c*) represents specifically an enhancement device. In (*a*) the substrate or body *B* is understood to be connected internally to the source. For a *p*-channel MOSFET (PMOS) the direction of the arrow is reversed. In the literature the lead *G* is sometimes shown connected to one end (toward the source) of the bar which represents the gate. In this book the lead *G* is shown connected to the center of the bar to emphasize the bidirectional nature of the MOSFET (the device is symmetrical with respect to *D* and *S*). For the sake of simplicity the arrowhead on the source lead is often omitted (if the channel type is clearly understood).

Inverter Loads

The circuit of Fig. 8-14*a* is analogous to the bipolar transistor inverter of Fig. 5-7. The diffusion resistor *R* would usually occupy an area about 20 times that of the active device *Q*. Hence the high-density advantage of MOSFETs would be lost, and this configuration is seldom used if large-scale integration is desired. In Fig. 8-14*b* and *c* a MOS device *Q*2 is used as a nonlinear resistance in place of the linear resistor *R* of Fig. 8-14*a*. Such configurations make possible MOSFET digital LSI systems which consist *entirely* of FETs and no other

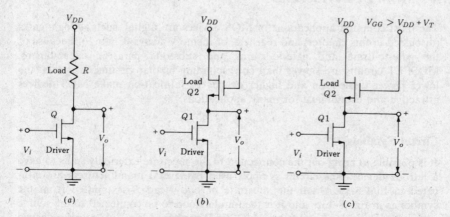

Figure 8-14 MOSFET inverters. (*a*) The load is a linear resistor *R*. The load is nonlinear and operates in the current saturation region in (*b*) and in the nonsaturation region in (*c*). Enhancement MOSFETs are used.

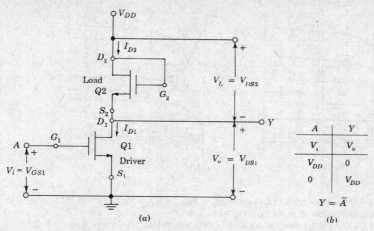

(a)   (b)

Figure 8-15 MOS inverter (NOT circuit). $Q2$ acts as a load for the driver $Q1$.

components such as diodes, resistors, or capacitors (except for small, parasitic capacitances).

Consider the enhancement inverter of Fig. 8-14b, which is redrawn in more detail in Fig. 8-15a. Device $Q1$ is the *driver* FET, whereas $Q2$ acts as its load resistance and is called the *load* FET. The nonlinear character of the load is brought into evidence as follows: Since the gate is tied to the drain, $V_{GS2} = V_{DS2}$. The drain characteristics of the type M116 (Siliconix) enhancement MOSFET is indicated in Fig. 8-16a, and the dashed curve represents the locus of the points $V_{GS2} = V_{DS2} = V_L$. This curve also gives I_{D2} versus V_L (for $V_{GS2} = V_{DS2}$), and its slope gives the incremental load conductance g_L of $Q2$ as a load. Clearly, the load resistance is nonlinear. Note that $Q2$ is always conducting (for $|V_{DS2}| > |V_T|$) regardless of whether $Q1$ is ON or OFF.

The incremental resistance is not a very useful parameter when considering large-signal (ON-OFF) digital operation. It is necessary to draw the *load curve* (corresponding to a *load line* with a constant resistance) on the volt-ampere characteristics of the driver FET $Q1$. The *load curve* is a plot of

$$I_{D1} = I_{D2} \quad \text{versus} \quad V_{DS1} = V_o = V_{DD} - V_L = 10 - V_{DS2} \quad (8\text{-}6)$$

where we have assumed a 10-V power supply. For a given value of $I_{D2} = I_{D1}$, we find $V_{DS2} = V_L$ from the dashed curve in Fig. 8-16a and then plot the locus of the values I_{D1} versus $V_o = V_{DS1}$ in Fig. 8-16b. For example, from Fig. 8-16a for $I_{D2} = 10$ mA, we find $V_{DS2} = 8$ V. Hence $I_{D1} = 10$ mA is located at $V_{DS1} = 10 - 8 = 2$ V on load curve A of Fig. 8-16b.

It is not necessary that $Q1$ and $Q2$ be fabricated with identical dimensions. If $Q2$ has a much higher resistance than $Q1$, load curve B of Fig. 8-16b is obtained.

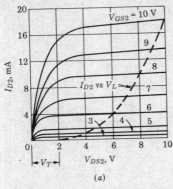

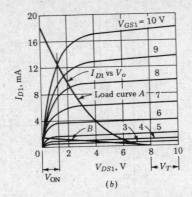

(a)   (b)

Figure 8-16 (a) $Q2$ acts as a load (I_{D2} versus $V_L = V_{DS2} = V_{GS2}$) for the driver $Q1$ of Fig. 8-15. Note that the load curve is in the current saturation region. (b) Load curve A is I_{D1} versus $V_o = V_{DS1}$ if $Q1$ and $Q2$ are identical MOSFETs (Siliconix M116). Load curve B results if $Q2$ has a much higher resistance than $Q1$.

The Transfer Characteristic

For each value of $V_{GS1} = V_i$ in Fig. 8-16b a value of $V_{DS1} = V_o$ is obtained from the load curve. A plot of V_o versus V_i, called the *transfer curve*, is plotted in Fig. 8-17 for load curves A and B.

Note that for $V_i = 0$ V the output differs from the supply voltage by the threshold voltage, $V_o = V_{DD} - V_T$. For $V_i = V_{DD}$ the output is V_{ON}. These values are indicated in the voltage truth table of Fig. 8-17b. For curve A, $V_{ON} \approx 1.4$ V whereas for B, $V_{ON} \approx 0.2$ V. Since the output swing is given by $V_{DD} - V_T - V_{ON}$, it is desirable to have a small value of V_{ON}. Curve B gives the larger output swing.

The FET resistance is proportional to the ratio L/w, where L is the channel length (the distance between source and drain) and w is the width [refer to Eq. (8-3)]. Hence, $Q2$ is fabricated with a much larger L and smaller w than $Q1$ in order to obtain curve B of Fig. 8-17a.

If we assume that $|V_{ON}|$ and $|V_T|$ are small compared with $|V_{DD}|$, we may take $V_{ON} = 0$ and $V_T = 0$. Subject to these approximations, it follows from Fig. 8-17b that, for $V_i = 0$, $V_o = V_{DD}$ and for $V_i = V_{DD}$, $V_o = 0$. These relationships define the ideal NOT circuit. For positive logic with $V(0) = 0$ and $V(1) = V_{DD}$, the logic truth table of Fig. 8-17c is satisfied.

A Nonsaturated Load

From the discussion in Sec. 8-4 it follows that a FET will operate in the ohmic region if $|V_{GS} - V_T| > V_{DS}$. The inverter of Fig. 8-14c shows $Q2$ connected in this manner, so that it represents a nonsaturated load. Consider the case where $V_{DD} = 10$ V and $V_{GG} = 16$ V. Then

$$V_{DS2} - V_{GS2} = V_{DD} - V_{GG} = -6 \quad \text{or} \quad V_{DS2} = V_{GS2} - 6 \quad (8\text{-}7)$$

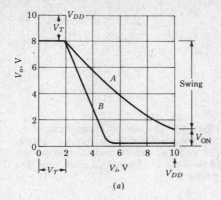

V_i	V_o
0	$V_{DD} - V_T$
V_{DD}	V_{ON}

(b)

A	Y
0	1
1	0

$Y = \bar{A}$

(c)

Figure 8-17 (*a*) The transfer characteristics of an NMOS enhancement inverter with a saturation load; (*A*) with $Q1$ and $Q2$ identical and (*B*) with the geometry of $Q2$ such that its resistance is much larger than that of $Q1$. (*b*) The voltage truth table. (*c*) The logic truth table and Boolean expression. (Note that A in the table refers to the logic input symbol of Fig. 8-15 and not to curve A).

The drain characteristics of Fig. 8-16 are repeated in Fig. 8-18. For each value of V_{GS2} indicated, V_{DS2} is calculated from Eq. (8-7). The current I_{D2} for each pair of values V_{GS2} and V_{DS2} is plotted versus V_{DS2}, as indicated by the dashed curve in Fig. 8-18*a*. The load curve A of Fig. 8-18*b* is a plot of I_{D1} versus V_{DS1} for

$$I_{D1} = I_{D2} \quad \text{and} \quad V_{DS1} = V_{DD} - V_{DS2} = 10 - V_{DS2} \qquad (8\text{-}8)$$

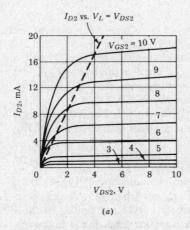

(a)

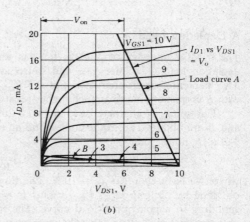

(b)

Figure 8-18 (*a*) $Q2$ acts as a load (I_{D2} versus V_{DS2}) for the driver $Q1$ of Fig. 8-14*c*, with $V_{DD} = 10$ V and $V_{GG} = 16$ V. Note that the load curve is in the nonsaturation region and is almost a straight line. (*b*) Load curve A is I_{D1} versus $V_o = V_{DS1}$ if $Q1$ and $Q2$ are identical devices (Siliconix M116). Load curve B results if $Q2$ has a much higher resistance than $Q1$.

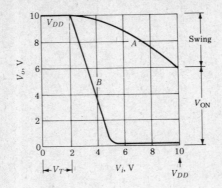

Figure 8-19 The transfer characteristic of an NMOS enhancement inverter with a nonsaturation load; (A) with $Q1$ and $Q2$ identical and (B) with the L/w ratio of $Q2$ much larger than that for $Q1$.

where the values of I_{D2} and V_{DS2} are obtained from the dashed curve of Fig. 8-18a. Note that the characteristic is almost linear and that it passes through the point $V_{DS1} = V_{DD} = 10$ V and $I_{D1} = 0$. If $Q2$ has a much higher resistance than $Q1$, then load curve B is obtained.

The transfer characteristics $V_o = V_{DS1}$ versus $V_i = V_{GS1}$, obtained from the load curves A and B are plotted in Fig. 8-19. Note that the output swing $V_{DD} - V_{ON}$ is relatively small (~ 4 V) for A. However, if $Q2$ is fabricated so that its resistance is much larger than that of $Q1$, the characteristic B results and since $V_{ON} \ll V_{DD}$, the swing is almost the full power-supply voltage V_{DD}. The nonsaturating-load circuit of Fig. 8-14c has the advantage over the saturating-load inverter of Fig. 8-14b in that the former output swing exceeds the latter by the threshold voltage V_T of the load. (Figures 8-17 and 8-19 should be compared.)

A Depletion Load

The NMOS inverter of Fig. 8-20a has an enhancement driver $Q1$ and a depletion load $Q2$. Since gate and source of $Q2$ are tied together, $V_{GS2} = 0$. Two depletion drain characteristics for $V_{GS2} = 0$ are indicated in Fig. 8-20b, with B corresponding to a much higher-resistance MOSFET than A.

The family of characteristics in Fig. 8-20c represent $Q1$. The load curves A and B are obtained from the corresponding enhancement characteristics of Fig. 8-20b by noting that

$$I_{D1} = I_{D2} \quad \text{and} \quad V_{DS1} = 10 - V_{DS2} \tag{8-9}$$

where we have chosen $V_{DD} = 10$ V. The transfer characteristic of Fig. 8-20d is obtained by plotting $V_o = V_{DS1}$ versus $V_i = V_{GS1}$, each pair of voltages corresponding to a point on a load curve. The transfer curves are similar to those obtained in Fig. 8-19 for an enhancement nonsaturating load. For a high-resistance FET the output swing is approximately equal to the power-supply voltage V_{DD}. In Fig. 8-14c two supply voltages are needed, whereas in the inverter of

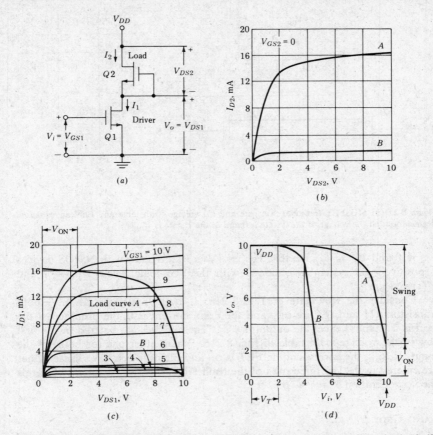

Figure 8-20 (*a*) An NMOS inverter with an enhancement driver $Q1$ and a depletion load $Q2$. (*b*) The drain characteristic of $Q2$ for $V_{GS2} = 0$; in case B device $Q2$ has a much higher resistance than MOSFET $Q1$. (*c*) The output characteristics for $Q1$ with the load curves (A and B) of $Q2$. (*d*) The transfer characteristics corresponding to A and B.

Fig. 8-20*a* only one voltage source is used. However, the latter circuit requires that both enhancement and depletion MOSFETs be fabricated on the same chip. Since very high packing densities can be obtained with the enhancement driver-depletion load configuration, this combination is used for LSI applications such as the microprocessor (Sec. 9-11).

8-8 MOSFET LOGIC GATES

The inverters discussed in the preceding section may be converted to NOR (NAND) gates by using multiple drivers in parallel (series). We shall assume that

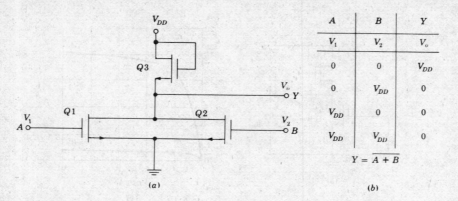

A	B	Y
V_1	V_2	V_o
0	0	V_{DD}
0	V_{DD}	0
V_{DD}	0	0
V_{DD}	V_{DD}	0

$$Y = \overline{A + B}$$

(a)   (b)

Figure 8-21 (a) MOSFET (positive) NOR gate and (b) voltage truth table and Boolean equation. (Remember that the voltage of the 0 state is 0 and of the 1 state is V_{DD}.)

$V_{ON} = 0$ and $V_T \ll V_{DD}$, so that $V_{DD} - V_T \approx V_{DD}$. Hence with NMOS devices a positive-logic system is obtained with the two logic levels $V(0) = 0$ and $V(1) = V_{DD}$.

Consider the NOR gate of Fig. 8-21a. If both inputs are at 0 V, both transistors $Q1$ and $Q2$ are OFF, and the drain current is 0, the drop across the load is 0 V and, hence, the output is V_{DD}. These values are entered in the first row of the voltage truth table of Fig. 8-21b. When either one (or both) of the inputs is V_{DD}, the corresponding FET is ON and the output is 0 V. These values are entered in the last three rows of the truth table. Clearly, the circuit performs the NOR operation $Y = \overline{A + B}$.

NAND Gate

The operation of the negative NAND gate of Fig. 8-22a can be understood if we realize that if either input V_1 or V_2 is at 0 V (the 0 state), the corresponding FET is OFF and the current is zero. Hence the voltage drop across the load FET is zero and the output $V_0 = V_{DD}$ (the 1 state). If both V_1 and V_2 are in the 1 state ($V_1 = V_2 = V_{DD}$), then both $Q1$ and $Q2$ are ON and the output is 0 V, or at the 0 state. These values are in agreement with the voltage truth table of Fig. 8-22b. If 1 is substituted for V_{DD} in Fig. 8-22b, this logic agrees with the truth table for a NAND gate, given by $Y = \overline{AB}$.

Note that only during one of the four possible input conditions is power delivered by the power supply for the NAND gate, whereas during three of the input possibilities power is taken by the NOR gate. Because of the high density of MOS devices on the same chip, it is important to minimize power consumption in LSI MOSFET systems. Typically, a three-input NAND gate uses about 16 mils^2 of chip area, whereas a single bipolar transistor needs about three times this area.

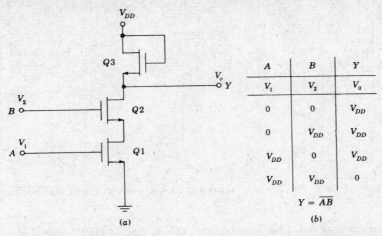

A	B	Y
V_1	V_2	V_0
0	0	V_{DD}
0	V_{DD}	V_{DD}
V_{DD}	0	V_{DD}
V_{DD}	V_{DD}	0

$$Y = \overline{AB}$$

(a)   (b)

Figure 8-22 (a) MOSFET (positive) NAND gate and (b) voltage truth table and Boolean expression.

The circuit of Fig. 8-21 may be considered to be a *negative* NAND gate, and that of Fig. 8-22 to be a *negative* NOR gate. A negative NAND gate may also be constructed with the topology of Fig. 8-22, provided that PMOS (*p*-channel) devices are used and that V_{DD} is replaced by $-V_{DD}$. Then $V(0) = 0$ and $V(1) = -V_{DD}$.

The gates discussed in this section are examples of direct-coupled transistor logic (DCTL), mentioned in Sec. 5-13. However, MOSFET DCTL circuits have none of the disadvantages (such as base-current "hogging") of bipolar DCTL gates.

Additional MOS Logic Functions

An AND gate is obtained by cascading a NAND and a NOT (inverter). An AND-OR-INVERT (AOI) is obtained by combining the series arrangement of the NAND with the parallel arrangement of the NOR. For example, the Boolean expression $Y = \overline{AB + CD}$ is satisfied (Prob. 8-17) with the circuit of Fig. 8-23, which corresponds to the TTL AOI gate of Fig. 6-2. All combinational logic in Chap. 6 can be implemented with MOS devices.

A 1-bit Memory

In Fig. 7-1 a 1-bit storage cell is shown constructed from two NOT gates with the output of the first connected to the input of the second and the output of the second tied to the input of the first. Such a latch built from MOSFETs is indicated in Fig. 8-24, where one NOT gate ($Q3$ and $Q4$) is interconnected as indicated above to the second NOT gate ($Q5$ and $Q6$). The two possible states of

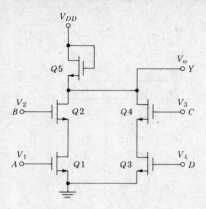

Figure 8-23 An AND-OR-INVERT gate ($Y = \overline{AB + CD}$).

this memory are $Q = 1$ (the 1 state) and $Q = 0$ (the 0 state). The input MOSFETs $Q1$ and $Q2$ allow a 1 or a 0 to be "written" or "stored" in the memory. For $S = 1$ and $R = 0$ a 1 is *set* into the cell, whereas for $R = 1$ and $S = 0$ the cell is *reset* or *cleared* to 0 (Prob. 8-22). Hence, the circuit of Fig. 8-24 is a bistable latch.

Assume that a MOSFET $Q7$ is added in series with $Q1$ and that $Q8$ is placed in series with $Q2$. If the inputs to $Q7$ and $Q8$ are excited by a pulse train Ck, then Fig. 8-24 augmented in this manner becomes a *clocked S-R* FLIP-FLOP analogous to Fig. 7-5 (Prob. 8-23).

A *J-K* master-slave FLIP-FLOP corresponding to Fig. 7-8 can be constructed entirely from MOSFETs. All the sequential systems (shift registers and counters) discussed in Chap. 7 may be implemented with MOS devices. A wide variety of digital MOSFET chips are available from the manufacturers listed in Appendix B-1.

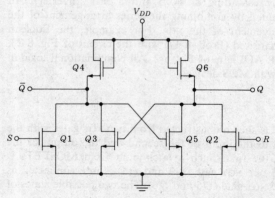

Figure 8-24 A 1-bit storage cell ($Q3$, $Q4$, $Q5$, and $Q6$). The state Q of the memory is 1 if $S = 1$ and $R = 0$, whereas $Q = 0$ if $S = 0$ and $R = 1$.

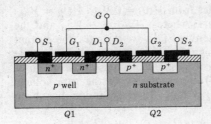

Figure 8-25 Cross section of complementary MOSFETs.

Q1   Q2

8-9 COMPLEMENTARY MOSFETs[15]

It is possible to construct p-channel and n-channel enhancement MOS devices on the same chip. Such devices are called *complementary* MOSFETs [abbreviated CMOS, COS/MOS (RCA), or McMOS (Motorola)]. The fabrication process starts with an n-type substrate into which a p-type "well" or "tub" is diffused. The NMOS $Q1$ is formed in this p-type well and the PMOS $Q2$ in the n-type body. The cross section of a CMOS is indicated in Fig. 8-25 (idealized, since lateral diffusion is ignored).

The CMOS Inverter

This circuit is shown in Fig. 8-26a. The driver is transistor $Q1$, which is the n-channel unit, and $Q2$ (the p-channel device) acts as the load. The two MOSFETs are in series with the drains tied together and the output taken from this node D. The two gates are connected and the input is applied to the

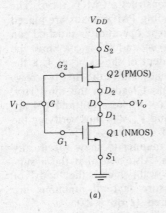

(a)

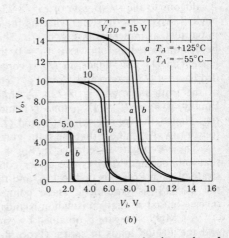

(b)

Figure 8-26 (a) Complementary MOS inverter. (b) The transfer characteristics for three values of V_{DD} and two values of temperature. *(Courtesy of Motorola Semiconductor Products, Inc.)*

common gate G. The input voltage V_i varies from $V(0) = 0$ V to $V(1) = V_{DD}$. When $V_i = 0$ then $V_{GS1} = 0$ and $Q1$ is OFF, whereas $V_{GS2} = -V_{DD}$ and the PMOS device $Q2$ is ON. However, since the two FETs are in series, the current in $Q2$ equals that in $Q1(I_D = 0)$, even though the gate voltage has a magnitude that nominally causes conduction. In other words, $Q2$ operates at the origin of the PMOS output characteristic (Fig. 8-8) corresponding to the gate voltage $V_{GS2} = -V_{DD}$. Since $V_{DS2} = 0$, then $V_o = V_{DD}$. Inverter action has been verified because $V_o = V(1)$ for $V_i = V(0)$.

Consider now that $V_i = V_{DD} = V_{GS1}$; then $Q1$ is ON but $Q2$ with $V_{GS2} = 0$ is OFF. Hence $I_{D1} = I_{D2} = 0$ and $Q1$ operates at the origin in Fig. 8-16b independent of V_{GS1}. Since the voltage across $Q1$ is zero, $V_o = 0$. Again the NOT property is obtained; $V_o = V(0)$ for $V_i = V(1)$. In either logic state $Q1$ or $Q2$ is OFF and the quiescent power dissipation is theoretically zero. In reality the standby power equals the product of the OFF leakage current and V_{DD} and equals a few nanowatts per gate.

The transfer characteristics of a CMOS INVERTER for three values of V_{DD} (5, 10, and 15 V) are shown in Fig. 8-26b. If the NMOS section had identical properties to the PMOS device, there would be complete symmetry between V_o and V_i. For $0 \leqslant V_i \leqslant V_T$, $Q1$ remains OFF and $V_o = V_{DD}$. For $V_i > V_T$, V_o decreases and falls rapidly to $V_{DD}/2$ for $V_i \approx V_{DD}/2$. For $V_{DD} \geqslant V_i > V_{DD} - V_T$, $Q2$ remains OFF, and $V_o = 0$. The transfer curves of Fig. 8-26b agree fairly well with these conclusions. Also note the relatively small temperature dependence of the inverter transfer characteristic.

CMOS Logic Functions

Analogous to the NAND gate of Fig. 8-22 we can construct a CMOS NAND. The NMOS drivers are in series whereas the corresponding PMOS loads are placed in parallel, as indicated in Fig. 8-27a. That NAND logic operation is satisfied can be demonstrated as follows: If $V_1 = 0$, then $Q1$ is OFF and $Q3$ is ON. Since the current in $Q1$ and, hence, in $Q3$ is zero (independent of the value of V_2), the drop across $Q3$ is zero and $V_o = V_{DD}$. This argument justifies the first two rows of the truth table of Fig. 8-22b. A similar argument leads to the third row. Finally, if $V_1 = V_2 = V_{DD}$, then $Q1$ and $Q2$ are ON whereas $Q3$ and $Q4$ are OFF. Hence, the voltage across $Q1$ and $Q2$ is zero and $V_o = 0$, thus verifying the fourth row. Again note that the static current is zero.

The NAND gate is redrawn in Fig. 8-27b in a manner to show specifically that enhancement devices are under consideration. Also note that the n substrate of each PMOS is tied to V_{DD} and that each p well (which is the "body" of the NMOS) is grounded. These connections ensure that all junctions are reverse-biased so that no isolation islands are required. (Prob. 8-27).

A CMOS NOR gate is obtained by connecting the NMOS drivers in parallel and the PMOS loads in series. (Prob. 8-30). From cross-coupled NOR gates a bistable latch can be constructed (Prob. 8-34). CMOS chips are now available to perform a wide variety of functions; for example, FLIP-FLOPS, counters, registers, decoders, etc.

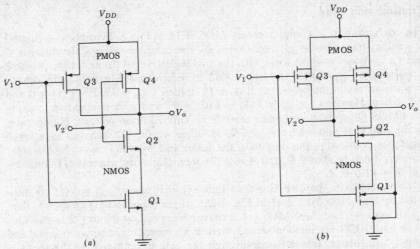

Figure 8-27 (*a*) A CMOS NAND gate. (*b*) An alternative drawing of this gate, showing the substrate connections. The *n* substrate of the PMOS is connected to the most positive supply voltage (V_{DD}) and the *p* well of the NMOS is tied to the most negative voltage (ground, for the above circuit).

CMOS Properties

The desirable features of CMOS gates are the following.

1. The quiescent (static) power dissipation is extremely small (a few nanowatts). Appreciable power is absorbed only when switching from one state to the other. At 1-MHz switching rate the dynamic power increases to a few milliwatts in a 50-pF load (about the same as in a Schottky low-power gate).
2. The noise immunity is better than 40 percent of V_{DD}. Note that, for $V_{DD} = 10$ V in the inverter transmission characteristic of Fig. 8-26*b*, a noise voltage of 4V superimposed upon $V(0) = 0$ reduces the output from $V(1) = 10$ V by only a fraction of a volt.
3. Propagation delay is about 50 ns per gate, allowing 10-MHz clock rates. Hence, CMOS is faster than MOS but slower than TTL logic.
4. The fan-out is very high, in excess of 50.
5. The logic swing is V_{DD}, independent of the fan-out.
6. A single power supply is required, and it can be a simple and inexpensive system (because of the small standby current).
7. If $V_{DD} = 5$ V, then CMOS is TTL compatible.
8. The temperature stability is excellent (Fig. 8-26*b*).

The above advantages are offset by the increased cost because of the additional processing steps required. Also the density of gates for a given chip area is decreased since CMOS requires that PMOS and NMOS devices appear in pairs. For example,[16] a four-input NAND gate requires about 50 mil^2 for CMOS, 30 mil^2 for TTL (low power, Schottky), 11 mil^2 for PMOS, and 5.6 mil^2 for I^{2}L (Sec. 9-13).

Transmission Gate

The configuration of complementary MOSFETs in Fig. 8-28a acts as a (digital or analog) transmission gate controlled by the complementary gate voltages C and $\bar{C}$. Consider positive logic with the two logic levels $V(0)$ and $V(1)$. Assume $C = 1$ so that $v_{G1} = V(1)$ and $v_{G2} = V(0)$, as indicated in Fig. 8-28b. (Ignore for now the values in the brackets.) If $A = V(1)$, then $v_{GS1} = V(1) - V(1) = 0$ and $Q1$ is OFF. However, $|v_{GS2}| = V(1) - V(0) > V_T$ and v_{GS2} is negative, causing the PMOS $Q2$ to conduct. Since there is no applied drain voltage, $Q2$ operates in the ohmic region where $v_{DS2} \approx 0$. In other words, $Q2$ behaves as a small resistance connecting the output to the input and $B = V(1) = A$. In a similar manner, it can be shown that, if $A = V(0)$, then $Q2$ is OFF whereas $Q1$ conducts and $B = V(0) = A$.

Now consider the case $C = 0$ so that $v_{G1} = V(0)$ and $v_{G2} = V(1)$, as indicated by the bracketed values in Fig. 8-28b. If the input is $V(1)$ as shown, then v_{GS1} is negative and the NMOS $Q1$ is OFF whereas $v_{GS2} = 0$ and $Q2$ is also OFF. Since both FETs are nonconducting there is an open circuit between input and output and, hence, transmission through the gate is inhibited. If the input is $V(0)$, it is again found that both devices are OFF. *In summary*, if $C = 1$, the gate transmits the input to the output so that $B = A$, whereas if $C = 0$ no transmission is possible.

The n well of the PMOS is tied to $V(1)$, the most positive voltage in the circuit and the p substrate of the NMOS is tied to $V(0)$, the most negative voltage. The symbol for the transmission gate is indicated in Fig. 8-28c. The control C is binary (it can have only one of two values) but the input v_i may be either digital, as discussed in the foregoing, or an analog signal whose instantaneous value must lie between $V(0)$ and $V(1)$. For example, if $V(0) = -5$ V and $V(1) = +5$ V, then a sinusoidal input signal (whose peak value does not exceed 5 V) appears at the output if $C = 1(v_{G1} = +5$ V) but is not transmitted through the gate if $C = 0$ ($v_{G1} = -5$ V) (Prob. 8-35).

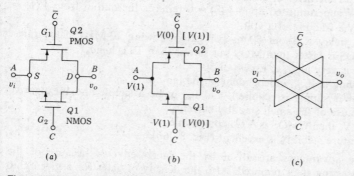

Figure 8-28 (*a*) A transmission gate using complementary MOSFETs. (*b*) The input voltage is considered first to be $V(1)$ and then $[V(0)]$. (*c*) The symbol for the transmission circuit.

REFERENCES

1. Hamilton, D. J., and W. G. Howard: "Basic Integrated Circuit Engineering," McGraw-Hill Book Company, New York 1975.
2. Carr, W. N., and J. P. Mize: "MOS/LSI Design and Applications," McGraw-Hill Book Company, New York, 1972.
3. Penney, W. M., and L. Lau, (eds.): "MOS Integrated Circuits," Van Nostrand Reinhold Company, New York, 1972.
4. Wallmark, J. T., and H. Johnson: "Field-Effect Transistors," Prentice-Hall Inc., Englewood Cliffs, N. J., 1966.
5. Sevin, L. J.: "Field-Effect Transistors," McGraw-Hill Book Company, New York, 1965.
   Yang, E. S.: "Fundamentals of Semiconductor Devices, " chaps. 7 and 8, Mc-Graw Hill Book Company, New York, 1978.
6. Millman, J., and H. Taub: "Pulse, Digital, and Switching Waveforms," sec. 17-20, McGraw-Hill Book Company, New York, 1965.
7. Ref. 4, p. 115.
8. Sevin, L. J., ref. 5, pp. 13–17.
9. Ref. 1, secs. 5-2 and 5-3.
10. Ref. 2, chap. 2.
    Ref. 3, chap. 3.
11. American Microsystems, Inc.: Guide to Standard MOS Products, 1975, pp. 2–4.
12. Allan, R.: Semiconductors: Toeing the (Microfine) Line, *IEEE Spectrum*, pp. 34 – 40, December, 1977.
13. Pashley, R., et al.: H-MOS Scales Traditional Devices to Higher Performance Level, *Electronics*, vol. 50, no. 17, pp. 94-99, August 18, 1977.
14. Ref. 1, chap. 14
    Ref. 2, chap. 4.
    Ref. 3, chap. 4.
15. Ref. 1, chap. 14
16. Texas Instruments, Inc.: "The Microprocessor Handbook," 1975, Fig. 3-26.

REVIEW QUESTIONS

8-1 (*a*) Sketch the basic structure of an *n*-channel junction field-effect transistor.
(*b*) Show the circuit symbol for the JFET.

8-2 (*a*) Draw a family of drain characteristics of an *n*-channel JFET.
(*b*) Explain the shape of these curves qualitatively.

8-3 How does the FET behave (*a*) for small values of $|V_{DS}|$? (*b*) for large $|V_{DS}|$?

8-4 (*a*) Define the *pinchoff voltage* V_P.
(*b*) Sketch the depletion region before and after pinch-off.

8-5 Sketch the cross-sectional view of an *n*-channel JFET.

8-6 Repeat Rev. 8-5 for a *p*-channel JFET.

8-7 Sketch the cross section of a *p*-channel enhancement MOSFET.

8-8 For the MOSFET in Rev. 8-7 draw (*a*) the drain characteristics and (*b*) the transfer curve.

8-9 Repeat Rev. 8-7*a* for an *n*-channel enhancement MOSFET.

8-10 List three advantages of *n*-channel over *p*-channel MOSFETs.

8-11 Repeat Rev. 8-7*a* for an *n*-channel depletion MOSFET.

8-12 Repeat Rev. 8-8 for an *n*-channel depletion MOSFET.

8-13 Describe an MNOS device.

8-14 (a) Describe the fabrication of a silicon-gate n-channel enhancement MOSFET.

8-15 (a) Describe the self-aligning property of a silicon-gate MOSFET.

(b) Why is this method of construction not possible with a metal gate.

8-16 Explain how it is possible to obtain depletion and enhancement MOSFETs on the same chip.

8-17 Explain how ion implantation is used to reduce the capacitances between gate and drain and also between gate and source.

8-18 (a) Draw the circuit of a MOS NOT circuit with the enhancement load operating in the saturation region.

(b) Obtain the volt-ampere characteristic of the load graphically.

8-19 For Rev. 8-18 sketch the transfer characteristic if (a) the two MOSFETs are identical; (b) the load FET has a much higher resistance.

8-20 Repeat Rev. 8-18 for a load which operates in the ohmic region.

8-21 Repeat Rev. 8-19 for an inverter with the load operating in the ohmic region.

8-22 Repeat Revs. 8-18 and 8-19 for a NOT gate with a depletion load.

8-23 Consider a MOSFET inverter with a load resistance much larger than the driver resistance. What is the high value of the output voltage and also the approximate value of the low output voltage for (a) a saturating load; (b) an ohmic load; and (c) a depletion load?

8-24 Sketch a two-input NOR gate and verify that it satisfies the Boolean NOR equation.

8-25 Repeat Rev. 8-24 for a two-input NAND gate.

8-26 Repeat Rev. 8-24 for an AND-OR-INVERT.

8-27 (a) Repeat Rev. 8-24 for an R-S latch.

(b) Verify that if $S = 1$ and $R = 0$, a one is set into the memory.

8-28 Sketch the cross section of complementary MOSFETs.

8-29 (a) Sketch the circuit of a CMOS inverter.

(b) Verify that this configuration satisfies the NOT operation.

(c) For $V_{DD} = 10$ V, $V_T = 2$ V, and matched FETs, sketch carefully the transfer characteristic.

8-30 Sketch the circuit of a two-input NAND CMOS gate and verify that it satisfies the Boolean NAND equation.

8-31 List five desirable properties of CMOS gates.

8-32 (a) Sketch the circuit of a transmission gate by using CMOS transistors.

(b) Explain the operation of this switch.

LARGE-SCALE INTEGRATION SYSTEMS

A chip containing over 1,000 components is referred to as an LSI and one with over 10,000 elements as a VLSI (very large-scale integration) system. There are reasonable expectations that in the 1980s chips with over one million components will be available commercially! This chapter describes large-system chips which are primarily "memories" of various types. The commercial importance of these microelectronic chips is evident from the fact that the estimated worldwide semiconductor memory consumption in 1978 exceeded 900 million dollars.†

Included are discussions of very long (over 1,000 bits) MOS shift-register serial memories, mask-programmable MOS read-only memories (ROMs with over 65,000 bits), erasable-programmable read-only memories (EPROMs with over 16,000 bits), programmable logic arrays, read/write memories (RAMs with over 65,000 bits), charge-coupled devices (CCDs with over a 65,000-bit memory), microcomputers, and integrated injection logic (I^2L).

9-1 DYNAMIC MOS SHIFT REGISTERS[1-3]

Very long shift registers (involving hundreds of bits) are impractical if built from FLIP-FLOPS as discussed in Sec. 7-4. Too much power is wasted and they are uneconomical because an excessive area of silicon is required. An alternative approach is to construct an LSI shift-register stage by cascading two dynamic

† Source: DATAQUIST, Inc.

MOS† inverters. A bit is stored temporarily by charging the parasitic capacitance between gate and substrate of a MOSFET. A dynamic NOT circuit is first described and then two such gates will be expanded into a 1-bit dynamic storage cell.

A Dynamic MOS Inverter

The circuit of Fig. 9-1 shows a dynamic MOS inverter which requires a clock waveform ϕ for proper operation. Positive logic using n-channel enhancement MOSFETs is assumed with a 0 state of 0 V and a 1 state of $V_{DD} = 10$ V. The capacitor C represents the parasitic capacitance (~ 0.5 pF) between the gate and substrate of the following MOS, fed by V_o.

When $\phi = 0$ V, then the gates of $Q2$ and $Q3$ are at 0 V and both these enhancement NMOS are OFF. The supply voltage is disconnected from the circuit and delivers essentially no power. When the clock is at 10 V, both $Q2$ and $Q3$ are ON and inversion of V_i takes place. For example, if $V_i = 0$ V, $Q1$ is OFF, C charges to V_{DD}‡ through $Q2$ in series with $Q3$, and $V_o = 10$ V. If, however, $V_i = 10$ V, $Q1$ is ON, C discharges to ground through $Q3$ and $Q1$, and $V_o = 0$ V. Note that $Q3$ is a bidirectional switch: Terminal 2 acts as the source when C charges to the supply voltage, whereas terminal 1 becomes the source when C discharges to ground.

The important features of MOSFETs for this dynamic inverter (and also for the shift register) are:

1. The MOS is a bidirectional switch.
2. The very high input resistance permits temporary data storage on the small gate-to-substrate capacitance of a MOS device.
3. The load FET may be turned off by a clock pulse to reduce power dissipation.

The inverter discussed above is called a *ratioed inverter*. The name derives from the fact that when the input is high and the clock is high, transistors $Q1$ and $Q2$ form a voltage divider between V_{DD} and ground. Therefore the output voltage V_o depends on the ratio of the ON resistance of $Q1$ and the effective load resistance of $Q2$ (typically, $< 1 : 5$). This ratio is related to the physical size of $Q1$ and $Q2$ and is often referred to as the *aspect ratio*.

Two-Phase Ratioed Memory Cell

Cascading two of the dynamic inverters of Fig. 9-1 allows each bit of information which is stored on the capacitance C of the first NOT gate to be transferred to the following inverter by applying a second clock pulse out of phase with the

---

† The following terms are used synonomously in this chapter: MOSFET, MOS, FET, MOST, NMOS, and transistor.

‡ It is assumed throughout this chapter that the threshold voltage V_T is much smaller than the supply voltage V_{DD} and that $V_{ON} = 0$ (Sec. 8-7).

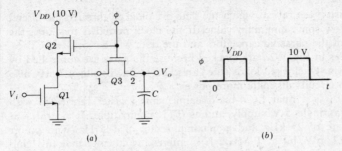

Figure 9-1 (*a*) Dynamic NMOS inverter. (*b*) Clock waveform ϕ.

first waveform. A typical MOS dynamic-shift-register stage is shown in Fig. 9-2*a* and the required two-phase clock waveforms are indicated in Fig. 9-2*b*. Each stage of the register requires six MOSFETs. The input V_i is the voltage on the gate capacitance C_1 of $Q1$, applied there by the previous stage (or by the input signal if this is the first stage of the shift register). When at $t = t_1$ the clock ϕ_1 goes positive (for NMOS devices), transistors $Q1$ and $Q2$ form an inverter and the bidirectional switch $Q3$ conducts. Hence, the complement of the level of C_1 is transferred to C_2. When ϕ_1 drops to 0 (at $t = t_2 +$), $Q2$ and $Q3$ are OFF and C_2 retains its charge as long as ϕ_1 remains at 0 V. However, at $t = t_3 +$, when $\phi_2 = V_{DD}$, then $Q4$ and $Q5$ act as an inverter and the switch $Q6$ is closed. Hence, the data stored on C_2 are inverted and deposited on C_3. The bit (a 1 or a 0) transferred to the output V_o is identical with that which was at the input V_i but delayed by an amount determined by the clock period. In other words the register stage in Fig. 9-2*a* is a 1-bit delay line. The combination $Q1Q2Q3$ can be called the *master inverter*, and $Q4Q5Q6$ the *slave section*. To retain data

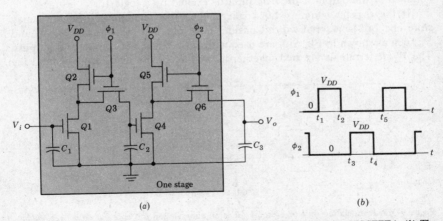

Figure 9-2 (*a*) A two-phase ratioed dynamic NMOS shift-register stage (six MOSFETs). (*b*) The two-phase clock waveforms ϕ_1 and ϕ_2.

stored in the register, the rate at which the data are clocked through the circuit must not fall below some minimum value. If the clock period is too long, the charge will leak of the parasitic capacitors and the data will be lost.

The load FETs in Fig. 9-2a are clocked because the gates are controlled by the clock waveform. Unclocked loads (the gates tied to fixed voltages) may also be used, but such circuits dissipate more power.

The Intel 2401 is a dual 1,024-bit dynamic shift register fabricated with NMOS. It uses a single 5-V supply and is TTL compatible. It operates at minimum data rate of 25 kHz and a maximum rate of 1 MHz, with power dissipation of 0.12 mW/bit at 1 MHz. It is interesting to note that this chip contains $2 \times 1,024 \times 6 = 12,288$ MOSFETs, exclusive of the control circuitry needed to convert it into a recirculating memory (Fig. 9-3).

Applications

Typical applications for MOS shift registers are as serial memories for calculators, cathode-ray tube displays, or communication equipment, as refresh or buffer memories, and as delay lines. A serial dynamic circulating shift-register memory is drawn in Fig. 9-3. The output of the register is returned to its input through an AND-OR combination as indicated. If the *write but not read* mode $W/\overline{R}$ is in the 1 state, the digital data at the *input* terminal are fed into the register. After a *clock* pulse cycle each bit is shifted to the right into the following stage, as explained in connection with Fig. 9-2. When the desired number of bits are entered sequentially into the register, the *recirculating* mode is commenced by changing $W/\overline{R}$ to the 0 state. In this mode further data are inhibited from entering the register and the bits stored in the memory are recirculated from the output back into the input of the shift register in synchronism with the clock waveform. Nondestructive reading of the data train is obtained at the output if the *read* input is excited by a logic 1.

If the register contains 1,024 stages, then this recirculating memory may store one 1,024-bit serial word. Consider now that four systems, S_0, S_1, S_2, and S_3, such as shown in Fig. 9-3, are used with *independent* data inputs and outputs. The $W/\overline{R}$ terminals are tied together as well as all the read terminals and the

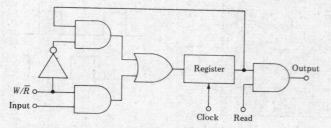

Figure 9-3 A recirculating shift register. ($W/\overline{R}$ is an abbreviation for *write but do not recirculate*). The AND, OR, and NOT gates are fabricated on the same chip with the register stages.

same clock synchronizes all systems. The resulting configuration is a serial memory which could be considered as storing 1,024 words, each containing 4 bits. All 4 bits in a particular word appear simultaneously: the LSB at the output of S_0 and the MSB at the output of S_3. One clock period later another 4-bit word can be read. To expand the system to n-bit words, it is necessary to use n-recirculating shift registers. If more words are required, then longer shift registers must be used.

If the desired use has been made of the data circulating in the memory of Fig. 9-3, then the W/R input is changed to a logic 1. This inhibits the bits from the last stage of the shift register from entering the first stage. In other words, the content of the memory is *erased* and new data may be simultaneously entered into the register.

Static MOS Shift Register

A "static" shift register is dc stable and can operate without a minimum clock rate. That is, it can store data indefinitely provided that power is supplied to the circuit. However, static shift-register cells are larger than the dynamic cells and consume more power. Consequently they are seldom used.

9-2 RATIOLESS SHIFT-REGISTER STAGES

It is pointed out in Sec. 9-1 that the load FET $Q2$ in Fig. 9-2 must have a much higher resistance than the driver $Q1$ in order for the low-state voltage V_{ON} to be close to zero. In Sec. 8-7 we emphasize that the FET resistance is proportional to L/w. Hence, $Q2$ must have a much larger channel length L and smaller width w than $Q1$. Consequently the inverter occupies more than the minimum possible area. Also, since the parasitic storage capacitance is charged through the load $Q2$ during a portion of the cycle, the high resistance of $Q2$ limits the speed of operation of the register. Both of these difficulties may be avoided by using a dynamic *ratioless* inverter, as indicated in Fig. 9-4a (where $Q1$ and $Q2$ may have identical geometries). Note that no power supply (dc) voltage is used in this inverter. The clock pulse ϕ (Fig. 9-1b for NMOS devices) must supply the required energy to this circuit and the power dissipation is proportional to the clocking frequency.

To understand the operation of the ratioless inverter consider first the case where $V_i = 0$. Then during the pulse the situation is as pictured in Fig. 9-4b. Since the gate voltage of $Q1$ is 0 and that of $Q2$ is V_{DD}, then (for NMOS enhancement devices) $Q1$ is OFF and $Q2$ is ON. Therefore C charges through $Q2$ to V_{DD}. At the end of the pulse ϕ falls to 0 and both MOSFETs remain OFF. Hence, with $V_i = 0$ (logic 0), the output $V_o = V_{DD}$ (logic 1) and an inversion has taken place.

Consider now that $V_i = V_{DD}$ and that $\phi = V_{DD}$ as shown in Fig. 9-4c. Both MOSFETs are ON and deliver current to C as indicated. Hence C is quickly

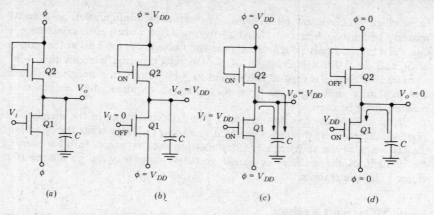

Figure 9-4 (a) A ratioless dynamic inverter using NMOS. (b) Input = logic 0 and $\phi = V_{DD}$. (c) Input = logic 1 and $\phi = V_{DD}$ (during pulse). (d) Input remains at logic 1 and $\phi = 0$ (after the pulse terminates).

charged to V_{DD}. Since $V_o = V_i = V_{DD}$, there is no inversion during the pulse. However, at the termination of the pulse when the clock voltage returns to 0, we have the situation depicted in Fig. 9-4d. Now the gate G_2 of $Q2$ is at 0 and $Q2$ is OFF, whereas G_1 of $Q1$ is at V_{DD} and $Q1$ is ON. Consequently C discharges to 0 V through $Q1$. Hence, shortly after the pulse ends, $V_o = 0$ while $V_i = V_{DD}$ indicating that a logical inversion has taken place.

Two-Phase Ratioless Dynamic Register Cell

If we cascade two inverters of the type shown in Fig. 9-4a through bidirectional transmission gates, the ratioless shift-register stage of Fig. 9-5 is obtained. The first inverter is powered by phase ϕ_1 and the second by phase ϕ_2, where these clock waveforms are drawn in Fig. 9-5b. At the beginning ($t = t_1 +$) of pulse ϕ_1 the switch $Q0$ closes and the voltage across C_0 (the input voltage to $Q1$) equals the input level V_i. By the inverter action described in connection with Fig. 9-4, the voltage across C_1 after the end of the pulse ϕ_1 (at $t = t_2 +$) corresponds to the complementary logic state of V_i. Since ϕ_1 is now at its low level, $Q0$ opens and V_i is retained on C_0 until the end of the period of ϕ_1 (at $t = t_5$).

At $t = t_3 +$ the second waveform ϕ_2 goes to its high level V_{DD} allowing transmission through $Q3$ and effectively placing C_1 and C_2 in parallel. If at $t = t_3 -$ the voltage on $C_1(C_2)$ is $V_1(V_2)$, then at $t = t_3 +$ the voltage V on C_2 (which must be the same as that on C_1) is found in Prob. 9-5 to be

$$V = \frac{C_1 V_1 + C_2 V_2}{C_1 + C_2} \tag{9-1}$$

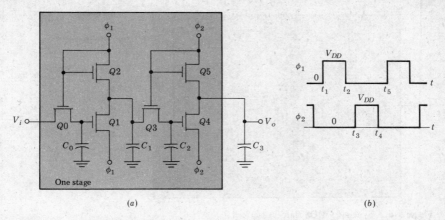

(a)   (b)

Figure 9-5 (a) A two-phase ratioless dynamic NMOS shift-register stage (six MOSFETs). (b) The two-phase clock waveforms ϕ_1 and ϕ_2.

If $C_1 \gg C_2$, note from Eq. (9-1) that $V \approx V_1$. In other words, pulse ϕ_2 causes the output voltage (across C_1) of the first inverter to appear at the input (across C_2) of the second NOT gate. Finally, by the inverter action described above, at the end of the pulse ϕ_2 (at $t = t_4 +$ and until $t = t_5$) the logic level V_o across C_3 is the complement of that across C_2, which, in turn, is the complement of that across C_0. Clearly, in one period of the clock the input level V_i has shifted through the stage to the output V_o, as it should in a 1-bit delay line or 1-bit shift register.

No dc power supply is used in Fig. 9-5, but the clocking waveforms must be capable of furnishing the heavy capacitive currents. Also in order to ensure that C_1 be much larger than C_2, additional area must be added to the chip for C_1. We can reduce the loading on the clock drivers by adding another transistor to each inverter as in Prob. 9-6. This modification results in an eight-MOSFET stage. A number of four-phase ratioless shift registers capable of operation at high speed are described in the literature.[1-3] Because of the large amount of chip area required for two-phase ratioless shift registers and the additional complications of four-phase clock drivers these systems are seldom used.

A Dynamic CMOS Shift-Register Stage

The static CMOS inverter indicated in Fig. 8-26 is discussed in Sec. 8-9. The CMOS transmission gate is shown in Fig. 8-28 and its operation is also explained in Sec. 8-9. We find in the foregoing discussion that interposing bidirectional transmission gates between inverters results in a dynamic shift register. Such a configuration using CMOS is indicated in Fig. 9-6. The transmission gates are labeled $T1$ and $T2$ and are controlled by complementary

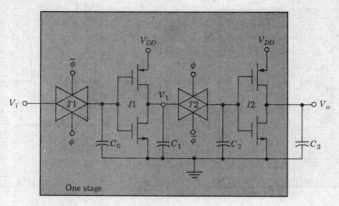

Figure 9-6 A dynamic CMOS shift-register cell.

clocks ϕ and $\bar{\phi}$. When $\phi = V_{DD}$ (corresponding to $C = V(1)$ in Fig. 8-28), $T1$ conducts whereas $T2$ acts as an open circuit. The CMOS inverters are marked $I1$ and $I2$.

The explanation of the operation of the register stage of Fig. 9-6 closely parallels that given in connection with Fig. 9-5. When $\phi = V_{DD}$ (logic 1), then $T1$ transmits and the input V_i appears across C_0. Because of the inverter action of $I1$, the complement of V_i appears across C_1 ($V_1 = \bar{V}_i$). On the next half cycle $\phi = 0$, $T1$ opens, C_0 retains the voltage V_i, and V_1 remains at the $\bar{V}_i$. Also when $\phi = 0$, $T2$ closes, putting C_2 in parallel with C_1, and $I2$ causes the voltage across C_3 to be the complement of that across C_2. Consequently at the end of a complete cycle $V_o = \bar{V}_1 = V_i$ and we have demonstrated that this cell behaves as a 1-bit delay line or register.

The CMOS stage consists of eight MOSFETs (or four complementary pairs). The power dissipation is very low since there are no dc current paths; power is used only for the transient charging of capacitors. From the explanation of the circuit given in the foregoing, it should be clear that the output voltage does not depend on the ratio of the resistances of any of the devices and, hence, ratioless operation is involved.

Another type of shift register (the charge-coupled device, CCD) is discussed in Sec. 9-8.

9-3 MOS READ-ONLY MEMORY[4]

The ROM is discussed in Sec. 6-9, where it is seen (Fig. 6-25) to consist of a decoder, followed by an encoder (memory matrix). Consider, for example, a 10-bit input code, resulting in $2^{10} = 1,024$ word lines, and with 4 bits per output code. The memory matrix for this system consists of $1,024 \times 4$ intersections, as indicated schematically in Fig. 9-7. The code conversion to be performed by the

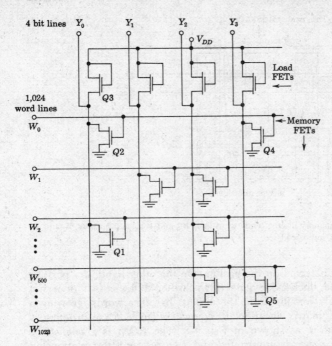

Figure 9-7 NMOS read-only-memory encoder. (Although there are a total of 1,024 word lines present, only 5 of these are indicated.)

ROM is permanently programmed during the fabrication process by using a custom-designed mask so as to construct or omit a MOS transistor at each matrix intersection. Such an encoder is indicated in Fig. 9-7, which shows how the memory FETs are connected between *word* and *bit* lines.

In Sec. 6-9 it is demonstrated that the relationship between the output bits Y and the word lines W is satisfied by the logic OR function. Consider, for example, that it is required by the desired code conversion that

$$\overline{Y}_0 = W_0 + W_2 \qquad\qquad \overline{Y}_1 = W_1$$

$$\overline{Y}_2 = W_1 + W_2 + W_{500} \qquad \overline{Y}_3 = W_0 + W_{500} \tag{9-2}$$

These relationships are satisfied by the connections in Fig. 9-7. The NOR gate for Y_0 of Eq. (9-2) is precisely that drawn in Fig. 8-21, with $Q3$ as the load FET and with signals W_0 and W_2 applied to the gates of $Q2$ and $Q1$, respectively.

The presence or absence of a MOS memory cell at a matrix intersection is determined during fabrication in the oxide-gate mask steps (Figs. 8-6b and c). If the MOSFET has a normal thin-oxide gate, its threshold voltage V_T is low; if the gate oxide is thick, then V_T is high. In response to a positive pulse on the word line, the low-threshold device will conduct and a logic 0 (because of

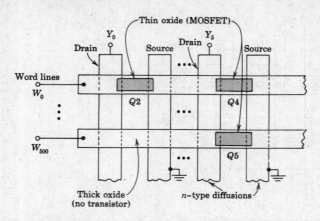

Figure 9-8 MOS read-only-memory matrix. (Only word lines W_0 and W_{500} and bit lines Y_0 and Y_3 corresponding to Fig. 9-7 are indicated.)

inverter action) will be detected on the bit line. On the other hand, if a positive pulse is applied to the thick-oxide gate (high-threshold device), it does not conduct; it is effectively missing from the circuit. In other words, growing a thick-oxide gate at a matrix location is equivalent to *not* constructing a MOSFET at this position, as shown in Fig. 9-8. The ROM is a *nonvolatile memory*, because if the power is interrupted and then restored, the relationship between input and output programmed into the ROM is not lost.

In a static ROM no clocks are needed and the output exists as long as the input address remains valid. Such ROMs are available in sizes from 1 kb to 65 kb (from Intel, Mostek, Texas Instruments, and other manufacturers), usually with four or eight output bits. Access times are in the range 0.5–1.5 μs (about an order of magnitude longer than for bipolar ROMs) and power dissipations extend from 0.2 to 1 W.

An example of a static ROM is the 16-kb (2,048 × 8) Intel 2316 in a 24-pin dual-in-line package. It uses *n*-channel MOSFETs (NMOS) and operates from a single 5-V supply so that inputs and outputs are TTL compatible. The two-dimensional addressing discussed in Sec. 6-10 is also used with LSI ROMs. The 2316 ROM is organized as indicated in Fig. 9-9. Eleven address bits are needed for 2,048 words. Note that seven of these address inputs (A_4 through A_{10}) are used for the X decoder to obtain 128 rows. The memory matrix is square with 128 columns, and these must be multiplexed onto 8 outputs (O_0 through O_7). This is accomplished with eight (16-to-1) selectors, using four-column-address inputs (A_0 through A_3). This organization is an extension of that shown in Fig. 6-26 for a 2-kb ROM (512 × 4), although the two diagrams (Figs. 9-9 and 6-26) are drawn somewhat differently.

The decoder in a static MOS ROM (Fig. 9-9) contains NAND gates which are static. Power dissipation as a result is relatively high. A dynamic ROM uses

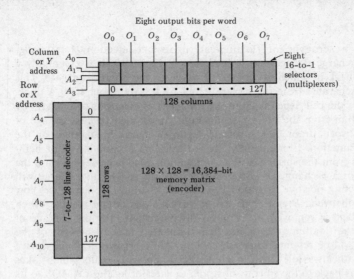

Figure 9-9 The organization of the Intel 2316A 16-kb ROM (2,048 words by 8 bits). Input and output buffers and chip-select circuitry are not indicated.

clocked or dynamic inverters in the decoder and/or load FETs and requires a minimum clock rate, since otherwise the information is lost. However, its power dissipation is lower than for a static ROM. Most commercial ROMs are static because of the advantages of requiring no clocks and of giving an output which remains valid as long as the input address is applied.

Increasing the number of bits per word (called *word expansion*) or increasing the number of words, with the same number of bits per word (called *address expansion*), is attained by interconnecting several ROM packages in the manner discussed in Sec. 6-10 for bipolar chips.

9-4 ERASABLE-PROGRAMMABLE READ-ONLY MEMORY (EPROM)

The bipolar ROM is programmed in the field by applying a sufficiently high voltage to blow out the fusible link in the ROM cell (Sec. 6-9). Clearly, the program of such a PROM cannot be changed because a burned out fuse is not repairable. This difficulty is overcome in a MOS ROM by using a novel device (the FAMOS described in the following) which can be programmed to act as a conductor, but which may be returned to its nonconducting state, if desired. In other words, such a MOS ROM is *reprogrammable*, a distinct advantage in a prototype system where the program is not yet fixed or in equipment requiring periodic updating of a ROM program.

The FAMOS Device[5]

The FAMOS (an acronym for *f*loating-gate *a*valanche-injection *m*etal-*o*xide-*s*emiconductor) charge-storage unit is indicated in Fig. 9-10. It is essentially a *p*-channel MOSFET in which no contact is made to the polysilicon gate. This floating gate is separated from the substrate by a SiO_2 layer about 1000 Å thick. The operation of the cell depends upon the transport of electrons to the gate by avalanche injection from the drain *p-n* junction. An applied reverse voltage of about 30 V between drain and substrate (source) causes breakdown of the drain-substrate junction. Consequently, there results an accumulation of high-energy electrons from the surface of the *p-n* junction avalanche region onto the floating gate. For a *p*-channel FAMOS device this stored negative charge will induce a conducting positive inversion layer in the channel. Hence a very low resistance exists between *S* and *D* so that a virtual short circuit results. When the applied voltage is removed, there is no discharge path for the leakage of electrons from the floating gate; it has been estimated that 70 percent of the induced charge will be retained *after 10 years* at 125 °C. However, if the device is illuminated with ultraviolet light, a photoelectric current results which takes the gate electrons back to the silicon substrate, thus restoring the FAMOS to its initial nonconducting condition (an open circuit between *S* and *D*).

The manner in which the device in Fig. 9-10*a* is inserted into a ROM encoder is illustrated in Fig. 9-10*b*. Between *every* intersection of the matrix a MOSFET is inserted with its gate connected to the word line and its drain tied to the bit line. The source of this memory FET is connected to the drain of the FAMOS unit whose source is grounded and whose gate is, of course, floating. In other words, in series with *each* MOSFET in the memory there is a FAMOS

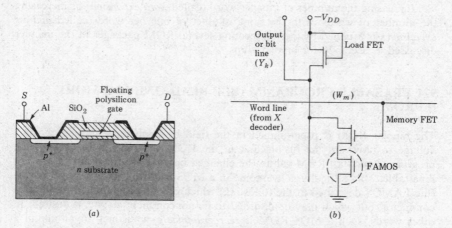

Figure 9-10 (*a*) A cross section of the *p*-channel FAMOS charge-storage cell. (*b*) The application of this cell in a ROM encoder.

device which acts as an open connection unless programmed to be conducting. For example, if the program calls for a connection between the mth word line W_m and the kth output line Y_k, then W_m is selected by the proper X address and -30 V is applied to Y_k. As explained above, the resulting avalanche breakdown in the FAMOS structure places electrons on the floating gate, rendering the channel conducting, and thereby connecting the source of the series MOSFET to ground. On the other hand, if no voltage is applied to Y_k when W_m is decoded, the FAMOS cell in Fig. 9-10b acts as an open circuit and the memory FET is effectively out of the circuit.

The Intel 1702 2-kb (256 $\times$ 8) EPROM uses PMOS devices. Initially, and after each erasure, all 2,048 bits are in the logic-0 state (output low = $-V_{DD}$). The chip is programmed by selectively introducing a logic 1 (output high = 0 V) by avalanching so that the FAMOS unit is conducting, as described in the foregoing. The 24-pin package has a transparent lid which allows the chip to be exposed to ultraviolet (UV) light to erase the bit pattern. Incidentally, such a reprogrammable ROM is referred to as a *read-mostly memory*.

The Stacked-Gate Memory Cell[6]

Engineers at Intel Corporation ingeniously have combined the memory FET and the FAMOS devices in Fig. 9-10b into a single transistor structure. As shown schematically in Fig. 9-11a, a second polysilicon gate (called the *select gate*) is added above the floating gate. The top gate (available externally) controls the *programming* (and also the *read*) *operations*, while the bottom gate stores charge almost indefinitely. To program the stacked-gate cell a relatively high voltage of 25 V (compared with the 5-V read voltage) is applied to the select gate and to the drain simultaneously. As a result of this excitation, some

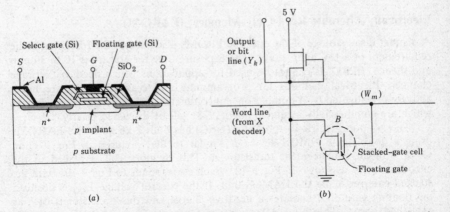

Figure 9-11 (*a*) A cross section of the NMOS stacked-gate structure. The total surface area is only about 1 mil². (*b*) The application of this cell in a RDM encoder. A depletion load is used.

of the electrons in the inversion layer acquire enough energy to be transported from the channel via the thin oxide (1000 Å) to the floating gate. This captured negative charge leaves the channel more positive and, hence, less conducting (since this is an NMOS device). As a result, the threshold voltage at the top gate is higher for a charged cell than for one which has no charge on the floating gate. Consequently, whereas a 5-V select-gate voltage turns on the enhancement MOSFET if the floating gate is uncharged, this double-gate cell remains non-conducting (for 5 V on the control gate) if the floating gate has stored electrons. The stacked-gate cell is positioned in the encoder matrix as indicated in Fig. 9-11b, corresponding to Fig. 9-10b for the FAMOS device.

The Intel 2716 16-kb (2,048 × 8) EPROM uses the stacked-gate memory cell and operates from a single 5-V power supply (although 25 V is also required for programming). Including the decode, address, drive, and sense circuitry the chip area is remarkably small (175 mil × 175 mil). The access time is 450 ns and the power dissipation is 525 mW. The time for programming all words is only 100 s. The erase time (with ultraviolet light) is 10 to 30 min. The 2716 package is organized into two 64 × 128 cell matrices. Since the 2716 EPROM is pin-compatible with the 2316 mask-programmed 16-kb ROM, designers can debug systems with the 2716 package and after the program is firm, they can order (for high-volume production) 2316 chips having the fixed desired program. Since these chips use a single 5-V supply, they are especially useful in designing the new 5-V microprocessors.

The Intel 2708 8-kb (1024 × 8) EPROM uses NMOS stacked-gate cells. This chip was available in 1975 and the 2716 in 1976. The 2708 uses three supply voltages +12 V, +5 V, and −5 V. Two 2708 IC's can now be replaced by one 2716 package, using a single 5-V supply.

In 1978 Texas Instruments produced a 32-kb EPROM and Intel expects to have available a 65-kb EPROM in 1979.

Electrically Alterable Read-Only Memory[7] (EAROM)

A distinct disadvantage of the erasable PROMs discussed in the foregoing is the requirement of a UV light source. The exposure time for erasure is 10 to 30 min and, hence, EPROMs cannot be used for applications which need fast storage changes. Electrical methods for programming and erasing ROMs are being actively investigated by a number of semiconductor companies, and two devices which are commercially available now (1978) will be discussed briefly.

The Nippon Electric Corporation (NEC) has built 2-kb and 8-kb EAROMs using a double-gate NMOS device B similar to that pictured in Fig. 9-11 in series with an enhancement transistor A. This memory cell is placed in the encoder matrix, as shown in Fig. 9-12, which corresponds to Fig. 9-10b with the stacked-gate replacing the FAMOS unit. If the control voltage V_{CG} is positive, the floating gate of B acquires a negative charge and this transistor is OFF, as explained above. This bit has been programmed to be logic 1. To erase the

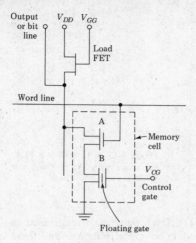

Figure 9-12 The memory cell in an EAROM consists of a conventional MOSFET in series with a double-gate n-channel device.

memory cell, V_{CG} is made sufficiently negative to induce positive charge to accumulate on the floating gate. Hence, the n-channel of B is now made conducting, and the memory cell stores a logic 0. The NEC EAROM μPD458 has an access time of about 1 μs and a programming or erasure time of approximately 1 s.

Another electrically alterable ROM[8] is the MNOS device (described in Sec. 8-6). This memory cell uses an extremely thin (~ 25 Å) silicon dioxide layer under a much thicker silicon nitride layer (~ 600 Å). Upon application of a high voltage (± 25 V) to the gate, tunneling (a quantum-mechanical process) takes place and, depending upon the polarity of the applied voltage, electrons or holes are trapped at the nitride-oxide interface (instead of in the floating gate of the FAMOS device). NCR, General Instrument Corporation, and Nitron have pioneered EAROMs using the aforementioned nitride cell, but most companies find it almost impossible to build such memories reliably. These devices are also designated EEROMs, which is an acronym for *e*lectrically *e*rasable *r*ead-*o*nly *m*emories.

9-5 PROGRAMMABLE LOGIC ARRAY (PLA)[9, 10]

In Sec. 9-3 we discuss the (Intel 2316A) 16-kb ($2,048 \times 8$) ROM, which has $M = 11$ inputs and $N = 8$ outputs. For every increase in M by 1, the number of bits is doubled. For example, if $M = 16$ and N remains at 8, then the number of words is $2^{16} = 2^5 \times 2^{11} = 32 \times 2,048 = 65,536$ and the number of bits is $65,536 \times 8 = 524,288$. This tremendous number of bits is not feasible in a single ROM chip and thirty-two 16-kb ROM packages are required, interconnected for address expansion (Sec. 6-10). This system implements $N = 8$ combinational

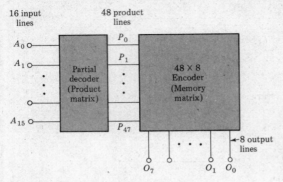

16 input lines   48 product lines

Figure 9-13 A $16 \times 48 \times 8$ PLA. (Compare with Fig. 6-25b for a ROM.)

logic equations in $M = 16$ variables (Sec. 6-11). Each equation is expressed in the sum-of-products canonical form. Each product contains 16 factors and there are a total of 65,536 product terms (or words).

Consider a subset of the foregoing ROM system. The number of inputs and outputs are unchanged ($M = 16$ and $N = 8$), but each sum contains only, say, 48 product terms instead of 65,536. These are called *partial products* of the input variables, because each product does *not* contain all 16 inputs (or the complements of these inputs). Such a system of combinational logic is referred to as a $16 \times 48 \times 8$ *programmable logic array* (PLA) and designates that there are 16 inputs, 8 outputs, and a total of 48 partial products, as indicated in Fig. 9-13.

The decoding array in the PLA of Fig. 9-13 consists of 48 AND gates. The output of each AND is a partial product term and the number of inputs to any gate is often small; the maximum equals the number of input data bits (16). The encoder matrix consists of eight OR gates whose outputs are the eight output functions of the PLA. The maximum number of inputs to any OR gate equals the number of product terms and is usually much smaller than this number (48). As an example, consider that two combinational logic equations (out of the eight implemented by the PLA of Fig. 9-13) are

$$O_0 = \bar{A}_3 A_0 + \bar{A}_9 A_4 \bar{A}_1 + A_{11} \bar{A}_{10} \bar{A}_7 A_3 + A_{13} + \bar{A}_{15} A_{14} \qquad (9\text{-}3)$$

$$O_1 = A_5 \bar{A}_4 A_0 + \bar{A}_9 A_4 \bar{A}_1 + A_{12} A_6 \qquad (9\text{-}4)$$

These two outputs use seven product terms because $\bar{A}_9 A_4 \bar{A}_1$ is common to both equations. The remaining $48 - 7 = 41$ terms are available for the other six outputs O_2 through O_7. One of the AND gates has one input, three have two inputs, one has three inputs, and one has four inputs. The OR gate for O_0 has five inputs and that for O_1 has three inputs.

The truth table for the foregoing equations is given in Table 9-1. Positive logic is used and each row represents one product term. If a data input is true (false), then a logic 1 (0) appears in the column representing this input. If a variable is missing in a product, then an X ("don't care") is placed in the column

Table 9-1 PLA truth table for Eqs. (9-3) and (9-4)

term	15	14	13	12	11	10	9	8	7	6	5	4	3	2	1	0	1	0
						Inputs											Outputs	
0	X	X	X	X	X	X	X	X	X	X	X	X	0	X	X	1	0	1
1	X	X	X	X	X	X	0	X	X	X	X	1	X	X	0	X	1	1
2	X	X	X	X	1	0	X	X	0	X	X	X	1	X	X	X	0	1
3	X	X	1	X	X	X	X	X	X	X	X	X	X	X	X	X	0	1
4	0	1	X	X	X	X	X	X	X	X	X	X	X	X	X	X	0	1
5	X	X	X	X	X	X	X	X	X	X	1	0	X	X	X	1	1	0
6	X	X	X	1	X	X	X	X	X	1	X	X	X	X	X	X	1	0

for the data input and product (row) under consideration. If the output O_k is 1 (0), it signifies that the product term represented by the row under consideration is present (absent) in the kth output function.

Programming a PLA

Table 9-1 (extended to cover 8 output functions and up to 48 product terms) becomes a *program table* for the $16 \times 48 \times 8$ PLA. The customer fills out this table to satisfy his combinational logic functions and the manufacturer makes a mask for the metalization so as to obtain the proper connections. For example, if an X appears at the pth data input and the rth product term, the aluminum connection from the pth input (nor from its complement) is *not* made to the rth AND gate. On the other hand, if the pth input is 1(0) for the rth term, the metalization is made from A_p ($\overline{A_p}$) to the rth AND gate. Similarly, if the kth output is 1(0) for the mth product line, aluminization connects (does not connect) the mth term to the input of the kth OR gate. An example of a mask-programmable logic array is the 6775 of Monolithic Memories, Inc. (or the DM 8575 of National Semiconductor) which has 14 inputs, 8 outputs, and 96 product terms ($14 \times 96 \times 8$). It is TTL compatible and the access time is approximately 50 ns.

    Field-programmable logic arrays[11] (FPLAs) are also available, such as the Signetics 82S100 ($16 \times 48 \times 8$), drawn in block-diagram form in Fig. 9-13. This bipolar chip uses (Schottky) diode AND gates (Fig. 5-5b) and emitter-follower OR gates (Fig. 6-24). This FPLA is indicated in Fig. 9-14, where each symbol $\times$ represents a Nichrome link which acts as a fuse. This system is programmed in the field by selectively blowing fuses so as to break the connections required in order to satisfy the program table, as explained in the preceding paragraph. The complement $\overline{A_k}$ of A_k is obtained by an inverter (not shown in Fig. 9-14). The complement of each output is also available on the chip, but this circuit is omitted from the figure for simplicity. The package is TTL compatible with either tristate or open-collector outputs (Fig. 5-24) and also includes chip-enable control.

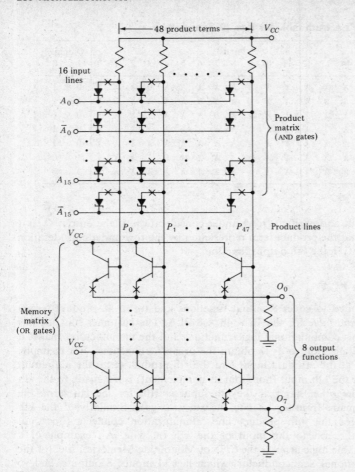

Figure 9-14 The 82S100 FPLA (16 × 48 × 8). *(Courtesy of Signetics.) Note:* The × represents a fusible link.

The PLA (or FPLA) is designed for the implementation of complex logic functions. A PLA can handle more data inputs and is more economical than a ROM. It is useful for the same type of applications as the ROM (Sec. 6-11), provided that the number of product terms needed is a small fraction of the total input combinations possible.

9-6 RANDOM-ACCESS MEMORY[12-14](RAM)

In computer, information processing, and control systems it is necessary to store digital data and to retrieve them as desired. In a semiconductor memory an

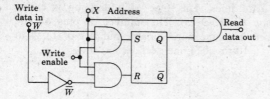

Figure 9-15 A 1-bit read/write memory.

array of storage cells is used, each cell holding 1 bit of data. In such a memory, as contrasted with a shift register, information can be put *randomly* into, or taken out of, each storage element as required. Hence, this system is called a *random-access memory* and abbreviated RAM. Since each bit can be read out of, or written into, each cell, the system is also called a *read/write* memory, to distinguish it from a read-only memory (ROM).†

Magnetic cores have been used as storage elements for many years but are rapidly being replaced by semiconductor memories. Monolithic RAMs are constructed by using microelectronic technology, and they employ either bipolar or MOS transistors for the storage and supporting circuits. One basic static storage device is the latch, or FLIP-FLOP, discussed in Sec. 7-1. Some of the advantages of semiconductor over core memories are lower cost, smaller size, lower power consumption, faster data access, and nondestructive reading of the array. On the other hand, RAMs have the disadvantage of being *volatile*, which means that all stored information is lost when the power supply fails.

The access time is the same for any bit in a RAM regardless of its location in the matrix. On the other hand, in a recirculating memory (the shift register arrangement of Fig. 9-3), the access time to any particular bit depends upon the position of the bit at the moment of accessing.

Linear Selection

To understand how the RAM operates we examine the simple 1-bit S-R FLIP-FLOP circuit shown in Fig. 9-15, with data input and output lines. From the figure we see that to read data out of or to write data into the cell, it is necessary to excite the *address line* ($X = 1$). To perform the write operation, the *write enable line* must also be excited. If the write input is a logic 1(0), then $S = 1(0)$ and $R = 0(1)$. Hence $Q = 1(0)$, and the data read out is 1(0), corresponding to that written in.

Suppose that we wish to have an 8-kb RAM organized as 1,024 words of 8 bits each. This system requires 10 addresses, 8 data in and 8 data out lines. A total of $1,024 \times 8 = 8,192$ storage cells must be used. Of this number, 8 cells are arranged in a horizontal line, all excited by the same address line. There are 1,024 such lines, each excited by a different address. In other words, addressing

---

† RAM and ROM terminology is confusing. As commonly used, RAM means a random-accessed read/write memory, whereas ROM designates a random-accessed read-only memory.

is provided by exciting 1 of 1,024 lines. This type of addressing is called *one-dimensional* or *linear selection* (Prob. 9-15). The number of pins on the package for addressing is reduced from the unreasonable number of 1,024 to only 10 by including on the chip a 10-to-1,024-line decoder.

Two-dimensional Addressing

A great economy of the number of NAND gates needed in the decoder mentioned above can be obtained (Prob. 9-16) by arranging the 1,024 memory elements in a rectangular 32×32 array, each cell storing 1 bit of one word. Eight such packages are required, one for each of the 8 bits in each word.

Each word is identified by a matrix number X-Y in a memory cell of the rectangular matrix. To read (or write into) a specific cell (say, 1-3), an X decoder identifies row 1 (X_1) and a Y decoder locates column 3 (Y_3). Such *two-dimensional addressing* (also called X-Y *addressing* or X-Y *selection*) is indicated in Fig. 9-16 for a 4-kb (64×64) RAM.

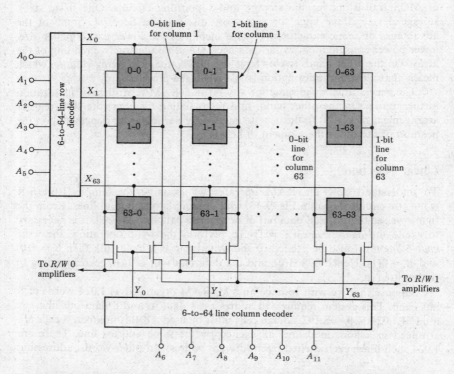

Figure 9-16 Organization of a 4,096-word-by-1-bit static MOS RAM. Each shaded square represents a 6-MOS cell (Fig. 9-20).

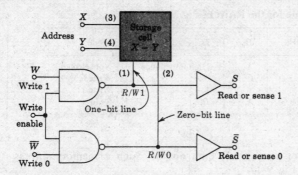

Figure 9-17 A basic storage cell can be constructed with complementary inputs and outputs and with the write and sense amplifiers meeting at a common node (1) for true data and (2) for complementary data.

Basic RAM Organization

In the 1-bit memory of Fig. 9-15 separate read and write leads are required. For either the bipolar or the MOS RAM it is possible to construct a FLIP-FLOP (as we demonstrate in Figs. 9-20 and 9-21) which has a common terminal for both writing and reading, such as terminals 1 and 2 in Fig. 9-17. This configuration requires the use not only of the write data W (write 1), but also of its complement $\overline{W}$ (write 0). At the cell terminal to which $W(\overline{W})$ is applied, there is

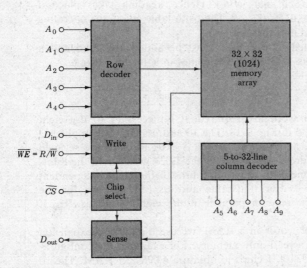

Figure 9-18 Organization of a 1,024-word-by-1-bit static RAM. (Courtesy of Mostek Corporation.)

Table 9-2. Truth table for the RAM of Fig. 9-18

$\overline{CS}$	$R/\overline{W}$ or $\overline{WE}$	D_{in}	D_{out}	Mode
1	X^*	X	High impedance	Not selected
0	0	0	0	Write 0
0	0	1	1	Write 1
0	1	X	D_{out}	Read

$^*X$ = don't care

obtained the read $R(\overline{R})$ or sense $S(\overline{S})$ data output. Such a memory unit is indicated schematically in Fig. 9-17.

The basic elements of which a RAM is constructed are indicated in Fig. 9-16. These include the rectangular array of storage cells, the X and Y decoders, the write amplifiers to drive the memory, and the sense amplifiers to detect (read) the stored digital information. The amplifiers labeled R/W 0 and R/W 1 are not indicated explicitly in Fig. 9-16 (but are drawn in Fig. 9-20).

The organization (also called a *functional diagram*) for a 1,024 word by 1-bit read/write memory is given in Fig. 9-18. Note that there are 32 rows and 32 columns in the memory array. Hence, each decoder has 5 inputs. The data input D_{in} (output D_{out}) corresponds to $W(S)$ in Fig. 9-17. The complements of D_{in}, D_{out}, and the addresses $A_0 \cdots A_9$ are generated on the chip. Both decoders are also fabricated on the chip, referred to as *on-chip decoding*. The terminal labeled $\overline{CS}$ is the chip-select input (sometimes designated CE for *chip enable*). If $CS = 1$, the chip is selected. The complement of the write-enable input is labeled $\overline{WE}$ or $R/\overline{W}$ (*read but not write*). Hence, reading takes place if $R/\overline{W} = 1$ and writing is done if $R/\overline{W} = 0$. The truth table giving the operating modes of this chip is indicated in Table 9-2.

To the 14 inputs indicated in Fig. 9-18 must be added a ground and a power-supply lead. Hence, this 1,024-kb RAM comes in a 16-pin package.

Memory Expansion

If 1,024 words of 8 bits per word are required, it is necessary to use eight packages like the one indicated in Fig. 9-18. The 10 address lines are applied in parallel to all eight packages, and with $CS = 1$ ($\overline{CS} = 0$) all chips are selected simultaneously. A specific address selects one of the 1,024 words, and the 8 bits of data are entered into the memory in parallel through the eight independent D_{in} terminals, with $R/\overline{W} = 0$. For this same address (and, hence, for the selected word), the 8 bits are read in parallel at the eight terminals D_{out} if $R/\overline{W} = 1$.

To expand the number of words in a RAM, we use the address-expansion organization of Fig. 6-28 for a read-only memory. For example, the configuration of Fig. 9-19 uses four $1,024 \times 1$ chips to obtain a $4,096 \times 1$ RAM. The 10 decoder inputs $A_0 \cdots A_9$ are applied to each chip in parallel, and an external

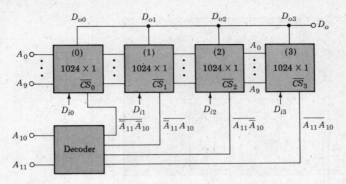

Figure 9-19 Four (1,024 words by 1 bit) RAMs organized to form a (4,096 words by 1 bit) read/write memory.

two-input (A_{10} and A_{11}) decoder is added with its four outputs going to the chip-selects $\overline{CS}$ of the four packages. The four data inputs are independent bits. The output circuit of each chip is a 3-state stage (Fig. 5-24) and the four outputs are OR-tied, as indicated in Fig. 9-19. The operation of the system is explained as follows. If $A_{11} = 1$ and $A_{10} = 0$, then $CS_2 = 1$ and chip 2 is enabled, whereas the other three packages present a high impedance at each output (Fig. 5-24). Hence, $D_o = D_{o2}$. As all possible addresses $A_9 \cdots A_1 A_0$ are applied (with $A_{11} = 1$ and $A_{10} = 0$) the 1,024 1-bit words stored in chip 2 appear at the output D_o. For all possible addresses of $A_{11} \cdots A_1 A_0$ a total of $4 \times 1,024 = 4,096$ 1-bit words are obtained at the output.

By using both types of expansion discussed in the previous two paragraphs, it is possible to increase simultaneously the number of words and the number of bits per word of a read/write memory (Prob. 9-19).

9-7 READ/WRITE MEMORY CELLS[12–14]

Storage cells in a RAM are fabricated with either bipolar or MOS components. The TTL cell is operated statically, whereas some MOS units are available which operate in a static mode and others in a dynamic mode. These various types of MOS memory cells are discussed in this section and a comparison among them is made.

Static MOS RAM

The MOS FLIP-FLOP of Fig. 8-24 is a 1-bit memory and is the basic storage cell for the static MOS RAM. In Fig. 9-20 $Q1$ through $Q4$ form such a bistable unit and MOSFETs $Q5$-$Q6$ form the gating network through which the interior node N_1 (N_2) is connected to the 0-bit (1-bit) data line. Cell (1-3) is indicated in Fig. 9-20. This six-transistor cell is inserted into the memory array in the manner

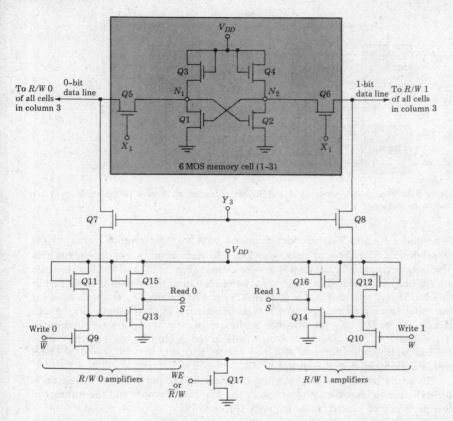

Figure 9-20 A storage cell (1-3) containing six *n*-channel MOSFETs. The X_1 and Y_3 address lines are shown. The write and read amplifiers are also included. *A logic 1 is stored if $Q2$ conducts.*

indicated in Fig. 9-16. The 0-bit and 1-bit lines are connected to every cell *in the same column*. To select a cell in a particular column (say, 3) it is necessary to address that column (Y_3). To select a cell in row 1, the row decoder must excite X_1. In other words, two-dimensional addressing is used to locate a specific cell (1-3).

Included in Fig. 9-20 are the read and write amplifiers for each of the data lines. Note that $Q17$ and $Q10$ ($Q9$) form an AND gate with inputs WE and W ($\overline{W}$), where WE = *write enable*, and W = *write* (also called *data input* D_{in}). The *sense S* or *read output* may also be labeled D_{out}.

It is desired to read cell 1-3. We must set X_1 and Y_3 to V_{DD} (logic 1 for NMOS). Assume that a 1 is stored in this cell ($Q2$ is ON and $Q1$ is OFF so that node N_2 is at 0 V and N_1 is at V_{DD}. In order to read, WE is set to 0. Then $Q17$ is OFF and hence $Q10$ ($Q9$) is nonconudcting, so that the 1-bit (0-bit) data line is tied to V_{DD} *through the load* $Q12$ ($Q11$). Consequently current flows into $Q2$ through $Q12$, $Q8$, and $Q6$ from V_{DD} (as well as through $Q4$ from V_{DD}) and the 1-bit line is effectively grounded. Hence, $Q14$ is OFF and $S = D_{out} = V_{DD}$ (logic 1). Since $Q1$ is OFF, no current flows in $Q3$, $Q5$, $Q7$, and $Q11$ in series, and the 0-bit data line is at V_{DD}, $Q13$ is ON, and $\bar{S} = 0$ V. We thus have correctly sensed that the 1-3 FLIP-FLOP stores a 1 (since $S = 1$ and $\bar{S} = 0$).

In order to write a 1 into the cell we address it ($X_1 = 1$ and $Y_3 = 1$), set $WE = 1$, $W = 1$, and $\bar{W} = 0$. Then $Q17$ and $Q10$ are ON and $Q9$ is OFF. Hence, the 1-bit line is grounded and the 0-bit line goes to V_{DD} through load $Q11$. Current now flows into the 1-bit line from V_{DD} through $Q4$, $Q6$, $Q8$, $Q10$, and $Q17$ to ground. Thus node N_2 is effectively grounded. This cuts off $Q1$, and N_1 rises to V_{DD}. Consequently $Q2$ is held ON and N_2 is maintained at 0. When the address is removed ($Q5$, $Q6$, $Q7$, and $Q8$ are OFF), $Q2$ is ON, $Q1$ is OFF, and a 1 has been written into the selected memory cell.

Static MOS RAMs with memory capacities of 1 kb and 4 kb are available from Intel, Mostek, Texas Instruments, and other vendors, and 16-kb RAMs are expected in 1979. Note that a 4-kb RAM using the 6-MOS cell of Fig. 9-20 has $6 \times 4,096 = 24,576$ MOSFETs in the memory array alone. It is clearly important to reduce the number of FETs per cell, and some of the configurations developed to accomplish this saving are described in the following.

Four-MOSFET Dynamic RAM Cell

The silicon area occupied by the six-transistor cell of Fig. 9-20 may be reduced by changing the load FETs $Q3$ and $Q4$ to clocked loads. In other words, the two cross-coupled inverters which form the latch are now dynamic inverters, as indicated in Fig. 9-21a. The gating excitation for each load is supplied by the word lines from the X decoder. MOSFETs $Q3$ and $Q4$ act simultaneously as loads and as row-selection transistors, thus reducing the cell from a six- to a four-device memory unit. If $X = 0$, $Q3$ and $Q4$ are OFF, and no information can be written into (or read out of) the cell. However, if $X = 1$, then $Q3$ and $Q4$ are ON, and the four transistors form a latch which can store a 1 ($Q2$ ON) or a 0 ($Q1$ ON).

As with the dynamic MOS shift register of Fig. 9-2, the information in the dynamic memory cell of Fig. 9-21a is stored on the parasitic capacitances C_1 and C_2 between gate and source of $Q1$ and $Q2$, respectively. If a 1 is stored, then C_2 (C_1) is charged to $V_{DD}(0)$, and if a 0 is written, the converse is true. Suppose that after the data are stored in the cell it is not accessed for a time T. The charge on the capacitors decreases during this interval because of the inevitable leakage currents. If T is too long the 1-state voltage may become small enough to be indistinguishable from the 0-V level and the information is lost. This same phenomenon is the reason why a dynamic shift register cannot be operated below a minimum operating frequency.

Clearly, some additional circuitry is required to refresh the stored data before the drop in capacitor voltage becomes excessive. Two transistors (Q and

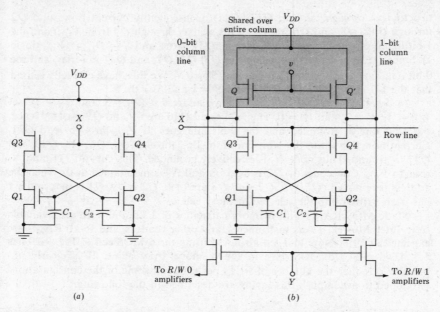

Figure 9-21 (*a*) A 4-MOSFET dynamic cell. A logic 1 is stored if the voltage across C_2 is V_{DD} so that $Q2$ is ON. (*b*) The memory unit is placed in a RAM organized as in Fig. 9-16. All FLIP-FLOPS in column Y are refreshed when a positive pulse v is applied to turn on Q and Q', provided that $X = 1$.

Q' in Fig. 9-21*b*) are added to restore *all of the* FLIP-FLOPS *in a given column*. The refresh waveform v is a pulse of less than 1 μs occurring about every 2 ms. All cells in a given row are refreshed simultaneously by addressing that row while v is high. Note that during the refresh interval, $Q3$ in series with Q forms the load for $Q1$, and Q' in series with $Q4$ acts as the load for $Q2$. If at the beginning of the refresh cycle the voltage across C_2 exceeds that across C_1 (~ 0), then $Q1$ is OFF and C_2 charges toward V_{DD} by the current in Q and $Q3$. The current charging C_1 through Q' and $Q4$ is smaller than that of C_2 because $Q2$ is conducting. Hence, C_2 rises rapidly to V_{DD} and the voltage across $Q2$ falls to zero, maintaining the voltage across C_1 at 0 V. In other words, because of the regenerative feedback action in the FLIP-FLOP the cell is restored to its initial state (a logic 1 in this case).

Note that the organization of the four-transistor cell into a RAM is the same as that for the 6-MOS cell of Fig. 9-20. The number of transistors saved in going from a 6- to a 4-MOS cell in a 4-kb square RAM, taking into account the MOSFETs which must be added to generate the refresh voltage v, is $2 \times 4,096 - 2 \times 64 = 8,064$. In addition to using much less chip area, the dynamic cell saves a great deal of power. The load devices conduct only during the refresh pulse and power is dissipated only during this very short interval.

One-MOSFET Dynamic RAM Cell[15]

In Fig. 9-21 the storage elements are capacitors but there is no fundamental reason to use a FLIP-FLOP to charge or discharge the capacitors. It should be possible to design a dynamic memory using a single capacitor and one transistor to act as a transmission gate to place electrons upon the capacitor or to remove the charge already stored there. This simplest of all RAM cells is indicated in Fig. 9-22 and is used in large (4-kb and 16-kb) dynamic RAMs commercially available. Complementary inputs and outputs are not required and, hence, the organization is that indicated in Fig. 9-16, except that only one bit (data) line is used to connect all cells in a column. As with the RAMs described in the foregoing discussions, only one cell in the memory is selected at a given time depending upon the X and Y addresses.

The cell is written into by applying the bit line voltage through the transistor to the capacitor C_1. Reading is done by connecting C_1 to the bit line through the gate and sensing the capacitor voltage level. One disadvantage of this simple cell is that the readout is destructive. This difficulty occurs because the transistor in the cell selected for reading places its storage capacitance C_1 in parallel with the capacitance C_2 of the data line. If V_1 is the voltage across C_1, the readout voltage V is given by Eq. (9-1) with $V_2 = 0$ or $V = C_1 V_1/(C_1 + C_2)$. Since many cells are connected to the column line, $C_2 \gg C_1$ and $V \ll V_1$. The stored information that is to be retained must therefore be regenerated after every read operation to its original level V_1. In order to increase the ratio C_1/C_2, n-channel two-layer polysilicon gate technology is used.

The capacitor C_1 also loses voltage because of leakage currents and, hence, additional circuitry must be added to refresh the stored information periodically, as was done in Fig. 9-21b. There is one refreshing amplifier on each data line.

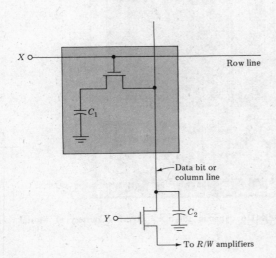

Figure 9-22 A 1-MOSFET dynamic memory cell is shown in the shaded block. It is organized into a RAM in the manner indicated in Fig. 9-16. Refresh circuitry is not shown.

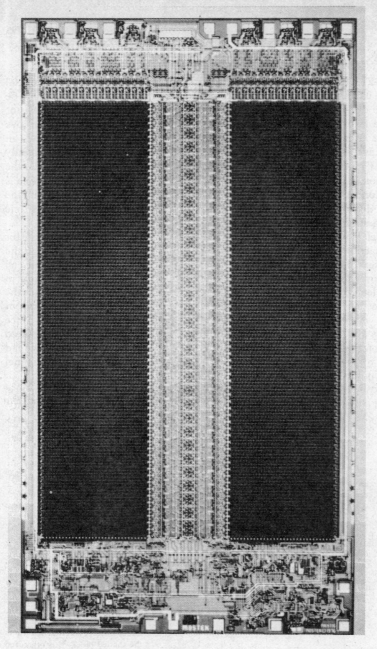

Figure 9-23 Photomicrograph of a MK4116 dynamic 16-kb × 1 RAM. (Courtesy of Mostek Corporation.)

A photomicrograph of the Mostek MK4116 dynamic 16,384-bit (16-kb × 1) RAM is shown in Fig. 9-23. The area of the entire chip is only 110 × 200 mils (2.8 × 5.1mm), corresponding to 1.34 mils^2/bit (including all auxiliary chip circuits). The memory is divided into two equal parts (easily identified in the figure), each with a 128 × 64 array containing a single transistor per cell. The rest of the chip consists of the decoders, the R/W amplifiers, the refresh circuitry, the clocks, etc. The access time is 200 ns, the active power is 460 mW, and the standby power is only 20 mW. This memory comes in an industry standard 16-pin package, made possible by eliminating 7 of the 14 addresses required to select 1 of 16,384 cells. In order to accomplish this pin reduction, the 14 address bits are multiplexed on 7 address pins. The two 7-bit X and Y words are latched onto the chip by two TTL clocks called *row address strobe* (RAS) and *column address strobe* (CAS). The 16 pins are the 7 addresses, D_{in}, D_{out}, $\overline{RAS}$, $\overline{CAS}$, $\overline{WE}$, ground, and three supply voltages +5, +12, and −5 V. The Intel 2116 or 2117 16-kb RAM is pin for pin compatible with the MK-4116 and so also is the TI-4070 chip. A 65-kb dynamic RAM is expected to be commercially available by 1979.

Comparison of Read/Write Memories

The most important characteristics of RAMs are storage capacity, cost per bit, physical size, power consumption, reliability, access time, and cycle time. (The minimum period for a RAM to complete a read or write operation is called the *cycle time*.)

A plot[16] of cost per bit versus access time is given in Fig. 9-24. Static memories have higher cost per bit (because each cell has more transistors than

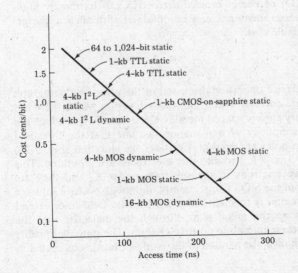

Figure 9-24 Cost per bit (1977) versus access time of the various RAM devices commercially available.[16] *(Courtesy of Electronics.)*

dynamic RAMs). However, for small memory systems they are favored because they need a minimum of controls (no refresh circuitry). Bipolar arrays are usually faster than MOS (ECL cells are the fastest) but generally require more power (Fairchild has the lead in bipolar RAMs). Read/write memories are volatile (the stored data are lost when power fails), except CMOS, which can operate from batteries. (The RCA 1-kb CMOS/SOS static RAM has a standby power of only 0.1 μW and access time of about 100 ns.) The densest memories use the 1-cell dynamic MOS devices and have the smallest cost per bit (about 0.1 cent/bit in 1978). Dynamic RAMs are organized in 1-bit word lengths and hence an 8-bit microprocessor (Sec. 9-11) requires eight chips. The I^2L devices indicated in Fig. 9-24 are discussed in Sec. 9-13.

Starting in 1978 HMOS and VMOS transistors (Sec. 8-6) were used in RAMs, resulting in decreased access times. For example, the Intel 2147 4-kb static RAM has an access time of only 45 ns.

9-8 CHARGE-COUPLED DEVICE[17, 18, 19](CCD)

Consider a MOS transistor with an extremely long channel and with many (perhaps 1,000) gates closely spaced between source and drain. Such an arrangement can be made to function as a shift register. Each gate electrode and the substrate form a MOS capacitor which can store charge. For example, if a logic 1 is introduced at the source, charge is stored by the capacitor nearest to the source, provided that an appropriate voltage is applied to this first electrode E_1. If now this voltage is removed from E_1 and simultaneously applied to E_2, the charge packet will move from E_1 to E_2. By repeating this process, charge is transferred from capacitor to capacitor, and hence this configuration is called a *charge-transfer device* (CTD) or *charge-coupled device* (CCD). Extremely high-density shift registers or serial memories can be obtained with such a charge-coupled device at a reasonable cost.

Basic CCD Operation[20]

To understand better the device operation discussed in the preceding paragraph consider an n-type substrate covered with a thin-oxide layer on which there has been deposited a row of very closely spaced metallic electrodes, five of which are shown in Fig. 9-25. For simplicity of explanation, assume that the threshold voltage is zero and that no holes are present. Consider the situation (Fig. 9-25) where the voltage on gate 3 is $-V$ and all other electrodes are grounded. This negative voltage repels free electrons in the substrate under E_3 and they are driven downward away from the SiO_2. Consequently, immobile positive ions are exposed and a depletion region is formed under E_3. Electric field lines extend from the negative charges on the metal plate through the dielectric into the depletion region and onto these immobile positive charges. The potential profile (the potential variation with distance parallel to the oxide surface) is indicated in

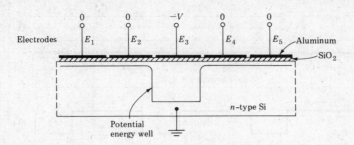

Figure 9-25 The simplest structure for a p-channel (n-substrate) CCD. A potential energy "well" is formed under gate 3 if this electrode is at a negative voltage and all other gates are at the substrate potential (ground).

Fig. 9-25. This plot also represents the potential-energy barrier ("well") for holes, the minority carriers. If a packet of holes is introduced in the region under E_3, these charges can move freely within the well, but cannot penetrate the potential-energy walls of the well (Sec. 1-2). In other words, as long as the voltage $- V$ is present, the positive charge cannot wander away but is trapped under E_3 near the channel surface.

We now consider how stored charge is moved from left to right down the channel, corresponding to shifting binary bits along this shift register. Consider the structure in Fig. 9-26a consisting of 10 plates where each third electrode is tied together. If at $t = t_1$ the applied voltages are $\phi_1 = - V$, $\phi_2 = \phi_3 = 0$, then as indicated in Fig. 9-26b, potential-energy wells (corresponding to Fig. 9-25) are formed under electrode 1, 4, 7, and 10. The plus signs indicate schematically that charge is stored near the surface under E_1, E_7, and E_{10} but not under E_4, indicating that the digital byte 1011 has been entered into the CCD. At a later time $t = t_2$, the voltage ϕ_2 changes to $- V$ but ϕ_1 and ϕ_2 maintain their previous values. The potential profile is thereby altered, as in Fig. 9-26c. The stored charge now is shared by two adjacent electrodes, owing to diffusion of the holes from the original well into the newly created empty well.

Shortly after the situation in Fig. 9-26c is established, $|\phi_1|$ starts to decrease and, at $t = t_3$, $\phi_1 = - V/2$, whereas ϕ_2 and ϕ_3 remain unchanged. The potential profile at t_3 is shown in Fig 9-26d. The fringing electric field caused by the potential difference between ϕ_1 and ϕ_2 causes the holes to move into the deeper well, as indicated. Finally at $t = t_4$, when $\phi_1 = 0$, $\phi_2 = - V$, and $\phi_3 = 0$, the potential profile is as shown in Fig. 9-26e. As a result of this sequence of voltage changes, the initial pattern of stored charge (1011) has transferred one electrode to the right, as is clear from a comparison of Fig. 9-26b and e.

The sequence just described represents transfer from one electrode to the next of the CCD shift register. Since three voltages are necessary, three-phase clocks are required. The waveforms ϕ_1, ϕ_2, and ϕ_3 necessary to conform with the profiles in Fig. 9-26 are given in Fig. 9-27. The times t_1, t_2, t_3, and t_4 in Fig. 9-26

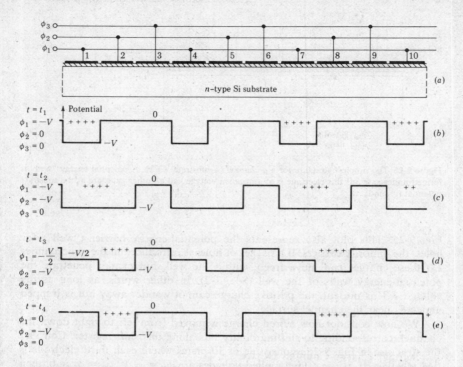

Figure 9-26 Illustrating charge transfer in a CCD. (a) Every third electrode is at the same potential, and three-phase voltages ϕ_1, ϕ_2, and ϕ_3 (Fig. 9-27) are applied. (b) through (e) The potential profile variations during one shift interval. The potential energy for positive charge is proportional to the potential and, hence, these curves also represent the potential-energy wells for holes.

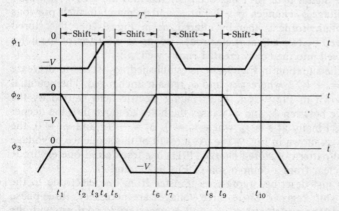

Figure 9-27 The three-phase excitation waveforms for the CCD of Fig. 9-26a. The potential profile of Fig. 9-26b corresponds to the instant t_1 of this figure, Fig. 9-26c corresponds to t_2, etc.

are also indicated in Fig. 9-27. Note that at t_1 in Fig. 9-27, $\phi_1 = - V$, $\phi_2 = 0$, and $\phi_3 = 0$, as in Fig. 9-26; at t_2, $\phi_1 = - V$, $\phi_2 = - V$, and $\phi_3 = 0$ in both figures, etc. The first transfer takes place between t_1 and t_4, the second between t_5 and t_6, the third between t_7 and t_8, the fourth between t_9 and t_{10}. Clearly, for every input cycle of period T, three shifts take place. In the interval between shifts (for example, between t_4 and t_5), the clock voltages remain constant and the potential profiles are unaltered.

Electrodes per Bit

From Fig. 9-26*b* or *e* it is clear that if a logic bit is latched under one electrode, no information can be stored below the next two electrodes. In other words, a storage cell or stage consists of three electrodes and 1 bit is stored in this cell. For this CCD the *electrodes per bit* is three ($E/B = 3$). The information is read at the output, say electrode 10, where at $t = t_1^-$ a 1 is observed. From Fig. 9-26 three shifts are required before the next bit (the 1 stored under electrode 7) can be sensed. Three transfers later the 0 under gate 4 appears at the output. Since three shifts take place in the period T, information must be read (or written) only once per cycle of the input waveform.

It has been assumed for simplicity that the threshold voltage V_T is negligible in the foregoing discussion. In reality, all levels marked 0 V in Figs. 9-26 and 9-27 should be at a voltage in excess of V_T in order for the electric field to penetrate into the channel and form the depletion region.

n-Channel Charge-Coupled Device

The CCD is a unipolar device because only charges of one sign are shifted longitudinally from electrode to electrode. The structure in Fig. 9-26 is referred to as a *p*-channel CCD because the transferred charge consists of holes (minority carriers) in the *n*-type silicon substrate. An *n*-channel device (*p*-type substrate) is excited by positive instead of negative voltages and the waveforms in Fig. 9-27 are inverted (change $- V$ to $+ V$). A positive gate voltage repels the mobile holes in the substrate and the depletion region contains immobile negative ions. The electric lines of force from a positively charged electrode end on these negative ions, giving rise to a positive voltage drop within the substrate. However, since potential energy U equals potential multiplied by the charge (Sec. 1-2) and the electronic charge q is negative, the potential-energy profiles for a three-phase *n*-channel CCD are those illustrated in Fig. 9-26*b* through *e*, multiplied by $|q|$. Of course, the plus signs should be changed to minus signs to indicate that electronic packets of electrons (a logic 1) are being transferred down the shift register.

Minimum and Maximum Operating Frequencies

Steady-state (dc) operation of a CCD is not possible. Thermally generated carriers become trapped in empty potential-energy wells and, in time, change the logic state from a 0 to a 1. This phenomenon, called the *dark-current effect*, sets a lower limit to the clock frequency (50 kHz to 1 MHz).

No steady-state power is required by a CCD cell since power is dissipated only in charging the effective cell capacitances. Consequently the upper limit of clocking frequency (1 to 10 MHz) may be determined by the maximum allowable power dissipation. Also, an increase in frequency reduces the efficiency of transfer of charge from one cell to the next. Hence, the upper frequency may be limited by the point at which transfer losses become unacceptable.

9-9 CCD STRUCTURES

Note that a charge-coupled device cannot be assembled from discrete components because a single continuous channel is required to provide the coupling between the depletion regions (the wells) under the electrodes. The gates in Fig. 9-26 must be separated by a very small distance (~ 1 μm) in order to supply this coupling. These narrow gaps are difficult to etch reliably because of mask defects, flaws in the photoemulsion, dust particles, etc. A number of alternative fabrication methods[21, 22] have been developed, with the use of metal electrodes, polysilicon gates, ion implantation, and combinations of these in order to avoid the abovementioned difficulties. One such polysilicon electrode structure for a three-phase n-channel CCD is indicated in Fig. 9-28. Each electrode has a 9-μm active length with a 3-μm overlap. The channel width w (the dimension perpendicular to the paper) is 17 μm. A large number of CCDs are fabricated in rows parallel to one another to cover the chip area. The separation between rows is 10 μm, resulting in an average area used for one three-electrode cell of 27×27 μm = 1.1 mils^2/bit (not including peripheral circuits).

Two-Phase CCD

For the simple planar electrodes of Fig. 9-26 or Fig. 9-28 it is necessary to use three-phase clocks to transfer the charge longitudinally in one direction only (Fig. 9-26). It is found that two-phase excitation is not possible, because it leads

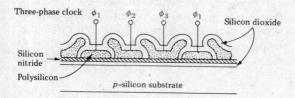

Figure 9-28 Electrode structure for a three-phase n-channel CCD. The electrodes are n-polysilicon (not aluminum). Note that each phase has a differently shaped electrode; this is a triple polysilicon structure. An MNOS device which has a low threshold voltage (Sec. 8-6) is indicated.

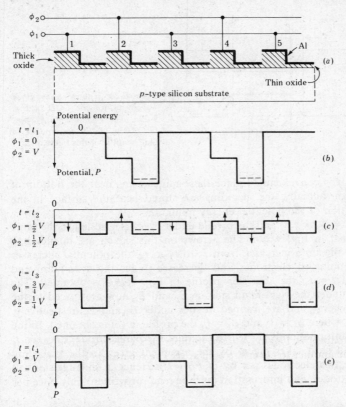

Figure 9-29 Illustrating charge transfer in a two-phase CCD. (a) The electrode structure. (b) through (e) The potential-energy profiles corresponding to the times indicated in Fig. 9-30.

to indeterminancy in the direction of shift (Prob. 9-20). However, two-phase clocking can be used if nonplanar electrodes, such as indicated schematically in Fig. 9-29a, are employed. The right-hand half of each electrode is over a thinner oxide layer than the left-hand section and, consequently, electric lines of force penetrate more deeply into the substrate at the right side of the metal. Hence, the depletion region and the potential-energy profile has the same step shape as the two-level electrodes. Alternate electrodes are tied together, resulting in a two-phase system whose clock waveforms ϕ_1 and ϕ_2 are given in Fig. 9-30.

Let us again assume that the threshold voltage is 0. The potential-energy profiles are drawn in Fig. 9-29 for the values of time t_1, t_2, t_3, and t_4, indicated in Fig. 9-30. At $t = t_1$, $\phi_1 = 0$ and $\phi_2 = V$, so that there is no barrier under E_1, and the step barrier under E_2 is as drawn in Fig. 9-29b. We assume that a logic 1 is stored under E_2 and E_4, and we have indicated the minority carriers (electrons)

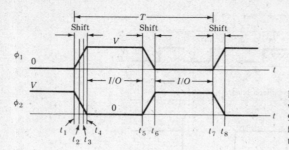

Figure 9-30 The two-phase clock waveforms for the CCD of Fig. 9-29a. The potential-energy profile of Fig. 9-29b corresponds to the instant t_1 of this figure, etc.

by minus signs. It is convenient to place these minus signs near the bottom of the well, although, actually, the electrons are stored near the surface in the longitudinal position of the potential energy minimum.

At $t = t_2$, $\phi_1 = \phi_2 = \frac{1}{2} V$, and the profile under every electrode has the same shape, as indicated in Fig. 9-29c. The arrows on this sketch are intended to indicate that, as the time increases from t_1 to t_2 to t_3, the potential increases under the odd-numbered electrodes and decreases under the even-numbered plates. Hence, at $t = t_3$, the stair-step profile in Fig. 9-29d is obtained. The electrons stored under the right-hand side of E_2 and E_4 now are forced to the lowest potential energy and are trapped in sites under E_3 and E_5, respectively. Finally, at $t = t_4$, where $\phi_1 = V$ and $\phi_2 = 0$, the profile in (e) is obtained. In the interval t_4-t_1 the information is shifted to the right by one electrode. Between t_5 and t_6 the second shift takes place. By the argument given in the preceding section, there are two electrodes per bit ($E/B = 2$). Hence, a shift-register cell contains two electrodes, and information must be read (or written) only once per

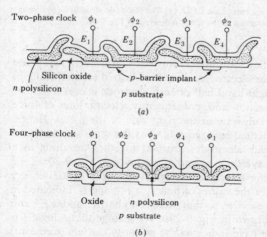

Figure 9-31 Polysilicon electrode structure for (a) a two-phase and (b) a four-phase n-channel CCD.

clock period in the interval $t_5 - t_4$ or $t_7 - t_6$, which is also called the input/output (I/O) interval.

An excellent implementation of the electrode structure of the two-phase CCD is given in Fig. 9-31a. Polysilicon electrode E_1 is shaped like its metallic counterpart of Fig. 9-29a, with thick oxide at the left. The p-type ions implanted (Sec. 8-6) under the left-hand side of E_2 in the p-type substrate supply the desired potential offset under this electrode. When a positive voltage is applied to E_2, the holes are repelled, leaving the high concentration of immobile negative charges from the implantation. Consequently the lines of force from the left-hand side of E_2 terminate on these negative ions and do not penetrate deeply into the substrate. Hence, the potential-energy profile is much closer to the surface on the left side than on the right side of E_2, as required.

Alternative Two-Phase Clocking Waveforms

The waveforms ϕ_1 and ϕ_2 in Fig. 9-30 are essentially symmetrical square waves with one waveform displaced half a period with respect to the other. If the two sections of the square wave have very different time intervals, then a pulse-type waveform results. One such possibility is indicated in Fig. 9-32a. Note that the *positive pulses are nonoverlapping*. It can be shown (Prob. 9-21) that the charge trapped under the surface of E_1 is shifted to the site under E_2 in the pulse width $t_5 - t_1$ of ϕ_2. Similarly this information is shifted from E_2 to E_3 by the positive pulse of ϕ_1 during the interval $t_7 - t_6$. In the remainder of the cycle no further shifting takes place and the data can be read (or written); in other words $t_8 - t_7$ is the I/O interval.

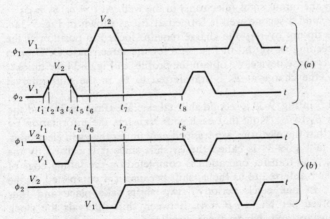

Figure 9-32 Pulse-type waveforms which may be used as clocks for an n-channel two-phase CCD. (a) Nonoverlapping narrow positive pulses. (b) Nonoverlapping narrow negative pulses (overlapping wide positive pulses).

Another possible two-phase clock is indicated in Fig. 9-32b, where a shift takes place during each *negative* pulse interval (Prob. 9-22). Note that now the *positive pulses overlap*.

Single-Phase Operation

The electrode structure of Fig. 9-31a may be used with a constant voltage $\phi_1 = V_3$ applied to the odd electrodes, and with the positive pulse waveform ϕ_2 of Fig. 9-32a as excitation for the even electrodes. The dc bias must be chosen so that $V_1 < V_3 < V_2$. It can be shown (Prob. 9-23) that charge is shifted from E_1 to E_2 in the interval when $\phi_2 > V_3$, and the information is shifted from E_2 to E_3 during the remainder of the cycle when $\phi_2 < V_3$. Hence, the memory cell consists of two electrodes per bit, as is the case for two-phase operation.

Four-Phase CCD

The polysilicon electrode structure for a four-phase shift register is indicated in Fig. 9-31b. All odd electrodes are over a thick oxide and the even electrodes are separated from the substrate by a thin oxide. Hence, for the same applied voltage the depletion region is deeper under the even gates. This electrode structure is used in the Intel 2416 (16-kb) CCD serial memory.[23] The shifting is accomplished with the clock waveforms of Fig. 9-33.

To follow the transfer of information along the channel of this CCD, the potential-energy profiles are drawn in Fig. 9-34. In Fig. 9-34a the electrode structure of Fig. 9-31b is drawn in a simplified form. At $t = t_1$ in Fig. 9-33, $\phi_1 = \phi_3 = \phi_4 = 0$ and $\phi_2 = V$. Hence, a potential-energy well exits under electrode 2 (E_2), as shown in Fig. 9-34b. We assume that a logic 1 is stored at this site, as indicated by the minus signs (electrons) in the well. At $t = t_2$, $\phi_1 = \phi_3 = 0$ and $\phi_2 = \phi_4 = V$, and a second well is formed at E_4, as shown in Fig. 9-34c. Because these wells do not overlap the charge remains locked in position at the surface of E_2. Proceeding as we did in Fig. 9-26 for the three-phase CCD or in Fig. 9-29 for the two-phase device, we obtain the profiles in Fig. 9-34 for times t_3 through t_6. Clearly, the charge at E_2 is transferred to E_4 in the shift interval $t_6 - t_1$.

Between t_7 and t_8 in Fig. 9-33 a second shift takes place (Prob. 9-24) and the charge moves from E_4 to E_6. Note that each shift transfers the information by two electrodes and that the electrons are trapped *only* under the even electrodes. The shift pictured in Fig. 9-34 is called the ϕ_3 shift since it is caused by the positive pulse of ϕ_3. The transfer operation is completed on the falling edge of ϕ_3. The next transfer is referred to as the ϕ_1 shift because it is completed on the falling edge of ϕ_1. In one clock period T two shifts take place and four electrodes are involved per bit ($E/B = 4$). Between shift intervals the clock voltages remain constant and, hence, these durations ($t_7 - t_6$ and $t_9 - t_8$) are used for write/read operations. They are labeled I/O in Fig. 9-33.

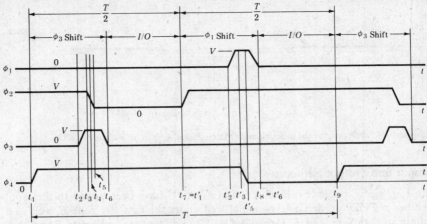

Figure 9-33 Four-phase clock waveforms used in the Intel 2416 CCD.

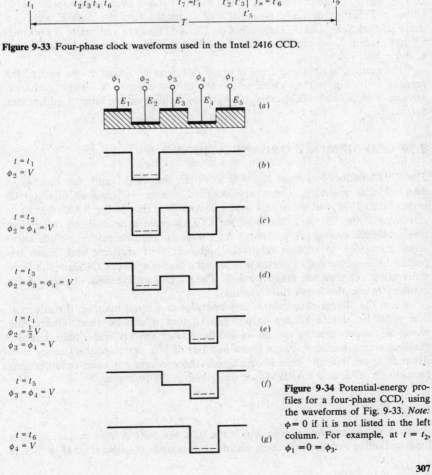

Figure 9-34 Potential-energy profiles for a four-phase CCD, using the waveforms of Fig. 9-33. *Note:* $\phi = 0$ if it is not listed in the left column. For example, at $t = t_2$, $\phi_1 = 0 = \phi_3$.

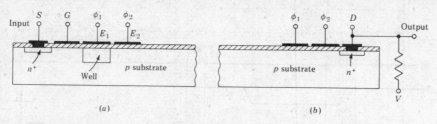

Figure 9-35 Structures for (a) injecting and (b) detecting charge in an n-channel CCD.

Input and Output Structures[17, 24, 27]

In Fig. 9-35a a source diffusion S and a gate G have been added to the input end of a CCD register. The potential well under the first electrode E_1 acts as a drain so that S, G, and E_1 form a MOSFET. Voltages are applied to S and to G so that current flows until the well fills with charge to the same potential as that of S.

The output is obtained from a drain diffusion D added at the end of the register as in Fig. 9-35b, which senses the output current. Voltage- or charge-sensing is obtained by fabricating an output amplifier on the chip or adding one externally.

9-10 CCD MEMORY ORGANIZATIONS[24, 25]

The CCD memory bridges the gap between the RAM and the fixed-head magnetic-disk memory. The charge-coupled memory is cheaper (about 0.02 cents/bit in 1978) but the access time is slower than that of the RAM, due to its serial operation. On the other hand the CCD is more expensive but faster than magnetic-bulk storage. Applications for charge-coupled memories include small serial memories to replace relatively high-cost shift registers and, more importantly, large memory systems in which the higher speed of CCDs make them more attractive than the magnetic disk. The CCD is an excellent display refresh memory in a cathode-ray-tube terminal.

Since the charge-coupled memory operates in a serial manner, the information must be shifted to the output port before it can be read. Hence, the worst-case access time to any bit (called the *latency time*) is longer than that of a random-access memory. For a given number of bits per chip, the latency time depends upon how the chip is organized. Three commonly used organizations (serpentine, SPS, and LARAM) are described in the following.

Serpentine

A synchronous organization in which the data are shifted from cell to cell in a long, snakelike manner in a recirculating shift-register configuration (Fig. 9-3) is

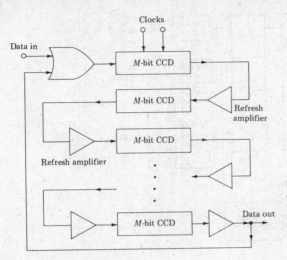

Figure 9-36 Serpentine or synchronous organization of a CCD memory ($M \approx 256$). The multiphase clock is applied simultaneously to all CCD sections (between refresh amplifiers).

indicated in Fig. 9-36. The transfer of charge from one cell to the next is quite efficient (~ 99.999 percent). The limitations imposed by transfer inefficiency and dark-current generation require that the stored charge must be replenished by refresh amplifiers, as in Fig. 9-36, after about every 100 to 300 cells. This configuration is the simplest of the three organizations to fabricate, and for a short CCD it is fast if a high-frequency clock is used. For example, for $f = 10$ MHz, the latency time for a 1,024-bit CCD is 102 μs. The higher the clock frequency, the greater is the power wasted, and a tradeoff must be reached between latency time and power dissipation.

Line-addressable Random-Access Memory (LARAM)

This is an organization optimized to give short access times. The LARAM consists of a number of short CCD recirculating memories operating in parallel, which share common input and output lines. A decoder is used to address any one of the registers at random; hence the designation *line-addressable random-access memory*.

The Intel 2416 is a 16-kb CCD in an 18-pin package, organized as 64 independent circulating shift registers of 256 bits each. A 6-to-64-line decoder can select any register in a random-access fashion. The LARAM is indicated schematically in Fig. 9-37. Random access and I/O operations are performed in a manner similar to that for a 64-bit RAM. For a 2-MHz clock the latency time is $\frac{256}{2} = 128$ μs. The *average* access time is half this amount, or 64 μs. The Intel 2464 is a 65,536-bit CCD organized in a similar manner to the 2416-chip CCD, but with 256 independent circulating registers of 256 bits each.

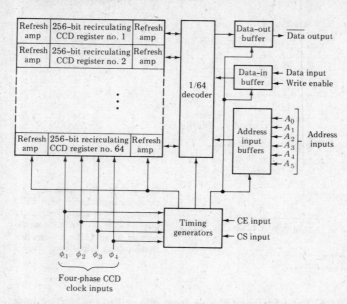

Figure 9-37 The Intel 2416 16,384-bit CCD memory organized as 64 recirculating shift registers of 256 bits each. The registers can be addressed randomly. The four-phase clock is obtained from a separate chip, #5244. (Courtesy of Intel Corporation.)

Serial-Parallel-Serial Organization (SPS)

This configuration is indicated in Fig. 9-38. The first N bits are shifted by the high-frequency (f_c) clock waveforms into a horizontal CCD register. The charges representing this byte are then transferred in parallel into the N vertical registers. The next N input bits are now fed into the horizontal shift-in CCD. Then this second byte is "dumped" into the vertical registers. In order to make room for the second N bits, the first byte must be shifted one cell down the vertical registers. Clearly, these vertical CCDs operate at a frequency f_c/N, which is much lower than f_c if N is large. The third byte of N bits is fed quickly into the input register and transferred to the slow, parallel CCDs. This process is repeated until the input register and all the parallel registers are full. At this time the first N bits (which are now at the bottom of the vertical CCDs) are transferred to the horizontal output register and then are shifted out serially at the same high frequency f_c at which they entered the system. The reason for the designation *serial-parallel-serial* should now be evident. Note that the output bits are fed back to the input so that a recirculating shift register is formed.

If the total number of bits in the memory is B, Fig. 9-38 represents a rectangular matrix of $M \times N = B$ cells. Hence, the number of bits in each vertical CCD is $M - 1$; one bit of the serial input register is part of a column of the matrix. Since the data appear at the output at a frequency f_c, the access time

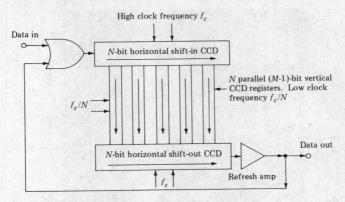

Figure 9-38 Serial-parallel-serial CCD organization.

is the same as that obtained from a serpentine-organized charge-coupled memory. This limits the maximum number of bits. For example, for $B = 4,096$ and $f_c = 5$ MHz, the latency time is 820 μs, with an average access time of half this value.

There are a number of advantages to the SPS architecture; low power, elimination of internal refresh amplifiers, low clock capacitances, and very high density. Consider the square array $M = N = 64$, $B = 4,096$, and $f_c = 5$ MHz. Of the 4,096 bits stored only $N = 64$ of these are shifted at the 5-MHz rate; the remaining bits are transferred at the much lower frequency, $f_c/M = \frac{5}{64} = 0.078$ MHz = 78 kHz. Consequently, the power dissipation, which decreases with decreasing clocking frequency, is greatly reduced. Note also that each bit moves through only $M + N = 128$ cells. Hence, only one refresh amplifier is needed, as indicated in Fig. 9-38. The high-frequency clock feeds the capacitance of only $2N = 128$ cells, whereas the low-frequency clocks feed the much larger capacitance of the remaining cells.

The Fairchild 65-kb CCD

A photomicrograph of the Fairchild F464 which has 65,536 bits in a standard 16-pin DIP is shown in Fig. 9-39. The memory is organized as sixteen 4,096-bit SPS square blocks (easily identified in the picture) with $M = N = 64$, as described in the preceding paragraph. Using a 4-bit code as input to an on-chip decoder, the 16 blocks are randomly accessible (LARAM) with each line containing 4,096 bits. The entire chip measures only 173 × 228 mils (4.4 × 5.8 mm), corresponding to 0.60 mil²/bit (including all the auxiliary chip circuits). This density is more than twice that of the 16,384-bit RAM of Fig. 9-23. The active power is 200 mW at 4 MHz.

The two-phase clock waveforms of Fig. 9-32a are used and each cell consists of two electrodes. Since the cell is twice as long as it is wide, 32 bits are

Figure 9-39 Photomicrograph of the F464, 65,536-bit CCD memory. (*Courtesy of Fairchild Semiconductor.*)

used in the serial registers, but there are 64 parallel CCD tracks, each of which is 63 bits long. The first 32 bits are loaded into the odd-numbered parallel registers from the 32 left-hand cell electrodes. The next 32 bits are shifted into the serial register, and these are loaded into the even-numbered tracks from the right-hand cell electrode. This arrangement is called *interlacing*. The parallel shifting frequency is now $f_c/32$ (not $f_c/64$). By 1980 262-kb CCDs are expected to be available.

In a two-phase system two electrodes are used to store one bit of information. At the expense of additional support circuitry complexity[25, 26] each electrode can effectively store one bit of information so that $E/B = 1$. This E/B *operation* is used in the Fairchild F464 to increase the packing density.

In addition to their use in high-density memories, charge-coupled devices are also used as analog-signal processors and as (line or area) image sensors.[17, 24] The latter two applications are outside the scope of this book.

9-11 MICROPROCESSOR AND MICROCOMPUTER[28–30]

A microprocessor is a single chip which contains the arithmetic, logic, and control circuitry of a general-purpose processing, control, and/or computing system. Such a combination, sometimes including a small amount of memory on the same chip, is the central processing unit (CPU) of the system, generally referred to simply as a *processor*. This CPU chip is available from many manufacturers (Appendix B-1) in word lengths of 4, 8, 12, or 16 bits. For example, the Intel 8080 is an 8-bit microprocessor which contains a static RAM array of six 16-bit registers for temporary storage in addition to the arithmetic logic unit (ALU, Sec. 6-3) and control sections. It has a cycle time of 2 μs.

All the technologies described in this book are used for microprocessors. The early types were PMOS, but now the dominant technology is NMOS. For very low-power applications, such as wrist-watch chips or military aerospace systems, CMOS is used. Higher speed systems use low-power Schottky TTL bipolar technology. I^2L has also been introduced into microprocessors.

Converting a Microprocessor into a Microcomputer

To be able to perform all the tasks required of a computer, the CPU must be augmented by additional memory, a clock, and peripheral interface adapters (PIAs) for input and output (I/O) devices. A block diagram of such a system is indicated in Fig. 9-40. The nonvolatile ROM chips store the program and tables. If the program is not fixed, then PROMs or EPROMs are used. Temporary storage resides in the random-access memory chips or, for large amounts of storage, flexible (floppy) magnetic disks or tape cassettes are employed. The input/output devices might be a keyboard, a cathode-ray-tube (CRT) display, paper or magnetic tape, a line printer, or a transducer which converts a physical measurement into an electrical digital signal. The number of auxiliary chips surrounding the microprocessor in Fig. 9-40 may be of the order of 10 to 80, and they are mounted on a printed-circuit (PC) board, often no larger than the size of this page. The equivalent number of transistors on such a board would probably exceed 100,000.

Single-Chip Microcomputer

By integrating the timing circuits, the memory (both ROM and RAM), the I/O, and the CPU, a microcomputer can be fabricated on a single chip. The Intel

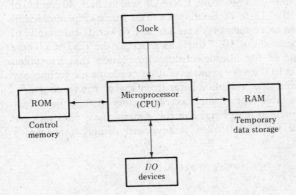

Figure 9-40 A microcomputer assembled from a microprocessor chip, ancillary chips, and input-output devices.

Figure 9-41 A photomicrograph of the 8748 microcomputer chip. (*Courtesy of Intel Corporation.*)

8048 or the 8748 (which uses an **EPROM**) is such a *computer on a chip*. A photomicrograph of the 8748 is shown in Fig. 9-41. This chip contains about 20,000 transistors and measures only 5.6 × 6.6 mm. This microcomputer sells for about $250 (in 1978). It has more computing capacity[31] than the first large-scale electron-tube computer built in 1946 (the ENIAC) which had 40 racks of equipment and occupied 10 × 13 m of floor space. The single-chip microcomputer is 20 times faster, has more memory, uses far less power, is thousands of times more reliable, and costs about 10^{-4} times as much as the ENIAC. These statistics indicate the scope of the microelectronics revolution that has taken place in the last decade due to the very rapid evolution of electronic technology of the integrated circuit†; and this revolution is far from having run its course.[31]

Interface chips are available from a number of vendors so that the microcomputer need not be devoted to peripheral control. Among these packages are a floppy-disk controller, a CRT controller, a keyboard display interface, a

† Read "A Short History of Electronics" starting on p. XIX.

matrix printer controller, and a communications interface; to name only a few.

The design of microcomputer-based instruments and commercial systems is considered in detail in Ref. 28. It is well known that in any computer (including the microcomputer) development time and costs for the software exceed those for the hardware.

Applications

The microcomputer will affect our daily lives in many ways by making possible the design and introduction into the marketplace of sophisticated products at reasonable prices. These may be characterized in a general way as "smart" instruments, control systems, and calculators. They will include (but certainly are not limited to the following): control and reduction of the exhaust-gas emission of an automobile, optimizing engine adjustments to improve gas mileage, safety devices in the car, electronic scales and sales terminals in supermarkets, improved control of traffic lights, temperature control in the home, improvements in control of kitchen appliances (food processors, microwave ovens, etc.), video and other electronic games and toys, distribution and control of information in the office (accounting, word processors, electronic mail, etc.), tremendous improvements in missile, satellite, and other space systems, numerous industrial process control and measurement systems, inventory control, electronic banking (bank terminals, check and credit card processing, etc.), communication's improvements, and so forth. This list of products is limited only by the imagination of the entrepreneur, the knowledge, experience, and ingenuity of the engineer and programmer, and the economics of the marketplace. It has been predicted that factory sales of microcomputers will reach one billion dollars in the early 1980s.

An important microprocessor advantage is that it shortens considerably the lengthy effort to design, debug, and manufacture a special LSI circuit. Furthermore, modifications and improvements can be made often by simple program changes even after the product has been sold, since this can be accomplished by reprogramming the PROM.

9-12 INTEGRATED-INJECTION LOGIC (I²L)[32-37]

The LSI chips discussed in this chapter are all MOS except the PLA. The reasons that MOSFETs dominate in LSI are:

1. A MOS transistor occupies a considerably smaller area than a bipolar junction transistor (BJT).
2. The MOSFET dissipates much less power than the BJT.
3. A BJT requires an isolation diffusion which wastes considerable chip "real estate" whereas a MOS needs no isolation.

4. A MOS logic gate has no resistors (a MOSFET is used as a load), whereas a bipolar circuit employs diffused resistors, and each such resistor usually requires even more silicon space than does a BJT. However, the logic delay of a BJT gate is much smaller than that of a MOS gate.

It clearly would be very desirable to be able to obtain the high speed of the BJT together with the high packing density and low power of the MOST. This goal was achieved almost simultaneously in 1972 by ingenious engineers at the IBM Laboratories,[32] Boeblingen, West Germany, and at the Philips Research Laboratories,[33] Eindhoven, The Netherlands. The former group called their logic family *merged-transistor logic* (MTL) and the latter designated their invention *integrated injection logic* (IIL or I^2L). These two approaches led to essentially the same LSI device, whose most popular acronym is I^2L. We now shall develop the basic I^2L inverter; the circuit configuration is new, but it is fabricated by using bipolar technology (Chap. 4).

Merging of Devices

If one semiconductor region is part of two or more devices, these components are said to be *merged*. This process may save considerable silicon area. We are already familiar with merging, although this designation was not used in the preceding pages. For example, in the evolution from the DTL gate to the TTL gate, the anodes of the three input diodes D plus the coupling diode $D1$ of Fig. 5-18 are merged into the base region of the transistor $Q1$ of Fig. 5-21. As a result of this merging a multiemitter transistor (Fig. 4-13) is formed. This multiemitter BJT, which is at the input to the TTL NAND gate (Fig. 5-21), is also used in the ROM encoder (Fig. 6-24b).

Another example of merging is supplied by the NMOS NOR gate of Fig. 8-21. If the fan-in is M, then M sources are grounded. Hence, a single n diffusion serves as the source for all M inputs. Note also that the drains of all input transistors are connected together. Consequently another n diffusion becomes the drain of all M input MOSFETs. The p-type substrate is the channel for all M input transistors. The MOS ROM encoder consists of NOR gates and, hence, this type of merging is visible in Fig. 9-8. Note from the top view of the chip of Fig. 9-8 that the output Y_3 consists of one drain, one source, and many gates (two of which are shown in Fig. 9-8).

Now we discuss the type of merging that leads to the basic I^2L structure. Direct-coupled transistor logic (DCTL) is the configuration used in MOS/LSI. Hence, let us reexamine DCTL with bipolar transistors, as indicated in Fig. 9-42, redrawn from Fig. 5-26. This logic system has a fan-in of 3 and a fan-out of 2. Consider the shaded block: it will be developed into the basic I^2L gate. Note that the bases of $Q4$ and $Q5$ are tied together and, hence, can be merged into a single p-type diffusion region. Similarly, since the emitters of $Q4$ and $Q5$ are at ground potential, they can be merged into a single n-type diffusion

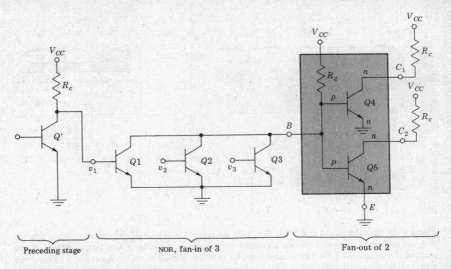

Figure 9-42 The DCTL gate of Fig. 5-26 redrawn.

volume. The result of this merging of $Q4$ and $Q5$ is a single transistor with two collectors, a cross section of which is pictured in Fig. 9-43a. Note that this standard bipolar transistor requires a space-wasting p^+ isolation diffusion; if an isolation island were not used, *the collectors of all transistors on the chip would be in a single n-epitaxial region,* and hence all collectors would be at the same potential. In Fig. 9-42 we note that not all the collectors are tied together but rather that *every emitter* is at the same (ground) potential. Consequently, no isolation is required if the collectors and emitters are interchanged. In Fig. 9-43b this is accomplished by building a multiple-emitter transistor and using it in the inverse mode of operation (Sec. 3-11). In other words, the two emitters (marked C_1 and C_2) are used as collectors and the collector of the normal transistor (labeled E) becomes the emitter E of the inverted device. In order to increase the current gain of the reverse transistor, a high-conductivity n^+ substrate is used, and an n-epitaxial layer is grown, as indicated in Fig. 9-43c. Since all emitters are grounded, the substrate acts as a *grounded plane* for the entire chip and only one emitter (ground) lead is necessary. Compare this inverted ("upside-down") transistor (Sec. 3-11) with the discrete transistor of Fig. 4-6b which is operated in the normal mode.

In Sec. 5-13 it is pointed out that either $Q4$ or $Q5$ in Fig. 9-26 may "hog" most of the base current because the input characteristics of these two discrete transistors are not identical. This current-hogging possibility is eliminated in the IC of Fig. 9-43 because the two transistors have been merged and share the same base-emitter junction.

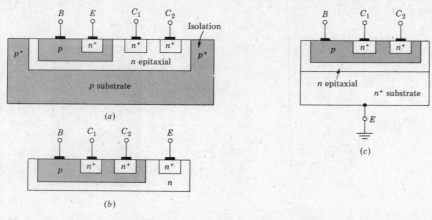

Figure 9-43 Transistors $Q4$ and $Q5$ of Fig. 9-42 are merged into a single two-collector transistor. (*a*) A cross section of the two-collector transistor. (The surface oxide layer and lateral diffusions are omitted, for simplicity.) (*b*) The p substrate and isolation diffusion are eliminated if a two-emitter transistor is fabricated, but used in the inverse mode. (*c*) An n^+ substrate serves as a grounded plane which is the single emitter for the entire chip.

Current Injection

In the shaded block of Fig. 9-42 there is a diffused resistor which wastes considerable power and takes so much silicon real estate that its use in LSI is prohibitive. Let us rethink the need for R_c, the collector resistor for the NOR-gate transistors $Q1$, $Q2$, and $Q3$ in Fig. 9-42. The effective load on these transistors is R_c in parallel with the input resistance of the fan-out stage. A large value of R_c improves the gain of the NOR inverters; so why not go to the extreme of infinite R_c and omit R_c altogether? In other words, let us connect the collectors of the NOR directly to the base B of the fan-out transistors. If this were done, there would be no biasing base current for $Q4$ and $Q5$. Consequently we have moved R_c from the collectors of $Q1$, $Q2$, and $Q3$ into the shaded block in Fig. 9-42 at the bases of $Q4$ and $Q5$. This position emphasizes that R_c is required *only* to supply the fan-out base currents.

It should now be clear that R_c may be eliminated, provided that it is replaced by a current source. The grounded-base p-n-p transistor Q in Fig. 9-44a acts as such a current source or *current injector*. *The resistor R_x is external to the chip.* Clearly

$$I_o = \frac{V_{CC} - V_{BE}}{R_x} \tag{9-5}$$

and the collector current of the p-n-p CB transistor (which is also the base current of the n-p-n transistor) is αI_o, where α (≈ 0.5) is the common-base current gain and the base-to-emitter voltage of the p-n-p transistor is V_{BE} (≈ 0.85 V).

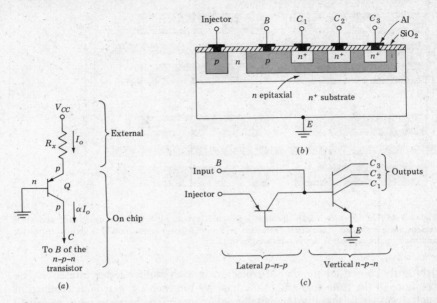

Figure 9-44 (*a*) A current source or injector is used to replace the resistor R_c in Fig. 9-42. (*b*) The *p-n-p* source is merged with the *n-p-n* transistor inverter. Only a *p* diffusion for the injector need be added to the monolithic structure of Fig. 9-43*c* to form the I^2L gate. The number of collectors depends upon the required logic. In this inverter three collectors are indicated. (*c*) The circuit model of the I^2L cell in (*b*).

Since the collector of the CB transistor is tied to the base of the fan-out transistors, these two elements can be merged. In other words, the *p* region in Fig. 9-43*c* can serve also as the collector of *Q*. Similarly the *n*-base of *Q* can be merged with the grounded emitter *E*. By adding a *p*-type emitter diffusion for *Q* to the structure in Fig. 9-43*c*, we obtain the complete I^2L gate of Fig. 9-44*b*. Note that the current injector *is a lateral p-n-p* transistor, whereas the multicollector inverter is a *vertical n-p-n* transistor. Since more than two collectors are sometimes required, we have extended the figure to include three outputs ($M = 3$) and have also indicated the surface oxide for the sake of completeness. The effective circuit model of the basic I^2L gate, corresponding to the structure shown in Fig. 9-44*b*, is indicated in Fig. 9-44*c*.

It must be emphasized that *all* injector currents are obtained from V_{CC} through a *single* external resistor R_x. A chip usually is fabricated with long *p*-type diffusion lines called *injector rails*. Each rail delivers base current to all the *n-p-n* transistors adjacent to it (within a diffusion length). A top view of a possible injector-logic-chip layout is shown in Fig. 9-45. A shaded rectangle represents an *n-p-n* transistor whose base (the transistor input) is shown as a small circle and whose collectors (the inverter outputs) are indicated by small squares. All *p* regions are shaded and all *n* regions are left white. We have

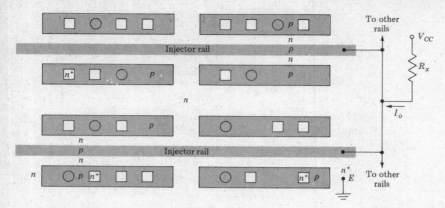

Figure 9-45 The top view of an injection logic chip layout. All p sections are shaded and all n regions are not shaded. Each circle represents an input (p base) connection of a vertical transistor and each square represents an output (n collector).

arbitrarily chosen the number of collectors in each multicollector transistor. The positions of the collectors and of the base lead in each I^2L gate is determined in such a way as to simplify the metallic interconnections between gates to satisfy the desired logic. The bottom left-hand inverter in Fig. 9-45 corresponds to the electrode arrangement in Fig. 9-44b. The emitter of each vertical transistor is the grounded n (unshaded) region. The organization in Fig. 9-45 shows only eight inverters, but it can, of course, be expanded both vertically and horizontally into a LSI system.

Summary

The basic I^2L unit consists of a p-n-p transistor which acts as a current injector to supply the base current for a multicollector n-p-n transistor inverter. This is a novel circuit configuration, but there is nothing new about the method of fabrication, which uses standard bipolar technology (Chap. 4). From Figs. 9-44b and 9-45 it is clear that the process starts with an n^+ substrate onto which an n epitaxial layer is grown. A p diffusion forms both the injector and the vertical transistor base regions. A second n diffusion forms the multiple collectors. A total of only four masks are required: two for the diffusions, the third to open the contact windows for bases, collectors, and injector rails, and the fourth to define the metalization interconnections. These numbers should be compared with the standard BJT which requires five masks and three diffusions. Note also that space-consuming isolation borders and diffused resistors are not required in I^2L chips. Consequently packing densities of over 300 gates/mm^2 are realized, corresponding to a cell size of 1.1×4.3 mils, which is comparable to (or somewhat smaller than) the area of a MOSFET gate. However, it must be pointed out that often more than four masking operations are required to provide isolated devices for I/O compatibility, to combine I^2L with other circuits, etc. These added fabrication steps decrease the packing density appreciably.

9-13 INJECTION LOGIC CIRCUITS

In this section we examine the operation of the (resistorless) I^2L inverter and show how to obtain the NAND, NOR, and FLIP-FLOP. Injection logic design is also illustrated.

Inverter

The inverter $Q1$ is loaded by $Q2$ in Fig. 9-46. Each transistor is biased by an injector current I_j. At the low logic level $V_i \approx 0$, $Q1$ is OFF, and the input signal V_i acts as a sink for I_j so that $I_{B1} = 0$ and $I_{C1} = 0$. Hence $I_{B2} = I_j$, and $Q2$ is ON so that $V_{BE2} \approx 0.75 \text{ V} = V_{CE1}$. On the other hand, if the input is high, $V_i \approx 0.75$ V, the base current I_{B1} increases above I_j and $Q1$ tends to saturate. Consequently, V_{CE1} drops very low (≈ 0 V). Now $Q2$ is cut off because $Q1$ acts as a sink for I_j of $Q2$, so that $I_{C1} = I_j$, reducing I_{B2} to zero. Clearly, an inversion has been performed by $Q1$ because $V_o = 0.75$ V for $V_i = 0$ and $V_o = 0$ V for $V_i = 0.75$ V. The logic swing is about 0.75 V, its exact value depending upon the bias current I_j.

Note that in saturation the collector current is I_j and the base current has about the same magnitude. Hence, a value of the CE current gain h_{FE} of only unity is required to cause saturation. A transistor operating in the inverse mode has a very much smaller value of h_{FE} than in the normal mode (~ 100). Nevertheless, a CE current gain in excess of unity (approximately 2 to 5) is easily obtained for the up-side-down transistor.

When an inverter is switched from one state to the other, the voltages of the transistor capacitances must change values, causing a propagation delay t_{pd}. The charging (and discharging) of these capacitances does not take place through resistors in an I^2L gate. The charging current is supplied by the injector. Large values of I_j result in small values of t_{pd}, but the penalty which must be paid for the decreased t_{pd} is an increased average-power dissipation P_{av}. In Sec. 5-15 the *delay-power product* $t_{pd}P_{av}$ is introduced. A value of $t_{pd}P_{av} = 1$ pJ is easily obtained. For $t_{pd} = 20$ ns this figure of merit corresponds to a power of 0.05

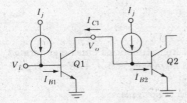

Figure 9-46 An I^2L inverter $Q1$ directly coupled to the following stage $Q2$.

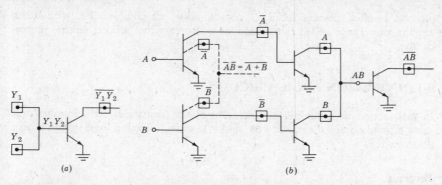

Figure 9-47 (*a*) NAND gate using wired-AND for internal (collector) logic variables. (*b*) NAND gate for externally applied logic variables (solid connections). The dashed portion of the circuit is a NOR gate. *Note:* The injectors have been omitted, for simplicity.

mW. Injection logic gates have been operated with $t_{pd} P_{av}$ below 0.1 pJ, a performance obtained in 1977 under laboratory conditions (not with a production device).

Another advantage of the I^2L configuration is that it can be operated over a wide range of speeds by simply altering the total injector current, by varying the one resistor R_x. The range of operation extends from about 1 nA to 1 mA. Thus after a chip is designed and built, the desired speed of operation may be selected by changing R_x.

NAND Gate

To obtain an AND gate in injector logic is extremely simple. In Fig. 9-47*a*, Y_1 is a logic variable at the output of an I^2L inverter and Y_2 is another variable at the collector of a second I^2L gate. Connecting Y_1 and Y_2 together yields $Y = Y_1 Y_2$ at the common node in Fig. 9-47*a*. This wired-AND logic is illustrated in Fig. 5-22 in connection with DTL logic and it is equally valid for I^2L logic (Prob. 9-27). If Y is applied to the input of an inverter, the output is the NAND function $\overline{Y_1 Y_2}$, as shown in Fig. 9-47*a*.

If $A(B)$ is an externally applied logic variable, then to obtain $A(B)$ *at the collector of an I^2L gate*, two cascaded inverters must be used, as indicated by the solid lines in Fig. 9-47*b*. Proceeding in this manner, the NAND function $\overline{AB}$ is obtained as shown in the figure.

NOR Gate

In Chap. 5 it is verified that all combinational logic functions can be generated by using only NAND gates.† From De Morgan's law (Sec. 5-7) $\overline{A + B} = \overline{A}\overline{B}$ and,

† Remember that a single input NAND gate is an inverter (a NOT gate).

hence, the NOR function is obtained from the wired-AND gate with inputs A and B. This procedure is illustrated by the dashed part of Fig. 9-47b. Note that the two input transistors have two collectors each, and the other three inverters each have one collector.

AND OR INVERT (AOI) Gate

As an example of an AOI gate consider the equality detector of Eq. (6-9):

$$E \equiv A\overline{B} + \overline{A}B = \overline{(A\overline{B})} \; \overline{(\overline{A}B)}$$

As expected, this function involves only NAND gates. Let us assume that the three functions $\overline{AB}$, $\overline{A+B}$, and E are to be available on a single chip. The layout for this chip is shown in Fig. 9-48, with the injector rails and the shading used in Fig. 9-45 omitted, for simplicity. This *interconnecting diagram* should be easy to follow if it is remembered that:

1. A rectangle represents an inverter.
2. A small circle is the input to the inverter.
3. Small squares indicate the multicollector outputs.
4. The logic at each inverter *output* is shown at the right of the rectangle.
5. AND logic is obtained by wiring two (or more) collectors together.

The digital comparator discussed in Sec. 6-4 has three outputs: E, $C \equiv A\overline{B}$, and $D \equiv \overline{A}B$. The outputs C and D are also generated by the interconnections in Fig. 9-48.

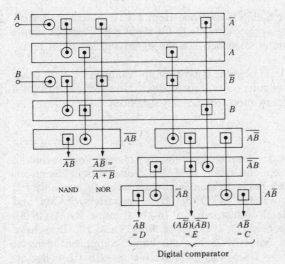

Digital comparator

Figure 9-48 An I²L interconnecting diagram to generate the NAND, NOR, and digital comparator logic functions C, E, and D for the two inputs A and B.

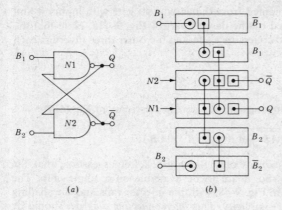

Figure 9-49 (*a*) The storage cell of Fig. 7-1*b*. (*b*) The I²L interconnecting diagram for this FLIP-FLOP latch.

(*a*)   (*b*)

FLIP-FLOP

Sequential circuits such as registors and counters discussed in Chap. 7 are based upon FLIP-FLOPs, which are easily implemented with I²L gates. The 1-bit storage cell or latch of Fig. 7-1*b* is repeated in Fig. 9-49*a*. The interconnecting diagram, obtained by following the foregoing rules, is drawn in Fig. 9-49*b*.

Compatability with TTL

The output of an I²L unit has an open collector. By tying the collector to $V_{CC} = 5$ V through a suitable resistance, output levels compatible with TTL are obtained.

If an input signal to an I²L chip is obtained from a TTL system, its logic-1 level is too high (~ 3 V) to be applied directly to the input n-p-n base. A simple resistive attenuator must be interposed between the signal and the input terminal.

Using standard bipolar technology it is possible to combine I²L with other bipolar logic and/or analog circuits on the same chip.[34] However, the beautiful simplicity of the four-mask-fabrication technique is sacrificed.

Applications

Injection logic is under active development by a number of companies including Bell Laboratories, Fairchild Semiconductor, IBM, Motorola, National Semiconductor, Philips, Signetics, and Texas Instruments. Commercially available I²L systems include digital wrist watches, A/D and D/A converters, RAMs, and microprocessors. Fairchild Semiconductor has a 16-kb dynamic RAM (1978). The TI 9900 (16-bit) microprocessor is available in NMOS and I²L. Both have the same speed and power, but NMOS has the larger component density. Innovations such as fabricating a Schottky diode clamp at each I²L collector, replacing the lateral with a vertical *p-n-p* injector, etc., will enhance injection logic performance. However, at the same time new developments will take place in MOS technology. There are many parameters which must be taken into consideration in the design of any system, such as speed, power, density, cost, reliability, and chip availability from at least two vendors. Consequently it is doubtful that either I²L or MOS systems will predominate the LSI market in the future. Predictions have been made in the literature that I²L actually will have a relatively small impact upon microelectronics. However, in research, development, and the marketplace predictions are often quite unreliable!

REFERENCES

1. Luecke, G., J. P. Mize, and W. N. Carr: "Semiconductor Memory Design and Applications," chap. 3, McGraw-Hill Book Company, New York, 1973.
2. Carr, W. N., and J. P. Mize: "MOS/LSI Design and Application," chap. 5, McGraw-Hill Book Company, New York, 1972.
3. Penney, W. M., and Lau, L.: "MOS Integrated Circuits," chap. 5, Van Nostrand Reinhold Company, New York, 1972.
4. Ref. 1, chap. 6. Terman, L. M.: MOSFET Memory Circuits, *Proc. IEEE*, vol. 59, no. 7, pp. 1044–1057, July, 1971.
5. Frohman-Bentchkowsky, D.: A Fully Decoded 2048-bit Electrically Programmable FAMOS Read Only Memory, *IEEE J. Solid-State Circuits*, vol. SC-6, pp. 301–306, October, 1971.
6. Greene, R., G. Perlegos, P. J. Salsbury, and W. L. Morgan: The Biggest Erasable PROM Yet Puts 16,384 Bits on a chip, *Electronics*, pp. 108–111, March 3, 1977.
7. Altman, A.: Memories, *Electronics* pp. 89–94, January 20, 1977.
   Kelley, J. W., and D. F. Millet: An Electrically Alterable ROM and It Doesn't Use Nitride, *Electronics*, pp. 101–104, December 9, 1976.
8. Frohman-Bentchkowsky, D.: The Metal-Nitride-Oxide-Silicon (MNOS) Transistor-Characteristics and Applications, *Proc. IEEE*, vol. 58, no. 8, pp. 1207–1219, August, 1970.
   Chang, J. J.: Theory of MNOS Memory Transistor, *IEEE Trans. Electron Devices*, vol. ED-24, no. 5, pp. 511–518, May 1977.
9. Hemel, A.: The PLA: a "Different Kind" of ROM, *Electronic Design*, vol. 1, pp. 78–84, January 5, 1976.
   Mrazek, D.: PLAs Replace ROMs for Logic Designs, *Electronic Design*, vol. 22, pp. 66–70, October 25, 1973.
10. Blakeslee, T. R.: "Digital Design with Standard MSI and LSI," sec. 4.9, John Wiley and Sons, Inc., 1975.
11. Cavlan, N.: "Signetics Field Programmable Logic Arrays," Brochure from Signetics, February, 1976.
12. Ref. 1, chaps. 4 and 5.
13. Hodges, D. A.: Microelectronic Memories, *Scientific American*, vol. 237, no. 3, pp. 130–145, September, 1977.
14. Taub, H., and D. Schilling, "Digital Integrated Electronics," chap. 12, McGraw-Hill Book Company, New York, 1977.
15. Ahlquist, C. N., J. R. Breivogel, J. T. Koo, J. L. McCollum, W. G. Oldham, and A. L. Renninger: A 16,384-Bit Dynamic RAM, *IEEE J. Solid-State Circuits*, vol. SC-11, no. 5, pp. 570–574, October, 1976.
    MOSTEK, 1977 Memory Products Catalog, sec. V on MK4027 and sec. 4 on MK4116.
16. Ref. 7, p. 82.
17. Sequin, C. H., and M. F. Tompsett: "Charge Transfer Devices," Academic Press, New York, 1975.

18. Special Issue on Charge Transfer Devices, *IEEE Trans. Electron Devices*, vol. ED-23, no. 2, pp. 71–126, February, 1976.

19. Kosonocky, W. F., and D. J. Sauer: The ABCs of CCDs, *Electronic Design*, April 12, 1975.

20. Boyle, W. S., and G. E. Smith: Charge-Coupled Devices—A New Approach to MIS Device Structures, *IEEE Spectrum*, pp. 18–27, July, 1971.

21. Ref. 17, chap. III.

22. Mohsen, A. M., and T. F. Retajczyk, Jr.: Fabrication and Performance of Offset-Mask Charge-Coupled Devices, *IEEE Trans. Electron Devices*, vol. ED-23, no. 2, pp. 248–256, February, 1976.

23. Chou, S.: Design of a 16,384-Bit Serial Charge-Coupled Memory Device, *IEEE Trans. Electronic Devices*, vol. ED-23, no. 2, pp. 78–86, February, 1976.

24. Kosonocky, W. F., and D. J. Sauer: Consider CCDs for a Wide Range of Uses, *Electronic Design*, March 15, 1976.
    Ref. 17, Chap. VII.

25. Terman, L. M., and L. G. Heller: Overview of CCD Memory, Ref. 18, pp. 72–77.
    Ref. 7, pp. 94–96.

26. Ref. 17, p. 247.

27. Ref. 17, pp. 132–207.

28. Peatman, J. B.: "Microcomputer-based Design," McGraw-Hill Book Company, New York, 1977.

29. Petritz, R. L.: The Pervasive Microprocessor: Trends and Prospects, *IEEE Spectrum*, vol. 14, no. 7, pp. 18–24, July 1977.

30. Toong, Hoo-Min D.: Microprocessors, *Scientific American*, vol. 237, no. 3, pp. 146–161, September, 1977.

31. Noyce, R. N.: Microelectronics, *Scientific American*, vol. 237, no. 3, p. 65, September, 1977.

32. Berger, H. H., and S. K. Wiedmann: Merged-Transistor Logic (MTL): A Low-Cost Bipolar Logic Concept, *IEEE J. Solid-State Circuits*, pp. 340–346, October, 1972.

33. Hart, K., and A. Slob: Integrated Injection Logic: A New Approach to LSI, *IEEE J. Solid-State Circuits*, pp. 346–351, October, 1972.

34. Berger, H. H., and S. K. Wiedmann: The Bipolar LSI Breakthrough, parts 1 and 2 in "Large Scale Integration" Electronics Book Series, McGraw-Hill Book Company, New York, pp. 38–49, 1976.

35. DeTroye, N. C.: Integrated Injection Logic-Present and Future, *IEEE J. Solid-State Circuits*, pp. 206–211, October, 1974.

36. Pedersen, R. A.: Integrated Injection Logic—A Bipolar LSI Technique, *Computer*, pp. 24–29, February, 1976.

37. Horton, R. L., J. Englade, and G. McGee: I²L Takes Bipolar Integration a Significant Step Forward, *Electronics*, pp. 83–90, February 6, 1975.

REVIEW QUESTIONS

9-1 (*a*) Draw the circuit for a *single-phase dynamic* MOS *inverter*.
    (*b*) Explain its operation.

9-2 List three important features of MOSFETs used in a dynamic inverter.

9-3 Explain what is meant by a *ratioed inverter*.

9-4 (*a*) Draw the circuit for one stage of a *two-phase ratioed dynamic* NMOS *shift register*.
    (*b*) Draw the clocking waveforms.
    (*c*) Explain the operation of the circuit.

9-5 (*a*) Draw the block diagram of a *recirculating shift register*.

(*b*) Explain its operation, including how to write into the register and how to read out nondestructively.

9-6 Explain how to obtain a serial memory which may store 512 words of 8 bits each.

9-7 (*a*) Sketch the circuit of a *ratioless* MOS *inverter*. Explain the operation (*b*) if $V_i = 0$ and (*c*) if $V_i = V_{DD}$.

9-8 Repeat Rev. 9-5 for one stage of a ratioless shift register.

9-9 (*a*) Draw one stage of a dynamic CMOS register.

(*b*) Explain its operation briefly.

9-10 (*a*) Draw the circuit of the *encoder* of a MOS read-only memory.

(*b*) Explain the operation of the circuit.

9-11 Explain how the MOS ROM is programmed.

9-12 Sketch, in block diagram form, the organization of a ROM having 512 words of 8 bits each. Use two-dimensional addressing. The encoder matrix is to have the same number of rows as columns. Give no explanation, but indicate the number of X addresses and the number of Y addresses.

9-13 (*a*) What does the acronym FAMOS mean?

(*b*) How is such a cell "programmed"?

(*c*) How is the cell erased?

9-14 (*a*) Show how a FAMOS cell is placed in an encoder.

(*b*) What purpose does the cell serve?

9-15 (*a*) What is a *stacked-gate memory cell*?

(*b*) How is such a cell programmed?

(*c*) How is the cell erased?

9-16 Repeat Rev. 9-14 for the stacked-gate cell.

9-17 Repeat Rev. 9-13 for the acronym EAROM.

9-18 Repeat Rev. 9-14 for the EAROM cell.

9-19 (*a*) What is a *programmable logic array* (PLA)?

(*b*) In particular what does a $12 \times 96 \times 4$ PLA designate?

(*c*) Show a block diagram for the PLA in (*b*).

9-20 (*a*) It is desired to mask program a PLA. Indicate the program table which must be given to the manufacturer for a $12 \times 48 \times 4$ PLA. Include only three of the product terms in the table.

(*b*) Indicate explicitly the logic equations satisfied by your simplified table.

9-21 (*a*) List four advantages of semiconductor random-access memories over core memories.

(*b*) What advantage does a core memory have over a RAM?

9-22 (*a*) Draw the block diagram of a 1-bit *read/write memory*.

(*b*) Explain its operation.

9-23 Explain *linear selection* in a *random-access memory* (RAM).

9-24 Repeat Rev. 9-23 for two-dimensional addressing.

9-25 (*a*) Draw in block-diagram form the basic elements of a RAM with two dimensional addressing used to store four words of 1 bit each.

(*b*) How is the system expanded to 3 bits/word?

(*c*) How is the system expanded to 25 words of 3 bits/word?

9-26 Indicate the organization (functional diagram) of a square static RAM array of 256 words by 1 bit.

9-27 Indicate how to expand the memory of a 256×1 RAM to $1,024 \times 1$. Give no explanation of the operation, but label the block diagram carefully and completely.

9-28 Explain how to expand a 256×1 RAM to 256×16.

9-29 (a) Sketch the circuit of a 6-MOSFET static storage cell.

(b) Explain its operation briefly.

9-30 How many transistors are saved in going from a static 6-MOS to a dynamic 4-MOS cell in a 16-kb RAM? Explain.

9-31 (a) Draw a 4-MOSFET dynamic RAM cell.

(b) Why is additional circuitry required to refresh the data stored in the cell?

9-32 (a) Show how the 4-MOSFET cell in Rev. 9-31 is placed into a random-access memory.

(b) Which devices are associated with a given column?

(c) Explain the function served by each MOSFET.

9-33 Draw a 1-MOSFET dynamic memory cell. Explain its operation briefly.

9-34 List five important characteristics of RAMs.

9-35 (a) Explain how a potential-energy well is formed under an electrode of a CCD.

(b) If the substrate is p-type, are electrons or holes captured in the well?

9-36 Consider a CCD with planar electrodes and using three-phase excitation.

(a) How many shifts of a charge packet take place in one cycle?

(b) What is meant by *electrodes per bit*?

(c) What is the value of E/B for this CCD?

9-37 What determines (a) the minimum frequency of operation of a CCD; (b) the maximum frequency?

9-38 (a) What are the approximate dimensions of a CCD electrode?

(b) What is the approximate area in square mils per bit?

9-39 (a) Sketch *schematically* the shape of the electrodes in a two-phase charge-coupled memory.

(b) Draw the excitation waveforms.

(c) How many shifts of a charge packet take place in each cycle?

9-40 For the CCD memory of Rev. 9-39 illustrate how the information is shifted down the register.

9-41 (a) Sketch *schematically* the shape of the electrodes in a four-phase CCD.

(b) How many shifts take place per excitation cycle?

(c) What is E/B?

9-42 With respect to cost per bit and latency time, how does the CCD compare with (a) the RAM and (b) the magnetic disk?

9-43 Describe with the aid of a block diagram the serpentine organization of a CCD memory.

9-44 Repeat Rev. 9-43 for the LARAM organization.

9-45 Repeat Rev. 9-43 for the SPS organization.

9-46 List three advantages of the SPS organization.

9-47 (a) Describe a microprocessor.

(b) How does a microcomputer differ from a microprocessor?

9-48 Give three reasons why MOSFETs dominate in LSI over BJTs.

9-49 Define *merging* of devices.

9-50 Draw a DCTL circuit. Place a border around all elements which are merged into an I^2L inverter.

9-51 (*a*) Explain why a collector resistor is not needed in DCTL and why it may be replaced by a current source.

(*b*) Indicate such a current injector.

(*c*) Which elements of the injector are external to the chip?

9-52 (*a*) Draw the cross section of an I^2L inverter including the current source.

(*b*) Show a circuit model for an I^2L unit.

9-53 Sketch a top view of an I^2L chip layout. Include injection rails.

9-54 Explain the operation of an I^2L inverter.

9-55 (*a*) A and B are available at the collectors of two I^2L inverters. Show how to obtain the NAND function $\overline{AB}$.

(*b*) Repeat part (*a*) if A and B are applied externally.

9-56 Show how to obtain the NOR function $\overline{A + B}$ in I^2L.

9-57 Show the interconnection diagram for a FLIP-FLOP latch in I^2L.

ANALOG CIRCUITS AND SYSTEMS

ANALOG DIODE CIRCUITS

The emphasis thus far in this book has been on digital circuits and systems, where the voltage ideally has only one of two values, $V(0)$ and $V(1)$. The study of analog circuits and systems begins with this chapter. An analog waveform is one whose voltage or current varies continuously with time, taking on all values between specified minimum and maximum amplitudes.

The piecewise linear diode model of Fig. 2-16 indicates that when this device is conducting it may be approximated by a battery V_γ in series with a small forward resistance R_f. When it is OFF, the diode behaves as a large reverse resistance R_r. This model and the method of analysis of diode circuits outlined in Sec. 2-12 are used in this chapter in the following applications: A simple rectifier, single-ended and double-ended clippers, voltage regulators, half-wave and full-wave rectifiers, including capacitor filters.

10-1 A SIMPLE APPLICATION

Consider a series circuit consisting of a diode D, a load resistor R_L, and a sinusoidal input $v_i = V_m \sin \alpha$, where $\alpha = \omega t$, $\omega = 2\pi f$, and f is the frequency of the input excitation. Assume that the piecewise linear model of Fig. 2-16 (with $R_r = \infty$) is valid. The current in the forward direction $(v_i > V_\gamma)$ may then be obtained from the equivalent circuit of Fig. 10-1a. We have

$$i = \frac{V_m \sin \alpha - V_\gamma}{R_L + R_f} \tag{10-1}$$

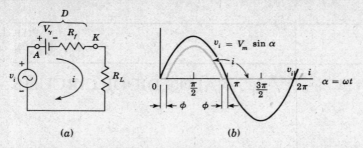

(a)   (b)

Figure 10-1 (a)The equivalent circuit of a diode D (in the ON state) in series with a load resistance R_L and a sinusoidal voltage v_i. (b) The input waveform v_i and the rectified current i.

for $v_i = V_m \sin \alpha \geqslant V_\gamma$ and $i = 0$ for $v_i < V_\gamma$. This waveform is plotted in Fig. 10-1b, where the cutin angle ϕ is given by

$$\phi = \arcsin \frac{V_\gamma}{V_m} \tag{10-2}$$

If, for example, $V_m = 2V_\gamma$, then $\phi = 30°$. For silicon (germanium),

$$V_\gamma = 0.6 \text{ V } (0.2 \text{ V}),$$

and hence a cutin angle of 30° is obtained for very small peak sinusoidal voltages; 1.2 V (0.4 V) for Si (Ge). On the other hand, if $V_m \geqslant 10$ V, then $\phi \leqslant 3.5°$ (1.2°) for Si (Ge) and the cutin angle may be neglected; the diode conducts essentially for a full half cycle. Such a rectifier is considered in more detail in Sec. 10-5.

Incidentally, the circuit of Fig. 10-1 may be used to charge a battery from an ac supply line. The battery V_B is placed in series with the diode D, and R_L is adjusted to supply the desired dc (average) charging current. The instantaneous current is given by Eq. (10-1), with V_B added to V_γ.

The Break Region

The piecewise linear approximation given in Fig. 2-16a indicates an abrupt discontinuity in slope at V_γ. Actually, the transition of the diode from the OFF condition to the ON state is not abrupt. Therefore the waveform transmitted through a clipper or a rectifier will not show an abrupt change of attenuation at a break point, but instead there will exist a *break* region, that is, a region over which the slope of the diode characteristic changes gradually from a very small to a very large value. We shall now estimate the range of voltage of this break region.

The break point is defined at the voltage V_γ, where the diode resistance changes discontinuously from the very large value R_r to the very small value R_f. Hence, let us arbitrarily define the break region as the voltage change over which the diode resistance is multiplied by some large factor, say 100. The incremental resistance $r \equiv dV/dI = 1/g$ is, from Eq. (2-7),

$$r = \eta \frac{V_T}{I_o} \epsilon^{-V/\eta V_T} \tag{10-3}$$

If $V_1(V_2)$ is the potential at which $r = r_1(r_2)$, then

$$\frac{r_1}{r_2} = \epsilon^{(V_2 - V_1)/\eta V_T} \tag{10-4}$$

For $r_1/r_2 = 100$, $\Delta V \equiv V_2 - V_1 = \eta V_T \ln 100 = 0.24$ V for Si ($\eta = 2$) and 0.12 V for Ge ($\eta = 1$) at room temperature. Note that the break region ΔV is only one- or two-tenths of a volt. If the input signal is large compared with this small range, then the piecewise linear volt-ampere approximation and models of Fig. 2-16 are valid.

10-2 CLIPPING (LIMITING) CIRCUITS

Clipping circuits are used to select for transmission that part of an arbitrary waveform which lies above or below some reference level. Clipping circuits are also referred to as voltage (or current) *limiters, amplitude selectors,* or *slicers.*

In the above sense, Fig. 10-1 is a clipping circuit, and input voltages below V_γ are *not* transmitted to the output, as is evident from the waveforms of Fig. 10-1b. Some of the more commonly employed clipping circuits are now to be described.

Consider the circuit of Fig. 10-2a. Using the piecewise linear model, the transfer characteristic of Fig. 10-2b is obtained, as may easily be verified. For example, if D is OFF, the diode voltage $v < V_\gamma$ and $v_i < V_\gamma + V_R$. However, if D is OFF, the circuit reduces to that in Fig. 10-3a, there is no current in R, and $v_o = v_i$. This argument justifies the linear portion (with unity slope) of the transmission characteristic extending from arbitrary negative values to $v_i = V_R + V_\gamma$. For v_i larger than $V_R + V_\gamma$, the diode conducts, and it behaves as a battery V_γ in series with a resistance R_f, as indicated in Fig. 10-3b. Hence the transfer characteristic is given by

$$v_i \leqslant V_R + V_\gamma \qquad v_o = v_i$$

$$v_i \geqslant V_R + V_\gamma \qquad v_o = v_i \frac{R_f}{R + R_f} + (V_R + V_\gamma) \frac{R}{R + R_f} \tag{10-5}$$

The second equation above is obtained by superposition (Sec. C-2), considering v_i as one voltage and $V_R + V_\gamma$ as a second independent source. Equation (10-5) verifies the linear portion of slope $R_f/(R_f + R)$ for $v_i > V_R + V_\gamma$ in the transfer curve. Note that the transmission characteristic is piecewise linear and continuous and has a break point at $V_R + V_\gamma$.

Figure 10-2b shows a sinusoidal input signal of amplitude large enough so that the signal makes excursions past the break point. The corresponding output

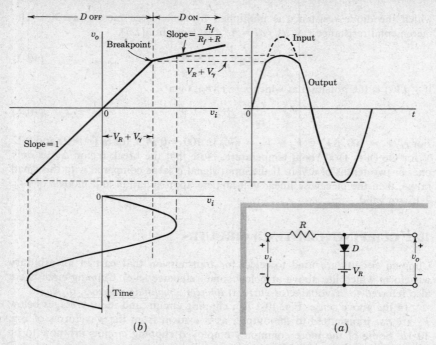

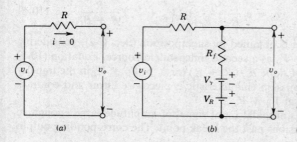

Figure 10-2 (a) A diode clipping circuit which transmits that part of the waveform more negative than $V_R + V_\gamma$. (b) The piecewise linear transmission characteristic of the circuit. A sinusoidal input and the clipped output are shown.

exhibits a suppression of the positive peak of the signal. If $R_f \ll R$, this suppression will be very pronounced, and the positive excursion of the output will be sharply limited at the voltage $V_R + V_\gamma$. The output will appear as though the positive peak had been "clipped off" or "sliced off." Often it turns out that $V_R \gg V_\gamma$, in which case one may consider that V_R itself is the limiting reference voltage.

Figure 10-3 Circuits from which to obtain the transfer characteristic of the clipper of Fig. 10-2. (a) The diode is OFF. (b) The diode is ON.

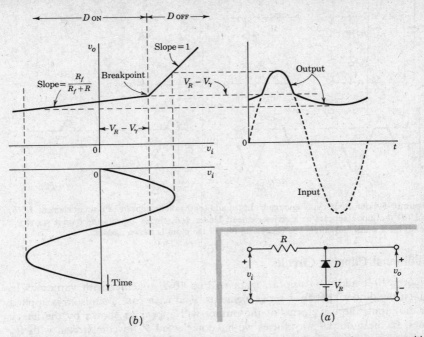

Figure 10-4 (*a*) A diode clipping circuit which transmits that part of the waveform more positive than $V_R - V_\gamma$. (*b*) The piecewise linear transmission characteristic of the circuit. A sinusoidal input and the clipped output are shown.

In Fig. 10-4*a* the clipping circuit has been modified in that the diode in Fig. 10-2*a* has been reversed. The corresponding piecewise linear representation of the transfer characteristic is shown in Fig. 10-4*b*. In this circuit, the portion of the waveform more positive than $V_R - V_\gamma$ is transmitted without attenuation, but the less positive portion is greatly suppressed.

In Figs. 10-2*b* and 10-4*b* we have assumed R_r arbitrarily large in comparison with R. If this condition does not apply, the transmission characteristics must be modified. The portions of these curves which are indicated as having unity slope must instead be considered to have a slope $R_r/(R_r + R)$.

In a transmission region of a diode clipping circuit we require that $R_r \gg R$, for example, that $R_r \doteq kR$, where k is a large number. In the attenuation region, we require that $R \gg R_f$, for example, that $R = kR_f$. From these two equations we deduce that $R = \sqrt{R_f R_r}$ and that $k = \sqrt{R_r/R_f}$. On this basis we conclude that it is reasonable to select R as the geometrical mean of R_r and R_f. And we note that the ratio R_r/R_f may well serve as a figure of merit for diodes used in the present application.

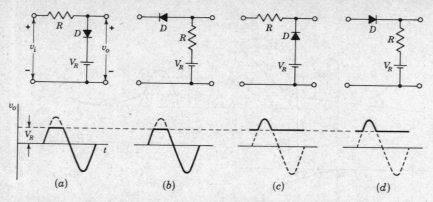

Figure 10-5 Four diode clipping circuits. In (a) and (c) the diode appears as a shunt element. In (b) and (d) the diode appears as a series element. Under each circuit appears the output waveform (solid) for a sinusoidal input. The clipped portion of the input is shown dashed.

Additional Clipping Circuits

Figures 10-2 and 10-4 appear again in Fig. 10-5, together with variations in which the diodes appear as series elements. If in each case a sinusoid is applied at the input, the waveforms at the output will appear as shown by the heavy lines. In these output waveforms we have neglected V_γ in comparison with V_R and we have assumed that the break region is negligible in comparison with the amplitude of the waveforms. We have also assumed that $R_r \gg R \gg R_f$. In two of these circuits the portion of the waveform transmitted is that part which lies below V_R; in the other two the portion above V_R is transmitted. In two the diode appears as an element in series with the signal lead; in two it appears as a shunt element. The use of the diode as a series element has the disadvantage that when the diode is OFF and it is intended that there be no transmission, fast signals or high-frequency waveforms may be transmitted to the output through the diode capacitance. The use of the diode as a shunt element has the disadvantage that when the diode is open (back-biased) and it is intended that there be transmission, the diode capacitance, together with all other capacitance in shunt with the output terminals, will round sharp edges of input waveforms and attenuate high-frequency signals. A second disadvantage of the use of the diode as a shunt element is that in such circuits the resistance R_s of the source which supplies V_R must be kept low. This requirement does not arise in circuits where V_R is in series with R, which is normally large compared with R_s.

10-3 CLIPPING AT TWO INDEPENDENT LEVELS

Diode clippers may be used in pairs to perform double-ended limiting at independent levels. A parallel, a series, or a series-parallel arrangement may be used. A parallel arrangement is shown in Fig. 10-6a. Figure 10-6b shows the

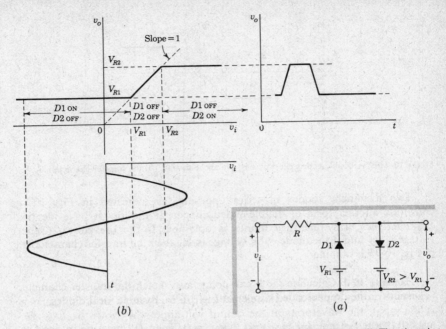

Figure 10-6 (*a*) A double-diode clipper which limits at two independent levels. (*b*) The piecewise linear transfer curve for the circuit in (*a*). The doubly clipped output for a sinusoidal input is shown.

piecewise linear and continuous input-output voltage curve for the circuit in Fig. 10-6*a*. The transfer curve has two break points, one at $v_o = v_i = V_{R1}$ and a second at $v_o = v_i = V_{R2}$, and has the following characteristics (assuming $V_{R2} > V_{R1} \gg V_\gamma$ and $R_f \ll R$):

Input v_i	Output v_o	Diode states	
$v_i \leqslant V_{R1}$	$v_o = V_{R1}$	$D1$ ON, $D2$ OFF	
$V_{R1} < v_i < V_{R2}$	$v_o = v_i$	$D1$ OFF, $D2$ OFF	(10-6)
$v_i \geqslant V_{R2}$	$v_o = V_{R2}$	$D1$ OFF, $D2$ ON	

The circuit of Fig. 10-6*a* is referred to as a *slicer* because the output contains a slice of the input between the two reference levels V_{R1} and V_{R2}.

The circuit is used as a means of converting a sinusoidal waveform into a square wave. In this application, to generate a symmetrical square wave, V_{R1} and V_{R2} are adjusted to be numerically equal but of opposite sign. The transfer characteristic passes through the origin under these conditions, and the waveform is clipped symmetrically top and bottom. If the amplitude of the sinusoidal waveform is very large in comparison with the difference in the reference levels, the output waveform will have been *squared*.

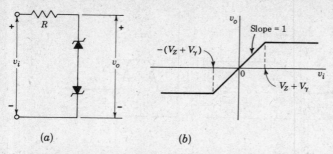

(a)

(b)

Figure 10-7 (a) A double-ended clipper using avalanche diodes; (b) the transfer characteristic.

Two avalanche diodes in series opposing, as indicated in Fig. 10-7a, constitute another form of double-ended clipper. If the diodes have identical characteristics, a symmetrical limiter is obtained. If the breakdown (Zener) voltage is V_Z and if the diode cutin voltage is V_γ, then the transfer characteristic of Fig. 10-7b is obtained.

Example 10-1 Calculate the break points and sketch the transfer characteristics for the double-ended clipper of Fig. 10-8a. Assume ideal diodes.

SOLUTION Assume $D1$ is OFF so that $i_1 = 0$. Then $D2$ must be ON, and the loop current i_2 is given by

$$i_2 = \frac{10 - 2.5}{10 + 5} = 0.5 \text{ mA}$$

The output voltage is

$$v_o = 10 - 5i_2 = 10 - (5)(0.5) = 7.5 \text{ V}$$

Since $D2$ is ON the voltage v_P at P equals $v_o = 7.5$ V and $D1$ is reverse-biased if $v_i < 7.5$ V. The first break point (B_1 in Fig. 10-8b) corresponds to $v_i = v_o = 7.5$ V.

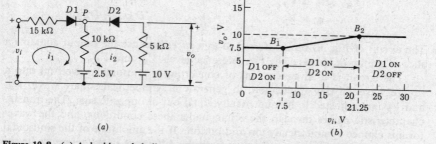

(a)

(b)

Figure 10-8 (a) A double-ended clipper. (b) The piecewise-linear transfer characteristic.

Now assume $D2$ is OFF, so that $i_2 = 0$ and $v_o = 10$ V. Under this circumstance $D1$ must be ON, and the loop current i_1 is given by

$$i_1 = \frac{v_i - 2.5}{15 + 10} = 0.04v_i - 0.1$$

and

$$v_P = 2.5 + 10i_1 = 1.5 + 0.4v_i$$

In order that $D2$ be reverse-biased, v_P must be at least 10 V. Hence,

$$1.5 + 0.4v_i \geqslant 10 \qquad \text{or} \qquad v_i \geqslant 21.25 \text{ V}$$

The second break-point (B_2 in Fig. 10-8b) corresponds to $v_i = 21.25$ V and $v_o = 10$ V. Between the two break-points both diodes are ON and the circuit consists only of sources and resistors. Such a network is linear and, hence, B_1 and B_2 are connected by a straight line, resulting in the broken-line transfer characteristic shown.

As a check on the foregoing calculations, assume that $D1$ and $D2$ are both ON. Proceeding as in Appendix C (Fig. C-5), the KVL loop equations are

$$- v_i + 25i_1 + 10i_2 + 2.5 = 0 \tag{10-7}$$

$$-10 + 15i_2 + 10i_1 + 2.5 = 0 \tag{10-8}$$

Eliminating i_1 from Eqs. (10-7) and (10-8) and solving for $v_o = 10 - 5i_2$, we obtain

$$v_o = \frac{2v_i + 67.5}{11} \tag{10-9}$$

This equation shows that v_o is indeed linearly related to v_i. Also if $v_i = 7.5$ V, Eq. (10-9) yields $v_o = 7.5$ V, whereas if $v_i = 21.25$ V, then $v_o = 10$ V, which are the break points found above.

Catching or Clamping Diodes

Consider that v_i and R in Fig. 10-6a represent Thévenin's circuit model, Sec. C-2, at the output of a device, such as an amplifier. In other words, R is the output resistance and v_i is the open-circuit output signal. In such a situation $D1$ and $D2$ are called *catching diodes*. The reason for this terminology should be clear from Fig. 10-9, where we see that $D1$ "catches" the output v_o and does not allow it to fall below $V_{R1} - V_\gamma$, whereas $D2$ "catches" v_o and does not permit it to rise above $V_{R2} + V_\gamma$.

Generally, whenever a node becomes connected through a low resistance (as through a conducting diode) to some reference voltage V_R, we say that the node has been clamped to V_R, since the voltage at that point in the circuit is unable to depart appreciably from V_R. In this sense the diodes in Fig. 10-9 are called *clamping diodes*.

A circuit for clamping the extremity of a periodic waveform to a reference voltage is considered in Sec. 10-8.

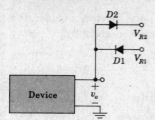

Figure 10-9 Catching diodes $D1$ and $D2$ limit the output excursion of the device between V_{R1} and V_{R2}.

10-4 A BREAKDOWN-DIODE VOLTAGE REGULATOR

A circuit using an avalanche or Zener diode as a voltage regulator is indicated in Fig. 10-10. The source V and resistor R are selected so that, initially, the diode is operating in the breakdown region. Here the diode voltage, which is also the voltage across the load R_L, is V_Z as in Fig. 2-12, and the diode current is I_Z. The diode will now regulate the load voltage against variations in load current and against variations in supply voltage V because, in the breakdown region, large changes in diode current produce only small changes in diode voltage. Moreover, as load current or supply voltage changes, the diode current will accommodate itself to these changes to maintain a nearly constant load voltage. The diode will continue to regulate until the circuit operation requires the diode current to fall to I_{ZK}, in the neighborhood of the knee of the diode volt-ampere curve of Fig. 2-12a. The upper limit on diode current is determined by the power-dissipation rating of the diode.

Example 10-2 (*a*) The avalanche diode regulates at 50 V over a range of diode currents from 5 to 40 mA. The supply voltage $V = 200$ V. Calculate R to allow voltage regulation from a load current $I_L = 0$ up to $I_{L(max)}$, the maximum possible value of I_L. What is $I_{L(max)}$? (*b*) If R is set as in part *a* and the load current is set at $I_L = 25$ mA, what are the limits between which V may vary without loss of regulation in the circuit?

SOLUTION (*a*) The current from the voltage source V is given by

$$I = \frac{V - V_Z}{R} \qquad (10\text{-}10)$$

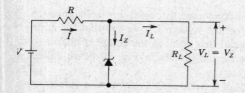

Figure 10-10 A Zener voltage regulator.

From Kirchhoff's current law (KCL) (Sec. C-1) the diode current is given by

$$I_Z = I - I_L \qquad (10\text{-}11)$$

As the load resistance R_L is varied so that $I_L = V_Z/R_L$ changes, we see from Eq. (10-10) that the current I remains constant. Hence, from Eq. (10-11) the diode current I_Z decreases with an increase in load current. Therefore, the maximum $I_Z = 40$ mA occurs at the minimum $I_L = 0$. From Eq. (10-11), $I = I_Z = 40$ mA. From Eq. (10-10)

$$R = \frac{200 - 50}{40} = 3.75 \text{ k}\Omega$$

Since the minimum Zener current (the value of I_{ZK} in Fig. 2-10a) is 5 mA, then from Eq. (10-11) the maximum load current is

$$I_{L(\text{max})} = I - I_{ZK} = 40 - 5 = 35 \text{ mA}$$

(*b*) At the minimum diode current $I = 5 + 25 = 30$ mA and from Eq. (10-10)

$$V = IR + V_Z = (30)(3.75) + 50 = 162.5 \text{ V}$$

At the maximum Zener current,

$$I = 40 + 25 = 65 \text{ mA} \qquad \text{and} \qquad V = (65)(3.75) + 50 = 293.8 \text{ V}$$

Hence, the source may vary between 162.5 and 293.8 V, and the output voltage will remain constant at 50 V and the load current will be constant at 25 mA.

10-5 RECTIFIERS

Almost all electronic circuits require a dc source of power. For portable low-power systems batteries may be used. More frequently, however, electronic equipment is energized by a *power supply*, a piece of equipment which converts the alternating waveform from the power lines into an essentially direct voltage. The study of ac-to-dc conversion is initiated in this section.

A Half-Wave Rectifier

A device, such as the semiconductor diode, which is capable of converting a sinusoidal input waveform (whose average value is zero) into a unidirectional (though not constant) waveform, with a nonzero average component, is called a *rectifier*. The basic circuit for half-wave rectification is shown in Fig. 10-11. Since in a rectifier circuit the input $v_i = V_m \sin \omega t$ has a peak value V_m which is very large compared with the cutin voltage V_γ of the diode, we assume in the following discussion that $V_\gamma = 0$. (The condition $V_\gamma \neq 0$ is treated in Sec. 10-1, and the current waveform is shown in Fig. 10-1b.) With the diode idealized to be a resistance R_f in the ON state and an open circuit in the OFF state, the current

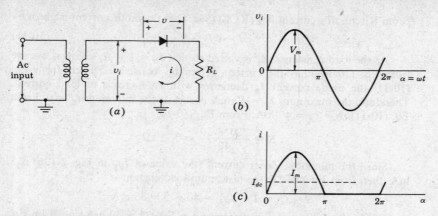

Figure 10-11 (a) Basic circuit of half-wave rectifier. (Although not necessary for circuit operation, the transformer is included for the sake of completeness since most power supplies have input transformers.) (b) Transformer sinusoidal secondary voltage v_i. (c) Diode and load current i.

i in the diode or load R_L is given by

$$i = I_m \sin \alpha \quad \text{if } 0 \leqslant \alpha \leqslant \pi$$
$$i = 0 \quad \text{if } \pi \leqslant \alpha \leqslant 2\pi \tag{10-12}$$

where $\alpha \equiv \omega t$ and

$$I_m \equiv \frac{V_m}{R_f + R_L} \tag{10-13}$$

The transformer secondary voltage v_i is shown in Fig. 10-11b, and the rectified current in Fig. 10-11c. Note that the output current is unidirectional. We now calculate this nonzero value of the average current.

A dc ammeter is constructed so that the needle deflection indicates the average value of the current passing through it. By definition, the average value of a periodic function is given by the area of one cycle of the curve divided by the base. Expressed mathematically,

$$I_{dc} = \frac{1}{2\pi} \int_0^{2\pi} i \, d\alpha \tag{10-14}$$

For the half-wave circuit under consideration, it follows from Eqs. (10-12) that

$$I_{dc} = \frac{1}{2\pi} \int_0^{\pi} I_m \sin \alpha \, d\alpha = \frac{I_m}{\pi} \tag{10-15}$$

Note that the upper limit of the integral has been changed from 2π to π since the instantaneous current in the interval from π to 2π is zero and so contributes nothing to the integral.

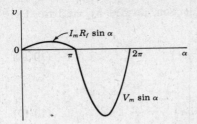

Figure 10-12 The voltage across the diode in Fig. 10-11.

The Diode Voltage

The dc (average) output voltage is clearly given as

$$V_{dc} = I_{dc}R_L = \frac{I_m R_L}{\pi} \tag{10-16}$$

However, the reading of a dc voltmeter placed across the diode is *not* given by $I_{dc}R_f$ because the diode cannot be modeled as a constant resistance, but rather it has two values: R_f in the ON state and ∞ in the OFF state.

A dc voltmeter reads the average value of the voltage across its terminals. Hence, to obtain V'_{dc} across the diode, the instantaneous voltage must be plotted as in Fig. 10-12 and the average value obtained by integration. Thus

$$V'_{dc} = \frac{1}{2\pi} \left(\int_0^\pi I_m R_f \sin \alpha \, d\alpha + \int_\pi^{2\pi} V_m \sin \alpha \, d\alpha \right)$$

$$= \frac{1}{\pi} (I_m R_f - V_m) = \frac{1}{\pi} \left[I_m R_f - I_m (R_f + R_L) \right]$$

where use has been made of Eq. (10-13). Hence

$$V'_{dc} = - \frac{I_m R_L}{\pi} \tag{10-17}$$

This result is negative, which means that if the voltmeter is to read upscale, its positive terminal must be connected to the cathode of the diode. From Eq. (10-16) the dc diode voltage is seen to be equal to the negative of the average voltage across the load resistor. This result is evidently correct because the sum of the dc voltages around the complete circuit must add up to zero.

The AC Current (Voltage)

A root-mean-square ammeter (voltmeter) is constructed so that the needle deflection indicates the effective, or rms, current (voltage). Such a "square-law" instrument may be of the thermocouple type. By definition, the effective or rms value squared of a periodic function of time is given by the area of one cycle of the curve, which represents the square of the function, divided by the base. Expressed mathematically,

$$I_{rms} = \left(\frac{1}{2\pi} \int_0^{2\pi} i^2 \, d\alpha \right)^{1/2} \qquad (10\text{-}18)$$

By use of Eqs. (10-12), it follows that

$$I_{rms} = \left(\frac{1}{2\pi} \int_0^{\pi} I_m^2 \sin^2\alpha \, d\alpha \right)^{1/2} = \frac{I_m}{2} \qquad (10\text{-}19)$$

The rms output voltage is given by $I_m R_L/2$.

Applying Eq. (10-18) to the *sinusoidal input voltage*, we obtain

$$V_{rms} = \frac{V_m}{\sqrt{2}} \qquad (10\text{-}20)$$

Regulation

The variation of dc output voltage as a function of dc load current is called *regulation*. The percentage regulation is defined as

$$\% \text{ regulation} \equiv \frac{V_{no \, load} - V_{load}}{V_{load}} \times 100\% \qquad (10\text{-}21)$$

where *no load* refers to zero current and *load* indicates the normal load current. For an ideal power supply the output voltage is independent of the load (the output current) and the percentage regulation is zero.

The variation of V_{dc} with I_{dc} for the half-wave rectifier is obtained as follows: From Eqs. (10-15) and (10-13),

$$I_{dc} = \frac{I_m}{\pi} = \frac{V_m/\pi}{R_f + R_L} \qquad (10\text{-}22)$$

Solving Eq. (10-22) for $V_{dc} = I_{dc}R_L$, we obtain

$$V_{dc} = \frac{V_m}{\pi} - I_{dc}R_f \qquad (10\text{-}23)$$

This result is consistent with the circuit model given in Fig. 10-13 for the dc voltage and current. Note that the rectifier circuit functions as if it were a constant (open-circuit) voltage source $V = V_m/\pi$ in series with an effective internal resistance (the *output resistance*) $R_o = R_f$. This model shows that V_{dc} equals V_m/π at no load and that the dc voltage decreases linearly with an increase in dc output current. In practice, the resistance R_s of the transformer secondary is in series with the diode, and in Eq. (10-23) R_s should be added to R_f. The best method of estimating the diode resistance is to obtain a regulation plot of V_{dc} versus I_{dc} in the laboratory. The negative slope of the resulting straight line gives $R_f + R_s$. Clearly, Fig. 10-13 represents a Thévenin's model, and hence a rectifier behaves as a linear circuit with respect to average current and voltage.

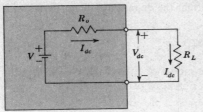

Figure 10-13 The Thévenin's model which gives the dc voltage and current for a power supply. For the half-wave circuit of Fig. 10-11, $V = V_m/\pi$ and $R_o = R_f$. For the full-wave circuit of Fig. 10-14, $V = 2V_m/\pi$ and $R_o = R_f$. For the full-wave rectifier with a capacitor filter (Sec. 10-7), $V_o = V_m$ and $R_o = 1/4fC$.

A Full-Wave Rectifier

The circuit of a full-wave rectifier is shown in Fig. 10-14a. This circuit is seen to comprise two half-wave circuits so connected that conduction takes place through one diode during one half of the power cycle and through the other diode during the second half of the cycle.

The current to the load, which is the sum of these two currents, $i = i_1 + i_2$, has the form shown in Fig. 10-14b. The dc and rms values of the load current and voltage in such a system are readily found to be

$$I_{dc} = \frac{2I_m}{\pi} \qquad I_{rms} = \frac{I_m}{\sqrt{2}} \qquad V_{dc} = \frac{2I_m R_L}{\pi} \qquad (10\text{-}24)$$

where I_m is given by Eq. (10-13) and V_m is the peak transformer secondary voltage from one end to the center tap. Note by comparing Eq. (10-24) with Eq. (10-16) that the dc output voltage for the full-wave connection is twice that for the half-wave circuit.

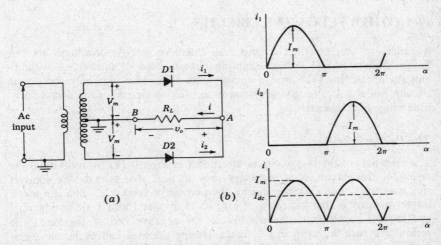

Figure 10-14 (a) A full-wave rectifier circuit. (b) The individual diode currents and the load current i. The output voltage is $v_o = iR_L$.

From Eqs. (10-13) and (10-24) we find that the dc output voltage varies with current in the following manner:

$$V_{\text{dc}} = \frac{2V_m}{\pi} - I_{\text{dc}}R_f \tag{10-25}$$

This expression leads to the Thévenin's dc model of Fig. 10-13, except that the internal (open-circuit) supply is $V = 2V_m/\pi$ instead of V_m/π.

Peak Inverse Voltage

For each rectifier circuit there is a maximum voltage to which the diode can be subjected. This potential is called the *peak inverse voltage* because it occurs during that part of the cycle when the diode is nonconducting. From Fig. 10-11 it is clear that, for the half-wave rectifier, the peak inverse voltage is V_m. We now show that, for a full-wave circuit, twice this value is obtained. At the instant of time when the transformer secondary voltage to midpoint is at its peak value V_m, diode $D1$ is conducting and $D2$ is nonconducting. If we apply KVL around the outside loop and neglect the small voltage drop across $D1$, we obtain $2V_m$ for the peak inverse voltage across $D2$. Note that this result is obtained without reference to the nature of the load, which can be a pure resistance R_L or a combination of R_L and some reactive elements which may be introduced to "filter" the ripple. We conclude that, *in a full-wave circuit, independently of the filter used, the peak inverse voltage across each diode is twice the maximum transformer voltage measured from midpoint to either end.*

Rectification of a sinusoid whose peak value is less than V_γ is discussed in Sec. 16-8.

10-6 OTHER FULL-WAVE CIRCUITS

A variety of other rectifier circuits find extensive use. Among these are the bridge circuit, several voltage-doubling circuits, and a number of voltage-multiplying circuits. The bridge circuit finds application not only for power circuits, but also as a rectifying system in rectifier ac meters for use over a fairly wide range of frequencies.

The Bridge Rectifier

The essentials of the bridge circuit are shown in Fig. 10-15. To understand the action of this circuit, it is necessary only to note that two diodes conduct simultaneously. For example, during the portion of the cycle when the transformer polarity is that indicated in Fig. 10-15, diodes 1 and 3 are conducting, and current passes from the positive to the negative end of the load. The conduction path is shown in the figure. During the next half cycle, the transformer voltage reverses its polarity, and diodes 2 and 4 send current through the load in the same direction as during the previous half cycle.

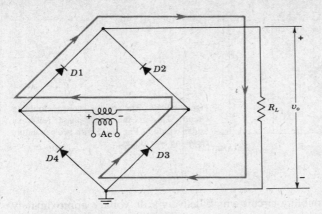

Figure 10-15 Full-wave bridge circuit.

The principal features of the bridge circuit are the following: The currents drawn in both the primary and the secondary of the supply transformer are sinusoidal, and therefore a smaller transformer may be used than for the full-wave circuit of the same output; a transformer without a center tap is used; and each diode has only transformer voltage across it on the inverse cycle. The bridge circuit is thus suitable for high-voltage applications.

The Rectifier Meter

This instrument, illustrated in Fig. 10-16, is essentially a bridge-rectifier system, except that no transformer is required. Instead, the voltage to be measured is applied through a multiplier resistor R to two corners of the bridge, a dc milliammeter being used as an indicating instrument across the other two corners. Since the dc milliammeter reads average values of current, the meter scale is calibrated to give rms values when a sinusoidal voltage is applied to the input terminals. As a result, this instrument will not read correctly when used with waveforms which contain appreciable harmonics.

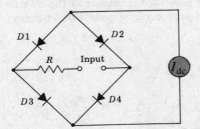

Figure 10-16 The rectifier voltmeter.

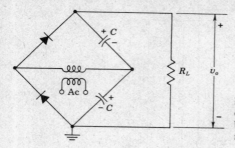

Figure 10-17 The bridge voltage-doubling circuit. This is the single-phase full-wave bridge circuit of Fig. 10-15 with two capacitors replacing two diodes.

Voltage Multipliers

A common voltage-doubling circuit which delivers a dc voltage approximately equal to twice the transformer maximum voltage at no load is shown in Fig. 10-17. This circuit is operated by alternately charging each of the two capacitors to the transformer peak voltage V_m, current being continually drained from the capacitors through the load. The capacitors also act to smooth out the ripple in the output.

10-7 CAPACITOR FILTERS

Filtering is frequently effected by shunting the load with a capacitor. The action of this system depends upon the fact that the capacitor stores energy during the conduction period and delivers this energy to the load during the inverse, or nonconducting, period. In this way, the time during which the current passes through the load is prolonged, and the ripple is considerably decreased. The ripple voltage is defined as the deviation of the load voltage from its average or dc value.

Consider the half-wave capacitive rectifier of Fig. 10-18. Suppose, first that the load resistance $R_L = \infty$. The capacitor will charge to the potential V_m, the transformer maximum value. Further, the capacitor will maintain this potential, for no path exists by which this charge is permitted to leak off, since the diode will not pass a negative current. The diode resistance is infinite in the inverse direction, and no charge can flow during this portion of the cycle. Consequently,

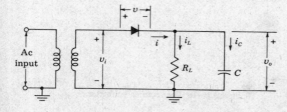

Figure 10-18 A half-wave capacitor-filtered rectifier.

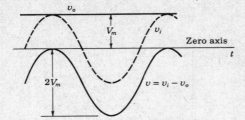

Figure 10-19 Voltages in a half-wave capacitor-filtered rectifier at no load. The output voltage v_o is a constant, indicating perfect filtering. The diode voltage v is negative for all values of time, and the peak inverse voltage is $2V_m$.

the filtering action is perfect, and the capacitor voltage v_o remains constant at its peak value, as is seen in Fig. 10-19.

The voltage v_o across the capacitor is, of course, the same as the voltage across the load resistor, since the two elements are in parallel. The diode voltage v is given by

$$v = v_i - v_o \tag{10-26}$$

We see from Fig. 10-19 that the diode voltage is always negative and that the peak inverse voltage is twice the transformer maximum. Hence the presence of the capacitor causes the peak inverse voltage to increase from a value equal to the transformer maximum when no capacitor filter is used to a value equal to twice the transformer maximum value when the filter is used.

Suppose, now, that the load resistor R_L is finite. Without the capacitor input filter, the load current and the load voltage during the conduction period will be sinusoidal functions of time. The inclusion of a capacitor in the circuit results in the capacitor charging in step with the applied voltage. Also, the capacitor must discharge through the load resistor, since the diode will prevent a current in the negative direction. Clearly, the diode acts as a switch which permits charge to flow into the capacitor when the transformer voltage exceeds the capacitor voltage, and then acts to disconnect the power source when the transformer voltage falls below that of the capacitor.

Output Voltage under Load

During the time interval when the diode in Fig. 10-18 is conducting, the transformer voltage is impressed directly across the load (assuming that the diode drop can be neglected). Hence, the output voltage is $v_o = V_m \sin \omega t$. During the interval when D is nonconducting, the capacitor discharges through the load with a time constant CR_L. The output waveform in Fig. 10-20 consists of portions of sinusoids (when D is ON) joined to exponential segments (when D is OFF). The point at which the diode starts to conduct is called the *cutin* point t_2, and that at which it stops conducting is called the *cutout point* t_1. These times are indicated in Fig. 10-20.

The cutout time is obtained from the expression (Prob. 10-37) for the current i in Fig. 10-18 when $v_o = V_m \sin \omega t$. Then the time for which $i = 0$ gives

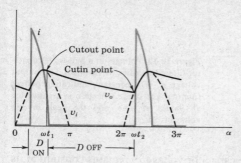

Figure 10-20 Theoretical sketch of diode current i and output voltage v_o in a half-wave capacitor-filtered rectifier.

the cutout angle ωt_1. The cutin point t_2 is obtained graphically by finding the time when the exponential portion of v_o in Fig. 10-20 intersects the curve $V_m \sin \omega t$ (in the following cycle). The validity of this statement follows from the fact that at an instant of time greater than t_2, the transformer voltage v_i (the sine curve) is greater than the capacitor voltage v_o (the exponential curve). Since the diode voltage is $v = v_i - v_o$, then v will be positive beyond t_2 and the diode will become conducting. Thus t_2 is the cutin point.

The use of a large capacitance to improve the filtering at a given load R_L is accompanied by a high-peak diode current I_m. For a specified average load current, i becomes more peaked and the conduction period decreases as C is made larger. It is to be emphasized that the use of a capacitor filter may impose serious restrictions on the diode, since the average current may be well within the current rating of the diode, and yet the peak current may be excessive.

Full-Wave Circuit

Consider a full-wave rectifier with a capacitor filter obtained by placing a capacitor C across R_L in Fig. 10-14. The analysis of this circuit requires a simple extension of that just made for the half-wave circuit. If in Fig. 10-20 a dashed half-sinusoid is added between π and 2π, the result is the dashed full-wave voltage in Fig. 10-21. The cutin point now lies between π and 2π, where the

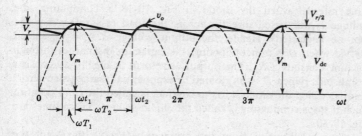

Figure 10-21 The approximate load-voltage waveform v_o in a full-wave capacitor-filtered rectifier.

exponential portion of v_o intersects this sinusoid. The cutout point is the same as that found for the halfwave rectifier.

Approximate Analysis

It is possible to obtain the dc output voltage for given values of the parameters ω, R_L, C, and V_m from the graphical construction indicated in Fig. 10-21. Such an analysis is involved and tedious. Hence we now present an approximate solution which is simple and yet sufficiently accurate for most engineering applications.

We assume that the output-voltage waveform of a full-wave circuit with a capacitor filter may be represented by the approximately piecewise linear curve shown in Fig. 10-21. For large values of C (so that $\omega C R_L \gg 1$) we note that $\omega t_1 \rightarrow \pi/2$ and $v_o \rightarrow V_m$ at $t = t_1$. Also, with C very large, the exponential decay can be replaced by a linear fall. If the total capacitor discharge voltage (the ripple voltage) is denoted by V_r, then from Fig. 10-21, the average value of the voltage is approximately

$$V_{dc} = V_m - \frac{V_r}{2} \tag{10-27}$$

It is necessary, however, to express V_r as a function of the load current and the capacitance. If T_2 represents the total nonconducting time, the capacitor, when discharging at the constant rate I_{dc}, will lose an amount of charge $I_{dc}T_2$. Hence the change in capacitor voltage is $I_{dc}T_2/C$, or

$$V_r = \frac{I_{dc}T_2}{C} \tag{10-28}$$

The better the filtering action, the smaller will be the conduction time T_1 and the closer T_2 will approach the time of half a cycle. Hence we assume that $T_2 = T/2 = 1/2f$, where f is the fundamental power-line frequency. Then

$$V_r = \frac{I_{dc}}{2fC} \tag{10-29}$$

and from Eq. (10-27),

$$V_{dc} = V_m - \frac{I_{dc}}{4fC} \tag{10-30}$$

This result is consistent with Thévenin's model of Fig. 10-13, with the open-circuit voltage $V = V_m$ and the effective output resistance $R_o = 1/4fC$.

The ripple is seen to vary directly with the load current I_{dc} and also inversely with the capacitance. Hence, to keep the ripple low and to ensure good regulation, very large capacitances (of the order of tens of microfarads) must be used. The most common type of capacitor for this rectifier application is the electrolytic capacitor. These capacitors are polarized, and care must be taken to insert them into the circuit with the terminal marked + to the positive side of the output.

The desirable features of rectifiers employing capacitor input filters are the small ripple and the high voltage at light load. The no-load voltage is equal, theoretically, to the maximum transformer voltage. The disadvantages of this system are the relatively poor regulation, the high ripple at large load currents, and the peaked currents that the diodes must pass.

An approximate analysis similar to that given above applied to the half-wave circuit shows that the ripple, and also the drop from no load to a given load, are double the values calculated for the full-wave rectifier.

10-8 ADDITIONAL DIODE CIRCUITS

Many applications depend upon the semiconductor diode besides those already considered in this chapter. We mention three others below.

Peak Detector

The half-wave capacitor-filtered rectifier circuit of Fig. 10-18 may be used to measure the peak value of an input waveform. Thus, for $R_L = \infty$, the capacitor charges to the maximum value V_{max} of v_i, the diode becomes nonconducting, and v_o remains at V_{max} (assuming an ideal capacitor with no leakage resistance shunting C). Refer to Fig. 10-19, where $V_{max} = V_m =$ the peak value of the input sinusoid. Improved peak detector circuits are given in Sec. 16-8.

In an AM radio the amplitude of the high-frequency wave (called the *carrier*) is varied in accordance with the audio information to be transmitted. This process is called *amplitude modulation*, and such an AM waveform is illustrated in Fig. 10-22. The audio information is contained in the envelope, shown dashed (the locus of the peak values) of the modulated waveform. The process of extracting the audio signal is called *detection*, or *demodulation*. If the input to Fig. 10-18 is the AM waveform shown in Fig. 10-22, the output v_o is the heavy-weight curve, provided that the time constant $R_L C$ is chosen properly; that is, $R_L C$ must be small enough so that, when the envelope decreases in magnitude, the voltage across C can fall fast enough to keep in step with the

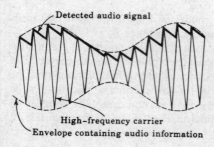

Detected audio signal

High-frequency carrier

Envelope containing audio information

Figure 10-22 An amplitude-modulated wave and the detected audio signal. (For ease of drawing, the carrier waveform is indicated triangular instead of sinusoidal and of much lower frequency than it really is, relative to the audio frequency.)

envelope, but $R_L C$ must not be so small as to introduce excessive ripple. The order of magnitude of the frequency of an AM radio carrier is 1,000 kHz, and the audio spectrum extends from about 20 Hz to 20 kHz. Hence there should be *at least* 50 cycles of the carrier waveform for each audio cycle. If Fig. 10-22 were drawn more realistically (with a much higher ratio of carrier to audio frequency), then clearly, the ripple amplitude of the demodulated signal would be very much smaller. This low-amplitude high-frequency ripple in v_o is easily filtered so that the smoothed detected waveform is an excellent reproduction of the audio signal. The capacitor-rectifier circuit of Fig. 10-18 therefore also acts as an *envelope demodulator*.

A Clamping Circuit

A function which must be frequently performed with a periodic waveform is the establishment of the recurrent positive or negative extremity at some constant reference level V_R. Such a clamping circuit is indicated in Fig. 10-23a. Assuming an ideal diode, the drop across the device is zero in the forward direction. Hence the output cannot rise above V_R and is said to be *clamped* to this level. If the input is sinusoidal with a peak value V_m and an average value of zero, then, as indicated in Fig. 10-23b, the output is sinusoidal, with an average value of $V_R - V_m$. This waveform is obtained subject to the following conditions: the diode parameters are $R_f = 0$, $R_r = \infty$, and $V_\gamma = 0$; the source impedance $R_s = 0$; and the time constant RC is much larger than the period of the signal. In practice these restrictions are not completely satisfied and the clamping is not perfect; the output voltage rises slightly above V_R, and the waveshape is a somewhat distorted version of the input.[1]

Digital-Computer Circuits

Since the diode is a binary device existing in either the ON or OFF state for a given interval of time, it is a very usual component in digital-computer applications. Such so-called "logic" circuits are discussed in Chap. 5 in conjunction with transistor binary applications.

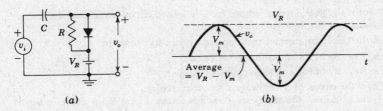

(a)   (b)

Figure 10-23 (a) A circuit which clamps to the voltage V_R. (b) The output voltage v_o for a sinusoidal input v_i.

REFERENCE

1. Millman, J., and H. Taub: "Pulse, Digital, and Switching Waveforms," chap. 3, McGraw-Hill Book Company, New York, 1965.

REVIEW QUESTIONS

10-1 Consider a diode D, a resistance R, and a signal source v_i in series. If $v_i = V_m \sin \omega t$ and if the diode is represented by its piecewise linear model, find the current as a function of time and plot.

10-2 What is meant by the *break region* of a diode?

10-3 Consider a series circuit consisting of a diode D, a resistance R, a reference battery V_R, and an input signal v_i. The output is taken across R and V_R in series. Draw the transfer characteristic if the anode of D is connected to the positive terminal of the battery. Use the piecewise linear diode model.

10-4 Repeat Rev. 10-3 with the anode of D connected to the negative terminal of the battery.

10-5 If v_i is sinusoidal and D is ideal (with $R_f = 0$, $V_\gamma = 0$, and $R_r = \infty$), find the output waveforms in (*a*) Rev. 10-3 and (*b*) Rev. 10-4.

10-6 Sketch the circuit of a *double-ended clipper* using ideal *p-n* diodes which limit the output between ± 10 V.

10-7 Repeat Rev. 10-6 using avalanche diodes.

10-8 (*a*) Draw a circuit which uses a breakdown diode to regulate the voltage across a load.

(*b*) Explain the circuit operation.

10-9 Define in words and as an equation: (*a*) dc current I_{dc}; (*b*) dc voltage V_{dc}; (*c*) ac current I_{rms}.

10-10 (*a*) Sketch the circuit for a full-wave rectifier.

(*b*) Derive the expression for (1) the dc current; (2) the dc load voltage; (3) the dc diode voltage; (4) the rms load current.

10-11 (*a*) Define *regulation*.

(*b*) Derive the regulation equation for a full-wave circuit.

10-12 Draw the Thévenin's model for a full-wave rectifier.

10-13 (*a*) Define *peak inverse voltage*.

(*b*) What is the peak inverse voltage for a full-wave circuit using ideal diodes?

(*c*) Repeat part (*b*) for a half-wave rectifier.

10-14 Sketch the circuit of a bridge rectifier and explain its operation.

10-15 Repeat Rev. 10-14 for a rectifier meter circuit.

10-16 Repeat Rev. 10-14 for a voltage-doubler circuit.

10-17 (*a*) Draw the circuit of a half-wave capacitive rectifier.

(*b*) At no load draw the steady-state voltage across the capacitor and also across the diode.

10-18 (*a*) Draw the circuit of a full-wave capacitive rectifier.

(*b*) Draw the output voltage under load. Indicate over what period of time the diode conducts. Make no calculations.

(*c*) Indicate the diode current waveform superimposed upon the output waveform.

10-19 (*a*) Consider a full-wave capacitor rectifier using a large capacitance *C*. Sketch the approximate output waveform.

(*b*) Derive the expression for the peak ripple voltage.

(*c*) Derive the Thévenin's model for this rectifier.

10-20 For a full-wave capacitor rectifier circuit, list (*a*) two advantages; (*b*) three disadvantages.

10-21 Describe (*a*) *amplitude modulation* and (*b*) *detection*.

LOW-FREQUENCY AMPLIFIERS

In a digital system a transistor operates in one of two states: it is either at cutoff or in saturation. On the other hand, when a transistor is used as an amplifier, it is biased so that it is operating in the active region. With no signal applied to the base, the collector current and voltage determine a point approximately in the center of the output characteristics. This zero-excitation operating point is called the *quiescent point* Q. Methods for establishing the BJT operating point are given. The quiescent collector current shifts with changes in temperature because the transistor parameters (β, I_{CO}, and V_{BE}) are functions of T. This increment in I_C is calculated.

The large-signal response of a transistor is obtained graphically. For small signals the transistor operates with reasonable linearity, and we inquire into small-signal linear models which represent the operation of the transistor in the active region. A detailed study of the bipolar transistor amplifier in its various configurations is made.

Very often, in practice, a number of stages are used in cascade to amplify a signal from a source, such as a phonograph pickup, to a level which is suitable for the operation of another transducer, such as a loudspeaker. We consider the problem of cascading a number of transistor amplifier stages at low frequencies, where the transistor internal capacitances may be neglected.

Methods of biasing the field-effect transistor are given, a small-signal model for the device is developed, and common-source and source-follower amplifiers are studied.

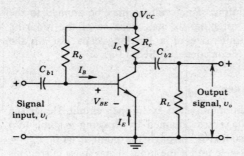

Figure 11-1 The fixed-bias circuit.

11-1 THE OPERATING POINT OF A BJT

The bipolar junction transistor functions most linearly when it is constrained to operate in its active region. To establish an operating point in this region it is necessary to provide appropriate direct potentials and currents, using external sources. Once an operating point Q is established, time-varying excursions of the input signal (base current, for example) should cause an output signal (collector voltage or collector current) of the same waveform. If the output signal is not a faithful reproduction of the input signal, for example, if it is clipped on one side, the operating point is unsatisfactory and should be relocated on the collector characteristics. The question now naturally arises as to how to choose the operating point. In Fig. 11-1 we show a common-emitter circuit. Figure 11-2 gives the output characteristics of the transistor used in Fig. 11-1. Note that even if we are free to choose R_c, R_L, R_b, and V_{CC}, we may not operate the transistor everywhere in the active region because the various transistor ratings limit the range of useful operation. These ratings [listed in the manufacturer's specification sheets (Appendix B-3)] are maximum collector dissipation $P_{C(\max)}$, maximum collector voltage $V_{C(\max)}$, maximum collector current $I_{C(\max)}$, and maximum emitter-to-base voltage $V_{EB(\max)}$. Figure 11-2 shows three of these bounds on typical collector characteristics.

Capacitive Coupling

A capacitor C_{b1} is used to couple the input signal to the transistor, as indicated in Fig. 11-1. In this diagram one end of v_i is at ground potential. The collector supply V_{CC} provides the biasing base current I_B and the quiescent collector current I_C. Under quiescent conditions (no input signal), C_{b1} (called a *blocking* capacitor) acts as an open circuit, because the reactance of a capacitor is infinite at zero frequency (dc). The capacitances C_{b1} and C_{b2} are chosen large enough so that, at the lowest frequency of excitation, their reactances are small enough so that they can be considered to be short circuits. These coupling capacitors block dc voltages but freely pass signal voltages. For example, the quiescent collector voltage does not appear at the output, but v_o is an amplified replica of the input signal v_i. The (ac or incremental) output signal voltage may be applied to the input of another amplifier without affecting its bias, because of the blocking capacitor C_{b2}. The effect of the finite size of a blocking capacitor on the frequency response of an amplifier is considered in Sec. 13-3.

The Static and Dynamic Load Lines

We noted above that under dc conditions C_{b2} acts as an open circuit. Hence the quiescent collector current and voltage are obtained by drawing a static (dc) load line corresponding to the resistance R_c through the point $I_C = 0$, $V_{CE} = V_{CC}$, as indicated in Fig. 11-2. If $R_L = \infty$ and if the input signal (base current) is large and symmetrical, we must locate the operating point Q_1 at the center of the load line. In this way the collector voltage and current may vary approximately symmetrically around the quiescent values V_C and I_C, respectively. If $R_L \neq \infty$, however, a *dynamic* (ac) load line must be drawn. Since we have assumed that, at the signal frequency, C_{b2} acts as a short circuit, the effective load R_L' at the collector is R_c in parallel with R_L. The dynamic load line must be drawn through the operating point Q_1 and must have a slope corresponding to $R_L' = R_c \| R_L$.

This ac load line is indicated in Fig. 11-2, where we observe that the input signal may swing a maximum of approximately 40 μA around Q_1 because if the base current decreases by more than 40 μA, the transistor is driven off.

If a larger input swing is available, then in order to avoid cutoff during a part of the cycle, the quiescent point must be located at a higher current. For example, by simple trial and error we locate Q_2 *on the dc load line* such that a line with a slope corresponding to the ac resistance R_L' and drawn through Q_2 gives as large an output as possible without too much distortion. In Fig. 11-2 the choice of Q_2 allows an input peak current swing of about 60 μA.

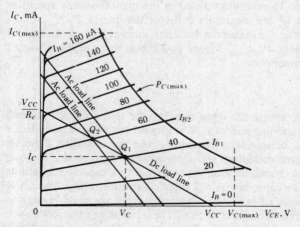

Figure 11-2 Common-emitter collector characteristics; ac and dc load lines.

The Fixed-Bias Circuit

The point Q_2 can be established by noting the required current I_{B2} in Fig. 11-2 and choosing the resistance R_b in Fig. 11-1 so that the base current is equal to I_{B2}. Therefore

$$I_B = \frac{V_{CC} - V_{BE}}{R_b} = I_{B2} \qquad (11\text{-}1)$$

The voltage V_{BE} across the forward-biased emitter junction is (Table 3-1, p. 79) approximately 0.7 V for a silicon transistor and 0.2 V for a germanium transistor in the active region. If V_{CC} is much larger than V_{BE}, we have

$$I_B \approx \frac{V_{CC}}{R_b} \qquad (11\text{-}2)$$

The current I_B is constant, and the network of Fig. 11-1 is called the *fixed-bias circuit*. In summary, we see that the selection of an operating point Q depends upon a number of factors. Among these factors are the ac and dc loads of the stage, the available power supply, the maximum transistor ratings, the peak signal excursions to be handled by the stage, and the tolerable distortion.

11-2 BIAS STABILITY

In the preceding section we examined the problem of selecting an operating point Q on the load line of the transistor. We now consider some of the problems of maintaining the operating point stable.

Let us refer to the biasing circuit of Fig. 11-1. In this circuit the base current I_B is kept constant since $I_B \approx V_{CC}/R_b$. Let us assume that the transistor of Fig. 11-1 is replaced by another of the same type. In spite of the tremendous strides that have been made in the technology of the manufacture of semiconductor devices, transistors of a particular type still come out of production with a wide spread in the values of some parameters. To provide information about this variability, a transistor data sheet (Appendix B-3), in tabulating parameter values, provides columns headed *minimum* and *maximum* (and often also *typical*).

In Sec. 3-6 we see that the spacing of the output characteristics will increase or decrease (for equal changes in I_B) as β increases or decreases. In Fig. 11-3 we have assumed that β is greater for the replacement transistor of Fig. 11-1, and since I_B is maintained constant at I_{B2} by the external biasing circuit, it follows that the operating point moves from Q_1 to Q_2. This new operating point may be completely unsatisfactory. Specifically, it is possible for the transistor to find itself in the saturation region. We now conclude that maintaining I_B constant will not provide operating-point stability as β changes. On the contrary, I_B should be allowed to change so as to maintain I_C and V_{CE} constant as β changes.

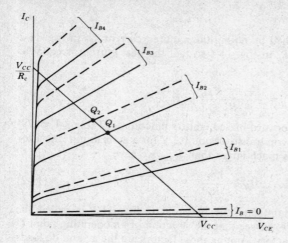

Figure 11-3 Graphs showing the collector characteristics for two transistors of the same type. The dashed characteristics are for a transistor whose β is much larger than that of the transistor represented by the solid curves.

Thermal Instability

A second very important cause for bias instability is a variation in temperature. In Sec. 3-7 we note that the reverse saturation current I_{CO}† changes greatly with temperature. Specifically, I_{CO} doubles for every 10°C rise in temperature. This fact may cause considerable practical difficulty in using a transistor as a circuit element. For example, the collector current I_C causes the collector-junction temperature to rise, which in turn increases I_{CO}. As a result of this growth of I_{CO}, I_C will increase [Eq. (3-12)], which may further increase the junction temperature, and consequently I_{CO}. It is possible for this succession of events to become cumulative, so that the ratings of the transistor are exceeded and the device burns out. This phenomenon is called *thermal runaway*.

Even if the drastic state of affairs described above does not take place, it is possible for a transistor which was biased in the active region to find itself in the saturation region as a result of this operating-point instability. To see how this may happen, we note that if $I_B = 0$, then, from Eq. (3-12), $I_C = I_{CO}(1 + \beta)$. As the temperature increases, I_{CO} increases, and even if we assume that β remains constant (actually, it also increases), it is clear that the $I_B = 0$ line in the CE output characteristics will move upward. The characteristics for other values of I_B will also move upward, and consequently the operating point will move if I_B is forced to remain constant. Hence the transistor could find itself almost in saturation at a temperature of $+100$°C, even though it would be biased in the middle of its active region at $+25$°C.

---

† Throughout this chapter I_{CBO} is abbreviated I_{CO} (Sec. 3-7).

11-3 SELF-BIAS OR EMITTER BIAS

A circuit which is used to establish a stable operating point is the self-biasing configuration of Fig. 11-4a. The current in the resistance R_e in the emitter lead causes a voltage drop which is in the direction to reverse-bias the emitter junction. Since this junction must be forward-biased, the base voltage is obtained from the supply through the R_1R_2 network.

The physical reason for an improvement in stability with this circuit is the following: If I_C tends to increase, say, because I_{CO} has risen as a result of an elevated temperature, the current in R_e increases. As a consequence of the increase in voltage drop across R_e, the base current is decreased. Hence I_C will increase less than it would have, had there been no self-biasing resistor R_e.

Analysis of the Self-Bias Circuit

If the circuit component values in Fig. 11-4a are specified, the quiescent point is found as follows:

If the circuit to the left between the base B and ground N terminals in Fig. 11-4a is replaced by its Thévenin equivalent, the two-mesh circuit of Fig. 11-4b is obtained, where

$$V \equiv \frac{R_2 V_{CC}}{R_2 + R_1} \qquad R_b \equiv \frac{R_2 R_1}{R_2 + R_1} \qquad (11\text{-}3)$$

Obviously, R_b is the effective resistance seen looking back from the base terminal. Kirchhoff's voltage law around the base circuit yields

$$V = I_B R_b + V_{BE} + (I_B + I_C)R_e \qquad (11\text{-}4)$$

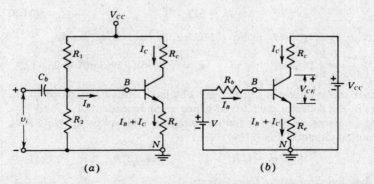

(a)   (b)

Figure 11-4 (a) A self-biasing circuit. (b) Simplification of the base circuit in (a) by the use of Thévenin's theorem.

An approximate solution is easily obtained if $I_B \ll I_C$ and if $I_B R_b \ll V$. Then

$$I_C \approx \frac{V - V_{BE}}{R_e} \tag{11-5}$$

If these approximations are not valid, but if β is known, then the calculation of the Q point may be carried out analytically as follows: In the active region the collector current is given by Eq. (3-12), namely,

$$I_C = \beta I_B + (1 + \beta)I_{CO} \tag{11-6}$$

Equations (11-4) and (11-6) can now be solved for I_B and I_C (since V_{BE} is known in the active region). Note that with this method the currents (in the active region) are determined by the base circuit and the values of β and I_{CO}.

Example 11-1 A silicon transistor is used in the circuit of Fig. 11-4a, with $V_{CC} = 22.5$ V, $R_c = 5.6$ kΩ, $R_e = 1$ kΩ, $R_2 = 10$ kΩ, and $R_1 = 90$ kΩ. For this transistor, $\beta = 50$. Find the Q point.

SOLUTION From Eqs. (11-3) we have

$$V = \frac{10 \times 22.5}{100} = 2.25 \text{ V} \qquad R_b = \frac{10 \times 90}{100} = 9.0 \text{ k}\Omega$$

From Table 3-1, $V_{BE} = 0.7$ V in the active region and Eq. (11-4) becomes

$$2.25 = 9.0I_B + 0.7 + (I_B + I_C)(1)$$

or
$$1.55 = 10I_B + I_C \tag{11-7}$$

For base currents large compared with the reverse saturation current ($I_B \gg I_{CO}$), it follows from Eq. (11-6) that

$$I_C = 51I_B \tag{11-8}$$

Substituting $I_B = 0.0196I_C$ in Eq. (11-7) for the base circuit yields

$$I_C = \frac{1.55}{1.196} = 1.296 \text{ mA} \qquad \text{and} \qquad I_B = (0.0196)(1.296) \text{ mA} = 25.40 \text{ } \mu\text{A}$$

Note that these values of current are obtained without reference to the voltages in the collector circuit, and hence are independent of R_c and V_{CC}.

The quiescent collector-to-emitter voltage is obtained by applying KVL to the output circuit of Fig. 11-4b.

$$V_{CE} = -I_C R_c + V_{CC} - (I_B + I_C)R_e \tag{11-9}$$

Using the numerical values found above,

$$V_{CE} = -(1.296)(5.6) + 22.5 - (1.296 + 0.026)(1) = 13.92 \text{ V}$$

11-4 STABILIZATION AGAINST VARIATIONS IN I_{CO}, V_{BE}, AND β

The sources of instability of I_C are essentially three.[1] These are the reverse saturation current I_{CO}, which doubles for every 10°C increase in temperature; the base-to-emitter voltage V_{BE}, which decreases at the rate of 2.5 mV/°C for both Ge and Si transistors; and β, which increases with temperature. Also β changes with transistor replacement.

We shall neglect the effect of the change of V_{CE} with temperature, because this variation is very small and we assume that the transistor operates in the active region, where I_C is approximately independent of V_{CE}.

Combining Eqs. (11-4) and (11-6) we obtain the following equation for the collector current as a function of I_{CO}, V_{BE}, and β:

$$I_C \left[\frac{R_b + R_e(1 + \beta)}{\beta} \right] = V - V_{BE} + (R_b + R_e) \frac{\beta + 1}{\beta} I_{CO} \quad (11\text{-}10)$$

This expression indicates clearly that I_C will vary with changes in one or more of the parameters β, V_{BE}, or I_{CO}.

Current Increment for a Change in β

Consider that a transistor whose current gain is β_1 is replaced by another whose gain is β_2. Find the resultant current change $\Delta I_C = I_{C2} - I_{C1}$, where $I_{C2}(I_{C1})$ corresponds to $\beta_2(\beta_1)$. If $\beta \gg 1$, then the right-hand side of Eq. (11-10) is essentially independent of β and, hence,

$$I_{C2} \left[\frac{R_b + R_e(1 + \beta_2)}{\beta_2} \right] = I_{C1} \left[\frac{R_b + R_e(1 + \beta_1)}{\beta_1} \right] \quad (11\text{-}11)$$

Solving Eq. (11-11) for I_{C2}/I_{C1} and subtracting unity from the result, we obtain (Prob. 11-5) for $I_{C2}/I_{C1} - 1$ (assuming $\beta \gg 1$ and with $\Delta\beta \equiv \beta_2 - \beta_1$),

$$\frac{I_{C2} - I_{C1}}{I_{C1}} = \frac{\Delta I_C}{I_{C1}} = \left(1 + \frac{R_b}{R_e} \right) \frac{M_2 \, \Delta\beta}{\beta_1 \beta_2} \quad (11\text{-}12)$$

where M is defined by

$$M \equiv \frac{1}{1 + R_b/[R_e(1 + \beta)]} \approx \frac{1}{1 + R_b/\beta R_e} \quad (11\text{-}13)$$

for $\beta \gg 1$. The symbol $M_2(M_1)$ corresponds to $\beta_2(\beta_1)$. It is clear that R_b/R_e should be kept small. Also, for a given spread in the value of β (say, $\beta_2/\beta_1 = 3$), a high-β circuit will be more stable than one using a lower-β transistor.

Example 11-2 A silicon transistor, used in the circuit of Fig. 11-4a, may have any value of β between 36 and 90 at a temperature of 25°C, and the leakage current I_{CO} has negligible effect on I_C at room temperature. Find

R_e, R_1, and R_2 subject to the following specifications: $R_c = 4 \text{ k}\Omega$, $V_{CC} = 20$ V; the nominal bias point is to be at $V_{CE} = 10$ V, $I_C = 2$ mA; and I_C should be in the range 1.75 to 2.25 mA as β varies from 36 to 90.

SOLUTION From the collector circuit (with $I_C \gg I_B$)

$$R_c + R_e = \frac{V_{CC} - V_{CE}}{I_C} = \frac{20 - 10}{2} = 5 \text{ k}\Omega$$

Hence $R_e = 5 - 4 = 1 \text{ k}\Omega$.

From Eqs. (11-12) and (11-13) we have (with $R_e = 1 \text{ k}\Omega$),

$$\frac{2.25 - 1.75}{1.75}\left(1 + \frac{R_b}{90}\right) = (1 + R_b)\frac{90 - 36}{(36)(90)}$$

Solving yields $R_b = 19.93 \text{ k}\Omega$.

To find R_1 and R_2 we use Eqs. (11-3), where V is given by Eq. (11-4). For $V_{BE} = 0.7$ V, $I_{CO} = 0$, $\beta_1 = 36$, $I_{C1} = 1.75$mA, and $I_B = I_C/\beta$

$$V = V_{BE} + \frac{R_b + R_e(1 + \beta)}{\beta} I_C$$

$$= 0.7 + \frac{19.93 + 37}{36}(1.75) = 3.467 \text{ V}$$

The same result should be obtained for V if we use $\beta_2 = 90$ and $I_{C2} = 2.25$ mA. Why?

From Eqs. (11-3), solving for R_1 and R_2, we find

$$R_1 = R_b \frac{V_{CC}}{V} = (19.93)\frac{20}{3.467} = 115.0 \text{ k}\Omega$$

$$R_2 = \frac{R_1 V}{V_{CC} - V} = \frac{(115.0)(3.467)}{20 - 3.467} = 24.11 \text{ k}\Omega$$

Current Increment for a Change in I_{CO}

From Eq. (11-10) with $\beta \gg 1$, it follows, *if β and V_{BE} remain constant,* that

$$\Delta I_C = \frac{R_b + R_e}{R_b/\beta + R_e} \Delta I_{CO} = \left(1 + \frac{R_b}{R_e}\right) M_1 \Delta I_{CO} \qquad (11\text{-}14)$$

Current Increment for a Change in V_{BE}

From Eq. (11-10) with $\beta \gg 1$, it follows, *if β and I_{CO} remain constant,* that

$$\Delta I_C = -\frac{\beta}{R_b + \beta R_e} \Delta V_{BE} = -\frac{M_1}{R_e} \Delta V_{BE} \qquad (11\text{-}15)$$

Table 11-1 Typical silicon transistor parameters

T, °C	-65	$+25$	$+175$
I_{CO}, nA	1.95×10^{-3}	1.0	33,000
β	25	55	100
V_{BE}, V	0.78	0.60	0.225

Table 11-2 Typical germanium transistor parameters

T, °C	-65	$+25$	$+75$
I_{CO}, µA	1.95×10^{-3}	1.0	32
β	20	55	90
V_{BE}, V	0.38	0.20	0.10

Total Current Increment

To obtain the total change in current over a specified temperature range due to simultaneous variations in β, I_{CO}, and V_{BE}, we add the individual increments found in Eqs. (11-12), (11-14), and (11-15). The fractional change in collector current is given by

$$\frac{\Delta I_C}{I_{C1}} = \left(1 + \frac{R_b}{R_e}\right)\frac{M_1 \Delta I_{CO}}{I_{C1}} - \frac{M_1 \Delta V_{BE}}{I_{C1} R_e} + \left(1 + \frac{R_b}{R_e}\right)\frac{M_2 \Delta \beta}{\beta_1 \beta_2} \quad (11\text{-}16)$$

where $M_1(M_2)$ corresponds to $\beta_1(\beta_2)$. Note that as T increases, $\Delta I_{CO}/I_{C1}$ and $\Delta\beta$ increase, whereas $\Delta V_{BE}/I_{C1}$ decreases. Hence all terms in Eq. (11-16) are positive for increasing T and negative for decreasing T.

We now examine in detail the order of magnitude of the terms of Eq. (11-16) for a silicon transistor over the entire range of temperature operation as specified by transistor manufacturers. This range usually is -65 to $+175$°C for silicon transistors and -65 to $+75$°C for germanium transistors.

Tables 11-1 and 11-2 show typical parameters of silicon and germanium transistors, each having the same β (55) at room temperature. For Si, I_{CO} is much smaller than for Ge. Note that I_{CO} doubles approximately every 10°C and $|V_{BE}|$ decreases by approximately 2.5 mV/°C.

Example 11-3 For the self-bias circuit of Fig. 11-4a, $R_e = 4.7$ kΩ, $R_b = 7.75$ kΩ, and $R_b/R_e = 1.65$. The collector supply voltage and R_c are adjusted to establish a collector current of 1.5 mA at 25°C. Determine the variation of I_C in the temperature range of -65 to $+175$°C when the silicon transistor of Table 11-1 is used.

SOLUTION Since R_b/R_e is known, we can find the percentage change in I_C using Eq. (11-16). At room temperature

$$M_1 = \frac{1}{1 + R_b/\beta_1 R_e} = \frac{1}{1 + 1.65/55} = 0.97 \approx 1$$

Since at 175°C, $\beta_2 = 100$, then M_2 is even closer to unity than M_1. Hence at $T = +175°C$, we shall assume $M_1 = M_2 = 1$. From Eq. (11-16) we have

$$\frac{\Delta I_C(+175°C)}{I_{C1}} = (1 + 1.65) \times \frac{33,000 \times 10^{-9}}{1.5 \times 10^{-3}} + \frac{0.6 - 0.225}{1.5 \times 4.7}$$

$$+ (1 + 1.65) \times \frac{100 - 55}{55 \times 100} = (5.83 + 5.32 + 2.17)\%$$

or the change in collector current is

$$\Delta I_C(+175°C) = 0.087 + 0.080 + 0.033 = 0.200 \text{ mA}$$

At $-65°C$, $M_2 = 1/(1 + 1.65/25) = 0.94$, and we shall take this small correction factor into account. From Eq. (11-16) we find

$$\frac{\Delta I_C(-65°C)}{I_{C1}} = -\frac{2.65 \times 10^{-9}}{1.5 \times 10^{-3}} - \frac{0.78 - 0.60}{1.5 \times 4.7} - \frac{2.65 \times (55 - 25) \times 0.94}{25 \times 55}$$

$$= (0 - 2.55 - 5.43)\%$$

or $\Delta I_C(-65°C) = 0 - 0.038 - 0.081 = -0.119 \text{ mA}$

Therefore, for the silicon transistor, the collector current will be approximately 1.70 mA at $+175°C$ and 1.38 mA at $-65°C$.

If the foregoing calculation is made for a germanium transistor by using the parameters in Table 11-2, the result (Prob. 11-10) is that the collector current is approximately 1.63 mA at $+75°C$ and $+1.34$ mA at $-65°C$.

Practical Considerations

The foregoing example illustrates the superiority of silicon over germanium transistors because, approximately, the same change in collector current is obtained for a much higher temperature change in the silicon transistor. In the above example the current change at the extremes of temperature is only about 10 percent. If R_e is smaller, the current instability is greater. For example, by using Eq. (11-16), we find for $R_e = 1$ kΩ and $R_b = 7.75$ kΩ that the 25°C collector current varies about 30 percent at $-65°C$ and $+175°C$ (Si) or at $-65°C$ and $+75°C$ (Ge). These numerical values illustrate why a germanium transistor is seldom used above 75°C, and a silicon device above 175°C. The importance of keeping R_e and β large is clear.

The change in collector current that can be tolerated in any specific application depends on design requirements, such as peak signal voltage required across R_c. We should also point out that the tolerance in bias resistors and supply voltages must be taken into account, in addition to the variation of β, I_{CO}, and V_{BE}.

Our discussion of stability and the results obtained are independent of R_c, and hence they remain valid for $R_c = 0$. If the output is taken across R_e, such a circuit is called an *emitter follower* or common-collector (CC) circuit (Sec. 11-10). If we have a direct-coupled emitter follower *driven from an ideal voltage source*, then $R_b = 0$ and this circuit can be used to a higher temperature than a similar circuit with $R_b \neq 0$. For the emitter follower, Eq. (11-16) reduces to

$$\frac{\Delta I_C}{I_{C1}} = \frac{\Delta I_{CO}}{I_{C1}} - \frac{\Delta V_{BE}}{I_{C1}R_e} + \frac{\Delta \beta}{\beta_1 \beta_2} \tag{11-17}$$

Example 11-4 Design the self-bias circuit of Fig. 11-4a using a Si transistor to meet the following specifications over the temperature range 25 to 145°C:

$$\frac{\Delta I_C}{I_C} \leqslant 15 \text{ percent}$$

V_{BE} at 25°C = 650 ± 50 mV V_{CC} = 20 V

β spread 150 to 600 at $I_C = 1$ mA and $T = 25°C$

Lowest β at 25°C = 150   highest β at 145°C = 1,200

I_{CO} at 25°C = 5 nA max I_{CO} at 145°C = 3.0 μA max

SOLUTION We shall assume for our design that each factor I_{CO}, β, and V_{BE} causes the same percentage change (5 percent) in I_C. We now proceed in the following steps:

1. Select R_b/R_e using the $\Delta \beta$ term of Eq. (11-16) and assuming $M \approx 1$.

$$\left(1 + \frac{R_b}{R_e}\right) \frac{\Delta \beta}{\beta_1 \beta_2} = \left(1 + \frac{R_b}{R_e}\right) \times \frac{1,200 - 150}{1,200 \times 150} = 0.05$$

or $$\frac{R_b}{R_e} = 7.56$$

Since the smallest β is 150, then

$$1 > M > \frac{1}{1 + R_b/\beta R_e} = \frac{1}{1 + 7.56/150} = 0.95$$

which justifies our assumption $M \approx 1$.

2. Select I_{C1}, considering the ΔI_{CO} term in Eq. (11-16).

$$\left(1 + \frac{R_b}{R_e}\right) \frac{\Delta I_{CO}}{I_{C1}} = 8.56 \frac{3 \times 10^{-6}}{I_{C1}} = 0.05$$

or $I_{C1} = 0.514$ mA. Use $I_C = 0.6$ mA.

3. Select $I_{C1}R_e$, considering the ΔV_{BE} term only in Eq. (11-16). Since V_{BE} changes -2.5 mV/°C, then $\Delta V_{BE} = -2.5 \times 120 = -300$ mV due to the temperature range. Since there is an uncertainty in V_{BE} at 25°C of ± 50 mV, the total increment is $\Delta V_{BE} = -300 - 100 = -400$ mV = -0.4 V. Hence

$$\frac{0.40}{I_{C1}R_e} = 0.05 \quad \text{or} \quad I_{C1}R_e = 8 \text{ V}$$

4. Since $I_{C1} = 0.6$ mA, then $R_e = 8/0.6 = 13.3$ kΩ. Also $R_b = (R_b/R_e)R_e$ $= 7.56 \times 13.3 = 100$ kΩ.

5. To determine the value of the biasing resistors R_1 and R_2, we must first find V. From Fig. 11-4b or Eq. (11-4), at 25°C and using an average value of $\beta = \frac{1}{2}(150 + 600) = 375$,

$$V = I_B R_b + V_{BE} + (I_B + I_C)R_e = \frac{0.6 \times 100}{375} + 0.65 + 8 = 8.81 \text{ V}$$

Solving Eqs. (11-3) for R_1 and R_2, we obtain

$$R_1 = R_b \frac{V_{CC}}{V} = 100 \times \frac{20}{8.81} = 227 \text{ k}\Omega$$

$$R_2 = \frac{R_1 V}{V_{CC} - V} = \frac{227 \times 8.81}{20 - 8.81} = 179 \text{ k}\Omega$$

The value of R_c is selected on the basis of the required small-signal gain and symmetric operation of the circuit.

The self-biasing configuration discussed in the preceding two sections is used for discrete transistors. It is not practical for an integrated transistor because of the following reasons:

1. R_1 and R_2 have such large values of resistance that they would require prohibitively large areas for fabrication.
2. The blocking capacitor C_B in Fig. 11-4 has a value of capacitance of the order of microfarads to obtain reasonable low-frequency response (Sec. 13-3). Integrated capacitors are limited to 200 pF (Sec. 4-11).
3. In order to increase the low-frequency gain, the emitter resistor must be bypassed with a capacitance even larger than C_b. The special techniques developed for stabilizing the operating point of an integrated transistor are discussed in Sec. 15-5. Biasing the FET is considered in Sec. 11-17.

We now consider the low-frequency behavior of a bipolar transistor. The discussion is valid for either a discrete or integrated device.

11-5 OUTPUT WAVEFORMS FOR A SINUSOIDAL INPUT

Consider a transistor in the CE configuration biased so that the quiescent collector current (voltage) is I_C (V_C). Assume that the input signal is a sinusoidal base current superimposed upon the quiescent current I_B. The output waveforms can then be obtained, in principle, from the load line and the collector characteristics. For example, assume that the transistor in Fig. 11-2 is biased at Q_1, where $I_B = 40$ μA, and that a sinusoidal base current of peak value 40 μA is applied. We can find the values of the instantaneous collector current i_C and voltage v_{CE}, corresponding to any given value of instantaneous base current i_B (for $0 \leqslant i_B \leqslant 80$ μA), at the intersection of the ac load line and the collector characteristics. This construction is carried out in Fig. 11-5. We observe that the collector-current and collector-voltage waveforms are not the same as the base-current waveform (the sinusoid of Fig. 11-5c) because the collector characteristics in the neighborhood of the load line in Fig. 11-2 are not parallel lines equally spaced for equal increments in base current. This change in waveform is known as *output nonlinear distortion*. In particular note that the peak increase in collector current from the quiescent point $Q = Q_1$ in Fig. 11-2 (and hence in Fig. 11-5) is smaller than the peak decrease in i_C.

The base current varies exponentially with base voltage (Fig. 3-11). Hence, for a sinusoidal i_B waveform, the input voltage waveform v_B may be very nonsinusoidal, as indicated in Fig. 11-5d. This change of waveshape is called *input nonlinear distortion*.

Since the *small-signal* low-frequency response of a transistor is linear, it can be obtained analytically rather than graphically. As a matter of fact, for very small signals the graphical technique used in connection with Fig. 11-2 would

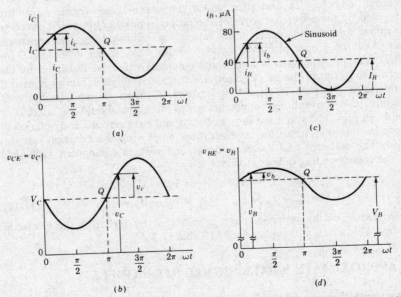

Figure 11-5 (*a*) Collector current and (*b*) collector voltage waveforms for the sinusoidal base current in (*c*). The corresponding base voltage waveform is shown in (*d*).

Table 11-3 Notation summarized

Notation	Base (collector) voltage with respect to emitter	Base (collector) current toward electrode from external circuit
Instantaneous total value	v_B (v_C)	i_B (i_C)
Quiescent value	V_B (V_C)	I_B (I_C)
Instantaneous value of varying component	v_b (v_c)	i_b (i_c)
Effective value of varying component (phasor, if a sinusoid)	V_b (V_c)	I_b (I_c)
Supply voltage (magnitude)	V_{BB} (V_{CC})	

require interpolation between the printed characteristics and would result in very poor accuracy. In the remainder of the chapter we assume small-signal (also called *incremental*) operation, obtain a linear circuit model for the transistor, and then analyze the network analytically.

Notation

At this point it is important to make a few remarks on transistor symbols. Specifically, instantaneous values of quantities which vary with time are represented by lowercase letters (i for current, v for voltage, and p for power). Maximum, average (dc), and effective, or root-mean-square (rms), values are represented by the uppercase letter of the proper symbol (I, V, or P). Average (dc) values and instantaneous total values are indicated by the uppercase subscript of the proper electrode symbol (B for base, C for collector, E for emitter). Varying components from some quiescent value are indicated by the lowercase subscript of the proper electrode symbol. A single subscript is used if the reference electrode is clearly understood. If there is any possibility of ambiguity, the conventional double-subscript notation should be used. For example, in Figs. 11-5a to d, we show collector and base currents and voltages in the common-emitter transistor configuration, employing the notation just described. The collector and emitter current and voltage component variations from the corresponding quiescent values are

$$i_c = i_C - I_C = \Delta i_C \qquad v_c = v_C - V_C = \Delta v_C$$
$$i_b = i_B - I_B = \Delta i_B \qquad v_b = v_B - V_B = \Delta v_B \qquad (11\text{-}18)$$

The *magnitude* of the supply voltage is indicated by repeating the electrode subscript. This notation is summarized in Table 11-3.

11-6 APPROXIMATE SMALL-SIGNAL BJT MODELS

A very simple approximate small-signal model for a bipolar junction transistor at low frequencies is indicated in Fig. 11-6. Since h_{fe} (Sec. 3-10) is the negative of

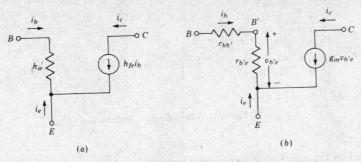

Figure 11-6 Two equivalent small-signal approximate transistor models.

the short-circuit current gain, this model is valid only for small values of load resistance (Sec. 11-15). In this approximate model the resistance between base and emitter is h_{ie} ohms. This simplified h-parameter model may be used in any configuration (CE, CB, or CC) by grounding the appropriate node. The signal is applied between the input terminal and ground, and the load R_L is placed between the output node and ground.

An alternative model equivalent to that of Fig. 11-6a is shown in Fig. 11-6b. The ohmic resistance in series with the base lead (the base-spreading resistance $r_{bb'}$ of Sec. 3-8) is shown explicitly in this model. In other words, h_{ie} is split into two resistors $r_{bb'}$ and $r_{b'e}$.

$$h_{ie} = r_{bb'} + r_{b'e} \tag{11-19}$$

Note that in Fig. 11-6a the current generator depends upon the input current whereas in Fig. 11-6b the current generator is proportional to the voltage between the internal point B' (not physically accessible) and the emitter E. The proportionality factor g_m is called the *mutual conductance* or *transconductance* of the transistor. In order for the two models to give identical results it is clear that

$$h_{fe}i_b = g_m v_{b'e} = g_m r_{b'e} i_b \tag{11-20}$$

or

$$h_{fe} = g_m r_{b'e} \tag{11-21}$$

Note that this equation is dimensionally correct, because conductance has the reciprocal dimension of resistance and h_{fe} is dimensionless (since it represents gain).

11-7 THE TRANSISTOR TRANSCONDUCTANCE

The parameter g_m depends only on the quiescent collector current and the temperature and is independent of the constructional features of the transistor. We now verify this statement: Figure 11-7 shows a p-n-p transistor with $v_{CE} = V_{CC} = $ constant. Hence, $\Delta v_{CE} = v_{ce} = 0$ which indicates no change in collector-to-emitter voltage for time-varying signals. In other words, the collector

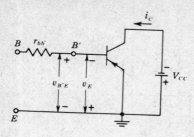

Figure 11-7 Pertaining to the derivation of g_m.

is short-circuited to the emitter *for incremental changes from the quiescent point.* In the active region the collector current is given by Eq. (3-3), namely (using the notation in Table 11-3),

$$i_C = I_{CO} - \alpha i_E$$

Since the short-circuit collector current in Fig. 11-6 is $g_m v_{b'e}$, the transconductance g_m is defined by

$$g_m \equiv \frac{i_c}{v_{b'e}} = \left. \frac{\partial i_C}{\partial v_{B'E}} \right|_{V_{CE}} = \alpha \frac{\partial i_E}{\partial v_{B'E}} = \alpha \frac{\partial i_E}{\partial v_E} \tag{11-22}$$

In the above we have assumed that α is independent of v_E. For a *p-n-p* transistor, $v_E = -v_{B'E}$, as shown in Fig. 11-7. If the emitter diode resistance is r_e, then $r_e = \partial v_E / \partial i_E$, and hence

$$g_m = \frac{\alpha}{r_e} \tag{11-23}$$

The dynamic resistance of a forward-biased diode is given in Eq. (2-8) as $V_T / I_E,$† where $V_T = kT/q$ [Eq. (1-35)], and I_E is the quiescent emitter current. Hence, using Eq. (3-3),

$$g_m = \frac{\alpha I_E}{V_T} = \frac{I_{CO} - I_C}{V_T} \tag{11-24}$$

For a *p-n-p* transistor I_C is negative. For an *n-p-n* transistor I_C is positive, but the foregoing analysis (with $v_E = +v_{B'E}$) leads to $g_m = (I_C - I_{CO})/V_T$. Hence, for either type of transistor, g_m is positive. Since $|I_C| \gg |I_{CO}|$, then g_m is given by

$$g_m \approx \frac{|I_C|}{V_T} \tag{11-25}$$

where, from Eq. (1-35), $V_T = T/11,600$. Equation (11-25) confirms that g_m *is*

---

† Since the recombination current in the emitter space-charge region does not reach the collector, the factor η in Eq. (2-8) is taken as unity in the calculation of g_m.

directly proportional to the quiescent collector current and inversely proportional to temperature. At room temperature

$$g_m = \frac{|I_C|(\text{mA})}{26} \qquad (11\text{-}26)$$

If $I_C = 1.3$ mA, $g_m = 0.05 \, \text{℧} = 50$ mA/V. For $I_C = 10$ mA, $g_m \approx 400$ mA/V.

11-8 LINEAR ANALYSIS OF A TRANSISTOR CIRCUIT

More accurate transistor models than those indicated in Fig. 11-6 are given in Sec. 11-15. In most practical cases sufficiently accurate values of voltage and current can be obtained by using the simplified model. Hence, we shall first apply this model to several amplifier circuits. A better "physical feeling" for the behavior of a transistor circuit can be obtained from the simple approximate solution than from a more laborious (albeit more exact) calculation.

There are many transistor circuits which do not consist simply of the CE, CB, or CC configurations. For example, a CE amplifier may have a feedback resistor from collector to base, or it may have an emitter resistor. Furthermore, a circuit may consist of several transistors which are interconnected in some manner. An analytic determination of the small-signal behavior of even relatively complicated amplifier circuits may be made by following these simple rules:

1. Draw the actual wiring diagram of the circuit neatly.
2. Mark the points B (base), C (collector), and E (emitter) on this circuit diagram. Locate these points as the start of the equivalent circuit. Maintain the same relative positions as in the original circuit.
3. Replace each transistor by its model.
4. Transfer all components (resistors, capacitors, and signal sources) from the network to the equivalent circuit.
5. Since we are interested only in changes from the quiescent values, replace each independent dc source by its internal resistance. The ideal voltage source is replaced by a short circuit, and the ideal current source by an open circuit.
6. Solve the resultant linear circuit for mesh or branch currents and node voltages by applying Kirchhoff's current and voltage laws (KCL and KVL).

It should be pointed out that the above procedure is not limited for use with the approximate model or for low frequencies. A basic restriction is that the voltages and currents are small enough so that linear operation results. In other words, over the signal excursion, the parameters in the model must remain essentially constant.

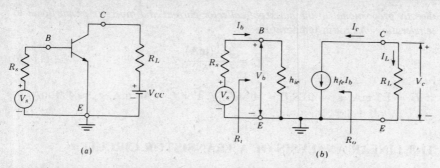

Figure 11-8 (a) The CE amplifier and (b) the transistor is replaced by the approximate h-parameter model.

11-9 THE COMMON-EMITTER (CE) AMPLIFIER

This configuration is shown in Fig. 11-8a, where for simplicity we have omitted the biasing resistors, coupling, and bypass capacitors. The effect of these components on the amplifier response is considered in subsequent sections. Figure 11-8b shows this CE stage with the transistor replaced by the approximate model of Fig. 11-6. Assuming sinusoidally varying voltages and currents, we can proceed with the analysis of this circuit with the use of the phasor notation to represent the sinusoidally varying quantities. The quantities of interest are *the current gain*, *the input resistance*, *the voltage gain*, and *the output resistance*.

The Current Gain, or Current Amplification A_I

For the transistor amplifier stage, A_I is defined as the ratio of output to input currents, or

$$A_I \equiv \frac{I_L}{I_b} = -\frac{I_c}{I_b} \tag{11-27}$$

From the circuit of Fig. 11-8 we have $I_c = h_{fe}I_b$. Hence

$$A_I = -h_{fe} \tag{11-28}$$

Note that subject to the approximate model, A_I equals the short-circuit current gain, independent of the load R_L.

The Input Resistance R_i

The resistance R_s in Fig. 11-8 represents the signal-source resistance. The resistance we see by looking into the transistor input terminals B and E is the amplifier *input resistance* R_i, or

$$R_i \equiv \frac{V_b}{I_b} = h_{ie} \tag{11-29}$$

Note that R_i equals the short-circuit input resistance.

The Voltage Gain, or Voltage Amplification, A_V

The ratio of output voltage V_c to input voltage V_b gives the voltage gain of the transistor, or

$$A_V \equiv \frac{V_c}{V_b} = \frac{I_L R_L}{I_b h_{ie}} \tag{11-30}$$

Combining Eqs. (11-27), (11-29), and (11-30) we obtain

$$A_V = \frac{A_I R_L}{R_i} \tag{11-31}$$

The above derivation indicates that Eq. (11-31) is a general expression for the voltage gain in terms of the current gain and the input resistance, and *it is valid independent of the transistor configuration and also of the transistor model used.*
For the CE circuit Eq. (11-31) becomes

$$A_V = \frac{-h_{fe} R_L}{h_{ie}} \tag{11-32}$$

The Voltage Amplification A_{Vs} Taking into Account the Resistance R_s of the Source

This overall voltage gain A_{Vs} is defined by

$$A_{Vs} \equiv \frac{V_c}{V_s} = \frac{V_c}{V_b} \frac{V_b}{V_s} = A_V \frac{V_b}{V_s} \tag{11-33}$$

From the equivalent input circuit of the amplifier, shown in Fig. 11-8,

$$V_b = \frac{V_s h_{ie}}{h_{ie} + R_s} \tag{11-34}$$

Then,

$$A_{Vs} = \frac{A_V h_{ie}}{h_{ie} + R_s} = \frac{A_I R_L}{h_{ie} + R_s} \tag{11-35}$$

where use has been made of Eq. (11-31). Note that, if $R_s = 0$, then $A_{Vs} = A_V$. Hence A_V is the voltage gain for an ideal voltage source (one with zero internal resistance). In practice, the quantity A_{Vs} is more meaningful than A_V since, usually, the source resistance has an appreciable effect on the overall voltage amplification. For example, if R_i is equal in magnitude to R_s, then $A_{Vs} = 0.5\,A_V$.

The Output Resistance

By definition, the output resistance R_o is obtained by setting the source voltage V_s to zero and the load resistance R_L to infinity and by driving the output terminals from a generator V_2. If the current drawn from V_2 is I_2, then

$$R_o \equiv \frac{V_2}{I_2} \quad \text{with} \quad V_s = 0 \text{ and } R_L = \infty \tag{11-36}$$

With $V_s = 0$ in Fig. 11-8, the input current I_b is zero. Hence, $I_2 = I_c = h_{fe} I_b = 0$ and

$$R_o = \frac{V_2}{0} = \infty$$

The CE stage, using the simplified h-parameter model, has infinite output resistance.

The Output Resistance R_o' Taking into Account the Resistance of the Load

This resistance is clearly the parallel combination of R_o and R_L.

$$R_o' = \frac{R_o R_L}{R_o + R_L} \tag{11-37}$$

For the circuit of Fig. 11-8, $R_o' = R_L$ because $R_o = \infty$.

The approximate solution for the CE configuration is summarized in the first column of Table 11-4, and numerical values are given in Table 11-5.

Table 11-4 Summary of approximate equations for $h_{oe}(R_e + R_L) \leqslant 0.1$†

	CE	CE with R_e	CC	CB
A_I	$-h_{fe}$	$-h_{fe}$	$1 + h_{fe}$	$\dfrac{h_{fe}}{1 + h_{fe}}$
R_i	h_{ie}	$h_{ie} + (1 + h_{fe})R_e$	$h_{ie} + (1 + h_{fe})R_L$	$\dfrac{h_{ie}}{1 + h_{fe}}$
A_V	$-\dfrac{h_{fe}R_L}{h_{ie}}$	$-\dfrac{h_{fe}R_L}{R_i} \approx -\dfrac{R_L}{R_e}$	$1 - \dfrac{h_{ie}}{R_i}$	$h_{fe}\dfrac{R_L}{h_{ie}}$
R_o	∞	∞	$\dfrac{R_s + h_{ie}}{1 + h_{fe}}$	∞
R_o'	R_L	R_L	$R_o \| R_L$	R_L

† $(R_i)_{CB}$ is an underestimation by less than 10 percent. All other quantities except R_o are too large in magnitude by less than 10 percent. h_{oe} is defined in Sec. 11-15.

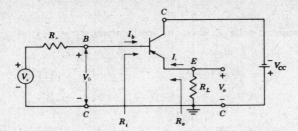

Figure 11-9 A common-collector, or emitter-follower, configuration.

11-10 THE EMITTER FOLLOWER

The circuit diagram of a common-collector (CC) transistor amplifier is given in Fig. 11-9. This configuration is also called the *emitter follower*, because its voltage gain is close to unity [Eq. (11-42)], and hence a change in base voltage appears as an equal change across the load at the emitter. In other words, the emitter *follows* the input signal. It is shown below that the input resistance R_i of an emitter follower is very high (hundreds of kilohms) and the output resistance R_o is very low (tens of ohms). Hence the most common use for the CC circuit is as a buffer stage which performs the function of resistance transformation (from high to low resistance) over a wide range of frequencies, with voltage gain close to unity. In addition, the emitter follower increases the power level of the signal.

Approximate Calculations for the Common-Collector Configuration

Figure 11-10 shows the simplified h-parameter model of Fig. 11-6 replacing the transistor in Fig. 11-9. Note that the collector is grounded with respect to the signal (because the supply V_{CC} must be replaced by a short-circuit, as indicated in rule 5, Sec. 11-8).

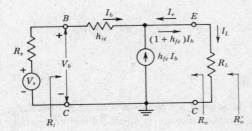

Figure 11-10 Simplified h-parameter model for the CC circuit.

Current Gain

If KCL is applied at the emitter node E, then the load current $I_L = - I_e = (1 + h_{fe})I_b$. Hence,

$$A_I = \frac{I_L}{I_b} = 1 + h_{fe} \tag{11-38}$$

Input Resistance

From Fig. 11-10 we obtain

$$V_b = I_b h_{ie} + (1 + h_{fe})I_b R_L \tag{11-39}$$

Hence,

$$R_i = \frac{V_b}{I_b} = h_{ie} + (1 + h_{fe})R_L \tag{11-40}$$

Note that $R_i \gg h_{ie}$ even if R_L is as small as 0.5 kΩ because $h_{fe} \gg 1$.

Voltage Gain

From Eqs. (11-31) and (11-38),

$$A_V = \frac{A_I R_L}{R_i} = \frac{(1 + h_{fe})R_L}{R_i} \tag{11-41}$$

From Eq. (11-40), $(1 + h_{fe})R_L = R_i - h_{ie}$ and, hence, Eq. (11-41) becomes

$$A_V = 1 - \frac{h_{ie}}{R_i} \tag{11-42}$$

Since $R_i \gg h_{ie}$, Eq. (11-42) indicates that the voltage gain of a CC amplifier is approximately unity (but slightly less than unity).

Output Resistance

From the definition of R_o given in Eq. (11-36) we must set $V_s = 0$ and $R_L = \infty$ and apply an external generator V_2 to the output terminals, as indicated in Fig. 11-11. From this figure we find

$$V_2 = - I_b(R_s + h_{ie}) \qquad \text{and} \qquad I_2 = -(1 + h_{fe})I_b$$

Hence,

$$R_o \equiv \frac{V_2}{I_2} = \frac{R_s + h_{ie}}{1 + h_{fe}} \tag{11-43}$$

Note that *the output resistance is a function of the source resistance R_s.* Because $h_{fe} \gg 1$, the output resistance of an emitter follower is small (ohms) compared with the input resistance which is large (tens or hundreds of kilohms).

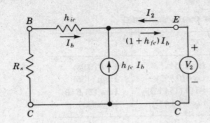

Figure 11-11 Network used to find the output resistance of a CC stage.

The output resistance R_o', taking the load into account, is obviously R_o in parallel with R_L. The formulas derived in this section are summarized in the third column of Table 11-4. Note that these results are independent of whether an *n-p-n* or a *p-n-p* transistor is under consideration.

11-11 THE COMMON-BASE (CB) AMPLIFIER

The circuit of Fig. 3-3 is a CB amplifier if a signal generator V_s (with internal resistance R_s) is added into the input (emitter circuit). Using the simplified h-parameter model of Fig. 11-6, the equivalent circuit of Fig. 11-12 is obtained for the CB amplifier. The current entering node E is I_e and that entering C is I_c. Since, from KCL, $I_b = -(I_e + I_c)$, the current in h_{ie} must be I_b, as indicated in Fig. 11-12. Applying the definitions of A_I, R_i, A_V, and R_o given in Sec. 11-9 to this figure, the formulas given in the fourth column of Table 11-4 are obtained (Prob. 11-23).

11-12 COMPARISON OF BJT AMPLIFIER CONFIGURATIONS

Numerical values for A_I, A_V, R_i, and R_o for the CE, CC, and CB amplifiers are given in Table 11-5 for $R_L = 3$ kΩ and $R_s = 3$ kΩ, based upon the simplified model using the parameters $h_{ie} = 1.1$ kΩ and $h_{fe} = 50$. These agree within better than 10 percent with the exact values using the four h-parameter model of Fig. 11-17, except for the two values of R_o marked ∞ in Table 11-5. The correct

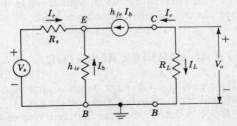

Figure 11-12 The model for the CB amplifier.

Table 11-5 Comparison of transistor configurations†

Quantity	CE	CC	CB
A_I	High (-50)	High (51)	Low (0.98)
A_V	High (-136)	Low (0.993)	High (136)
R_i	Medium (1,100 Ω)	High (154 kΩ)	Low (21.6 Ω)
R_o	High (∞)	Low (80.4 Ω)	High (∞)

† $h_{ie} = 1.1$ kΩ, $h_{fe} = 50$, $R_L = R_s = 3$ kΩ

values for output impedance are 45.5 kΩ and 1.75 MΩ for the CE and CB configurations, respectively. However, R_o', taking the 3-kΩ load into account, is within 10 percent of 3 kΩ (for the CE and CB stages), using either the two- or the four-parameter model.

The CE Configuration

From Table 11-4 we see that only the CE stage is capable of both a voltage gain and a current gain greater than unity. This configuration is the most versatile and useful of the three connections. Note that the magnitudes of R_i and R_o lie between those for the CB and CC configurations (refer to the exact values of R_o quoted in the preceding paragraph).

The CB Configuration

For the common-base stage, A_I is less than unity, A_V is high (equal to that of the CE stage), R_i is the lowest, and R_o is the highest of the three configurations. The CB stage has few applications. It is sometimes used to match a very-low-imped-ance source, to drive a high-impedance load, or as a noninverting amplifier with a voltage gain greater than unity.

The CC Configuration

For the common-collector stage, A_I is high (approximately equal to that of the CE stage), A_V is less than unity (but close to unity), R_i is the highest, and R_o is the lowest of the three configurations. This circuit finds wide application as a buffer stage between a high-impedance source and a low-impedance load.

11-13 THE CE AMPLIFIER WITH AN EMITTER RESISTOR

We see from Table 11-4 that the voltage gain of a CE stage depends upon h_{fe}. This transistor parameter depends upon temperature, aging, and operating point. Moreover, h_{fe} may vary widely from device to device, even for the same type of transistor. It is often necessary to stabilize the voltage amplification of each stage, so that A_V will become essentially independent of h_{fe}. A simple and effective way to obtain voltage-gain stabilization is to add an emitter resistor R_e to a CE stage, as indicated in the circuit of Fig. 11-13. This stabilization is a result of the feedback provided by the emitter resistor. The general concept of feedback is discussed in Chap. 12.

We show in this section that the presence of R_e has the following effects on the amplifier performance, in addition to the beneficial effect on bias stability discussed in Sec. 11-3. It leaves the current gain A_I essentially unchanged; it increases the input resistance by $(1 + h_{fe})R_e$; it increases the output resistance; and under the condition $(1 + h_{fe})R_e \gg h_{ie}$, it stabilizes the voltage gain, which becomes essentially equal to $-R_L/R_e$ (and thus is independent of the transistor).

The Approximate Solution

An approximate analysis of the circuit of Fig. 11-13a can be made using the simplified model of Fig. 11-6 as shown in Fig. 11-13b.

The current gain is, from Fig. 11-13b,

$$A_I = \frac{-I_c}{I_b} = \frac{-h_{fe}I_b}{I_b} = -h_{fe} \tag{11-44}$$

The current gain equals the short-circuit value and is unaffected by the addition of R_e.

The input resistance, as obtained from inspection of Fig. 11-13b, is

$$R_i = \frac{V_i}{I_b} = h_{ie} + (1 + h_{fe})R_e \tag{11-45}$$

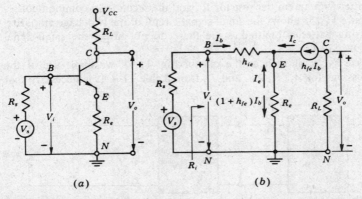

(a)   (b)

Figure 11-13 (a) Common-emitter amplifier with an emitter resistor. (b) Approximate small-signal equivalent circuit.

The input resistance is augmented by $(1 + h_{fe})R_e$, and may be very much larger than h_{ie}. Hence an emitter resistance greatly increases the input resistance.

The voltage gain is

$$A_V = \frac{A_I R_L}{R_i} = \frac{-h_{fe} R_L}{h_{ie} + (1 + h_{fe})R_e} \tag{11-46}$$

Clearly, the addition of an emitter resistance greatly reduces the voltage amplification. This reduction in gain is often a reasonable price to pay for the improvement in stability. We note that, if $(1 + h_{fe})R_e \gg h_{ie}$, and since $h_{fe} \gg 1$, then

$$A_V \approx \frac{-h_{fe}}{1 + h_{fe}} \frac{R_L}{R_e} \approx \frac{-R_L}{R_e} \tag{11-47}$$

Subject to the above approximations, A_V is completely stable (if stable resistances are used for R_L and R_e), since it is independent of all transistor parameters.

The output resistance of the transistor alone (with R_L considered external) is infinite for the approximate circuit of Fig. 11-13, just as it was for the CE amplifier of Sec. 11-9 with $R_e = 0$. Hence the output resistance of the stage, including the load, is R_L.

The above expressions are included in the second column of Table 11-4.

11-14 CASCADING TRANSISTOR AMPLIFIERS

When the amplification of a single transistor is not sufficient for a particular purpose, or when the input or output impedance is not of the correct magnitude for the intended application, two or more stages may be connected in cascade; i.e., the output of a given stage is connected to the input of the next stage, as shown in Fig. 11-14. In the circuit of Fig. 11-15a the first stage is connected common-emitter with an emitter resistor R_e, and the second is a common-collector stage. Figure 11-15b shows the small-signal circuit of the two-stage amplifier, with the biasing batteries omitted, since these do not affect the small-signal calculations.

To analyze a circuit such as the one of Fig. 11-15, we make use of the general expressions for A_I, R_i, A_V, and R_o from Table 11-4. It is necessary that

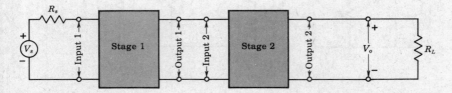

Figure 11-14 Two cascaded stages.

we have available the *h* parameters for the specific transistors used in the circuit.

The voltage gain A_V of a multistage amplifier from the input of the first to the output of the last stage equals the product of the voltage gains of each stage. This statement is easily verified by reference to Fig. 11-14 or 11-15*b*. Thus,

$$A_V \equiv \frac{V_o}{V_1} = \frac{V_o}{V_2}\frac{V_2}{V_1} = A_{V2}A_{V1} \tag{11-48}$$

The overall voltage gain A_{Vs}, taking the source resistance into account, is given by

$$A_{Vs} \equiv \frac{V_o}{V_s} = \frac{V_o}{V_1}\frac{V_1}{V_s} = A_V \frac{R_{i1}}{R_{i1} + R_s} \tag{11-49}$$

because there is an effective resistance R_{i1} between base and emitter of the first stage.

The total current gain A_I is *not* equal to the product of the current gains of the individual stages because the output current of one stage is not equal to the input current of the following stage. For example, in Fig. 11-15*b* we see that

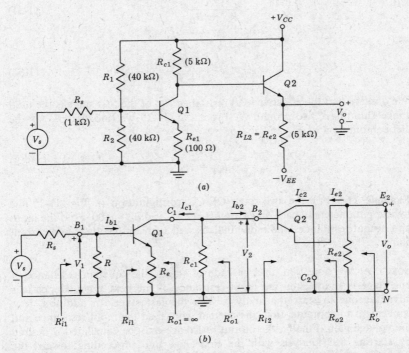

(a)

(b)

Figure 11-15 (*a*) Common-emitter–common-collector amplifier (*b*) Approximate small-signal circuit of the CE-CC amplifier ($R \equiv R_1 \| R_2$).

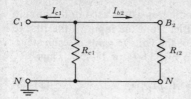

Figure 11-16 Relating to the calculation of overall current gain.

$I_{b2} \neq - I_{c1}$. The correct expressions for A_I is given by

$$A_I \equiv - \frac{I_{e2}}{I_{b1}} = - \frac{I_{e2}}{I_{b2}} \frac{I_{b2}}{I_{c1}} \frac{I_{c1}}{I_{b1}} = - A_{I2} \frac{I_{b2}}{I_{c1}} A_{I1} \qquad (11\text{-}50)$$

From Fig. 11-16 we can obtain the ratio I_{b2}/I_{c1}. Since $- I_{c1}$ passes through the parallel combination of R_{c1} and R_{i2},

$$V_{B_2 N} = - \frac{I_{c1} R_{c1} R_{i2}}{R_{c1} + R_{i2}} = I_{b2} R_{i2}$$

then,

$$\frac{I_{b2}}{I_{c1}} = - \frac{R_{c1}}{R_{c1} + R_{i2}} \qquad (11\text{-}51)$$

Hence,

$$A_I = A_{I2} A_{I1} \frac{R_{c1}}{R_{i2} + R_{c1}} \qquad (11\text{-}52)$$

We observe that the effective load resistance R_{L1} of the first stage is the total resistance shunting C_1 to ground N. From Fig. 11-15 we find that R_{L1} is the parallel combination of R_{c1} and R_{i2}, or

$$R_{L1} = \frac{R_{c1} R_{i2}}{R_{c1} + R_{i2}} \qquad (11\text{-}53)$$

Example 11-5 For the two-stage CE-CC configuration in Fig. 11-15 the hybrid parameters of each stage are $h_{ie} = 2 \text{ k}\Omega$ and $h_{fe} = 100$. Find the input and output resistances and individual, as well as overall, voltage and current gains.

SOLUTION We note that, in a cascade of stages, the collector resistance of one stage is shunted by the input resistance of the next stage. Hence it is advantageous to start the analysis with the last stage. In addition, it is convenient to compute, first, the current gain, then the input resistance and the voltage gain. Finally, the output resistance may be calculated if desired by starting this analysis with the first stage and proceeding toward the output stage.

For the CC output stage we have, from Table 11-4

$$A_{I2} = 1 + h_{fe} = 101$$

$$R_{i2} = h_{ie} + (1 + h_{fe})R_L = 2 + (101)(5) = 507 \text{ k}\Omega$$

$$A_{V2} = 1 - \frac{h_{ie}}{R_{i2}} = 1 - \frac{2}{507} = 0.996$$

Note the high input resistance of the CC stage and that its voltage gain is close to unity

For the CE stage with an emitter resistor we find from Table 11-4,

$$A_{I1} = \frac{-I_{c1}}{I_{b1}} = -h_{fe} = -100$$

$$R_{i1} = h_{ie} + (1 + h_{fe})R_{e1} = 2 + (101)(0.1) = 12.1 \text{ k}\Omega$$

The effective load on the first stage, its voltage gain, and the output resistance are

$$R_{L1} = \frac{R_{c1}R_{i2}}{R_{c1} + R_{i2}} = \frac{(5)(507)}{512} = 4.95 \text{ k}\Omega$$

$$A_{V1} = \frac{A_{I1}R_{L1}}{R_{i1}} = \frac{-(100)(4.95)}{12.1} = -40.9$$

$$R'_{o1} = R_{c1} = 5 \text{ k}\Omega$$

Since R'_{o1} is the effective source resistance for $Q2$, then, from Table 11-4,

$$R_{o2} = \frac{h_{ie} + R'_{o1}}{1 + h_{fe}} = \frac{2,000 + 5,000}{101} = 69.3 \ \Omega$$

$$R'_{o2} = \frac{R_{o2}R_{L2}}{R_{o2} + R_{L2}} = \frac{(69.3)(5,000)}{5,069} = 68.4 \ \Omega = R'_o$$

The voltage and current gains of the cascade are, from Eqs. (11-48) and (11-50)

$$A_V \equiv \frac{V_o}{V_1} = A_{V1}A_{V2} = (-40.9)(0.996) = -40.75$$

$$A_I \equiv \frac{-I_{e2}}{I_{b1}} = A_{I1}A_{I2}\frac{R_{c1}}{R_{c1} + R_{i2}} = (-100)(101)\frac{5}{5 + 507} = -98.6$$

Alternatively, A_V may be computed from

$$A_V = A_I \frac{R_{L2}}{R_{i1}} = \frac{(-98.6)(5)}{12.1} = -40.74$$

The biasing resistors R_1 and R_2 have had no effect upon the above calculations. They do influence the calculation of the overall voltage gain

$$A_{Vs} \equiv \frac{V_o}{V_s} = A_V \frac{R_{i1}'}{R_{i1}' + R_s} = (-40.74) \frac{7.539}{8.539} = -35.97$$

where from Fig. 11-15 we see that

$$R_{i1}' = R \| R_{i1} = (40\|40)\|12.1 = \frac{(20)(12.1)}{32.1} = 7.539 \text{ k}\Omega$$

Note that whereas the input resistance of the cascaded amplifier is the input resistance of the first stage R_{i1}, the resistance seen by the signal source (V_s in series with R_s) is R_{i1}'.

Choice of the Transistor Configuration in a Cascade

It is important to note that the previous calculations of input and output resistances and voltage and current gains are applicable for any connection of the cascaded stages. They could be CE, CC, CB, or combinations of all three possible connections.

Consider the following question: Which of the three possible connections must be used in cascade if maximum voltage gain is to be realized? For the intermediate stages, the common-collector connection is not used because the voltage gain of such a stage is less than unity. Hence it is not possible (without a transformer) to increase the overall voltage amplification by cascading common-collector stages.

Grounded-base (CB) coupled stages also are seldom cascaded because the voltage gain of such an arrangement is approximately the same as that of the output stage alone. This statement may be verified as follows: The voltage gain of a stage equals its current gain times the effective load resistance R_L divided by the input resistance R_i. The effective load resistance R_L is the parallel combination of the actual collector resistance R_c and (except for the last stage) the input resistance R_i of the following stage. This parallel combination is certainly less than R_i, and hence, for identical stages, the effective load resistance is less than R_i. The maximum current gain is less than unity (Table 11-4). Hence the voltage gain of any stage (except the last, or output, stage) is less than unity.

Since the short-circuit current gain h_{fe} of a common-emitter stage is much greater than unity, it is possible to increase the voltage amplification by cascading such stages. We may now state that *in a cascade the intermediate transistors should be connected in a common-emitter configuration.*

The choice of the input stage may be decided by criteria other than the maximization of voltage gain. For example, the amplitude or the frequency response of the transducer V_s may depend upon the impedance into which it operates. Some transducers require essentially open-circuit or short-circuit operation. In many cases the common-collector or common-base stage is used at the input because of impedance considerations, even at the expense of voltage or current gain.

The output stage is selected also on the basis of impedance considerations. Since a CC stage has a very low output resistance, it is often used for the last stage if it is required to drive a low impedance (perhaps capacitive) load.

11-15 ACCURATE SMALL-SIGNAL BJT MODELS

The approximate models of Fig. 11-6 ignore two physical phenomena: (1) The output resistance of a BJT is not infinite and (2) there is internal feedback between output and input (the Early base-modulation effect of Sec. 3-5). The first effect can be taken into account by placing a conductance h_{oe} between C and E in Fig. 11-6a to represent the nonzero slope of the CE output characteristics.

The second effect mentioned above is taken into account in Fig. 11-6a by adding a voltage generator $h_{re}v_c$ which places a voltage source into the input circuit. The polarity of this feedback voltage is such that it opposes the applied signal; that is, it has the effect of reducing the gain over what it would be if this reverse feedback were missing. The complete h-parameter model is indicated in Fig. 11-17. This model defines the h parameters as follows:

$$h_{ie} = \frac{v_b}{i_b}\bigg|_{v_c=0} = \text{input resistance with output short-circuited (ohms).}$$

$$h_{re} = \frac{v_b}{v_c}\bigg|_{i_b=0} = \text{fraction of output voltage at input with input open-circuited, or,}$$
more simply, reverse-open-circuit voltage amplification (dimensionless).

$$h_{fe} = \frac{i_c}{i_b}\bigg|_{v_c=0} = \text{negative of current transfer ratio (or current gain) with output}$$
short-circuited. (Note that the current into a load across the output port would be the negative of i_c.) This parameter is usually referred to, simply, as *the short-circuit current gain* (dimensionless).

$$h_{oe} = \frac{i_c}{v_c}\bigg|_{i_b=0} = \text{output conductance with input open-circuited (mhos or A/V).}$$

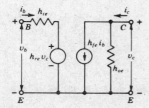

Figure 11-17 The four h-parameter small-signal model for the BJT.

Expressions for the current gain A_I, the input resistance R_i, the voltage gain A_V, and the output resistance R_o are given in Prob. 11-33 for the CE configuration in terms of the four h parameters.

Typical values of h parameters for a transistor operating at room temperature and at an emitter current $I_E = 1.3$ mA are (Appendix B-3):

$$h_{ie} = 2.1 \text{ k}\Omega \qquad h_{re} = 10^{-4}$$

$$\text{(11-54)}$$

$$h_{fe} = 100 \qquad h_{oe} = 10^{-5} \text{ A/V} \qquad \text{or} \qquad \frac{1}{h_{oe}} = 100 \text{ k}\Omega$$

Some manufacturers supply curves of the h parameters as a function of T and also of I_C.

Since $h_{re} \ll 1$, it is often possible to neglect the reverse voltage generator $h_{re}v_c$ and to use a model using only the three h parameters, h_{ie}, h_{fe}, and h_{oe}. It turns out[2] that h_{re} and h_{oe} both may be neglected, provided that the inequality

$$h_{oe}(R_L + R_e) \leqslant 0.1 \qquad \text{(11-55)}$$

is satisfied. For $h_{oe} = 10^{-5}$ as in Eq. (11-54), $R_L + R_e$ must not exceed 10 kΩ. The errors in A_I, R_i, A_V, and R_o' are then less than 10 percent (Table 11-4).

The Hybrid-π Model[3]

The finite output resistance of the transistor is represented by r_{ce} between C and E in Fig. 11-6b. The Early effect is taken into account in this model by applying the feedback from the output to the input through a resistor $r_{b'c}$ between the collector C and the internal base node B'. The complete low-frequency network (using phasor notation $V_{b'e}$) is drawn in Fig. 11-18 and is known as the *hybrid-π* or *Giacoletto* model.

Typical magnitudes for the hybrid-π parameters operating at room temperature for $I_C = 1.3$ mA are:

$$r_{bb'} = 100 \ \Omega \qquad r_{b'e} = 2 \text{ k}\Omega \qquad r_{b'c} = 20 \text{ M}\Omega \qquad r_{ce} = 200 \text{ k}\Omega \quad \text{(11-56a)}$$

and from Eq. (11-26)

$$g_m = \frac{|I_C| \text{mA}}{26} = \frac{1.3}{26} \text{ A/V} = 50 \text{ mA/V} \qquad \text{(11-56b)}$$

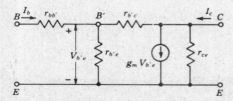

Figure 11-18 The hybrid-π model at low frequencies.

We now demonstrate that all the resistive components in the hybrid-π model can be obtained from the h parameters in the CE configuration.

The Input Conductance $g_{b'e}$

From Eqs. (11-56) we see that $r_{b'c} \gg r_{b'e}$. Hence I_b flows into $r_{b'e}$ and $V_{b'e} \approx I_b r_{b'e}$. The short-circuit collector current is given by

$$I_c = g_m V_{b'e} \approx g_m I_b r_{b'e} = h_{fe} I_b \tag{11-57}$$

or $$r_{b'e} = \frac{h_{fe}}{g_m} = \frac{h_{fe} V_T}{|I_C|} \qquad \text{or} \qquad g_{b'e} = \frac{g_m}{h_{fe}} \tag{11-58}$$

Note that, over the range of currents for which h_{fe} remains fairly constant, $r_{b'e}$ *is directly proportional to temperature and inversely proportional to current.* Observe in Fig. 3-14 that at both very low and very high currents, $h_{fe} \approx h_{FE}$ decreases. Note that Eq. (11-58) is identical with Eq. (11-21) obtained from the approximate model.

The Feedback Conductance $g_{b'c}$

With the input open-circuited, h_{re} is defined as the reverse voltage gain, or from Fig. 11-18 with $I_b = 0$,

$$h_{re} = \frac{V_{b'e}}{V_{ce}} = \frac{r_{b'e}}{r_{b'e} + r_{b'c}} \tag{11-59}$$

or $$r_{b'e}(1 - h_{re}) = h_{re} r_{b'c}$$

Since $h_{re} \ll 1$, then to a good approximation

$$r_{b'e} = h_{re} r_{b'c} \qquad \text{or} \qquad g_{b'c} = h_{re} g_{b'e} \tag{11-60}$$

Since $h_{re} \approx 10^{-4}$, Eq. (11-60) verifies that $r_{b'c} \gg r_{b'e}$.

The Base-spreading Resistance $r_{bb'}$

The input resistance with the output short-circuited is h_{ie}. Under these conditions $r_{b'e}$ is in parallel with $r_{b'c}$. Using Eq. (11-60), we have $r_{b'e} \| r_{b'c} \approx r_{b'e}$, and hence,

$$h_{ie} = r_{bb'} + r_{b'e} \tag{11-61}$$

or $$r_{bb'} = h_{ie} - r_{b'e} \tag{11-62}$$

Incidentally, note from Eqs. (11-58) and (11-61) that the short-circuit input resistance h_{ie} varies with current and temperature in the following manner:

$$h_{ie} = r_{bb'} + \frac{h_{fe} V_T}{|I_C|} \approx \frac{h_{fe} V_T}{|I_C|} \tag{11-63}$$

since usually $r_{b'e} \gg r_{bb'}$.

The Output Conductance g_{ce}

With the input open-circuited, this conductance is defined as h_{oe}. For $I_b = 0$, we have from Fig. 11-18

$$I_c = \frac{V_{ce}}{r_{ce}} + \frac{V_{ce}}{r_{b'c} + r_{b'e}} + g_m V_{b'e} \tag{11-64}$$

With $I_b = 0$, we have, from Eq. (11-59), $V_{b'e} = h_{re} V_{ce}$, and from Eq. (11-64), we find

$$h_{oe} \equiv \frac{I_c}{V_{ce}} = \frac{1}{r_{ce}} + \frac{1}{r_{b'c}} + g_m h_{re} \tag{11-65}$$

where we made use of the fact that $r_{b'c} \gg r_{b'e}$. If we substitute Eqs. (11-58) and (11-60) in Eq. (11-65), we have

$$h_{oe} = g_{ce} + g_{b'c} + g_{b'e} h_{fe} \frac{g_{b'c}}{g_{b'e}}$$

or

$$g_{ce} = h_{oe} - (1 + h_{fe}) g_{b'c} \tag{11-66}$$

Summary

If the CE h parameters at low frequencies are known at a given collector current I_C, the conductances or resistances in the hybrid-π circuit are calculable from the following five equations in the order given:

$$g_m = \frac{|I_C|}{V_T}$$

$$r_{b'e} = \frac{h_{fe}}{g_m} = \frac{h_{fe} V_T}{|I_C|} \quad \text{or} \quad g_{b'e} = \frac{g_m}{h_{fe}}$$

$$r_{bb'} = h_{ie} - r_{b'e} \tag{11-67}$$

$$r_{b'c} = \frac{r_{b'e}}{h_{re}} \quad \text{or} \quad g_{b'c} = \frac{h_{re}}{r_{b'e}}$$

$$g_{ce} = h_{oe} - (1 + h_{fe}) g_{b'c} = \frac{1}{r_{ce}}$$

For the typical h parameters in Eq. (11-54), at $I_C = 1.3$ mA and room temperature, we obtain the component values listed in Eq. (11-56).

At high frequencies the h parameters must be considered to be complex functions of f. Hence the h-parameter model is not useful for analyzing high-frequency amplifiers. The high-frequency behavior of a transistor can easily be taken into consideration in the hybrid-π model as follows: The emitter diffusion capacitance is added between E and B' and the collector transition capacitance is placed between C and B'. This augmented hybrid-π model is used to study high-frequency amplifiers in Chap. 13. However, at low frequencies it is usually more convenient to use the h-parameter than the hybrid-π model.

11-16 HIGH-INPUT-RESISTANCE BJT CIRCUITS

For some applications the need arises for an amplifier with a very high input resistance. Consider an emitter follower with parameter values given in Eq. (11-54) and with a 5-kΩ emitter resistance. Since $h_{oe}R_e = (10^{-5})(5 \times 10^3) = 0.05 < 0.1$, the relationship Eq. (11-55) is satisfied. We may use Table 11-4 for the input-resistance formula and we find $R_i = 507$ kΩ. Suppose we wish to obtain an even higher value for R_i. We might use a much larger resistance for R_e, say, $R_e = 500$ kΩ. Now what is the value of R_i? Since $h_{oe}R_e = (10^{-5})(500 \times 10^3) = 5 \nless 0.1$, Eq. (11-55) is not satisfied and we must analyze the common-collector amplifier by using all four h parameters.

> **Example 11-6** Find expressions for (a) A_I, (b) R_i, and (c) A_V for an emitter follower by taking h_{oe} and h_{re} into account.

SOLUTION (a) The CC equivalent circuit of Fig. 11-10 modified to take all four h parameters into account is shown in Fig. 11-19. The parallel combination R of R_L and $1/h_{oe}$ is

$$R \equiv \frac{R_L(1/h_{oe})}{R_L + 1/h_{oe}} = \frac{R_L}{1 + h_{oe}R_L} \tag{11-68}$$

Since $V_{ec} = (1 + h_{fe})I_b R$ and $I_L = V_{ec}/R_L$,

$$A_I \equiv \frac{I_L}{I_b} = \frac{(1 + h_{fe})R}{R_L} = \frac{1 + h_{fe}}{1 + h_{oe}R_L} \tag{11-69}$$

(b) The input resistance $R_i \equiv V_b/I_b$ is obtained by applying KVL to the outside mesh.

$$V_b = h_{ie}I_b + h_{re}V_{ce} + V_{ec} = h_{ie}I_b + V_{ec}(1 - h_{re})$$

Since $h_{re} \ll 1$, it may be neglected so that

$$V_b = h_{ie}I_b + (1 + h_{fe})I_b R$$

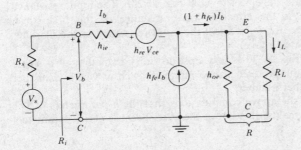

Figure 11-19 The equivalent circuit for the CC configuration using all four h parameters.

and $R_i = V_b/I_b$ is given by

$$R_i = h_{ie} + (1 + h_{fe})R = h_{ie} + \frac{(1 + h_{fe})R_L}{1 + h_{oe}R_L} \qquad (11\text{-}70)$$

(c) The voltage gain is given by

$$A_V = \frac{V_{ec}}{V_b} = \frac{(1 + h_{fe})I_b R}{I_b R_i} = (1 + h_{fe})\frac{R}{R_i} \qquad (11\text{-}71)$$

Since the voltage gain of an emitter follower is close to unity, let us evaluate the difference between unity and A_V. Thus,

$$1 - A_V = \frac{R_i - (1 + h_{fe})R}{R_i} = \frac{h_{ie}}{R_i} \qquad (11\text{-}72)$$

where use was made of Eq. (11-70).

Note that h_{re} does not appear in any of these equations or in that for R_o [Eq. (11-73)].

In Prob. 11-44 it is shown that the output conductance $G_o = 1/R_o$ is given by

$$G_o = \frac{1 + h_{fe}}{R_s + h_{ie}} + h_{oe} \qquad (11\text{-}73)$$

For $R_L = 507$ kΩ, $h_{ie} = 2.1$ kΩ, $h_{fe} = 100$, and $h_{oe} = 10^{-5}$ Ω, Eq. (11-70) gives $R_i = 8.44$ MΩ. Incidentally, for the limiting case $R_L \to \infty$, $R_i \to (1 + h_{fe})/h_{oe} = 10.1$ MΩ which is the maximum possible value of input resistance for an emitter follower with the above values of the h parameters.

To obtain $R_i = 8.44$ MΩ requires a 507-kΩ emitter resistance which is an unreasonably large value, especially for an integrated resistor. To circumvent this difficulty the circuit shown in Fig. 11-20a, called the Darlington connection, is used.† Note that two transistors form a composite pair, the input resistance of the second transistor constituting the emitter load for the first. More specifically, the Darlington circuit consists of two cascaded emitter followers with infinite emitter resistance in the first stage, as shown in Fig. 11-20b. The input resistance R_{i2} of the second stage is the effective load resistance of the first stage, $R_L = R_{i2} = 507$ kΩ. Hence, for $R_e = 5$ kΩ, the input resistance of the Darlington pair is 8.44 MΩ (indicated above).

The voltage gain of Fig. 11-20 is less close to unity than that of a single emitter follower because the product of two numbers, each of which is smaller than 1, is smaller than either number.

We have assumed in the above computations that the h parameters of $Q1$ and $Q2$ are identical. In reality, this is not the case, because the h parameters

---

† For many applications the field-effect transistor (Chap. 8), with its extremely high input resistance, would be preferred to the Darlington pair.

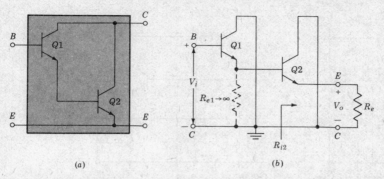

Figure 11-20 (a) Darlington pair. Some vendors package this device as a single composite transistor with only three external leads. (b) The Darlington circuit drawn as two cascaded CC stages.

depend on the quiescent conditions of $Q1$ and $Q2$. Since the emitter current of $Q1$ is the base current of $Q2$, the quiescent current of the first stage is much smaller than that of the second. From Fig. 3-14 we see that h_{fe} does not vary drastically with current, and hence we have assumed $h_{fe1} = h_{fe2} = h_{fe}$ in all the previous equations in this chapter. The symbol h_{oe}, which increases rapidly with current, refers to $Q1$ in these equations. The short-circuit input resistance varies approximately inversely with collector current [Eq. (11-63)]. Since the current in $Q2$ is $1 + h_{fe}$ times the current in $Q1$, then $h_{ie1} \approx (1 + h_{fe})h_{ie2}$. Values of R_i, A_V and R_o using this relationship are calculated in Prob. 11-45. The output resistance of a Darlington circuit is less than that of a single CC stage if $R_s > h_{ie2}$ but may be greater than that of an emitter follower if $R_s < h_{ie2}$ (Prob. 11-46).

A major drawback of the Darlington transistor pair is that the leakage current of the first transistor is amplified by the second. Hence the overall leakage current may be high and a Darlington connection of three or more transistors is usually impractical.

The composite transistor pair of Fig. 11-20a can, of course, be used as a common-emitter amplifier. The advantage of this pair would be a very high overall h_{fe}, nominally equal to the product of the CE short-circuit current gains of the two transistors. In fact, Darlington integrated transistor pairs are commercially available with h_{fe} as high as 30,000.

The Biasing Problem

In discussing the Darlington transistor pair, we have emphasized its value in providing high input resistance. However, we have oversimplified the problem by disregarding the effect of the biasing arrangement used in the circuit. Figure 11-21a shows a typical biasing network of resistors R_1, R_2, and R_e, if discrete transistors are used. The input resistance R_i' of this stage consists of $R_i \| R'$, where $R' \equiv R_1 \| R_2$. Since $R_i \gg R'$, then $R_i' \approx R'$, and we no longer have the high input resistance of the emitter follower of Fig. 11-9.

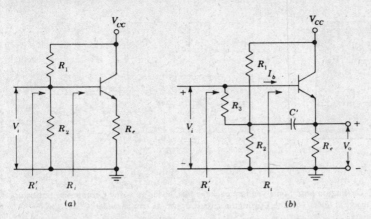

Figure 11-21 (a) A BJT self-biasing circuit. (b) The bootstrap principle increases the effective value of R_3 and R_i'.

Assume that the input circuit is modified as in Fig. 11-21b by the addition of R_3 but with $C' = 0$ (that is, for the moment, ignore the presence of C'). Now R' is increased to $R_3 + R_1 \| R_2$. However, since R_i is usually much greater than R', it is seen that $R_i' \approx R'$, which may be a few hundred kilohms at most.

To overcome the decrease in the input resistance due to the biasing network, the input circuit of Fig. 11-21b is modified by the addition of C' between the emitter and the junction of R_1 and R_2. The capacitance C' is chosen large enough to act as a short circuit at the lowest frequency under consideration. Hence the bottom of R_3 is effectively connected to the output (the emitter), whereas the top of R_3 is at the input (the base). Since the input voltage is V_i and the output voltage is $V_o = A_V V_i$, the voltage across R_3 is $V_3 = V_i - A_V V_i$, the current in R_3 is $I_3 = V_3/R_3$, and the effective resistance R_{eff} due to the biasing arrangement R_1, R_2, and R_3 is

$$R_{\text{eff}} = \frac{V_i}{I_3} = \frac{V_i}{(1 - A_V)V_i/R_3} = \frac{R_3}{1 - A_V} \qquad (11\text{-}74)$$

Since, for an emitter follower, A_V approaches unity, R_{eff} becomes extremely large. For example, with $A_V = 0.995$ and $R_3 = 100$ kΩ, we find $R_{\text{eff}} = 20$ MΩ. Note that the quiescent base current passes through R_3, and hence that a few hundred kilohms is probably an upper limit for R_3.

The above effect, when $A_V \to + 1$, is called *bootstrapping*. The term arises from the fact that, if one end of the resistor R_3 changes in voltage, the other end of R_3 moves through the same potential difference; it is as if R_3 were "pulling itself up by its bootstraps." The input resistance of the CC amplifier as given by Eq. (11-42) is $R_i = h_{ie}/(1 - A_V)$. Since this expression is of the form of Eq. (11-74), here is an example of bootstrapping of the resistance h_{ie} which appears between base and emitter.

Biasing an IC Emitter Follower

The self-biasing arrangements in Fig. 11-21 cannot be used with integrated devices for the reasons given in Sec. 11-4. This difficulty can be avoided if a second power supply is used. In Fig. 11-21a omit R_1 and R_2 and tie the bottom of R_e to a negative supply $-V_{EE}$, as in Fig. 11-15a. The input signal V_i is applied between the base and ground. If the average value of V_i is V_{DC} and if the desired quiescent collector current is I_C, then the value of V_{EE} is given by

$$V_{EE} = -V_{DC} + V_{BE} + I_C R_e$$

where $V_{BE} \approx 0.7$ V. The resistance R_e is chosen from Eq. (11-17) to limit the increment in collector current with a variation in V_{BE} to a reasonable value. This configuration may be used for a discrete as well as an integrated CC transistor.

11-17 BIASING THE FIELD EFFECT TRANSISTOR

The selection of an appropriate operating point (I_D, V_{GS}, V_{DS}) for a FET amplifier stage is determined by considerations similar to those given to bipolar transistors, as discussed in Sec. 11-1. These considerations are output voltage swing, distortion, power dissipation, voltage gain, and drift of drain current. In most cases it is not possible to satisfy all desired specifications simultaneously. In this section we examine several biasing circuits for field-effect devices.

Source Self-Bias

The configuration shown in Fig. 11-22 can be used to bias junction FET devices or depletion-mode MOS transistors. For a specified drain current I_D, the corresponding gate-to-source voltage V_{GS} can be obtained from the plotted drain or transfer characteristics. Since the gate current (and, hence, the voltage drop across R_g) is negligible, the source resistance R_s can be found as the ratio of V_{GS} to the desired I_D.

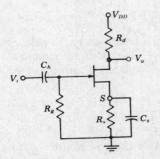

Figure 11-22 A FET self-bias circuit.

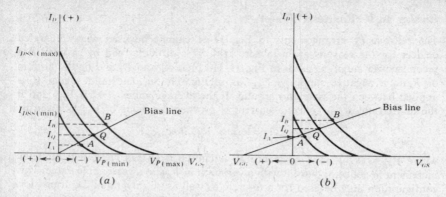

Figure 11-23 Maximum and minimum transfer curves for an *n*-channel FET. The drain current must lie between I_A and I_B. The bias line can be drawn through the origin for the current limits indicated in (*a*), but this is not possible for the currents specified in (*b*).

Biasing Against Device Variation

FET manufacturers usually supply information on the maximum and minimum values of I_{DSS} and V_P at room temperature. They also supply data to correct these quantities for temperature variations. The transfer characteristics for a given type of *n*-channel FET may appear as in Fig. 11-23*a*, where the top and bottom curves are for extreme values of temperature and device variation. Assume that, on the basis of considerations previously discussed, it is necessary to bias the device at a drain current which will not drift outside of $I_D = I_A$ and $I_D = I_B$. Then the bias line $V_{GS} = -I_D R_s$ must intersect the transfer characteristics between the points A and B, as indicated in Fig. 11-23*a*. The slope of the bias line is determined by the source resistance R_s. For any transfer characteristic between the two extremes indicated, the current I_Q is such that $I_A < I_Q < I_B$, as desired.

Consider the situation indicated in Fig. 11-23*b*, where a line drawn to pass between points A and B does not pass through the origin. This bias line satisfies the equation

$$V_{GS} = V_{GG} - I_D R_s$$

Such a bias relationship may be obtained by adding a fixed bias to the gate in addition to the source self-bias, as indicated in Fig. 11-24*a*. A circuit requiring only one power supply and which can satisfy this bias equation is shown in Fig. 11-23*b*. For this circuit

$$V_{GG} = \frac{R_2 V_{DD}}{R_1 + R_2} \qquad R_g = \frac{R_1 R_2}{R_1 + R_2}$$

We have assumed that the gate current is negligible. It is also possible for V_{GG} to

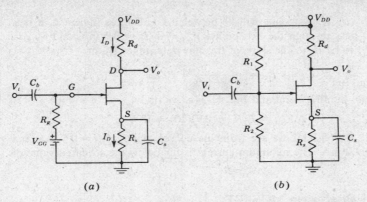

Figure 11-24 (*a*) Biasing an FET with a fixed-bias V_{GG} in addition to self-bias through R_S. (*b*) A single power-supply configuration which is equivalent to the circuit in (*a*).

fall in the reverse-biased region so that the line in Fig. 11-23*b* intersects the axis of abscissa to the right of the origin. Under these circumstances two separate supply voltages must be used.

Example 11-7 FET 2N3684 is used in the circuit of Fig. 11-24*b*. For this *n*-channel device the manufacturer specifies $V_{P(min)} = -2$ V, $V_{\dot{P}(max)} = -5$ V, $I_{DSS(min)} = 1.6$ mA, and $I_{DSS(max)} = 7.05$ mA. The extreme transfer curves are plotted in Fig. 11-25. It is desired to bias the circuit so that $I_{D(min)} = 0.8$ mA $= I_A$ and $I_{D(max)} = 1.2$ mA $= I_B$ for $V_{DD} = 24$ V. Find (*a*) V_{GG} and R_s, and (*b*) the range of possible values in I_D if $R_s = 3.3$ kΩ and $V_{GG} = 0$.

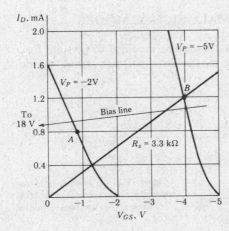

Figure 11-25 Extreme transfer curves for the 2N3684 field-effect transistor. (*Courtesy of Union Carbide Corporation.*)

SOLUTION (a) The bias line will lie between A and B, as indicated, if it is drawn to pass through the two points $V_{GS} = 0$, $I_D = 0.9$ mA, and $V_{GS} = -4$ V, $I_D = 1.1$ mA. The slope of this line determines R_g, or

$$R_S = \frac{4 - 0}{1.1 - 0.9} = 20 \text{ k}\Omega$$

Then, from the first point and Fig. 11-24a, we find

$$V_{GG} = I_D R_s = (0.9)(20) = 18 \text{ V}$$

(b) If $R_s = 3.3$ kΩ, we see from the curves that $I_{D(\min)} = 0.4$ mA and $I_{D(\max)} = 1.2$ mA. The minimum current is far below the specified value of 0.8 mA.

Biasing the Enhancement MOSFET

The self-bias technique of Fig. 11-22 cannot be used to establish an operating point for the enhancement-type MOSFET because the voltage drop across R_s is in a direction to reverse-bias the gate, and a forward gate bias is required. The circuit of Fig. 11-26a can be used, and for this case we have $V_{GS} = V_{DS}$, since no current flows through R_f. If for reasons of linearity in device operation or maximum output voltage it is desired that $V_{GS} \neq V_{DS}$, then the circuit of Fig. 11-26b is suitable. We note that $V_{GS} = [R_1/(R_1 + R_f)]V_{DS}$. Both circuits discussed here offer the advantages of dc stabilization through the feedback introduced with R_f. However, the input resistance is reduced because R_f corresponds to an equivalent resistance $R_{\text{eff}} = R_f/(1 - A_V)$ shunting the amplifier input [Eq. (11-74) is valid. Why?].

Finally, note that the circuit of Fig. 11-24b is often used with the enhancement MOSFET. The dc stability introduced in Fig. 11-26 through the feedback resistor R_f is then missing and is replaced by the dc feedback through R_s.

11-18 THE JFET OR MOSFET SMALL-SIGNAL MODEL

The hybrid-π model for a BJT (Fig. 11-18) may be modified to be valid for a field-effect transistor. The input ohmic resistance is extremely small, the dy-

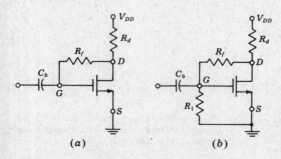

(a)   (b)

Figure 11-26 (a) Drain-to-gate bias circuit for enhancement-mode MOS transistors. (b) Improved version of (a).

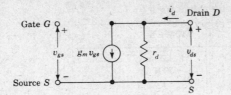

Figure 11-27 The low-frequency small-signal model for a JFET or MOSFET.

namic input resistance is exceedingly large, and there is no resistive feedback from output to input. Hence we may set $r_{bb'} = 0$, $r_{b'e} = \infty = r_{b'c}$. If these changes are made in Fig. 11-18 and if the symbol r_d is used in place of r_{ce}, the very simple low-frequency FET model of Fig. 11-27 results. Applying KCL to the output node of Fig. 11-27, we obtain

$$i_d = g_m v_{gs} + \frac{1}{r_d} v_{ds} \tag{11-75}$$

where

$$g_m \equiv \frac{i_d}{v_{gs}} \bigg|_{v_{ds}=0} = \frac{\partial i_D}{\partial v_{GS}} \bigg|_{V_{DS}} \tag{11-76}$$

is the *mutual conductance*, or *transconductance*. It is also often designated by y_{fs} or g_{fs} and called the *(common-source) forward transadmittance*. The second parameter r_d in Eq. (11-75) is the *drain (or output) resistance*, and is defined by

$$r_d \equiv \frac{v_{ds}}{i_d} \bigg|_{v_{gs}=0} = \frac{\partial v_{DS}}{\partial i_D} \bigg|_{V_{GS}} \tag{11-77}$$

The reciprocal of r_d is the drain conductance g_d. It is also designated by y_{os} and g_{os} and called the (common-source) output conductance.

An *amplification factor* μ for a FET may be defined by

$$\mu \equiv -\frac{v_{ds}}{v_{gs}} \bigg|_{i_d=0} = -\frac{\partial v_{DS}}{\partial v_{GS}} \bigg|_{I_D} \tag{11-78}$$

We can verify that μ, r_d, and g_m are related by

$$\mu = r_d g_m \tag{11-79}$$

by setting $i_d = 0$ in Eq. (11-75).

An expression for g_m may be obtained from Eq. (8-5) rewritten in the following form:

$$I_{DS} = I_{DSS} \left(1 - \frac{V_{GS}}{V_P}\right)^2 \tag{11-80}$$

where I_{DS} is the drain current in the saturation region, V_P is the pinchoff voltage for a depletion FET or the threshold voltage V_T for an enhancement device, and I_{DSS} is the value of current for $V_{GS} = 0$. From Eq. (11-80),

$$g_m = \frac{\partial I_{DS}}{\partial V_{GS}} \bigg|_{V_{DS}} = g_{mo} \left(1 - \frac{V_{GS}}{V_P}\right) = \frac{2}{|V_P|} (I_{DSS} I_{DS})^{1/2} \tag{11-81}$$

Table 11-6 Range of parameter values for a FET

Parameter	JFET	MOSFET
g_m	0.1 – 10 mA/V	0.1 – 20 mA/V or more
r_d	0.1 – 1 M	1 – 50 K
C_{ds}	0.1 – 1 pF	0.1 – 1 pF
C_{gs}, C_{gd}	1 – 10 pF	1 – 10 pF
r_{gs}	$> 10^{10}\Omega$	$> 10^{10}\Omega$
r_{gd}	$> 10^{10}\Omega$	$> 10^{14}\Omega$

where g_{mo} is the value of g_m for $V_{GS} = 0$, and is given by

$$g_{mo} = \frac{-2I_{DSS}}{V_P} \qquad (11\text{-}82)$$

Since I_{DSS} and V_P are of opposite sign, g_{mo} is always positive. Note that the transconductance varies as the square root of the drain current. The relationship connecting g_{mo}, I_{DSS}, and V_P has been verified experimentally.[4] Since g_{mo} can be measured and I_{DSS} can be read on a dc milliammeter placed in the drain lead (with zero gate excitation), Eq. (11-82) gives a method for obtaining V_P. The linear relationship between g_m and V_{GS} is found experimentally to be approximately true.

The drain current (and also g_m) has a negative temperature coefficient of resistance because the mobility decreases with increasing temperature.[5] Since this majority-carrier current decreases with T (unlike the bipolar transistor whose minority-carrier current increases with T), the troublesome phenomenon of *thermal runaway* (Sec. 11-2) is not encountered with field-effect transistors.

The high-frequency model is obtained by adding the capacitances between drain and source C_{ds}, between gate and source C_{gs}, and also between gate and drain C_{gd} to Fig. 11-18 (Sec. 13-14). The order of magnitude of the parameters for a JFET and MOSFET are given in Table 11-6. Note that the transconductance values in Table 11-6 are much smaller than those for a BJT (50 mA/V at 1.3 mA and 500 mA/V at 13 mA).

11-19 THE LOW-FREQUENCY COMMON-SOURCE AND COMMON-DRAIN AMPLIFIERS

The common-source (CS) stage is indicated in Fig. 11-28a and the common-drain (CD) configuration in Fig. 11-28b. The former is analogous to the bipolar transistor CE amplifier, and the latter to the CC stage. We shall analyze both of these circuits simultaneously by considering the generalized configuration in Fig. 11-29a. For the CS stage the output is v_{o1} taken at the drain and $R_s = 0$. The signal-source resistance is unimportant since it is in series with the gate, which

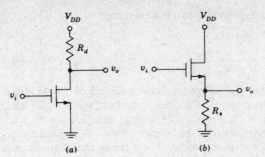

Figure 11-28 (*a*) The CS and (*b*) the CD configurations.

draws negligible current. No biasing arrangements are indicated (Sec. 11-17), but it is assumed that the stage is properly biased for linear operation.

Replacing the FET by its low-frequency small-signal model of Fig. 11-27, the equivalent circuit of Fig. 11-29*b* is obtained. Applying KVL to the output circuit yields

$$i_d R_d + (i_d - g_m v_{gs})r_d + i_d R_s = 0 \tag{11-83}$$

From Fig. 11-29*b* the voltage from *G* to *S* is given by

$$v_{gs} = v_i - i_d R_s \tag{11-84}$$

Combining Eqs. (11-83) and (11-84) and remembering that $\mu = r_d g_m$ [Eq. (11-79)], we find

$$i_d = \frac{\mu v_i}{r_d + R_d + (\mu + 1)R_s} \tag{11-85}$$

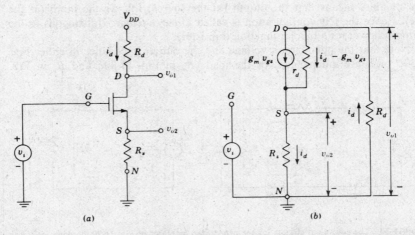

Figure 11-29 (*a*) A generalized FET amplifier configuration. (*b*) The small-signal equivalent circuit.

The CS Amplifier with an Unbypassed Source Resistance

Since $v_{o1} = -i_d R_d$, then

$$v_{o1} = \frac{-\mu v_i R_d}{r_d + R_d + (\mu + 1)R_s} \tag{11-86}$$

From Eq. (11-86) we obtain the Thévenin's equivalent circuit of Fig. 11-30a "looking into" the drain node (to ground). The open-circuit voltage is $-\mu v_i$, and the output resistance is $R_o = r_d + (\mu + 1)R_s$. The voltage gain is $A_V = v_{o1}/v_i$. The minus sign in Eq. (11-86) indicates that the output is 180° out of phase with the input. If R_s is bypassed with a large capacitance or if the source is grounded, the above equations are valid with $R_s = 0$. Under these circumstances,

$$A_V = \frac{v_{o1}}{v_i} = \frac{-\mu R_d}{r_d + R_d} = -g_m R'_d \tag{11-87}$$

where $\mu = r_d g_m$ [Eq. (11-79)] and $R'_d = R_d \| r_d$.

The CD Amplifier with a Drain Resistance

Since $v_{o2} = i_d R_s$, then from Eq. (11-85)

$$v_{o2} = \frac{\mu v_i R_s}{r_d + R_d + (\mu + 1)R_s} = \frac{\left[\mu v_i/(\mu + 1)\right]R_s}{(r_d + R_d)/(\mu + 1) + R_s} \tag{11-88}$$

From Eq. (11-88) we obtain the Thévenin's equivalent circuit of Fig. 11-30b "looking into" the source node (to ground). The open-circuit voltage is $\mu v_i/(\mu + 1)$, and the output resistance is $R_o = (r_d + R_d)/(\mu + 1)$. The voltage gain is $A_V = v_{o2}/v_i$. Note that there is no phase shift between input and output. If $R_d = 0$ and if $(\mu + 1)R_s \gg r_d$, then $A_V \approx \mu/(\mu + 1) \approx 1$ for $\mu \gg 1$. A voltage gain of unity means that the output (at the source) follows the input (at the gate). Hence the CD configuration is called a *source follower* (analogous to the *emitter* follower for a bipolar junction transistor).

Note that the open-circuit voltage and the output impedance in either Fig. 11-30a or b are independent of the load (R_d in Fig. 11-30a and R_s in Fig.

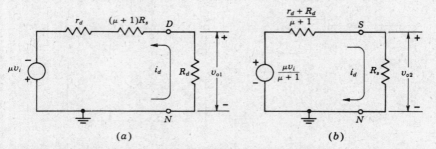

Figure 11-30 The equivalent circuits for the generalized amplifier of Fig. 11-29 "looking into" (a) the drain and (b) the source. Note that $\mu = r_d g_m$.

11-30*b*). These restrictions must be satisfied if the networks in Fig. 11-30 are to represent the true Thévenin equivalents of the amplifier in Fig. 11-29.

11-20 THE JFET AS A VOLTAGE-CONTROLLED RESISTOR[6] (VCR)

In most linear applications of field-effect transistors the device is operated in the constant-current portion of its output characteristics. We now consider FET transistor operation in the region before pinch-off, where V_{DS} is small. In this region the FET is useful as a voltage-controlled resistor; i.e., the drain-to-source resistance is controlled by the bias voltage V_{GS}. In such an application the FET is also referred to as a *voltage-controlled resistor* (VCR), a *voltage-variable resistor* (VVR), or *voltage-dependent resistor* (VDR).

Figure 11-31*a* shows the low-level bidirectional characteristics of a JFET. The slope of these characteristics gives r_{ds} as a function of V_{GS}. Figure 11-31*a* has been extended into the third quadrant to give an idea of device linearity around $V_{DS} = 0$.

Variation of r_{ds} with V_{GS} is plotted in Fig. 11-31*b* for a JFET. The dynamic range of r_{ds} is shown to be greater than 100 : 1, although for best control a range of 10 : 1 is usually used. Siliconix Inc. offers a family of both *n*- and *p*-channel FETs intended for use as voltage-controlled resistors, with values (at $V_{GS} = 0$) ranging from about 20 to 8,000 Ω.

Applications of the VCR

Since the FET operated as described above acts like a variable passive resistor, it finds applications in many areas where this property is useful, such as for

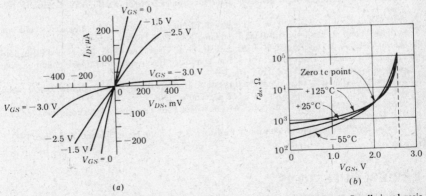

Figure 11-31 (*a*) Low-voltage drain characteristics for an *n*-channel JFET. (*b*) Small-signal resistance variation with applied gate voltage for a VCR JFET for three values of temperature. The zero temperature-coefficient (tc) voltage is $V_{GS} \approx 2.0$ V. (*Courtesy of Siliconix, Inc.*)

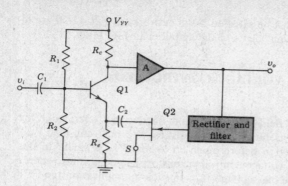

Figure 11-32 AGC amplifier using the FET as a voltage-controlled resistor.

small-signal attenuators, voltage-tuned filters, and amplifier-gain control. The VCR, for example, can be used to vary voltage gain of a multistage amplifier A as the signal level is increased. This action is called *automatic gain control* (AGC). A typical arrangement is shown in Fig. 11-32. The signal is taken at a high-level point, rectified, and filtered to produce a dc voltage proportional to the output-signal level. This voltage is applied to the gate of $Q2$, thus causing the ac resistance between the drain and source to change, as shown in Fig. 11-31b. We thus may cause the gain of transistor $Q1$ to decrease as the output-signal level increases. The dc bias conditions of $Q1$ are not affected by $Q2$ since $Q2$ is isolated from $Q1$ by means of capacitor C_2.

REFERENCES

1. Hunter, L. P.: "Handbook of Semiconductor Electronics," McGraw-Hill Book Company, New York, 1970.
2. J. Millman, and C. C. Halkias: "Integrated Electronics; Analog and Digital Circuits and Systems," chap. 8, McGraw-Hill Book Company, New York, 1972.
3. Searle, C. L., A. R. Boothroyd, E. J. Angelo, Jr., P. E. Gray, and D. O. Pederson: "Elementary Circuit Properties of Transistors," vol. 3, Semiconductor Electronics Education Committee, John Wiley & Sons, Inc., New York, 1964.
   Giacoletto, L. J.: Study of *p-n-p* Alloy Junction Transistors from DC through Medium Frequencies, *RCA Rev.*, vol. 15, no. 4, pp. 506–562, December, 1954.
4. Sevin, L. J.: "Field-Effect Transistors," McGraw-Hill Book Company, New York, 1965, p. 23.
5. Ref. 4, p. 34.
6. Capella, D.: FETs as Voltage-Controlled Resistors, Application Note, Siliconix, Inc., Feb. 1973.

REVIEW QUESTIONS

11-1 What ratings limit the range of operation of a transistor?

11-2 Why is capacitive coupling used to connect a signal source to an amplifier?

11-3 For a capacitively coupled load, is the dc load larger or smaller than the ac load? Explain.

11-4 (a) Draw a fixed-bias circuit.

(b) Explain why the circuit is unsatisfactory if the transistor is replaced by another of the same type.

11-5 Discuss *thermal instability*.

11-6 (a) Draw a self-bias circuit.

(b) Explain qualitatively why such a circuit is an improvement on the fixed-bias circuit, as far as stability is concerned.

11-7 List the three sources of instability of a collector current.

11-8 How does the designer minimize the percentage variations in I_C due to (a) variations in I_{CO} and V_{BE}; (b) variations in β?

11-9 Over what temperature range can a transistor be used if it is (a) silicon and (b) germanium?

11-10 A transistor is excited by a large sinusoidal base current whose magnitude exceeds the quiescent value I_B for $0 < \omega t < \pi$ and is less than I_B for $\pi < \omega t < 2\pi$. Is the magnitude of the collector-current variation from the quiescent current greater at $\omega t = \pi/2$ or $3\pi/2$? Explain your answer with the aid of a graphical construction.

11-11 (a) Draw two approximate small-signal models for a BJT.

(b) Give dimensions of each parameter.

11-12 Find the relationships between the parameters in the two models of Rev. 11-11.

11-13 (a) How is g_m related to the collector current and to the temperature?

(b) Evaluate g_m for $I_C = 5$ mA.

11-14 Draw the circuit of a CE transistor configuration and give its h-parameter approximate model.

11-15 Repeat Rev. 11-14 for the CC configuration.

11-16 Repeat Rev. 11-14 for the CB configuration.

11-17 Using the approximate h-parameter model, obtain the expression for a CE circuit for (a) A_I; (b) R_i; (c) A_V; (d) R_o.

11-18 Repeat Rev. 11-17 for the emitter-follower circuit.

11-19 Repeat Rev. 11-17 for the CE circuit with an emitter resistor.

11-20 Repeat Rev. 11-17 for the emitter follower with a collector resistor.

11-21 Which of the configurations (CB, CE, CC) has the (a) highest R_i; (b) lowest R_i; (c) highest R_o; (d) lowest R_o; (e) lowest A_V; (f) lowest A_I.

11-22 Draw a CE (first) stage cascaded with a CC (second) stage. In terms of A_{V1}, A_{V2}, A_{I1}, and A_{I2} derive the expression for (a) the resultant voltage gain A_V; (b) the resultant current gain A_I.

11-23 It is desired to have a high-gain amplifier with high input resistance and low output resistance. If a cascade of four stages is used, what configuration should be used for each stage? Explain.

11-24 (a) Draw the h-parameter small-signal low-frequency BJT model, using all four parameters.

(b) What are the dimensions of each parameter?

11-25 Define as an equation and also in words the parameter (a) h_{ie}, (b) h_{fe}, (c) h_{oe}, and (d) h_{re}.

11-26 Explain how to obtain from the output characteristic (a) h_{fe} and (b) h_{oe}.

11-27 Draw the hybrid-π small-signal low-frequency BJT model containing four resistances.

11-28 (a) Draw a Darlington emitter follower.

(b) Explain why the input impedance is higher than that of a single-stage emitter follower.

11-29 (a) Indicate the circuit of an emitter follower with biasing resistors R_1 and R_2. Show that the input resistance is reduced because of these biasing resistors.

(b) Add a bootstrapped resistor R_3 and explain how this increases the input resistance.

11-30 (a) Draw two biasing circuits for a JFET or a depletion-type MOSFET.

(b) Explain under what circumstances each of these two arrangements should be used.

11-31 Draw two biasing circuits for an enhancement-type MOSFET.

11-32 Define (a) *transconductance* g_m, (b) *drain resistance* r_d, and (c) *amplification factor* μ of an FET.

11-33 Give the order of magnitude of g_m, r_d, and μ for a MOSFET.

11-34 Show the small-signal model of a FET at low frequencies.

11-35 (a) Draw the circuit of a FET amplifier with a source resistance R_s and a drain resistance R_d.

(b) What is the Thévenin's equivalent circuit looking into the drain at low frequencies?

11-36 Repeat Rev. 11-35 looking into the source.

11-37 (a) Sketch the circuit of a source follower. At low frequencies what is (b) the maximum value of the voltage gain? (c) the order of magnitude of the output resistance?

11-38 (a) Sketch the circuit of a CS amplifier.

(b) Derive the expression for the voltage gain at low frequencies.

(c) What is the maximum value of A_V?

11-39 (a) How is a JFET used as a voltage-variable resistance? (b) Explain.

FEEDBACK AMPLIFIER CHARACTERISTICS

In this chapter we introduce the concept of feedback and show how to modify the characteristics of an amplifier by combining a portion of the output signal with the external signal. Many advantages are to be gained from the use of negative (degenerative) feedback, and these are studied. Examples of feedback amplifier circuits at low frequencies are given, but the frequency response of feedback amplifiers is deferred to Chap. 14.

12-1 CLASSIFICATION OF AMPLIFIERS

Before proceeding with the concept of feedback, it is useful to classify amplifiers into four broad categories,[1] as either *voltage, current, transconductance*, or *transresistance amplifiers*. This classification is based on the magnitudes of the input and output impedances of an amplifier relative to the source and load impedances, respectively.

Voltage Amplifier

Figure 12-1 shows a Thévenin's equivalent circuit of a two-port network which represents an amplifier. If the amplifier input resistance R_i is large compared with the source resistance R_s, then $V_i \approx V_s$. If the external load resistance R_L is large compared with the output resistance R_o of the amplifier, then $V_o \approx A_v V_i \approx A_v V_s$. This amplifier provides a voltage output proportional to the voltage input, and *the proportionality factor is independent of the magnitudes of the source and load resistances*. Such a circuit is called a *voltage amplifier*. An ideal voltage amplifier must have infinite input resistance R_i and zero output resistance R_o. The symbol A_v in Fig. 12-1 represents V_o/V_i, with $R_L = \infty$, and hence represents *the open-circuit voltage amplification, or gain*.

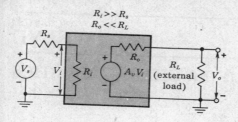

Figure 12-1 Thévenin's equivalent circuits of a voltage amplifier.

Current Amplifier

An ideal current amplifier[1] is defined as an amplifier which provides an output current proportional to the signal current, and *the proportionality factor is independent of R_s and R_L*. An ideal current amplifier must have zero input resistance R_i and infinite output resistance R_o. In practice, the amplifier has low input resistance and high output resistance. It drives a low-resistance load $(R_o \gg R_L)$, and is driven by a high-resistance source $(R_i \ll R_s)$. Figure 12-2 shows Norton's equivalent circuit of a current amplifier. Note that $A_i \equiv I_L/I_i$, with $R_L = 0$, representing *the short-circuit current amplification, or gain*. We see that if $R_i \ll R_s$, $I_i \approx I_s$, and if $R_o \gg R_L$, $I_L \approx A_i I_i \approx A_i I_s$. Hence the output current is proportional to the signal current. The characteristics of the four ideal amplifier types are summarized in Table 12-1.

Table 12-1 Ideal amplifier characteristics

Parameter	Amplifier type			
	Voltage	Current	Transconductance	Transresistance
R_i	∞	0	∞	0
R_o	0	∞	∞	0
Transfer characteristic	$V_o = A_v V_s$	$I_L = A_i I_s$	$I_L = G_m V_s$	$V_o = R_m I_s$
Reference	Fig. 12-1	Fig. 12-2	Fig. 12-3	Fig. 12-4

Transconductance Amplifier

The ideal transconductance amplifier[1] supplies an output current which is proportional to the signal voltage, independently of the magnitudes of R_s and

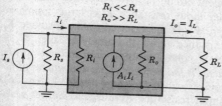

Figure 12-2 Norton's equivalent circuits of a current amplifier.

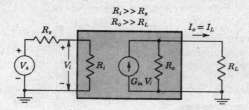

$R_i \gg R_s$
$R_o \gg R_L$

$I_o = I_L$

Figure 12-3 A transconductance amplifier is represented by a Thévenin's equivalent in its input circuit and a Norton's equivalent in its output circuit.

R_L. This amplifier must have an infinite input resistance R_i and infinite output resistance R_o. A practical transconductance amplifier has a large input resistance $(R_i \gg R_s)$ and hence must be driven by a low-resistance source. It presents a high output resistance $(R_o \gg R_L)$ and hence drives a low-resistance load. The equivalent circuit of a transconductance amplifier is shown in Fig. 12-3. From the figure we see that $V_i \approx V_s$ for $R_i \gg R_s$ and $I_o \approx G_m V_i \approx G_m V_s$ if $R_o \gg R_L$. Note that $G_m = I_o / V_i$ for $R_L = 0$. Hence G_m is *the short-circuit mutual or transfer conductance*.

Transresistance Amplifier

Finally, in Fig. 12-4, we show the equivalent circuit of an amplifier which ideally supplies an output voltage V_o in proportion to the signal current I_s independently of R_s and R_L. This amplifier is called a *transresistance amplifier*. For a practical transresistance amplifier we must have $R_i \ll R_s$ and $R_o \ll R_L$. Hence the input and output resistances are low relative to the source and load resistances. From Fig. 12-4 we see that if $R_s \gg R_i$, $I_i \approx I_s$, and if $R_o \ll R_L$, $V_o \approx R_m I_i \approx R_m I_s$. Note that $R_m \equiv V_o / I_i$, with $R_L = \infty$. In other words, R_m is the *open-circuit mutual or transfer resistance*.

12-2 THE FEEDBACK CONCEPT[2]

In the preceding section we summarize the properties of four basic amplifier types. For each one of these amplifiers we may sample the output voltage or current by means of a suitable sampling network and apply this signal to the input through a feedback two-port network, as shown in Fig. 12-5. At the input the feedback signal is combined with the external (source) signal through a mixer network and is fed into the amplifier proper.

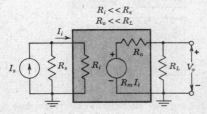

$R_i \ll R_s$
$R_o \ll R_L$

I_i

R_o

I_s R_s R_i R_L V_o

$R_m I_i$

Figure 12-4 A transresistance amplifier is represented by a Norton's equivalent in its input circuit and a Thévenin's equivalent in its output circuit.

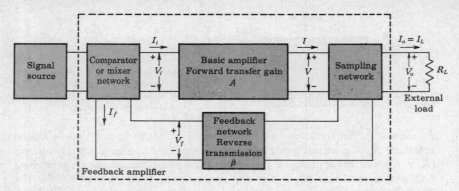

Figure 12-5 Representation of any single-loop feedback connection around a basic amplifier. The transfer gain A may represent A_V, A_I, G_M, or R_M.

Signal Source

This block in Fig. 12-5 is either a signal voltage V_s in series with a resistor R_s (a Thévenin's representation as in Fig. 12-1) or a signal current I_s in parallel with a resistor R_s (a Norton's representation as in Fig. 12-2).

Feedback Network

This block in Fig. 12-5 is usually a passive two-port network which may contain resistors, capacitors, and inductors. Most often it is simply a resistive configuration.

Sampling Network

Two sampling blocks are shown in Fig. 12-6. In Fig. 12-6a the output voltage is sampled by connecting the feedback network *in shunt* across the output. This

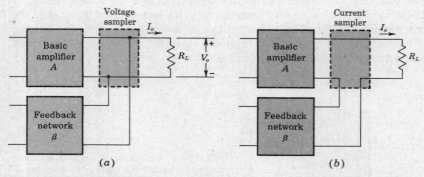

Figure 12-6 Feedback connections at the output of a basic amplifier, sampling the output (*a*) voltage and (*b*) current.

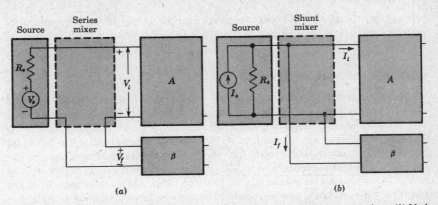

Figure 12-7 Feedback connections at the input of a basic amplifier. (*a*) Series comparison. (*b*) Node mixing.

type of connection is referred to as *voltage*, or *node*, *sampling*. Another feedback connection which samples the output current is shown in Fig. 12-6*b*, where the feedback network is connected *in series* with the output. This type of connection is referred to as *current*, or *loop*, *sampling*. Other sampling networks are possible.

Comparator, or Mixer, Network

Two mixing blocks are shown in Fig. 12-7. Figure 12-7*a* and *b* shows the simple and very common *series* (*loop*) *input* and *shunt* (*node*) *input* connections, respectively. A differential amplifier (Sec. 15-2) is often also used as the mixer. Such an amplifier has two inputs and gives an output proportional to the difference between the signals at the two inputs.

Transfer Ratio, or Gain

The symbol A in Fig. 12-5 represents the ratio of the output signal to the input signal of the basic amplifier. The transfer ratio V/V_i is the voltage amplification, or the *voltage gain*, A_V. Similarly, the transfer ratio I/I_i is the current amplification, or *current gain*, A_I for the amplifier. The ratio I/V_i of the basic amplifier is the *transconductance* G_M, and V/I_i is the *transresistance* R_M. Although G_M and R_M are defined as the ratio of two signals, one of these is a current and the other is a voltage waveform. Hence the symbol G_M or R_M does not represent an amplification in the usual sense of the word. Nevertheless, it is convenient to refer to each of the four quantities A_V, A_I, G_M, and R_M as a *transfer gain of the basic amplifier without feedback* and to use the symbol A to represent any one of these quantities.

The symbol A_f is defined as the ratio of the output signal to the input signal of the amplifier configuration of Fig. 12-5 and is called the *transfer gain of the amplifier with feedback*. Hence A_f is used to represent any one of the four ratios $V_o/V_s \equiv A_{Vf}$, $I_o/I_s \equiv A_{If}$, $I_o/V_s \equiv G_{Mf}$, and $V_o/I_s \equiv R_{Mf}$. The relationship between the transfer gain A_f with feedback and the gain A of the ideal amplifier without feedback is derived in Sec. 12-3 [Eq. (12-4)].

Advantages of Negative Feedback

When any increase in the output signal results in a feedback signal into the input in such a way as to cause a decrease in the output signal, the amplifier is said to have *negative feedback*. The usefulness of negative feedback lies in the fact that, in general, any of the four basic amplifier types discussed in Sec. 12-1 may be improved by the proper use of negative feedback. For example, the normally high input resistance of a voltage amplifier can be made higher, and its normally low output resistance can be lowered. Also, the transfer gain A_f of the amplifier with feedback can be stabilized against variations of the h or hybrid-π transistor parameters. Another important advantage of the proper use of negative feedback is the significant improvement in the frequency response and in the linearity of operation of the feedback amplifier compared with that of the amplifier without feedback.

It should be pointed out that all the advantages mentioned above are obtained at the expense of the gain A_f with feedback, which is lowered in comparison with the transfer gain A of an amplifier without feedback. Also, under certain circumstances, discussed in Chap. 14, a negative-feedback amplifier may become unstable and break into oscillations. The precautions which must be taken to avoid this undesirable effect are given in Chap. 15.

12-3 THE TRANSFER GAIN WITH FEEDBACK

Any one of the output connections of Fig. 12-6 may be combined with any of the input connections of Fig. 12-7 to form the feedback amplifier of Fig. 12-5. The analysis of the feedback amplifier can then be carried out by replacing each active element (a BJT or a FET) by its small-signal model and by writing Kirchhoff's loop, or nodal, equations. That approach, however, does not place in evidence the main characteristics of feedback.

As a first step toward a method of analysis which emphasizes the benefits of feedback, consider Fig. 12-8, which represents a generalized feedback amplifier. The basic amplifier of Fig. 12-8 may be a voltage, transconductance, current, or transresistance amplifier connected in a feedback configuration, as indicated in Fig. 12-9. The four topologies indicated in this figure are referred to as (1) *voltage-series* or *node-loop feedback*, (2) *current-series* or *loop-loop feedback*, (3) *current-shunt* or *loop-node feedback*, and (4) *voltage-shunt* or *node-node feedback*. In Fig. 12-8, the source resistance R_s is considered to be part of the amplifier, and the *transfer gain A (A_V, G_M, A_I, R_M) includes the effect of the loading of the β network (as well as R_L) upon the amplifier.* The input signal X_s, the output signal X_o, the feedback signal X_f, and the difference signal X_d, each represents either a

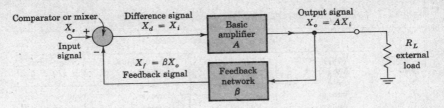

Figure 12-8 Schematic representation of a single-loop feedback amplifier.

voltage or a current. These signals and also the ratios A and β are summarized in Table 12-2. The symbol indicated by the circle in Fig. 12-8 represents a mixing, or comparison, network, whose output is the sum of the inputs, taking the sign shown at each input into account. Thus

$$X_d = X_s - X_f = X_i \tag{12-1}$$

Since X_d represents the difference between the applied signal and that fed back to the input, X_d is called the *difference*, *error*, or *comparison*, *signal*.

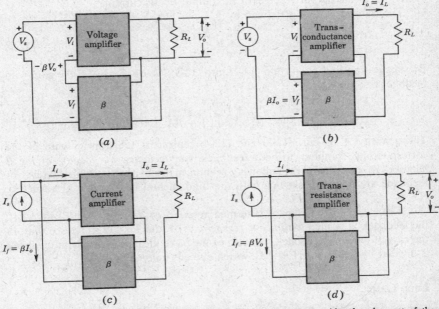

Figure 12-9 Feedback-amplifier topologies. The source resistance is considered to be part of the amplifier. (*a*) Voltage amplifier with voltage-series feedback. (*b*) Transconductance amplifier with current-series feedback. (*c*) Current amplifier with current-shunt feedback. (*d*) Transresistance amplifier with voltage-shunt feedback.

Table 12-2 Voltage and current signals in feedback amplifiers

Signal or ratio	Type of feedback			
	Voltage-series Fig. 12-9a	Current-series Fig. 12-9b	Current-shunt Fig. 12-9c	Voltage-shunt Fig. 12-9d
X_o	Voltage	Current	Current	Voltage
X_s, X_f, X_d	Voltage	Voltage	Current	Current
A	A_V	G_M	A_I	R_M
β	V_f/V_o	V_f/I_o	I_f/I_o	I_f/V_o

The reverse transmission factor β is defined by

$$\beta \equiv \frac{X_f}{X_o} \tag{12-2}$$

The factor β is often a positive or a negative real number, but in general, β is a complex function of the signal frequency. (This symbol should not be confused with the symbol β used previously for the CE short-circuit current gain.) The symbol X_o is the output voltage, or the output (load) current.

The transfer gain A is defined by

$$A \equiv \frac{X_o}{X_i} \tag{12-3}$$

By substituting Eqs. (12-1) and (12-2) into (12-3), we obtain for A_f the gain with feedback,

$$A_f \equiv \frac{X_o}{X_s} = \frac{A}{1 + \beta A} \tag{12-4}$$

The quantity A in Eqs. (12-3) and (12-4) represents the transfer gain of the corresponding amplifier without feedback, but *including the loading of the β network*, R_L and R_s. In the following section many of the desirable features of feedback are deduced, starting with the fundamental relationship given in Eq. (12-4).

If $|A_f| < |A|$, the feedback is termed *negative*, or *degenerative*. If $|A_f| > |A|$, the feedback is termed *positive*, or *regenerative*. From Eq. (12-4) we see that, in the case of negative feedback, the gain of the basic ideal amplifier with feedback is divided by the factor $|1 + \beta A|$, which exceeds unity.

Loop Gain

The signal X_d in Fig. 12-8 is multiplied by A in passing through the amplifier, is multiplied by β in transmission through the feedback network, and is multiplied by -1 in the mixing or differencing network. Such a path takes us from the input terminals around the loop consisting of the amplifier and feedback network back to the input; the product $-A\beta$ is called the *loop gain*, or *return ratio*. The difference between unity and the loop gain is called the *return difference* $D = 1 + A\beta$. Also, the amount of feedback introduced into an amplifier is often expressed in decibels by the definition

$$N = \text{dB of feedback} = 20 \log \left| \frac{A_f}{A} \right| = 20 \log \left| \frac{1}{1 + A\beta} \right| \qquad (12\text{-}5)$$

If negative feedback is under consideration, N will be a negative number.

Fundamental Assumptions

Three conditions must be satisfied for the feedback network of Fig. 12-8 in order that Eq. (12-4) be true and that the expressions for input and output resistances (derived in Secs. 12-5 and 12-6) be valid.

1. The input signal is transmitted to the output through the amplifier A and *not* through the β network. In other words, if the amplifier is deactivated (say, set $A = 0$ by reducing h_{fe} or g_m for a transistor to zero), the output signal must drop to zero.

    This first assumption is equivalent to the statement that *the system has rendered the β block unilateral*, so that it does *not* transmit a signal from input to output. This condition is often not satisfied exactly because β is a passive bilateral network. It is, however, approximately valid for practical feedback connections, as we shall verify in each of the feedback amplifiers to be considered.

2. The feedback signal is transmitted from the output to the input through the β block, and *not* through the amplifier. In other words, *the basic amplifier is unilateral from input to output* and the reverse transmission is zero. Note that the amplifiers in Figs. 12-1 to 12-4 satisfy this unilateral condition (such is not the case, for example, with a transistor amplifier at low frequencies if $h_{re} \neq 0$).

3. *The reverse transmission factor β of the feedback network is independent of the load and the source resistances R_L and R_s.*

For each topology studied we shall point out the approximations involved.

12-4 GENERAL CHARACTERISTICS OF NEGATIVE-FEEDBACK AMPLIFIERS[3]

Since negative feedback reduces the transfer gain, why is it used? The answer is that many desirable characteristics are obtained for the price of gain reduction. We now examine some of the advantages of negative feedback.

Desensitivity of Transfer Amplification

The variation due to aging, temperature, replacement, etc., of the circuit components and the BJT or the FET characteristics is reflected in a corresponding lack of stability of the amplifier transfer gain. The fractional change in amplification with feedback divided by the fractional change without feedback is called the *sensitivity* of the transfer gain. If Eq. (12-4) is differentiated with respect to A, the absolute value of the resulting equation is

$$\left| \frac{dA_f}{A_f} \right| = \frac{1}{|1 + \beta A|} \left| \frac{dA}{A} \right| \tag{12-6}$$

Hence the sensitivity is $1/|1 + \beta A|$. If, for example, the sensitivity is 0.1, the percentage change in gain with feedback is one-tenth the percentage variation in amplification if no feedback is present. The reciprocal of the sensitivity is called the *desensitivity D*, or

$$D \equiv 1 + \beta A \tag{12-7}$$

The fractional change in gain without feedback is divided by the desensitivity D when feedback is added. [In passing, note that the desensitivity is another name for the *return difference*, and that the amount of feedback is $-20 \log D$ (Eq. 12-5).] For an amplifier with 20 dB of negative feedback, $D = 10$, and hence, for example, a 5 percent change in gain without feedback is reduced to a 0.5 percent variation after feedback is introduced.

Note from Eq. (12-4) that the transfer gain is divided by the desensitivity after feedback is added. Thus

$$A_f = \frac{A}{D} \tag{12-8}$$

In particular, if $|\beta A| \gg 1$, then

$$A_f = \frac{A}{1 + \beta A} \approx \frac{A}{\beta A} = \frac{1}{\beta} \tag{12-9}$$

and *the gain may be made to depend entirely on the feedback network*. The worst offenders with respect to stability are usually the active devices (transistors) involved. If the feedback network contains only stable passive elements, the improvement in stability may indeed be pronounced.

Since A represents A_V, G_M, A_I, or R_M, then A_f represents the corresponding transfer gains with feedback: A_{Vf}, G_{Mf}, A_{If}, or R_{Mf}. *The topology determines which transfer ratio* (Table 12-2) *is stabilized*. For example, for voltage-series feedback, Eq. (12-9) signifies that $A_{Vf} \approx 1/\beta$, and it is the voltage gain which is stabilized. For current-series feedback, Eq. (12-9) is $G_{Mf} \approx 1/\beta$, and hence, for this topology, it is the transconductance gain which is desensitized. Similarly, it follows from Eq. (12-9) that the current gain is stabilized for current-shunt feedback ($A_{If} \approx 1/\beta$) and the transresistance gain is desensitized for voltage-shunt feedback ($R_{Mf} \approx 1/\beta$).

Feedback is used to improve stability in the following way: Suppose an amplifier of gain A_1 is required. We start by building an amplifier of gain

$A_2 = DA_1$, in which D is a large number. Feedback is now introduced to divide the gain by the factor D. The stability will be improved by the same factor D, since both gain and instability are divided by the desensitivity D. If now the instability of the amplifier of gain A_2 is not appreciably larger than the instability of an amplifier of gain without feedback equal to A_1, this procedure will have been useful. It often happens as a matter of practice that amplifier gain may be increased appreciably without a corresponding loss of stability. For example, the voltage gain of a transistor may be increased by increasing the collector resistance R_c.

Frequency Distortion

It follows from Eq. (12-9) that if the feedback network does not contain reactive elements, the overall gain is not a function of frequency. Under these circumstances a substantial reduction in frequency and phase distortion (Sec. 13-1) is obtained. The frequency response of feedback amplifiers is analyzed in the following chapter.

If a frequency-selective feedback network is used, so that β depends upon frequency, the amplification may depend markedly upon frequency. For example, it is possible to obtain an amplifier with a high-Q bandpass characteristic by using a feedback network which gives little feedback at the center of the band and a great deal of feedback on both sides of this frequency.

Nonlinear Distortion

Suppose that a large amplitude signal is applied to a stage of an amplifier so that the operation of the device extends slightly beyond its range of linear operation, and as a consequence the output signal is slightly distorted. Negative feedback is now introduced, and the input signal is increased by the same amount by which the gain is reduced, so that the output-signal amplitude remains the same. For simplicity, let us consider that the input signal is sinusoidal and that the distortion consists, simply, of a second-harmonic signal generated within the active device. We assume that the second-harmonic component, in the absence of feedback, is equal to B_2. Because of the effects of feedback, a component B_{2f} actually appears in the output. To find the relationship that exists between B_{2f} and B_2, it is noted that the output will contain the term $-A\beta B_{2f}$, which arises from the component $-\beta B_{2f}$ that is fed back to the input. Thus the output contains two terms: B_2, generated in the transistor, and $-A\beta B_{2f}$, which represents the effect of the feedback. Hence

$$B_2 - A\beta B_{2f} = B_{2f}$$

or

$$B_{2f} = \frac{B_2}{1 + \beta A} = \frac{B_2}{D} \tag{12-10}$$

Since A and β are generally functions of the frequency, they must be evaluated at the second-harmonic frequency.

The signal X_s to the feedback amplifier may be the actual signal externally available, or it may be the output of an amplifier preceding the feedback stage or stages under consideration. To multiply the input to the feedback amplifier by the factor $|1 + A\beta|$, it is necessary either to increase the nominal gain of the preamplifying stages or to add a new stage. If the full benefit of the feedback amplifier in reducing nonlinear distortion is to be obtained, these preamplifying stages must not introduce additional distortion, because of the increased output demanded of them. Since, however, appreciable harmonics are introduced only when the output swing is large, most of the distortion arises in the last stage. The preamplifying stages are of smaller importance in considerations of harmonic generation.

It has been assumed in the derivation of Eq. (12-10) that the small amount of additional distortion that might arise from the second-harmonic component fed back from the output to the input is negligible. This assumption leads to little error. Further, it must be noted that the result given by Eq. (12-10) applies only in the case of small distortion. The principle of superposition has been used in the derivation, and for this reason it is required that the device operate approximately linearly.

Reduction of Noise

By employing the same reasoning as that in the discussion of nonlinear distortion, it can be shown that the noise introduced in an amplifier is divided by the factor D if feedback is employed. If D is much larger than unity, this would seem to represent a considerable reduction in the output noise. However, as noted above, for a given output the amplification of the preamplifier for a specified overall gain must be increased by the factor D. Since the noise generated is independent of the signal amplitude, there may be as much noise generated in the preamplifying stage as in the output stage. Furthermore, this additional noise will be amplified, as well as the signal, by the feedback amplifier, so that the complete system may actually be noisier than the original amplifier without feedback. If the additional gain required to compensate what is lost because of the presence of inverse feedback can be obtained by a readjustment of the circuit parameters rather than by the addition of an extra stage, a definite reduction will result from the presence of the feedback. In particular, the hum introduced into the circuit by a poorly filtered power supply may be decreased appreciably.

12-5 INPUT RESISTANCE[4]

We now discuss qualitatively the effect of the topology of a feedback amplifier upon the input resistance. *If the feedback signal is returned to the input in series with the applied voltage* (regardless of whether the feedback is obtained by sampling the output current or voltage), *it increases the input resistance*. Since the feedback voltage V_f opposes V_s, the input current I_i is less than it would be if V_f

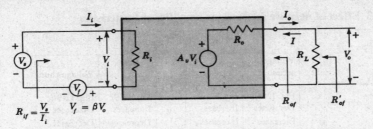

Figure 12-10 Voltage-series feedback circuit used to calculate input and output resistances.

were absent. Hence the input resistance with feedback $R_{if} \equiv V_s/I_i$ (Fig. 12-10) is greater than the input resistance without feedback R_i. We show below that, for this topology, $R_{if} = R_i(1 + \beta A) = R_i D$.

Negative feedback in which the feedback signal is returned to the input in shunt with the applied signal (regardless of whether the feedback is obtained by sampling the output current or voltage) *decreases the input resistance*. Since $I_i = I_s - I_f$ (Fig. 12-11), then the current I_i (for a fixed value of I_s) is decreased from what it would be if there were no feedback current. Hence, $R_{if} \equiv V_i/I_s = I_i R_i/I_s$ (Fig. 12-11) is decreased because of this type of feedback. We show below that, for this topology, $R_{if} = R_i/(1 + \beta A) = R_i/D$.

Table 12-3 summarizes the characteristics of the four types of negative-feedback configurations: *For series comparison, $R_{if} > R_i$, whereas for shunt mixing, $R_{if} < R_i$.*

Voltage-Series Feedback

We now obtain R_{if} quantitatively. The topology of Fig. 12-9a is indicated in Fig. 12-10, with the amplifier replaced by its Thévenin's model. In this circuit A_v (corresponding to A_{vs} in Chap. 11) represents the open-circuit voltage gain *taking R_s into account*. Since throughout the discussion of feedback amplifiers we shall consider R_s to be part of the amplifier, we shall drop the subscript s on the

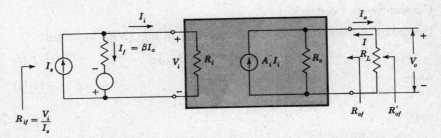

Figure 12-11 Current-shunt feedback circuit used to calculate input and output resistances.

Table 12-3 Effect of negative feedback on amplifier characteristics

	Type of feedback			
	Voltage-series	Current-series	Current-shunt	Voltage-shunt
Reference	Fig. 12-9a	Fig. 12-9b	Fig. 12-9c	Fig. 12-9d
R_{of}	Decreases	Increases	Increases	Decreases
R_{if}	Increases	Increases	Decreases	Decreases
Improves characteristics of	Voltage amplifier	Transconductance amplifier	Current amplifier	Transresistance amplifier
Desensitizes	A_{Vf}	G_{Mf}	A_{If}	R_{Mf}
Bandwidth	Increases	Increases	Increases	Increases
Nonlinear distortion	Decreases	Decreases	Decreases	Decreases

transfer gain and input impedance (A_v instead of A_{vs}, R_i instead of R_{is}, R_{if} instead of R_{ifs}, G_m instead of G_{ms}, etc.).

From Fig. 12-10 the input impedance with feedback is $R_{if} = V_s/I_i$. Also

$$V_s = I_i R_i + V_f = I_i R_i + \beta V_o \tag{12-11}$$

and

$$V_o = \frac{A_v V_i R_L}{R_o + R_L} = A_V I_i R_i \tag{12-12}$$

where

$$A_V \equiv \frac{V_o}{V_i} = \frac{A_v R_L}{R_o + R_L} \tag{12-13}$$

From Eqs. (12-11) and (12-12)

$$R_{if} = \frac{V_s}{I_i} = R_i(1 + \beta A_V) \tag{12-14}$$

Whereas A_v represents the open-circuit voltage gain without feedback, Eq. (12-13) indicates that A_V is the voltage gain without feedback taking the load R_L into account. Therefore

$$A_v = \lim_{R_L \to \infty} A_V \tag{12-15}$$

Current-Series Feedback

Proceeding in a similar manner for the topology of Fig. 12-9b, we obtain

$$R_{if} = R_i(1 + \beta G_M) \tag{12-16}$$

where

$$G_m = \lim_{R_L \to 0} G_M \tag{12-17}$$

and

$$G_M \equiv \frac{I_o}{V_i} = \frac{G_m R_o}{R_o + R_L} \tag{12-18}$$

Note that G_m is the short-circuit transconductance, whereas G_M is the transconductance without feedback taking the load into account. Note that Eqs. (12-14) and (12-16) confirm that for series mixing $R_{if} > R_i$.

Current-Shunt Feedback

The topology of Fig. 12-9c is indicated in Fig. 12-11, with the amplifier replaced by its Norton's model. In this circuit A_i represents the short-circuit current gain *taking R_s into account*. From Fig. 12-11

$$I_s = I_i + I_f = I_i + \beta I_o \qquad (12\text{-}19)$$

and

$$I_o = \frac{A_i R_o I_i}{R_o + R_L} = A_I I_i \qquad (12\text{-}20)$$

where

$$A_I \equiv \frac{I_o}{I_i} = \frac{A_i R_o}{R_o + R_L} \qquad (12\text{-}21)$$

From Eqs. (12-19) and (12-20)

$$I_s = (1 + \beta A_I) I_i \qquad (12\text{-}22)$$

From Fig. 12-11, $R_{if} = V_i/I_s$ and $R_i = V_i/I_i$. Using Eq. (12-22), we obtain

$$R_{if} = \frac{V_i}{(1 + \beta A_I) I_i} = \frac{R_i}{1 + \beta A_I} \qquad (12\text{-}23)$$

Whereas A_i represents the short-circuit current gain, Eq. (12-21) indicates that A_I *is the current gain without feedback taking the load R_L into account*. Therefore

$$A_i = \lim_{R_L \to 0} A_I \qquad (12\text{-}24)$$

Voltage-Shunt Feedback

Proceeding in a similar manner for the topology of Fig. 12-9d, we obtain

$$R_{if} = \frac{R_i}{1 + \beta R_M} \qquad (12\text{-}25)$$

where

$$R_M \equiv \frac{V_o}{I_i} = \frac{R_m R_L}{R_o + R_L} \qquad (12\text{-}26)$$

Note that R_m is the open-circuit transresistance, whereas R_M is the transresistance without feedback taking the load into account. Therefore

$$R_m = \lim_{R_L \to \infty} R_M \qquad (12\text{-}27)$$

Note that Eqs. (12-23) and (12-25) confirm that for shunt comparison $R_{if} < R_i$. The expressions for R_{if} are summarized in Table 12-4.

12-6 OUTPUT RESISTANCE[4]

We now discuss qualitatively the effect of the topology of a feedback amplifier upon the output resistance. Negative feedback which samples the output *voltage*, regardless of how this output signal is returned to the input, tends to *decrease the output resistance*. For example, if R_L increases so that V_o increases, the effect of feeding this voltage back to the input in a degenerative manner (negative feedback) is to cause V_o to increase less than it would if there were no feedback. Hence the output voltage tends to remain constant as R_L changes, which means that $R_{of} \ll R_L$. This argument leads to the conclusion that this type of feedback (sampling the output voltage) reduces the output resistance.

By reasoning similar to that given above, negative feedback which samples the output *current* will tend to hold this current constant. Hence an output current source is created ($R_{of} \gg R_L$), and we conclude that this type of sampling connection increases the output resistance.

In summary (Table 12-3): *For voltage sampling, $R_{of} < R_o$, whereas for current sampling, $R_{of} > R_o$.*

Voltage-Series Feedback

We now obtain quantitatively the resistance with feedback R_{of} looking into the output terminals but with R_L disconnected. To find R_{of} we must remove the external signal (set $V_s = 0$ or $I_s = 0$), let $R_L = \infty$, impress a voltage V across the output terminals, and calculate the current I delivered by V. Then $R_{of} \equiv V/I$. From Fig. 12-10 we find (with V_o replaced by V)

$$I = \frac{V - A_v V_i}{R_o} = \frac{V + \beta A_v V}{R_o} \tag{12-28}$$

because, with $V_s = 0$, $V_i = -V_f = -\beta V$. Hence

$$R_{of} \equiv \frac{V}{I} = \frac{R_o}{1 + \beta A_v} \tag{12-29}$$

Note that R_o is divided by the desensitivity factor $1 + \beta A_v$, which contains the open-circuit voltage gain A_v (*not* A_V).

The output resistance with feedback R'_{of} which includes R_L as part of the amplifier is given by R_{of} in parallel with R_L, or

$$R'_{of} = \frac{R_{of} R_L}{R_{of} + R_L} = \frac{R_o R_L}{1 + \beta A_v} \frac{1}{R_o/(1 + \beta A_v) + R_L} = \frac{R_o R_L}{R_o + R_L + \beta A_v R_L}$$

$$= \frac{R_o R_L/(R_o + R_L)}{1 + \beta A_v R_L/(R_o + R_L)} \tag{12-30}$$

Since $R'_o = R_o \| R_L$ is the output resistance without feedback but *with R_L considered as part of the amplifier*, and using Eq. (12-13) relating A_V to A_v, we obtain

$$R'_{of} = \frac{R'_o}{1 + \beta A_V} \qquad (12\text{-}31)$$

Note that R'_o is now divided by the desensitivity factor $1 + \beta A_V$ which contains the voltage gain A_V that takes R_L into account.

Voltage-Shunt Feedback

Proceeding as outlined above, we obtain for this topology

$$R_{of} = \frac{R_o}{1 + \beta R_m} \qquad \text{and} \qquad R'_{of} = \frac{R'_o}{1 + \beta R_M} \qquad (12\text{-}32)$$

Note that Eqs. (12-30) and (12-32) confirm that for voltage sampling $R_{of} < R_o$.

Current-Shunt Feedback

From Fig. 12-11 we find (with V_o replaced by V)

$$I = \frac{V}{R_o} - A_i I_i \qquad (12\text{-}33)$$

With $I_s = 0$, $I_i = -I_f = -\beta I_o = +\beta I$. Hence

$$I = \frac{V}{R_o} - \beta A_i I \qquad \text{or} \qquad I(1 + \beta A_i) = \frac{V}{R_o} \qquad (12\text{-}34)$$

$$R_{of} = \frac{V}{I} = R_o(1 + \beta A_i) \qquad (12\text{-}35)$$

Note that R_o is multiplied by the desensitivity factor $1 + \beta A_i$ which contains the short-circuit current gain A_i (*not* A_I).

The output resistance R'_{of} which includes R_L as part of the amplifier is not given by $R'_o(1 + \beta A_I)$, as one might thoughtlessly expect. We shall now find the correct expression for R'_{of}.

$$R'_{of} = \frac{R_{of}R_L}{R_{of} + R_L} = \frac{R_o(1 + \beta A_i)R_L}{R_o(1 + \beta A_i) + R_L}$$

$$= \frac{R_oR_L}{R_o + R_L} \frac{1 + \beta A_i}{1 + \beta A_i R_o/(R_o + R_L)} \qquad (12\text{-}36)$$

Using Eq. (12-21) and with $R'_o = R_o \| R_L$, we obtain

$$R'_{of} = R'_o \frac{1 + \beta A_i}{1 + \beta A_I} \qquad (12\text{-}37)$$

For $R_L = \infty$, $A_I = 0$ and $R'_o = R_o$, so that Eq. (12-37) reduces to

$$R'_{of} = R_o(1 + \beta A_i) = R_{of}$$

in agreement with Eq. (12-35).

Table 12-4 Feedback amplifier analysis

Characteristic	(1) Voltage series	(2) Current series	(3) Current shunt	(4) Voltage shunt
Feedback signal X_f	Voltage	Voltage	Current	Current
Sampled signal X_o	Voltage	Current	Current	Voltage
Input circuit: set†	$V_o = 0$	$I_o = 0$	$I_o = 0$	$V_o = 0$
Output circuit: set†	$I_i = 0$	$I_i = 0$	$V_i = 0$	$V_i = 0$
Signal source	Thévenin	Thévenin	Norton	Norton
$\beta = X_f/X_o$	V_f/V_o	V_f/I_o	I_f/I_o	I_f/V_o
$A = X_o/X_i$	$A_V = V_o/V_i$	$G_M = I_o/V_i$	$A_I = I_o/I_i$	$R_M = V_o/I_i$
$D = 1 + \beta A$	$1 + \beta A_V$	$1 + \beta G_M$	$1 + \beta A_I$	$1 + \beta R_M$
A_f	A_V/D	G_M/D	A_I/D	R_M/D
R_{if}	$R_i D$	$R_i D$	R_i/D	R_i/D
R_{of}	$\dfrac{R_o}{1 + \beta A_v}$	$R_o(1 + \beta G_m)$	$R_o(1 + \beta A_i)$	$\dfrac{R_o}{1 + \beta R_m}$
$R'_{of} = R_{of}\|R_L$	$\dfrac{R'_o}{D}$	$R'_o\dfrac{1 + \beta G_m}{D}$	$R'_o\dfrac{1 + \beta A_i}{D}$	$\dfrac{R'_o}{D}$

† This procedure gives the basic amplifier circuit without feedback but taking the loading of β, R_L, and R_s into account.

Current-Series Feedback

Proceeding as outlined above, we obtain for this topology

$$R_{of} = R_o(1 + \beta G_m) \qquad \text{and} \qquad R'_{of} = R'_o \frac{1 + \beta G_m}{1 + \beta G_M} \tag{12-38}$$

Note that Eqs. (12-35) and (12-38) confirm that, for current sampling, $R_{of} > R_o$. The expressions for R_{of} and R'_{of} are summarized in Table 12-4. The above derivations do *not* assume that the network is resistive. Hence, if A or β is a function of frequency, R should be changed to Z in Table 12-4. Then $Z_{if}(Z_{of})$ gives the input (output) impedance with feedback.

12-7 METHOD OF ANALYSIS OF A FEEDBACK AMPLIFIER

The first step in the analysis is to identify the topology. The *input loop* is defined as the mesh containing the applied signal voltage V_s and either (*a*) the base-to-emitter region of the first bipolar transistor, or (*b*) the gate-to-source region of the first FET in the amplifier, or (*c*) the section between the two inputs of a differential amplifier (Sec. 15-2). The mixing or comparison is identified as *series* if, in the input circuit, there is a circuit component γ in series with V_s and if γ is connected to the output (the portion of the system containing the load). If this condition is true, the voltage across γ is the feedback signal $X_f = V_f$ (Figs. 12-9*a* and *b*).

If the above condition for series mixing is not satisfied, we must test for shunt comparison. The *input node* is defined as either (*a*) the base of the first bipolar transistor, or (*b*) the gate of the first FET, or (*c*) the inverting terminal of a differential amplifier. A Norton's source is now used for the external excitation so that the current signal I_s enters the input node. The mixing is identified as *shunt* if there is a connection between the input node and the output circuit. The current in this connection is the feedback signal $X_f = I_f$ (Figs. 12-9*c* and *d*).

In summary, since $X_i = X_s - X_f$, then series mixing is obtained if the feedback signal subtracts from the externally applied signal as a voltage in the input loop, and shunt mixing is present if the feedback signal subtracts from the applied excitation as a current at the input node.

The sampled output quantity can be either a voltage or a current. The output node at which the output voltage V_o (with respect to ground) is taken must be specified in each application. This voltage V_o appears across the load resistor (often designated by R_L) and the output current I_o is the current in R_L (Fig. 12-9). Tests for the type of sampling are the following.

1. Set $V_o = 0$ (that is, set $R_L = 0$). If X_f becomes zero, the original system exhibited *voltage sampling*.
2. Set $I_o = 0$ (that is, set $R_L = \infty$). If X_f becomes zero, *current sampling* was present in the original amplifier.

The Amplifier without Feedback

It is desirable to separate the feedback amplifier into two blocks, the basic amplifier A and the feedback network β, because with a knowledge of A and β, we can calculate the important characteristics of the feedback system, namely, A_f, R_{if}, and R_{of}. The basic amplifier configuration *without feedback but taking the loading of the β network into account* is obtained[5] by applying the following rules:

To find the input circuit:

1. Set $V_o = 0$ for voltage sampling. In other words, short-circuit the output node.
2. Set $I_o = 0$ for current sampling. In other words, open-circuit the output loop.

To find the output circuit:

1. Set $V_i = 0$ for shunt comparison. In other words, short-circuit the input node (so that none of the feedback current enters the amplifier input).
2. Set $I_i = 0$ for series comparison. In other words, open-circuit the input loop (so that none of the feedback voltage reaches the amplifier input).

These procedures ensure that the feedback is reduced to zero without altering the loading on the basic amplifier. These rules are included in Table 12-4.

Outline of Analysis

To find A_f, R_{if}, and R_{of} the following steps are carried out:

1. Identify the topology as indicated above. These tests will determine whether X_f (and also X_o) is a voltage or a current.
2. Draw the basic amplifier circuit without feedback, following the rules listed above.
3. Use a Thévenin's source if X_f is a voltage and a Norton's source if X_f is a current.
4. Replace each active device by the proper model (for example, the h-parameter model at low frequencies or the hybrid-π model for a transistor at high frequencies (Sec. 11-15).
5. Indicate X_f and X_o on the circuit obtained by carrying out steps 2, 3, and 4. Evaluate $\beta = X_f/X_o$.
6. Evaluate A by applying KVL and KCL to the equivalent circuit obtained after step 4.
7. From A and β, find D, A_f, R_{if}, R_{of}, and R'_{of}.

Table 12-4 summarizes the above procedure and should be referred to when carrying out the analyses of the feedback circuits discussed in the following sections. We shall consider only the low-frequency response in this chapter and reserve the high-frequency analysis for Chap. 14.

12-8 VOLTAGE-SERIES FEEDBACK

Two examples of the voltage-series topology are considered in this section: the FET common-drain amplifier (source follower) and the bipolar transistor common-collector amplifier (emitter follower). A transistor two-stage voltage-series feedback configuration is given in the following section.

The FET Source Follower

The circuit is given in Fig. 12-12a. The load resistance is $R_L = R$. Since the input loop contains a component R which is connected to the output (V_o is across R), this is a case of series mixing. The feedback signal X_f is the voltage V_f across R. The type of sampling is found by setting $V_o = 0$. Since this operation causes V_f to be zero, voltage sampling is under consideration. Hence this is a case of voltage-series feedback, and we must refer to the first topology in Table 12-4.

We must now draw the basic amplifier without feedback. To find the input circuit, set $V_o = 0$, and hence V_s appears directly between G and S. To find the output circuit, set $I_i = 0$ (the input loop is opened), and hence R appears only in the output loop. Following these rules we obtain Fig. 12-12b. If the FET is replaced by its low-frequency model of Fig. 11-27, the result is Fig. 12-12c. Note that the positive side of V_f across R must be at node S so that V_f opposes V_s

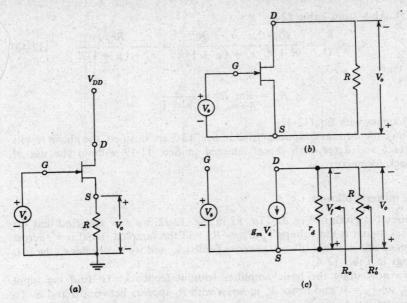

Figure 12-12 (*a*) The source-follower. (*b*) The amplifier without feedback and (*c*) the FET replaced by its small-signal low-frequency model.

around the input loop, as in Fig. 12-9*a*. From Fig. 12-12*c*, V_f and V_o are equal, and $\beta \equiv V_f/V_o = 1$.

This topology stabilizes voltage gain. A_V is calculated by inspection of Fig. 12-12*c*. Since without feedback $V_i = V_s$, then

$$A_V = \frac{V_o}{V_i} = \frac{g_m V_s r_d R}{(r_d + R)V_s} = \frac{\mu R}{r_d + R} \tag{12-39}$$

where $\mu = g_m r_d$ from Eq. (11-79).

$$D = 1 + \beta A_V = 1 + \frac{\mu R}{r_d + R} = \frac{r_d + (1 + \mu)R}{r_d + R} \tag{12-40}$$

$$A_{Vf} = \frac{A_V}{D} = \frac{\mu R}{r_d + (1 + \mu)R} \tag{12-41}$$

The input impedance of an FET is infinite, $R_i = \infty$, and hence $R_{if} = R_i D = \infty$.

We are interested in finding the output resistance seen looking into the FET source S. Hence R is considered as an external load R_L. From Table 12-4

$$R_{of} = \frac{R_o}{1 + \beta A_v} = \frac{r_d}{1 + \mu} \tag{12-42}$$

because $R_o = r_d$ from Fig. 12-12*c*, $\beta = 1$, and $A_v = \lim_{R \to \infty} A_V = \mu$ from Eq.

(12-15). Also, from Table 12-4

$$R'_{of} = \frac{R'_o}{D} = \frac{Rr_d}{R + r_d} \frac{r_d + R}{r_d + (\mu + 1)R} = \frac{Rr_d}{r_d + (\mu + 1)R} \qquad (12\text{-}43)$$

Note that

$$R_{of} = \lim_{R \to \infty} R'_{of} = \frac{r_d}{\mu + 1}$$

which agrees with Eq. (12-42).

Since the three assumptions listed in Sec. 12-3 are satisfied, the above results are exact and agree with those obtained in Sec. 11-19 without the use of feedback formulas.

The Emitter Follower

The circuit is given in Fig. 12-13a. As in Fig. 12-12, we now also find that the feedback signal is the voltage V_f across R_e, and the sampled signal is V_o across R_e. Hence this is a case of voltage-series feedback, and we must refer to the first topology in Table 12-4.

We now draw the basic amplifier without feedback. To find the input circuit, set $V_o = 0$, and hence V_s in series with R_s appears between B and E. To

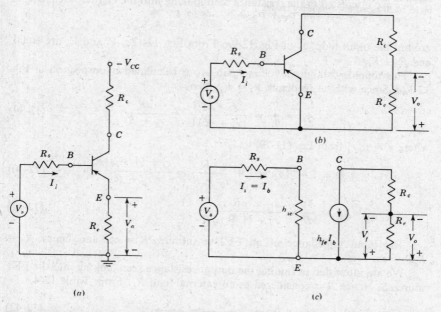

Figure 12-13 (a) An emitter follower. (b) The amplifier without feedback and (c) the transistor replaced by its approximate low-frequency model.

find the output circuit, set $I_i = I_b = 0$ (the input loop is opened), and hence R_e appears only in the output loop. Following these rules, we obtain the circuit of Fig. 12-13b. If the transistor is replaced by its low-frequency approximate model of Fig. 11-6a, the result is Fig. 12-13c. From this figure $V_o = V_f$ and $\beta = V_f/V_o = 1$.

This topology stabilizes the voltage gain. A_V is calculated by inspection of Fig. 12-13c. Since R_s is considered as part of the amplifier, then $V_i = V_s$, and

$$A_V = \frac{V_o}{V_i} = \frac{h_{fe}I_b R_e}{V_s} = \frac{h_{fe}R_e}{R_s + h_{ie}} \tag{12-44}$$

$$D = 1 + \beta A_V = 1 + \frac{h_{fe}R_e}{R_s + h_{ie}} = \frac{R_s + h_{ie} + h_{fe}R_e}{R_s + h_{ie}} \tag{12-45}$$

$$A_{Vf} = \frac{A_V}{D} = \frac{h_{fe}R_e}{R_s + h_{ie} + h_{fe}R_e} \tag{12-46}$$

For $h_{fe}R_e \gg R_s + h_{ie}$, $A_{Vf} \approx 1$, as it should be for an emitter follower.

The input resistance without feedback is $R_i = R_s + h_{ie}$ from Fig. 12-13c. Hence

$$R_{if} = R_i D = (R_s + h_{ie})\frac{R_s + h_{ie} + h_{fe}R_e}{R_s + h_{ie}} = R_s + h_{ie} + h_{fe}R_e \tag{12-47}$$

We are interested in the resistance seen looking into the emitter. Hence R_e is considered as an external load. From Table 12-4

$$R_{of} = \frac{R_o}{1 + \beta A_v} = \frac{\infty}{\infty} \tag{12-48}$$

because, from Fig. 12-13c, we are looking into a current source $R_o = \infty$ and $A_v = \lim\limits_{R \to \infty} A_V = \infty$ from Eq. (12-15). The indeterminacy in Eq. (12-48) may be resolved by first evaluating R'_{of} and then going to the limit $R_e \to \infty$. Thus, since $R'_o = R_e$,

$$R'_{of} = \frac{R'_o}{D} = \frac{R_e(R_s + h_{ie})}{R_s + h_{ie} + h_{fe}R_e} \tag{12-49}$$

and

$$R_{of} = \lim\limits_{R_e \to \infty} R'_{of} = \frac{R_s + h_{ie}}{h_{fe}} \tag{12-50}$$

Note that the feedback desensitizes voltage gain with respect to changes in h_{fe} and that it increases the input resistance and decreases the output resistance.

The foregoing expressions for A_{Vf}, R_{if}, and R_{of} are based on the assumption of zero forward transmission through the feedback network. Since there is such forward transmission because the input current passes through R_e in Fig. 12-13a, these expressions are only approximately true. In this example we have in effect neglected the base current which flows in R_e compared with the collector current. The more exact answers are obtained in Sec. 11-10, and they differ from those given above only in that h_{fe} must be replaced by $h_{fe} + 1$.

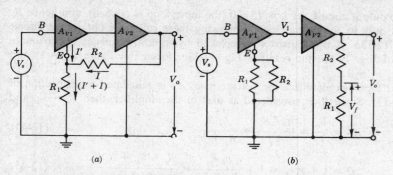

(a)

(b)

Figure 12-14 (a) Voltage-series feedback pair. (b) Equivalent circuit, without external feedback, but including the loading of R_2.

12-9 A VOLTAGE-SERIES FEEDBACK PAIR

Figure 12-14 shows two cascaded stages whose voltage gains are A_{V1} and A_{V2}, respectively. The output of the second stage is returned through the feedback network $R_1 R_2$ in opposition to the input signal V_s. From the same reasoning used in connection with the feedback amplifiers of Sec. 12-8, it follows that Fig. 12-14 is a case of voltage-series negative feedback. According to Table 12-3, we should expect the input resistance to increase, the output resistance to decrease, and the voltage gain to be stabilized (desensitized).

The first basic assumption listed in Sec. 12-3 is not strictly satisfied for the circuit of Fig. 12-14a because I' represents transmission through the feedback network from input to the output. We shall neglect I' compared with I on the realistic assumption that the current gain of the second stage is much larger than unity. Under these circumstances very little error is made in using the feedback formulas developed in this chapter.

The input of the basic circuit without feedback is found by setting $V_o = 0$ (Table 12-4), and hence R_2 appears in parallel with R_1. The output of the basic amplifier without feedback is found by opening the input loop (set $I_i = 0$ so that $I' = 0$), and hence R_1 is placed in series with R_2. Following these rules results in Fig. 12-14b, to which has been added the series feedback voltage V_f across R_1 in the output circuit. Since the feedback voltage must be proportional to the sampled voltage V_o, then V_f is indicated across the R_1 in the *output* circuit and not the R_1 in the input mesh. Clearly,

$$\beta = \frac{V_f}{V_o} = \frac{R_1}{R_1 + R_2} \tag{12-51}$$

Second-Collector to First-Emitter Feedback Pair

The circuit of Fig. 12-15 shows a two-stage amplifier which makes use of voltage-series feedback by connecting the second collector to the first emitter

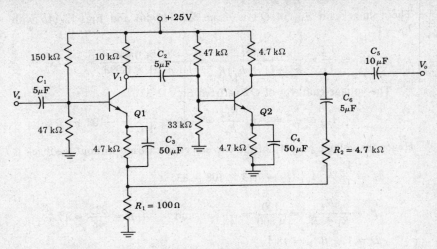

Figure 12-15 Second-collector to first-emitter feedback pair.

through the voltage divider $R_1 R_2$. Capacitors C_1, C_2, C_5, and C_6 are dc blocking capacitors, and capacitors C_3 and C_4 are bypass capacitors for the emitter bias resistors. All these capacitances represent negligible reactances at the frequencies of operation of this circuit. For this amplifier the voltage gain A_{Vf} is given approximately by $1/\beta$, and is thus stabilized against temperature changes and transistor replacement. A more accurate determination of A_{Vf}, as well as a calculation of input and output resistance, is given in the following illustrative problem.

Example 12-1 Calculate A_{Vf}, R_{of}, and R_{if} for the amplifier of Fig. 12-15. Assume $R_s = 0$, $h_{fe} = 50$, $h_{ie} = 1.1$ kΩ, $h_{re} = h_{oe} = 0$, and identical transistors.

SOLUTION We first calculate the overall voltage gain without feedback from $A_V = A_{V1}A_{V2}$. The effective load R'_{L1} of transistor $Q1$ is

$$R'_{L1} = 10\|47\|33\|1.1 \text{ k}\Omega = 943 \ \Omega$$

From Fig. 12-14b we see that the effective load R'_{L2} of transistor $Q2$ is the collector resistance $R_{c2} = 4.7$ kΩ in parallel with $R_1 + R_2 = 4.8$ kΩ,

$$R'_{L2} = 4.7\|4.8 = 2.37 \text{ k}\Omega$$

From Fig. 12-14b we see that the effective emitter impedance R_e of $Q1$ is $R_1\|R_2$ or

$$R_e = R_1\|R_2 = 0.1\|4.7 \text{ k}\Omega = 0.098 \text{ k}\Omega = 98 \ \Omega$$

The voltage gain A_{V1} of $Q1$ is, from Eq. (11-46) and Fig. 12-14b with $V_i = V_s$,

$$A_{V1} \equiv \frac{V_1}{V_i} = \frac{-h_{fe}R'_{L1}}{h_{ie} + (1 + h_{fe})R_e} = \frac{-50 \times 0.943}{1.1 + 51 \times 0.098} = -7.73$$

The voltage gain A_{V2} of $Q2$ is, from Eq. (11-32),

$$A_{V2} \equiv \frac{V_o}{V_1} = -h_{fe}\frac{R'_{L2}}{h_{ie}} = -50 \times \frac{2.37}{1.1} = -108$$

Hence the voltage gain A_V of the two stages in cascade without feedback is

$$A_V \equiv \frac{V_o}{V_i} = A_{V1}A_{V2} = 7.73 \times 108 = 835$$

$$\beta = \frac{R_1}{R_1 + R_2} = \frac{100}{4,800} = \frac{1}{48} \quad \text{and} \quad A_V\beta = \frac{835}{48} = 17.4$$

$$D = 1 + \beta A_V = 18.4$$

$$A_{Vf} = \frac{A_V}{D} = \frac{835}{18.4} = 45.4$$

This value is to be compared with the approximate solution (based upon $A_V \to \infty$) given by $A_{Vf} = 1/\beta = 48$.

The input resistance without external feedback is, from Eq. (11-45),

$$R_i \equiv h_{ie} + (1 + h_{fe})R_e = 1.1 + 51 \times 0.098 = 6.10 \text{ k}\Omega$$

Hence, from Table 12-4,

$$R_{if} = R_i D = 6.10 \times 18.4 = 112 \text{ k}\Omega$$

The output resistance without feedback is $R'_o = R'_{L2} = 2.37$ kΩ. Hence, from Table 12-4

$$R'_{of} = \frac{R'_o}{D} = \frac{2.37}{18.4} \text{ k}\Omega = 129 \ \Omega$$

It is interesting to note that there is internal (local) feedback in the first stage of Fig. 12-14b because the R_1R_2 parallel combination acts as an emitter resistor. This first stage is an example of current-series feedback, which is analyzed in the next section.

12-10 CURRENT-SERIES FEEDBACK

Consider the feedback amplifier in Fig. 12-16a. From the discussion of the emitter follower of Fig. 12-13, it is clear that series comparison is involved and $X_f = V_f$ is the feedback voltage across R_e. To test for the type of sampling, set $V_o = 0$. This grounding of the collector node does not eliminate the collector current or emitter current (due to the input signal), and hence the feedback

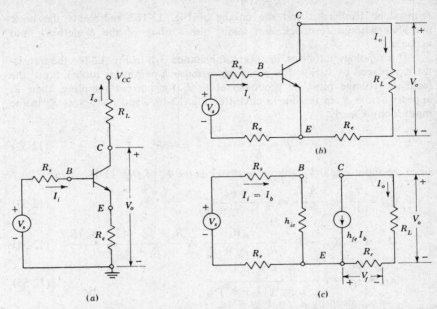

Figure 12-16 (*a*) Amplifier with an unbypassed emitter resistance as an example of current-series feedback. (*b*) The amplifier without feedback, but including the loading of R_e. (*c*) The approximate *h*-parameter model used for the transistor in (*b*).

voltage across R_e does not become zero. Therefore, the circuit does *not* exhibit voltage sampling (Sec. 12-7). Now set $I_o = 0$ so that the collector current is reduced to zero. Now the voltage across R_e is $X_f = 0$, which indicates (Sec. 12-7) that this amplifier samples the output current. Hence Fig. 12-16*a* is a case of current-series feedback.

In passing, note that although I_o is proportional to V_o, it is *not* correct to conclude that this is a voltage-series feedback. Thus, if the sampled signal is taken as the voltage V_o, then

$$\beta = \frac{V_f}{V_o} = \frac{-I_o R_e}{I_o R_L} = -\frac{R_e}{R_L}$$

Since β is now a function of the load R_L, the third basic assumption given in Sec. 12-3 is violated.

We must refer to the second topology in Table 12-4. The input circuit of the amplifier without feedback is obtained by setting $I_o = 0$. Hence R_e must appear in the input side. Similarly, the output circuit is obtained by opening the input loop ($I_i = 0$), and this places R_e also in the output side. The resulting equivalent circuit is given in Fig. 12-16*b*. No ground can be indicated in this circuit because to do so would again couple the input to the output via R_e; that is, it would reintroduce feedback. And the circuit of Fig. 12-16b represents the basic amplifier without feedback, but taking the loading of the β network into account.

This topology stabilizes the transconductance G_M. In Fig. 12-16c the transistor is replaced by its low-frequency *approximate* h-parameter model. Since the feedback voltage must be proportional to I_o for current sampling, then V_f appears across R_e in the *output* circuit of Fig. 12-16c (and not across R_e in the input loop). Clearly,

$$\beta = \frac{V_f}{I_o} = \frac{-I_o R_e}{I_o} = -R_e \tag{12-52}$$

Since the input signal V_i without feedback is the V_s of Fig. 12-16c, then

$$G_M = \frac{I_o}{V_i} = \frac{-h_{fe}I_b}{V_s} = \frac{-h_{fe}}{R_s + h_{ie} + R_e} \tag{12-53}$$

$$D = 1 + \beta G_M = 1 + \frac{h_{fe}R_e}{R_s + h_{ie} + R_e} = \frac{R_s + h_{ie} + (1 + h_{fe})R_e}{R_s + h_{ie} + R_e} \tag{12-54}$$

$$G_{Mf} = \frac{G_M}{D} = \frac{-h_{fe}}{R_s + h_{ie} + (1 + h_{fe})R_e} \tag{12-55}$$

Note that if $(1 + h_{fe})R_e \gg R_s + h_{ie}$, and since $h_{fe} \gg 1$, then $G_{Mf} \approx -1/R_e$, in agreement with $G_{Mf} \approx 1/\beta$. If R_e is a stable resistor, the transconductance gain with feedback is stabilized (desensitized). The load current is given by

$$I_o = G_{Mf}V_s = \frac{-h_{fe}V_s}{R_s + h_{ie} + (1 + h_{fe})R_e} \approx -\frac{V_s}{R_e} \tag{12-56}$$

Under the conditions $(1 + h_{fe})R_e \gg R_s + h_{ie}$ and $h_{fe} \gg 1$, *the load current is directly proportional to the input voltage, and this current depends only upon R_e, and not upon any other circuit or transistor parameter.* As an example, consider that this circuit is used as the driver for the deflection current I_o in a magnetic cathode-ray oscilloscope. The load is then the deflection yoke impedance, which is essentially an inductance whose reactance is proportional to frequency. Yet, from Eq. (12-56) the load current is independent of the characteristics of the yoke. If a deflection which varies linearly with time is desired, it is only necessary to generate a voltage waveform V_s which increases linearly with time (we are assuming that the deflection of the spot on the tube face is proportional to the yoke current).

The voltage gain is given by

$$A_{Vf} = \frac{I_o R_L}{V_s} = G_{Mf}R_L = \frac{-h_{fe}R_L}{R_s + h_{fe} + (1 + h_{fe})R_e} \tag{12-57}$$

Subject to the approximations made above, $A_{Vf} \approx -R_L/R_e$ and the voltage gain is stable if R_L and R_e are stable resistors.

From Fig. 12-16c, we see that $R_i = R_s + h_{ie} + R_e$. Hence, from Table 12-4,

$$R_{if} = R_i D = R_s + h_{ie} + (1 + h_{fe})R_e \qquad (12\text{-}58)$$

Because R_s is considered to be part of the amplifier, it appears above as a component of the input resistance.

Since $R_o = \infty$, then $R_{of} = R_o(1 + \beta G_m) = \infty$. Hence $R'_{of} = R_L \| R_{of} = R_L$. An alternative derivation is to use the expression in Table 12-4, namely,

$$R'_{of} = R'_o \frac{1 + \beta G_m}{1 + \beta G_M}$$

Since G_m represents the short-circuit transconductance, then $G_m = \lim_{R_L \to 0} G_M$ [Eq. (12-17)]. However, from Eq. (12-53), G_M is independent of R_L, and hence $G_m = G_M$ and $R'_{of} = R'_o = R_L$.

The above results agree exactly with those derived in Sec. 8-15 because all three assumptions listed in Sec. 12-3 are satisfied. Note, in particular, that if the amplifier is deactivated (say, $h_{fe} = 0$), then $I_o = 0$, which means that none of the input signal appears at the output via the feedback block. Hence the first condition is satisfied, even though the β network is simply a resistor R_e. Note that for this topology it is *not* necessary to assume that the base current is negligible compared with the collector current.

Example 12-2 The circuit of Fig. 12-16a is to have an overall transconductance gain of -1 mA/V, a voltage gain of -4, and a desensitivity of 50. If $R_s = 1$ kΩ, $h_{fe} = 150$, and $r_{bb'}$ is negligible, find (a) R_e, (b) R_L, (c) R_{if}, and (d) the quiescent collector current I_C at room temperature.

SOLUTION (a) $G_{Mf} = \dfrac{G_M}{D} = \dfrac{G_M}{50} = -1$ mA/V

or $\qquad\qquad\qquad\qquad\qquad G_M = -50$ mA/V

Since $\beta = -R_e$, then

$$D = 1 + \beta G_M = 1 + 50R_e = 50$$

or $\qquad\qquad\qquad\qquad R_e = 0.98\ \text{k}\Omega \approx 1\ \text{k}\Omega$

(b) $A_{Vf} = G_{Mf} R_L$

or $\qquad\qquad\qquad\qquad R_L = \dfrac{A_{Vf}}{G_{Mf}} = \dfrac{-4}{-1} = 4\ \text{k}\Omega$

(c) From Eq. (12-53)

$$G_M = -50 = \frac{-h_{fe}}{R_s + h_{ie} + R_e} = \frac{-150}{1 + h_{ie} + 1}$$

or
$$h_{ie} = 1 \text{ k}\Omega$$
$$R_i = R_s + h_{ie} + R_e = 3 \text{ k}\Omega$$
$$R_{if} = R_i D = (3)(50) = 150 \text{ k}\Omega$$

(d) From Eqs. (11-19), (11-21), and (11-25)

$$h_{ie} = r_{bb'} + r_{b'e} \approx \frac{h_{fe}}{g_m} = \frac{h_{fe} V_T}{I_C}$$

or
$$I_C = \frac{h_{fe} V_T}{h_{ie}} = \frac{(150)(0.026)}{1} = 3.90 \text{ mA}$$

12-11 CURRENT-SHUNT FEEDBACK

Figure 12-17 shows two transistors in cascade with feedback from the second emitter to the first base through the resistor R'. From the discussion in Sec. 12-7 it follows that shunt mixing is used in this circuit and that X_f is the feedback current I_f in R' which connects the input node to the output circuit. The excitation is now expressed by a Norton's circuit consisting of a current $I_s = V_s / R_s$ entering the input node shunted by the resistor R_s from the base of $Q1$ to ground. To find the type of sampling we set $V_o = 0$. This procedure does *not* reduce I_o and, hence, the emitter current of $Q2$ to zero. Therefore some fraction of I_o flows through R'. Since I_f is not reduced to zero, this amplifier does not exhibit voltage sampling. However, if we set $I_o = 0$ then the feedback current from the output is $I_f = 0$, indicating that current sampling is involved. These arguments show that the network of Fig. 12-17 is a current-shunt feedback amplifier.

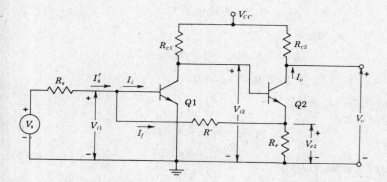

Figure 12-17 Second-emitter to first-base feedback pair. (The input blocking capacitor and the biasing resistors are not indicated.)

We now verify that the feedback is degenerative (negative). The voltage V_{i2} is much larger than V_{i1} because of the voltage gain of $Q1$. Also, V_{i2} is $180°$ out of phase with V_{i1}. Because of emitter-follower action, V_{e2} is only slightly smaller than V_{i2}, and these voltages are in phase. Hence V_{e2} is larger in magnitude than V_{i1} and is $180°$ out of phase with V_{i1}. If the input signal increases so that I'_s increases, I_f also increases, and $I_i = I'_s - I_f$ is smaller than it would be if there were no feedback. This action is characteristic of *negative* feedback.

The Amplifier without Feedback

We must refer to the third topology in Table 12-4. The input circuit of the amplifier without feedback is obtained by setting $I_o = 0$. Neglecting I_{b2}, the emitter of $Q2$ is open-circuited ($I_{e2} \approx 0$). Consequently, R' is placed in series with R_e from base to emitter of $Q1$. The output circuit is found by short-circuiting the input node (the base of $Q1$). This places R' in parallel with R_e at E_2. The resultant equivalent circuit is given in Fig. 12-18. Since the feedback signal is a current, the source is represented by a Norton's equivalent circuit with $I_s = V_s/R_s$.

The feedback signal is the current I_f in the resistor R', which is in the output circuit. From Fig. 12-18, with $I_{b2} < I_{c2} = |I_o|$.

$$\beta = \frac{I_f}{I_o} = \frac{R_e}{R' + R_e} \tag{12-59}$$

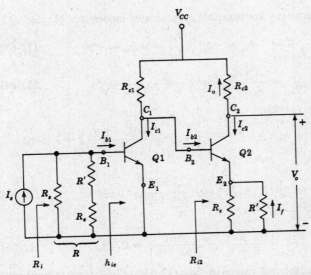

Figure 12-18 The amplifier of Fig. 12-17 without feedback, but including the loading of R'.

From Table 12-3 we expect the input resistance to be low, the output resistance to be high, and the transfer (current) gain A_{If} to be stabilized. From Eqs. (12-9) and (12-59),

$$A_{If} \equiv \frac{I_o}{I_s} \approx \frac{1}{\beta} = \frac{R' + R_e}{R_e} \tag{12-60}$$

and hence we have verified that A_{If} is desensitized provided that R' and R_e are stable resistances. The voltage gain with feedback is

$$A_{Vf} = \frac{V_o}{V_s} = \frac{I_o R_{c2}}{I_s R_s} = A_{If} \frac{R_{c2}}{R_s} \approx \frac{R' + R_e}{R_e} \frac{R_{c2}}{R_s} = \frac{R_{c2}}{\beta R_s} \tag{12-61}$$

Note that if R_e, R', R_{c2}, and R_s are stable elements, then A_{Vf} is stable (independent of the transistor parameters, the temperature, or supply-voltage variations).

Example 12-3 The circuit of Fig. 12-17 has the following parameters: $R_{c1} = 3$ kΩ, $R_{c2} = 500$ Ω, $R_e = 50$ Ω, $R' = R_s = 1.2$ kΩ, $h_{fe} = 50$, $h_{ie} = 1.1$ kΩ, and $h_{re} = h_{oe} = 0$. Find (a) A_{Vf}; (b) R_{if}; (c) the resistance seen by the voltage source; and (d) the output resistance.

SOLUTION (a) Since the current gain is stabilized, we first calculate A_{If} from A_I. We can then obtain A_{Vf} from A_{If}. Referring to Fig. 12-18,

$$A_I = \frac{-I_{c2}}{I_s} = \frac{-I_{c2}}{I_{b2}} \frac{I_{b2}}{I_{c1}} \frac{I_{c1}}{I_{b1}} \frac{I_{b1}}{I_s} \tag{12-62}$$

Using the low-frequency approximate h-parameter models for $Q1$ and $Q2$,

$$\frac{-I_{c2}}{I_{b2}} = -h_{fe} = -50 \qquad \frac{I_{c1}}{I_{b1}} = +h_{fe} = +50 \tag{12-63}$$

$$\frac{I_{b2}}{I_{c1}} = \frac{-R_{c1}}{R_{c1} + R_{i2}} = \frac{-3}{3 + 3.55} = -0.458 \tag{12-64}$$

because, from Eq. (11-45),

$$R_{i2} = h_{ie} + (1 + h_{fe})(R_e \| R') = 1.1 + (51)\left(\frac{0.05 \times 1.20}{1.25} \right) = 3.55 \text{ k}\Omega$$

If R is defined by

$$R \equiv R_s \| (R' + R_e) = \frac{(1.2)(1.25)}{1.2 + 1.25} = 0.612 \text{ k}\Omega \tag{12-65}$$

then from Fig. 12-18

$$\frac{I_{b1}}{I_s} = \frac{R}{R + h_{ie}} = \frac{0.61}{0.61 + 1.1} = 0.358 \tag{12-66}$$

Substituting the numerical values in Eqs. (12-63), (12-64), and (12-66) into

Eq. (12-62) yields

$$A_I = (-50)(-0.458)(50)(0.358) = +410$$

$$\beta = \frac{R_e}{R' + R_e} = \frac{50}{1,250} = 0.040$$

$$D = 1 + \beta A_I = 1 + (0.040)(410) = 17.4$$

$$A_{If} = \frac{A_I}{D} = \frac{410}{17.4} = 23.6$$

$$A_{Vf} = \frac{V_o}{V_s} = \frac{-I_{c2}R_{c2}}{I_s R_s} = \frac{A_{If}R_{c2}}{R_s} = \frac{(23.6)(0.5)}{1.2} = 9.83$$

The approximate expression of Eq. (12-61) yields

$$A_{Vf} \approx \frac{R_{c2}}{\beta R_s} = \frac{0.5}{(0.040)(1.2)} = 10.4$$

which is in error by 6 percent.

(b) From Fig. 12-18, the input impedance without feedback seen by the current source is, using Eq. (12-65),

$$R_i = R \| h_{ie} = \frac{(0.612)(1.1)}{1.71} = 0.394 \text{ k}\Omega$$

and from Table 12-4 the resistance R_{if} with feedback *seen by the current source* I_s is

$$R_{if} = \frac{R_i}{D} = \frac{394}{17.4} = 22.6 \ \Omega$$

(c) Note that the input resistance is quite small, as predicted. If the resistance looking to the right of R_s (from base to emitter of $Q1$) in Fig. 12-17 is R'_{if}, then $R_{if} = R'_{if} \| R_s$, or

$$22.6 = \frac{1,200 R'_{if}}{1,200 + R'_{if}}$$

which yields $R'_{if} = 23.0 \ \Omega$. Hence from Fig. 12-17, the resistance with feedback seen by the voltage source V_s is

$$R_s + R'_{if} = 1,200 + 23.0 \ \Omega = 1.22 \text{ k}\Omega$$

(d) If R_{c2} is considered as an external load, then R_o is the resistance seen looking into the collector of $Q2$. Since $h_{oe} = 0$, then $R_o = \infty$. From Table 12-4, $R_{of} = R_o(1 + \beta A_i) = \infty$.

From the calculations in part a, we note that A_I is independent of the load $R_L = R_{c2}$. Hence $A_i = \lim\limits_{R_{c2} \to 0} A_I = A_I$. Since $R'_o = R_o \| R_{c2} = R_{c2}$, then from Table 12-4,

$$R'_{of} = R'_o \frac{1 + \beta A_i}{1 + \beta A_I} = R'_o = R_{c2} = 500 \ \Omega$$

R'_{of} may also be calculated from Eq. (C-12) as the ratio of the open-circuit voltage V to the short-circuit current I. Since for $h_{oe} = 0$, I_o is independent of R_{c2}, then $I = I_o$. Hence,

$$R'_{of} = \frac{V}{I} = \frac{V_o}{I_o} = \frac{I_o R_{c2}}{I_o} = R_{c2} = 500 \, \Omega$$

which agrees with the value found in Table 12-4.

12-12 VOLTAGE-SHUNT FEEDBACK

Figure 12-19a shows a common-emitter stage with a resistor R' connected from the output to the input. We first show that this configuration conforms to voltage-shunt topology, and then obtain approximate expressions for transresistance and the voltage gain with feedback.

By the same argument used in connection with Fig. 12-17, it is clear that shunt mixing takes place in Fig. 12-19 and that X_f is I_f in R'. If we set $V_o = 0$, the feedback current from the output node is reduced to zero, indicating that voltage sampling is invoked. Hence the configuration is that of a voltage-shunt feedback amplifier. From Table 12-3 we expect the transfer gain (the transresistance) $A_f = R_{Mf}$ to be desensitized and both the input and output resistances to be low.

The Amplifier without Feedback

We must refer to the fourth topology in Table 12-4. The input circuit of the amplifier without feedback is obtained by short-circuiting the output node ($V_o = 0$). This places R' from base to emitter of the transistor. The output

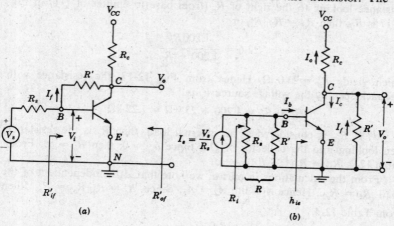

Figure 12-19 (a) Voltage-shunt feedback. (b) The amplifier without feedback, but including the loading of R'.

circuit is found by short-circuiting the input node ($V_i = 0$), thus connecting R' from collector to emitter. The resultant equivalent circuit is given in Fig. 12-19b. Since the feedback signal is a current, the source is represented by a Norton's equivalent with $I_s = V_s/R_s$.

The feedback signal is the current I_f in the resistor R' which is in the output circuit. From Fig. 12-19b

$$\beta = \frac{I_f}{V_o} = -\frac{1}{R'} \tag{12-67}$$

which verifies that I_f is proportional to V_o, as expected for voltage sampling.

From Eq. (12-9), for the amplifier with feedback,

$$R_{Mf} \equiv \frac{V_o}{I_s} \approx \frac{1}{\beta} = -R' \tag{12-68}$$

Note that the transresistance equals the negative of the feedback resistance from output to input of the transistor and is stable if R' is a stable resistance.

The voltage gain with feedback is

$$A_{Vf} = \frac{V_o}{V_s} = \frac{V_o}{I_s R_s} \approx \frac{1}{\beta R_s} = -\frac{R'}{R_s} \tag{12-69}$$

where use is made of Eq. (12-68). Note that if R' and R_s are stable elements, then A_{Vf} is stable (independent of the transistor parameters, the temperature, and supply-voltage variations).

Example 12-4 The circuit of Fig. 12-19 has the following parameters: $R_c = 4$ kΩ, $R' = 40$ kΩ, $R_s = 10$ kΩ, $h_{ie} = 1.1$ kΩ, $h_{fe} = 50$, and $h_{re} = h_{oe} = 0$. Find (a) A_{Vf}, (b) R'_{if}, and (c) R'_{of}.

SOLUTION (a) Since the transresistance is stabilized, we first calculate R_{Mf} from R_M. Define R'_c and R by

$$R'_c \equiv R_c \| R' = \frac{(4)(40)}{44} = 3.64 \text{ k}\Omega \tag{12-70}$$

and

$$R \equiv R_s \| R' = \frac{(10)(40)}{50} = 8 \text{ k}\Omega \tag{12-71}$$

From Fig. 12-19b

$$R_M = \frac{V_o}{I_s} = \frac{-I_c R'_c}{I_s} = \frac{-h_{fe} I_b R'_c}{I_s} = \frac{-h_{fe} R'_c R}{R + h_{ie}} \tag{12-72}$$

$$R_M = \frac{(-50)(3.64)(8)}{8 + 1.1} = -160 \text{ k}\Omega$$

$$\beta = -\frac{1}{R'} = -\frac{1}{40} = -0.025 \text{ mA/V}$$

$$D = 1 + \beta R_M = 1 + 0.025 \times 160 = 5.00$$

For the amplifier with feedback

$$R_{Mf} = \frac{R_M}{D} = \frac{-160}{5.00} = -32.0 \text{ k}\Omega$$

$$A_{Vf} = \frac{V_o}{V_s} = \frac{V_o}{I_s R_s} = \frac{R_{Mf}}{R_s} \qquad (12\text{-}73)$$

or $$A_{Vf} = \frac{-32.0}{10} = -3.20$$

(b) From Fig. 12-19b

$$R_i = \frac{Rh_{ie}}{R + h_{ie}} = \frac{(8)(1.1)}{9.1} = 0.967 \text{ k}\Omega = 967 \text{ }\Omega$$

From Table 12-4

$$R_{if} = \frac{R_i}{D} = \frac{967}{5.00} = 193 \text{ }\Omega$$

Note that the input resistance is quite small, as predicted.

If the input resistance looking to the right of R_s (from base to emitter in Fig. 12-19a) is R'_{if}, then $R_{if} = R'_{if} \| R_s$. Solving, we find $R'_{if} = 196 \text{ }\Omega$. The impedance seen by the voltage source V_s is $R_s + R'_{if} = 10.2 \text{ k}\Omega$.

(c) R'_{of} can be calculated from the formula in Table 12-4. The output resistance, taking R_c into account but neglecting feedback, is, from Fig. 12-19b, $R'_o = R_c \| R' = R'_c = 3.64 \text{ k}\Omega$. From Table 12-4

$$R'_{of} = \frac{R'_o}{D} = \frac{3.64}{5.00} \text{ k}\Omega = 728 \text{ }\Omega$$

It is instructive to examine the approximate expression for the voltage gain given in Eq. (12-69).

$$A_{Vf} \approx \frac{-R'}{R_s} = -\frac{40}{10} = -4.00$$

which differs from the value of -3.20 by about 22 percent. This approximate formula leads to the erroneous conclusion that A_{Vf} increases without limit as $R_s \to 0$. The difficulty arises because Eq. (12-69) is valid only if $\beta R_M \gg 1$. However, $R_M \to 0$ as $R_s \to 0$ (Prob. 12-29).

The voltage-shunt feedback topology of Fig. 12-19 used with a high-gain multistage amplifier [for which Eq. (12-69) is valid] is called an *operational amplifier*. This configuration is the most important analog building block. The operational amplifier is discussed in detail in Chaps. 15, 16, and 17.

REFERENCES

1. Jennings, R. R.: Negative Feedback in Voltage Amplifiers, *Electro-technol. (New York)*, vol. 70, pp. 80–83, December, 1962.

Jennings, R. R.: Negative Feedback in Current Amplifier, *ibid.*, vol. 72, pp. 100–103, July, 1963.

Jennings, R. R.: Negative Feedback in Transconductance and Transresistance Amplifiers, *ibid.,* vol. 74, pp. 37–41, July, 1964.

2. Bode, H. W.: "Network Analysis and Feedback Amplifier Design," D. Van Nostrand Company, Inc., Princeton, N. J., 1945.

3. Uzunoglu, V.: "Semiconductor Network Analysis and Design," chap. 8, McGraw-Hill Book Company, New York, 1964.

Ghausi, M. S.: "Principles and Design of Linear Active Circuits," chap. 4, McGraw-Hill Book Company, New York, 1965.

Thornton, R. D., et al.: "Multistage Transistor Circuits," vol. 5, chap. 3, Semiconductor Electronics Education Committee, John Wiley & Sons, Inc., New York, 1965.

Hakim, S. S.: "Junction Transistor Circuit Analysis," John Wiley & Sons, Inc., New York, 1962.

4. Uzunoglu, V.: Feedback and Impedance Levels in Transistor Circuits, *Electron. Equipment Eng.,* pp. 42–43, July, 1962.

Blecher, F. H.: Design Principles for Single Loop Transistor Feedback Amplifiers, *IRE Trans. Circuit Theory*, vol. CT-4, p. 145, September, 1957.

Blackman, R. B.: Effect of Feedback on Impedance, *Bell System Tech. J.*, vol. 22, no. 3, p. 269, October, 1943.

5. Gray, P. E., and Searle, C. L.: "Electronic Principles," chap. 18, John Wiley & Sons, Inc., New York, 1969.

REVIEW QUESTIONS

12-1 (*a*) Draw the equivalent circuit for a voltage amplifier.

(*b*) For the ideal amplifier, what are the values of R_i and R_o?

(*c*) What are the dimensions of the transfer gain?

12-2 Repeat Rev. 12-1 for a current amplifier.

12-3 Repeat Rev. 12-1 for a transconductance amplifier.

12-4 Repeat Rev. 12-1 for a transresistance amplifier.

12-5 Draw a feedback amplifier in block-diagram form. Identify each block and state its function.

12-6 (*a*) What are the four possible topologies of a feedback amplifier?

(*b*) Identify the output signal X_o and the feedback signal X_f for each topology (either as a current or voltage).

(*c*) Identify the transfer gain A for each topology (for example, give its dimensions).

(*d*) Define the feedback factor β.

12-7 (*a*) What is the relationship between the transfer gain with feedback A_f and that without feedback A?

(*b*) Define *negative feedback*.

(*c*) Define *positive feedback*.

(*d*) Define the amount of feedback in decibels.

12-8 State the three fundamental assumptions which are made in order that the expression $A_f = A/(1 + A\beta)$ be satisfied exactly.

12-9 (*a*) Define *desensitivity D*.

(*b*) For large values of D, what is A_f? What is the significance of this result?

12-10 List five characteristics of an amplifier which are modified by negative feedback.

12-11 State whether the input resistance R_{if} is increased or decreased for each topology.

12-12 Repeat Rev. 12-11 for the output resistance R_{of}.

12-13 Explain how to determine whether the comparison in a feedback amplifier is series or shunt.

12-14 Explain how to determine whether the sampling in a feedback amplifier is voltage or current.

12-15 List the procedures to follow to obtain the basic amplifier configuration without feedback but taking the loading of the β network into account.

12-16 List the steps required to carry out the analysis of a feedback amplifier.

12-17 Draw the circuit of a single-stage voltage-series feedback amplifier.

12-18 Repeat Rev. 12-17 for a two-stage amplifier.

12-19 Find A_f for a source follower using the feedback method of analysis.

12-20 Repeat Rev. 12-19 for an emitter follower.

12-21 Draw a circuit of a current-series feedback amplifier.

12-22 Repeat Rev. 12-19 for a CE stage with an unbypassed emitter resistor.

12-23 Draw the circuit of a feedback pair with current-shunt topology.

12-24 Draw the circuit of a voltage-shunt feedback amplifier.

FREQUENCY RESPONSE OF AMPLIFIERS

Frequently, the need arises for amplifying a signal with a minimum of distortion. Under these circumstances the active devices involved must operate linearly. In the analysis of such circuits the first step is the replacement of the actual circuit by a linear model. Thereafter it becomes a matter of circuit analysis to determine the distortion produced by the transmission characteristics of the linear network.

The frequency range of the amplifiers discussed in this chapter extends from a few cycles per second (hertz), or possibly from zero, up to some tens of megahertz. The original impetus for the study of such wideband amplifiers was supplied because they were needed to amplify the pulses occurring in a television signal. Therefore such amplifiers are often referred to as *video amplifiers*.

In this chapter, then, we consider the following problem: Given a low-level input waveform which is not necessarily sinusoidal but may contain frequency components from a few hertz to a few megahertz, how can this signal be amplified with a minimum of distortion?

The response of both bipolar and field-effect single- and multistage amplifiers are discussed.

13-1 FREQUENCY DISTORTION

The application of a low-level sinusoidal signal to the input of an amplifier will result in a sinusoidal output waveform. However, for a nonsinusoidal excitation the output waveshape is not an exact replica of the input signal because the input components of different frequencies are amplified differently. In a bipolar transistor or a FET this distortion may be caused by the internal device capacitances, or it may arise because the associated circuit (for example, the coupling components or the load) is reactive. Under these circumstances, the gain A is a complex number whose magnitude and phase angle depend upon the frequency of the impressed signal. A plot of gain (phase) vs. frequency of an amplifier is called the *amplitude (phase) frequency-response characteristic*.

Fidelity Considerations

A criterion which may be used to compare one amplifier with another with respect to fidelity of reproduction of the input signal is suggested by the following considerations: Any arbitrary waveform of engineering importance may be resolved into a Fourier spectrum. If the waveform is periodic, the spectrum will consist of a series of sines and cosines whose frequencies are all integral multiples of a fundamental frequency. The fundamental frequency is the reciprocal of the time which must elapse before the waveform repeats itself. If the waveform is not periodic, the fundamental period extends in a sense from a time $-\infty$ to a time $+\infty$. The fundamental frequency is then infinitesimally small; the frequencies of successive terms in the Fourier series differ by an infinitesimal amount rather than by a finite amount; and the Fourier series becomes instead a Fourier integral. In either case the spectrum includes terms whose frequencies extend, in the general case, from zero frequency to infinity.

Consider a sinusoidal signal of angular frequency ω represented by $V_m \sin(\omega t + \phi)$. If the voltage gain of the amplifier has a magnitude A and if the signal suffers a phase change (lead angle) θ, then the output will be

$$AV_m \sin(\omega t + \phi + \theta) = AV_m \sin\left[\omega\left(t + \frac{\theta}{\omega}\right) + \phi\right]$$

Therefore, *if the amplification A is independent of frequency and if the phase shift θ is proportional to frequency (or is zero), then the amplifier will preserve the form of the input signal, although the signal will be shifted in time by an amount θ/ω.*

This discussion suggests that the extent to which an amplifier's amplitude response is not uniform, and its time delay is not constant with frequency, may serve as a measure of the lack of fidelity to be anticipated in it. In principle, it is not necessary to specify both amplitude and delay response since, for most practical circuits, the two are related and, one having been specified, the other is uniquely determined. However, in particular cases, it may well be that either the time-delay or amplitude response is the more sensitive indicator of frequency distortion.

Frequency-Response Characteristics

For an amplifier stage the frequency characteristics may be divided into three regions: There is a range, called the *midband frequencies*, over which the amplification is reasonably constant and equal to A_o and over which the delay is also quite constant. The idealized amplifiers of Chap. 11 and 12 (where all capacitances have been omitted) belong to this midband classification. For the present discussion we assume that the midband gain is normalized to unity, $A_o = 1$. In the second (low-frequency) region, below midband, external coupling capacitors or bypass capacitors (Fig. 13-8) play the important role in determining the frequency response. An amplifier stage at low frequencies behaves like the simple high-pass circuit of Fig. 13-1. The response decreases with decreasing frequency, and the output approaches zero at zero frequency (dc). In the third (high-frequency) region, above midband, shunt capacitors within the device model or across the load play the prominent role in determining the frequency response. The circuit often behaves like the simple low-pass network of Fig. 13-2, and the response decreases with increasing frequency. The total frequency characteristic, indicated in Fig. 13-3 for all three regions, will now be discussed.

Low-Frequency Response

In terms of the complex variable $s = j\omega = j2\pi f$, the reactance† of a capacitor is $1/sC$. Hence, from the circuit of Fig. 13-1, we find

$$V_o = \frac{R_1}{R_1 + 1/sC_1} V_i = \frac{s}{s + 1/R_1 C_1} V_i \qquad (13\text{-}1)$$

The voltage transfer ratio, or voltage gain (in the low-frequency region), $A_L = V_o/V_i$ becomes, in terms of the frequency $f = s/2\pi j$,

$$A_L(f) = \frac{1}{1 - j(f_L/f)} \qquad (13\text{-}2)$$

where

$$f_L \equiv \frac{1}{2\pi R_1 C_1} \qquad (13\text{-}3)$$

The magnitude $|A_L|$ and the phase lead θ_L of the gain are given by

$$|A_L(f)| = \frac{1}{\sqrt{1 + (f_L/f)^2}} \qquad \theta_L = \arctan \frac{f_L}{f} \qquad (13\text{-}4)$$

At the frequency $f = f_L$, $A_L = 1/\sqrt{2} = 0.707$, whereas in the midband region ($f \gg f_L$), $A_L \to 1$. Hence f_L is that frequency at which the gain has fallen to 0.707 times its midband value A_o. This drop in signal level corresponds to a decibel‡ reduction of 20 log $(1/\sqrt{2})$, or 3 dB. Accordingly, f_L is referred to as the *low 3-dB frequency*. From Eq. (13-3) we see that f_L is that frequency for which the resistance R_1 equals the capacitive reactance $1/2\pi f_L C_1$.

---

† A discussion of reactive circuits and the complex frequency s is given in Secs. C-3 and C-4.
‡ The voltage gain A expressed in decibels is given by 20 log A.

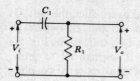

Figure 13-1 A high-pass RC circuit used to calculate the low-frequency response of an amplifier.

High-Frequency Response

In the high-frequency region, above the midband, the amplifier stage can often be approximated by the simple low-pass circuit of Fig. 13-2. In terms of the complex variable s, we find

$$V_o = \frac{1/sC_2}{R_2 + 1/sC_2} V_i = \frac{1}{1 + sR_2C_2} V_i \qquad (13\text{-}5)$$

For real frequencies ($s = j\omega = j2\pi f$) we obtain for the magnitude $|A_H(f)|$ and for the phase lead angle θ_H of the gain in the high-frequency region,

$$|A_H(f)| = \frac{1}{\sqrt{1 + (f/f_H)^2}} \qquad \theta_H = -\arctan \frac{f}{f_H} \qquad (13\text{-}6)$$

where

$$f_H \equiv \frac{1}{2\pi R_2 C_2} \qquad (13\text{-}7)$$

Since at $f = f_H$ the gain is reduced to $1/\sqrt{2}$ times its midband value, then f_H is called the *high 3-dB frequency*. It also represents that frequency at which the resistance R_2 equals the capacitive reactance $1/2\pi f_H C_2$. In the foregoing expressions θ_L and θ_H represent the angle by which the output *leads* the input, neglecting the initial 180° phase shift through the amplifier if $A_o < 0$. The frequency dependence of the gains in the high- and low-frequency range is indicated in Fig. 13-3. Such characteristics, called *Bode plots*, are discussed in detail in Sec. 14-10.

Bandwidth

The frequency range from f_L to f_H is called the *bandwidth* of the amplifier stage. We may anticipate in a general way that a signal, all of whose frequency components of appreciable amplitude lie well within the range f_L to f_H, will pass through the stage without excessive distortion. This criterion must be applied, however, with caution.

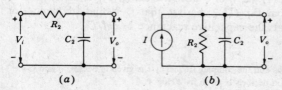

(a)   (b)

Figure 13-2 (*a*) A low-pass RC circuit used to calculate the high-frequency response of an amplifier. (*b*) The Norton's equivalent of the circuit in (*a*), where $I = V_i/R_2$.

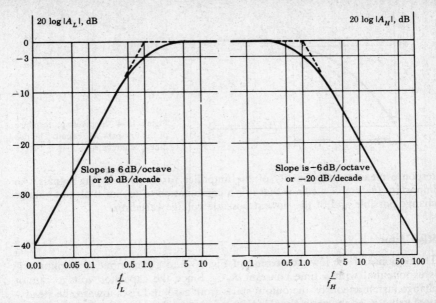

Figure 13-3 A semi-log plot of the amplitude frequency-response (Bode) characteristic of an RC-coupled amplifier. The dashed curve is the idealized Bode plot. *Note*: An octave is a factor of 2 in frequency.

13-2 STEP RESPONSE OF AN AMPLIFIER

An alternative criterion of amplifier fidelity is the response of the amplifier to a particular input waveform. Of all possible available waveforms, the most generally useful is the step voltage. In terms of a circuit's response to a step, the response to an arbitrary waveform may be written in the form of the superposition integral. Another feature which recommends the step voltage is the fact that this waveform is one which permits small distortions to stand out clearly. Additionally, from an experimental viewpoint, we note that excellent pulse (a short step) and square-wave (a repeated step) generators are available commercially.

As long as an amplifier can be represented by a single pole [Eq. (13-5)], the correlation between its frequency response and the output waveshape for a step input is that given below. Quite generally, even for more complicated amplifier circuits, there continues to be an intimate relationship between the distortion of the leading edge of a step and the high-frequency response. Similarly, there is a close relationship between the low-frequency response and the distortion of the flat portion of the step. We should, of course, expect such relationships, since the high-frequency response measures essentially the ability of the amplifier to respond faithfully to rapid variations in signal, whereas the low-frequency

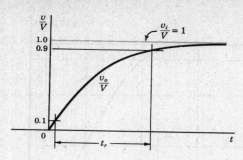

Figure 13-4 Step-voltage response of the low-pass RC circuit. The rise time t_r is indicated.

response measures the fidelity of the amplifier for slowly varying signals. An important feature of a step is that it is a combination of the most abrupt voltage change possible and of the slowest possible voltage variation.

Rise Time

The response of the low-pass circuit of Fig. 13-2 to a step input of amplitude V is exponential with a time constant R_2C_2. Since the capacitor voltage cannot change instantaneously, the output starts from zero and rises toward the steady-state value V, as shown in Fig. 13-4. The output is given by (Sec. 5-5)

$$v_o = V(1 - \epsilon^{-t/R_2C_2}) \tag{13-8}$$

The time required for v_o to reach one-tenth of its final value is readily found to be $0.1R_2C_2$, and the time to reach nine-tenths its final value is $2.3R_2C_2$. The difference between these two values is called the *rise time* t_r of the circuit and is shown in Fig. 13-4. The time t_r is an indication of how fast the amplifier can respond to a discontinuity in the input voltage. We have, using Eq. (13-7),

$$t_r = 2.2R_2C_2 = \frac{2.2}{2\pi f_H} = \frac{0.35}{f_H} \tag{13-9}$$

Note that the rise time is inversely proportional to the upper 3-dB frequency. For an amplifier with 1 MHz bandpass, $t_r = 0.35\ \mu\text{s}$.

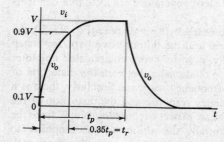

$$f_H = 1/t_p.$$

Figure 13-5 Pulse response of a low-pass RC circuit for the case $f_H = 1/t_p$.

Consider a pulse of width t_p. What must be the high 3-dB frequency f_H of an amplifier used to amplify this signal without excessive distortion? A reasonable answer to this question is: *Choose f_H equal to the reciprocal of the pulse width*, $f_H = 1/t_p$. From Eq. (13-9) we then have $t_r = 0.35t_p$. Using this relationship, the (shaded) pulse in Fig. 13-5 becomes distorted into the (solid) waveform, which is clearly recognized as a pulse.

Tilt or Sag

If a step of amplitude V is impressed on the high-pass circuit of Fig. 13-1, the output is

$$v_o = V\epsilon^{-t/R_1C_1} \qquad (13\text{-}10)$$

For times t which are small compared with the time constant R_1C_1, the response is given by

$$v_o \approx V\left(1 - \frac{t}{R_1C_1}\right) \qquad (13\text{-}11)$$

From Fig. 13-6 we see that the output is tilted, and the percent tilt, or sag, in time t_1 is given by

$$P \equiv \frac{V - V'}{V} \times 100\% = \frac{t_1}{R_1C_1} \times 100\% \qquad (13\text{-}12)$$

It is found[1] that this same expression is valid for the tilt of each half cycle of a symmetrical square wave of peak-to-peak value V and period T provided that we set $t_1 = T/2$. If $f = 1/T$ is the frequency of the square wave, then, using Eq. (13-3), we may express P in the form

$$P = \frac{T}{2R_1C_1} \times 100 = \frac{1}{2fR_1C_1} \times 100 = \frac{\pi f_L}{f} \times 100\% \qquad (13\text{-}13)$$

Note that the tilt is directly proportional to the lower 3-dB frequency. If we wish to pass a 50-Hz square wave with less than 10 percent sag, then f_L must not exceed 1.6 Hz.

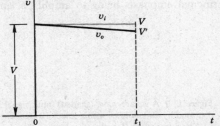

Figure 13-6 The response v_o exhibits a tilt when a step v_i is applied to a high-pass *RC* circuit.

Square-Wave Testing

An important experimental procedure (called *square-wave testing*) is to observe with an oscilloscope the output of an amplifier excited by a square-wave generator. It is possible to improve the response of an amplifier by adding to it certain circuit elements, which then must be adjusted with precision. It is a great convenience to be able to adjust these elements and to see simultaneously the effect of such an adjustment on the amplifier output waveform. The alternative is to take data, after each successive adjustment, from which to plot the amplitude and phase responses. Aside from the extra time consumed in this latter procedure, we have the problem that it is usually not obvious which of the attainable amplitude and phase responses corresponds to optimum fidelity. On the other hand, the step response gives immediately useful information.

It is possible, by judicious selection of two square-wave frequencies, to examine individually the high-frequency and low-frequency distortion. For example, consider an amplifier which has a high-frequency time constant of 1 μs and a low-frequency time constant of 0.1 s. A square wave of half period equal to several microseconds, on an appropriately fast oscilloscope sweep, will display the rounding of the leading edge of the waveform and will not display the tilt. At the other extreme, a square wave of half period approximately 0.01 s on an appropriately slow sweep will display the tilt, and not the distortion of the leading edge. Such a waveform is indicated in Fig. 13-7.

It should *not* be inferred from the above comparison between steady-state and transient response that the phase and amplitude responses are of no importance at all in the study of amplifiers. The frequency characteristics are useful for the following reasons. In the first place, much more is known generally about the analysis and synthesis of circuits in the frequency domain than in the time domain, and for this reason the design of coupling networks is often done on a frequency-response basis. Second, it is often possible to arrive at least at a qualitative understanding of the properties of a circuit from a study of the steady-state response in circumstances where transient calculations are extremely cumbersome. Third, compensating an amplifier against unwanted oscillations (Chap. 15) is accomplished in the frequency domain. Finally, it happens occasionally that an amplifier is required whose characteristics are specified on a frequency basis, the principal emphasis being to amplify a sine wave.

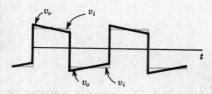

Figure 13-7 A square-wave (shaded) input signal is distorted by an amplifier with a low 3-dB frequency f_L. The output (solid) waveform shows a tilt where the input is horizontal.

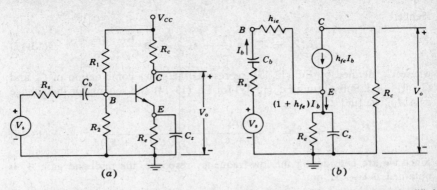

Figure 13-8 (*a*) An amplifier with a bypassed emitter resistor; (*b*) the low-frequency simplified *h*-parameter model of the circuit in (*a*).

13-3 EFFECT OF COUPLING AND EMITTER BYPASS CAPACITORS ON LOW-FREQUENCY RESPONSE

If an emitter resistor R_e is used for self-bias in an amplifier and if it is desired to avoid the degeneration, and hence the loss of gain due to R_e, we might attempt to bypass this resistor with a very large capacitance C_z. The circuit is indicated in Fig. 13-8*a*. It is shown in the following that the effect of this capacitor and of the coupling capacitor C_b is to affect adversely the low-frequency response.

In this analysis we assume that $R_1 \| R_2 \gg R_s$ and that the load R_c is small enough so that the simplified low-frequency model of Fig. 11-6 is valid. The equivalent circuit subject to these assumptions is shown in Fig. 13-8*b*.

The output voltage V_o is given by

$$V_o = - I_b h_{fe} R_c = - \frac{V_s h_{fe} R_c}{R_s + h_{ie} + Z_b + Z_e'} \tag{13-14}$$

where $Z_b = 1/j\omega C_b$ and Z_e' is defined by

$$Z_e' \equiv (1 + h_{fe}) \frac{R_e}{1 + j\omega C_z R_e} \tag{13-15}$$

Assume that C_z is chosen large enough so that its reactance is small compared with the resistance R_e at all frequencies of interest. Thus,

$$\frac{1}{\omega C_z} \ll R_e \quad \text{and} \quad \omega C_z R_e \gg 1 \tag{13-16}$$

From Eq. (13-15), $Z_e' \approx (1 + h_{fe})/j\omega C_z$. This equation signifies that the emitter capacitance C_z is reflected into the base circuit as a capacitance $C_z/(1 + h_{fe})$.

Hence,

$$j\omega(Z_b + Z_e') = \frac{1}{C_b} + \frac{1 + h_{fe}}{C_z} \equiv \frac{1}{C_1} \tag{13-17}$$

where C_1, defined by Eq. (13-17), represents the series combination of C_b and $C_z/(1 + h_{fe})$. Substituting Eq. (13-17) for Eq. (13-14) and solving for the voltage gain A_{Vs}, we find

$$A_{Vs} = \frac{V_o}{V_s} = -\frac{h_{fe}R_c}{R_s + h_{ie}} \frac{1}{1 - j/\omega C_1(R_s + h_{ie})} \tag{13-18}$$

Since we are considering the low-frequency response, the midband gain A_o is obtained as $\omega \to \infty$, or

$$A_o = \frac{-h_{fe}R_c}{R_s + h_{ie}} \tag{13-19}$$

Hence,

$$\frac{A_{Vs}}{A_o} = \frac{1}{1 - j(f_L/f)} \tag{13-20}$$

where the low 3-dB frequency f_L is given by

$$f_L \equiv \frac{1}{2\pi C_1(R_s + h_{ie})} \tag{13-21}$$

As predicted in Sec. 13-1, the low-frequency gain corresponds to the circuit of Fig. 13-1 [compare Eq. (13-21) with Eq. (13-2)]. For the configuration of Fig. 13-8 the total capacitance C_1 corresponds to the series combination of C_b and $C_z/(1 + h_{fe})$ and the total resistance R_1 is the sum of R_s and h_{ie}. The 3-dB frequency f_L is that frequency for which the reactance of C_1 equals the resistance $R_s + h_{ie} = R_1$

Example 13-1 (a) It is desired to have a low 3-dB frequency of not more than 10 Hz for the amplifier of Fig. 13-8. If $R_s = h_{ie} = 1$ kΩ and $h_{fe} = 100$, what minimum value of C_b is required, assuming that $C_z = C_b$? (b) Repeat part (a) if $C_z = 100 \, C_b$. (c) Calculate C_1 so that we may reproduce a 50-Hz square wave with a tilt of less than 10 percent. Calculate C_b if $C_z = C_b$.

SOLUTION (a) From Eq. (13-21),

$$C_1 = \frac{1}{2\pi f_L(R_s + h_{ie})} = \frac{10^6}{2\pi 10 \times 2,000} \, \mu F = 7.96 \, \mu F$$

If $C_z = C_b$, then from Eq. (13-17),

$$\frac{1}{C_b} + \frac{101}{C_b} = \frac{1}{7.96} \qquad C_b = 812 \, \mu F = C_z$$

(*b*) From Eq. (13-17)

$$\frac{1}{C_b} + \frac{101}{100C_b} = \frac{1}{7.96} \qquad C_b = 16.0 \ \mu F \qquad C_z = 1,600 \ \mu F$$

(*c*) From Eq. (13-13)

$$P = \frac{100}{2f(R_s + h_{ie})C_1}$$

$$C_1 = \frac{100 \times 10^6}{2 \times 50 \times 2,000 \times 10} \ \mu F = 50 \ \mu F$$

From Eq. (13-17),

$$\frac{1}{C_b} + \frac{101}{C_b} = \frac{1}{50} \qquad C_b = 5,100 \ \mu F = C_z$$

The capacitance values calculated in Example 13-1 are extremely large. Fortunately, with discrete transistors, it is possible to obtain physically small electrolytic capacitors having such high capacitance values at the low voltages at which transistors operate. Since the capacitances required for good low-frequency response are far larger than those obtainable in integrated form, *cascaded integrated stages must be direct-coupled* (Chap. 15).

Electrolytic capacitors are often used as emitter or source bypass capacitors because they offer the greatest capacitance per unit volume. It is important to note that these capacitors have a series resistance which arises from the conductive losses in the electrolyte. This resistance, typically 1 to 20 Ω, must be taken into account in computing the midband gain of the stage.

In the foregoing discussion the reasonable assumption is made that the reactance of C_z is much smaller than R_e. If this inequality is not satisfied, then the effect of C_z on the frequency response is considered in Prob. 13-11.

The low-frequency analysis of a FET amplifier with a source resistor R_s bypassed with a capacitor C_s is considered in Probs. 13-14 and 13-15.

13-4 THE *RC*-COUPLED AMPLIFIER

A cascaded arrangement of common-emitter (CE) transistor stages is shown in Fig. 13-9. The output Y_1 of one stage is coupled to the input X_2 of the next stage via a blocking capacitor C_b which is used to keep the dc component of the output voltage at Y_1 from reaching the input X_2. The collector circuit resistor is R_c, the emitter resistor R_e, and the resistors R_1 and R_2 are used to establish the bias. The bypass capacitor, used to prevent loss of amplification due to negative feedback (Chap. 12), is C_z in the emitter circuit. Also present are junction capacitances, to be taken into account when we consider the high-frequency response, which is limited by their presence. In any practical mechanical arrangement of the amplifier components, there are also capacitances associated

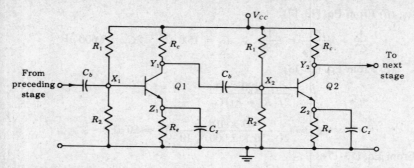

Figure 13-9 A cascade of common-emitter (CE) transistor stages.

with device sockets and the proximity to the chassis of components (for example, the body of C_b) and signal leads. These stray capacitances are also considered later. We assume that the active device operates linearly, so that small-signal models are used throughout this chapter.

Low-Frequency Response of an RC-coupled Stage

The calculation of the gain and 3-dB frequency of an intermediate stage $Q2$ is easily obtained from the analysis of the preceding section. The source of excitation for $Q2$ is the signal at the output Y_1 of Q_1. Hence, the source resistance R_s for $Q2$ is the resistance seen from Y_1 to ground, which is R_c in parallel with $1/h_{oe}$. Usually $R_c \ll 1/h_{oe}$ and $R_s \approx R_c$. Therefore, for any stage (except the first) in a cascaded amplifier, Eqs. (13-19) and (13-21) are valid with R_s replaced by R_c.

In the remainder of this chapter we consider the transistor at high frequencies. First a model for the transistor is found and the high-frequency response of a single stage is obtained. Then the high-frequency behavior of cascaded stages is studied.

13-5 THE HYBRID-π TRANSISTOR MODEL AT HIGH FREQUENCIES

As mentioned in Sec. 11-15 the hybrid-π model of Fig. 11-18 is valid at high frequencies provided that a capacitor C_c is placed across the collector junction (between C and B') and a capacitor C_e is added across the emitter junction (between E and B'). This high-frequency model is indicated in Fig. 13-10.

The collector-junction capacitance $C_c = C_{b'c}$ is the measured CB output capacitance with the input open ($I_E = 0$), and is usually specified by manufacturers as C_{ob}. Since in the active region the collector junction is reverse-biased,

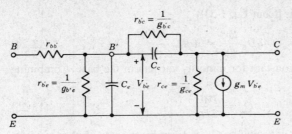

Figure 13-10 The high-frequency hybrid-π model for a transistor.

then C_c is the transition capacitance and varies as V_{CB}^{-n}, where n is $\frac{1}{2}$ or $\frac{1}{3}$ for an abrupt or graded junction, respectively.

The capacitance C_e represents the sum of the emitter diffusion capacitance C_{De} and the emitter-junction capacitance C_{Te}. For a forward-biased emitter junction, C_{De} is usually much larger than C_{Te}, and hence

$$C_e = C_{De} + C_{Te} \approx C_{De} = C_{ib} \tag{13-22}$$

where C_{ib} is the measured CB input capacitance with the output open ($I_C = 0$).

We shall now show that C_{De} is proportional to the emitter bias current I_E and is almost independent of temperature.

The Diffusion Capacitance

Refer to Fig. 13-11, which represents the injected hole concentration vs. distance in the base region of a p-n-p transistor. The base width W is assumed to be small compared with the diffusion length L_B of the minority carriers. Since the collector is reverse-biased, the injected charge concentration p' at the collector junction is essentially zero. If $W \ll L_B$, then p' varies almost linearly from the value $p'(0)$ at the emitter to zero at the collector, as indicated in Fig. 13-11. The stored base charge Q_B is the average concentration $p'(0)/2$ times the volume of the base WA (where A is the base cross-sectional area) times the electronic charge q; that is,

$$Q_B = \tfrac{1}{2} p'(0) A W q \tag{13-23}$$

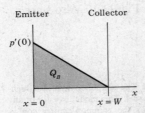

Figure 13-11 Minority-carrier charge distribution in the base region.

The diffusion current is [from Eq. 1-33]

$$I = -AqD_B \frac{dp'}{dx} = AqD_B \frac{p'(0)}{W} \tag{13-24}$$

where D_B is the diffusion constant for minority carriers in the base. Combining Eqs. (13-23) and (13-24),

$$Q_B = \frac{IW^2}{2D_B} \tag{13-25}$$

The static emitter diffusion capacitance C_{De} is given by the rate of change of Q_B with respect to emitter voltage V, or

$$C_{De} = \frac{dQ_B}{dV} = \frac{W^2}{2D_B} \frac{dI}{dV} = \frac{W^2}{2D_B} \frac{1}{r_e} \tag{13-26}$$

where $r_e \equiv dV/dI = V_T/I_E$ is the emitter-junction incremental resistance. From Eq. (11-23) with $\alpha = 1$

$$C_{De} = \frac{W^2 I_E}{2D_B V_T} = g_m \frac{W^2}{2D_B} \tag{13-27}$$

which indicates that *the diffusion capacitance is proportional to the emitter bias current I_E*. It is found experimentally that C_{De} varies as T^n, where for silicon $n = +0.5 (+0.7)$ for electrons (holes) and for germanium $n = -0.34 (+0.33)$ for electrons (holes).

C_e is determined from a measurement of f_T, the frequency at which the CE short-circuit current gain drops to unity. We verify in Sec. 13-7 that

$$C_e \approx \frac{g_m}{2\pi f_T} \tag{13-28}$$

Giacolletto[2] has shown that the network components of Fig. 13-10 are independent of frequency provided that $f \ll 3f_T$. Hence, the hybrid-π model of Fig. 13-10 is valid for frequencies up to approximately $f_T/3$. The relationships between the hybrid-π resistances and the h-parameters are given in Eq. (11-67).

Simplified Hybrid-π Model

The resistor r_{ce} is in parallel with the load resistor R_L in the CE configuration and usually $r_{ce} \gg R_L$. Hence, r_{ce} may be deleted from Fig. 13-10. Also since $r_{b'c}$

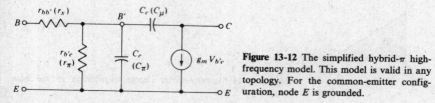

Figure 13-12 The simplified hybrid-π high-frequency model. This model is valid in any topology. For the common-emitter configuration, node E is grounded.

is very much larger than the other resistances in the network, it may also be deleted. With these deletions the simplified but accurate high-frequency model of Fig. 13-12 is obtained. In addition to the notation in Fig. 13-10 we have also indicated alternative symbols (r_x, r_π, C_π, and C_μ) which are sometimes used in the literature. The order of magnitude of these parameters are:

$$r_{bb'} = r_x = 100\ \Omega \qquad r_\pi = r_{b'e} = 1\ k\Omega$$

$$C_c = C_\mu = 3\ pF \qquad C_e = C_\pi = 100\ pF \tag{13-29}$$

g_m is given by Eq. (11-56b) and $g_m = 50$ mA/V at 1.3 mA. For this transistor $h_{fe} = g_m r_{b'e} = 50$ at 1.3 mA.

13-6 VARIATIONS OF HYBRID-π PARAMETERS[3]

Table 13-1 summarizes the dependence of g_m, $r_{b'e}$, $r_{bb'}$, C_e, C_c, h_{fe}, and h_{ie} on the collector current magnitude $|I_C|$, the collector-to-emitter voltage magnitude $|V_{CE}|$, and the temperature. The conclusions in the table are based on Eqs. (11-67), (13-22), and (13-27). We must also recall that increasing $|V_{CE}|$ decreases the effective base width. The dependence of $r_{bb'}$ on $|I_C|$ and temperature requires some explanation. The decrease of $r_{bb'}$ with $|I_C|$ is due to conductivity modulation of the base with increasing collector current. On the contrary, $r_{bb'}$ increases with increasing temperature because the mobility of majority and minority carriers decreases, and this results in reduced conductivity. The increase of h_{fe} with temperature has been determined experimentally, whereas the increase with $|V_{CE}|$ is due to the decrease of the base width and the reduction in recombination which increases the transistor alpha.

No entry in Table 13-1 means that the particular dependence varies with the absolute value of $|I_C|$, $|V_{CE}|$, or T in a complicated fashion.

Table 13-1 Dependence of parameters upon current, voltage, and temperature

Parameter	Variation with increasing:		
	$|I_c|$	$|V_{CE}|$	T
g_m	$|I_C|$	Independent	$1/T$
$r_{bb'}$	Decreases		Increases
$r_{b'e}$	$1/|I_C|$	Increases	Increases
C_e	$|I_C|$	Decreases	
C_c	Independent	Decreases	Independent
h_{fe}	See Fig. 3-14	Increases	Increases
h_{ie}	$1/|I_C|$	Increases	Increases

13-7 THE CE SHORT-CIRCUIT CURRENT GAIN

Consider a single-stage CE transistor amplifier, or the last stage of a cascade. The load R_L on this stage is the collector-circuit resistor, so that $R_c = R_L$. In this section we assume that $R_L = 0$, whereas the circuit with $R_L \neq 0$ is analyzed in the next section. To obtain the *frequency response* (the gain as a function of frequency) of the transistor amplifier, we use the hybrid-π model of Fig. 13-12. Representative values of the circuit components are specified in Eq. (13-29) for a transistor intended for use at high frequencies. We use these values as a guide in making simplifying assumptions.

The approximate equivalent circuit from which to calculate the short-circuit current gain is shown in Fig. 13-13. A current source furnishes a sinusoidal input current of magnitude I_i, and the load current is I_L. We have neglected the current delivered directly to the output through C_c. We see shortly that this approximation is justified.

The load current is $I_L = -g_m V_{b'e}$, where

$$V_{b'e} = \frac{I_i}{g_{b'e} + j\omega(C_e + C_c)} \tag{13-30}$$

The current amplification under short-circuited conditions is

$$A_i = \frac{I_L}{I_i} = \frac{-g_m}{g_{b'e} + j\omega(C_e + C_c)} \tag{13-31}$$

Using $h_{fe} = g_m/g_{b'e}$ given in Eqs. (11-67), we have

$$A_i = \frac{-h_{fe}}{1 + j(f/f_\beta)} \qquad |A_i| = \frac{h_{fe}}{\left[1 + (f/f_\beta)^2\right]^{1/2}} \tag{13-32}$$

where

$$f_\beta = \frac{g_{b'e}}{2\pi(C_e + C_c)} = \frac{1}{h_{fe}} \frac{g_m}{2\pi(C_e + C_c)} \tag{13-33}$$

At $f = f_\beta$, $|A_i|$ is equal to $1/\sqrt{2} = 0.707$ of its low-frequency value h_{fe}. The frequency range up to f_β is referred to as the *bandwidth* of the circuit. Note that the value of A_i at $\omega = 0$ is $-h_{fe}$, in agreement with the definition of $-h_{fe}$ as the low-frequency short-circuit CE current gain.

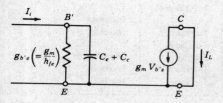

Figure 13-13 Approximate equivalent circuit for the calculation of the short-circuit CE current gain.

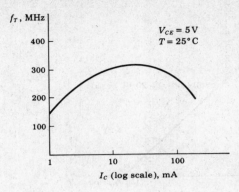

Figure 13-14 Variation of f_T with collector current.

The Parameter f_T

We now introduce f_T, which is defined as the *frequency at which the short-circuit common-emitter current gain attains unit magnitude*. Since $h_{fe} \gg 1$, we have, from Eqs. (13-32) and (13-33), that f_T is given by

$$f_T \approx h_{fe}f_\beta = \frac{g_m}{2\pi(C_e + C_c)} \approx \frac{g_m}{2\pi C_e} \qquad (13\text{-}34)$$

since $C_e \gg C_c$. From Eq. (13-34), $f_T \approx 80$ MHz and $f_\beta \approx 1.6$ MHz for our typical transistor. The parameter f_T is an important high-frequency characteristic of a transistor. Like other transistor parameters, its value depends on the operating conditions of the device. Typically, the dependence of f_T on collector current is as shown in Fig. 13-14.

Since $f_T \approx h_{fe}f_\beta$, this parameter may be given a second interpretation. It represents the *short-circuit current gain-bandwidth product*; that is, for the CE configuration with the output shorted, f_T is the product of the low-frequency current gain and the upper 3-dB frequency. It is to be noted from Eq. (13-34) that there is a sense in which gain may be sacrificed for bandwidth, and vice versa. Thus, if two transistors are available with equal f_T, the transistor with lower h_{fe} will have a correspondingly larger bandwidth.

In Fig. 13-15, A_i expressed in decibels (i.e., $20 \log |A_i|$) is plotted against frequency on a logarithmic frequency scale. When $f \ll f_\beta$, $A_i \approx -h_{fe}$, and A_i (dB) approaches asymptotically the horizontal line A_i (dB) $= 20 \log h_{fe}$. When $f \gg f_\beta$, $|A_i| \approx h_{fe}f_\beta/f = f_T/f$, so that A_i (dB) $= 20 \log f_T - 20 \log f$. Accordingly, A_i (dB) $= 0$ dB at $f = f_T$. And for $f \gg f_\beta$, the plot approaches as an asymptote a straight line passing through the point $(f_T, 0)$ and having a slope which causes a decrease in A_i (dB) of 6 dB per octave (f is multiplied by a factor of 2, and $20 \log 2 = 6$ dB), or 20 dB per decade. The intersection of the two asymptotes occurs at the "corner" frequency $f = f_\beta$, where A_i is down by 3 dB. Hence f_β is also called the 3-dB frequency for the short-circuit current gain.

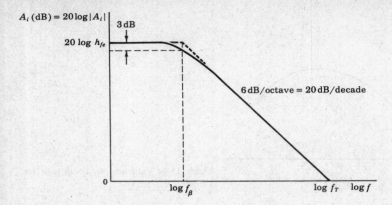

Figure 13-15 The short-circuit CE current gain vs. frequency (plotted on a log-log scale).

Earlier, we neglected the current delivered directly to the output through C_c. Now we may see that this approximation is justified. The magnitude of this current is $\omega C_c V_{b'e}$, whereas the current due to the controlled generator is $g_m V_{b'e}$. The ratio of currents is $\omega C_c / g_m$. At the highest frequency of interest f_T, we have, from Eq. (13-34), using the typical values of capacitance in Eq. (13-29),

$$\frac{\omega C_c}{g_m} = \frac{2\pi f_T C_c}{g_m} = \frac{C_c}{C_e + C_c} \approx 0.03$$

Measurement of f_T

The frequency f_T is often inconveniently high to allow a direct experimental determination of f_T. However, a procedure is available which allows a measurement of f_T at an appreciably lower frequency. We note from Eq. (13-32) that, for $f \gg f_\beta$, we may neglect the unity in the denominator and write $|A_i| f \approx h_{fe} f_\beta = f_T$ from Eq. (13-34). Accordingly, at some particular frequency f_1 (say, f_1 is five or ten times f_β), we measure the gain $|A_{i1}|$. The parameter f_T may be calculated now from $f_T = f_1 |A_{i1}|$. The experimentally determined value of f_T is used to calculate the value of C_e in the hybrid-π circuit [Eq. (13-34)].

13-8 THE GENERALIZED VOLTAGE-GAIN FUNCTION

Before we proceed to obtain the frequency response $A(s)$ of a single-stage amplifier loaded by a resistor R_L and then the response of a multistage amplifier, let us make a few general observations about the form of $A(s)$.

The high-frequency response of an amplifier is assumed in Fig. 13-2 to be determined by a single time constant $R_2 C_2$. In reality, a multistage amplifier may contain many energy-storage elements (capacitors), and the transfer function will be given by an equation of the form

$$A_V(s) = \frac{V_o(s)}{V_i(s)} = \frac{K(s - s_{z1})(s - s_{z2})(s - s_{z3})}{(s - s_{p1})(s - s_{p2})(s - s_{p3})(s - s_{p4})} \qquad (13\text{-}35)$$

The values of s for which $A(s) = 0$ are called the *zeros* of the transfer function. In Eq. (13-35) we have assumed three zeros, s_{z1}, s_{z2}, and s_{z3}. The values of s for which $A_V(s) = \infty$ are called the *poles* of the transfer function. There are four poles $s_{p1}, \ldots, s_{p4}$ in the amplifier represented by Eq. (13-35). The simple high-pass circuit of Fig. 13-1 has one zero $s_z = 0$ and one pole $s_p = -1/R_1C_1$, as is evident from Eq. (13-1). Equation (13-5) shows that the low-pass circuit of Fig. 13-2 has no zeros and one pole at $s_p = -1/R_2C_2$. Note that frequency f_p of a pole has a magnitude $s_p/2\pi$.

Determination of the Number of Poles and Zeros

The number of poles in a transfer function is equal to the number of independent energy-storing elements in the network. In electronic amplifiers the storing components are almost exclusively capacitors. An *independent* capacitor is one to which you can assign an arbitrary voltage, independent of all other capacitor voltages. For example, two capacitors in parallel are *not* independent because the voltage across the first must be the same as that across the second. Similarly, two capacitors in series are *not* independent because the stored charge Q is the same in each component and the voltage across a capacitor C is Q/C. Also, if a network loop can be traversed by passing only through capacitors, then not all of these C's are independent (because the sum of the voltages around the closed path must be zero).

The number of zeros in a transfer function is determined from a knowledge of the number of poles and the behavior of the network as $s \to \infty$. In Eq. (13-35), $A_V(s) \to 1/s$, as $s \to \infty$, because there is one pole more than the number of zeros. In general, if $A_V(s) \to 1/s^m$ as $s \to \infty$ *then the number of zeros is m less than the number of poles.*

The behavior of a network as $s \to \infty$ is usually obtained by inspection since the voltage across a capacitor is $1/sC$. For example, in Fig. 13-2,

$$A_V = \frac{V_0}{V_i} \to \frac{1}{s} \qquad \text{as} \quad s \to \infty$$

Hence, the number of zeros is one less than the number of poles. Since the circuit contains only one capacitor, A_V must contain one pole and (by the above argument) no zero. This conclusion is confirmed in Eq. (13-5).

In Fig. 13-1 there is also only one capacitor and, hence, the transfer function has one pole. Since the drop across C_1 approaches zero as $s \to \infty$, $V_o \to V_i$ as $s \to \infty$. Hence, $A_V = V_0/V_i \to 1 = 1/s^0$ as $s \to \infty$ and the number of zeros equals the number of poles. This conclusion is verified by Eq. (13-1) which has one zero and one pole.

The Dominant Pole

Consider a two-pole transfer function where f_{p1} is much smaller than f_{p2}. Then a frequency-response plot such as that in Fig. 13-3 indicates that the upper 3-dB frequency is given approximately by f_{p1}. If $f_{p2} = 4f_{p1}$, an exact plot indicates (Prob. 13-28) that the 3-dB frequency is less than 6 percent smaller than f_{p1}. We conclude that *if a transfer function has several poles determining the high-frequency response, if the smallest of these is f_{p1} and if each other pole is at least two octaves higher, then the amplifier behaves essentially as a single-time-constant circuit whose 3-dB frequency is f_{p1}.* The frequency f_{p1} is called the *dominant pole*. Note that *the high 3-dB frequency of a single-pole circuit is given by $f_H = |s_p|/2\pi = 1/(2\pi\tau)$, where τ is the time constant of the circuit.*

13-9 SINGLE-STAGE CE TRANSISTOR AMPLIFIER RESPONSE

In Sec. 13-7 we assume that the transistor is driven from an ideal current source, that is, a source of infinite resistance. We now remove this restriction and consider that the source V_s has a finite resistance R_s. The equivalent circuit from which to obtain the response is shown in Fig. 13-16. This circuit is obtained from Fig. 13-12 by adding V_s in series with R_s between B and E and by including a load resistor R_L between C and E. Since the circuit has two independent capacitors, the transfer function must have two poles. As $s \to \infty$, points B' and C become short-circuited together because of C_c, and the output falls toward zero as $1/s$ owing to the shunting action of C_e. From the discussion in the preceding section there must be one zero less than the number of poles. Therefore we expect the transfer function to have two poles and one zero.

We can obtain the zero by inspection from Fig. 13-16. If for some value of s, say s_0, $V_o = 0$, there is no current in R_L. Hence the current $sC_cV_{b'e}$ in C_c must equal the dependent generator current $g_mV_{b'e}$. Therefore the zero is given by $sC_cV_{b'e} = g_mV_{b'e}$. Hence, $s = g_m/C_c \equiv s_0$.

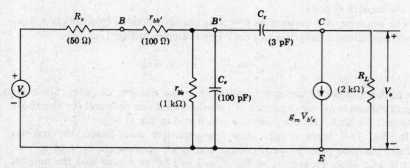

Figure 13-16 Equivalent circuit for the frequency analysis of a CE amplifier stage driven from a voltage source.

The Transfer Function

We wish to calculate the transfer function V_o/V_s as a function of the complex-frequency (or Laplace transform) variable s. Introducing $R_s' + r_{bb'} = 1/G_s'$, and noting that the admittance of a capacitor C is sC, we obtain in Sec. C-4 the following KCL equations at nodes B' and C, respectively (the so-called nodal equations):

$$G_s' V_s = \left[G_s' + g_{b'e} + s(C_e + C_c) \right] V_{b'e} - sC_c V_o \qquad (13\text{-}36)$$

$$0 = (g_m - sC_c) V_{b'e} + \left(\frac{1}{R_L} + sC_c \right) V_o \qquad (13\text{-}37)$$

Solving Eqs. (13-36) and (13-37), we find

$$\frac{V_o}{V_s} = \frac{-G_s' R_L (g_m - sC_c)}{s^2 C_e C_c R_L + s\left[C_e + C_c + C_c R_L (g_m + g_{b'e} + G_s') \right] + G_s' + g_{b'e}} \qquad (13\text{-}38)$$

The above equation is of the form

$$A_{Vs} \equiv \frac{V_o}{V_s} = \frac{K_1(s - s_0)}{(s - s_1)(s - s_2)} \qquad (13\text{-}39)$$

We thus see that the transfer function of the single CE transistor at high frequencies has one zero $s_0 = g_m/C_c$ and two poles s_1 and s_2 as predicted above. These poles are calculated by finding the roots of the denominator of Eq. (13-38). Also from this equation it follows that $K_1 = G_s' R_L C_c / C_e C_c R_L = G_s'/C_e$.

For the numerical values indicated in Fig. 13-16 and with $g_m = 50$ mA/V, we find

$$K_1 = 6.67 \times 10^7 \qquad\qquad s_0 = 1.67 \times 10^{10} \text{ rad/s}$$

$$s_1 = -1.75 \times 10^7 \text{ rad/s} \qquad s_2 = -7.31 \times 10^8 \text{ rad/s}$$

The magnitude and the phase of the transfer function [obtained from Eq. (13-38) with $s = j\omega = j2\pi f$] are plotted in Fig. 13-17. The 3-dB frequency is found to be 2.8 MHz.

Since $s_2/s_1 = 73.1/1.75 = 41.8 \gg 4$, s_1 is a dominant pole. The dominant-pole frequency is

$$f_D = \frac{|s_1|}{2\pi} = \frac{1.75 \times 10^7}{2\pi} \quad \text{Hz} = 2.78 \text{ MHz}$$

in agreement with the value of f_H found graphically.

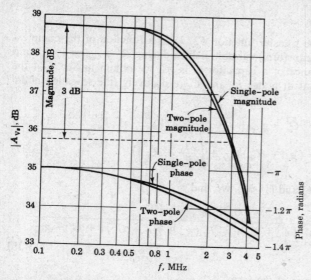

Figure 13-17 Magnitude in decibels and phase response for the two-pole transfer function of a CE amplifier stage. Also included is the response for the single-pole approximation.

Approximate Analysis

We can obtain a very simple approximate expression for the transfer function by applying Miller's theorem to the circuit of Fig. 13-16. Proceeding as in Sec. C-4, we obtain the circuit of Fig. 13-18, with $K \equiv V_{ce}/V_{b'e}$. This circuit is still rather complicated because it has two independent time constants, one associated with the input circuit and one associated with the output. We now show that in a practical situation the output time constant is negligible in comparison with the input time constant, and may be ignored.

Since $|K| \gg 1$, the output capacitance is C_c and the output time constant is

$$C_c R_L = 3 \times 10^{-12} \times 2 \times 10^3 = 6 \times 10^{-9}\ s = 6\ \text{ns}$$

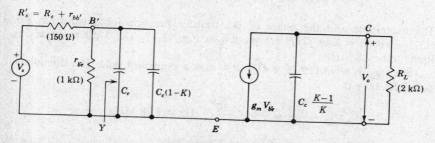

Figure 13-18 The equivalent circuit of the CE amplifier stage, using the Miller effect.

Neglecting C_c, it follows that $K = -g_m R_L$ and the input capacitance is

$$C \equiv C_e + C_c(1 + g_m R_L) = 100 + (3)(101) = 403 \text{ pF} \qquad (13\text{-}40)$$

The input loop resistance is

$$R \equiv R'_s \| r_{b'e} = \frac{1}{G'_s + g_{b'e}} = \frac{1}{1/150 + 1/1,000} = 130 \ \Omega \qquad (13\text{-}41)$$

and the input time constant is $(130)(403)$ ps $= 52$ ns. Since this time constant is almost nine times as large as the output time constant, we assume that the bandpass of the amplifier is determined by the input time constant alone. Neglecting the output time constant and using $K = -g_m R_L$, the transfer function obtained from Fig. 13-18 is

$$A_{Vs} = \frac{V_o}{V_s} = \frac{-g_m R_L G'_s}{G'_s + g_{b'e} + sC} \qquad (13\text{-}42)$$

This equation indicates that we have obtained a single-pole approximation for the transfer function.

$$A_{Vs} = \frac{A_{Vso}}{1 + jf/f_H} \qquad (13\text{-}43)$$

where the midband gain is

$$A_{Vso} = -g_m R_L G'_s R \qquad (13\text{-}44)$$

and the high 3-dB frequency is

$$f_H = \frac{G'_s + g_{b'e}}{2\pi C} = \frac{1}{2\pi RC} \qquad (13\text{-}45)$$

The numerical value of f_H is $1/[2\pi(130)(403 \times 10^{-12})]$ Hz $= 3.04$ MHz. The magnitude of the voltage gain as a function of frequency is

$$|A_{Vs}| = \frac{|A_{Vso}|}{\left[1 + (f/f_H)^2\right]^{1/2}} \qquad (13\text{-}46)$$

and the phase-lead angle is

$$\theta_1 = -\pi - \arctan \frac{f}{f_H} \qquad (13\text{-}47)$$

The phase angle $-\pi$ results from the fact that A_{Vso} is negative. The magnitude and phase as given by Eqs. (13-46) and (13-47) are plotted in Fig. 13-17. A comparison of the two-pole with the one-pole curves shows that the value obtained with the more exact two-pole model is $f_H = 2.8$ MHz. The error in the single-pole value of 3.0 MHz is about 7 percent, and hence it is not necessary to use the more complicated two-pole solution.

The Miller Input Impedance[4]

In the Miller capacitance C we used the low-frequency value of $|K| = g_m R_L$. Since this is the maximum value of K, too large a value of C is used, and we should expect to obtain too small a value of f_H. However, as noted above, the single-pole approximation gives too large a value of f_H (3.0 MHz instead of the correct value of 2.8 MHz). This apparent anomaly is resolved if we take into account the frequency dependence of K in calculating the input admittance Y_i. From Fig. 13-18

$$Y_{b'e} = Y_i = j\omega \left[C_e + C_c (1 - K) \right] \qquad (13\text{-}48)$$

Since $|K| \gg 1$ even at the 3-dB frequency, the output circuit consists of a capacitor C_c in parallel with R_L. Hence

$$K = \frac{V_o}{V_{b'e}} = \frac{-g_m}{j\omega C_c + 1/R_L} = \frac{-g_m R_L}{1 + j\omega C_c R_L} \qquad (13\text{-}49)$$

and

$$Y_i = j\omega \left[C_e + C_c \left(1 + \frac{g_m R_L}{1 + j\omega C_c R_L} \right) \right] \qquad (13\text{-}50)$$

If we consider the input to consist of a capacitor C_i in parallel with a resistor R_i, then

$$Y_i = j\omega C_i + \frac{1}{R_i} \qquad (13\text{-}51)$$

From Eq. (13-50) it follows that

$$C_i = C_e + C_c + \frac{g_m R_L C_c}{1 + \omega^2 C_c^2 R_L^2} \qquad (13\text{-}52)$$

and

$$R_i = \frac{1}{g_m} \left(1 + \frac{1}{\omega^2 C_c^2 R_L^2} \right) \qquad (13\text{-}53)$$

At the frequency $f_H = 3.0 \times 10^6$ Hz,

$$\omega^2 C_c^2 R_L^2 = \left[(2\pi)(3 \times 10^6)(3 \times 10^{-12})(2 \times 10^3) \right]^2 = 0.0128$$

Hence, for $0 \leqslant f \leqslant f_H$, C_i remains essentially constant at its zero frequency value of $C = C_e + C_c(1 + g_m R_L)$, whereas R_i decreases from infinity to $[1/(50 \times 10^{-3})](1 + 1/0.0128) = 1,590\ \Omega$. This value is comparable with $r_{b'e} = 1,000\ \Omega$, and hence the Miller input resistance reduces the value of $V_{b'e}/V_s$, and hence also V_o/V_s. This explains why the bandwidth obtained with the two-pole exact transfer function is more conservative (smaller by approximately 7 percent, as seen from Fig. 13-17) than the value obtained from the single-pole approximation. For most applications the approximate single-pole transfer function is sufficiently accurate for bandwidth calculations.

13-10 THE GAIN-BANDWIDTH PRODUCT

By using the approximate single-pole transfer function obtained in Eq. (13-43), it is found that the gain-bandwidth product for voltage amplification is

$$|A_{Vso}f_H| = \frac{g_m}{2\pi C} \frac{R_L}{R_s + r_{bb'}} = \frac{f_T}{1 + 2\pi f_T C_c R_L} \frac{R_L}{R_s + r_{bb'}} \quad (13\text{-}54)$$

The quantities f_H and A_{Vso}, which characterize the transistor stage, depend on both R_L and R_s. The form of this dependence, as well as the order of magnitude of these quantities, may be seen in Fig. 13-19. Here f_H has been plotted as a function of R_L, up to $R_L = 2,000\ \Omega$, for several values of R_s. The topmost f_H curve in Fig. 13-19 for $R_s = 0$ corresponds to ideal-voltage-source drive. The voltage gain ranges from zero at $R_L = 0$ to 90.9 at $R_L = 2,000\ \Omega$. Note that a source impedance of only $100\ \Omega$ reduces the bandwidth by a factor of about 1.8. The bottom curve has $R_s = \infty$ and corresponds to the ideal current source. The voltage gain is zero for all R_L if $R_s = \infty$. For any R_L the bandwidth is highest for lowest R_s. The voltage gain-bandwidth product increases with increasing R_L and decreases with increasing R_s. Even if we know the gain-bandwidth product at a particular R_s and R_L, we cannot use the product to determine the improvement, say, in bandwidth corresponding to a sacrifice in gain. For if

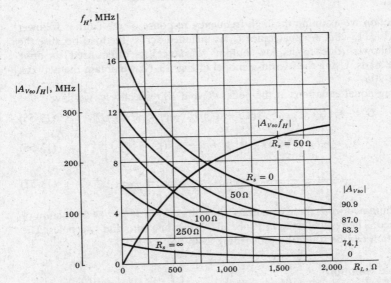

Figure 13-19 Bandwidth f_H as a function of R_L, with source resistance as a parameter, for an amplifier consisting of one CE transistor whose parameters are given in Eq. (13-29). Also, the gain-bandwidth product for a 50-Ω source is plotted. The tabulated values of $|A_{V_{so}}|$ correspond to $R_L = 2,000\ \Omega$ and to the values of R_s on the curves.

we change the gain by changing R_s or R_L or both, generally, the gain-bandwidth product will no longer be the same as it had been.

Summary

The high-frequency response of a transistor amplifier is obtained in this chapter in terms of the transistor parameters g_m, $r_{b'e}$, $r_{bb'}$, C_e, and C_c. We shall now show that these may be obtained from the four independent parameters h_{fe}, f_T, h_{ie}, and C_c.

From the operating current I_C and the temperature T, the transconductance is obtained as $g_m = |I_C|/V_T$ and is independent of the particular device under consideration. Knowing g_m, we can find, from Eqs. (11-67) and (13-28),

$$r_{b'e} = \frac{h_{fe}}{g_m} \qquad r_{bb'} = h_{ie} - r_{b'e} \qquad C_e \approx \frac{g_m}{2\pi f_T}$$

If R_s and R_L are given, all quantities in Eq. (13-38) are known. We have therefore verified that the frequency response may be determined from the four parameters h_{fe}, f_T, h_{ie}, and $C_c = C_{ob}$. Hence these four are usually specified by manufacturers of high-frequency transistors.

13-11 EMITTER FOLLOWER AT HIGH FREQUENCIES

In this section we examine the high-frequency response of the emitter follower shown in Fig. 13-20a. A capacitance C_L is included across the load because the emitter follower (due to its low output resistance) is often used to drive capacitive loads. Using the hybrid-π model of Fig. 13-12, we obtain the network of Fig. 13-20b.

Writing nodal equations at the nodes B' and E, respectively, we have

$$G_s'V_s = [G_s' + g_{b'e} + s(C_c + C_e)]V_i' - (g_{b'e} + sC_e)V_e \qquad (13\text{-}55)$$

$$0 = -(g + sC_e)V_i' + \left[g + \frac{1}{R_L} + s(C_e + C_L)\right]V_e \qquad (13\text{-}56)$$

where $\qquad G_s' \equiv \dfrac{1}{R_s + r_{bb'}} \qquad$ and $\qquad g \equiv g_m + g_{b'e} \qquad (13\text{-}57)$

If V_i' is eliminated from these equations, the voltage gain V_e/V_s as a function of s is obtained. The result, given in Prob. 13-37, has one zero and two poles. The exact solution can be found by proceeding as in Sec. 13-9.

Single-Pole Solution

We can obtain a very simple approximate expression for the transfer function by applying Miller's theorem to the circuit of Fig. 13-20b. With $K \equiv V_e/V_i'$ we obtain the circuit of Fig. 13-21.

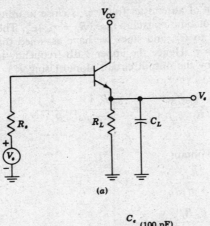

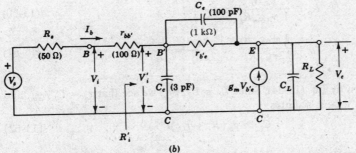

Figure 13-20 (a) Emitter follower. (b) High-frequency equivalent circuit of emitter follower.

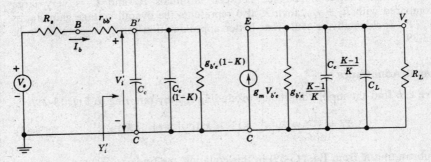

Figure 13-21 The equivalent circuit of the emitter follower, using Miller's theorem.

The low-frequency gain of an emitter follower is close to unity: $K \approx 1$ and $1 - K \approx 0$. Hence the input time constant $\tau_i \approx (R_s + r_{bb'})C_c$. The output time constant τ_o is proportional to C_L, and since we have assumed that the load is highly capacitive, then $\tau_o \gg \tau_i$. Hence the upper 3-dB frequency is determined, to a good approximation, by the output circuit alone. Using $K = 1$, we obtain

$$V_e = \frac{g_m V_{b'e}}{1/R_L + j\omega C_L} = \frac{g_m R_L (V_i' - V_e)}{1 + j\omega C_L R_L} \tag{13-58}$$

Solving for $V_e/V_i' = K$, we obtain

$$K = \frac{K_o}{1 + jf/f_H} \tag{13-59}$$

where

$$K_o \equiv \frac{g_m R_L}{1 + g_m R_L} \approx 1 \tag{13-60}$$

and

$$f_H \equiv \frac{1 + g_m R_L}{2\pi C_L R_L} \approx \frac{g_m}{2\pi C_L} = \frac{f_T C_e}{C_L} \tag{13-61}$$

and f_T is given by Eq. (13-34). Since $f_H = 1/2\pi\tau_o$, we see that $\tau_o = C_L/g_m$, and the condition $\tau_o \gg \tau_i$ requires

$$C_L \gg g_m(R_s + r_{bb'})C_c \tag{13-62}$$

For the parameter values in Fig. 13-20 and $g_m = 50$ mA/V, this condition is $C_L \gg (0.05)(150)(3) = 23$ pF.

Since the input impedance between terminals B' and C is very large compared with $R_s + r_{bb'}$, then K also represents the overall voltage gain $A_{Vs} \equiv V_e/V_s$. Incidentally, a somewhat better approximation for f_H is given in Prob. 13-39.

Input Admittance

We can find the input admittance (excluding $r_{bb'}$) by referring to Fig. 13-21.

$$Y_i' = \frac{I_b}{V_i'} = j\omega\big[C_c + (1 - K)C_e\big] + (1 - K)g_{b'e} \tag{13-63}$$

Substituting K from Eq. (13-59) in this equation, we find

$$Y_i' = j2\pi f C_c + (g_{b'e} + j2\pi f C_e)\frac{1 - K_o + jf/f_H}{1 + jf/f_H} \tag{13-64}$$

Since $K_o \approx 1$, the numerator of Eq. (13-64) is affected by frequency at a much lower value of f than is the denominator. Hence, for $f < f_H$, Eq. (13-64) can be written, with K_o replaced by unity,

$$Y_i' \approx j2\pi f C_c + (g_{b'e} + j2\pi f C_e)\frac{jf}{f_H}$$

$$\approx j2\pi f C_c + jg_{b'e}\frac{f}{f_H} - 2\pi f^2 \frac{C_e}{f_H} \qquad (13\text{-}65)$$

where the last term represents a negative resistance which is a function of frequency. Thus, the input impedance consists of a capacitance shunted by a negative resistance, and if the source resistance R_s contains some inductance in series with it, it is possible for the circuit to sustain undesirable oscillations. One way to remedy this condition is to use a small resistance in series with R_s so that the net resistance seen by the source is positive.

13-12 HIGH-FREQUENCY RESPONSE OF TWO CASCADED CE TRANSISTOR STAGES[5]

Since there is interaction between CE cascaded transistor stages, the analysis of a multistage amplifier is complicated and tedious. Fortunately, it is possible to make certain approximations in the analysis, and thus reduce the complexity of bandwidth calculations while keeping the error under approximately 20 percent.

Figure 13-9 shows two CE transistors in cascade. For high-frequency calculations each transistor is replaced by its small-signal hybrid-π model, as indicated in Fig. 13-22. We have included a voltage source V_s with $R_s = 50\ \Omega$ and have assumed that capacitors C_b and C_z represent short circuits for high frequencies. The base-biasing resistors R_1 and R_2 in Fig. 13-9 are assumed to be large compared with R_s. The symbol R_{L1} represents the parallel combination of R_1, R_2, and collector-circuit resistance R_c of the first stage. The symbol R_{L2} is the total load resistance of the second stage. A complete stage is included in each shaded block.

The network can be described by four nodal equations. If

$$R_s' \equiv R_s + r_{bb'} = 1/G_s',\ G_{L1} = 1/R_{L1},\ G_{L2} = 1/R_{L2},\ \text{and}\ g_{bb'} = 1/r_{bb'}$$

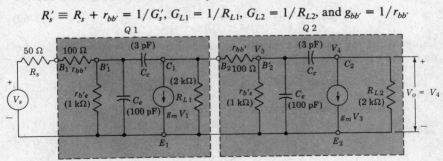

Figure 13-22 Two-stage interacting CE amplifier ($g_m = 50$ mA/V).

these equations are

$$G'_s V_s = \left[G'_s + g_{b'e} + s(C_e + C_c) \right] V_1 - sC_c V_2$$

$$0 = (g_m - sC_c)V_1 + (G_{L1} + g_{bb'} + sC_c)V_2 - g_{bb'}V_3 \qquad (13\text{-}66)$$

$$0 = -g_{bb'}V_2 + \left[g_{b'e} + g_{bb'} + s(C_e + C_c) \right] V_3 - sC_c V_4$$

$$0 = (g_m - sC_c)V_3 + (G_{L2} + sC_c)V_4$$

We can find the transfer gain V_o/V_s from these equations by recalling Cramer's rule,

$$A_{Vs} \equiv \frac{V_4}{V_s} = \frac{G'_s \Delta_{14}}{\Delta} = \text{transfer function} \qquad (13\text{-}67)$$

where Δ is the determinant of the set of equations, and Δ_{14} is the minor formed by removing the first row and fourth column of the complete determinant. Thus we see that the poles of the transfer function are given by the zeros of the determinant $\Delta = 0$, and the zeros of the transfer function are given by the zeros of the minor $\Delta_{14} = 0$. So much effort is required to find the poles and zeros that it is advisable to use a computer. There are several computer programs for finding roots of determinants.

The transfer function V_4/V_s of Fig. 13-22 must have four poles since the network contains four independent energy storage elements. In addition, the C_e capacitance for each of the two transistors provides a short circuit to ground as $s \to \infty$, and thus $V_4/V_s \to 1/s^2$ as $s \to \infty$. Hence (Sec. 13-8) we must have two zeros in addition to the poles. The values of the zeros for the circuit of Fig. 13-22 can be found by inspection. If, for some value of s, say s_5, V_4 is zero, the current fed through C_c must be equal to $g_m V_3$. Hence

$$s_5 C_c V_3 = g_m V_3 \qquad (13\text{-}68)$$

and the zero is $s_5 = g_m/C_c$. Similarly, for $Q1$, the zero is $s_6 = g_m/C_c$. The transfer function now has the form

$$A_{Vs} = \frac{K(s - s_5)(s - s_6)}{(s - s_1)(s - s_2)(s - s_3)(s - s_4)} \qquad (13\text{-}69)$$

Using one of the standard computer programs to solve $\Delta = 0$, we find for the numerical values given in Fig. 13-22 that the poles are given by

$$s_1 = -0.00342 \times 10^9 \text{ rad/s} \qquad s_2 = -0.0670 \times 10^9 \text{ rad/s}$$

$$s_3 = -0.680 \times 10^9 \text{ rad/s} \qquad s_4 = -4.21 \times 10^9 \text{ rad/s}$$

The zeros are

$$s_5 = s_6 = \frac{g_m}{C_c} = 16.65 \times 10^9 \text{ rad/s}$$

The program used to obtain the poles and zeros is CORNAP.[6] This computer

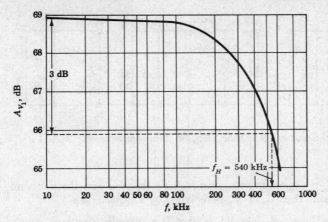

Figure 13-23 Computer-obtained frequency response of a two stage amplifier.

program is a network analysis routine that finds the state equations, the transfer function, and the frequency response of a general linear active network. The frequency response of the amplifier shown in Fig. 13-22 as obtained using CORNAP is plotted in Fig. 13-23. From the frequency-response curve we can read the high 3-dB frequency of the amplifier as $f_H = 540$ kHz.

The two transistors in Fig. 13-22 were assumed to have identical parameters. This condition simplifies the notation somewhat, but is not an essential restriction. The general method of analysis outlined above is equally valid for nonidentical transistors.

Dominant Pole

In the above example we observe that one of the poles, $s_1 = -0.00342 \times 10^9$ rad/s, is much lower than the other poles and zeros, and hence is the dominant pole. From our discussion in Sec. 13-8, we recognize that since all other poles or zeros are at least two octaves away, the upper 3-dB frequency for the amplifier is given, approximately, by the dominant pole

$$f_H \approx f_1 = \frac{-s_1}{2\pi} = \frac{0.00342 \times 10^9}{2\pi} \text{ Hz} = 545 \text{ kHz}$$

which is essentially the same value read from the curve.

Two-Stage-Cascade Simplified Analysis

If a computer is not available to help with the computations, we must make simplified assumptions in order to proceed with the analysis. We follow the method outlined in Sec. 13-9.

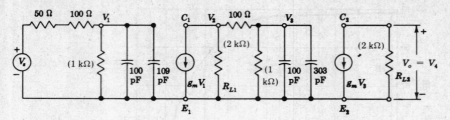

Figure 13-24 Two-stage interacting CE amplifier using the Miller approximation.

The effect of C_c is approximated using Miller's theorem and the midband value of the stage gain. Thus C_c of $Q2$ is replaced by a capacitance

$$C_c(1 + g_m R_{L2}) = 3(1 + 50 \times 2) = 303 \text{ pF}$$

across the input of $Q2$. Similarly, C_c of $Q1$ is replaced by

$$C_c[1 + g_m R_{L1}\|(r_{b'e} + r_{bb'})] = 3(1 + 50 \times 0.709) = 109 \text{ pF}$$

across the input of $Q1$. The circuit is now considerably simplified since, as shown in Fig. 13-24, there are only two independent capacitors in the network.

The transfer function obtained (Prob. 13-40) from Fig. 13-24 is

$$A_{Vs} \equiv \frac{V_4}{V_s} = \frac{2{,}810}{\left(1 + j\dfrac{f}{5.83 \times 10^5}\right)\left(1 + j\dfrac{f}{5.83 \times 10^6}\right)} \tag{13-70}$$

Clearly, we have a dominant pole at $f_1 = 583$ kHz, and thus this is the approximate high 3-dB frequency of the two-stage cascade.

We note that the simplified analysis yields a value for the bandwidth which is higher by 8 percent than the 540 kHz obtained by using the exact method.

13-13 MULTISTAGE CE AMPLIFIER AT HIGH FREQUENCIES

Regardless of the number of stages in an amplifier chain, the general method of solution is that given in the preceding section. Of course, the larger the number of stages, the greater is the computational complexity. We shall now outline the analysis of the three-stage CE cascade. Draw the network with each transistor replaced by its small-signal model for high-frequency analysis. Calculate and plot the transfer function V_o/V_s as a function of frequency. If we write nodal equations, we obtain a system of six equations in six unknowns similar to Eqs. (13-66). Clearly, due to the six independent capacitors (C_e and C_c for each stage), the transfer function must have six poles. In addition, three zeros must also be present, because the three emitter capacitors cause $V_o \rightarrow 1/s^3$ as $s \rightarrow \infty$.

Thus the transfer function is of the form

$$\frac{V_o}{V_s} = K \frac{(s - s_7)(s - s_8)(s - s_9)}{(s - s_1)(s - s_2)(s - s_3)(s - s_4)(s - s_5)(s - s_6)} \tag{13-71}$$

The zeros of this equation can be obtained by inspection of this particular network, as we explained in the preceding section. To obtain the poles, we must solve for the zeros of the network determinant, and since this determinant is of order six by six, the amount of computational labor required is prohibitive without the aid of a digital computer.

Using the program CORNAP, not only are the poles, zeros, and the gain constant K obtained, but the magnitude, phase, and the derivative of the phase with respect to frequency (this is the delay introduced by the amplifier) are also computed.

Approximate Analysis

We can perform a simplified analysis of the 3-stage amplifier by replacing C_{c1}, C_{c2}, and C_{c3} by the Miller capacitances. The resulting approximate network has three independent capacitances and we obtain (Prob. 13-41) the following transfer function (with f expressed in megahertz):

$$\frac{V_o}{V_s} = - \frac{90.5 \times 10^4}{\left(1 + j\dfrac{f}{0.582}\right)\left(1 + j\dfrac{f}{1.12}\right)\left(1 + j\dfrac{f}{5.86}\right)} \tag{13-72}$$

Since two of the poles are closer together than two octaves, this amplifier does not have a dominant pole. We must use an amplitude-response plot to deduce the upper 3-dB frequency. From this plot it is found that $f_H = 480$ kHz, whereas 420 kHz is obtained by computer analysis of the exact six-pole transfer function.

13-14 THE COMMON-SOURCE AMPLIFIER AT HIGH FREQUENCIES

The small-signal high-frequency model of an FET is obtained from the low-frequency model of Fig. 11-27 by adding three capacitors: C_{gs} between gate and source, C_{gd} between gate and drain, and C_{ds} between drain and source. Hence, the CS amplifier of Fig. 13-25a has the small-signal high-frequency equivalent circuit indicated in Fig. 13-25b. The output voltage V_o between D and S is easily found with the aid of the theorem in Sec. C-2, namely, $V_o = IZ$, where I is the short-circuit current and Z is the impedance seen between the terminals. To find Z, the independent generator V_i is (imagined) short-circuited, so that $V_i = 0$, and hence there is no current in the dependent generator $g_m V_i$. We then note that Z is the parallel combination of the impedances corresponding to Z_L, r_d,

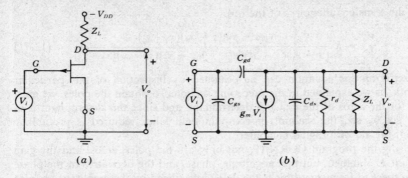

Figure 13-25 (a) The common-source amplifier circuit; (b) small-signal equivalent circuit at high frequencies. (The biasing network is not indicated.)

C'_{ds}, and C_{gd}. Hence

$$Y = \frac{1}{Z} = Y_L + g_d + Y_{ds} + Y_{gd} \tag{13-73}$$

where $Y_L = 1/Z_L$ = admittance corresponding to Z_L
$g_d = 1/r_d$ = conductance corresponding to r_d
$Y_{ds} = j\omega C_{ds}$ = admittance corresponding to C_{ds}
$Y_{gd} = j\omega C_{gd}$ = admittance corresponding to C_{gd}

The current in the direction from D to S in a zero-resistance wire connecting the output terminals is

$$I = -g_m V_i + V_i Y_{gd} \tag{13-74}$$

The amplification A_V with the load Z_L in place is given by

$$A_V = \frac{V_o}{V_i} = \frac{IZ}{V_i} = \frac{I}{V_i Y} \tag{13-75}$$

or from Eqs. (13-73) and (13-74)

$$A_V = \frac{-g_m + Y_{gd}}{Y_L + g_d + Y_{ds} + Y_{gd}} \tag{13-76}$$

At low frequencies the FET capacitances can be neglected and hence

$$Y_{ds} = Y_{gd} = 0$$

Under these conditions Eq. (13-76) reduces to

$$A_V = \frac{-g_m}{Y_L + g_d} = \frac{-g_m r_d Z_L}{r_d + Z_L} = -g_m Z'_L \tag{13-77}$$

where $Z'_L \equiv Z_L \| r_d$. This equation agrees with Eq. (11-87), with Z_L replaced by R_d.

Input Admittance

An inspection of Fig. 13-25b reveals that the gate circuit is not isolated from the drain circuit, but rather that they are connected by C_{gd}. From Miller's theorem (Sec. C-4), this admittance may be replaced by $Y_{gd}(1 - K)$ between G and S, and by $Y_{gd}(1 - 1/K)$ between D and S, where $K = A_V$. Hence the input admittance is given by

$$Y_i = Y_{gs} + (1 - A_V)Y_{gd} = G_i + j\omega C_i \qquad (13\text{-}78a)$$

where the input resistance is $R_i = 1/G_i$ and the input capacitance in parallel with R_i is C_i.

Input Capacitance (Miller Effect)

Consider an FET with a drain-circuit resistance R_d. From the previous discussion it follows that within the audio-frequency range, the gain is given by the expression $A_V = - g_m R_d' \equiv K$, where R_d' is $R_d \| r_d$. In this case Eq. (13-78a) becomes

$$\frac{Y_i}{j\omega} = C_i = C_{gs} + (1 + g_m R_d')C_{gd} \qquad (13\text{-}78b)$$

This increase in input capacitance C_i over the capacitance from gate to source is called the *Miller effect*.

The input capacitance is important in the operation of cascaded amplifiers. In such a system the output from one stage is used as the input to a second amplifier. Hence the input impedance of the second stage acts as a shunt across the output of the first stage and R_d is shunted by the capacitance C_i. Since the reactance of a capacitor decreases with increasing frequencies, the resultant output impedance of the first stage will be correspondingly low for the high frequencies. This will result in a decreasing gain at the higher frequencies.

Example 13-2 A MOSFET has a drain-circuit resistance R_d of 100 kΩ and operates at 20 kHz. Calculate the voltage gain of this device as a single stage, and then as the first transistor in a cascaded amplifier consisting of two identical stages. The MOSFET parameters are $g_m = 1.6$ mA/V, $r_d = 44$ kΩ, $C_{gs} = 3.0$ pF, $C_{ds} = 1.0$ pF, and $C_{gd} = 2.8$ pF.

SOLUTION

$$Y_{gs} = j\omega C_{gs} = j2\pi \times 2 \times 10^4 \times 3.0 \times 10^{-12} = j3.76 \times 10^{-7} \, \text{℧}$$

$$Y_{ds} = j\omega C_{ds} = j1.26 \times 10^{-7} \, \text{℧}$$

$$Y_{gd} = j\omega C_{gd} = j3.52 \times 10^{-7} \, \text{℧}$$

$$g_d = \frac{1}{r_d} = 2.27 \times 10^{-5} \, \text{℧}$$

$$Y_d = \frac{1}{R_d} = 10^{-5} \, \text{℧} = Y_L$$

$$g_m = 1.60 \times 10^{-3} \, \text{℧}$$

The gain of a one-stage amplifier is given by Eq. (13-76):

$$A_V = \frac{-g_m + Y_{gd}}{Y_d + g_d + Y_{ds} + Y_{gd}} = \frac{-1.60 \times 10^{-3} + j3.52 \times 10^{-7}}{3.27 \times 10^{-5} + j4.78 \times 10^{-7}}$$

It is seen that the j terms (arising from the interelectrode capacitances) are negligible in comparison with the real terms. If these are neglected, then $A_V = -48.8$. This value can be checked by using Eq. (13-77), which neglects interelectrode capacitances. Thus

$$A_V = \frac{-g_m r_d R_d}{R_d + r_d} = \frac{-1.6 \times 44 \times 100}{100 + 44} = -48.9 = -g_m R_d'$$

Since the gain is a real number, the input impedance consists of a capacitor whose value is given by Eq. (13-78b):

$$C_i = C_{gs} + (1 + g_m R_d')C_{gd} = 3.0 + (1 + 49)(2.8) = 143 \text{ pF}$$

Consider now a two-stage amplifier, each stage consisting of a FET operating as above. The gain of the second stage is that just calculated. However, in calculating the gain of the first stage, it must be remembered that *the input impedance of the second stage acts as a shunt on the output of the first stage.* Thus the drain load now consists of a 100-kΩ resistance in parallel with 143 pF. To this must be added the capacitance from drain to source of the first stage since this is also in shunt with the drain load. Furthermore, any stray capacitances due to wiring should be taken into account. For example, for every 1-pF capacitance between the leads going to the drain and gate of the second stage, 50 pF is effectively added across the load resistor of the first stage! This clearly indicates the importance of making connections with very short direct leads in high-frequency amplifiers. Let it be assumed that the input capacitance, taking into account the various factors just discussed, is 200 pF. Then the load admittance is

$$Y_L = \frac{1}{R_d} + j\omega C_i = 10^{-5} + j2\pi \times 2 \times 10^4 \times 200 \times 10^{-12}$$

$$= 10^{-5} + j2.52 \times 10^{-5} \text{ ℧}$$

The gain is given by Eq. (13-77):

$$A_V = \frac{-g_m}{g_d + Y_L} = \frac{-1.6 \times 10^{-3}}{2.27 \times 10^{-5} + 10^{-5} + j2.52 \times 10^{-5}}$$

$$= -30.7 + j23.7 = 38.8 \underline{/143.3°}$$

Thus the effect of the capacitances has been to reduce the magnitude of the amplification from 48.8 to 38.8 and to change the phase angle between the output and input from 180 to 143.3°.

If the frequency were higher, the gain would be reduced still further. For example, this circuit would be useless as a video amplifier, say, to a few megahertz, since the gain would then be less than unity.

Output Admittance

For the common-source amplifier of Fig. 13-25a the output impedance is obtained by "looking into the drain" with the input voltage set equal to zero. If $V_i = 0$ in Fig. 13-25b, we see r_d, C_{ds}, and C_{gd} in parallel. Hence the output admittance with Z_L considered external to the amplifier is given by

$$Y_o = g_d + Y_{ds} + Y_{gd} \tag{13-79}$$

13-15 THE COMMON-DRAIN AMPLIFIER AT HIGH FREQUENCIES

The source-follower configuration is given in Fig. 13-26a. Its equivalent circuit with the FET replaced by its high-frequency model is shown in Fig. 13-26b.

Voltage Gain

The output voltage V_o can be found from the product of the short-circuit current and the impedance between terminals S and N. We now find for the voltage gain

$$A_V = \frac{(g_m + j\omega C_{gs})R_s}{1 + (g_m + g_d + j\omega C_T)R_s} \tag{13-80}$$

$$C_T \equiv C_{gs} + C_{ds} + C_{sn} \tag{13-81}$$

where C_{sn} represents the capacitance from source to ground. At low frequencies the gain reduces to

$$A_V \approx \frac{g_m R_s}{1 + (g_m + g_d)R_s} \tag{13-82}$$

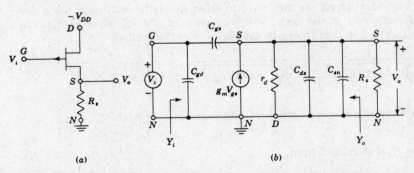

Figure 13-26 (*a*) The source-follower; (*b*) small-signal high-frequency equivalent circuit. (The biasing network is not indicated.)

Note that the amplification is positive and has a value less than unity. If $g_m R_s \gg 1$, then $A_V \approx g_m/(g_m + g_d) = \mu/(\mu + 1)$.

Input Admittance

The source follower offers the important advantage of lower input capacitance than the CS amplifier. The input admittance Y_i is obtained by applying Miller's theorem to C_{gs}. We find

$$Y_i = j\omega C_{gd} + j\omega C_{gs}(1 - A_V) \approx j\omega C_{gd} \qquad (13\text{-}83)$$

because $A_V \approx 1$. Hence, $C_i \approx C_{gd}$.

Output Admittance

The output admittance Y_o, with R_s considered external to the amplifier, is given by

$$Y_o = g_m + g_d + j\omega C_T \qquad (13\text{-}84)$$

where C_T is given by Eq. (13-81). At low frequencies the output resistance R_o is

$$R_o = \frac{1}{g_m + g_d} \approx \frac{1}{g_m} \qquad (13\text{-}85)$$

since $g_m \gg g_d$. For $g_m = 2$ mA/V, then $R_o = 500\ \Omega$.

The source follower is used for the same applications as the emitter follower, those requiring high input impedance and low output impedance.

13-16 BANDPASS OF CASCADED STAGES

The high 3-dB frequency for n cascaded stages is f_H^* and equals the frequency for which the overall voltage gain falls 3 dB to $1/\sqrt{2}$ of its midband value. To obtain the overall transfer function of *noninteracting* stages, the transfer gains of the individual stages are multiplied together. Hence, if each stage has a dominant pole and if the high 3-dB frequency of the ith stage is f_{Hi}, where $i = 1, 2, \ldots, n$, then f_H^* can be calculated from the product

$$\frac{1}{\sqrt{1 + (f_H^*/f_{H1})^2}} \cdots \frac{1}{\sqrt{1 + (f_H^*/f_{Hi})^2}} \cdots \frac{1}{\sqrt{1 + (f_H^*/f_{Hn})^2}} = \frac{1}{\sqrt{2}}$$

$$(13\text{-}86)$$

For n stages with identical upper 3-dB frequencies we have

$$f_{H1} = f_{H2} = \cdots = f_{Hi} = \cdots = f_{Hn} = f_H$$

Thus f_H^* is calculated from

$$\left[\frac{1}{\sqrt{1 + (f_H^*/f_H)^2}} \right]^n = \frac{1}{\sqrt{2}}$$

to be

$$\frac{f_H^*}{f_H} = \sqrt{2^{1/n} - 1} \tag{13-87}$$

For example, for $n = 2$, $f_H^*/f_H = 0.64$. Hence two cascaded stages, each with a bandwidth $f_H = 10$ kHz, have an overall bandwidth of 6.4 kHz. Similarly, three cascaded 10-kHz stages give a resultant upper 3-dB frequency of 5.1 kHz, etc.

If the low 3-dB frequency for n identical noninteracting cascaded stages is f_L^*, then, corresponding to Eq. (13-87), we find

$$\frac{f_L^*}{f_L} = \frac{1}{\sqrt{2^{1/n} - 1}} \tag{13-88}$$

We see that a cascade of stages has a lower f_H and a higher f_L than a single stage, resulting in a shrinkage in bandwidth.

If the amplitude response for a single stage is plotted on log-log paper, the resulting graph will approach a straight line whose slope is 6 dB per octave both at the low and at the high frequencies, as indicated in Fig. 13-3. For an n-stage amplifier it follows that the amplitude response falls $6n$ dB per octave, or equivalently, $20n$ dB per decade.

Interacting Stages

If in a cascade of stages the input impedance of one stage is low enough to act as an appreciable shunt on the output impedance of the preceding stage, then it is no longer possible to isolate stages. Under these circumstances individual 3-dB frequencies for each stage cannot be defined. However, when the overall transfer function of the cascade is obtained, it is found to contain n poles (in addition to k zeros). If the pole frequencies are $f_1, \ldots, f_2, \ldots, f_n$, then the high 3-dB frequency of the entire cascade f_H^* is given by Eq. (13-86) (with f_{Hi} replaced by f_i), provided that the zero frequencies are much higher than the pole frequencies (Prob. 13-50).

If the cascade has a dominant pole f_D which is much smaller than all other poles, all terms in the product in Eq. (13-86) may be neglected except the first. It then follows that $f_H = f_D$, or the high 3-dB frequency equals the dominant-pole frequency. (From here on we shall drop the asterisk on f_H^*.)

Consider now the situation discussed in Sec. 13-8, where there is a dominant frequency f_D, a second pole whose frequency is only two octaves away, and all other poles are at very much higher frequencies. Then Eq. (13-86) becomes

$$\frac{1}{\sqrt{1 + (f_H/f_D)^2}} \frac{1}{\sqrt{1 + (f_H/4f_D)^2}} = \frac{1}{\sqrt{2}} \tag{13-89}$$

Since we expect the 3-dB frequency to be approximately equal to the dominant frequency, substitute $f_H = f_D$ into the second term in Eq. (13-89) to obtain

$$1 + \left(\frac{f_H}{f_D}\right)^2 = \frac{2}{1 + \left(\frac{1}{4}\right)^2} \tag{13-90}$$

or
$$f_H = 0.94f_D \qquad (13\text{-}91)$$

This calculation verifies that the high 3-dB frequency is less than 6 percent smaller than the dominant frequency provided that the next higher pole frequency is at least two octaves away.

If the pole frequencies are not widely separated, the result of Prob. 13-51 indicates that f_H is given (to within 10 percent) by

$$\frac{1}{f_H} = 1.1\sqrt{\frac{1}{f_1^2} + \frac{1}{f_2^2} + \cdots + \frac{1}{f_n^2}} \qquad (13\text{-}92)$$

If this equation is applied to the case considered above, $f_1 = f_D$ and $f_2 = 4f_D$ and all other poles much higher, the result is $f_H = 0.88f_D$, in close agreement with Eq. (13-91).

Step Response of Noninteracting Stages

If the rise time of isolated individual cascaded stages is $t_{r1}, t_{r2}, \ldots, t_{rn}$ and if the input waveform rise time is t_{ro}, then, corresponding to Eq. (13-92) for the resultant upper 3-dB frequency, we have that the output-signal rise time t_r is given (to within 10 percent) by

$$t_r = 1.1\sqrt{t_{ro}^2 + t_{r1}^2 + t_{r2}^2 + \cdots + t_{rn}^2} \qquad (13\text{-}93)$$

If, upon application of a voltage step, one circuit produces a tilt of P_1 percent and if a second stage gives a tilt of P_2 percent, the effect of cascading these two noninteracting circuits is to produce a tilt of $P_1 + P_2$ percent. This result applies only if the individual tilts and the combined tilt are small enough so that in each case the waveform falls approximately linearly with time.

REFERENCES

1. Millman, J., and H. Taub: "Pulse, Digital, and Switching Waveforms," sec. 2-1, McGraw-Hill Book Company, New York, 1965.
2. Giacoletto, L. J.: Study of p-n-p Alloy Junction Transistors from DC through Medium Frequencies, *RCA Rev.*, vol. 15, no. 4, pp. 506–562, December, 1954.
   Searle, C. L., et al.: Ref. 1, vol. 3, chap. 3.
3. Gray, P. E., and C. L. Searle: "Electronic Principles," pp. 373–380, 421-424, John Wiley & Sons, Inc., New York, 1969.
4. Cherry, E. M., and D. E. Hooper: "Amplifying Devices and Low-pass Amplifier Design," pp. 337–343, John Wiley & Sons, Inc., New York, 1968.
5. Thornton, R. D., C. L. Searle, D. O. Pederson, R. B. Adler, and E. J. Angelo, Jr.: "Multistage Transistor Circuits," SEEC Series, vol. 5, chaps. 1, 2, John Wiley & Sons, Inc., New York, 1965.
6. Pottle, C.: A Textbook Computerized State Space Network Analysis Algorithm, *Cornell Univ. Elec. Eng. Res. Lab. Rept.*, September, 1968.

REVIEW QUESTIONS

13-1 Under what conditions does an amplifier preserve the form of the input signal?
13-2 (*a*) Define the *frequency-response magnitude characteristic* of an amplifier.

(b) Sketch a typical response curve.

(c) Indicate the high and low 3-dB frequencies.

(d) Define *bandwidth*.

13-3 (a) Sketch the high-frequency *step response* of a low-pass single-pole amplifier.

(b) Define the *rise time* t_r.

(c) What is the relationship between t_r and the high 3-dB frequency f_H?

13-4 (a) The input to a low-pass amplifier is a pulse of width t_p. Sketch the output waveshape.

(b) What must be the relationship between t_p and f_H in order to amplify the pulse without excessive distortion?

13-5 (a) Sketch the response of an amplifier to a low-frequency square wave.

(b) Define *tilt*.

(c) How is the tilt related to the low 3-dB frequency f_L?

13-6 (a) Sketch the circuit of a single-stage amplifier properly biased and indicate the two capacitors which affect the low-frequency response.

(b) Explain qualitatively why this amplifier behaves approximately as a single-time-constant circuit.

(c) What is the effective capacitance in the circuit?

13-7 (a) Sketch two *RC*-coupled CE transistor stages.

(b) Show the low-frequency model for one stage.

(c) What is the expression for f_L?

13-8 Draw the small-signal high-frequency model of a transistor.

13-9 (a) What is the physical origin of the two capacitors in the hybrid-π model?

(b) What is the order of magnitude of each capacitance?

13-10 (a) Draw the simplified high-frequency hybrid-π model.

(b) What is the order of magnitude of each resistance in this model?

13-11 How does g_m vary with (a) $|I_C|$; (b) $|V_{CE}|$; (c) T?

13-12 (a) How does C_e vary with $|I_C|$ and $|V_{CE}|$?

(b) How does C_c vary with $|I_C|$ and $|V_{CE}|$?

13-13 Derive the expression for the CE short-circuit current gain A_i as a function of frequency.

13-14 (a) Define f_β.

(b) Define f_T.

(c) What is the relationship between f_β and f_T?

13-15 (a) Indicate the form of the voltage gain of a multistage amplifier having two zeros and three poles.

(b) Under what conditions will this amplifier possess a dominant pole?

13-16 (a) Consider a CE stage with a resistive load R_L. Using Miller's theorem, what is the midband input capacitance?

(b) Assuming the output time constant is small compared with the input time constant, what is the high 3-dB frequency f_H for the voltage gain?

13-17 For a CE amplifier with a load R_L and source R_s sketch curves (as a function of R_L) of (a) f_H; (b) A_{Vso}; and (c) $|A_{Vso}f_H|$.

13-18 In terms of what four parameters is the high-frequency response of a CE stage obtained?

13-19 (a) Draw the circuit of an emitter follower with a capacitive load (C_L in parallel with R_L) and with a source resistance R_s.

(b) Sketch the high-frequency equivalent circuit.

(c) Explain qualitatively why this circuit may oscillate.

13-20 (a) Outline the general method for obtaining the high-frequency response of two interacting transistor amplifier stages.

(b) Outline an approximate method of solution.

13-21 (a) For a cascade of n transistor CE stages, how many poles will the voltage transfer function have? Explain.

(b) Repeat for the number of zeros.

13-22 (a) Sketch the small-signal high-frequency circuit of a CS amplifier.

(b) Derive the expression for the voltage gain.

13-23 (a) From the circuit of Rev. 13-22, derive the input admittance.

(b) What is the expression for the input capacitance in the audio range?

13-24 What specific capacitance has the greatest effect on the high-frequency response of a cascade of FET amplifiers? Explain.

13-25 Repeat Rev. 13-22 for a source-follower circuit.

13-26 Repeat Rev. 13-23 for a CD amplifier.

13-27 Derive the expression for the high 3-dB frequency f_H^* of n identical noninteracting stages in terms of f_H for one stage.

13-28 (a) Is f_H^* for two stages greater or smaller than f_H for a single stage? Explain.

(b) Repeat for f_L^* versus f_L.

13-29 (a) Give an approximate expression relating f_H^* and the 3-dB frequencies of n nonidentical stages.

(b) For two identical stages, what is f_H^*/f_H? Repeat for three stages.

13-30 Give an approximate relationship between the output rise time t_r, the rise time t_{ro} of an input signal, and the rise times of n nonidentical stages.

FEEDBACK AMPLIFIER FREQUENCY RESPONSE

The frequency-response characteristics of multipole feedback amplifiers are studied. It is shown that single-pole and double-pole closed-loop amplifiers are inherently stable. The step response of these amplifiers is obtained. A feedback amplifier with more than two poles can become unstable and break into oscillation if too much feedback is applied.

14-1 EFFECT OF FEEDBACK ON AMPLIFIER BANDWIDTH

The transfer gain of an amplifier employing feedback is given by Eq. (12-4), namely,

$$A_f = \frac{A}{1 + \beta A} \tag{14-1}$$

If $|\beta A| \gg 1$, then

$$A_f \approx \frac{A}{\beta A} = \frac{1}{\beta}$$

and from this result we conclude that the transfer gain may be made to depend entirely on the feedback network β. However, it is now important to consider the fact that even if β is constant, the gain A is not, since it depends on frequency. This means that at certain high or low frequencies, $|\beta A|$ will not be much larger than unity. To study the effect of feedback on bandwidth we shall assume in this section that the transfer gain A without feedback is given by a single-pole transfer function. In subsequent sections we consider multipole transfer functions.

Single-Pole Transfer Function

The gain A of a single-pole amplifier is given by

$$A = \frac{A_o}{1 + j(f/f_H)} \tag{14-2}$$

where A_o (real and negative) is the midband gain without feedback, and f_H is the high 3-dB frequency. The gain with feedback is given by Eq. (14-1), or using Eq. (14-2),

$$A_f = \frac{A_o/[1 + j(f/f_H)]}{1 + \beta A_o/[1 + j(f/f_H)]} = \frac{A_o}{1 + \beta A_o + j(f/f_H)}$$

By dividing numerator and denominator by $1 + \beta A_o$, this equation may be put in the form

$$A_f = \frac{A_{of}}{1 + j(f/f_{Hf})} \tag{14-3}$$

where $\qquad A_{of} \equiv \dfrac{A_o}{1 + \beta A_o} \qquad$ and $\qquad f_{Hf} \equiv f_H(1 + \beta A_o) \tag{14-4}$

We see that the *midband amplification with feedback* A_{of} equals the midband amplification without feedback A_o divided by $1 + \beta A_o$. Also, the *high 3-dB frequency with feedback* f_{Hf} equals the corresponding 3-dB frequency without feedback f_H multiplied by the same factor, $1 + \beta A_o$. The gain-frequency product has not been changed by feedback because, from Eqs. (14-4),

$$A_{of}f_{Hf} = A_o f_H \tag{14-5}$$

By starting with Eq. (13-2) for the low-frequency gain of a single RC-coupled stage and proceeding as above, we can show that the *low 3-dB frequency with feedback* f_{Lf} is decreased by the same factor as is the gain, or

$$f_{Lf} = \frac{f_L}{1 + \beta A_o} \tag{14-6}$$

For an audio or video amplifier, $f_H \gg f_L$, and hence the bandwidth is $f_H - f_L \approx f_H$. Under these circumstances, Eq. (14-5) may be interpreted to mean that the gain-bandwidth product is the same with or without feedback. Figure 14-1a is a plot of $|A|$ and $|A_f|$ versus frequency, whereas in Fig. 14-1b we show the Bode plots (Sec. 14-10) of both A and A_f in decibels vs. log f (compare with Fig. 13-3). The intersection of the two curves in Fig. 14-1b takes place at the frequencies $f = f_{Lf}$ and $f = f_{Hf}$ on the low end and high end, respectively. To verify this we equate the magnitude of Eq. (14-4) to the magnitude of Eq. (14-2) and solve for the value of f. Thus, if $f \gg f_H$ at the intersection point, we have

$$20 \log \left| \frac{A_o}{1 + \beta A_o} \right| = 20 \log \frac{|A_o|}{\sqrt{1 + (f/f_H)^2}} \approx 20 \log \frac{|A_o|}{f/f_H}$$

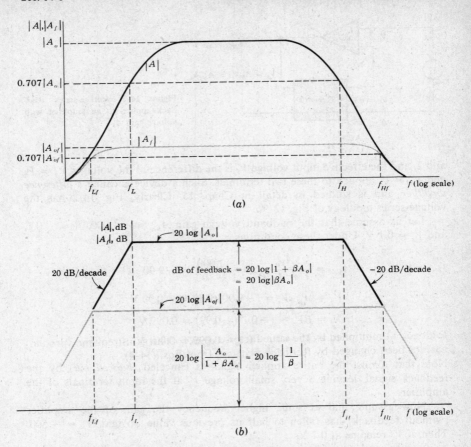

Figure 14-1 (*a*) Transfer gain is decreased and bandwidth is increased for an amplifier using negative feedback. (*b*) Idealized Bode plot.

or
$$f = f_H(1 + \beta A_o) = f_{Hf} \tag{14-7}$$

Similarly, we can show that the intersection at the low-frequency end occurs at $f = f_{Lf}$.

Bandwidth Improvement

Equations (14-4) and (14-6) show how the upper and lower 3-dB frequencies are affected by negative feedback. We may obtain a physical feeling of the mechanism by which feedback extends bandwidth by considering the voltage-series feedback circuit of Fig. 14-2. The amplifier A_V has two input terminals 1

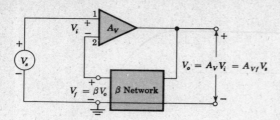

$$V_o = A_V V_i = A_{Vf} V_s$$

and 2, and the effective input voltage V_i is the difference in the voltages $V_1 = V_s$ and $V_2 = V_f$ applied to these two terminals. Such a device is called a *difference amplifier* and is studied in detail in Chap. 15. Clearly, Fig. 14-2 has the voltage-series topology of Fig. 12-9a.

Let us assume that the midband voltage gain $A_{Vo} = -1,000$, $\beta = -0.1$, and $V_s = 0.1$ V. Under these conditions

$$A_{Vof} = \frac{A_{Vo}}{1 + A_{Vo}\beta} = \frac{-1,000}{1 + 100} = -9.90$$

$$V_o = A_{Vof}V_s = (-9.90)(0.1) = -0.99 \text{ V}$$

$$V_f = \beta V_o = (-0.1)(-0.99) = 0.099 \text{ V}$$

$$V_i = V_s - V_f = 0.1 - 0.099 = 0.001 \text{ V}$$

Note that almost the entire applied signal is canceled (*bucked out*) by the feedback signal, leaving a very small voltage V_i at the input terminals of the amplifier.

Now assume that at some higher frequency the gain of the amplifier (without feedback) has fallen to half its previous value, so that $A_V = -500$. Then, if V_s remains at 0.1 V,

$$A_{Vf} = \frac{A_V}{1 + A_V\beta} = \frac{-500}{1 + 50} = -9.80$$

$$V_o = A_{Vf}V_s = (-9.80)(0.1) = -0.98 \text{ V}$$

$$V_f = (-0.1)(-0.98) = 0.098 \text{ V}$$

$$V_i = 0.1 - 0.098 = 0.002 \text{ V}$$

Note that although the base amplifier gain has been halved, the amplification with feedback has changed by only 1 percent. In the second case, V_i has doubled to compensate for the drop in A_V. There exists a self-regulating action so that, if the open-loop voltage gain falls (as a function of frequency), the feedback voltage also falls. Therefore less of the input voltage is bucked out, permitting more voltage to be applied to the amplifier input, and V_o remains almost constant.

Step Response

The transient behavior is that discussed in Sec. 13-2 for a single-pole transfer function. The output for a pulse input is given in Fig. 13-5, and for a square-wave input in Fig. 13-7.

14-2 DOUBLE-POLE TRANSFER FUNCTION WITH FEEDBACK

Let us consider a circuit where the basic amplifier gain (without feedback) has two poles on the negative real axis at $s_1 = -\omega_1$ and $s_2 = -\omega_2$ (ω_1 and ω_2 are positive), as shown in Fig. 14-3. If the midband gain is A_o, then the transfer gain is given by

$$A = \frac{A_o}{(1 - s/s_1)(1 - s/s_2)} = \frac{A_o}{(1 + s/\omega_1)(1 + s/\omega_2)} \tag{14-8}$$

If this expression for A is substituted into Eq. (14-1), we obtain for A_f, the transfer gain with feedback,

$$A_f = \frac{A_o\omega_1\omega_2}{s^2 + (\omega_1 + \omega_2)s + \omega_1\omega_2(1 + \beta A_o)} \tag{14-9}$$

or

$$A_f = \frac{A_{of}}{(s/\omega_o)^2 + (1/Q)(s/\omega_o) + 1} \tag{14-10}$$

where A_{of}, the midband gain with feedback, is given by Eq. (14-4) and ω_o and Q

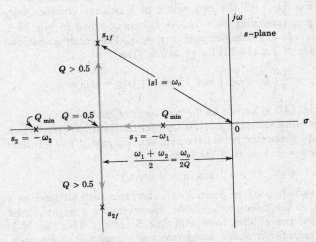

Figure 14-3 Root locus of the two-pole transfer function in the $s = \sigma + j\omega$ plane. The value Q_{min} corresponds to $\beta A_o = 0$.

are defined by

$$\omega_o \equiv \sqrt{\omega_1 \omega_2 (1 + \beta A_o)} \qquad Q \equiv \frac{\omega_o}{\omega_1 + \omega_2} \qquad (14\text{-}11)$$

The poles of A_f are

$$\frac{s}{\omega_o} = -\frac{1}{2Q} \pm \frac{1}{2} \sqrt{\frac{1}{Q^2} - 4} \qquad (14\text{-}12)$$

or

$$s = -\frac{\omega_1 + \omega_2}{2} \pm \frac{\omega_1 + \omega_2}{2} \sqrt{1 - 4Q^2} \qquad (14\text{-}13)$$

where the value $\omega_o/Q = \omega_1 + \omega_2$ from Eq. (14-11) is used. For negative feedback we shall assume βA_o to be real and positive. Hence the minimum value of Q is obtained for $\beta A_o = 0$ or $Q_{min} = \sqrt{\omega_1 \omega_2} /(\omega_1 + \omega_2)$. Substituting Q_{min} into Eq. (14-13) yields, after a little algebraic manipulation, the two values $s_{1f} = -\omega_1$ and $s_{2f} = -\omega_2$. Clearly, this result is correct, since if $\beta A_o = 0$, the poles with feedback (s_{1f} and s_{2f}) must coincide with the poles ($s_1 = -\omega_1$ and $s_2 = -\omega_2$) of the basic amplifier before feedback is added.

Root Locus[1, 2]

The movement of the poles in the s plane ($s = \sigma + j\omega$) as the feedback is increased is indicated in Fig. 14-3. The poles start at $-\omega_1$ and $-\omega_2$ at Q_{min} and move toward each other along the negative real axis as Q is increased until at $Q = 0.5$ the poles coincide. This behavior follows from Eq. (14-13), which shows that the values of s are real for $Q < 0.5$, and at $Q = 0.5$ the poles coincide at the value $-\frac{1}{2}(\omega_1 + \omega_2)$. For $Q > 0.5$ the two values of s become complex conjugates with the real part remaining at $-\frac{1}{2}(\omega_1 + \omega_2)$, as sketched in Fig. 14-3. Incidentally, the magnitude of a complex pole is $|s| = \omega_o$. This result is obtained by taking the magnitude of Eq. (14-12) for $Q \geqslant 0.5$. Thus

$$\left| \frac{s}{\omega_0} \right|^2 = \frac{1}{4Q^2} + \frac{1}{4}\left(4 - \frac{1}{Q^2}\right) = 1 \qquad \text{or} \qquad |s_{1f}| = |s_{2f}| = \omega_o$$

The shaded path in Fig. 14-3 is known as the *root locus* of the poles.

Note that for all positive values of βA_o the transfer function has poles which remain in the left-hand s plane (the poles have negative real parts). The transient response due to a pole $s = \sigma + j\omega$ is of the form

$$\epsilon^{st} = \epsilon^{(\sigma + j\omega)t} = \epsilon^{\sigma t}(\cos \omega t + j \sin \omega t)$$

If the real part of s (which is σ) is positive, $\epsilon^{\sigma t}$ increases without limit as t increases, indicating that the circuit is unstable. However, for σ negative, the transient which depends upon $\epsilon^{\sigma t}$ dies down with time. Since $\sigma < 0$ for Fig. 14-3, *the negative feedback amplifier is stable; independent of the amount of feedback.* This statement is true for a single-pole or double-pole transfer gain, but may not be true if three or more poles are present (Sec. 14-3).

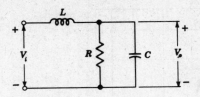

Figure 14-4 A circuit model for a two-pole feedback amplifier.

Circuit Model

We now demonstrate that the network in Fig. 14-4 is the analog of the two-pole feedback amplifier. The transfer function is found to be

$$\frac{V_o(s)}{V_i(s)} = \frac{1}{s^2 LC + s(L/R) + 1} \tag{14-14}$$

Introducing

$$\omega_o \equiv \frac{1}{\sqrt{LC}} \qquad Q \equiv \frac{R}{\omega_o L} = R\sqrt{\frac{C}{L}} \tag{14-15}$$

leads to

$$\frac{V_o(s)}{V_i(s)} = \frac{1}{(s/\omega_o)^2 + (1/Q)(s/\omega_o) + 1} = \frac{A_f}{A_{of}} \tag{14-16}$$

where the second equality follows from Eq. (14-10). Clearly, Fig. 14-4 is a circuit model of a two-pole feedback amplifier in the sense that both have the same frequency, phase, and transient response; *the transfer gain of the amplifier is A_{of} times the transfer function of the network.* Physical meanings can now be given to the symbols ω_o and Q, introduced in connection with the feedback amplifier. From Eqs. (14-15) it is evident that

ω_o = undamped ($R = \infty$) resonant angular frequency of oscillation

Q = quality factor (Q) at the resonant frequency

Frequency Response

If in Eq. (14-16) s is replaced by $j\omega$, then the magnitude of this expression gives the frequency response of the two-pole amplifier with feedback. It is convenient to use the *damping factor k* in place of Q. These are related by

$$k \equiv \frac{1}{2Q} \tag{14-17}$$

Thus, from Eqs. (14-16) and (14-17), we obtain

$$\left|\frac{A_f}{A_{of}}\right| = \frac{1}{\sqrt{\left[1 - (\omega/\omega_o)^2\right]^2 + 4k^2(\omega/\omega_o)^2}} \tag{14-18}$$

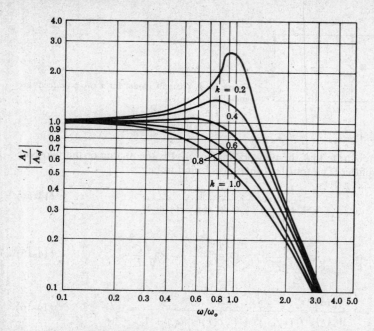

Figure 14-5 Normalized plot of the frequency response of a two-pole amplifier with feedback $(k = 1/2Q)$.

The peaks of this function are obtained by setting the derivative of the quantity under the square-root sign equal to zero. We find that a peak occurs at

$$\omega = \omega_o \sqrt{1 - 2k^2} \tag{14-19}$$

and the magnitude of the peak is given by

$$\left| \frac{A_f}{A_{of}} \right|_{\text{peak}} = \frac{1}{2k\sqrt{1 - k^2}} \tag{14-20}$$

Note that if $2k^2 > 1$ or $k > 0.707$ or $Q < 0.707$, the frequency response will not exhibit a peak. A plot of the normalized frequency response is given in Fig. 14-5.

Step Response

It has been proved in this section that regardless of how much negative feedback is employed, a two-pole amplifier remains stable (its poles are always in the left-half s plane). However, if the loop gain βA_o is too large, the transient response of the amplifier may be entirely unsatisfactory.

For example, in Fig. 14-6 there is indicated one possible response to a voltage step. Note that the output overshoots its final value by 37 percent and

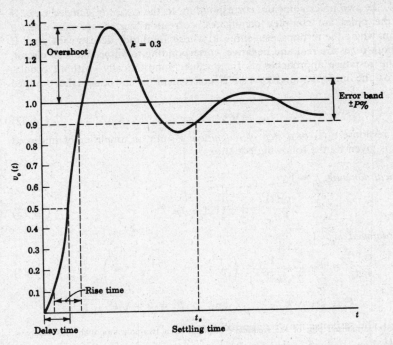

Figure 14-6 The step response of a two-pole feedback amplifier for a damping factor $k = 0.3$.

oscillates before settling down to the steady-state voltage. For most applications such a violent response is not acceptable.

The important parameters of the waveform are indicated in Fig. 14-6 and are defined as follows:

Rise time = time for waveform to rise from 0.1 to 0.9 of its steady-state value
Delay time = time for waveform to rise from 0 to 0.5 of its steady-state value
Overshoot = peak excursion above the steady-state value
Damped period = time interval for one cycle of oscillation
Settling time = time for response to settle to within $\pm P$ percent of the steady-state value (P specified for a particular application, say $P = 0.1$)

Analytical expressions for the response of the amplifier to a step of amplitude V is obtained by setting $V_i(s) = V/s$ into Eq. (14-16) and solving for the inverse Laplace transform. Recalling from Eq. (14-17) that $Q = 1/2k$, the poles, given in Eq. (14-12), can be put into the form

$$s = -k\omega_o \pm \omega_o\sqrt{k^2 - 1} \qquad (14\text{-}21)$$

If $k = 1$, the two poles coincide, corresponding to the *critically damped* case. If $k < 1$, the poles are complex conjugates, corresponding to an *underdamped* condition, where the response is a sinusoid whose amplitude decays with time. If $k > 1$, both poles are real and negative, corresponding to an *overdamped* circuit, where the response approaches its final value monotonically (without oscillation). For the underdamped case it is convenient to introduce the damped frequency

$$\omega_d \equiv \sqrt{1 - k^2}\, \omega_o \tag{14-22}$$

and the response $v_o(t)$ *to a step of magnitude* V into an amplifier of midband gain A_{of} is given by the following equations:

Critical damping, $k = 1$:

$$\frac{v_o(t)}{VA_{of}} = 1 - (1 + \omega_o t)\epsilon^{-\omega_o t} \tag{14-23}$$

Overdamped, $k > 1$:

$$\frac{v_o(t)}{VA_{of}} = 1 - \frac{1}{2\sqrt{k^2 - 1}}\left(\frac{1}{k_1}\epsilon^{-k_1 \omega_o t} - \frac{1}{k_2}\epsilon^{-k_2 \omega_o t}\right) \tag{14-24}$$

where $\quad k_1 \equiv k - \sqrt{k^2 - 1} \quad$ and $\quad k_2 \equiv k + \sqrt{k^2 - 1}$

If $4k^2 \gg 1$, the response may be approximated by

$$\frac{v_o(t)}{VA_{of}} \approx 1 - \epsilon^{-\omega_o t/2k} \tag{14-25}$$

Underdamped, $k < 1$:

$$\frac{v_o(t)}{VA_{of}} = 1 - \left(\frac{k\omega_o}{\omega_d}\sin \omega_d t + \cos \omega_d t\right)\epsilon^{-k\omega_o t} \tag{14-26}$$

These equations are plotted in Fig. 14-7 using the normalized coordinates $x \equiv t/T_o$ and $y \equiv v_o(t)/VA_{of}$, where $T_o \equiv 2\pi/\omega_o$ is the undamped period. If the derivative of Eq. (14-26) is set equal to zero, the positions $x = x_m$ and magnitudes $y = y_m$ of the maxima and minima are obtained. The results are

$$x_m = \frac{\omega_o t_m}{2\pi} = \frac{m}{2(1 - k^2)^{1/2}} \qquad y_m = \frac{v_o(t_m)}{VA_{of}} = 1 - (-1)^m \epsilon^{-2\pi k x_m} \tag{14-27}$$

where m is an integer. The maxima occur for odd values of m, and the minima are obtained for even values of m. By using Eq. (14-27) the waveshape of the underdamped output may be sketched very rapidly. From Eq. (14-27) it follows that the *overshoot* is given by $\exp[-\pi k m/(1 - k^2)^{1/2}]$.

Note that for heavy damping (k large or Q small) the rise time t_r is very long. As k is decreased (Q or βA_o increased), t_r decreases. For the critically

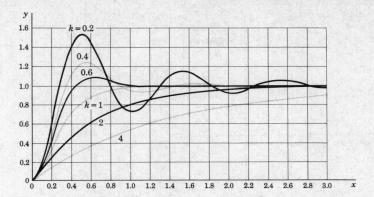

Figure 14-7 The response of a two-pole feedback amplifier to an input step [$y \equiv v_o(t)/VA_{of}$ and $x \equiv \omega_o t/2\pi = t/T_o$].

damped case we find from Fig. 14-7 that $t_r = 0.53T_o = 3.33/\omega_o$. If the feedback is increased so that $k < 1$, the rise time is decreased further, but this improvement is obtained at the expense of a ringing (oscillatory) response which may be unacceptable for some applications. Often $k \geqslant 0.707$ ($Q \leqslant 0.707$) is specified as a satisfactory response (corresponding to an overshoot of 4.3 percent or less).

14-3 THREE-POLE TRANSFER FUNCTION WITH FEEDBACK

In the previous two cases of single-pole and double-pole transfer functions we verify that the feedback-amplifier transfer function always has poles which lie in the left-hand plane. We now consider a three-pole transfer function and find that if the loop gain is sufficiently large, the poles of the feedback amplifier move into the right-hand plane, and thus the circuit becomes unstable.

Let us assume the open-loop gain to be given by

$$A(s) = \frac{A_o}{(1 + s/\omega_1)(1 + s/\omega_2)(1 + s/\omega_3)} \tag{14-28}$$

Using Eq. (14-1) for the gain with feedback, we find

$$A_f(s) = \frac{A_{of}}{(s/\omega_o)^3 + a_2(s/\omega_o)^2 + a_1(s/\omega_o) + 1} \tag{14-29}$$

where A_{of} is the midband gain with feedback and ω_o, a_2, and a_1 are given in Prob. 14-17 The stability of the feedback amplifier is determined by the poles of its transfer function. The techniques for the construction of the root locus of Eq. (14-29) are given in Ref. 1. The general shape of the root locus is shown in Fig. 14-8. It is clear that the poles start at $-\omega_1$, $-\omega_2$, and $-\omega_3$ when $\beta A_o = 0$. As

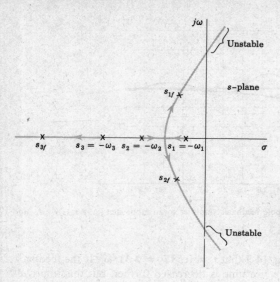

Figure 14-8 Root locus of the three-pole transfer function in the s plane. The poles without feedback ($\beta A_o = 0$) are s_1, s_2, and s_3, whereas the poles after feedback is added are s_{1f}, s_{2f}, and s_{3f}.

βA_o is increased one pole, s_{3f} increases in magnitude but always remains on the negative real axis, while the other two poles, s_{2f} and s_{1f}, approach each other, and then become complex conjugates when they break away from the real axis. The real part of s_{2f} and s_{1f} is negative when the roots coincide. However, as βA_o increases beyond this critical value, the real part of s_2 and s_1 becomes less negative. It is shown in Ref. 2 that if $\omega_1 = 0.1 \times 10^8$, $\omega_2 = 0.7 \times 10^8$, and $\omega_3 = 1.8 \times 10^8$ rad/s, the complex conjugate poles s_{2f} and s_{1f} move into the right-hand plane when $\beta A_o = 31$. We conclude that a three-pole amplifier can become unstable if sufficient negative feedback is applied to it.

14-4 APPROXIMATE ANALYSIS OF A MULTIPOLE FEEDBACK AMPLIFIER

In the general case, the response of a feedback system with three or more poles is so complicated that a computer must be used to obtain a solution. However, if the open-loop poles are widely separated, a simple approximate method of analysis is possible. We describe this technique now.

Assume that we have obtained the poles $|s_1| < |s_2| < |s_3| \cdots < |s_n|$ by removing the feedback. This can be done by measuring in the laboratory the corner frequencies of the Bode plot (Sec. 14-10), or the poles may be calculated. We wish to determine the closed-loop response. In the preceding section we note that, for a three-pole amplifier, the effect of adding feedback is to bring the poles s_1 and s_2 closer together along the real axis and to separate further the poles s_2 and s_3 (Fig. 14-8). Hence, if the second and third poles without feedback are at least two octaves apart ($s_3/s_2 \geqslant 4$), they will be separated even further apart after feedback is added ($s_{3f}/s_{2f} > 4$). *The response of an amplifier with three (or more) poles is determined approximately by the two lowest poles, s_1 and s_2, provided that $|s_3/s_2| \geqslant 4$.*

The above conclusion is consistent with the fact that *for a set of widely separated poles, the higher poles will remain almost fixed for moderate amounts of feedback.*[2] Hence we can use the theory in Sec. 14-2 for a double-pole transfer function with feedback to describe a multipole amplifier with widely separated poles. In other words, k is calculated for the first two poles, and then the frequency response is found from Fig. 14-5 and the transient response from Fig. 14-7.

It is convenient to introduce n as the ratio of the first two open-loop poles:

$$n = \frac{s_2}{s_1} = \frac{\omega_2}{\omega_1} > 1 \tag{14-30}$$

and then from Eq. (14-11) the Q is given in terms of n, and the desensitivity $1 + \beta A_o$ by

$$Q = \frac{\sqrt{n(1 + \beta A_o)}}{n + 1} = \frac{1}{2k} \tag{14-31}$$

Dominant Pole

If the first two open-loop poles are widely separated and if the desensitivity, and hence Q, is not too great, the first two closed-loop poles may be more than two octaves apart. Under these circumstances the response is simply that of a single (dominant) pole, and is discussed in Sec. 14-1.

We now determine the maximum value of Q for which a dominant pole exists. From Eq. (14-13), with $n = \omega_2/\omega_1$, the two lowest poles with feedback are

$$s_{1f} = \frac{-\omega_1(n + 1)}{2} \left(1 - \sqrt{1 - 4Q^2}\right)$$

$$s_{2f} = \frac{-\omega_1(n + 1)}{2} \left(1 + \sqrt{1 - 4Q^2}\right) \tag{14-32}$$

Setting $s_{2f}/s_{1f} \geqslant 4$, we find $Q \leqslant 0.4 = Q_{max}$. *For the poles of an amplifier with feedback to be separated by at least two octaves, Q must be no larger than 0.4 (or $Q^2 < 0.16$).* Incidentally, if Q increases to 0.5, the two poles coincide.

If $Q < 0.4$, *the dominant pole is s_{1f} in Eq. (14-32).* If $4Q^2 \ll 1$, this equation yields (Prob. 14-23), for the upper 3-dB frequency,

$$f_{Hf} \approx f_1\left(\frac{n}{n + 1}\right)(1 + \beta A_o)(1 + Q^2) = f_1(n + 1)Q^2(1 + Q^2) \tag{14-33}$$

In the following sections we analyze a number of feedback amplifiers by the approximate method outlined above and compare the results with exact computer solutions.

14-5 VOLTAGE-SHUNT FEEDBACK AMPLIFIER—FREQUENCY RESPONSE

We see in the preceding sections that negative feedback in general increases the bandwidth of the transfer function stabilized by the specific type of feedback (topology) used in a circuit. For example, voltage-series feedback increases the upper 3-dB frequency f_H of A_V to the value $f_{Hf} = (1 + \beta A_V)f_H$ after feedback is employed, provided A_V is a single-pole transfer function. We shall now examine specific feedback-amplifier circuits and obtain their frequency response. We start with a *voltage-shunt feedback amplifier*.

Figure 12-19a shows a common-emitter stage with a resistance R' connected from collector to base. Clearly, this is a case of voltage-shunt feedback, and we expect the bandwidth of the transresistance $R_{Mf} = V_o/I_s$ to be improved due to the feedback through R'. The example in Sec. 12-12 deals with this amplifier at low frequencies. Let us now examine the frequency response of the transresistance R_{Mf} and of the voltage gain $A_{Vf} = V_o/V_s$. The circuit of Fig. 14-9a is the same as Fig. 12-19a, with the transistor replaced by the hybrid-π equivalent circuit. The loading effects of the β network on the basic amplifier without feedback are included in Fig. 14-9b, which is equivalent to Fig. 12-19b. The voltage source is represented by its Norton's equivalent current source $I_s = V_s/R_s$. Since Fig. 14-9b is equivalent to Fig. 13-16, then from Eq. (13-38) we find the transfer function R_M without feedback:

$$R_M = \frac{V_o}{I_s} = \frac{- R_c' R G_1(g_m - sC_c)}{s^2 C_e C_c R_c' + s\left[C_e + C_c + C_c R_c'(g_m + g_{b'e} + G_1) \right] + G_1 + g_{b'e}}$$

$$(14\text{-}34)$$

where the following abbreviations are introduced:

$$R_c' \equiv R_c \| R' = 4\|40 = 3.64 \text{ k}\Omega$$

$$R \equiv R_s \| R' = 10\|40 = 8.00 \text{ k}\Omega$$

$$R_1 \equiv R + r_{bb'} = 8.10 \text{ K} \equiv \frac{1}{G_1}$$

We assume that $C_e = 100$ pF, $C_c = 3$ pF, $r_{bb'} = 100\ \Omega$, and to be consistent with the low-frequency h parameters, we have from Eqs. (11-19) and (11-21)

$$r_{b'e} = h_{ie} - r_{bb'} = 1{,}100 - 100 = 1{,}000\ \Omega$$

and

$$g_m = g_{b'e} h_{fe} = 50 \text{ mA/V}$$

With these numerical values, R_M is given by

$$R_M = \frac{V_o}{I_s} = 0.985 \times 10^{10}\, \frac{s - 16.7 \times 10^9}{(s + 600 \times 10^6)(s + 1.70 \times 10^6)} \qquad (14\text{-}35)$$

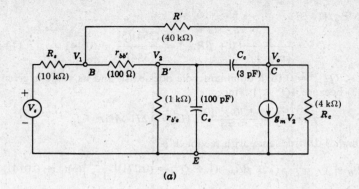

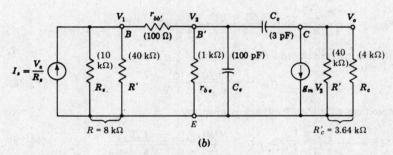

Figure 14-9 (*a*) CE amplifier with voltage-shunt feedback; (*b*) the *A* network including the loading of *R'*.

Dominant Pole

We see from Eq. (14-35) that we have a double-pole transfer function. However, the zero and one of the poles are much higher than $s_1 = -1.70 \times 10^6$ rad/s, and thus s_1 is a dominant open-loop pole. The question now arises whether the amplifier will have a dominant pole after feedback is applied to it. Thus we must check to see if $Q \leqslant Q_{max} = 0.4$.

From Eq. (14-35), with $s = j2\pi f = 0$ (or from Sec. 12-12), we obtain the midband transresistance

$$R_{Mo} = -1.60 \times 10^5 \ \Omega \tag{14-36}$$

Also from Eq. (12-67),

$$\beta = -1/R' = -2.50 \times 10^{-5} \quad \text{and} \quad 1 + \beta R_{Mo} = 5$$

Since

$$n = \frac{\omega_2}{\omega_1} = \frac{600}{1.70} = 353$$

then from Eq. (14-31)

$$Q^2 = \frac{n}{(n+1)^2}(1 + \beta R_{Mo}) = \frac{(353)(5)}{(354)^2} = 0.0141 \qquad (14\text{-}37)$$

Since $Q^2 < Q_{max}^2 = 0.16$, a dominant pole does exist, and its value is given by Eq. (14-33) because $4Q^2 \ll 1$. Since

$$f_1 = \frac{\omega_1}{2\pi} = \frac{1.70 \times 10^6}{2\pi} \text{ Hz} = 0.271 \text{ MHz} = f_H$$

then the high 3-dB frequency with feedback is

$$f_{Hf} = f_H \frac{n}{n+1}(1 + \beta R_{Mo})(1 + Q^2) = (0.271)\left(\frac{353}{354}\right)(5)(1 + 0.0141)$$

$$= 1.37 \text{ MHz}$$

Since $R_{Mof} = \dfrac{R_{Mo}}{1 + \beta R_{Mo}} = \dfrac{-1.60 \times 10^5}{5} = -3.20 \times 10^4 \ \Omega$

then, with f in megahertz and R_{Mf} in ohms,

$$R_{Mf} = \frac{-3.20 \times 10^4}{1 + j(f/1.37)} \qquad (14\text{-}38)$$

The voltage gain with feedback is given by

$$A_{Vf} = \frac{V_o}{V_s} = \frac{V_o}{I_s R_s} = \frac{R_{Mf}}{R_s} = \frac{-3.20}{1 + j(f/1.37)} \qquad (14\text{-}39)$$

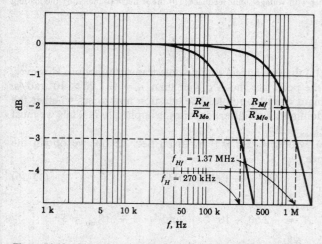

Figure 14-10 Computer-obtained frequency response of normalized transfer functions $|R_M/R_{Mo}|$ and $|R_{Mf}/R_{Mfo}|$ of the amplifier shown in Fig. 14-9a. $R_{mo} = 104$ dB and $R_{Mfo} = 90.1$ dB. Due to feedback the bandwidth is increased from 0.27 to 1.37 MHz.

From Eqs. (14-38) and (14-39) we see that R_{Mf} and A_{Vf} have the same upper 3-dB frequency.

In Fig. 14-9a we show the amplifier with feedback. Exact analysis of the circuit shown in Fig. 14-9b, using the CORNAP computer program,[3] yields the same values for the zero and two poles as in Eq. (14-35) for R_M. Computer analysis of the circuit shown in Fig. 14-9a yields for R_{Mf} two poles (in radians per second),

$$s_{1f} = -8.60 \times 10^6 \qquad s_{2f} = -5.95 \times 10^8$$

and one zero,

$$s_{3f} = 15.3 \times 10^9$$

Note that $s_{2f}/s_{1f} \gg 4$, verifying our conclusion of a dominant pole. The frequency responses for R_M and R_{Mf} are plotted in Fig. 14-10. From these curves we find $f_H = 270$ kHz and $f_{Hf} = 1.37$ MHz, in excellent agreement with the values obtained with the dominant-pole approximate analysis.

14-6 CURRENT-SERIES FEEDBACK AMPLIFIER— FREQUENCY RESPONSE

A common-emitter stage with an unbypassed emitter resistance represents a case of current-series feedback. The low-frequency operation of such a stage was studied in Sec. 12-10. In Fig. 14-11a the stage is shown with biasing and power supplies omitted. We shall obtain the frequency response of this amplifier using feedback concepts. Clearly, the transfer function which is stabilized is the transconductance $G_M = I_o/V_s$, and since $V_o = I_o R_c$, then $A_V = R_c G_M$ is also stabilized if R_c is a stable resistance. By referring to Fig. 12-16a and b, we construct the network shown in Fig. 14-11b, which is used for the calculation of the open-loop gain, taking the loading of the β network into account. Note that R_e appears in both input and output loops and that the transistor is replaced by its hybrid-π model. From this figure we find for the transfer function $G_M = I_o/V_s$ without feedback

$$G_M = \frac{-G_s'(g_m - sC_c)}{s^2 C_e C_c R_L + s\left[C_e + C_c + C_c R_L(g_m + g_{b'e} + G_s')\right] + G_s' + g_{b'e}}$$

$$(14\text{-}40)$$

where we make use of Eq. (13-38) and introduce

$$R_L \equiv R_e + R_c = 1.1 \text{ k}\Omega$$

and

$$G_s' \equiv \frac{1}{R_s + r_{bb'} + R_e} = \frac{1}{250} = 4 \times 10^{-3} \text{ A/V}$$

With $g_m = 50 \times 10^{-3}$ A/V, we obtain

$$G_M = -3.67 \times 10^4 \frac{167 \times 10^8 - s}{(s + 0.182 \times 10^8)(s + 8.45 \times 10^8)} \qquad (14\text{-}41)$$

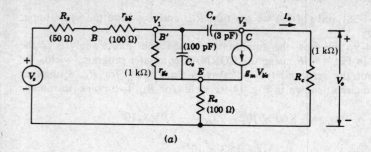

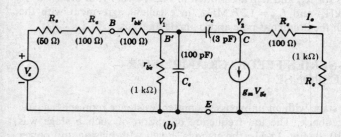

Figure 14-11 (a) Current-series feedback amplifier at high frequencies. (b) The basic amplifier without feedback but taking the loading of the β network (R_e) into account.

Dominant Pole

We now check to see if $s = -0.182 \times 10^8$ rad/s results in a dominant pole after the loop is closed. The transconductance G_{Mo} at low frequencies is found by substituting $s = 0$ in Eq. (14-41) or by using Eq. (12-53). We find

$$G_{Mo} = \frac{-3.67 \times 167 \times 10^{-4}}{0.182 \times 8.45} = -4 \times 10^{-2} \, \text{A/V} \qquad (14\text{-}42)$$

Since from Sec. 12-10, $\beta = -R_e = -100 \, \Omega$, then $1 + \beta G_{Mo} = 1 + 4 = 5$. With

$$n = \frac{\omega_2}{\omega_1} = \frac{8.45}{0.182} = 46.4$$

then from Eq. (14-31)

$$Q^2 = \frac{n}{(n+1)^2}(1 + \beta G_{Mo}) = \frac{46.4 \times 5}{(47.4)^2} = 0.103$$

Since $Q^2 < Q_{max}^2 = 0.16$, a closed-loop dominant pole does exist. The open-loop dominant-pole high 3-dB frequency is

$$f_H = \frac{0.182 \times 10^8}{2\pi} \, \text{Hz} = 2.90 \, \text{MHz}$$

and the closed-loop dominant frequency is given by Eq. (14-32):

$$f_{Hf} = \frac{f_H}{2}(n+1)(1 - \sqrt{1 - 4Q^2})$$

$$= \frac{2.90 \times 47.4}{2}(1 - \sqrt{1 - 0.412}) = 16.0 \text{ MHz}$$

Since $\quad G_{Mof} = \dfrac{G_{Mo}}{1 + \beta G_{Mo}} = \dfrac{-4 \times 10^{-2}}{5} = -8 \times 10^{-3} \text{ A/V} \qquad (14\text{-}43)$

then, with f in megahertz and G_{Mf} in amperes per volt,

$$G_{Mf} = \frac{-8 \times 10^{-3}}{1 + j(f/16.0)} \qquad (14\text{-}44)$$

The voltage gain $A_{Vf} = R_c G_{Mf} = 1,000 G_{Mf}$ has the same value of upper 3-dB frequency as the transconductance gain. The upper 3-dB frequency is $f_{Hf} = 16.0$ MHz. An exact computer analysis[3] of the circuit shown in Fig. 14-11b yields the same value of the zero and the two poles as in Eq. (14-41) for G_M.

For the circuit of Fig. 14-11a we expect two poles. Since as $s \rightarrow \infty$ the capacitors C_e and C_c act as short circuits and the current $g_m V_{b'e} \rightarrow 0$, Fig. 14-11$a$ becomes a resistive network and G_{mf} approaches a constant. Since $G_{mf} \rightarrow 1/s^0$ as $s \rightarrow \infty$, the number of zeros equals the number of poles. A computer analysis confirms these conclusions and yields two poles (in radians per second),

$$s_{1f} = -1.025 \times 10^8 \qquad s_{2f} = -7.65 \times 10^8$$

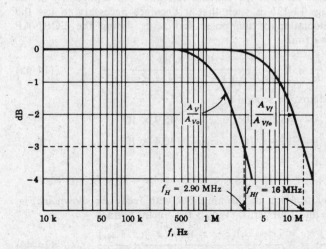

Figure 14-12 Computer-obtained frequency response of the amplifier shown in Fig. 14-11a. Due to feedback the bandwidth improvement is from 2.90 to 16.0 MHz.

and two zeros,

$$s_{3f} = 10.2 \times 10^8 \qquad s_{4f} - 16.3 \times 10^8$$

The frequency responses for A_{Vf} and A_V are plotted in Fig. 14-12, from which we read $f_{Hf} = 16.0$ MHz and $f_H = 2.90$ MHz, which are identical with the values obtained using the dominant-pole approximation.

14-7 CURRENT-SHUNT FEEDBACK PAIR—FREQUENCY RESPONSE

In Sec. 12-11 the low-frequency response of a current-shunt feedback pair shown in Fig. 12-17 is analyzed. This amplifier stabilizes the current gain $A_{If} = I_o/I_s$, and since from Eq. (12-61) we have $A_{Vf} = A_{If}R_{c2}/R_s$, then the voltage gain is also stabilized if R_{c2} and R_s are stable resistors. In other words, A_{If} and A_{Vf} have the same dependence on frequency. Figure 12-17 is redrawn in Fig. 14-13*a* for high-frequency calculations using the hybrid-π equivalent circuit, with $g_m = 50$ mA/V, $C_e = 100$ pF, $C_c = 3$ pF, $r_{bb'} = 100\ \Omega$, and $r_{b'e} = 1$ kΩ.

We desire to obtain the frequency response of A_{If} of the feedback amplifier shown in Fig. 14-13*a*, with $R_s = R' = 1.2$ kΩ, $R_{c2} = 500\ \Omega$, $R_{c1} = 3$ kΩ, and $R_e = 50\ \Omega$. These are the same parameters used in the example in Sec. 12-11.

The gain A_I without feedback, but with the loading effects of the feedback network included, can be obtained from the network of Fig. 14-13*b*. Since this network has four independent capacitors, we expect the transfer function for current gain to have four poles. In addition, the presence of the C_e capacitor from node V_2 to ground will cause the gain to go to zero as $1/s$ as $s \to \infty$; thus we must also have three zeros in the transfer function. The complexity of the network shown in Fig. 14-13*b* is such that it becomes necessary to use the computer for the calculation of the poles and zeros. Using the CORNAP computer program,[3] we find

$$A_I = \frac{I_o}{I_s} = K' \frac{(s - s_5)(s - s_6)(s - s_7)}{(s - s_1)(s - s_2)(s - s_3)(s - s_4)} \tag{14-45}$$

where the poles in radians per second are

$$s_1 = -46.2 \times 10^5 \qquad s_2 = -45.9 \times 10^6$$

$$s_3 = -11.4 \times 10^8 \qquad s_4 = -30.4 \times 10^8$$

and the zeros are

$$s_5 = 16.65 \times 10^9 \qquad s_6 = 15.4 \times 10^8 \qquad s_7 = -22.55 \times 10^8$$

The exact transfer function of the amplifier with feedback shown in Fig. 14-13*a* has four poles and four zeros. (Why?) It is of the form

$$A_{If} = \frac{I_o}{I_s} = K \frac{(s - s_{5f})(s - s_{6f})(s - s_{7f})(s - s_{8f})}{(s - s_{1f})(s - s_{2f})(s - s_{3f})(s - s_{4f})} \tag{14-46}$$

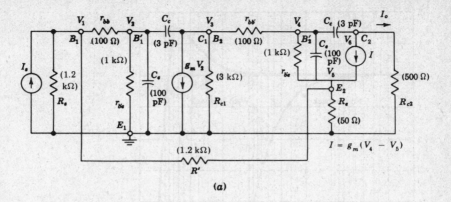

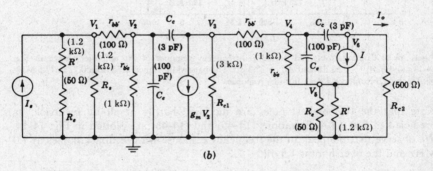

Figure 14-13 (a) Current-shunt feedback pair at high frequencies. (b) The amplifier without feedback but taking the loading of R' into account.

Using the CORNAP program,[3] we find the poles (in radians per second)

$$s_{1f} = -29.2 \times 10^6 + j5.40 \times 10^7$$

$$s_{2f} = -29.2 \times 10^6 - j5.40 \times 10^7$$

$$s_{3f} = -11.5 \times 10^8$$

$$s_{4f} = -30.2 \times 10^8$$

and the zeros

$$s_{5f} = 18.35 \times 10^8 - j9.75 \times 10^8$$

$$s_{6f} = 18.35 \times 10^8 + j9.75 \times 10^8$$

$$s_{7f} = -21.5 \times 10^8$$

$$s_{8f} = -7.40 \times 10^9$$

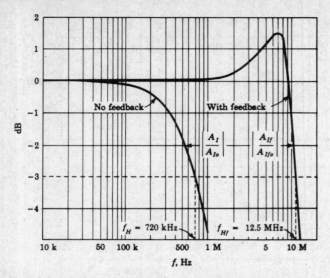

Figure 14-14 Computer-obtained normalized-frequency response of the amplifier shown in Fig. 14-13a and b. A_{Io} = 52.2 dB and A_{Ifo} = 27.2 dB. The bandwidth is increased from 0.72 to 12.5 MHz but feedback introduces a peak into the response.

Note that the two highest poles are modified hardly at all by feedback, as predicted in Sec. 14-4. Equations (14-45) and (14-46) are plotted in Fig. 14-14. We observe that the peak in the frequency response with feedback comes at 7.0 MHz and the overshoot is 1.5 dB.

Example 14-1 (a) Calculate the two lowest poles of the amplifier with feedback shown in Fig. 14-13a, using the fact that poles s_1 and s_2 in Eq. (14-45) are much lower than poles s_3 and s_4 and that $|s_3/s_2| > 4$. (b) Compute the frequency at which the frequency-response peaks, and find the overshoot in decibels.

SOLUTION (a) We approximate the transfer function of Eq. (14-45), using the lowest two poles, and refer to our discussion of the two-pole transfer function in Sec. 14-2. Thus

$$A_I = \frac{I_o}{I_s} \approx \frac{K''}{(s + 46.2 \times 10^5)(s + 45.9 \times 10^6)}$$

where $K'' = -K's_5s_6s_7/s_3s_4$.

The same example is considered at low frequencies in Sec. 12-11, where we find $1 + \beta A_{Io} = 17.4$. Since $n = 459/46.2 = 9.92$, we find from Eq. (14-31)

$$Q^2 = \frac{n}{(n+1)^2}(1 + \beta A_{Io}) = \frac{9.92 \times 17.4}{(10.92)^2} = 1.44 \quad \text{or} \quad Q = 1.20$$

Since $Q > 0.4$, a dominant pole does *not* exist. From Eq. (14-31) $k = 1/2Q$ $= 1/2.40 = 0.417$ and $k^2 = 0.174$.

The poles of the amplifier with feedback can be found using Eq. (14-13) or Eq. (14-32). Hence s_{1f} and s_{2f} are given by

$$-(46.2 \times 10^5)\left(\frac{10.92}{2}\right)(1 \mp j\sqrt{5.76 - 1}\,) = (-25.3 \pm j55.2) \times 10^6 \text{ rad/s}$$

as compared with the values $(-29.2 \pm j54.0) \times 10^6$ rad/s obtained using the computer and the exact analysis.

(b) The frequency-response peak occurs at the frequency $\omega = \omega_o \sqrt{1 - 2k^2}$, given by Eq. (14-19), where

$$\omega_o = Q(\omega_1 + \omega_2) = 1.20(45.9 + 4.62) \times 10^6 = 60.05 \times 10^6$$

as given by Eq. (14-11). Thus the response peaks at

$$f_{\text{peak}} = \frac{60.05}{6.28} \sqrt{1 - 2 \times 0.174} = 7.7 \text{ MHz}$$

The magnitude of the peak is given by Eq. (14-20), or at the peak we have

$$\left|\frac{A_f}{A_{of}}\right| = \frac{1}{2k\sqrt{1 - k^2}} = \frac{1}{2 \times 0.417\sqrt{1 - 0.174}} = 1.32$$

or $20 \log 1.32 = 2.4$ dB.

The exact analysis required the use of a computer and gave $f_{\text{peak}} = 7.0$ MHz and an overshoot of 1.5 dB.

14-8 VOLTAGE-SERIES FEEDBACK PAIR— FREQUENCY RESPONSE

In this section we examine the frequency response of the voltage-series feedback pair of Fig. 12-15. This amplifier is analyzed for low frequencies in Sec. 12-9. In Fig. 14-15a we show the amplifier prepared for high-frequency analysis, where the hybrid-π equivalent circuit is used to represent transistors $Q1$ and $Q2$. The same values of transistor parameters are assumed as in the preceding section. The 6.6-kΩ load of $Q1$ represents the parallel combination of $R_{c1} = 10$ kΩ and the biasing resistors of 47 kΩ and 33 kΩ for $Q2$. In the analysis, we need, as before, the open-loop gain, with the loading effects of the feedback network taken into account. Figure 14-15b shows this network. From Sec. 12-9 we know that this type of feedback stabilizes the voltage gain $A_V = V_o/V_s$. Thus we expect to increase the bandwidth of the voltage-gain transfer function. The network of Fig. 14-15b is similar to the network of Fig. 14-13b, except for the interchange of the unbypassed emitter resistance from the second stage to the first stage. Hence the expression for the voltage gain A_V is similar to A_I in Eq. (14-45); that is, it has three zeros and four poles.

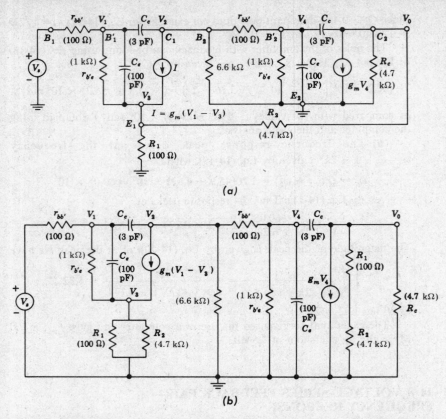

Figure 14-15 (a) Voltage-series feedback pair at high frequencies. (b) The basic amplifier without feedback but taking the loading of the β network into account.

Using the CORNAP program,[3] we find from Fig. 14-15b that the poles (in radians per second) are

$$s_1 = -24.4 \times 10^5 \qquad s_2 = -26.8 \times 10^7$$

$$s_3 = -6.45 \times 10^8 \qquad s_4 = -26.3 \times 10^8$$

and the zeros are

$$s_5 = -16.4 \times 10^8 \qquad s_6 = 10.3 \times 10^8 \qquad s_7 = 16.6 \times 10^9$$

Approximate Solution

Since s_1 is much smaller in magnitude than all other poles or zeros, we test to see if $f_1 = -s_1/2\pi = 390$ kHz results in a dominant pole after feedback. We find

(Prob. 14-29) that $4Q^2 = 0.67$, which slightly exceeds $4Q^2_{max} = 0.64$. This means that the second pole will not be quite two octaves away. If we ignore the presence of this second pole, the approximate dominant-pole solution of the closed-loop amplifier is

$$A_{Vf} = \frac{45.4}{1 + j(f/8.95)} \tag{14-47}$$

where f is in megahertz.

Exact Solution

The exact transfer function A_{Vf} obtained from Fig. 14-15a, using the CORNAP program,[3] has three zeros and four poles.

The poles (in radians per second) are

$$s_{1f} = -6.55 \times 10^7 \qquad s_{2f} = -17.4 \times 10^7$$
$$s_{3f} = -6.90 \times 10^8 \qquad s_{4f} = -26.2 \times 10^8$$

and the zeros are

$$s_{5f} = -17.0 \times 10^8 \qquad s_{6f} = 11.9 \times 10^8 \qquad s_{7f} = 8.10 \times 10^9$$

Note that the two highest poles are modified only slightly by feedback.

The open-loop and closed-loop frequency responses obtained with the aid of the computer are plotted in Fig. 14-16, from which we find the upper 3-dB

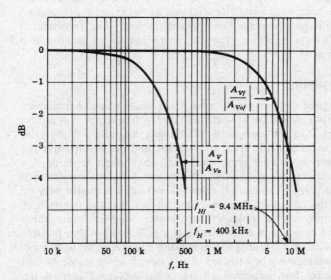

Figure 14-16 Computer-obtained normalized-frequency response of the amplifiers shown in Fig. 14-15a and b. $A_{Vo} = 58.4$ dB and $A_{Vof} = 33.1$ dB. Due to feedback the bandwidth increases from 0.40 to 9.4 MHz.

frequency without feedback to be 400 kHz, and the corresponding frequency of the amplifier with feedback is 9.4 MHz, which should be compared with the approximate value of 390 kHz and 8.95 MHz.

14-9 STABILITY[4]

Negative feedback for which $|1 + A\beta| > 1$ has been considered in some detail in the foregoing sections. If $|1 + A\beta| < 1$, the feedback is termed *positive*, or *regenerative*. Under these circumstances, the resultant transfer gain A_f will be greater than A, the nominal gain without feedback, since $|A_f| = |A|/|1 + A\beta| > |A|$. Regeneration as an effective means of increasing the amplification of an amplifier was first suggested by Armstrong.[5] Because of the reduced stability of an amplifier with positive feedback, this method is seldom used.

To illustrate the instability in an amplifier with positive feedback, consider the following situation: No signal is applied, but because of some transient disturbance, a signal X_o appears at the output terminals. A portion of this signal, $-\beta X_o$, will be fed back to the input circuit, and will appear in the output as an increased signal $-A\beta X_o$. If this term just equals X_o, then the spurious output has regenerated itself. In other words, if $-A\beta X_o = X_o$ (that is, if $-A\beta = 1$), the amplifier will oscillate. Hence, if an attempt is made to obtain large gain by making $|A\beta|$ almost equal to unity, there is the possibility that the amplifier may break out into spontaneous oscillation. This would occur if, because of variation in supply voltages, aging of transistors, etc., $-A\beta$ becomes equal to unity. There is little point in attempting to achieve amplification at the expense of stability. In fact, because of all the advantages enumerated in Sec. 12-2, feedback in amplifiers is almost always negative. However, combinations of positive and negative feedback are used.

The Condition for Stability

If an amplifier is designed to have negative feedback in a particular frequency range but breaks out into oscillation at some high or low frequency, it is useless as an amplifier. Hence, in the design of a feedback amplifier, it must be ascertained that the circuit is stable at *all* frequencies, and not merely over the frequency range of interest. In the sense used here, the system is stable if a transient disturbance results in a response which dies out. A system is unstable if a transient disturbance persists indefinitely or increases until it is limited only by some nonlinearity in the circuit. In Sec. 14-2 it is shown that the question of stability involves a study of the poles of the transfer function since these determine the transient behavior of the network. If a pole exists with a positive real part, this will result in a disturbance increasing exponentially with time. Hence the condition which must be satisfied, if a system is to be stable, is that the poles of the transfer function must all lie in the left-hand half of the complex-frequency plane. If the system without feedback is stable, the poles of A do lie in the left-hand half plane. It follows from Eq. (14-1), therefore, that *the stability condition requires that the zeros of $1 + A\beta$ all lie in the left-hand half of the complex-frequency plane.*

It should be clear from the foregoing discussion that *no oscillations are possible if the magnitude of the loop gain* $|A\beta|$ *is less than unity when the phase angle of* $A\beta$ *is* 180°. This condition is sought for in practice to ensure that the amplifier will be stable.

Consider, for example, a three-pole transfer function given by Eq. (14-28). To be specific, we can assume that this gain represents three cascaded stages, each with a dominant pole due to shunt capacitance. To simplify the discussion, consider that the amplifier stages are noninteracting, and that the poles are equal; $\omega_1 = \omega_2 = \omega_3$. (The general case of nonidentical poles is considered in the next section.) There is a definite maximum value of the feedback fraction $|\beta|$ allowable for stable operation. To see this, note that at low frequencies there is a 180° phase shift in each stage and 540°, or equivalently 180°, for the three stages. In other words, the midband gain A_o in Eq. (14-28) is negative. Since we are considering negative feedback, then $1 + \beta A_o > 1$ and β must be negative (it is assumed to be real). At high frequencies there is a phase shift due to the shunting capacitances, and at the frequency for which the phase shift per stage is 60°, the total phase shift of A is zero, and of $A\beta$ is 180°. If the magnitude of the gain at this frequency is called A_{60}, then β must be chosen so that $A_{60}|\beta|$ is less than unity, if the possibility of oscillations is to be avoided. Similarly, because of the phase shift introduced by the blocking capacitors between stages, there is a low frequency for which the phase shift per stage is also 60°, and hence there is the possibility of oscillation at this low frequency also, unless the maximum value of β is restricted as outlined above.

Gain Margin

The gain margin is defined as the value of $|A\beta|$ in decibels at the frequency at which the phase angle of $A\beta$ is 180°. If the gain margin is negative, this gives the decibel rise in open-loop gain, which is theoretically permissible without oscillation. If the gain margin is positive, the amplifier is potentially unstable. This definition is illustrated in Fig. 14-17.

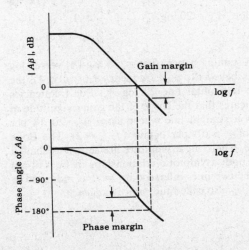

Figure 14-17 Bode plots relating to the definitions of gain and phase margins. Depending upon the feedback topology, A equals A_V, A_I, G_M, or R_M.

Phase Margin

The phase margin is 180° minus the magnitude of the angle of $A\beta$ at the frequency at which $|A\beta|$ is unity (zero decibels). The magnitudes of these quantities give an indication of how stable an amplifier is. For example, a linear amplifier of good stability requires gain and phase margins of at least 10 dB and 50°, respectively. This definition is illustrated in Fig. 14-17.

14-10 BODE PLOTS[5]

Gain and phase margins are obtained from Bode plots. These Bode characteristics can be constructed quickly and fairly accurately by approximating the curves by piecewise linear regions. These interconnected straight-line characteristics are referred to as *idealized Bode plots*.

Single-Pole Transfer Function

A one-pole transfer gain is of the form given in Eq. (13-5), or with $s = j2\pi f$ and with f_p equal to the frequency of the pole,

$$A(jf) = \frac{A_o}{1 + jf/f_p} \qquad (14-48)$$

The magnitude in decibels is defined by

$$|A|(\text{dB}) \equiv 20 \log |A| = 20 \log |A_o| - 20 \log \sqrt{1 + \left(\frac{f}{f_p}\right)^2} \qquad (14-49)$$

or

$$|A|(\text{dB}) = \begin{cases} 20 \log |A_o| & \text{if } \dfrac{f}{f_p} \ll 1 \\[2ex] 20 \log |A_o| - 20 \log \dfrac{f}{f_p} & \text{if } \dfrac{f}{f_p} \gg 1 \end{cases} \qquad (14-50)$$

These equations are plotted on semilog paper (that is, 20 log $|A|$ versus log f/f_p) in Fig. 14-18. For frequencies below the pole frequency $f_p(f/f_p < 1)$, the characteristic is asymptotic to the horizontal line 20 log A_o. For frequencies above $f_p(f/f_p > 1)$, Eq. (14-50) indicates that the curve of the gain magnitude in decibels asymptotically approaches a straight line whose slope is -20 dB per decade of frequency $(f/f_p = 10)$, or -6 dB per octave $(f/f_p = 2)$. The Bode characteristic is drawn in Fig. 14-18, where the asymptotes are shown as shaded straight lines. Note that for $f = f_p$, both asymptotic relationships in Eq. (14-50) yield $|A|(\text{dB}) = 20 \log A_o$. Hence, the two lines intersect at $f = f_p$ as indicated in Fig. 14-18 and the 3-dB frequency f_p is also called the *corner frequency*.

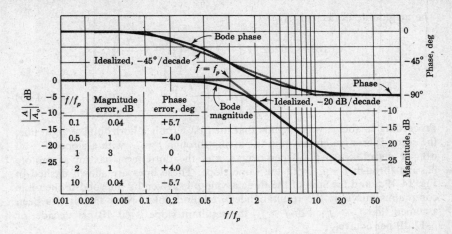

Figure 14-18 Bode plots of a single-pole low-pass amplifier. The piecewise linear approximations are shown shaded.

Since A_o and f_p are known, the two asymptotes can be easily located. The true Bode characteristic can then be sketched in simply by noting the deviations listed in the table given in Fig. 14-18. For example, the true response curve is 3 dB below the idealized Bode plot at $f = f_p$; is 1 dB below at $f = 0.5f_p$; etc.

The phase-shift angle θ of the single-pole transfer function is given by

$$\theta = -\arctan \frac{f}{f_p} \tag{14-51}$$

For $f \ll f_p$, $\theta \to 0°$, and for $f \gg f_p$, $\theta \to -90°$. At $f = f_p$, $\theta = -45°$. In view of these facts, a piecewise linear approximation to the phase characteristic is constructed as follows: The two asymptotes are the horizontal lines $\theta = 0°$ for $0 \leqslant f/f_p \leqslant 0.1$ and $\theta = -90°$ for $f/f_p \geqslant 10$. These are joined by a line of slope $-45°$ per decade passing through $\theta = -45°$ at $f = f_p$. In other words this line passes through the points $\theta = 0°$, $f = 0.1f_p$ and $\theta = -90°$, $f = 10f_p$. This broken-line Bode phase characteristic is indicated shaded in Fig. 14-18. The table shows that the idealized Bode plot differs from the true characteristic by less than 6° everywhere.

Two-Pole Transfer Function

A transfer gain with one pole at f_{p1} and a second at f_{p2} is given by

$$A(jf) = \frac{A_o}{\left[1 + j(f/f_{p1})\right]\left[1 + j(f/f_{p2})\right]} \tag{14-52}$$

The magnitude in decibels is

$$|A|(\text{dB}) = 20 \log |A_o| - 20 \log \sqrt{1 + \left(\frac{f}{f_{p1}}\right)^2}$$

$$- 20 \log \sqrt{1 + \left(\frac{f}{f_{p2}}\right)^2} \qquad (14\text{-}53)$$

Proceeding as above, we conclude that the first term is a horizontal straight line, the second term has an asymptote passing through $f = f_{p1}$ with a slope of -20 dB per decade, or -6 dB per octave, and the third term has an asymptote passing through $f = f_{p2}$ with the same slope. These lines are shown dashed in Fig. 14-19a, and the sum of the three asymptotes, given by the solid-broken-line continuous curve, is the resultant idealized Bode plot. Note that it has been assumed that $f_{p2} > f_{p1}$. For $f > f_{p2}$ the resultant slope is 40 dB per decade, or -12 dB per octave.

The phase response is given by

$$\theta = -\arctan \frac{f}{f_{p1}} - \arctan \frac{f}{f_{p2}} \qquad (14\text{-}54)$$

The linearized Bode plot is obtained by considering each term in Eq. (14-54)

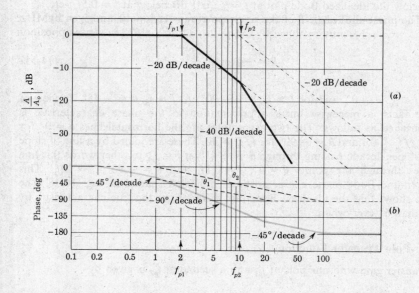

Figure 14-19 (a) The idealized Bode amplitude plots for a two-pole transfer function. The individual asymptotes for each pole are shown dashed, and the resultant is drawn as a solid continuous broken-line plot. (b) Phase-response Bode plot for (a).

separately and proceeding as in Fig. 14-19b. The contribution to the phase θ by each pole is indicated separately by the dashed lines. For example, the curve marked θ_2 corresponds to the pole at f_{p2}. Consistent with the above discussion $\theta_2 = 0$ for $f \leqslant 0.1f_{p2}$ and $\theta = -90°$ for $f \geqslant 10f_{p2}$, and θ_2 decreases by $45°$ per decade for $0.1f_{p2} \leqslant f \leqslant 10f_{p2}$. At $f = f_{p2}$, $\theta_2 = -45°$. The resultant phase at any frequency is the sum of the two dashed curves at that frequency and is indicated by the shaded-broken-line plot. (Figure 14-19 is drawn for the special case where $f_{p2} = 5f_{p1}$.)

The true (not linearized) Bode plot may be obtained by using the table in Fig. 14-18. The two individual curves corresponding to break points at f_{p1} and f_{p2} are drawn and then are added to obtain the resultant Bode amplitude (Prob. 14-31). The Bode phase response is obtained in an analogous manner by using the table in Fig. 14-18 to correct the piecewise linear phase lines.

Three-Pole Amplifier

To illustrate the stability problem in greater detail than in the preceding section, consider an amplifier with voltage gain

$$A_V = \frac{-10^3}{\left(1 + j\dfrac{f}{1}\right)\left(1 + j\dfrac{f}{10}\right)\left(1 + j\dfrac{f}{50}\right)} \tag{14-55}$$

where f is in megahertz. The three poles are at $f_1 = 1$, $f_2 = 10$, and $f_3 = 50$ MHz. If we sample the output voltage V_o and return a fraction βV_o in series opposition to the input voltage, the amplifier exhibits voltage-series feedback. For negative feedback, β must be a real negative number, and we assume it is independent of frequency. *This amplifier will oscillate when* $-A_V|\beta| = 1/180°$, *or when the magnitude of the gain* $|A_V|$ *is equal to* $|1/\beta|$ *and the phase of* $-A_V$ *is* $180°$.

In Fig. 14-20 we show the ideal Bode plot of $|A_V|$ and the phase of $-A_V$ versus frequency. The breaks in the magnitude plot occur at f_1, f_2, and f_3, and the slope of the lines increases by 20 dB per decade (6 dB per octave) after each break, as discussed above. The phase curve is more complicated, and hence the contribution to the phase θ by each pole is indicated separately by the dashed lines. For example, the curve marked θ_2 corresponds to the pole at f_2. Consistent with the discussion of Fig. 14-19, $\theta_2 = 0$ for $f \leqslant 0.1f_2$ and $\theta_2 = -90°$ for $f \geqslant 10f_2$, and θ_2 decreases by $45°$ per decade for $0.1f_2 \leqslant f \leqslant 10f_2$. At $f = f_2 = 10$ MHz, $\theta_2 = -45°$. The resultant phase at any frequency is the sum of the three dashed curves at that frequency, and is indicated by the shaded line plot.

We see from Fig. 14-20 that the phase of $-A_V$ is $180°$ and the magnitude of A_V is 26 dB at $f = 22$ MHz. The amplifier will oscillate at this frequency if the magnitude of the loop gain equals unity, i.e., if $|A_V\beta| = 1$. Hence,

$$20 \log |A_V| + 20 \log |\beta| = 0 \quad \text{or} \quad 20 \log |A_V| = 20 \log \left|\frac{1}{\beta}\right| \tag{14-56}$$

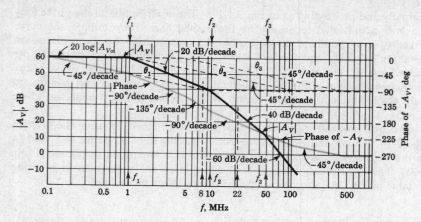

Figure 14-20 The open-loop voltage gain $|A_V|$ (solid lines) and phase of $-A_V$ (shaded lines) of a three-pole amplifier. The dashed lines are the phase contributions from each pole.

Expressed in decibels, if

$$\left| \frac{1}{\beta} \right| = |A_V| = 26 \text{ dB}$$

the network will oscillate at 22 MHz.

If we desire to maintain a phase margin of 45°, then the phase of $-A_V$ is $-180 + 45 = -135°$, which occurs in Fig. 14-20 at $f = 8$ MHz and $|A_V| = 42$ dB. From Fig. 14-17 the loop gain is 0 dB at the phase margin or $|A_V\beta| = 1$. Hence if

$$\left| \frac{1}{\beta} \right| = |A_V| = 42 \text{ dB}$$

the phase margin will be 45° at $f = 8$ MHz. Note that

$$20 \log |A_{Vo}\beta| = 20 \log |A_{Vo}| - 20 \log \left| \frac{1}{\beta} \right|$$

$$= 20 \log |A_{Vo}| - 20 \log |A_V| \qquad (14\text{-}57)$$

The midband loop gain for a phase margin θ is the difference (in dB) *between the gain A_{Vo} at zero frequency and the gain A_V at the frequency where the phase margin is θ.* Hence, for the amplifier whose Bode plots are given in Fig. 14-20, the midband loop gain for a phase margin of 45° is

$$|A_{Vo}\beta| = 60 - 42 = 18 \text{ dB}$$

This value of loop gain is the maximum allowed since, if $|\beta A_{Vo}| > 18$ dB, the phase margin will be less than 45°. (This statement should be checked by the reader by referring to Fig. 14-20.)

We now summarize what has been learned from the above discussion. The amplifier without feedback is stable and the phase margin for $|A_{Vo}\beta| = 0$ dB is

180° (from Fig. 14-20 with $|A_V| = |A_{Vo}|$ the phase is 0° corresponding to 180° phase margin). As the amount of feedback is increased the phase margin decreases and reaches 45° for $|A_{Vo}\beta| = 18$ dB. If we continue to increase the feedback, the phase margin becomes 0° at $|A_{Vo}\beta| = 60 - 26 = 34$ dB and the network breaks out into oscillations at a frequency of 22 MHz.

In some applications the desensitivity $(1 + A_{Vo}\beta)$ obtained at a phase margin of 45° may not be sufficient, and then compensation is used (Sec. 15–10) to increase the maximum loop gain while maintaining the same phase margin. From the foregoing discussion it should be clear that an unstable feedback system will function as an oscillator (Chap. 17).

REFERENCES

1. Gray, P. E., and C. L. Searle: "Electronic Principles: Physics, Models, and Circuits," p. 677, John Wiley & Sons, Inc., New York, 1969.
2. Thornton, R. D., C. L. Searle, D. O. Peterson, R. B. Adler, and E. J. Angelo, Jr.: "Multistage Transistor Circuits," SEEC Committee Series, vol. 5, pp. 108-118, John Wiley & Sons, Inc., 1965.
3. Pottle, C.: A Textbook Computerized State Space Network Analysis Algorithm, *Cornell Univ. Elec. Eng. Res. Lab. Rep.*, September, 1968.
4. Bode, H. W.: "Network Analysis and Feedback Amplifier Design," chap. 7, D. Van Nostrand Company, Inc., New York, 1956.
5. Ref. 4, chap. 15.

REVIEW QUESTIONS

14-1 Consider a feedback amplifier with a single-pole transfer function.

(*a*) What is the relationship between the high 3-dB frequency with and without feedback?

(*b*) Repeat part (*a*) for the low 3-dB frequency.

(*c*) Repeat part (*a*) for the gain-bandwidth product.

14-2 Consider a feedback amplifier with a double-pole transfer function.

(*a*) Without proof sketch the locus of the poles in the *s* plane after feedback.

(*b*) Why is this amplifier stable, independent of the amount of negative feedback?

14-3 (*a*) Indicate (without proof) a circuit having the same transfer function as the double-pole feedback amplifier.

(*b*) Sketch the step response of this amplifier for both the underdamped and overdamped condition.

14-4 For an underdamped two-pole amplifier response, define (*a*) rise time; (*b*) delay time; (*c*) overshoot; (*d*) damped period; (*e*) settling time.

14-5 (*a*) Sketch (without proof) the root locus of the poles of a three-pole amplifier after feedback is added.

(*b*) Indicate where the amplifier becomes unstable.

14-6 Consider a multipole amplifier with $|s_1| < |s_2| < |s_3| \cdots < |s_n|$. Under what circumstances is the response with feedback determined by (*a*) s_1 and s_2 and (*b*) s_1 alone?

14-7 Explain in words (without equations) how to obtain the frequency response of a voltage-shunt *feedback amplifier* using feedback concepts. Consider a single stage with a resistor R' between collector and base.

14-8 Repeat Rev. 14-7 for a current-series feedback amplifier. Consider a CE stage with an emitter resistor R_e.

14-9 (a) Define *positive* feedback.

    (b) What is the relationship between A_f and A for positive feedback?

14-10 State the stability condition in terms of Bode plots.

14-11 Consider an amplifier consisting of three identical noninteracting stages. Explain why a definite maximum value of $|\beta|$ exists before oscillations take place.

14-12 Define with the aid of graphs; (a) gain margin (b) phase margin.

14-13 (a) For a low-pass single-pole amplifier, sketch the Bode magnitude plot and its piecewise linear approximation.

    (b) Repeat for the Bode phase plot.

    (c) What are the slopes of the idealized Bode plots?

    (d) What are the corner frequencies, in both plots?

14-14 Repeat Rev. 14-13 for a two-pole transfer function.

14-15 Consider a three-stage amplifier.

    (a) What is the phase margin for zero feedback?

    (b) As the amount of feedback is increased, what happens to the phase margin?

    (c) How much feedback must be applied to cause the system to oscillate?

OPERATIONAL AMPLIFIER CHARACTERISTICS

The operational amplifier[1-4] (abbreviated OP AMP) is a direct-coupled high-gain amplifier to which feedback is added to control its overall response characteristic. It is used to perform a wide variety of linear functions (and also some nonlinear operations) and is often referred to as the *basic linear* (or more accurately, *analog*) *integrated circuit*. Its many applications are detailed in the following two chapters.

The integrated operational amplifier has gained wide acceptance as a versatile, predictable, and economic system building block. It offers all the advantages of monolithic integrated circuits: small size, high reliability, reduced cost, temperature tracking, and low offset voltage and current (which terms are defined carefully, later in this chapter).

Analog design techniques are described and a detailed analysis of the several stages in an OP AMP is made. Experimental methods for measuring OP AMP parameters are given. The frequency response and methods of compensation are discussed.

15-1 THE BASIC OPERATIONAL AMPLIFIER

The schematic diagram of the OP AMP is shown in Fig. 15-1a, and the equivalent circuit in Fig. 15-1b. A large number of operational amplifiers have a differential input, with voltages V_2 and V_1 applied to the inverting and noninverting terminals, respectively. The gain between the output V_o and V_1 is positive

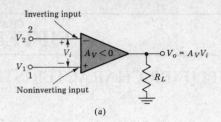

(a)

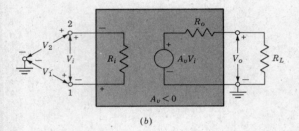

(b)

Figure 15-1 (a) Basic operational amplifier. (b) Low-frequency circuit model of an operational amplifier ($V_i = V_2 - V_1$). The open-circuit voltage gain is A_v and the gain under load is A_V.

(noninverting) whereas the amplification V_o/V_2 is negative (inverting). A single-ended amplifier may be considered as a special case where one of the input terminals is grounded. Nearly all OP AMPS have only one output terminal.

Ideal Operational Amplifier

The ideal OP AMP has the following characteristics:

1. Input resistance $R_i = \infty$.
2. Output resistance $R_o = 0$.
3. Voltage gain $A_v = -\infty$.
4. Bandwidth $= \infty$.
5. $V_o = 0$ when $V_1 = V_2$ independent of the magnitude of V_1.
6. Characteristics do not drift with temperature.

In Fig. 15-2a we show the ideal operational amplifier with feedback impedances (Z and Z') and the + terminal grounded. This is the basic inverting circuit. This topology represents *voltage-shunt feedback*, and is discussed in Sec. 12-12. The circuit of Fig. 15-2a is a generalization of Fig. 12-19a, with the single transistor replaced by the multistage OP AMP and the resistors R_s and R' replaced by impedances Z and Z', respectively. From Eq. (12-69) the voltage gain A_{Vf} with feedback is given by

$$A_{Vf} = -\frac{Z'}{Z} \qquad (15\text{-}1)$$

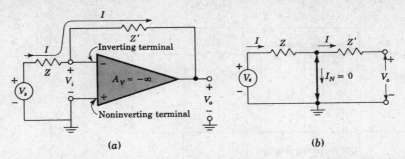

Figure 15-2 (*a*) Inverting operational amplifier with added voltage-shunt feedback. (*b*) Virtual ground in the operational amplifier.

For shunt (nodal) mixing the input resistance with feedback R_{if} is very low (Table 12-3), and for voltage sampling the output resistance with feedback R_{of} is very small.

An instructive alternative proof of Eq. (15-1) is obtained as follows: Since $R_i \to \infty$, the current I through Z also passes through Z', as indicated in Fig. 15-2a. In addition, we note that $V_i = V_o/A_V \to 0$ as $|A_V| \to \infty$, so that the inverting terminal is effectively grounded. Hence

$$A_{Vf} = \frac{V_o}{V_s} = \frac{-IZ'}{IZ} = -\frac{Z'}{Z}$$

in agreement with Eq. (15-1).

The operation of the circuit may now be described in the following terms: At the input to the amplifier proper there exists a *virtual ground* or *short circuit*. The term "virtual" is used to imply that, although the feedback from output to input through Z' serves to keep the voltage V_i at zero, no current actually flows into this short circuit. The situation is depicted in Fig. 15-2b, where the virtual ground is represented by the heavy double-headed arrow. This figure does not represent a physical circuit, but it is a convenient mnemonic aid from which to calculate the output voltage for a given input signal. This symbolism is very useful in connection with the analysis of analog systems discussed in Chap. 16. In summary: for an *ideal* OP AMP:

1. *The current to each input is zero.*
2. *The voltage between the two input terminals is zero.*

Practical Inverting Operational Amplifier

Equation (15-1) is valid only if the voltage gain is infinite. It is sometimes important to consider a physical amplifier which does not satisfy these restrictions. In Fig. 15-3 the amplifier in Fig. 15-2a is replaced by its small-signal model, with $|A_v| \neq \infty$, $R_i \neq \infty$, and $R_o \neq 0$. The symbol A_v is the *open-circuit* (*unloaded*) *voltage gain*. From the nodal equations at the output and input

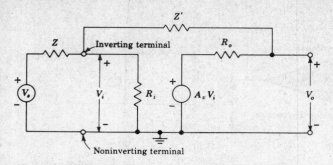

Figure 15-3 The model of the OP AMP of Fig. 15-1*b* used in the circuit of Fig. 15-2*a*.

terminals in Fig. 15-3 the following expression for the *closed-loop gain* is obtained (Prob. 15-1).

$$A_{Vf} = \frac{-Y}{Y' - (1/A_V)(Y' + Y + Y_i)} \tag{15-2}$$

where the Y's are the admittances corresponding to the Z's (for example, $Y' = 1/Z'$) and where the voltage gain $A_V \equiv V_o/V_i$, taking the loading of Z' into account, is given by

$$A_V = \frac{A_v + R_o Y'}{1 + R_o Y'} \tag{15-3}$$

Note that if $R_o = 0$ or $Y' = 0$ ($Z' = \infty$), the loading is effectively removed and $A_V = A_v$. Also observe that as $|A_v| \to \infty$, then $|A_V| \to \infty$ and

$$A_{Vf} \to -\frac{Y}{Y'} = -\frac{Z'}{Z}$$

in agreement with Eq. (15-1).

Noninverting Operational Amplifier

Very often there is a need for an amplifier whose output is equal to, and in phase with, the input, and in addition $R_i = \infty$ and $R_o = 0$, so that the source and load are in effect isolated. An emitter follower approximates these specifications. More ideal characteristics can be obtained by using an operational amplifier having a noninverting terminal for signals and an inverting terminal for the feedback voltage, as shown in Fig. 15-4. This configuration is that of a *voltage-series feedback* amplifier (Sec. 12-8) with the feedback voltage $V_f = V_2$. For $I_2 = 0$ the feedback factor is

$$\beta = \frac{V_2}{V_o} = \frac{Z}{Z + Z'}$$

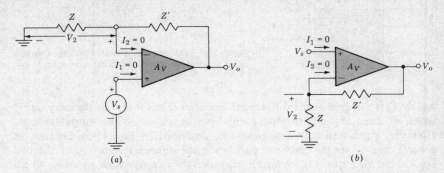

Figure 15-4 (*a*) A noninverting operational amplifier (*b*) An alternative presentation of the same system.

If $A_V\beta \gg 1$, then from Eq. (12-9),

$$A_{Vf} \approx \frac{1}{\beta} = \frac{Z + Z'}{Z} = 1 + \frac{Z'}{Z} \tag{15-4}$$

From Table 12-3 we expect extremely high input resistance and low output resistance. Problem 15-5 gives A_{Vf} if $|A_V| \neq \infty$, $R_i \neq \infty$, and $R_o \neq 0$.

The result obtained in Eq. (15-4) may be achieved without recourse to feedback theory. Since by rule 2 on p. 525, the potentials of the two inputs are equal, $V_s = V_2$ and, from Fig. 15-4,

$$A_{Vf} \equiv \frac{V_o}{V_s} = \frac{V_o}{V_2} = \frac{Z + Z'}{Z}$$

Note that, if $Z' \ll Z$ (for example, if Z is an open circuit and/or Z' is a short circuit), then $A_{Vf} = 1$. This configuration is a *voltage follower* ($V_o = V_s$), having a very high input resistance and a very low output resistance.

15-2 THE DIFFERENTIAL AMPLIFIER

The function of a differential amplifier (abbreviated DIFF AMP) is, in general, to amplify the difference between two signals. The need for DIFF AMPS arises in many physical measurements where response from dc to many megahertz is required. It is also the basic stage of an integrated operational amplifier with differential inputs.

Figure 15-5 represents a linear active device with two input signals v_1, v_2 and one output signal v_o, each measured with respect to ground. In an ideal DIFF AMP the output signal v_o should be given by

$$v_o = A_d(v_1 - v_2) \tag{15-5}$$

Figure 15-5 The output is a linear function of v_1 and v_2 for an ideal differential amplifier; $v_o = A_d(v_1 - v_2)$

where A_d is the gain of the differential amplifier. Thus it is seen that any signal which is common to both inputs will have no effect on the output voltage. However, a practical DIFF AMP cannot be described by Eq. (15-5), because, in general, the output depends not only upon the *difference signal* v_d of the two signals, but also upon the average level, called the *common-mode signal* v_c, where

$$v_d \equiv v_1 - v_2 \quad \text{and} \quad v_c \equiv \tfrac{1}{2}(v_1 + v_2) \qquad (15\text{-}6)$$

For example, if one signal is $+50 \ \mu\text{V}$ and the second is $-50 \ \mu\text{V}$, the output will not be exactly the same as if $v_1 = 1,050 \ \mu\text{V}$ and $v_2 = 950 \ \mu\text{V}$, even though the difference $v_d = 100 \ \mu\text{V}$ is the same in the two cases.

The Common-Mode Rejection Ratio

The foregoing statements are now clarified, and a figure of merit for a difference amplifier is introduced. The output of Fig. 15-5 can be expressed as a linear combination of the two input voltages

$$v_o = A_1 v_1 + A_2 v_2 \qquad (15\text{-}7)$$

where $A_1(A_2)$ is the voltage amplification from input 1(2) to the output under the condition that input 2(1) is grounded. From Eqs. (15-6),

$$v_1 = v_c + \tfrac{1}{2}v_d \quad \text{and} \quad v_2 = v_c - \tfrac{1}{2}v_d \qquad (15\text{-}8)$$

If these equations are substituted in Eq. (15-7), we obtain

$$v_o = A_d v_d + A_c v_c \qquad (15\text{-}9)$$

where $\qquad A_d \equiv \tfrac{1}{2}(A_1 - A_2) \quad \text{and} \quad A_c \equiv A_1 + A_2 \qquad (15\text{-}10)$

The voltage gain for the difference signal is A_d, and that for the common-mode signal is A_c. We can measure A_d directly by setting $v_1 = -v_2 = 0.5$ V, so that $v_d = 1$ V and $v_c = 0$. Under these conditions the measured output voltage v_o gives the gain A_d for the difference signal [Eq. (15-9)]. Similarly, if we set $v_1 = v_2 = 1$ V, then $v_d = 0$, $v_c = 1$ V, and $v_o = A_c$. The output voltage now is a direct measurement of the common-mode gain A_c.

Clearly, we should like to have A_d large, whereas ideally, A_c should equal zero. A quantity called the *common-mode rejection ratio*, which serves as a figure of merit for a DIFF AMP, is defined by

$$\text{CMRR} \equiv \rho \equiv \left| \frac{A_d}{A_c} \right| \qquad (15\text{-}11)$$

From Eqs. (15-9) and (15-11) we obtain an expression for the output in the following form:

$$v_o = A_d v_d \left(1 + \frac{1}{\rho} \frac{v_c}{v_d}\right) \tag{15-12}$$

From this equation we see that the amplifier should be designed so that ρ is large compared with the ratio of the common-mode signal to the difference signal. For example, if $\rho = 1{,}000$, $v_c = 1$ mV, and $v_d = 1$ μV, the second term in Eq. (15-12) is equal to the first term. Hence, for an amplifier with a common-mode rejection ratio of 1,000, a 1-μV difference of potential between the two inputs gives the same output as a 1-mV signal applied with the same polarity to both inputs.

> **Example 15-1** (*a*) Consider the situation given on p. 528, where the first set of signals is $v_1 = +50$ μV and $v_2 = -50$ μV and the second set is $v_1 = 1{,}050$ μV and $v_2 = 950$ μV. If the common-mode rejection ratio is 100, calculate the percentage difference in output voltage obtained for the two sets of input signals. (*b*) Repeat part *a* if $\rho = 10{,}000$.
>
> SOLUTION (*a*) In the first case, $v_d = 100$ μV and $v_c = 0$, so that, from Eq. (15-12), $v_o = 100 A_d$ μV.
> In the second case, $v_d = 100$ μV, the same value as for the first case, but now $v_c = \frac{1}{2}(1{,}050 + 950) = 1{,}000$ μV, so that, from Eq. (15-12),
>
> $$v_o = 100 A_d \left(1 + \frac{10}{\rho}\right) = 100 A_d \left(1 + \frac{10}{100}\right) \mu\text{V}$$
>
> These two measurements differ by 10 percent.
> (*b*) For $\rho = 10{,}000$, the second set of signals results in an output
>
> $$v_o = 100 A_d (1 + 10 \times 10^{-4}) \,\mu\text{V}$$
>
> whereas the first set of signals gives an output $v_o = 100 A_d$ μV. Hence the two measurements now differ by only 0.1 percent.

15-3 THE EMITTER-COUPLED DIFFERENTIAL AMPLIFIER[5, 6]

The operational amplifier must exhibit gain down to zero frequency. Such direct-coupled (dc) amplifiers cannot use blocking capacitors, since these would reduce the amplification to zero at $f = 0$. Large bypass capacitors (Sec. 13-3) must be avoided with monolithic circuits.

In a dc amplifier any change in the value of a circuit parameter (due to a temperature change, for example) results in an output-voltage variation even if the input remains constant. Various techniques have been devised for minimizing such drifts in output, and some of these are described in this chapter. The differential amplifier of Fig. 15-6 is well suited as the input stage for an OP AMP;

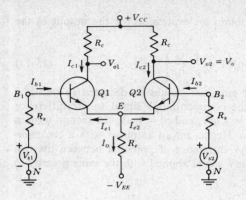

Figure 15-6 Symmetrical emitter-coupled difference amplifier.

its drift is low because of its symmetrical IC construction, it can be designed to have high input resistance, it has two inputs (an inverting and a noninverting terminal), a high CMRR, and many other of its properties approach the ideal characteristics listed in Sec. 15-1.

If the emitter resistance R_e is large, then common-mode rejection ratio is large for the circuit of Fig. 15-6. This statement can be justified as follows: If $V_{s1} = V_{s2} = V_s$, then from Eqs. (15-6) and (15-9) we have $V_d = V_{s1} - V_{s2} = 0$ and $V_o = A_c V_s$. However, if $R_e = \infty$, $I_o \approx 0$, and because of the symmetry of Fig. 15-6, we obtain $I_{e1} = I_{e2} = 0$. If $I_{b2} \ll I_{c2}$, then $I_{c2} \approx I_{e2}$, and it follows that $V_o = 0$. Hence the common-mode gain A_c becomes very small, and the common-mode rejection ratio is very large for a very large value of R_e and a symmetrical circuit.

The difference-mode gain A_d can be obtained by setting $V_{s1} = -V_{s2} = V_s/2$. From the symmetry of Fig. 15-6, we see that, if $V_{s1} = -V_{s2}$, then $I_{e1} = -I_{e2}$ and $I_o = 0$. Hence the drop across R_e is zero, and E is grounded for small-signal operation. Under these conditions the circuit of Fig. 15-7a can be

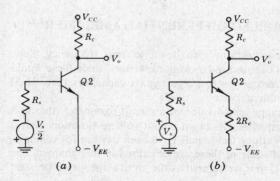

(a)   (b)

Figure 15-7 Equivalent circuit for a symmetrical differential amplifier used to determine (a) the difference gain A_d and (b) the common-mode gain A_c.

used to obtain A_d. Hence from Eq. (11-35) and Eq. (11-28) we get

$$A_d = \frac{V_o}{V_s} = \frac{1}{2} \frac{h_{fe} R_c}{R_s + h_{ie}} \qquad (15\text{-}13)$$

provided $h_{oe} R_c \ll 1$.

The common-mode gain A_c can be evaluated by setting $V_{s1} = V_{s2} = V_s$. If R_e is replaced by two resistances of value $2R_e$ in parallel, one of these to the left of node E and the other to the right of this node, then Fig. 15-6 is not altered. The circuit is now symmetrical with respect to an imaginary line passing through $+ V_{cc}$, E, and $- V_{EE}$. Bisecting the network along this line, the right-hand half circuit is as indicated in Fig. 15-7b. Using the approximate h-parameter model, the value of A_c is given (Sec. 11-13) by

$$A_c = \frac{- h_{fe} R_c}{R_s + h_{ie} + (1 + h_{fe}) 2 R_e} \qquad (15\text{-}14)$$

From Eqs. (15-13) and (15-14) it is seen that the CMRR $= A_d / A_c$ increases without limit as $R_e \to \infty$. However, it must be recalled that the approximate model which was used to derive Eq. (15-14) is not valid [Eq. (11-55)] when $h_{oe}(2R_e + R_c) > 0.1$. The more accurate expression for A_c for large R_e is given in Prob. 15-10a.

There are practical limitations on the magnitude of R_e because of the quiescent dc voltage across it; the emitter supply V_{EE} must become larger as R_e is increased, in order to maintain the quiescent current at its proper value. If the operating currents of the transistors are allowed to decrease, this will lead to higher h_{ie} values and lower values of h_{fe}. Both of these effects will tend to decrease the common-mode rejection ratio.

Differential Amplifier Supplied with a Constant Current

Frequently, in practice, R_e is replaced by a transistor circuit, as in Fig. 15-8, in which R_1, R_2, and R_3 can be adjusted to give the same quiescent conditions for $Q1$ and $Q2$ as the original circuit of Fig. 15-6. This modified circuit of Fig. 15-8 presents a very high effective emitter resistance R_e for the two transistors $Q1$ and $Q2$. In Prob. 15-10c it is verified that R_e is hundreds of kilohms even if R_3 is as small as 1 kΩ.

We now verify that transistor $Q3$ acts as an approximately constant-current source, subject to the condition that the base current of $Q3$ is negligible. Applying KVL to the base circuit of $Q3$, we have

$$I_3 R_3 + V_{BE3} = V_D + (V_{EE} - V_D) \frac{R_2}{R_1 + R_2} \qquad (15\text{-}15)$$

where V_D is the diode voltage. Hence

$$I_O \approx I_3 = \frac{1}{R_3} \left(\frac{V_{EE} R_2}{R_1 + R_2} + \frac{V_D R_1}{R_1 + R_2} - V_{BE3} \right) \qquad (15\text{-}16)$$

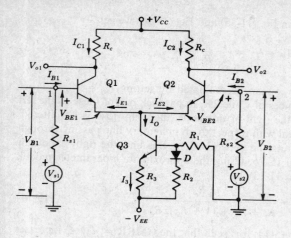

Figure 15-8 Differential amplifier with constant-current stage in the emitter circuit. Nominally, $R_{s1} = R_{s2}$.

If the circuit parameters are chosen so that

$$\frac{V_D R_1}{R_1 + R_2} = V_{BE3} \qquad (15\text{-}17)$$

then

$$I_O = \frac{V_{EE} R_2}{R_3 (R_1 + R_2)} \qquad (15\text{-}18)$$

Since this current is independent of the signal voltages V_{s1} and V_{s2}, then $Q3$ acts to supply the DIFF AMP consisting of $Q1$ and $Q2$ with the constant current I_O.

The above result for I_O has been rendered independent of temperature because of the added diode D. Without D the current would vary with temperature because V_{BE3} decreases approximately 2.5 mV/°C (Sec. 3-8). The diode has this same temperature dependence, and hence the two variations cancel each other and I_O does not vary appreciably with temperature. Since the cutin voltage V_D of a diode has approximately the same value as the base-to-emitter voltage V_{BE3} of a transistor, then Eq. (15-17) cannot be satisfied with a single diode. Hence *two diodes in series are used for* V_D.

Consider that $Q1$ and $Q2$ are identical and that $Q3$ is a true constant-current source. Under these circumstances we can demonstrate that the common-mode gain is zero. Assume that $V_{s1} = V_{s2} = V_s$, so that from the symmetry of the circuit, the collector current I_{c1} (the increase over the quiescent value for $V_s = 0$) in $Q1$ equals the current I_{c2} in $Q2$. However, since the total current increase $I_{c1} + I_{c2} = 0$ if $I_O = $ constant, then $I_{c1} = I_{c2} = 0$ and $A_c = V_{o2}/V_s = - I_{c2}R_c/V_s = 0$.

Practical Considerations[6]

In some applications the choice of V_{s1} and V_{s2} as the input voltages is not realistic because the resistances R_{s1} and R_{s2} represent the output resistances of the voltage generators V_{s1} and V_{s2}. In such a case we use as input voltages the base-to-ground voltages V_{b1} and V_{b2} of $Q1$ and $Q2$, respectively.

The differential amplifier is often used in dc applications. It is difficult to design dc amplifiers using transistors because of drift due to variations of h_{FE}, V_{BE}, and I_{CBO} with temperature. A shift in any of these quantities changes the output voltage and cannot be distinguished from a change in input-signal voltage. Using the techniques of integrated circuits (Chap. 4), it is possible to construct a DIFF AMP with $Q1$ and $Q2$ having almost identical characteristics. For example, base-to-emitter voltage matching to within 1 mV is possible. Because of the proximity of the devices on the chip, any parameter changes due to temperature and power-supply variations tend to cancel so that the output variation is small.

Differential amplifiers may be cascaded to obtain larger amplifications for the difference signal. Outputs V_{o1} and V_{o2} are taken from each collector (Fig. 15-8) and are coupled directly to the two bases, respectively, of the next stage.

Finally, the differential amplifier may be used as an emitter-coupled phase inverter. For this application the signal is applied to one base, whereas the second base is not excited (but is, of course, properly biased). The output voltages taken from the collectors are equal in magnitude and 180° out of phase.

15-4 TRANSFER CHARACTERISTICS OF A DIFFERENTIAL AMPLIFIER

It is important to examine the transfer characteristic[7] (I_C versus $V_{B1} - V_{B2}$) of the DIFF AMP of Fig. 15-8 to understand its advantages and limitations. We first consider this circuit qualitatively. When V_{B1} is below the cutoff point of $Q1$, all the current I_O flows through $Q2$ (assume for this discussion that V_{B2} is constant). As V_{B1} carries $Q1$ above cutoff, the current in $Q1$ increases, while the current in $Q2$ decreases, and the sum of the currents in the two transistors remain constant and equal to I_O. The total range ΔV_O over which the output can follow the input is $R_C I_O$ and is therefore adjustable through an adjustment of I_O.

From Fig. 15-8 we have

$$I_{E1} + I_{E2} = -I_O \tag{15-19}$$

$$V_{B1} - V_{B2} = V_{BE1} - V_{BE2} \tag{15-20}$$

The emitter current I_E of each transistor is related to the voltage V_{BE} by a diode volt-ampere characteristic of the form

$$I_E = I_S \epsilon^{V_{BE}/V_T} \tag{15-21}$$

where V_T is defined in Eq. (1-35).

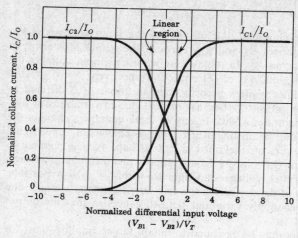

Figure 15-9 Transfer characteristics of the basic differential-amplifier circuit. Compare with Fig. 5-27.

If we assume that $Q1$ and $Q2$ are matched, it follows from Eqs. (15-19) to (15-21) that

$$I_{C1} \approx -I_{E1} = \frac{I_O}{1 + \exp\left[-(V_{B1} - V_{B2})/V_T\right]} \qquad (15\text{-}22)$$

and I_{C2} is given by the same expression with V_{B1} and V_{B2} interchanged. The transfer characteristics described by Eq. (15-22) for the normalized collector currents I_{C1}/I_O (and I_{C2}/I_O) are shown in Fig. 15-9, where the abscissa is the normalized differential input $(V_{B1} - V_{B2})/V_T$.

If Eq. (15-22) is differentiated with respect to $V_{B1} - V_{B2}$, we have the transconductance g_{md} of the DIFF AMP with respect to the differential input voltage, or

$$\frac{dI_{C1}}{d(V_{B1} - V_{B2})} = g_{md} = \frac{I_O}{4V_T} \qquad (15\text{-}23)$$

where g_{md} is evaluated at $V_{B1} = V_{B2}$. This equation indicates that, for the same value of I_O, the effective transconductance of the differential amplifier is one-fourth that of a single transistor [Eq. (11-25)]. An alternative proof of Eq. (15-23) is given in Prob. 15-12.

The following conclusions can be drawn from the transfer curves of Fig. 15-9:

1. The differential amplifier is a very good limiter, since when the input $(V_{B1} - V_{B2})$ exceeds $\pm 4V_T$ ($\approx \pm 100$ mV at room temperature), very little further increase in the output is possible.

2. The slope of these curves defines the transconductance, and it is clear that g_{md} starts from zero, reaches a maximum of $I_O/4V_T$ when $I_{C1} = I_{C2} = \frac{1}{2} I_O$, and again approaches zero.

3. The value of g_{md} is proportional to I_O [Eq. (15-23)]. Since the output voltage change V_{o2} is given by

$$V_{o2} = g_{md} R_c \Delta (V_{B1} - V_{B2}) = g_{md} R_c (V_{b1} - V_{b2}) \qquad (15\text{-}24)$$

it is possible to change the differential gain by varying the value of the current I_O. This means that automatic gain control (AGC) is possible with the DIFF AMP.

4. The transfer characteristics are linear in a small region around the operating point where the input varies approximately $\pm V_T (\pm 26$ mV at room temperature). In Prob. 15-17 we show that it is possible to increase the region of linearity by inserting two equal resistors R_e in series with the emitter leads of $Q1$ and $Q2$. This current-series feedback added to each transistor results in a smaller value of g_{md}. Reasonable values for R_e are 50-100 Ω, since for large values, A_d is reduced too much. The insertion of R_e also increases the input resistance.

15-5 OPERATIONAL AMPLIFIER DESIGN TECHNIQUES[1, 3, 8, 9]

The integrated OP AMP is fabricated in many different and complex IC configurations, depending upon the imagination and ingenuity of the designer. The system on the chip usually consists of four cascaded blocks. As indicated in Fig. 15-10, the first stage is a DIFF AMP, the second block of one or more stages supplies additional amplification, the third stage is a buffer, and the last block is the output driver. The buffer is usually an emitter follower whose high input resistance prevents loading down the high-gain stage. This buffer together with the circuitry associated with it and the output stage also act as a level shifter so that the output voltage is zero (approximately) for zero input. The final stage has low output resistance and supplies the desired large-signal output current or voltage. Power-supply voltages of ± 15 V are common, with rated undistorted outputs of ± 12 V and peak outputs of ± 14 V approximately.

We shall indicate a few of the design techniques commonly used in many of the excellent OP AMPs commercially available. The input stage is invariably a DIFF AMP because:

1. It has two inputs (inverting and noninverting).
2. It has a common-mode rejection ratio which is very high.
3. The signals are direct-coupled to the inputs.
4. Because of the symmetry and temperature-tracking of the small IC structure, the output drift voltage is extremely small.

In addition, the input bias currents can be made small enough to be neglected in many applications, as we now demonstrate.

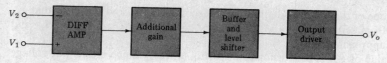

Figure 15-10 The usual OP AMP configuration.

Input Resistance

The DIFF AMP of Fig. 15-8 consist of $Q1$ and $Q2$, with $Q3$ used as a constant-current source to provide high common-mode rejection. The differential input resistance R_{id} to the total input signal $V_{s1} - V_{s2}$ is $2h_{ie}$, provided $R_s = 0$ and $h_{oe}R_c \leqslant 0.1$. This statement follows from the fact that, since $Q3$ acts as a constant current I_O, the emitters of $Q1$ and $Q2$ are floating. Hence the resistance between the two inputs 1 and 2 is $h_{ie1} + h_{ie2} = 2h_{ie}$. If input 2 is grounded, then input 1 is loaded by $2h_{ie}$.

If we neglect $r_{bb'}$ compared with $r_{b'e}$, then

$$h_{ie} \approx r_{b'e} = \frac{h_{fe}}{g_m} = \frac{h_{fe}V_T}{|I_C|} \tag{15-25}$$

where use is made of Eqs. (11-19), (11-21), and (11-25). If we assume that $I_{C1} = I_{C2} = 0.5$ mA and $h_{fe} = 100$ for $Q1$ and $Q2$, the differential input resistance is

$$R_{id} = 2h_{ie} = \frac{2 \times 100 \times 26}{0.5} \ \Omega = 10.4 \ \text{k}\Omega$$

If the biasing currents are reduced from 500 μA to 10 μA, then R_{id} is increased to 520 kΩ. If even this resistance is too small for the applied signal source, it can be increased by modifying the input circuit. For example, some IC OP AMPs have a Darlington pair (Sec. 11-16) in place of $Q1$ and another in place of $Q2$. Another modification is to add a matched discrete FET differential stage at the input, or preferably to fabricate a FET differential pair on the same chip with the rest of the OP AMP. Input resistances of the order of 10^{12} Ω are possible with such JFET inputs (Texas Instruments, TL080 family).

Widlar[8] has designed OP AMPs (National Semiconductor Corp. LM108, for example) using supergain transistors (Sec. 4-6) in the input stage (current gains of 5,000 can be obtained at 1-μA collector current). For this transistor we find from Eq. (15-25),

$$h_{ie} = \frac{5,000 \times 26 \times 10^{-3}}{10^{-6}} \ \Omega = 130 \ \text{M}\Omega$$

which is very high indeed for a bipolar transistor.

Current Sources

In Sec. 15-3 the constant-current source I_O obtained by using $Q3$ as connected in Fig. 15-8 is discussed in detail. We have indicated that such a constant-current source leads to a very high value of the common-mode rejection ratio. The

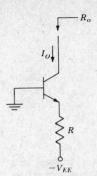

Figure 15-11   A constant-current source $[I_O \approx (V_{EE} - V_{BE})/R]$

current source I_O (in its essential form) is indicated in Fig. 15-11. Since constant current independent of voltage changes implies an extremely large resistance, we shall assume that $R_o \to \infty$.

The open-circuit voltage gain of an OP AMP should be as large as possible and, as indicated in Fig. 15-10, cascaded gain stages must be used. However, the larger the number of stages the greater is the phase shift introduced, and the more susceptable the amplifier becomes to breaking out into oscillations, Sec. 14-9. Some OP AMPs use three gain stages, but the more stable designs employ only two gain-producing amplifiers. The gain of a CE stage is $A_V = - h_{fe}R_c/h_{ie}$ $\approx - g_m R_c$ if the approximate models of Fig. 11-6 are used. To obtain high gain, a large value of collector resistance R_c must be used. However, this is impractical for two reasons: (1) A large diffused resistance requires a prohibitive amount of chip area, and (2) for a given quiescent collector current, the larger R_c the greater is the quiescent drop across this resistor and, hence, unrealistically large power supplies would be necessary. These difficulties are circumvented by using a current source such as shown in Fig. 15-11 (or current repeaters, discussed in Fig. 15-12 and Fig. 15-14) for the load in place of R_c. If we consider R_o to be infinite, then $R_c = R_o = \infty$ and the more exact model of Fig. 11-18 must be used. In Prob. 11-35 it is found $A_v \approx - g_m r_{ce}$, where r_{ce} is the internal transistor resistance between collector and emitter. At a quiescent current of 1.3 mA, $g_m = 50$ mA/V and $r_{ce} = 200$ kΩ [Eq. (11-56)], $A_v = 10,000$ for a single stage with an ideal-current load.

A Biasing Technique and Current Repeaters

It has already been emphasized that the self-bias circuit of Fig. 11-4 is not practical with IC fabrication because of the large resistances and capacitances which are required. Hence the biasing technique shown in Fig. 15-12 has been developed for monolithic circuits. In Fig. 15-12a the transistor $Q1$ is connected as a diode across the base-to-emitter junction of $Q2$ whose collector current is to be temperature-stabilized. From KCL at the junction of the two bases

$$I_1 = I_{C1} + I_{B1} + I_{B2} \tag{15-26}$$

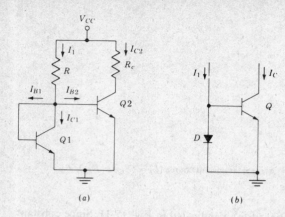

Figure 15-12 (*a*) A biasing technique for linear integrated circuits. (*b*) The circuit as a current repeater ($I_C \approx I_1$).

If transistors $Q1$ and $Q2$ are identical and have the same V_{BE}, their collector currents will be equal. Hence $I_{C2} = I_{C1} = I_C$. Even if the two transistors are not identical, experiments have shown that this biasing scheme gives collector-current matching between the biasing and operating transistors typically better than 5 percent and is stable over a wide temperature range. Since $I_B = I_C/\beta$ then from Eq. (15-26)

$$I_C = \frac{\beta I_1}{\beta + 2} \tag{15-27}$$

If $\beta \gg 2$, then $I_C \approx I_1 = (V_{CC} - V_{BE})/R$. Since the variation of V_{BE} with temperature is usually small compared with V_{CC}, then I_C is almost independent of T.

The representation in Fig. 15-12*b* is equivalent to that in Fig. 15-12*a*. Transistor $Q1$ with base tied to collector is a diode and is represented by D. The current I_1 (equal to the diode current, if the base current of Q can be neglected) is equal to the transistor collector current. This configuration is called a *current-repeater* or a *current-mirror* circuit.

If it is required that N transistors be biased with the same current, the circuit of Fig. 15-12 may be expanded as in Fig. 15-13*a*. For identical transistors $I_{C1} = I_{C2} = \cdots I_{CN} = I_C$. From KCL at the common-base node we obtain (with $I_C = \beta I_B$)

$$I_C = \frac{\beta I_1}{\beta + N + 1} \tag{15-28}$$

For $\beta \gg N + 1$, $I_C \approx I_1$. Since the N transistor bases are connected together and all their emitters are grounded, these N transistors may be fabricated as a single transistor with N collectors, as shown in Fig. 15-13*b*. The cross section of such a multiple-collector transistor is indicated in Fig. 9-43*a*. If the collector areas are not equal, then the collector currents will not be identical but will be proportional to the collector areas and to I_1 (Prob. 15-18).

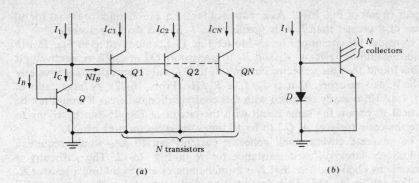

Figure 15-13 (a) Repeated-current sources. (All bases are tied together.) (b) An implementation using a diode and a multiple-collector transistor, obtained by *merging* the N transistors (Sec. 9-12).

Considerable error may result in assuming that $I_C = I_1$ in the current repeaters of Fig. 15-12 and Fig. 15-13 if β is small, as it would be for a lateral p-n-p transistor. This error can be greatly reduced by adding an emitter follower to supply the base currents, as indicated in Fig. 15-14a. In Prob. 15-19 we find that for identical transistors

$$I_2 = \frac{I_1(\beta)(\beta+1)}{(\beta)(\beta+1)+2} \tag{15-29}$$

and the ratio I_2/I_1 is much less dependent upon β than is the case with the

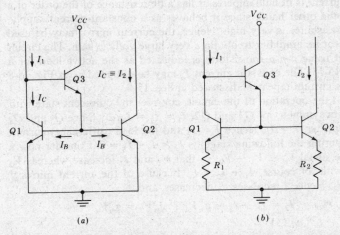

Figure 15-14 (a) An improved version of the current repeater ($I_2 \approx I_1$). (b) A current repeater with magnification ($I_2/I_1 \approx R_1/R_2$).

simpler circuit of Fig. 15-12. As a matter of fact, Eq. (15-29) indicates that for all values of β greater than 5, I_2 is smaller than I_1 by less than 6 percent.

Consider the circuit of Fig. 15-14a (or Fig. 15-12) modified as in Fig. 15-14b by adding emitter resistors R_1 and R_2 to transistors $Q1$ and $Q2$, respectively. It is now found that the collector currents I_1 and I_2 vary approximately inversely as the added resistors; that is $I_2/I_1 \approx R_1/R_2$ (Prob. 15-22). A large ratio of I_2/I_1 (say 10) is easily obtained with this configuration, whereas it would not be practical to obtain the same result with the circuit of Fig. 15-14a by trying to make the collector area of $Q2$ to be 10 times the collector area of $Q1$.

To generate a very small collector current, say 10 μA, would require a prohibitively large diffused resistance for R in Fig. 15-12. This difficulty is overcome by choosing R so that I_1 is in milliamperes and by adding a resistor R_2 into the emitter of $Q2$. This degeneration will cause I_{C2} to become a small fraction of I_{C1} so that the desired value of 10 μA is obtained. The same result may be obtained with the circuit of Fig. 15-14b by omitting the resistor in the emitter of $Q1$; that is, $R_1 = 0$. The required value of R_2 is found in Prob. 15-21 to be given by

$$R_2 = \frac{V_T}{I_{C2}} \ln \frac{I_{C1}}{I_{C2}} \tag{15-30}$$

15-6 ANALOG DESIGN TECHNIQUES (CONTINUED)[1, 3, 8, 9]

Active Loads

Since the quiescent voltage across a current repeater is a fraction of the supply voltage and the current is in milliamperes, it has a dc resistance of the order of a few kilohms. On the other hand, since it behaves as a constant-current supply, its dynamic (ac) resistance is very high. Hence, the current mirror may be used as an active load for an amplifier to obtain a very large voltage gain. The circuit of Fig. 15-12 (but using p-n-p transistors) is indicated as the active load for a DIFF AMP in Fig. 15-15. The constant current I_o may be obtained as in Fig. 15-8 or from one of the current repeaters discussed in Sec. 15-5.

To understand the operation of the circuit, consider the quiescent state with $V_1 = V_2 = 0$. From symmetry of $Q1$ and $Q2$, $I_1 = I_2 = I_o/2$.† Since Q_3 and Q_4 form a current repeater, $I = I_1$. Hence $I = I_2$ and the load (difference) current I_d (the current entering the following stage) is $I_d = I - I_2 = 0$. Consider now a positive difference signal $V_d \equiv V_1 - V_2$, so that V_1 and I_1 increase whereas V_2 and I_2 decrease with, of course, $I_1 + I_2 = I_o$. Because of the current mirror I remains equal to I_1. Hence I increases, I_2 decreases, and

$$I_d = I - I_2 = I_1 - I_2 = g_m V_1 - g_m V_2 = g_m V_d$$

---

† Base currents are neglected compared with collector currents in the circuits discussed in this section.

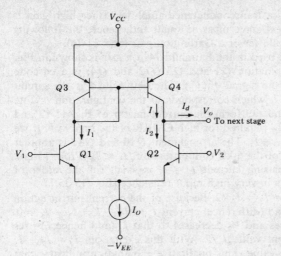

Figure 15-15 A DIFF AMP with an active load $Q3$-$Q4$.

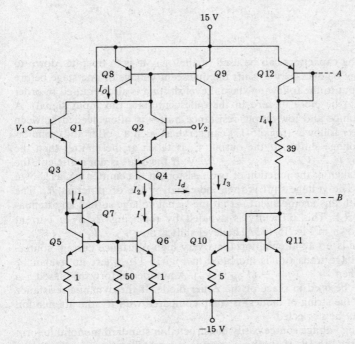

Figure 15-16 A low-current DIFF AMP (μA 741) with an active load $Q5$-$Q6$-$Q7$. All resistance values are in kilohms. (Lines A and B connect with the output section of this OP AMP shown in Fig. 15-20.)

so that Fig. 15-15 behaves as a transconductance amplifier. Since high gain is desired, the loading of the next stage must be small and, hence, the following circuit is usually an emitter follower or a Darlington pair (Sec. 11-16).

An alternative active load used in the Fairchild 741 OP AMP is shown in Fig. 15-16. The n-p-n-p-n-p combination $Q1$ and $Q3$ ($Q2$ and $Q4$) is a cascode arrangement whose input is the signal $V_1(V_2)$. The active load is the circuit consisting of $Q5$, $Q6$, and $Q7$, which corresponds to the configuration of Fig. 15-14b. The transistors $Q8$ and $Q9$ form the current mirror of Fig. 15-12 and provide the constant current I_O required for high CMRR of the DIFF AMP. If we neglect base currents, then $I_O \approx I_3$. The arrangement $Q10$ and $Q11$ is another current repeater, but because of the 5-kΩ emitter resistor, $I_3 \ll I_4$ (Prob. 15-26). Consequently the constant biasing current $I_o = I_3$ is small (of the order of microamperes) and this results in very high input resistance [Eq. (15-25)].

In the quiescent state $I_1 = I_2 = I_O/2$. Because of the current-mirror action I always equals I_1, and the load (difference) current $I_d = I_2 - I = I_2 - I_1 = 0$. Consider now that V_1 increases and V_2 decreases so that I_1 (and hence I) rises and I_2 falls from the quiescent value $I_o/2$. With this excitation, $I_d = I_2 - I_1$ changes from 0 to a negative value. This qualitative explanation indicates that the circuit functions as a transconductance amplifier because I_d is proportional to $V_1 - V_2$.

Level Shifting

Since no coupling capacitors can be used (if the OP AMP is to operate down to dc; $f = 0$), it may be necessary to shift the quiescent voltage of one stage before applying its output to the following stage. Level shifting is also required in order for the output to be close to zero in the quiescent state (no input signal). A high-input-resistance and low-output-resistance buffer is often needed between stages. An emitter follower (Fig. 15-17) can serve as such a buffer and, simultaneously, as a voltage shifter. If the output V_o is taken at the emitter then the change in level is $V_o - V_i = - V_{BE} \approx - 0.7$ V. If this shift is not sufficient, the output may be taken at the junction of two resistors in the emitter leg, as shown in Fig. 15-17a. The voltage shift is then increased by the drop across R_1. The disadvantage with this arrangement is that the signal voltage suffers an attenuation $R_2/(R_1 + R_2)$. This difficulty is avoided by replacing R_2 by a current source I_O, as shown in Fig. 15-17b. The level shift is now $V_o - V_i = - (V_{BE} + I_O R_1)$, and there is no ac attenuation for a very high resistance current source.

Another voltage translator is indicated in Fig. 15-17c where an avalanche diode is used. Then $V_o - V_i = - (V_{BE} + V_Z)$. A number of forward-biased p-n diodes may also be used in place of the Zener diode. If the dynamic resistance of the V_Z (or of the string of diodes) is small compared with R_2, the attenuation of the signal may be neglected.

An interesting voltage source easily constructed in standard monolithic form is indicated in Fig. 15-18. If the base current is negligible compared with the

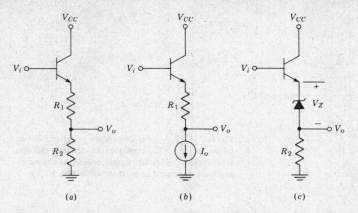

Figure 15-17 Level shifters using an emitter-follower buffer.

current in R_3 and in R_4, the circuit acts as a "V_{BE} multiplier" because

$$V = \frac{V_{BE}}{R_4}(R_4 + R_3) = V_{BE}\left(1 + \frac{R_3}{R_4}\right) \qquad (15\text{-}31)$$

This voltage source may be used in place of R_1 in Fig. 15-17a (or, equivalently, to replace V_Z in Fig. 15-17c) or in any branch of a network where a voltage translation is desired. Since Eq. (15-31) is valid independently of I, its dynamic resistance is ideally zero ($R_o \equiv \Delta V/\Delta I = 0/\Delta I = 0$). A better approximation for R_o is given in Prob. 15-24.

Output Stages

An output stage must be capable of supplying the load current and must have a low output resistance. A common configuration for the output stage of an OP

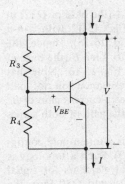

Figure 15-18 A voltage source V which is a multiple of V_{BE}.

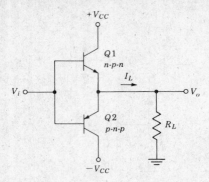

Figure 15-19 A complementary emitter-follower output stage. (This configuration is also called a class B push-pull power amplifier, and its characteristics are discussed in more detail in Sec. 18-5.)

AMP is the emitter follower with complementary transistors as shown in Fig. 15-19. If the input signal V_i goes positive, the n-p-n transistor $Q1$ acts as a source to supply current to the load R_L and the p-n-p transistor $Q2$ is cut off. On the other hand, if V_i becomes negative, $Q1$ is cut off and $Q2$ acts as a sink to remove current from the load, that is, to decrease I_L.

There is a fundamental difficulty with the circuit of Fig. 15-19, because the output remains zero until the input $|V_i|$ exceeds $V_{BE(\text{cutin})} \approx 0.5$ V. This phenomenon is called *crossover distortion*. This distortion can be eliminated by applying a bias voltage V, slightly larger than $2V_{BE(\text{cutin})} = 1$ V between the two bases, so that a small current flows in the transistors in the quiescent state. The output portion of the μA 741 OP AMP is shown in Fig. 15-20. The block marked V is the V_{BE} multiplier source of Fig. 15-18, and it supplies approximately 1.1 V potential difference between the bases of the complementary emitter followers $Q14$ and $Q15$. (Some OP AMPs use two p-n diodes in series in place of the V_{BE} multiplier.) The small (25 Ω) emitter resistors help to stabilize the quiescent base currents and to improve the small-signal gain linearity. The signal voltage is applied to this output configuration at point P, which is the ac junction of the bases of the complementary transistors.

Note that line A (the base of $Q13$) connects with line A (the base of $Q12$) in Fig. 15-16. Hence, $Q12$-$Q13$ form a current mirror (Fig. 15-12 with p-n-p transistors; equivalent to the $Q3$-$Q4$ combintation of Fig. 15-15). Hence, the current I_4 is constant and equal to the value I_4 in Fig. 15-16. Line B (the base of $Q16$) connects with line B (the output of the DIFF AMP of Fig. 15-16). Note that $Q16$ and $Q17$ form a Darlington pair whose high input resistance will not load the first stage appreciably. The Darlington CE amplifier has a current gain of β^2 and the second-stage load is r_{ce}. The resultant overall voltage gain of both stages of the 741 OP AMP has a typical value of 200,000.

An alternative output stage to that considered above is a totem-pole configuration similar to that used with digital circuits (Fig. 5-23). Such a topology is used in the Motorola MC1530 OP AMP, and the essential elements for signal operation are indicated in Fig. 15-21. Note again that $D4$ and $Q10$ form the

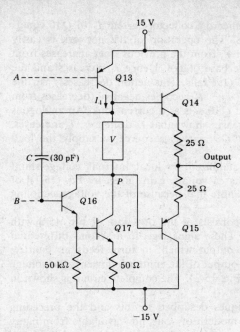

Figure 15-20 The essential configuration of the output stage ($Q14$ and $Q15$) of the μA 741, including also the second stage of amplification ($Q16$, $Q17$, $Q13$, and the constant current load I_4). The function of the compensating capacitor C is explained in Sec. 15-12. (Overload circuitry which limits the short-circuit output current to 25 mA is not indicated in this diagram.)

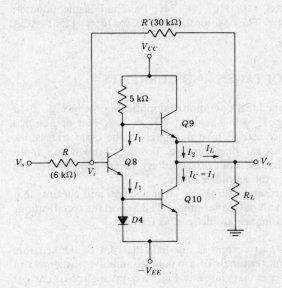

Figure 15-21 The totem-pole output stage of the MC 1530. Note that $I_L = I_2 - I_1$.

current repeater of Fig. 15-12, and hence the collector current I_C of $Q10$ equals the current I_1 in the emitter of $Q8$. The operation of the network is easily understood qualitatively. If the signal V_s from the previous stage increases from its quiescent value, so does V_i at the base of $Q8$. Hence I_1 increases and the collector voltage of $Q8$ (the base of $Q9$) falls. This action decreases I_2 in the emitter of $Q9$. The load current $I_L = I_2 - I_1$ becomes negative (decreases from its previous quiescent zero value) and $Q10$ acts as a current sink. An analogous argument for the case where V_s decreases shows that I_1 decreases, I_2 increases, and $I_L = I_2 - I_1$ is positive so that $Q9$ acts as a source to supply the load current.

We observe that this stage is stabilized by local internal voltage-shunt feedback. Hence, this output stage has a voltage gain [given by Eq. (15-1)] of $A_V = -R'/R = -\frac{30}{6} = -5$. Also note that because of the voltage sampling the output resistance is very small.

The complete OP AMP MC1530 consists of the DIFF AMP of Fig. 15-8 with double-ended outputs V_{o1} and V_{o2}. These signals go to a second differential amplifier stage with a single-ended output which, in turn, feeds an emitter-follower buffer and level shifter. The output of the emitter follower is the voltage V_s which excites the output stage of Fig. 15-21. The complete circuit is shown in Prob. 15-25.

The basic analog design techniques described in this and the preceding section are employed in most of the modern OP AMPs now available from many manufacturers (Appendix B-1). We mention a few of these: National Semiconductor, LM 101, 102, 107, 108, 110, 112, 118, 741; Fairchild Semiconductor, μA 702, 709, 710, 726, 741, 776; Motorola, MC1530, 1556, 1558; RCA CA3130, 3160.

15-7 OFFSET ERROR VOLTAGES AND CURRENTS[10]

In Sec. 15-1 we observe that the ideal operational amplifier shown in Fig. 15-1a is perfectly balanced, that is, $V_o = 0$ when $V_1 = V_2 = 0$. A real operational amplifier exhibits an unbalance caused by a mismatch of the input transistors. This mismatch results in unequal bias currents flowing through the input terminals, and also requires that an input offset voltage be applied between the two input terminals to balance the amplifier output.

In this section we are concerned with the dc error voltages and currents that can be measured at the input and output terminals. The important specifications used to describe OP AMP performance are defined as follows:

Input bias current The input bias current is one-half the sum of the separate currents entering the two input terminals of a balanced amplifier, as shown in Fig. 15-22. Since the input stage is that shown in Fig. 15-8, the input bias current is $I_B \equiv (I_{B1} + I_{B2})/2$ when $V_o = 0$.

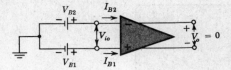

Figure 15-22 Input bias currents I_{B1} and I_{B2} and offset voltage V_{io}.

Input offset current The input offset current I_{io} is the difference between the separate currents entering the input terminals of a balanced amplifier. As shown in Fig. 15-22, we have $I_{io} \equiv I_{B1} - I_{B2}$ when $V_o = 0$.

Input offset current drift The input offset current drift $\Delta I_{io}/\Delta T$ is the ratio of the change of input offset current to the change of temperature.

Input offset voltage The input offset voltage V_{io} is that voltage which must be applied between the input terminals to balance the amplifier, as shown in Fig. 15-13a.

Input offset voltage drift The input offset voltage drift $\Delta V_{io}/\Delta T$ is the ratio of the change of input offset voltage to the change in temperature.

Output offset voltage The output offset voltage is the difference between the dc voltages present at the two output terminals (or at the output terminal and ground for an amplifier with one output) when the two input terminals are grounded.

Input common-mode range The common-mode input-signal range for which a differential amplifier remains linear.

Input differential range The maximum difference signal that can be applied safely to the OP AMP input terminals.

Output voltage range The maximum output swing that can be obtained without significant distortion (at a given load resistance).

Full-power bandwidth The maximum frequency at which a sinusoid whose size is the output voltage range is obtained.

Power supply rejection ratio The power supply rejection ratio (PSRR) is the ratio of the change in input offset voltage to the corresponding change in one power supply voltage, with all remaining power supply voltages held constant.

Slew rate The slew rate is the time rate of change of the closed-loop amplifier output voltage under large-signal conditions.

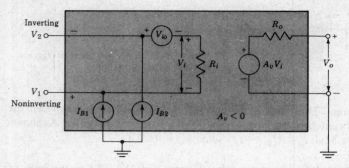

Figure 15-23 The model of an OP AMP taking offset voltage and bias currents into account.

Table 15-1 Typical parameters of monolithic operational amplifier at 25°C

Input offset voltage V_{io}	5 mV
Input offset current I_{io}	20 nA
Input bias current I_B	100 nA
Common-mode rejection ratio ρ	100 dB
PSRR	20 μV/V
I_{io} drift	0.1 nA/°C
V_{io} drift	5 μV/°C
Slew rate	1 V/μs
Unity gain frequency	1 MHz
Full-power bandwidth	50 kHZ
Open-loop difference gain A_d	100,000
Open-loop output resistance R_o	100 Ω
Open-loop input resistance R_i	1 MΩ
R_i for a JFET input stage	10^{12} Ω

The idealized model of Fig. 15-1 must be modified as in Fig. 15-23 to include the offset voltage and the bias currents. The various parameters of a typical monolithic operational amplifier are given in Table 15-1. The specification sheet for the LM741 OP AMP is given in Appendix B-7.

Example 15-2 (a) The inverting and the noninverting OP AMPs (with no applied signal voltages) have the same configuration (Fig. 15-24a). Assuming negligible input offset voltage, find the output dc voltage V_o due to the input bias current by assuming $I_{B1} = I_{B2} \equiv I_B$. Use the parameter values in Table 15-1. (b) How can the effect of the bias current be eliminated, so that $V_o = 0$? (c) With the circuit amended as in part (b), calculate V_o if $I_{B1} - I_{B2} = I_{io} \neq 0$. (d) If $I_{io} = 0$, what is V_o due to a nonzero value of V_{io}? (e) If $I_{io} \neq 0$ and $V_{io} \neq 0$, find V_o.

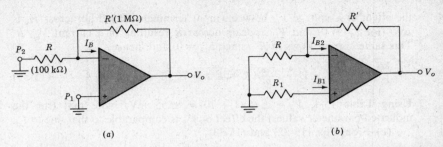

(a)   (b)

Figure 15-24 Illustrative example. Note that if P_1 in (a) is disconnected from ground and a signal voltage is applied to P_1, a noninverting OP AMP is under consideration. On the other hand, if P_2 is removed from ground and a signal is applied to P_2, this configuration represents an inverting OP AMP.

SOLUTION (a) As mentioned in Sec. 15-1, for very large values of A_o there exists a short circuit between the two input terminals. Hence there is no current in R. The current I_B must exist in R' and hence $V_o = I_B R'$.

Using the value $I_B = 100$ nA in Table 15-1,

$$V_o = 100 \times 10^{-9} \times 10^6 = 0.1 \text{ V} = 100 \text{ mV}$$

(b) Add a resistor R_1 between the noninverting terminal and ground, as indicated in Fig. 15-24b. If $V_o = 0$, then R and R' are in parallel and the voltage from the inverting terminal to ground is $-I_{B2}R_\parallel$. Since there is zero voltage between input terminals, $-I_{B2}R_\parallel$ must equal $-I_{B1}R_1$ or (for $I_{B1} = I_{B2}$)

$$R_1 = R_\parallel = \frac{RR'}{R + R'} = \frac{100 \times 1,000}{1,100} = 90.9 \text{ k}\Omega$$

If $I_{B1} \neq I_{B2}$, we must choose $I_{B1}R_1 = I_{B2}R_\parallel$.

(c) In Fig. 15-24b set $I_{B2} = I_{B1} - I_{io}$. In part (b) it is demonstrated that due to I_{B1} entering both the inverting and noninverting terminals, the output is $V_o = 0$. Applying superposition (Sec. C-2) to the two current sources I_{B1} and I_{io}, we may now set $I_{B1} = 0$ and find the effect of I_{io}. Since the drop across R_1 is $I_{B1}R_1 = 0$ and the two input terminals are at the same potential, the drop across R is 0 and the current R is also 0. Hence, I_{io} flows in R' and

$$V_o = -I_{io}R' \qquad (15\text{-}32)$$

For the numerical values given above,

$$V_o = -20 \times 10^{-9} \times 10^6 \text{ V} = -20 \text{ mV}$$

The sign of V_o is not significant because I_{io} may be positive or negative. Two alternative derivations of Eq. (15-32) are indicated in Prob. 15-27.

(d) If $I_{io} = 0$, then $I_{B1} = I_{B2}$ and, from part (b), $V_o = 0$. Hence, we may assume that the bias currents in Fig. 15-24b are zero and consider only the effect of a voltage V_{io} between input terminals. The drop across R_1 is zero (for $I_{B1} = 0$) and V_{io} appears across R resulting in a current V_{io}/R. This same current flows in R' (since $I_{B2} = 0$) and, hence,

$$V_o = \frac{V_{io}}{R}(R + R') = V_{io}\left(1 + \frac{R'}{R}\right) \tag{15-33}$$

Using Table 15-1, $V_o = \pm (5)(1 + 10) = \pm 55$ mV. Note that (for the indicated parameter values) the effect of V_{io} is comparable to that due to I_{io}.

(e) From Eqs. (15-32) and (15-33)

$$V_o = -I_{io}R' + V_{io}\left(1 + \frac{R'}{R}\right) \tag{15-34}$$

If all resistance values are divided by a factor F, the output due to V_{io} is not altered, whereas the component of V_o caused by I_{io} is divided by F. The inverting and also the noninverting gains depend only upon resistance ratios and, hence, are independent of the factor F.

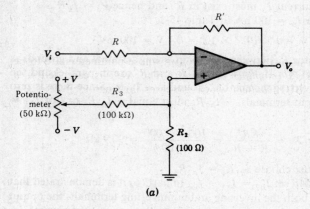

(a)

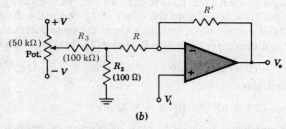

(b)

Figure 15-25 Universal output offset-voltage balancing circuits for (a) inverting and (b) noninverting operational amplifiers. These OP AMPs may include a resistor $R_1 = R \| R'$ in the positive input terminal (Fig. 15-24b) in order to minimize the output offset due to the bias currents. (See Fig. 16-7 for balancing of a DIFF AMP.)

Universal Balancing Techniques

When we use an operational amplifier, it is often necessary to balance the offset voltage. This means that we must apply a small dc voltage in the input so as to cause the dc output voltage to become zero. The techniques shown here allow offset-voltage balancing without regard to the internal circuitry of the amplifier. The circuit shown in Fig. 15-25a supplies a small voltage effectively in series with the noninverting input terminal in the range $\pm V[R_2/(R_3 + R_2)] = \pm 15$ mV if ± 15-V supplies are used and $R_3 = 100$ kΩ, $R_2 = 100$ Ω. This circuit is useful for balancing inverting amplifiers even when the feedback element R' is a capacitor or a nonlinear element. If the operational amplifier is used as a noninverting amplifier, the circuit of Fig. 15-25 is used for balancing the offset voltage.

15-8 MEASUREMENT OF OPERATIONAL AMPLIFIER PARAMETERS[11]

In this section we describe practical methods of measuring some of the important parameters of operational amplifiers. Specifically, we examine (1) input offset voltage V_{io}, (2) input bias current I_B and input offset current I_{io}, (3) open-loop voltage gain A_V, (4) common-mode rejection ratio, and (5) slew rate. In the circuits discussed in this section the OP AMP whose parameters are to be determined is labeled AUT (*amplifier under test*). The AUT is cascaded with another OP AMP labeled BUF (*buffer*) which increases the open-loop gain and also allows the output voltage of the AUT to be adjusted to any desired value. The input offset voltage of the buffer in Fig. 15-26 is balanced out by means of the arrangement of Fig. 15-25 applied to the inverting terminal. Since the BUF is within the feedback loop, the potential difference between its input terminals is zero. By neglecting the bias current of the BUF, it follows that $V_o = - V'$ since $V_B = 0$. Hence, the output of the AUT is always equal to the negative of V', which can be set at any desired value from an external voltage supply.

The system in Fig. 15-26 may oscillate if not properly compensated (Sec. 15-10). A capacitor across R' will normally stabilize the loop (Sec. 15-13).

Input Offset Voltage V_{io}

For this measurement set $V' = 0$ so that $V_o = 0$. Both switches S_1 and S_2 are closed. From the circuit model in Fig. 15-23, if $V_o = 0$, then $V_i = 0$ and V_{io} appears between the inverting and noninverting terminals. In other words, V_{io} of the AUT is across R and the corresponding current V_{io}/R (which is much larger than the bias current) also passes through the feedback resistor R'. Hence,

$$V = \frac{V_{io}}{R} (R + R') = 1{,}001 V_{io} \approx 10^3 V_{io} \equiv V_3 \qquad (15\text{-}35)$$

From the meter reading V_3 in volts we obtain V_{io} in millivolts. Note that V_{io} is

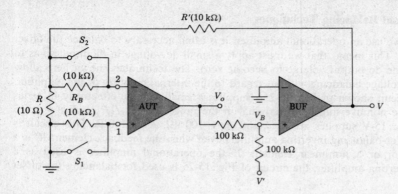

Figure 15-26 System for measuring V_{io}, I_B, and $A_v = A_d$.

measured with the output of the AUT set at zero, as required by the definition of the *input offset voltage*.

The power supply rejection ratio is obtained by repeating the V_{io} measurement for two values of the supply voltage V_{CC}. Then the PSRR is calculated from $\Delta V_{io}/\Delta V_{CC}$, where ΔV_{io} (ΔV_{CC}) represents the difference in the two input offset (power supply) voltages.

Input Bias Current

Switch S_1 in Fig. 15-26 is opened, S_2 is closed, and $V' = 0$ for this measurement. The voltage across R is now, from Fig. 15-23, $V_{io} - R_B I_{B1}$ and

$$V = \frac{R + R'}{R} (V_{io} - R_B I_{B1}) \approx 10^3 (V_{io} - 10^4 I_{B1}) \equiv V_4 \quad (15\text{-}36)$$

From Eqs. (15-35) and (15-36), it follows that

$$-I_{B1} = (V_4 - V_3)10^{-7} \text{ A} = 100(V_4 - V_3) \text{ nA} \quad (15\text{-}37)$$

If S_2 is left open but S_1 is closed and $V' = 0$, then I_{B2} is obtained by proceeding as above and $+I_{B2}$ is given by Eq. (15-37). The bias current is given by $I_B = \frac{1}{2}(I_{B1} + I_{B2})$ and the offset current by $I_{io} = I_{B1} - I_{B2}$.

Open-Loop Differential Voltage Gain $A_v = A_d$

The open-loop gain is defined as the ratio of the output voltage to the differential voltage input signal. A direct measurement of A_v based upon this definition is extremely difficult. It is essential that the effect of the input offset voltage and current in the *open-loop* amplifier be canceled almost exactly, since otherwise the high amplification of the unbalanced input will result in output saturation of the amplifier (whereas it should be operating in its linear region). If an output of, say, 10 V is desired then, with $A_d = 100,000$, an accurately

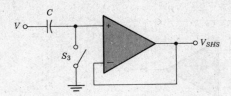

Figure 15-27 A sample-hold-subtract system.

calibrated input signal of 0.1 mV is required. With such small signals, noise voltages may be troublesome. All of these difficulties are circumvented by using the AUT in the closed-loop arrangement of Fig. 15-26.

Switches S_1 and S_2 are closed and V' is set to the recommended output voltage, say, -10 V. Then $V_o = -V' = +10$ V. Since the output resistance of the AUT is very small compared with its load of 100 kΩ, then from Fig. 15-23, $A_v V_i = V_o$. The voltage across the resistor R between the input terminals of the AUT is $V_{io} + V_i$. Hence

$$V = \frac{R + R'}{R}(V_{io} + V_i) \approx 10^3 \left(V_{io} + \frac{V_o}{A_v} \right) \equiv V_5 \qquad (15\text{-}38)$$

Subtracting Eq. (15-35) from (15-38) yields (for $V_o = 10$ V)

$$A_v = \frac{10^3 V_o}{V_5 - V_3} = \frac{10^4}{V_5 - V_3} \qquad (15\text{-}39)$$

If V' is adjusted to $+10$ V and the above procedure is repeated, then A_v for an output of $V_o = -10$ V is obtained. If the voltage gain A_V under load is desired, it is only necessary to place the required load resistor R_L from V_o to ground while carrying out the above measurements.

For $A_v = 100,000$, $V_5 - V_3 = 0.1$ V, and very poor accuracy is obtained since two large and almost equal numbers must be subtracted. This difficulty is overcome as follows. The subtractions required in Eq. (15-39), and also in Eq. (15-37), may be performed electronically by use of the circuit of Fig. 15-27. The OP AMP is a noninverting unity-gain follower (Fig. 15-4 with $Z = \infty$ and $Z' = 0$), with very high input resistance. The capacitor C will store the measured voltage V_3. The input to this circuit [called a *sample-hold-subtract* (SHS) configuration] is the output V of the BUF of Fig. 15-26. The experimental procedure is as follows: S_3, S_2, and S_1 are closed and $V' = 0$, so that $V = V_3$ is stored upon the high-quality capacitor C. Now S_3 is opened and the procedure outlined above for measuring A_v (or I_{B1}) is followed. Then V is V_5 (or V_4) and $V_{\text{SHS}} = V_5 - V_3$ (or $V_4 - V_3$).

Common-Mode Rejection Ratio

This ratio is defined in Eq. (15-11) as CMRR $\equiv |A_d/A_c| \equiv \rho$, where A_d is the differential gain and A_c is the common-mode gain. The circuit for measuring the CMRR is that of Fig. 15-26 with switches S_1 and S_2 closed, $V' = 0$, and a signal

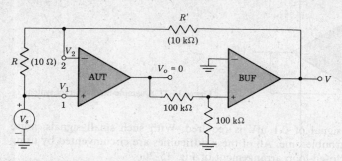

Figure 15-28 Measurement of the common-mode rejection ratio $\rho = |A_d/A_c|$ of AUT.

voltage V_s inserted between the noninverting terminal and ground. These modifications result in the network of Fig. 15-28. From Eq. (15-9) applied to the AUT, with $V_o = 0$,

$$V_o = A_d V_d + A_c V_c = 0 \tag{15-40}$$

To obtain V_d and V_c we first obtain V_1 and V_2 in Fig. 15-28. Clearly $V_1 = V_s$. Using superposition

$$V_2 = V_s \frac{R'}{R + R'} + V \frac{R}{R + R'} \approx V_s + V \frac{R}{R'} \tag{15-41}$$

since $R' \gg R$. The difference voltage V_d is the voltage $-V_i$ across R_i. If we take the input offset voltage into account (Fig. 15-23) and if use is made of Eqs. (15-41) and (15-35) with $R' \gg R$,

$$V_d = V_1 - V_2 - V_{io} = -\frac{VR}{R'} - V_{io} = -\frac{R}{R'}(V + V_3) \tag{15-42}$$

and

$$V_c = \tfrac{1}{2}(V_1 + V_2) = V_s + \frac{VR}{2R'} \tag{15-43}$$

Substituting Eqs. (15-42) and (15-43) into Eq. (15-40) yields

$$-A_d \frac{R}{R'}(V + V_3) + A_c\left(V_s + \frac{VR}{2R'}\right) = 0 \tag{15-44}$$

Since $A_d \gg A_c$, the fourth term in this equation may be neglected compared with the first term. Hence, if the measured value of V is designated by V_6,

$$\rho \frac{R}{R'}(V_6 + V_3) = V_s \tag{15-45}$$

For $\rho = 10^5$, $R/R' = 10^{-3}$, and $V_s = 10$ V we find that $V_6 + V_3 = 0.1$ V. For $V_{io} = 5$ mV, $V_1 = 5$ V. Hence $V_6 = -4.9$ V, and very poor accuracy is obtained from this measurement since two large and almost equal voltages $|V_6|$ and $|V_3|$ must be subtracted. This difficulty is overcome by changing the input to a new value V_s' and measuring the new value of V, called V_6'. Then, corresponding to Eq. (15-45), we have

$$\rho \frac{R}{R'} (V_6' + V_3) = V_s' \tag{15-46}$$

Subtracting Eq. (15-45) from Eq. (15-46) eliminates V_3 and yields

$$\rho = \frac{R'}{R} \frac{V_s' - V_s}{V_6' - V_6} \tag{15-47}$$

If $V_s' = 5$ V, $V_s = -5$ V, $\rho = 10^5$, and $R'/R = 10^3$ we obtain $V_6' - V_6 = 0.1$ V. However, this subtraction can now be done electronically with the sample-hold-subtract circuit of Fig. 15-27. The switch S_3 is closed for the V_6 measurement and opened for the V_6' measurement.

Slew Rate[12]

The maximum rate of change of the output voltage when supplying the rated output is defined in Sec. 15-7 as the slewing rate. This rate dV_o/dt can be measured using the noninverting circuit of Fig. 15-4, with $Z = \infty$ and $Z' = 0$, since this usually represents the worst case. If the amplifier has a single-ended input, then the circuit of Fig. 15-2a is used, with $Z = R = 1$ K and $Z' = R' = 10$ kΩ. The input V_s is a high-frequency square wave, and the slopes with respect to time of the leading and trailing edges of the output signal are measured. It is common to specify the slower of the two rates as the slew rate of the device.

15-9 FREQUENCY RESPONSE OF OPERATIONAL AMPLIFIERS

The typical OP AMP consists of two or three gain stages, a buffer-level-shifter stage and an output stage as indicated in Fig. 15-10. The general method of obtaining the high-frequency response of such a chain of interacting stages is presented in Secs. 13-12 and 13-13. The computational complexity is so great that computer-aided analysis is required. The response may also be obtained by laboratory measurements.

The open-loop gain of the OP AMP has a transfer function with several poles and with zeros at much higher frequencies than the poles. Experimentally, the poles can be found from the amplitude response curve (the plot of the magnitude of gain in decibels versus $\log f$). Tangent to this curve are drawn straight lines whose slopes are 0, -20 dB per decade, -40 dB per decade, ..., as indicated in Figs. 14-19 and 14-20. The pole frequencies $f_1, f_2, \ldots$ are then obtained from the *corner* frequencies, the values of f at which adjacent lines intersect.

The poles and zeros of $A(jf)$ are sometimes specified by the manufacturer in data sheets provided with commercial operational amplifiers. In Fig. 15-29 we show the open-loop gain and phase response of a typical OP AMP (μA702A), using the straight-line Bode approximation. We see that the transfer function has

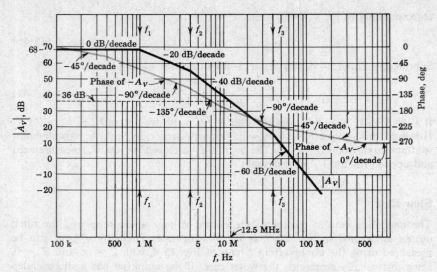

Figure 15-29 Open-loop gain and phase-shift characteristics of the μA702A.

three poles, one at 1 MHz, a second at 4 MHz, and a third at 40 MHz. For the MC1530 the manufacturer gives the first three poles at 1, 6, and 22 MHz.

Stability of an OP AMP

For the inverting OP AMP of Fig. 15-2a, with $Z = R$, $Z' = R'$, and $R_i = \infty$, we obtain from Eq. (15-2)

$$A_{Vf} = \frac{V_o}{V_s} = \frac{R'}{R + R'} \frac{A_V}{1 - RA_V/(R + R')} \tag{15-48}$$

This equation may also be obtained by using the feedback concepts of Chap. 12, as we now demonstrate.

The topology corresponds to voltage-shunt feedback, and it is found (Prob. 15-33) that

$$\beta = -\frac{1}{R'} \qquad R_M = \frac{A_V R R'}{R + R'} \qquad A_{Vf} = \frac{R_{Mf}}{R} \tag{15-49}$$

Using the feedback formula [Eq. (12-4)]

$$R_{Mf} = \frac{R_M}{1 + \beta R_m}$$

and Eqs. (15-49), the expression for A_{Vf} in Eq. (15-48) is obtained. From Eqs. (15-49)

$$\beta R_M = -\frac{R A_V}{R + R'} \tag{15-50}$$

The stability of an amplifier (for example, the gain margin or the frequency of oscillation) is obtained from Bode plots of the magnitude and phase of $\beta A = \beta R_M$ versus frequency (Fig. 14-17). Since βR_M is proportional to A_V, it is customary to plot $|A_V|$ and the phase of $-A_V$ versus frequency as the Bode characteristics from which to determine stability (even though R_M is the transfer gain stabilized in a voltage-shunt feedback configuration).

The noninverting amplifier of Fig. 15-4 corresponds to *voltage-series feed-back* with $A = -A_V (A_V < 1)$ and $\beta = R/(R + R')$. Hence,

$$\beta A = -\frac{R A_V}{R + R'} \tag{15-51}$$

which is the same expression for the loop gain as in Eq. (15-50) for the inverting OP AMP. For the noninverting amplifier, we find from Eq. (12-4)

$$A_{Vf} = \frac{V_o}{V_s} = \frac{-A_V}{1 - [R/(R + R')]A_V} \tag{15-52}$$

It is important to point out that for negative feedback the gain A_V represents a negative real number at low frequencies. Hence we observe that if the product $[R/(R + R')]|A_V|$ becomes unity when the phase shift of $-A_V$ reaches 180°, the amplifier will oscillate. From the open-loop gain and phase shift of the μA702A shown in Fig. 15-29, we find that at the frequency of $f = 12.5$ MHz, where the phase shift of $-A_V$ equals 180°, the magnitude of A_V is 36 dB. From Eq. (12-5) the amount of negative feedback introduced into the amplifier is $20 \log R_M - 20 \log R_{Mf}$. Since from Eq. (15-49) R_M is proportional to A_V, then if the amount (in decibels) of feedback added exceeds $68 - 36 = 32$ dB, the amplifier will oscillate.

We now calculate the minimum closed-loop gain A_{Vf} for which the circuit becomes unstable.

$$20 \log \frac{R}{R + R'} + 20 \log |A_V| = 20 \log 1 = 0 \tag{15-53}$$

$$20 \log \frac{R}{R + R'} = -36 \quad \text{and} \quad \frac{R + R'}{R} = 63$$

Since the closed-loop gain of the inverting amplifier is $A_{Vf} \approx -R'/R$, we see that, if the low-frequency gain is less than 62 in magnitude, the circuit will oscillate. Since the closed-loop gain of a noninverting OP AMP is $A_{Vf} = 1 + R'/R$, then if feedback is added to the μA702A OP AMP to reduce its gain below 63, it will break out into oscillations. This amplifier is clearly unsuitable as a voltage follower (unless compensation is added, Sec. 15-10).

For 45° phase margin we find, from Fig. 15-29, $20 \log |A_V| = 50$ dB, and from Eq. (15-53) we obtain $20 \log[R/(R + R')] = -50$ dB, or $R'/R = 316$. We see that the low-frequency closed-loop voltage gain of the inverting feedback amplifier cannot be less than 316 in magnitude for a phase margin of at least 45°.

15-10 COMPENSATION[13]

From Eq. (14-57) we have that the maximum magnitude of loop gain for the μA702A OP AMP (Fig. 15-29) for a 45° phase margin is $|A\beta| = |R_M\beta| = 68 - 50 = 18$ dB. If the desensitivity afforded by this loop gain is inadequate for the desired performance of the OP AMP, then a capacitor or RC combination is added (called *compensation*) so as to modify the gain and phase characteristics. Compensation is used so as to allow an increase in the amount of feedback (loop gain) while maintaining the same phase margin. In other words, compensation reshapes the magnitude and phase plots of βA so that $|\beta A| = |RA_V/(R + R')| < 1$ when the phase of $-A_V$ is $-180°$. For a 45° phase margin $|\beta A|$ will just reach unity (0 dB) when the phase of βA is $-135°$. There are three general methods of accomplishing this goal.

1. *Dominant-pole compensation.* This method inserts an extra pole into the transfer function at a lower frequency than the existing poles. Such a circuit introduces a *phase lag* into the amplifier.
2. *Pole-zero (lag) compensation.* This technique adds both a pole f_p and a zero f_z to the transfer function. The zero is chosen so as to cancel the lowest pole $f_z = f_1 > f_p$. The compensation leaves the number of poles unchanged, but the first pole f_1 has effectively shifted to a higher frequency f_p where it introduces a lag in phase.
3. *Lead compensation.* The amplifier or feedback network is modified so that a zero is added to the transfer function, thereby increasing the phase (without appreciably changing the magnitude of the gain). This added zero improves the phase margin.

15-11 DOMINANT-POLE COMPENSATION

The amplifier is modified by adding a dominant pole, that is, a pole much smaller in magnitude than all other poles in the forward transfer function. Consequently, the loop gain drops to 0 dB with a slope of 20 dB per decade at a frequency where the poles of A_V contribute negligible phase shift.

Suppose we modify A_V of Fig. 15-29 by adding a dominant pole so that the new forward transfer function becomes

$$A'_V = \frac{1}{1 + j(f/f_d)} A_V \qquad (15\text{-}54)$$

where $f_d \ll 1$ MHz. This can be accomplished by a simple RC network placed in the forward amplifier, as shown in Fig. 15-30, or by connecting a capacitance C from a suitable high-resistance point to ground. The capacitance is chosen such that it creates a dominant pole in A'_V low enough in frequency so that the magnitude of the loop gain becomes less than unity at a frequency where the amplifier introduces negligible phase shift. Since the capacitor adds a phase shift smaller than 90°, the circuit will be stable.

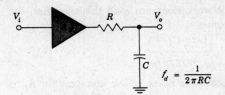

$$f_d = \frac{1}{2\pi RC}$$

Figure 15-30 Dominant-pole compensation.

Usually, f_d is selected so that A'_V passes through 0 dB at the first pole f_1 of the uncompensated A_V. At the frequency f_1 the phase due to this pole is $-45°$ and the phase of $f_d \ll f_1$ is $-90°$. Since the total phase is $-135°$ when $|A'_V| = 1$, the phase margin is $45°$ and occurs at the first uncompensated pole f_1. The frequency f_d is found graphically by having A'_V pass through 0 dB at the frequency f_1 with a slope of -20 dB per decade. The size of the capacitor C is determined from f_d since $C = 1/2\pi Rf_d$.

Dominant-pole compensation reduces the open-loop bandwidth drastically, and the other methods of compensation described in the following sections must be used if we desire better bandwidth. Of course, narrowbanding due to dominant-pole compensation improves the noise immunity of the system, since noise-frequency components outside of the bandwidth are eliminated.

In some applications, as in a power supply design, adding a large capacitance from the output to ground to obtain a dominant pole is extremely advantageous since then the output is not affected by load transients.

15-12 POLE-ZERO COMPENSATION

In this type of compensation the forward transfer function A_V is altered by adding both a pole and a zero, with the zero at a higher frequency than the pole. Figure 15-31 shows the forward amplifier with the pole-zero network cascaded with it. The transfer function of the compensating network is found to be

$$\frac{V_3}{V_2} = \frac{1 + j(f/f_z)}{1 + j(f/f_p)} \tag{15-55}$$

where

$$f_z = \frac{1}{2\pi R_c C_c} \qquad f_p = \frac{1}{2\pi (R_y + R_c)C_c} \tag{15-56}$$

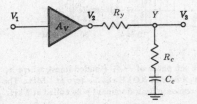

Figure 15-31 Pole-zero compensation.

If we assume that the compensation network does not load the amplifier, the modified forward transfer function becomes

$$A'_V = \frac{V_3}{V_1} = A_V \frac{1 + j(f/f_z)}{1 + j(f/f_p)} \tag{15-57}$$

Let us assume that the amplifier has three poles whose voltage gain A_V is given by Eq. (14-55) and whose Bode plots are shown in Fig. 14-20. Let us also select the zero frequency f_z to be equal to 1 MHz, which is the lowest pole of A_V, so as to have pole-zero cancellation. The forward transfer function now becomes (with f expressed in megahertz)

$$A'_V = \frac{-10^3}{\left(1 + j\dfrac{f}{f_p}\right)\left(1 + j\dfrac{f}{10}\right)\left(1 + j\dfrac{f}{50}\right)} \tag{15-58}$$

If we set the pole at $f_p = 0.2$ MHz, the Bode plots of the magnitude of A'_V and phase of $-A'_V$ are drawn in Fig. 15-32. We see that the phase margin is 45° for $A'_V = 29$ dB and, hence, the maximum loop gain for 45° phase margin is $60 - 29 = 31$ dB. This value should be compared with the uncompensated value of only 18 dB obtained in Sec. 14-10 from Fig. 14-20.

In Fig. 15-32 we have also plotted $|A'_V|$ for dominant-pole compensation. The dominant pole f_d is chosen so that $\log |A'_V| = 0$ (or $\log |A'_V| = 1$) at the first pole $f_1 = 1$ MHz of the uncompensated amplifier. Since the rolloff is 20 dB/decade then in two decades (from 10 kHz to 1 MHz) $|A'_V|$ changes by 40 dB, as indicated in Fig. 15-32. If the dominant-pole-compensation line were extended to the left another decade, it would rise another 20 dB to $|A'_{Vo}| = 60$ dB. Hence, f_d is one decade lower than 10 kHz, or $f_d = 1$ kHz. Notice how pole-zero cancellation has resulted in an improvement of bandwidth. For

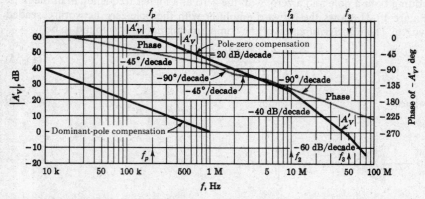

Figure 15-32 The open-loop gain $|A'_V|$ (solid lines) and the phase of $-A'_V$ (shaded lines), where A'_V [Eq. (15-58)] represents A_V of Eq. (14-55) augmented by a pole at 200 kHz and a zero at 1 MHz. The solid line between 10 kHz and 1 MHz is the amplitude response for a dominant pole added at 1 kHz.

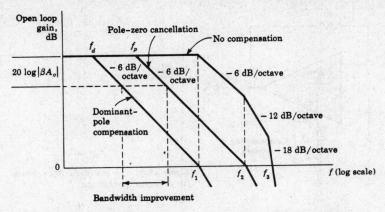

Figure 15-33 Comparison of dominant-pole and pole-zero compensation techniques.

example, for a loop-gain of 20 dB ($|A_V'| = 40$ dB), the closed-loop amplifier with pole-zero compensation has a bandwidth of 2 MHz, whereas the amplifier with dominant-pole compensation has a bandpass of only 10 kHz. The corresponding bandwidth of the uncompensated OP AMP, as obtained from Fig. 14-20, is 10 MHz.

Consider again the Bode plots in Fig. 15-32 for the pole-zero-compensation configuration. This OP AMP *cannot* be used as a unity-gain amplifier, because to do so would require 60 dB of feedback so that $|A_V'| = 1$. However, the amplifier becomes unstable and oscillates (at $f \approx 25$ MHz) if the loop gain exceeds 50 dB. If f_p is chosen so $|A_V'|$ passes through the second pole $f_2 = 10$ MHz and $A_V' = 0$, then the OP AMP used as a voltage follower is possible, if $f_3 \gg f_2$. Why?

A comparison of the dominant-pole and pole-zero compensation techniques is given in Fig. 15-33. The dominant pole is selected so that the compensated forward transfer function goes through 0 dB at the first pole f_1 of the uncompensated response. For the pole-zero compensation the zero is chosen equal to f_1, while the pole is selected so that the compensated forward transfer function goes through 0 dB at the second pole of the uncompensated transfer function. The bandwidth improvement is also shown in this figure.

The $R_y R_c C_c$ combination for pole-zero compensation need not be placed at the output of the amplifier. As indicated in Fig. 15-34, the series combination of R_c and C_c may be connected to any node Y of the amplifier, where R_y is the resistance seen by the signal at the point Y. Two other methods of obtaining pole-zero compensation will now be discussed.

Modification of Open-Loop Input Impedance

It is found that the amplifier slew rate (or the maximum time rate of the output swing at high frequencies) increases if the RC network is connected at a point

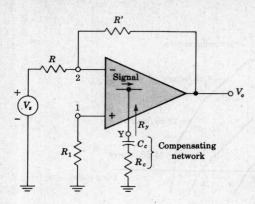

Figure 15-34 The use of an RC network for pole-zero cancellation.

where the signal swing is small, and thus only a small current is required to charge the compensating capacitor. Since the input terminals of an OP AMP are virtually short-circuited together, it is advantageous to place the compensating $R_c C_c$ network between terminals 1 and 2 instead of from point Y to ground in Fig. 15-34. Thus the network is connected in parallel with the operational amplifier input impedance Z_i. Let us assume that $R' > R_o$ and $|Z_c| \ll Z_i$, where $Z_c = R_c + 1/j\omega C_c$. The compensated gain without feedback is found in Prob. 15-41 to be given by,

$$A'_V(\omega) = A_V(\omega) \frac{1 + j\omega R_c C_c}{1 + j\omega C_c (R_c + R\|R' + R_1)} \frac{R'}{R + R'} \qquad (15\text{-}59)$$

This equation is of the same form as Eq. (15-57), and thus this procedure allows compensation by means of pole-zero cancellation.

It should be emphasized again that for the inverting amplifier it is the transresistance gain R_M which is stabilized. However, since R_M is proportional to the voltage gain [Eq. (15-49)] then Bode plots of A'_V are used to determine stability for any type of compensation configuration.

Miller-Effect Compensation

Adding a feedback capacitance around an intermediate stage (say, stage 2) of the operational amplifier is another method of providing phase compensation by means of pole-zero cancelation. Due to the Miller effect (Sec. C-4), a response zero is developed in the first stage coincident with the pole of stage 2, as we now demonstrate. Consider a differential-amplifier input stage followed by a second stage across which is the compensating capacitance C_f, as indicated in Fig. 15-35a. *In the absence of C_f, we assume that the input stage has a dominant pole f_1*, and hence its voltage gain is approximated by

$$A_{V1} = \frac{V_2}{V_i} = \frac{A_{Vo1}}{1 + j(f/f_1)} \qquad (15\text{-}60)$$

In Fig. 15-35b is shown the small-signal model of the input DIFF AMP which results in the transfer function of Eq. (15-60). In this circuit R_L is the effective load resistance and C_L the effective load capacitance from collector to ground of the input stage, g_{md} is the differential transconductance, and C_M is the Miller capacitance (which is zero for $C_f = 0$). Solving for V_2/V_i from this circuit, we obtain (with $C_M = 0$) Eq. (15-60), with

$$A_{Vo1} = g_{md}R_L \quad \text{and} \quad f_1 = \frac{1}{2\pi R_L C_L} \tag{15-61}$$

We assume that the second stage also has a dominant pole f_2 (with C_f in place), and hence its voltage gain is approximated by

$$A_{V2} = \frac{V_3}{V_2} = \frac{A_{Vo2}}{1 + j(f/f_2)} \tag{15-62}$$

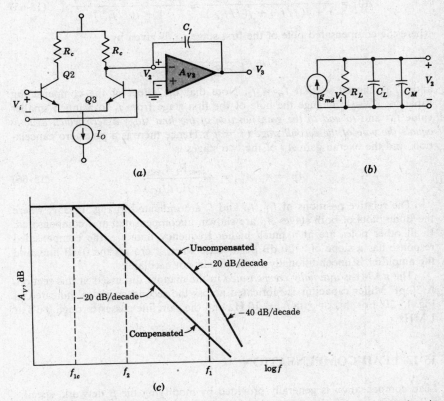

(a)

(b)

(c)

Figure 15-35 (a) Differential input stage with Miller-effect compensation. (b) Small-signal equivalent circuit of first stage. (c) Overall response of the two stages.

From Fig. 15-35b, with $C_M \neq 0$, we obtain the voltage gain A_{V1} of the first stage after the compensating capacitor C_f has been added.

$$A'_{V1} = \frac{g_{md}R_L}{1 + j2\pi fR_L(C_L + C_M)} \tag{15-63}$$

For negative feedback we assume $A_{Vo2} < 0$. The Miller effect indicates that $C_M = (1 - A_{V2})C_f$, and using Eq. (15-62), we find that

$$C_L + C_M \approx \frac{-A_{Vo2}C_f}{1 + jf/f_2} \tag{15-64}$$

This result is obtained in Prob. 15-43 if use is made of the inequality $|A_{Vo2}C_f| \gg C_L + C_f$.

From Eqs. (15-63) and (15-64),

$$A'_{V1} = \frac{g_{md}R_L[1 + j(f/f_2)]}{1 + j(f/f_2) + j(f/f_{1C})} \approx \frac{g_{md}R_L[1 + j(f/f_2)]}{1 + j(f/f_{1C})} \tag{15-65}$$

where the compensated pole of the first stage f_{1C} is given by

$$f_{1C} \equiv \frac{-1}{2\pi R_L A_{Vo2}C_f}$$

and C_f is chosen so that $f_{1C} \ll f_2$. Note that the effect of the compensating capacitor C_f is to change the pole of the first stage from f_1 to a much smaller value f_{1C} and *to add to the gain function of the first stage a zero which exactly equals the pole of the second stage* ($f_z = f_2$). Hence there is a pole-zero cancelation, and the overall gain A_V of the two stages is

$$A_V = A'_{V1}A_{V2} = \frac{g_{md}R_L A_{Vo2}}{1 + j(f/f_{1C})} \tag{15-66}$$

The relative positions of f_{1C}, f_1, and f_2 are indicated in Fig. 15-35c, where the Bode plots of both stages A_V are shown, uncompensated and compensated. If all other poles are at a much higher frequency than f_{1c}, the compensated response has a slope of -20 dB per decade when it crosses the 0-dB line, and the amplifier is unconditionally stable (it can not oscillate).

The μA741 is *internally compensated* in the manner discussed in this section. A 30-pF Miller capacitor is fabricated across the second stage, as indicated in Fig. 15-20. For this OP AMP $f_{1c} \approx 10$ Hz and the gain line passes through 0 dB at 1 MHz.

15-13 LEAD COMPENSATION

Lead compensation is generally provided by modifying the β network, specifically, by shunting resistor R' with a capacitance C', as shown in Fig. 15-36, so that the new loop gain will have an added positive phase shift in the frequency

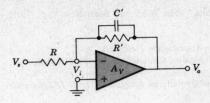

Figure 15-36 Lead-compensated operational amplifier.

range near the unity-loop-gain crossover point. Equation (15-50) gives the loop gain for the uncompensated amplifier. If we substitute for R' an impedance Z', which is the parallel combination of R' and C', Eq. (15-50) becomes

$$\beta R_M = - \frac{RA_V}{R + Z'} = \frac{-RAA_V}{R + R'} \tag{15-67}$$

where we find (Prob. 15-44) that A is given by

$$A \equiv \frac{1 + j(f/f_z)}{1 + j(f/f_p)} \tag{15-68}$$

$$f_z \equiv \frac{1}{2\pi C'R'} \qquad \text{and} \qquad f_p \equiv \frac{R + R'}{R} f_z \tag{15-69}$$

Since $f_p \gg f_z$, then in the neighborhood of the zero, $A \approx 1 + j(f/f_z)$, which is the transfer function of a lead network. An alternative system for obtaining lead compensation is given in Prob. 15-46.

As already mentioned, some OP AMPs are internally compensated (for example, Fairchild's μA741, National Semiconductor's LM741, LM107, and LM112, and Motorola's MC1558). If an OP AMP is to be compensated by adding external capacitance (and also perhaps resistance), the proper pins must be available on the chip. Often the manufacturer suggests the type of compensation and the useful range of C and R values.

REFERENCES

1. Roberge, K. R.: "Operational Amplifiers: Theory and Practice," John Wiley and Sons, Inc., New York, 1975.
2. Stout, D. F., and M. Kaufman: "Handbook of Operational Amplifier Circuit Design," McGraw-Hill Book Company, New York, 1976.
3. Hnatek, E. R.: "Applications of Linear Integrated Circuits," John Wiley and Sons, Inc., New York, 1975.
4. Tobey, G. E., Graeme, J. G., and L. P. Huelsman: "Operational Amplifiers: Design and Applications," McGraw-Hill Book Company, New York, 1971.
   Graeme, J. G.: "Applications of Operational Amplifiers: Third Generation Techniques," McGraw-Hill Book Company, New York, 1973.
5. Hamilton, D. J., and W. G. Howard: "Basic Integrated Circuit Engineering," sec. 9-4, McGraw-Hill Book Company, New York, 1975.
   Grebene, A. B.: "Analog Integrated Circuit Design," McGraw-Hill Book Company, New York, 1977, chap. 5.

Connelly, J. A. (ed.): "Analog Integrated Circuits," John Wiley and Sons, New York, 1975, Chap. 3.

Ref. 1, chap. 7.

6. Giacoletto, L. J.: "Differential Amplifiers," Wiley-Interscience, New York, 1970.

7. Radio Corporation of America: "RCA Linear Integrated Circuits," pp. 28–43, Harrison, N.J., 1967.

8. Widlar, R. J.: Design Techniques for Monolithic Operational Amplifiers, *IEEE J. Solid-state Circuits*, vol. SC-4, pp. 184-191, August, 1969.

9. Hamilton, D. J.: Ref. 5, chaps. 9 and 10.

Grebene, A. B.: Ref. 5, chaps. 4 and 5.

Connelly, J. A.: Ref. 5, chap. 3.

10. Ref. 2, chaps. 1 and 2.

11. Wojslaw, C. F.: Use OP AMPs with Greater Confidence, *Electronic Design*, no. 6, March 16, 1972.

Hamilton, D. J.: Ref. 5, sec. 11-6.

Connelly, J. A.: Ref. 5, sec. 5.8.

12. Hearn, W. E.: Fast Slewing Monolithic Operational Amplifier, *IEEE J. Solid-State Circuits*, vol. SC-6, pp. 20-24, February, 1971.

13. Thornton, R. D., C. L. Searle, D. O. Peterson, R. B. Adler, and E. J. Angelo, Jr.: "Multistage Transistor Circuits," SEEC Committee Series, vol. 5, pp. 108-118, John Wiley and Sons, Inc., 1965.

Ref. 1, chaps. 5 and 13.

REVIEW QUESTIONS

15-1 (*a*) Draw the schematic block diagram of the basic OP AMP with inverting and noninverting inputs.

(*b*) Indicate its equivalent circuit.

15-2 List six characteristics of the ideal OP AMP.

15-3 (*a*) Draw the schematic diagram of an ideal inverting OP AMP with voltage-shunt feedback impedances Z and Z'.

(*b*) Indicate the virtual-ground model for calculating the gain.

15-4 For the OP AMP of Rev. 15-3, assume finite A_v and R_i and nonzero R_o. Draw the equivalent circuit for this nonideal OP AMP.

15-5 (*a*) Draw the schematic diagram of an ideal noninverting OP AMP with voltage-series feedback.

(*b*) Derive the expression for the voltage gain.

15-6 (*a*) Define an ideal DIFF AMP.

(*b*) Define difference signal v_d and common-mode signal v_c.

(*c*) Define *common-mode rejection ratio*.

15-7 (*a*) Draw the circuit of an emitter-coupled DIFF AMP.

(*b*) Explain why the CMRR $\to \infty$ for a symmetrical circuit with $R_e \to \infty$.

15-8 (*a*) Draw the equivalent circuit from which to calculate A_d for the emitter-coupled DIFF AMP.

(*b*) Repeat for A_c.

15-9 (*a*) Why is R_e in an emitter-coupled DIFF AMP replaced by a constant-current source?

(*b*) Draw such a circuit.

(c) Explain why the network replacing R_e acts as an approximately constant current I_O.

(d) Explain how I_O is made to be independent of temperature.

15-10 Explain why the CMRR is infinite if a true constant-current source is used in a symmetrical emitter-coupled DIFF AMP.

15-11 (a) Sketch the transfer characteristics of a DIFF AMP.

(b) Over what differential voltage is the DIFF AMP a good limiter?

(c) Over what differential voltage is the transfer characteristic quite linear?

(d) How does the transconductance vary (qualitatively) with differential voltage?

(e) Explain why AGC is possible with the DIFF AMP.

15-12 (a) Draw an IC OP AMP in block-diagram form.

(b) Identify each block by function.

15-13 (a) What is the differential input resistance of an emitter-coupled DIFF AMP in terms of the parameters of the low-frequency approximate transistor model?

(b) How does this resistance depend upon the quiescent current?

15-14 Indicate two methods for obtaining very high OP AMP input resistance.

15-15 (a) Draw a two-transistor circuit for biasing a linear IC.

(b) Explain why this circuit acts as a current repeater if $I_B \ll I_C$.

15-16 Draw a circuit for obtaining the same current I_C in each of three different transistors. This current I_C is to equal the current I_1 which is obtained in a different branch of the network.

15-17 (a) Draw a three-transistor version of a current mirror.

(b) Explain how to modify the circuit to obtain current magnification.

15-18 What are the advantages of an active over a passive load?

15-19 (a) Sketch a DIFF AMP configuration which uses an active load.

(b) Explain qualitatively how the circuit functions.

15-20 (a) Show two forms of emitter-follower level shifters, one using a constant-current source and the other using a Zener diode.

(b) What is the expression for the shift in voltage in each circuit?

15-21 (a) Draw the circuit of a transistor used as a level shifter.

(b) What is this configuration called?

15-22 (a) Draw the simplest form of a complementary emitter-follower output stage.

(b) This circuit exhibits *crossover distortion*. Explain this difficulty and indicate how to modify the circuit to eliminate this distortion.

15-23 (a) Sketch a totem-pole output stage using *n-p-n* transistors.

(b) Explain qualitatively the operation of the circuit.

15-24 Define (a) *input bias current*; (b) *input offset current*; (c) *input offset voltage*; (d) *input offset voltage drift*; and (e) *output offset voltage*.

15-25 Define (a) *input common-mode range*; (b) *full-power bandwidth*; (c) *power-supply rejection ratio*; and (d) *slew rate*.

15-26 Give the order of magnitudes of the following OP AMP parameters: (a) V_{io}, (b) I_B, (c) I_{io}, (d) A_v, (e) CMRR.

15-27 Show the model of an OP AMP taking into account I_{B1}, I_{B2}, V_{io}, R_i, R_o, and A_v.

15-28 (a) Derive the expression for the output voltage V_o of an inverting OP AMP due to a bias current I_B.

(b) Explain carefully how to reduce V_o to zero if $I_{B1} = I_{B2}$.

(c) If $I_{B1} \neq I_{B2}$, give the expression (without proof) for V_o with the circuit modified as in part (b).

15-29 Derive the expression for V_o due to V_{io}.

15-30 Show the balancing arrangement for (a) an inverting and (b) a noninverting OP AMP.

15-31 Show the circuit and explain how to measure V_{io}.

15-32 Repeat Rev. 15-31 for I_{io}.

15-33 Repeat Rev. 15-31 for $A_v = A_d$.

15-34 (a) Draw the circuit of a *sample-hold-subtract* system.

(b) What is this circuit used for in connection with the measurements of I_{B1} and A_d? Explain.

15-35 How is the *slew rate* measured?

15-36 Explain how the poles of an OP AMP may be determined experimentally.

15-37 An inverting OP AMP has the configuration of a voltage-shunt feedback amplifier and hence the loop gain is $-\beta R_M$. Explain why it is possible to discuss stability in terms of the magnitude $|A_V|$ and the phase of $-A_V$.

15-38 (a) Discuss dominant-pole compensation.

(b) What are the advantages and disadvantages of this method?

15-39 (a) Explain carefully what is meant by pole-zero compensation.

(b) With this method of compensation are the number of zeros (and poles) of the transfer gain increased or decreased?

15-40 (a) Sketch ideal graphs of the open-loop gain in decibels versus $\log f$ for an uncompensated amplifier, this amplifier with dominant-pole compensation, and the original amplifier with pole-zero compensation. State the assumptions you are making.

15-41 Indicate three methods of implementing pole-zero compensation. Give no explanations.

15-42 (a) A capacitor is added between the input and output of the second stage of an OP AMP. Explain, without proof, what type of compensation this represents.

(b) Before the capacitor is added, assume that each of the two stages has a dominant pole. After the compensation, what happens to these two poles?

15-43 (a) Explain how to obtain *lead compensation* in an inverting OP AMP.

(b) Are the number of zeros (and poles) of the transfer function increased or decreased? Justify the term *lead compensation* as applied to the configuration in part (a).

OPERATIONAL AMPLIFIER SYSTEMS

Many analog systems (both linear and nonlinear) are constructed with the OP AMP or DIFF AMP as the basic building block. These IC's augmented by a few external discrete components, either singly or in combination, are used in the following *linear* systems: analog computers, voltage-to-current and current-to-voltage converters, dc instrumentation amplifiers, voltage followers, and active filters.

Among the *nonlinear* system configurations discussed in this chapter are: Precision ac/dc converters, peak detector, sample-and-hold systems, multiplexer, logarithmic amplifiers, analog multipliers, analog-to-digital and digital-to-analog converters.

16-1 BASIC OPERATIONAL AMPLIFIER APPLICATIONS[1-4]

An OP AMP may be used to perform many mathematical operations. This feature accounts for the name *operational amplifier*. Some of the basic applications are given in this section. Consider the ideal OP AMP of Fig. 15-2a, which is repeated for convenience in Fig. 16-1a. Recalling (Sec. 15-1) that the equivalent circuit of Fig. 16-1b has a virtual ground (which takes no current), it follows that the voltage gain is given by Eq. (15-1), namely,

$$A_{Vf} = \frac{V_o}{V_s} = -\frac{Z'}{Z} \tag{16-1}$$

Based upon this equation we can readily obtain an *analog inverter*, a *scale changer*, a *phase shifter*, and an *adder*.

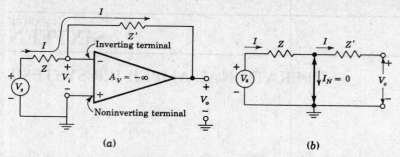

Figure 16-1 (*a*) Inverting operational amplifier with voltage-shunt feedback. (*b*) Virtual ground in the OP AMP.

Sign Changer, or Inverter

If $Z = Z'$ in Fig. 16-1, then $A_{Vf} = -1$, and the sign of the input signal has been changed. Hence such a circuit acts as a phase inverter. If two such amplifiers are connected in cascade, the output from the second stage equals the signal input without change of sign. Hence the outputs from the two stages are equal in magnitude but opposite in phase, and such a system is an excellent *paraphase amplifier*.

Scale Changer

If the ratio $Z'/Z = k$, a real constant, then $A_{Vf} = -k$, and the scale has been multiplied by a factor $-k$. Usually, in such a case of multiplication by a constant, -1 or $-k$, Z and Z' are selected as precision resistors.

Phase Shifter

Assume that Z and Z' are equal in magnitude but differ in angle. Then the operational amplifier shifts the phase of a sinusoidal input voltage while at the same time preserving its amplitude. Any phase shift from 0 to 360° (or ±180°) may be obtained.

Adder or Summing Amplifier

The arrangement of Fig. 16-2 may be used to obtain an output which is a linear combination of a number of input signals. Since a virtual ground exists at the OP AMP input, then

$$i = \frac{v_1}{R_1} + \frac{v_2}{R_2} + \cdots + \frac{v_n}{R_n}$$

and

$$v_o = -R'i = -\left(\frac{R'}{R_1} v_1 + \frac{R'}{R_2} v_2 + \cdots + \frac{R'}{R_n} v_n \right) \qquad (16\text{-}2a)$$

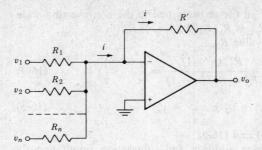

Figure 16-2 Operational inverting adder, or summing amplifier.

If $R_1 = R_2 = \cdots = R_n$, then

$$v_o = -\frac{R'}{R_1}(v_1 + v_2 + \cdots + v_n) \qquad (16\text{-}2b)$$

and the output is proportional to the sum of the inputs.

Many other methods may, of course, be used to combine signals. The present method has the advantage that it may be extended to a very large number of inputs requiring only one additional resistor for each additional input. The result depends, in the limiting case of large amplifier gain, only on the resistors involved, and because of the virtual ground, there is a minimum of interaction between input sources.

Noninverting Summing

An adder whose output is a linear combination of the inputs without a change of sign is obtained by using the noninverting amplifier. In Fig. 16-3 we show such a summer. From Eq. (15-4) the output is given by

$$v_o = \left(1 + \frac{R'}{R}\right)v_+ = \left(\frac{R + R'}{R}\right)v_+ \qquad (16\text{-}3)$$

where the voltage at the noninverting terminal v_+ is found by using superposition. For example, the contribution to v_+ due to v_2' is $v_2'R_{p2}'/(R_2' + R_{p2}')$, where

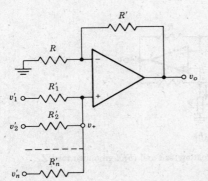

Figure 16-3 A noninverting OP AMP adder.

R'_{p2} is the parallel combination of all the resistors tied to the noninverting node *with the exception of* R'_2; that is, $R'_{p2} = R'_1 \| R'_3 \| R'_4 \| \cdots \| R'_n$.

For n equal resistors each of value R'_2,

$$\frac{R'_{p2}}{R'_2 + R'_{p2}} = \frac{R'_2/(n-1)}{R'_2 + R'_2/(n-1)} = \frac{1}{n} \tag{16-4}$$

and

$$v_+ = \frac{1}{n}(v'_1 + v'_2 + \cdots + v'_n) \tag{16-5}$$

The output is given by Eqs. (16-3) and (16-5).

It is possible to perform analog addition and subtraction simultaneously with a single OP AMP by replacing the resistor R in Fig. 16-3 by the n input voltages and resistors of Fig. 16-2. Again superposition is used to find the contribution to v_o from any of the input voltages. It should be emphasized that, when one of the voltages $v_1, v_2, \ldots, v_n$ is under consideration, then the positive input terminal is effectively grounded (if the bias current is negligible). Similarly when one of the voltages $v'_1, v'_2, \ldots, v'_n$ is under consideration, then R in Fig. 16-3 represents the parallel combination of $R_1, R_2, \ldots, R_n$.

Voltage-to-Current Converter (Transconductance Amplifier)

Often it is desirable to convert a voltage signal to a proportional output current. This is required, for example, when we drive a deflection coil in a television tube. If the load impedance has neither side grounded (if it is floating), the simple circuit of Fig. 16-2 with R' replaced by the load impedance Z_L is an excellent *voltage-to-current converter*. For a single input $v_1 = v_s(t)$, the current in Z_L is

$$i_L = \frac{v_s(t)}{R_1} \tag{16-6}$$

Note that i is independent of the load Z_L, because of the virtual ground of the operational amplifier input. Since the same current flows through the signal source and the load, it is important that the signal source be capable of providing this load current. On the other hand, the amplifier of Fig. 16-4a

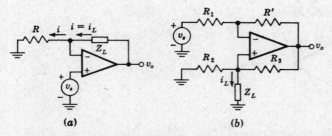

Figure 16-4 Voltage-to-current converter for (*a*) a floating load and (*b*) a grounded load Z_L.

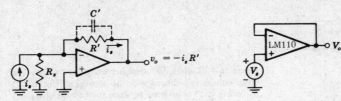

Figure 16-5 Current-to-voltage converter.   **Figure 16-6** A voltage follower, $V_o = V_s$.

requires very little current from the signal source due to the very large input resistance seen by the noninverting terminal.

If the load Z_L is grounded, the circuit of Fig. 16-4b can be used. In Prob. 16-6 we show that if $R_3/R_2 = R'/R_1$, then

$$i_L(t) = -\frac{v_s(t)}{R_2} \tag{16-7}$$

Current-to-Voltage Converter (Transresistance Amplifier)

Photocells and photomultiplier tubes give an output current which is independent of the load. The circuit in Fig. 16-5 shows an operational amplifier used as a current-to-voltage converter. Due to the virtual ground at the amplifier input, the current in R_s is zero and i_s flows through the feedback resistor R'. Thus the output voltage v_o is $v_o = -i_s R'$. It must be pointed out that the lower limit on current measurement with this circuit is set by the bias current of the inverting input. It is common to parallel R' with a capacitance C' to reduce high-frequency noise and the possibility of oscillations. The current-to-voltage converter makes an excellent current-measuring instrument since it is an ammeter with zero voltage across the meter.

DC Voltage Follower

The simple configuration of Fig. 16-6 approaches the ideal *voltage follower*. Because the two inputs are tied together (virtually), then $V_o = V_s$ and *the output follows the input*. The LM110 (National Semiconductor Corporation) is specifically designed for voltage-follower usage and its output is internally connected to the inverting output. It has extremely high input resistance (10^6 MΩ), very low input current (1 nA), very low output resistance (0.75 Ω), a bandwidth of 10 MHz, and a gain of 0.9997. The LM110 may oscillate if the source resistance is too small and it may be necessary to add about 10 kΩ in series with the source.

16-2 DIFFERENTIAL (INSTRUMENTATION) AMPLIFIERS[1-6]

The differential-input single-ended-output instrumentation amplifier is often used to amplify inputs from transducers which convert a physical parameter and

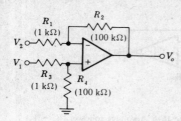

Figure 16-7 Differential amplifier using one OP AMP. The offset voltage is balanced with the network of Fig. 15-25; R_2 of Fig. 5-25 is inserted between the bottom of R_4 and ground.

its variations into an electric signal. Such transducers are strain-gauge bridges, thermocouples, etc. The circuit shown in Fig. 16-7 is very simple and uses only one OP AMP. To find V_o we use the superposition theorem. If we set $V_1 = 0$, then, neglecting the bias current, the voltage v_+ at the noninverting terminal is zero and the standard inverting configuration results. Hence $V_o = -(R_2/R_1)V_2$. On the other hand, if $V_2 = 0$ then $V_+ = [R_4/(R_3 + R_4)]V_1$ and V_o is given by Eq. (16-3). By superposition

$$V_o = -\frac{R_2 V_2}{R_1} + \frac{R_4}{R_3 + R_4}\,\frac{R_1 + R_2}{R_1}\,V_1$$

$$= -\frac{R_2}{R_1}\left[V_2 - \frac{1}{R_3/R_4 + 1}\left(\frac{R_1}{R_2}+1\right)V_1\right] \tag{16-8}$$

If $R_1/R_2 = R_3/R_4$, then

$$V_o = \frac{R_2}{R_1}(V_1 - V_2) \tag{16-9}$$

If the signals V_1 and V_2 have source resistances R_{s1} and R_{s2}, then these resistances add to R_3 and R_1, respectively. Note that the signal source V_1 sees a resistance $R_3 + R_4 = 101$ kΩ. If $V_1 = 0$, the inverting input is at ground potential and hence V_2 is loaded by $R_1 = 1$ kΩ. If this is too heavy a load for the transducer, a high-resistance buffer may be used preceding each input in Fig. 16-7. The resulting system in Fig. 16-8 of three OP AMPs represents a dc instrumentation amplifier with very high input resistance and improved common-mode rejection ratio. (Since 2, 3, or 4 OP AMPs are available on a single chip, the cost of this configuration is low.)

It is easy to demonstrate that the gain of each buffer $A1$ and $A2$ is unity for a common-mode voltage but is high for a difference signal. Because there is almost zero voltage between amplifier input terminals, the top of R is at a voltage V_2 and the bottom of this resistor is at V_1. If a common-mode signal is under consideration, $V_1 = V_2$ and the voltage across R is zero. Hence, there is no current in R or R'. Consequently $V_2' = V_2$ and $V_1' = V_1$, and the buffers act as unity-gain amplifiers. However, if $V_1 \neq V_2$, there is current in R and R' and $V_2' - V_1' > V_2 - V_1$. Therefore the differential gain and the CMRR of the

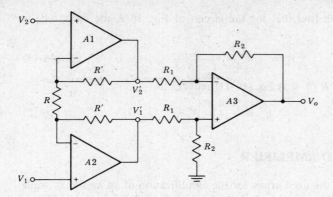

Figure 16-8 An improved instrumentation amplifier.

two-stage system have been increased over the single-stage circuit of Fig. 16-7. Continuing this analysis (Prob. 16-16), we obtain

$$V_o = \left(1 + \frac{2R'}{R}\right) \frac{R_2}{R_1} (V_1 - V_2) \qquad (16\text{-}10)$$

Note that the difference gain may be varied by using an adjustable resistance for R.

The system consisting of only $A1$, $A2$, R, and R' is an amplifier with a double-ended output (*a differential-output amplifier*). Clearly, $V_2' - V_1' = (1 + 2R'/R)(V_2 - V_1)$.

Bridge Amplifier

A differential amplifier is often used to amplify the output from a transducer bridge, as shown in Fig. 16-9. Nominally, the four arms of the bridge have equal resistances R. However, one of the branches has a resistance which changes to $R + \Delta R$ with temperature or some other physical parameter. The goal of the measurement is to obtain the fractional change δ of the resistance value of the active arm, or $\delta = \Delta R / R$.

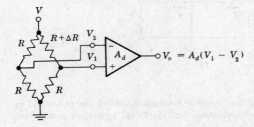

Figure 16-9 Differential bridge amplifier.

In Prob. 16-19 we find that for the circuit of Fig. 16-9, the output V_o is given by

$$V_o = - \frac{A_d V}{4} \frac{\delta}{1 + \delta/2} \tag{16-11}$$

For small changes in R ($\delta \ll 1$) Eq. (16-11) reduces to

$$V_o = - \frac{A_d V}{4} \delta \tag{16-12}$$

16-3 AC-COUPLED AMPLIFIER

In some applications the need arises for the amplification of an ac signal, while any dc signal present must be blocked. A very simple and stable ac amplifier is shown in Fig. 16-10a, where capacitor C blocks the dc component of the input signal and together with the resistor R sets the low-frequency 3-dB response for the overall amplifier.

The output voltage V_o as a function of the complex variable s is found from the equivalent circuit of Fig. 16-10b (where the double-ended heavy arrow represents the virtual ground) to be

$$V_o = - IR' = - \frac{V_s}{R + 1/sC} R'$$

and

$$A_{Vf} = \frac{V_o}{V_s} = - \frac{R'}{R} \frac{s}{s + 1/RC} \tag{16-13}$$

From Eq. (16-13) we see that the low 3-dB frequency is

$$f_L = \frac{1}{2\pi RC} \tag{16-14}$$

The high-frequency response is determined by the frequency characteristics of the operational amplifier A_V and the amount of voltage-shunt feedback present (Sec. 14-5). The midband gain is, from Eq. (16-1), $A_{Vf} = - R'/R$.

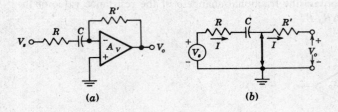

(a)   (b)

Figure 16-10 (a) AC-coupled feedback amplifier. Potential feedback instability can be avoided by judicious selection of R, R^1, and C (lag compensation, Sec. 15-10). (b) Equivalent circuit when $|A_V| = \infty$.

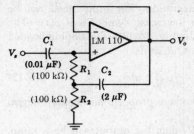

Figure 16-11 AC voltage follower. (*Courtesy of National Semiconductor Corporation.*)

AC Voltage Follower

The ac voltage follower is used to provide impedance buffering, that is, to connect a signal source with high internal source resistance to a load of low impedance, which may even be capacitive. In Fig. 16-11 is shown a practical high-input impedance ac voltage follower using the LM110 operational amplifier. We assume that C_1 and C_2 represent short circuits at all frequencies of operation of this circuit. Resistors R_1 and R_2 are used to provide RC coupling and allow a path for the dc input current into the noninverting terminal. In the absence of the bootstrapping capacitor C_2, the ac signal source would see an input resistance of only $R_1 + R_2 = 200$ kΩ. Since the LM110 is connected as a voltage follower, the voltage gain A_V between the output terminal and the noninverting terminal is very close to unity (0.9997). Thus, from Miller's theorem in Sec. C-4, the input resistance the source sees becomes, approximately, $R_1/(1 - A_V)$, which is measured to be 12.5 MΩ at 100 Hz and increases to 117 MΩ at 1 kHz, and 322 MΩ at 10 kHz.

16-4 ANALOG INTEGRATION AND DIFFERENTIATION[1-4]

The analog integrator is very useful in many applications which require the generation or processing of analog signals. If, in Fig. 16-1, $Z = R$ and a capacitor C is used for Z', as in Fig. 16-12, we can show that the circuit

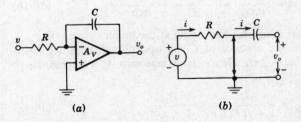

Figure 16-12 (*a*) Operational integrator. (*b*) Equivalent circuit of an ideal integrator.

performs the mathematical operation of integration. The input need not be sinusoidal, and hence is represented by the lowercase symbol $v = v(t)$. (The subscript s is now omitted, for simplicity.) In Fig. 16-12b, the double-headed arrow represents a virtual ground. Hence $i = v/R$, and

$$v_o = - \frac{1}{C} \int i \, dt = - \frac{1}{RC} \int v \, dt \qquad (16\text{-}15)$$

The amplifier therefore provides an output voltage proportional to the integral of the input voltage.

If the input voltage is a constant, $v = V$, then the output will be a ramp, $v_o = - Vt/RC$. Such an integrator makes an excellent sweep circuit for a cathode-ray-tube oscilloscope (Sec. 17-6), and is called a *Miller integrator*, or *Miller sweep*.[7]

DC Offset and Bias Current

The input stage of the operational amplifier used in Fig. 16-12 is usually a DIFF AMP. The dc input offset voltage V_{io} appears across the amplifier input, and this voltage will be integrated and will appear at the output as a linearly increasing voltage. The input bias current will also flow through the feedback capacitor, charging it and producing an additional linearly increasing component of the output voltage. These two effects (error sources) cause a continually increasing output until the amplifier reaches its saturation point. We see then that a limit is set on the feasible integration time by the above error components. The effect of the bias current can be minimized by increasing the feedback capacitor C while simultaneously decreasing the value of R for a given value of the time constant RC.

Finite Gain and Bandwidth

The integrator supplies an output voltage proportional to the integral of the input voltage, provided the operational amplifier shown in Fig. 16-12a has infinite gain $|A_V| \to \infty$ and infinite bandwidth. The voltage gain as a function of the complex variable s is, from Eq. (16-1),

$$A_{Vf}(s) = \frac{V_o(s)}{V(s)} = - \frac{Z'}{Z} = - \frac{1}{RCs} \qquad (16\text{-}16)$$

and it is clear that the ideal integrator has a pole at the origin.

Let us assume that in the absence of C the operational amplifier has a dominant pole at f_1, or $s_1 \equiv - 2\pi f_1$. Hence its voltage gain A_v is approximated by

$$A_v = \frac{A_{vo}}{1 + j(f/f_1)} = \frac{A_{vo}}{1 - s/s_1} \qquad (16\text{-}17)$$

If we further assume that $R_o = 0$ in Fig. 15-3, then $A_v = A_V$. Substituting Eq.

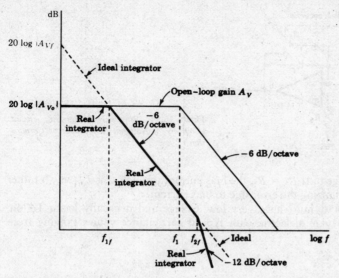

Figure 16-13 Bode magnitude plots of open-loop OP AMP gain A_V, ideal integrator and real integrator. Note that $f_{1f} = 1/2\pi RC|A_{Vo}|$ and $f_{2f} = A_{Vo}s_1/2\pi$.

(16-17) in Eq. (15-2) with $R_i = \infty$ and using $|A_{Vo}| \gg 1$, $|A_{Vo}|RC \gg 1/|s_1|$, we find (Prob. 16-21)

$$A_{Vf} = -\frac{s_1}{RC} A_{Vo} \frac{1}{(s + A_{Vo}s_1)(s - 1/RCA_{Vo})} \qquad (16\text{-}18)$$

where A_{Vo} is a negative number and represents the low-frequency voltage gain of the operational amplifier.

The foregoing transfer function has two poles on the negative real axis as compared with one pole at the origin for the ideal integrator. In Fig. 16-13 we show the Bode plots of Eqs. (16-16) to (16-18). We note that the response of the real integrator departs from the ideal at both low and high frequencies. At high frequencies the integrator performance is affected by the finite bandwidth $(-s_1/2\pi)$ of the operational amplifier, while at low frequencies the integration is limited by the finite gain of the OP AMP.

Practical Circuit

A practical integrator must be provided with an external circuit to introduce initial conditions, as shown in Fig. 16-14. When switch S is in position 1, the input is zero and capacitor C is charged to the voltage V, setting an initial condition of $v_o = V$. When switch S is in position 2, the amplifier is connected as an integrator and its output will be V plus a constant times the time integral

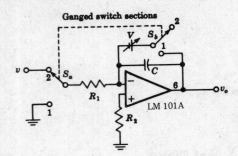

Figure 16-14 Practical integrator circuit. (*Courtesy of National Semiconductor Corporation.*)

of the input voltage v. If $R_2 = R_1$, the bias current through C is I_{io} (why?) rather than I_B, thus minimizing the error due to bias current.

The capacitor C must have very low leakage and it usually has a Teflon, polystyrene, or Mylar dielectric with typical capacitance values ranging from 0.001 to 10 μF.

Differentiator

If Z is a capacitor C and if $Z' = R$, we see from the equivalent circuit of Fig. 16-15 that $i = C\, dv/dt$ and

$$v_o = - Ri = - RC\frac{dv}{dt} \qquad (16\text{-}19)$$

Hence the output is proportional to the time derivative of the input. If the input signal is $v = \sin \omega t$, then the output will be $v_o = - RC\omega \cos \omega t$. Thus the magnitude of the output increases linearly with increasing frequency, and the differentiator circuit has high gain at high frequencies. This results in amplification of the high-frequency components of amplifier noise, and the noise output may completely obscure the differentiated signal.

The General Case

In the important cases considered above, Z and Z' have been simple elements such as a single R or C. In general, they may be any series or parallel combinations of R, L, or C. Using the methods of Laplace transform analysis, Z and Z' can be written in their operational form as $Z(s)$ and $Z'(s)$, where s is the

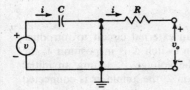

Figure 16-15 Equivalent circuit of an ideal operational differentiator.

complex-frequency variable. In this notation the reactance of an inductor is written formally as Ls and that of a capacitor as $1/sC$. The current $I(s)$ is then $V(s)/Z(s)$, and the output is

$$V_o(s) = -\frac{Z'(s)}{Z(s)} V(s) \tag{16-20}$$

The amplifier thus solves this operational equation.

16-5 ELECTRONIC ANALOG COMPUTATION[8]

The OP AMP is the fundamental building block in an electronic analog computer. As an illustration, let us consider how to program the differential equation

$$\frac{d^2v}{dt^2} + K_1 \frac{dv}{dt} + K_2v - v_1 = 0 \tag{16-21}$$

where v_1 is a given function of time, and K_1 and K_2 are real positive constants.

We begin by assuming that d^2v/dt^2 is available in the form of a voltage. Then, by means of an integrator, a voltage proportional to dv/dt is obtained. A second integrator gives a voltage proportional to v. Then an adder (and scale changer) gives $-K_1(dv/dt) - K_2v + v_1$. From the differential equation (16-21), this equals d^2v/dt^2, and hence the output of this summing amplifier is fed to the input terminal, where we had assumed that d^2v/dt^2 was available in the first place.

The procedure outlined above is carried out in Fig. 16-16. The voltage d^2v/dt^2 is assumed to be available at an input terminal. The integrator (1) has a time constant $RC = 1$ s, and hence its output at terminal 1 is $-dv/dt$. This voltage is fed to a similar integrator (2), and the voltage at terminal 2 is $+v$. The voltage at terminal 1 is fed to the inverter and scale changer (3), and its output at terminal 3 is $+K_1(dv/dt)$. This same operational amplifier (3) is used as an adder. Hence, if the given voltage $v_1(t)$ is also fed into it as shown, the output at terminal 3 also contains the term $-v_1$, or the net output is $+K_1(dv/dt) - v_1$. Scale changer-adder (4) is fed from terminals 2 and 3, and hence delivers a resultant voltage $-K_2v - K_1(dv/dt) + v_1$ at terminal 4. By Eq. (16-21) this must equal d^2v/dt^2, which is the voltage that was assumed to exist at the input terminal. Hence the computer is completed by connecting terminal 4 to the input terminal. (This last step is omitted from Fig. 16-16 for the sake of clarity of explanation.)

The specified initial conditions (the value of dv/dt and v at $t = 0$) must now be inserted into the computer. We note that the voltages at terminals 1 and 2 in Fig. 16-16 are proportional to dv/dt and v, respectively. Hence initial conditions are taken care of (as in Fig. 16-14) by applying the correct voltages V_1 and V_2 across the capacitors in integrators 1 and 2, respectively.

The solution is obtained by opening switches S_1 and S_2 and simultaneously closing S_3 (by means of relays) at $t = 0$ and observing the waveform at terminal

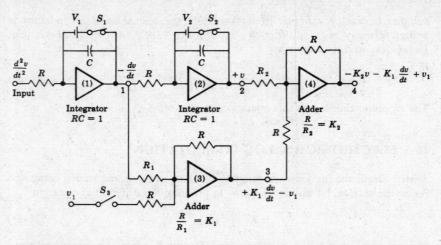

Figure 16-16 A block diagram of an electronic analog computer. At $t = 0$, S_1 and S_2 are opened and S_3 is closed. Each OP AMP input is as in Fig. 16-14.

2. If the derivative dv/dt is also desired, its waveform is available at terminal 1. The indicator may be a cathode-ray tube (with a triggered sweep) or a recorder or, for qualitative analysis with slowly varying quantities, a high-impedance voltmeter.

The solution of Eq. (16-21) can also be obtained with a computer which contains differentiators instead of integrators. However, integrators are almost invariably preferred over differentiators in analog computer applications, for the following reasons: Since the gain of an integrator decreases with frequency whereas the gain of a differentiator increases nominally linearly with frequency, it is easier to stabilize the former than the latter with respect to spurious oscillations. As a result of its limited bandwidth, an integrator is less sensitive to noise voltages than a differentiator. Furthermore, if the input waveform changes rapidly, the amplifier of a differentiator may overload. Finally, as a matter of practice, it is convenient to introduce initial conditions in an integrator.

16-6 ACTIVE FILTERS[1, 9, 10]

Consider the ideal low-pass-filter response shown in Fig. 16-20a. In this plot all signals within the band $0 \leqslant f \leqslant f_o$ are transmitted without loss, whereas inputs with frequencies $f > f_o$ give zero output. It is known that such an ideal characteristic is unrealizable with physical elements, and thus it is necessary to approximate it. An approximation for an ideal low-pass filter is of the form

$$\frac{A_V(s)}{A_{Vo}} = \frac{1}{P_n(s)} \tag{16-22}$$

where $P_n(s)$ is a polynomial in the variable s with zeros in the left-hand plane. Active filters permit the realization of arbitrary left-hand poles for $A_V(s)$, using the operational amplifier as the active element and only resistors and capacitors for the passive elements. The fact that no inductors are required is an important advantage of practical filter design. The use of inductors should always be avoided, if possible, because they are bulky, heavy, and nonlinear; they generate stray magnetic fields; and they may dissipate considerable power.

Since commercially available OP AMPs have unity gain-bandwidth products as high as 100 MHz, it is possible to design active filters up to frequencies of several MHz. The limiting factor for full-power response at those high frequencies is the slew rate (Sec. 15-7) of the operational amplifier. (Commercial integrated OP AMPs are available with slew rates as high as 500 V/μs.)

Butterworth Filter[11]

A common approximation of Eq. (16-22) uses the Butterworth polynomials $B_n(s)$ for $P_n(s)$, where the magnitude of $B_n(\omega)$ is given by

$$|B_n(\omega)| = \sqrt{1 + \left(\frac{\omega}{\omega_o}\right)^{2n}} \tag{16-23}$$

The Butterworth response for various values of n is plotted in Fig. 16-17. Note that the magnitude of A_V is down 3 dB at $\omega = \omega_o$ for all n and is monotonically decreasing. The larger the value of n, the more closely the curve approximates the ideal low-pass response of Fig. 16-20a on p. 586.

If we normalize the frequency by assuming $\omega_o = 1$ rad/s, then Table 16-1 gives the Butterworth polynomials for n up to 8. Note that for n even, the

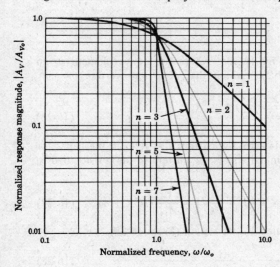

Figure 16-17 Butterworth low-pass-filter frequency response.

Table 16-1 Normalized Butterworth polynomials

n	Factors of polynomial $B_n(s)$
1	$(s + 1)$
2	$(s^2 + 1.414s + 1)$
3	$(s + 1)(s^2 + s + 1)$
4	$(s^2 + 0.765s + 1)(s^2 + 1.848s + 1)$
5	$(s + 1)(s^2 + 0.618s + 1)(s^2 + 1.618s + 1)$
6	$(s^2 + 0.518s + 1)(s^2 + 1.414s + 1)(s^2 + 1.932s + 1)$
7	$(s + 1)(s^2 + 0.445s + 1)(s^2 + 1.247s + 1)(s^2 + 1.802s + 1)$
8	$(s^2 + 0.390s + 1)(s^2 + 1.111s + 1)(s^2 + 1.663s + 1)(s^2 + 1.962s + 1)$

polynomials are the products of quadratic forms, and for n odd, there is present the additional factor $s + 1$. The *damping factor* k is defined as one-half the coefficient of s in each quadratic factor in Table 16-1. For example, for $n = 4$, there are two damping factors, namely, $0.765/2 = 0.383$ and $1.848/2 = 0.924$.

From the table and Eq. (16-22) we see that the typical second-order Butterworth filter transfer function is of the form

$$\frac{A_V(s)}{A_{Vo}} = \frac{1}{(s/\omega_o)^2 + 2k(s/\omega_o) + 1} \tag{16-24}$$

where $\omega_o = 2\pi f_o$ is the high-frequency 3-dB point. Similarly, the first-order filter is

$$\frac{A_V(s)}{A_{Vo}} = \frac{1}{s/\omega_o + 1} \tag{16-25}$$

Practical Realization

Consider the circuit shown in Fig. 16-18a, where the active element is an operational amplifier whose midband gain is that of a noninverting OP AMP, namely,

$$A_{Vo} = \frac{V_o}{V_i} = \frac{R_1 + R_1'}{R_1} \tag{16-26}$$

Since the amplifier input current is zero, the current $I = sCV_i$ in the capacitor connected to the noninverting terminal equals the current in the resistor tied to this terminal. Hence,

$$V' = I\left(R + \frac{1}{sC}\right) = sCV_i\left(R + \frac{1}{sC}\right) = \frac{V_o}{A_{Vo}}(sCR + 1) \tag{16-27}$$

If we apply KCL to the node labeled V', we obtain (Prob. 16-35),

$$\frac{A_V(s)}{A_{Vo}} = \frac{1}{(RCs)^2 + (3 - A_{Vo})(RCs) + 1} \tag{16-28}$$

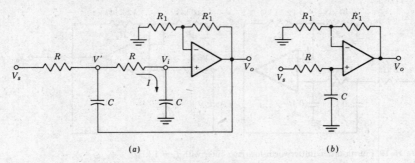

Figure 16-18 (*a*) Second-order low-pass section. (*b*) First-order low-pass section.

Note that Eqs. (16-24) and (16-28) are of the same form. Equating the coefficients of s^2 in these two equations, we obtain

$$\omega_o = \frac{1}{RC} \tag{16-29}$$

and matching the coefficients of s in these equations yields

$$2k = 3 - A_{Vo} \qquad \text{or} \qquad A_{Vo} = 3 - 2k \tag{16-30}$$

We are now in a position to synthesize even-order Butterworth filters by cascading prototypes of the form shown in Fig. 16-18*a*. We must select the gain A_{Vo} of each operational amplifier to satisfy Eq. (16-30), where the damping factors are obtained from Table 16-1.

To realize odd-order filters, it is necessary to cascade the first-order filter of Eq. (16-25) with second-order sections such as indicated in Fig. 16-18*a*. The first-order prototype of Fig. 16-18*b* has the transfer function of Eq. (16-25) for arbitrary A_{Vo} provided that ω_o is given by Eq. (16-29). For example, a third-order Butterworth active filter consists of the circuit in Fig. 16-18*a* in cascade with the circuit of Fig. 16-18*b*, with R and C chosen so that $RC = 1/\omega_o$, with A_{Vo} in Fig. 16-18*a* selected to give $k = 0.5$ (Table 16-1, $n = 3$), and A_{Vo} in Fig. 16-18*b* chosen arbitrarily.

Example 16-1 Design a fourth-order Butterworth low-pass filter with a cutoff frequency of 1 kHz.

SOLUTION We cascade two second-order prototypes as shown in Fig. 16-19. For $n = 4$ we have from Table 16-1 and Eq. (16-30)

$$A_{Vo1} \equiv A_{V1} = 3 - 2k_1 = 3 - 0.765 = 2.235$$

and $$A_{Vo2} \equiv A_{V2} = 3 - 2k_2 = 3 - 1.848 = 1.152$$

From Eq. (16-26), $A_{V1} = (R_1 + R_1')/R_1$. If we arbitrarily choose $R_1 = 10 \text{ k}\Omega$, then for $A_{V1} = 2.235$, we find $R_1' = 12.35 \text{ k}\Omega$, whereas for $A_{V2} = 1.152$, we

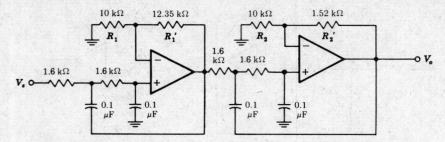

Figure 16-19 Fourth-order Butterworth low-pass filter with $f_o = 1$ kHz.

find $R_2' = 1.520$ kΩ and $R_2 = 10$ kΩ. To satisfy the cutoff-frequency requirement, we have, from Eq. (16-29), $f_o = 1/2\pi RC$. We arbitrarily choose a
convenient value of capacitance, say, $C = 0.1$ μF and find that $R = 1.6$ kΩ.
Figure 16-19 shows the complete fourth-order low-pass design.

The Butterworth filter possesses maximum flatness in the pass band,
whereas a Chebyshev frequency-response characteristic possesses ripples (increases and decreases in gain) within the pass frequency range. However, the
Chebyshev response has the sharpest cutoff characteristics of all filters. The
design of a Chebyshev filter is carried out in the same manner as described
above, except that a table of Chebyshev polynomials is used in place of Table
16-1.

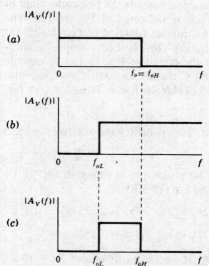

Figure 16-20 Ideal filter characteristics. (a) Low-
pass; (b) high-pass; and (c) bandpass.

High-Pass Prototype

An idealized high-pass-filter characteristic is indicated in Fig. 16-20b. The high-pass second-order filter is obtained from the low-pass second-order prototype of Eq. (16-24) by applying the transformation

$$\left.\frac{s}{\omega_o}\right|_{\text{low-pass}} \rightarrow \left.\frac{\omega_o}{s}\right|_{\text{high-pass}} \tag{16-31}$$

Thus, interchanging R's and C's in Fig. 16-18a results in a second-order high-pass active filter.

Bandpass Filter

A second-order bandpass prototype is obtained by cascading a low-pass second-order section whose cutoff frequency is f_{oH} with a high-pass second-order section whose cutoff frequency is f_{oL}, provided $f_{oH} > f_{oL}$, as indicated in Fig. 16-20c.

Band-reject Filter

Figure 16-21 shows that a band-reject filter is obtained by paralleling a high-pass section whose cutoff frequency is f_{oL} with a low-pass section whose cutoff frequency is f_{oH}. Note that for band-reject characteristics it is required that $f_{oH} < f_{oL}$.

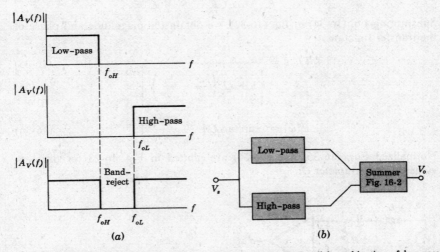

Figure 16-21 (a) Ideal band-reject-filter frequency response. (b) Parallel combination of low-pass and high-pass filters results in a band-reject filter.

16-7 ACTIVE RESONANT BANDPASS FILTERS[1, 12]

The idealized bandpass filter of Fig. 16-20c has a constant response for $f_{oL} < f < f_{oH}$ and zero gain outside this range. An infinite number of Butterworth sections are required to obtain this filter response. A very simple approximation to a narrowband characteristic is obtained using a single LC resonant circuit. Such a bandpass filter has a response which peaks at some center frequency f_o and drops off with frequency on both sides of f_o. A basic prototype for a resonant filter is the second-order section shown in Fig. 16-22, whose transfer function we now derive. Then we show how to obtain the same frequency response by using an OP AMP in combination with a network of resistors and capacitors, but no inductors.

If we assume that the amplifier provides a gain $A_o = V_o/V_i$ which is positive and constant for all frequencies, we find

$$A_V(j\omega) = \frac{V_o}{V_s} = \frac{V_o V_i}{V_i V_s} = \frac{RA_o}{R + j(\omega L - 1/\omega C)} \tag{16-32}$$

The *center*, or *resonant*, frequency $f_o = \omega_o/2\pi$ is defined as that frequency at which the inductance resonates with the capacitance; in other words, the inductive and capacitive reactances are equal (in magnitude), or

$$\omega_o^2 = \frac{1}{LC} \tag{16-33}$$

It is convenient to define the *quality factor Q* of this circuit by

$$Q \equiv \frac{\omega_o L}{R} = \frac{1}{\omega_o CR} = \frac{1}{R}\sqrt{\frac{L}{C}} \tag{16-34}$$

Substituting Eq. (16-34) in Eq. (16-32), we obtain the magnitude and phase of the transfer function

$$|A_V(j\omega)| = \frac{A_o}{\left[1 + Q^2\left(\frac{\omega}{\omega_o} - \frac{\omega_o}{\omega}\right)^2\right]^{1/2}} \tag{16-35}$$

$$\theta(\omega) = -\arctan Q\left(\frac{\omega}{\omega_o} - \frac{\omega_o}{\omega}\right) \tag{16-36}$$

Normalized Eqs. (16-35) and (16-36) are plotted in Fig. 16-23 for different values of the parameter Q.

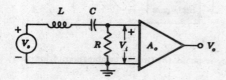

Figure 16-22 A resonant circuit.

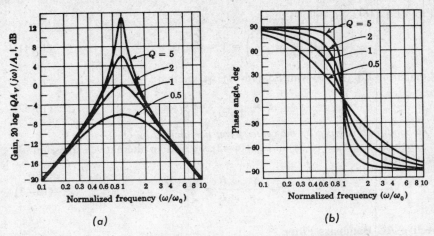

Figure 16-23 The bandpass characteristics of a tuned circuit. (*a*) Amplitude; (*b*) phase response.

Geometric Symmetry

In the $|A_V(j\omega)|$ curves of Fig. 16-23 it is seen that, for every frequency $\omega' < \omega_o$, there exists a frequency $\omega'' > \omega_o$ for which $|A_V(j\omega)|$ has the same value. We now show that these frequencies have ω_o as their geometric mean; that is, $\omega_o^2 = \omega'\omega''$.

Setting $|A_V(j\omega')| = |A_V(j\omega'')|$, we obtain

$$\frac{\omega'}{\omega_o} - \frac{\omega_o}{\omega'} = -\left(\frac{\omega''}{\omega_o} - \frac{\omega_o}{\omega''} \right) \tag{16-37}$$

where the minus sign is required outside the parentheses because $\omega' < \omega_o < \omega''$. From Eq. (16-37) we find

$$\omega_o^2 = \omega'\omega'' \tag{16-38}$$

Bandwidth

Let $\omega_1 < \omega_o$ and $\omega_2 > \omega_o$ be the two frequencies on either side of ω_o for which the gain drops by 3 dB from its value A_o at ω_o. Then the bandwidth is defined by

$$B \equiv \frac{\omega_2 - \omega_1}{2\pi} = \frac{1}{2\pi}\left(\omega_2 - \frac{\omega_o^2}{\omega_2} \right) \tag{16-39}$$

where use is made of Eq. (16-38). The frequency ω_2 is found by setting

$$\left| \frac{A_V(j\omega)}{A_o} \right| = \frac{1}{\sqrt{2}} \tag{16-40}$$

From Eq. (16-35) it follows that

$$Q\left(\frac{\omega_2}{\omega_o} - \frac{\omega_o}{\omega_2}\right) = 1 = \frac{Q}{\omega_o}\left(\omega_2 - \frac{\omega_o^2}{\omega_2}\right) \tag{16-41}$$

Comparing Eq. (16-39) with Eq. (16-41), we see that

$$B = \frac{1}{2\pi}\frac{\omega_o}{Q} = \frac{f_o}{Q} \tag{16-42}$$

The bandwidth is given by the center frequency divided by Q.

Substituting Eq. (16-34) in Eq. (16-42), we find an alternative expression for B, namely,

$$B = \frac{1}{2\pi}\frac{\omega_o R}{\omega_o L} = \frac{1}{2\pi}\frac{R}{L} \tag{16-43}$$

Active *RC* Bandpass Filter

The general form for the second-order bandpass filter is obtained if we let $s = j\omega$ in Eq. (16-32).

$$A_V(s) = \frac{RA_o}{R + sL + 1/sC} = \frac{(R/L)A_o s}{s^2 + s(R/L) + 1/LC} \tag{16-44}$$

Substituting Eqs. (16-33) and (16-34) into (16-44) yields

$$A_V(s) = \frac{(\omega_o/Q)A_o s}{s^2 + (\omega_o/Q)s + \omega_o^2} \tag{16-45}$$

The transfer function of Eq. (16-45) obtained from the *RLC* circuit shown in Fig. 16-22 can be implemented with the multiple-feedback circuit of Fig. 16-24, which uses two capacitors, three resistors, and one OP AMP, but *no inductors*. Because of the virtual ground at the input of the OP AMP the voltage across R_3 is V_o and the current I_3 in R_3 is V_o/R_3. Assuming that the bias current is negligible, V' is given by

$$V' = -\frac{I_3}{sC} = -\frac{V_o}{sCR_3} \tag{16-46}$$

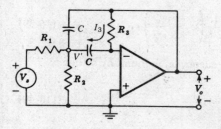

Figure 16-24 An active resonant filter without an inductance.

Applying KCL to the node marked V', we show in Prob. 16-42 that

$$\frac{V_o(s)}{V_s} = \frac{-s/R_1C}{s^2 + (2/R_3C)s + 1/R'R_3C^2} \tag{16-47}$$

where $R' = R_1 \| R_2$, or

$$R' \equiv \frac{R_1R_2}{R_1 + R_2} \tag{16-48}$$

Note that Eqs. (16-45) and (16-47) each has a zero at the origin in addition to two poles. Equating the coefficients of s in the numerators of these equations, we obtain

$$R_1C = \frac{Q}{\omega_o(-A_o)} \tag{16-49}$$

Equating the coefficients of s in the denominators of Eqs. (16-47) and (16-45) yields

$$\frac{R_3C}{2} = \frac{Q}{\omega_o} \tag{16-50}$$

Equating the constant terms in the denominators of Eqs. (16-47) and (16-45), we find

$$R'R_3C^2 = \frac{1}{\omega_o^2} \tag{16-51}$$

Dividing this equation by Eq. (16-50) eliminates R_3 and yields the following equation for R':

$$2R'C = \frac{1}{\omega_o Q} \tag{16-52}$$

Since we have only three independent relationships [Eqs. (16-49), (16-50), and (16-52)] from which to determine the four parameters, R_1, R_3, R', and C, one of these (usually C) is chosen arbitrarily.

Example 16-2 Design a second-order bandpass filter with a midband voltage gain $-A_o = 50$ (34 dB), a center frequency $f_o = 160$ Hz, and a 3-dB bandwidth $B = 16$ Hz.

SOLUTION From Eq. (16-42) we see that the required $Q = 160/16 = 10$. The center angular frequency is $\omega_o = 2\pi f_o = 2\pi \times 160 \approx 1{,}000$ rad/s. Assume $C = 0.1 \ \mu$F. From Eq. (16-49)

$$R_1 = \frac{Q}{-A_o\omega_oC} = \frac{10}{50 \times 10^3 \times 0.1 \times 10^{-6}} \ \Omega = 2 \text{ k}\Omega$$

From Eq. (16-50),

$$R_3 = \frac{2Q}{\omega_oC} = \frac{20}{1{,}000 \times 0.1 \times 10^{-6}} \ \Omega = 200 \text{ k}\Omega$$

From Eq. (16-52),

$$R' = \frac{1}{2C\omega_o Q} = \frac{1}{2 \times 0.1 \times 10^{-6} \times 1,000 \times 10} = 500 \ \Omega$$

Finally, from Eq. (16-48),

$$R_2 = \frac{R_1 R'}{R_1 - R'} = \frac{2,000 \times 500}{2,000 - 500} = 667 \ \Omega$$

If the above specifications were to be met with the RLC circuit of Fig. 16-22, an unreasonably large value of inductance would be required (Prob. 16-47).

Controlled driving-point impedance circuits are discussed in Probs. 16-29 through 16-31. These include the *negative impedance converter* and the *gyrator*. The latter allows an inductor to be simulated with two OP AMPs, resistors, and a capacitor.

All the systems discussed thus far in this chapter have operated linearly. The remainder of this chapter is concerned with nonlinear OP AMP functions.

16-8 PRECISION AC/DC CONVERTERS[1, 13]

If a sinusoid whose peak value is less than the threshold or cutin voltage V_γ (~ 0.6 V) is applied to the rectifier circuit of Fig. 2-13, we see that the output is zero for all times. In order to be able to rectify millivolt signals, it is clearly necessary to reduce V_γ. By placing the diode in the feedback loop of an OP AMP, the cutin voltage is divided by the open-loop gain A_V of the amplifier. Hence V_γ is virtually eliminated and the diode approaches the ideal rectifying component. If in Fig. 16-25a the input v_i goes positive by at least V_γ/A_V, then v' exceeds V_γ and D conducts. Because of the virtual connection between the noninverting and inverting inputs (due to the feedback with D ON), $v_o \approx v_i$. Therefore the circuit acts as a voltage follower for positive signals (in excess of approximately $0.6/10^5$ V $= 60$ μV). When v_i swings negatively, D is OFF and no current is delivered to the external load except for the small bias current of the LM101A and the diode reverse saturation current.

Precision Limiting

By modifying the circuit of Fig. 16-25a, as indicated in Fig. 16-25b, an almost ideal limiter is obtained. If $v_i < V_R$, then v' is positive and D conducts. As explained above, under these conditions the output equals the voltage at the noninverting terminal, or $v_o = V_R$. If $v_i > V_R$, then v' is negative, D is OFF, and $v_o = v_i R_L/(R_L + R) \approx v_i$ if $R \ll R_L$. In summary: The output follows the input for $v_i > V_R$ and v_o is clamped to V_R if v_i is less than V_R by about 60 μV. When D is reverse-biased in Fig. 16-25a or b, a large differential voltage may appear

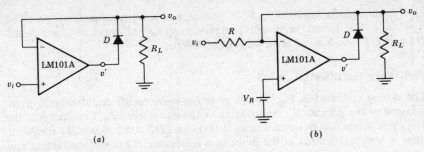

Figure 16-25 (*a*) A precision diode. (*b*) A precision clamp.

between the inputs, and the OP AMP must be able to withstand this voltage. Also note that when $v_i > V_R$, the OP AMP saturates because the feedback through D is missing.

Fast Half-Wave Rectifier

By adding R' and $D2$ to Fig. 16-25*b* and setting $V_R = 0$, we obtain the circuit of Fig. 16-26*a*. If v_i goes negative, $D1$ is ON, $D2$ is OFF, and the circuit behaves as an inverting OP AMP, so that $v_o = -(R'/R)v_i$. If v_i is positive, $D1$ is OFF and $D2$ is ON. Because of the feedback through $D2$, a virtual ground exists at the input and $v_o = 0$. If v_i is a sinusoid, the circuit performs half-wave rectification.

The principal limitation of this circuit is the slew rate of the OP AMP. As the input passes through zero, the OP AMP output v' must change as quickly as possible from $+0.6$ to -0.6 V (or vice versa) in order for the conduction to switch very rapidly from one diode to the other. If the slew rate is 1 V/μs, then this switching time is 1.2 μs. Hence, 1.2 μs must be a small fraction of the period of the input sinusoid.

An alternative noninverting configuration to that in Fig. 16-26*a* is to ground the left-hand side of R and to impress v_i at the noninverting terminal. The output now has a value of $(R + R')/R$ times the input for positive voltages and

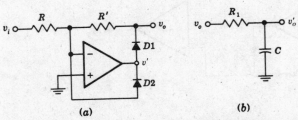

Figure 16-26 (*a*) A half-wave rectifier. (*b*) A low-pass filter which can be cascaded with the circuit in (*a*) to obtain an average detector.

$v_o = v_i$ for negative inputs if $R_L \gg R'$. Hence, half-wave rectification is obtained if $R' \gg R$. For either the inverting or noninverting half-wave rectifier, diodes $D1$ and $D2$ may *both* be reversed (Prob. 16-49).

Full-Wave Rectifier[14]

The system indicated in Fig. 16-27a gives full-wave rectification without inversion and with a gain R/R_1, controllable by the one resistor R_1. Consider first the half cycle where v_i is positive. Then $D1$ is ON and $D2$ is OFF. Since $D1$ conducts, there is a virtual ground at the input to $A1$. Because $D2$ is nonconducting and since there is no current in the R which is connected to the noninverting input to $A2$, then $v_1 = 0$. Hence, the system consists of two OP-AMPS in cascade, with the gain of $A1$ equal to $-R/R_1$ and the gain of $A2$ equal to $-R/R = -1$. The result is

$$v_o = +\frac{R}{R_1}v_i > 0 \qquad \text{for} \qquad v_i > 0 \tag{16-53}$$

Consider now the half cycle where v_i is negative. Then $D1$ is OFF and $D2$ is ON, as indicated in Fig. 16-27b. Because of the virtual ground at the input to $A2$,

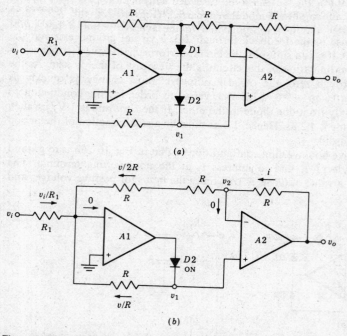

(a)

(b)

Figure 16-27 (a) Full-wave rectifier system. (b) During the half cycle when v_i is negative, $D1$ is nonconducting and $D2$ is ON, as indicated. Note that $v_1 = v_2 \equiv v$ and that $i = v/2R$.

$v_2 = v_1 \equiv v$. Since the input terminals of $A1$ are at the same (ground) potential, the currents coming to the inverting terminal of $A1$ are as indicated in the figure. From KCL at this node,

$$\frac{v_i}{R_1} + \frac{v}{2R} + \frac{v}{R} = 0 \qquad \text{or} \qquad v = -\frac{2}{3}\frac{R}{R_1}v_i \qquad (16\text{-}54)$$

The output voltage is $v_o = iR + v$, where the current i equals $v/2R$ because the inverting terminal of $A2$ takes no current. Hence,

$$v_o = \frac{v}{2R}R + v = \tfrac{3}{2}v = -\frac{R}{R_1}v_i > 0 \qquad \text{for} \qquad v_i < 0 \qquad (16\text{-}55)$$

where use is made of Eq. (16-54). Note that the sign of v_o is positive in Eq. (16-55) because v_i is negative in this half cycle. Since v_o in Eq. (16-55) equals v_o in Eq. (16-54), the outputs for the two half cycles are identical, thus verifying that the system performs full-wave rectification (with a gain of R/R_1). Note that for any input waveform, v_o is proportional to the absolute value of the input $|v_i|$.

Active Average Detector

Consider the circuit of Fig. 16-26a to be cascaded with the low-pass filter of Fig. 16-26b. If v_i is an amplitude-modulated carrier (Fig. 10-22), the R_1C filter removes the carrier and v_o' is proportional to the average value of the audio signal. In other words, this configuration represents an *average detector*.

Active Peak Detector

If a capacitor is added at the output of the precision diode of Fig. 16-25a, with $R_L = \infty$, a peak detector results. The capacitor in Fig. 16-28a will hold the output at $t = t'$ to the most positive value attained by the input v_i prior to t', as indicated in Fig. 16-28b. This operation follows from the fact that if $v_i > v_o$, the voltage at the noninverting terminal exceeds that at the inverting terminal and the OP AMP output v' is positive, so that D conducts. The capacitor is then

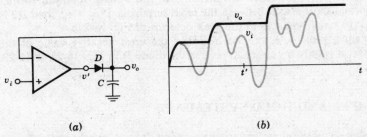

(a)   (b)

Figure 16-28 (a) A positive peak detector. (b) An arbitrary input waveform v_i and the corresponding output v_o.

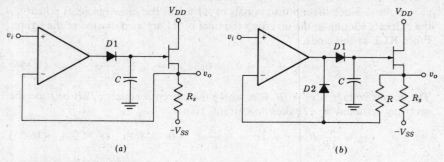

Figure 16-29 Improved versions of a peak detector ($R \gg R_s$).

charged through D (by the output current of the amplifier) to the value of the input because the circuit is a voltage follower. When v_i falls below the capacitor voltage, the OP AMP output goes negative and the diode becomes reverse-biased. Thus the capacitor charges until its voltage equals the most positive value of the input. To reset the circuit, a low-leakage switch such as a MOSFET gate must be placed across the capacitor.

The bias current of the OP AMP is integrated by the capacitor. Also, if the output is loaded, then C discharges through the load. Both of these difficulties are avoided by modifying the system with a source follower, as indicated in Fig. 16-29. By connecting the inverting terminal to the load at the output, v_o is forced to equal the peak value of v_i, as desired (but the capacitor voltage differs from v_o by the gate-to-source voltage of the FET). This network is a special case of a sample-and-hold circuit, and the capacitor-leakage-current considerations given in the following section also apply to this configuration. For an ideal capacitor, the voltage across C in the hold position changes only because of the very small FET input current and the diode reverse current.

If the input v_i falls below the output v_o, the OP AMP will saturate (the maximum input differential range may also be exceeded). To prevent this difficulty another diode is added to the circuit, as indicated in Fig. 16-29b. If now $v_i < v_o$, then $D2$ conducts and the OP AMP is a voltage follower, so that there is a virtual short circuit between the input terminals. If $v_i > v_o$, then $D2$ is cut off and the circuit reduces to the peak detector of Fig. 16-29a.

To obtain a peak detector that measures the most negative value of the input voltage, it is only necessary to reverse the diode D in Fig. 16-28 or 16-29. Why?

16-9 SAMPLE-AND-HOLD SYSTEMS[1, 15]

A typical data-acquisition system receives signals from a number of different sources and transmits these signals in suitable form to a computer or a communication channel. A multiplexer (Sec. 16-10) selects each signal in

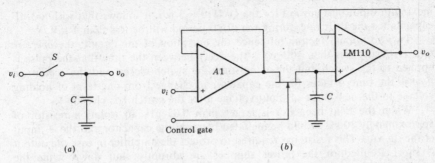

Figure 16-30 Sample-and-hold circuit. (*a*) Schematic; (*b*) practical. (The negative terminal of $A1$ may be connected to the output v_o instead of to the output of $A1$.)

sequence, and then the analog information is converted into a constant voltage over the gating-time interval by means of a *sample-and-hold system*. The constant output of the sample-and-hold may then be converted to a digital signal by means of an analog-to-digital (A/D) converter (Sec. 16-13) for digital transmission.

A sample-and-hold circuit in its simplest form is a switch S in series with a capacitor, as in Fig. 16-30a. The voltage across the capacitor tracks the input signal during the time T_g when a logic control gate closes S, and holds the instantaneous value attained at the end of the interval T_g when the control gate opens S. The switch[16] may be a relay (for very slow waveforms), a sampling diode-bridge gate, a bipolar transistor switch, a FET controlled by a gating-signal voltage, or a CMOS transmission gate (Fig. 8-28). The FET makes an excellent chopper because its *offset voltage* is zero whereas that of a bipolar transistor is several millivolts.

The configuration shown in Fig. 16-30b is one of the simplest practical sample-and-hold systems. A positive pulse at the gate of the n-channel FET will turn the switch ON, and the holding capacitor C will charge to the instantaneous value of the input voltage with a time constant $(R_o + r_{DS(ON)})C$, where R_o is the very small output resistance of the input OP AMP voltage follower and $r_{DS(ON)}$ is the ON resistance of the FET (Sec. 8-2). In the absence of a positive pulse, the switch is turned OFF and the capacitor is isolated from any load through the LM 110 OP AMP. Thus it will hold the voltage impressed upon it. It is recommended that a capacitor with polycarbonate, polyethylene, polystyrene, Mylar, or Teflon dielectric be used. Most other capacitors do not retain the stored voltage, due to a polarization phenomenon[17] which causes the stored voltage to decay with a time constant of several seconds.

Dielectrics other than those mentioned above also exhibit a phenomenon called *dielectric absorption*, which causes a capacitor to "remember" a fraction of its previous charge (if there is a change in capacitor voltage). Even if the polarization and absorption effects do not occur, the OFF current of the switch (~ 1 nA) and the bias current of the OP AMP will flow through C. Since the maximum input bias current for the LM110 is 3 nA, it follows that with a 3-μF capacitance the drift rate during the HOLD period will be less than 1 mV/s.

Two additional factors influence the operation of the circuit: the *aperture time* (typically less than 100 ns) is the delay between the time that the pulse is applied to the switch and the actual time the switch closes, and the *acquisition time* is the time it takes for the capacitor to change from one level of holding voltage to the new value of input voltage after the switch has closed.

When the hold capacitor is larger than 0.05 μF, an isolation resistor of approximately 10 kΩ should be included between the capacitor and the + input of the OP AMP. This resistor is required to protect the amplifier in case the output is short-circuited or the power supplies are abruptly shut down while the capacitor is charged.

If R_o and $r_{DS(ON)}$ were negligibly small, the acquisition time would be limited by the maximum current I which the input OP AMP follower can deliver. The capacitor voltage then changes at a peak rate of $dv_c/dt = I/C$. Since the short-circuit current of an OP AMP is limited (25 mA for the 741 chip), an external complementary emitter follower is used to increase the current available to charge (or discharge) C extremely rapidly. Such an arrangement is indicated in Fig. 16-31 between the sampling switch and the capacitor. Note that the input OP AMP is no longer operated as a follower, but its negative-input terminal now is connected to the output v_o. This connection is analogous to that shown in Fig. 16-29 for the improved peak detector, and it ensures that $v_o = v_i$ during the *sampling interval*. In the *hold interval* v_o remains at the value which v_i attained at the end of the sampling time, except for the very small changes in voltage across C due to the bias current of the output OP AMP and the leakage current of the switch and emitter follower. The larger the value of the capacitance C, the smaller is the drift in voltage during the hold mode. However, the smaller the capacitance C, the smaller is the acquisition time and, hence, the greater is the fidelity with which the output follows the input during the sampling mode. Furthermore, the holding capacitor creates an additional pole which must be

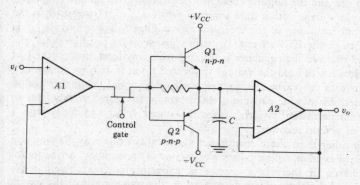

Figure 16-31 An improved sample-and-hold configuration. (The complementary emitter follower is discussed in detail in Sec. 18-5.)

reckoned with when considering loop transmission and stability. Hence, the value of C must be chosen as a compromise between these three conflicting requirements, depending upon the application.

A sample-and-hold (S-H) system is available on a single monolithic chip (for example, Harris Semiconductor, HA-2420 or National Semiconductor, LF 198), with the storage capacitor added externally. The inverting terminal of $A1$ is available at an external pin and, hence, this chip may be used to build either a noninverting or an inverting S-H system which exhibits gain, if the usual external resistors are added (Prob. 16-52).

16-10 ANALOG MULTIPLEXER AND DEMULTIPLEXER[18]

As indicated in Fig. 6-17b a *multiplexer* selects one out of N sources and transmits the (analog) signals to a single transmission line. Of all the switches (mentioned in the preceding section) which are available to feed the input signals to the output channel, the best performance is obtained with the CMOS transmission gate (Fig. 8-28). If dielectric isolation is used in the fabrication of this gate, then typically a leakage current of only 1 nA at $+125°C$ with a switching time of 250 ns is obtainable. Large arrays of such CMOS gates are available for this application.

A block diagram of a 16-input analog commutator is indicated in Fig. 16-32. Time-division multiplexing results if the complementary MOSFET switch $S1$

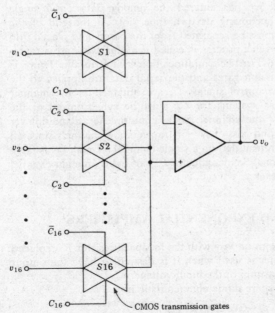

Figure 16-32 A 16-input analog multiplexer using CMOS gates.

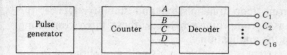

Figure 16-33 A system for generating the digital control voltages C_k required in a multiplexer system. (The inverters to obtain $\overline{C}_k$ are not shown.)

closes (i.e., it is in its low-resistance state) for a time T, switch $S2$ closes for the second interval T, $S3$ transmits for the third period T, and so forth. In Fig. 16-32 the symbol C_k ($k = 1, 2, \ldots, 16$) represents the digital control voltage and $\overline{C}_k$ is its complementary value obtained from an inverter (not shown). If C_k equals binary 1, the CMOS gate transmits the analog signal v_k to the output, but if C_k is binary 0, no transmission is allowed.

The block diagram for obtaining the required digital control voltages for the analog multiplexer of Fig. 16-32 is indicated in Fig. 16-33. The control C_k is the output of the kth line of a 4-to-16-line decoder (Sec. 6-6). The four address lines A, B, C, and D are the outputs from a binary counter which is excited by a pulse generator. If the time interval between pulses is T, time-division multiplexing is obtained with the system of Figs. 16-33 and 16-32 (corresponding to parallel-to-serial conversion of digital data, discussed in Sec. 6-7).

Analog Demultiplexer

The multiplexer described above has entered the analog data on a single channel, each analog signal occupying its own time slot. At the end of the transmission line, each signal must be separated from the others and placed into an individual channel. This reverse process is called *demodulation* and is represented schematically in Fig. 6-17a. The multiposition switch in this figure is replaced by N CMOS transmission gates and the serial data are applied to the input of all these gates. The control signals C_k are obtained in the manner indicated in Fig. 16-33. These systems for C_k must be synchronized at the sending and receiving ends of the channel. Such a multiplexer/demultiplexer system saves the size, weight, and cost of $N - 1$ transmission channels since all the analog signals have been transmitted on a single channel (N may be as large as several hundred). The National Semiconductor CD4051M is an eight-channel analog multiplexer/demultiplexer.

16-11 LOGARITHMIC AND EXPONENTIAL AMPLIFIERS[1, 19]

In Fig. 16-34 there is indicated an OP AMP with the feedback resistor R' replaced by the diode $D1$. This amplifier is used when it is desired to have the output voltage proportional to the logarithm of the input voltage.

From Eq. (2-3) the volt-ampere diode characteristic is

$$I_f = I_o(\epsilon^{V_f/\eta V_T} - 1) \approx I_o\epsilon^{V_f/\eta V_T}$$

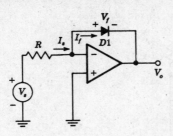

Figure 16-34 Logarithmic amplifier. The input voltage V_s must be positive.

provided that $V_f/\eta V_T \gg 1$ or $I_f \gg I_o$. Hence

$$V_f = \eta V_T(\ln I_f - \ln I_o) \tag{16-56}$$

Since $I_f = I_s = V_s/R$ due to the virtual ground at the amplifier input, then

$$V_o = -V_f = -\eta V_T\left(\ln \frac{V_s}{R} - \ln I_o\right) \tag{16-57}$$

Logarithmic Amplifier Using Matched Transistors

We note from Eq. (16-57) that V_o is temperature dependent due to the scale factor ηV_T and to the saturation current I_o. The factor η, whose value normally depends on the diode current, can be eliminated by replacing the diode with a grounded-base transistor. Another important advantage of using a transistor in place of a diode is that the exponential relationship between current and voltage extends over a much wider voltage range for a transistor than a diode. By augmenting Fig. 16-34 with a second matched transistor it is possible to eliminate from the expression for V_o the reverse saturation current I_o (which doubles for every 10°C rise in temperature). The final system shown in Fig. 16-35 includes an output noninverting OP AMP with a gain of 60.

We now derive the logarithmic expression for V_o. For the present discussion ignore the high-resistance-balancing-potentiometer arrangements. From Eq. (15-21) it follows for matched transistors with $I_{S1} = I_{S2}$ and with $I_B \ll I_C$, that the plus input to $A2$ is at a voltage

$$V \equiv V_{BE2} - V_{BE1} = V_T \ln I_{C2} - V_T \ln I_{C1} = -V_T \ln \frac{I_{C1}}{I_{C2}} \tag{16-58}$$

Since V equals the small difference in the base-emitter voltages of $Q2$ and $Q1$, we neglect V compared with the reference voltage V_R. Then, since $I_{B2} \ll I_{C2}$ and because of the virtual ground at the input of $A1$, it follows that

$$I_{C2} = \frac{V_R}{R_2} \quad \text{and} \quad I_{C1} = \frac{V_s}{R_1} \tag{16-59}$$

Since $A2$ is a noninverting OP AMP, $V_o = V(R_3 + R_4)/R_3$. Combining this equation with Eqs. (16-58) and (16-59) gives

$$V_o = -V_T \frac{R_3 + R_4}{R_3} \ln\left(\frac{V_s}{R_1} \frac{R_2}{V_R}\right) \tag{16-60}$$

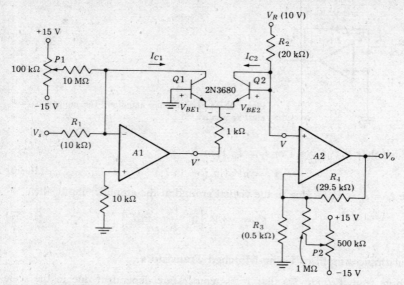

Figure 16-35 An improved logarithmic amplifier. Each OP AMP is type 709 and both are mounted in the same package. (*Courtesy of Texas Instruments, Inc.*)

For the parameter values indicated in Fig. 16-35 and at room temperature with $V_T = 0.0259$ V, $V_o = -3.58 \log(0.2 V_s)$.

Experimentally, it is found that Eq. (16-60) is satisfied over a dynamic range of four decades, from input voltages of approximately 2 mV to 20 V. Beyond 20 V the higher values of transistor currents passing through the ohmic collector and base resistances give a component of linear voltage drop, which leads to a departure from the logarithmic relationship. Below an input voltage of about 2 mV, the input current becomes comparable to the bias current and the logarithmic dependence of V_o on V_s is no longer valid.

The potentiometer $P1$ is used to balance out the offset voltage of $A1$; that is, with $V_s = 0$, $P1$ is varied until $V' \approx 0$ (less than 50 μV). The system is nulled as follows: With $V_s = V_R R_1 / R_2 = 5$ V, potentiometer $P2$ is varied until $V_o = 0$, thus satisfying Eq. (16-60).

Note from Eq. (16-60) that the slope of the characteristic

$$\frac{dV_o}{d(\ln V_s)} = -V_T \frac{R_3 + R_4}{R_3} \tag{16-61}$$

This result has been verified experimentally. Since V_T is proportional to temperature, R_3 should be chosen as a temperature-sensitive resistance. If R_3 increases linearly with T, the slope in Eq. (16-61) can be made to be quite constant as the temperature changes.

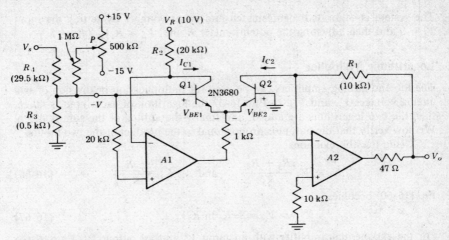

Figure 16-36 An exponential amplifier, using 709 OP AMPS. (*Courtesy of Texas Instruments, Inc.*)

Exponential (Antilog) Amplifier

This system is depicted in Fig. 16-36 and should be compared with that of Fig. 16-35. In the exponential amplifier the feedback current I_{C1} is constant and is derived from the reference voltage V_R, whereas I_{C2} depends upon the input signal. In the logarithmic amplifier the converse is true.

Because of the virtual ground at the inputs to $A1$ and $A2$, the collector and base of $Q1$ are at the same potential $-V = V_{BE1} - V_{BE2}$. Neglecting V relative to V_R,

$$I_{C1} = \frac{V_R}{R_2} \quad \text{and} \quad I_{C2} = \frac{V_o}{R_1} \tag{16-62}$$

From the input attenuator it is clear that

$$- V = \frac{R_3 V_s}{R_3 + R_4} = V_T \ln \frac{I_{C1}}{I_{C2}} \tag{16-63}$$

where use is made of Eq. (16-58). Substituting the currents I_{C1} and I_{C2} from Eq. (16-62) into Eq. (16-63) we obtain,

$$V_s = - V_T \frac{R_3 + R_4}{R_3} \ln\left(\frac{V_o}{R_1} \frac{R_2}{V_R} \right) \tag{16-64}$$

Note that this equation becomes identical with Eq. (16-60) if V_s and V_o are interchanged. Hence, V_o is proportional to the antilog or exponential of V_s. From Eq. (16-64),

$$V_o = \frac{R_1 V_R}{R_2} \exp\left(- \frac{V_s}{V_T} \frac{R_3}{R_3 + R_4} \right) \tag{16-65}$$

The system is calibrated for mismatch and offset voltages by setting the input $V_s = 0$ and then adjusting the potentiometer P until $V_o = R_1 V_R / R_2 = 5$ V.

Logarithmic Multiplier

The log and antilog amplifiers can be used for multiplication or division of two analog voltages V_{s1} and V_{s2}. In Fig. 16-37 the logarithm of each input is taken, then the two logarithms are added, and finally the antilog of the sum is taken. We now verify that the output is proportional to the product of the two inputs.

Using the abbreviations

$$K_1 \equiv V_T \frac{R_3 + R_4}{R_3} \quad \text{and} \quad K_2 \equiv \frac{R_2}{R_1 V_R} \tag{16-66}$$

Eq. (16-60) becomes

$$V_o = -K_1 \ln K_2 V_s \tag{16-67}$$

For the exponential amplifier with an input V_s' and an output V_o', Eq. (16-65) may be written in the form

$$V_o' = \frac{1}{K_2} \epsilon^{-V_s'/K_1} \tag{16-68}$$

Using this notation the output V_o of the summing OP AMP in Fig. 16-37 is

$$V_o = -K_1 \ln K_2 V_{s1} - K_1 \ln K_2 V_{s2} = -K_1 \ln K_2^2 V_{s1} V_{s2} \tag{16-69}$$

Since V_o is the input to the antilog amplifier, $V_o = V_s'$, and from Eqs. (16-68) and (16-69)

$$V_o' = \frac{1}{K_2} \exp(\ln K_2^2 V_{s1} V_{s2}) = K_2 V_{s1} V_{s2} \tag{16-70}$$

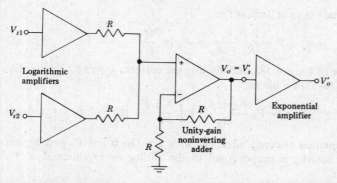

Figure 16-37 Logarithmic multiplier of two analog signals. (The adder of Fig. 16-3 is used.)

We show in Prob. 16-53 that it is possible to raise the input V_s to an arbitrary power by cascading log and antilog amplifiers.

The input signals can be divided if we subtract the logarithm of V_{s1} from that of V_{s2} and then take the antilog. We must point out that the logarithmic multiplier or divider is useful for unipolar inputs only. This is often called one-quadrant operation. Other techniques[19] are available for the accurate multiplication of two signals, one of which is described below.

Differential Amplifier Multiplier

From Eqs. (15-24) and (15-23) we observe that the output voltage of a differential amplifier depends on the current source I_O. If V_{s1} is applied to one input and V_{s2} is used to vary I_O, as in Fig. 16-38, the output will be proportional to the product of the two signals $V_{s1}V_{s2}$. The device AD 532S manufactured by Analog Devices, Inc., is a completely pretrimmed (no external trim networks are required) monolithic multiplier/divider with basic accuracy of ± 1 percent and bandwidth of 1 MHz. As a multiplier, the AD 532S has the transfer function $XY/10$ and as a divider $+10Z/X$. The X, Y, and Z input levels are ± 10 V for multiplication and the output is ± 10 V at 5 mA. As a divider, operation is restricted to two quadrants (where X is negative) only.

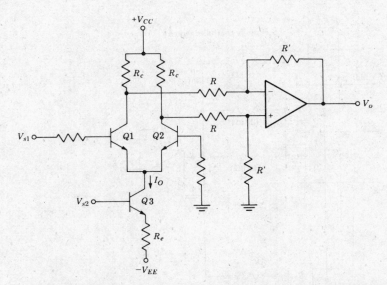

Figure 16-38 Variable transconductance multiplier ($V_o = KV_{s1}V_{s2}$). Note that the transistor $Q3$ configuration is a current source with I_O proportional to V_{s2}; in other words, it functions as a voltage-to-current converter.

16-12 DIGITAL-TO-ANALOG (D/A) CONVERTERS[1, 20-24]

Many systems accept a digital word as an input signal and translate or convert it to an analog voltage or current. These systems are called *digital-to-analog*, or *D/A, converters* (or DACs). The digital word is presented in a variety of codes, the most common being pure binary or binary-coded-decimal (BCD).

The output V_o of an N-bit D/A converter is given by the following equation:

$$V_o = (2^{N-1}a_{N-1} + 2^{N-2}a_{N-2} + \cdots + 2^2a_2 + 2^1a_1 + a_0)V$$

$$= \left(a_{N-1} + \frac{1}{2}a_{N-2} + \frac{1}{4}a_{N-3} + \cdots + \frac{1}{2^{N-2}}a_1 + \frac{1}{2^{N-1}}a_0\right)2^{N-1}V \quad (16\text{-}71)$$

where V is a proportionality factor determined by the system parameters and where the coefficients a_n represent the binary word and $a_n = 1(0)$ if the nth bit is $1(0)$. The voltage V_R is a stable reference voltage used in the circuit. The most significant bit (MSB) is that corresponding to a_{N-1}, and its weight is $2^{N-1}V$, while the least significant bit (LSB) corresponds to a_0, and its weight is $2^0V = V$.

Consider, for example, a 5-bit word ($N = 5$) so that Eq. (16-71) becomes

$$V_o = (16a_4 + 8a_3 + 4a_2 + 2a_1 + a_0)V \quad (16\text{-}72)$$

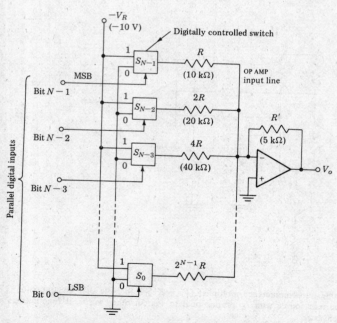

Figure 16-39 D/A converter with binary weighted resistors.

For simplicity, assume $V = 1$. Then, if $a_0 = 1$ and all other a's are zero, we have $V_o = 1$. If $a_1 = 1$ and all other a's are zero, we obtain $V_o = 2$. If $a_0 = a_1 = 1$ and all other a's are zero, $V_o = 2 + 1 = 3$ V, etc. Clearly, V_o is an analog voltage proportional to the digital input.

A D/A converter is indicated schematically in Fig. 16-39. The blocks $S_0, S_1, S_2, \ldots, S_{N-1}$ in Fig. 16-39 are electronic switches which are digitally controlled. For example, when a 1 is present on the MSB line, switch S_{N-1} connects the resistor R to the reference voltage $-V_R$; conversely, when a 0 is present on the MSB line, the switch connects the resistor to the ground line. Thus the switch is a single-pole double-throw (SPDT) electronic switch. The operational amplifier acts as a current-to-voltage converter (Sec. 16-1). We see that if the MSB is 1 and all other bits are 0, the current through the resistor R is $-V_R/R$ and the output is $V_R R'/R$. Similarly, the output of the LSB (if $N = 5$) becomes $V_o = V_R R'/16R$. If all five bits are 1, the output becomes

$$V_o = \left(1 + \tfrac{1}{2} + \tfrac{1}{4} + \tfrac{1}{8} + \tfrac{1}{16}\right)\frac{V_R R'}{R} = (16 + 8 + 4 + 2 + 1)\frac{V_R R'}{16R} \quad (16\text{-}73)$$

which agrees with Eq. (16-71) if $V = V_R R'/16R$. This argument confirms that the analog voltage V_o is proportional to the digital input.

Many implementations are possible for the digitally controlled switches of Fig. 16-39, two of which are indicated in Fig. 16-40. A totem-pole MOSFET driver in Fig. 16-40a feeds each resistor connected to the OP AMP input. The two complementary gate inputs Q and $\bar{Q}$ come from a MOSFET S-R FLIP-FLOP or register which holds the digital information to be converted to an analog number. Let us assume that logic 1 corresponds to -10 V and logic 0 corresponds to 0 V (negative logic). A 1 on the bit line sets the FLIP-FLOP at $Q = 1$ and $\bar{Q} = 0$, and thus transistor $Q1$ is ON, connecting the resistor R_1 to the reference voltage $-V_R$, while transistor $Q2$ is kept OFF. Similarly, a 0 at the input bit line will connect the resistor to the ground terminal.

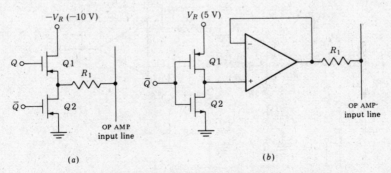

Figure 16-40 Two implementations of the digitally controlled switch (Fig. 16-39) of a DAC. (a) A totem-pole and (b) a CMOS inverter (Fig. 8-26) configuration. The resistance R_1 depends upon the bit under consideration. For example for the $N - 3$ bit, $R_1 = 4R$ in Fig. 16-39.

An excellent alternative single-pole double-throw electronic switch is that shown in Fig. 16-40b. This configuration consists of a CMOS inverter feeding an OP AMP follower which drives R_1 from a very low output resistance. A positive logic system is indicated with $V(1) = V_R = +5$ V and $V(0) = 0$ V. The complement $\overline{Q}$ of the bit $Q = a_n$ under consideration is applied to the input. Hence, if $a_n = 1$, then $\overline{Q} = 0$, the output of the inverter is logic 1, and 5 V is applied to R_1. On the other hand, if the nth is a binary 0, $\overline{Q} = 1$ and the output of the inverter is 0 V, so that R_2 is connected to ground. This confirms the proper operation of the circuit in Fig. 16-40b as a SPDT switch.

The accuracy and stability of the DAC in Fig. 16-39 depend primarily on the absolute accuracy of the resistors and the tracking of each other with temperature. Since all resistors are different and the largest is $2^{N-1}R$, where R is the smallest resistor, their values become excessively large, and it is very difficult and expensive to obtain stable, precise resistances of such values. For example, for a 10-digit DAC the largest resistance is 5.12 MΩ if the smallest is 10 kΩ. Also, the voltage drop across such a large resistance due to the bias current would affect the accuracy. On the other hand, if the largest resistance is a reasonable value, the smallest may become comparable to the output resistance of the switch, again affecting the accuracy. The ladder-type converter described in the following avoids these difficulties.

A Ladder-type D/A Converter

A circuit utilizing twice the number of resistors in Fig. 16-39 for the same number of bits (N) but of values R and $2R$ only is shown in Fig. 16-41. The ladder used in this circuit is a current-splitting device, and thus the ratio of the resistors is more critical than their absolute value. We observe from the figure

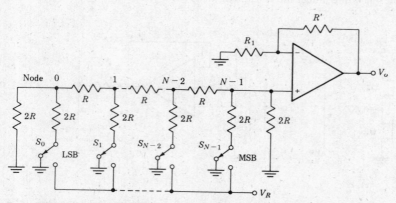

Figure 16-41 D/A converter using an R, $2R$ ladder.

that at any of the ladder nodes the resistance is $2R$ looking to the left or the right or toward the switch.

For example, to the left of node 0 there is $2R$ to ground; to the left of node 1 there is the parallel combination of two $2R$ resistors to ground in series with R, for a total resistance of $R + R = 2R$, and so forth. Hence, if any switch, say $N - 2$, is connected to V_R, the resistance seen by V_R is $2R + 2R \| 2R = 3R$ and the voltage at node $N - 2$ is $(V_R/3R)R = V_R/3$.

Consider now that MSB is logic 1 so that the voltage at node $N - 1$ is $\frac{1}{3} V_R$, the output is

$$V_o = \frac{V_R}{3} \frac{R_1 + R'}{R_1} \equiv V' \qquad (16\text{-}74)$$

Similarly, when the second MSB bit $(N - 2)$ is binary 1 and all other bits are logic 0, the output voltage at node $N - 2$ is $V_R/3$, but at node $N - 1$ the voltage is half this value, because of the attenuation due to the resistance R between the nodes and the resistance R from node $N - 1$ to ground. Hence, $V_o = \frac{1}{2} V'$ for the second most significant bit $(N - 2)$. In a similar manner (Prob. 16-55) it can be shown that the third MSB gives an output $\frac{1}{4} V'$, and so forth. Clearly, the output is of the form of Eq. (16-71) with $V' = 2^{N-1}V$.

Because of the stray capacitance from the nodes to ground, there is a propagation delay time from left to right down the ladder network. When switch S_0 closes, the propagation delay is much longer than when the MSB switch closes. Hence, when the digital voltage changes, a transient waveform will appear at the output before V_o settles down to its proper value. These transients are avoided by using an inverted-ladder DAC (Prob. 16-56).

Hybrid Systems DAC 371I-10 is a 10-bit, $1\mu s$-current-output D/A converter in a standard 16-pin dual-in-line package which contains switches, ladder, and reference.

Multiplying D/A Converter

A D/A converter which uses a varying analog signal V_a instead of a fixed reference voltage is called a *multiplying D/A converter*. From Eq. (16-71) we see that the output is the product of the digital word and the analog voltage $V_a (= 2^{N-1}V)$ and its value depends on the binary word (which represents a number smaller than unity). This arrangement is often referred to as a *programmable attenuator* because the output V_o is a fraction of the input V_a and the attenuator setting can be controlled by a computer.

16-13 ANALOG-TO-DIGITAL (A/D) CONVERTERS[20-24]

It is often required that data taken in a physical system be converted into digital form. Such data would normally appear in electrical analog form. For example, a temperature difference would be represented by the output of a thermocouple, the strain of a mechanical member would be represented by the electrical unbalance of a strain-gage bridge, etc. The need therefore arises for a device that converts analog information into digital form. A very large number of such devices have been invented. We shall consider the four most popular systems: (1) The counting analog-to-digital converter (ADC); (2) the successive-approximation ADC; (3) the parallel-comparator ADC; and (4) the dual-slope or ratiometric ADC.

The Counting A/D Converter

This system will be explained with reference to Fig. 16-42a. The *clear* pulse resets the counter to the zero count. The counter then records in binary form the number of pulses from the clock line. The clock is a source of pulses equally spaced in time. Since the number of pulses counted increases linearly with time, the binary word representing this count is used as the input of a D/A converter whose output is the staircase waveform shown in Fig. 16-42b. As long as the

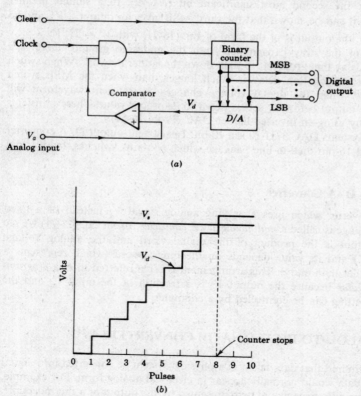

Figure 16-42 (a) A/D converter using a counter; (b) D/A output staircase waveform.

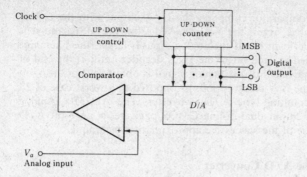

Figure 16-43 A tracking or servo counting A/D converter.

analog input V_a is greater than V_d, the comparator (which is a high-gain DIFF AMP, Sec. 15-4) has an output which is high and the AND gate is open for the transmission of the clock pulses to the counter. When V_d exceeds V_a, the comparator output changes to the low value and the AND gate is disabled. This stops the counting at the time when $V_a \approx V_d$ and the counter can be read out as the digital word representing the analog input voltage.

If the analog voltage varies with time, it is not possible to convert the analog data continuously, but it will be necessary that the input signal be sampled at intervals. If the maximum value of the analog voltage is represented by n pulses and if the period of the clock is T seconds, the minimum interval between samples (the conversion time) is nT seconds.

An improved version of the counting ADC, called a *tracking* or *servo converter*, is obtained by using an UP-DOWN counter (Sec. 7-6, Fig. 7-18). This modification of the system of Fig. 16-42a is indicated in Fig. 16-43. Neither a start command (a clear pulse) nor an AND gate is now used. However, an UP-DOWN counter is now required and the comparator output feeds the UP-DOWN control of the counter. To understand the operation of the system, assume initially the output of the DAC is less than the analog input V_a. Then the positive comparator output causes the counter to read UP. The D/A converter output increases with each clock pulse until it exceeds V_a. The UP-DOWN control line changes state so that it now counts DOWN (but by only one count, LSB). This causes the control to change to UP and the count increases by one LSB. This process keeps repeating so that the digital output bounces back and forth by ± 1 LSB around the correct value. The conversion time is small for small changes in the sampled analog signal and, hence, this system can be used effectively as a tracking A/D converter.

Successive-Approximation A/D Converter

Instead of a binary counter, as shown in Fig. 16-43, this system uses a programmer. The programmer sets the most significant bit (MSB) to 1, with all other bits to 0, and the comparator compares the D/A output with the analog signal. If the D/A output is larger, the 1 is removed from the MSB, and it is tried in the next most significant bit. If the analog input is larger, the 1 remains in that bit. Thus a 1 is tried in each bit of the D/A decoder until, at the end of the process, the binary equivalent of the analog signal is obtained. For an N-bit system, the conversion time is N clock periods as opposed to a worst case of 2^N pulse intervals for the counting-type A/D converter. The AD7570 (Analog Devices Co.), which is a 28-pin dual-in-line CMOS package, is an 8-bit A/D converter which makes use of the successive-approximation technique.

The Parallel-Comparator A/D Converter

This system is by far the fastest of all converters. Its operation is easily understood if reference is made to the 3-bit A/D converter of Fig. 16-44. The

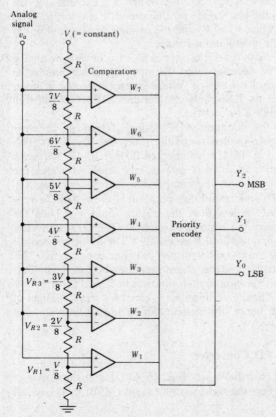

Figure 16-44 A 3-bit parallel-comparator A/D converter.

Table 16-2 The truth table for the A/D converter of Fig. 16-44

Inputs							Outputs		
W_7	W_6	W_5	W_4	W_3	W_2	W_1	Y_2	Y_1	Y_0
0	0	0	0	0	0	0	0	0	0
0	0	0	0	0	0	1	0	0	1
0	0	0	0	0	1	1	0	1	0
0	0	0	0	1	1	1	0	1	1
0	0	0	1	1	1	1	1	0	0
0	0	1	1	1	1	1	1	0	1
0	1	1	1	1	1	1	1	1	0
1	1	1	1	1	1	1	1	1	1

analog voltage v_a is applied simultaneously to a bank of comparators with equally spaced thresholds (reference voltages $V_{R1} = V/8$, $V_{R2} = 2V/8$, etc.). This type of processing is called *bin conversion*, because the analog input is sorted into a given voltage range or "voltage bin" determined by the thresholds of two adjacent comparators. Note that the comparator outputs W take on a very distinctive pattern: low output (logic 0) for all comparators with thresholds *above* the input voltage and high output (logic 1) for each comparator whose threshold is *below* the analog input. For example, if $\frac{2}{8}V < v_a < \frac{3}{8}V$, then $W_1 = 1$, $W_2 = 1$, and all other W's are 0. For this situation the digital output should be 2 ($Y_2 = 0$, $Y_1 = 1$, $Y_0 = 0$), which is interpreted to mean an input analog voltage between $\frac{2}{8}V$ and $\frac{3}{8}V$.

The truth table with inputs W and outputs Y is given in Table 16-2. A comparison with Table 6-3 shows that the logic is that of a 3-bit priority encoder. The X's in Table 6-3 are all replaced by 1s. The column labeled W_0 in Table 6-3 is missing in Table 16-2 because, if $v_a < \frac{1}{8}V$ then W_1 through W_7 are all 0, and the output is zero ($Y_2 = 0$, $Y_1 = 0$, $Y_0 = 0$).

Conversion time is limited only by the speed of the comparator and of the priority encoder. Using an Advanced Micro Devices AMD 686A comparator and a TI-147 priority-encoder conversion, delays of the order of 20 ns can be obtained.

An obvious drawback of this technique is the complexity of the hardware. The number of comparators needed is $2^N - 1$, where N is the desired number of bits (seven comparators for the 3-bit converter of Fig. 16-44). Hence, the number of comparators approximately doubles for each added bit. Also the larger the N, the more complex is the priority encoder.

Dual-Slope or Ratiometric A/D Converter

This widely used system is depicted in Fig. 16-45. Consider unipolar operation with $V_a > 0$ and $V_R < 0$. Initially S_1 is open, S_2 is closed, and the counter is cleared. Then at $t = t_1$, S_1 connects V_a to the integrator and S_2 opens. The sampled (and hence constant) analog voltage V_a is now integrated for a fixed

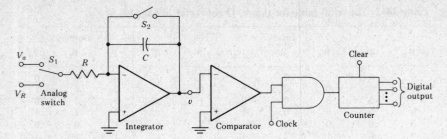

Figure 16-45 Schematic diagram of a dual-slope ADC.

number n_1 of clock pulses. If the clock period is T, the integration takes place for a definite known time $T_1 = n_1 T$, and the waveform v at the output of the integrator is indicated in Fig. 16-46.

If an N-stage ripple counter is used and if $n_1 = 2^N$, then at time t_2 (the end of the integration of V_a) all FLIP-FLOPs in the counter read 0. This is clearly indicated in the waveform chart of Fig. 7-14 for a four-stage ripple counter where, after $n_1 = 2^4 = 16$ counts, $Q_0 = 0$, $Q_1 = 0$, $Q_2 = 0$, and $Q_3 = 0$. In other words, the counter automatically resets itself to zero at the end of the interval T_1. Note also from Fig. 7-14 that at the 2^N the pulse, the state of Q_{N-1} (MSB), changes from 1 to 0 for the first time. This change of state can be used as the control signal for the analog switch or transmission gate (Fig. 8-28).

Because of the counter operation described in the preceding paragraph, the reference voltage V_R is automatically connected to the input of the integrator at $t = t_2$, at which time the counter reads zero. Since V_R is negative, the waveform v has the positive slope shown in Fig. 16-46. We have assumed that $|V_R| > V_a$, so that the integration time T_2 is less than T_1, as indicated. As long as v is negative, the output of the comparator is positive and the AND gate allows clock pulses to be counted. When v falls to zero, at $t = t_3$, the AND gate is inhibited and no further clock pulses enter the counter.

We now show that the reading of the counter at time t_3 is proportional to the analog input voltage. The value of v at t_3 is given by

$$v = -\frac{1}{RC} \int_{t_1}^{t_2} V_a \, dt - \frac{1}{RC} \int_{t_2}^{t_3} V_R \, dt = 0 \tag{16-75}$$

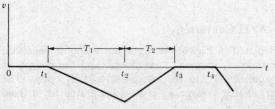

Figure 16-46 The waveform at the integrator output of Fig. 16-45.

Since V_a and V_R are constant,

$$V_a(t_2 - t_1) + V_R(t_3 - t_2) = 0 \quad \text{or} \quad V_a T_1 + V_R T_2 = 0 \quad (16\text{-}76)$$

If the number of pulses accumulated in the interval T_2 is n_2, then $T_2 = n_2 T$. Since $T_1 = n_1 T = 2^N T$, then from Eq. (16-76),

$$V_a = \frac{T_2 |V_R|}{T_1} = \frac{n_2 |V_R|}{n_1} = n_2 \frac{|V_R|}{2^N} \quad (16\text{-}77)$$

Since $|V_R|$ and N are constant, we have verified that V_a is proportional to the counter reading n_2. Note that this result is independent of the time constant RC.

The system includes automatic logic sequencing (not shown in Fig. 16-45), which clears the counter between t_3 and t_4, takes a new sample of the analog voltage, and moves S_1 back to V_a at t_4, so that the process is repeated; thus a new reading of V_a is obtained each $t_3 = t_1 + T_1 + T_2$ seconds. This technique can be very accurate; six-digit digital voltmeters employ such signal processing. The counter feeds a decoder/lamp driver so that the output is visible on Nixie tubes (Sec. 6-6). For each cycle of operation a new voltage reading is obtained.

The dual-slope system is inherently noise-immune because of input signal integration, i.e., the ubiquitous 60-Hz interference can be all but eliminated by choosing the integration time to be an integral number of power line periods. This statement also brings to light the obvious disadvantage of the system, namely, the conversion time is long since $\frac{1}{60}$s $\approx$ 16 ms. Such a dual-slope A/D converter can be obtained in various degrees of user complexity. Motorola Semiconductor provides a three-chip system: MC1505 analog subsystem, MC14435 digital subsystem, and MC14511 decoder/driver for display.

REFERENCES

1. Stout, D. F., and M. Kaufman: "Handbook of Operational Amplifier Circuit Design," McGraw-Hill Book Company, New York, 1976.
2. Wait, J. T., L. P. Huelsman, and G. A. Korn: "Introduction to Operational Amplifier Theory and Applications," chap. 1, McGraw-Hill Book Company, New York, 1975.
3. Graeme, J. G., G. E. Tobey, and L. P. Huelsman: "Operational Amplifiers, Design and Applications," chap. 6, McGraw-Hill Book Company, New York, 1971.
4. Connelly, J. A.: "Analog Integrated Circuits," sec. 5.3, John Wiley and Sons, Inc., New York, 1975.
5. Roberge, J. K.: "Operational Amplifiers, Theory and Practice," sec. 11.4.1, John Wiley and Sons, Inc., New York, 1975.
6. Giacoletto, L. J.: "Differential Amplifiers," Wiley-Interscience, New York, 1970.
7. Millman, J., and H. Taub: "Pulse, Digital, and Switching Waveforms," pp. 536–548, McGraw-Hill Book Company, New York, 1965.
8. Korn, G. A., and T. M. Korn: "Electronic Analog and Hybrid Computers," McGraw-Hill Book Company, New York, 1964.
   Ref. 5, sec. 12.3.
9. Hilburn, J. L., and D. E. Johnson: "Manual of Active Filter Design," McGraw-Hill Book Company, New York, 1973.
   Huelsman, P. L.: "Active Filters," McGraw-Hill Book Company, New York, 1970.
10. Ref. 2, chap. 4.
    Ref. 3, chap. 8.
    Ref. 5, sec. 12.4.

11. Kuo, F. F.: "Network Analysis and Synthesis," John Wiley & Sons, Inc., New York, 1962.
12. Stremler, F. G.: "Design of Active Bandpass Filters," *Electronics*, vol. 44, no. 12, pp. 86-89, June 7, 1971.
13. Dobkin, R. C.: "Linear Brief 8," National Semiconductor Corporation, August, 1969.
    Ref. 2, chap. 3.
    Ref. 3, chap. 7.
    Ref. 5, sec. 11.5.
14. Norris, B. (ed.): "Digital Integrated Circuits and Operational-Amplifier and Optoelectronic Circuit Design," pp. 140-141, McGraw-Hill Book Company, New York, 1976.
15. Ref. 2, sec. 5.8.
    Ref. 3, sec. 9.4.
    Ref. 4, sec. 6.3.
    Ref. 5, sec. 11.6.
16. Ref. 2, chap 5.
    Ref. 7, chap. 17.
17. Dow, Jr., P. C.: An Analysis of Certain Errors in Electronic Differential Analyzers: Capacitor Dielectric Absorption, *IRE Trans. Electronic Computers*, pp. 17–22, March, 1958.
18. Ref. 4, chap. 7.
19. Ref. 2, chap. 3.
    Ref. 3, chap. 7.
    Ref. 5, sec. 11.5.
    Ref. 14, chap. 15.
20. Ref. 2, chap. 6.
    Ref. 3, chap. 9.
    Ref. 4, chap. 8.
21. Hnatek, E. R.: "A User's Handbook of D/A and A/D Converters," John Wiley and Sons, Inc., New York, 1976.
22. Taub, H., and D. Schilling: "Digital Integrated Electronics," chap. 14, McGraw-Hill Book Company, New York, 1977.
23. Schmid, H.: "Electronic Analog/Digital Converters," Van Nostrand Company, New York, 1970.
24. Analog Devices Company: "Nonlinear Circuits Handbook," Norwood, Mass., 1974.
    "Analog-Digital Conversion Handbook," Norwood, Mass., 1972.

REVIEW QUESTIONS

16-1 Indicate an OP AMP connected as (*a*) an *inverter*, (*b*) a *scale changer*, (*c*) a *phase shifter*.

16-2 Indicate an OP AMP connected (*a*) as inverting adder and (*b*) as a noninverting summer.

16-3 Draw the circuit of a *voltage-to-current converter* if the load is (*a*) floating and (*b*) grounded.

16-4 Draw the circuit of a *current-to-voltage converter*. Explain its operation.

16-5 Draw the circuit of a dc *voltage follower* and explain its operation.

16-6 (*a*) Draw the circuit of a dc differential amplifier using a single OP AMP.
   (*b*) What are the resistances seen by each input source?

16-7 Draw the system of a very-high-input-resistance instrumentation amplifier whose gain is controlled by an adjustable resistance.

16-8 Draw the circuit of an ac *voltage follower* having very high input resistance. Explain its operation.

16-9 Draw the circuit of an OP AMP integrator and indicate how to apply the initial condition. Explain its operation.

16-10 Sketch the idealized characteristics for the following filter types: (*a*) low-pass, (*b*) high-pass, (*c*) bandpass, and (*d*) band-rejection.

16-11 Draw the prototype for a low-pass active-filter section of (*a*) first order, (*b*) second order, and (*c*) third order.

16-12 (*a*) Obtain the frequency response of an *RLC* circuit in terms of ω_o and Q.
    (*b*) Verify that the bandwidth is given by f_o/Q.
    (*c*) What is meant by an active resonant bandpass filter?

16-13 Sketch the circuit of a precision (*a*) diode and (*b*) clamp and explain their operation.

16-14 (*a*) Sketch the circuit of a fast half-wave rectifier and explain its operation.
    (*b*) How is this circuit converted into an *average detector*?

16-15 Sketch the circuit of a *peak detector* and explain its operation.

16-16 Sketch the circuit of a peak detector in which the external load does not discharge the storage capacitor.

16-17 (*a*) Sketch a sample-and-hold system with very high input resistance and very low output resistance.
    (*b*) Explain the operation of this system.

16-18 (*a*) What limits the acquisition time in a sample-and-hold configuration?
    (*b*) Sketch a system for minimizing the acquisition time.

16-19 (*a*) Sketch an analog multiplexer system.
    (*b*) How are the switches implemented?

16-20 Draw the block diagram from which to obtain the gating signals for time-division multiplexing.

16-21 (*a*) Sketch the circuit of a *logarithmic amplifier* using one OP AMP and explain its operation.
    (*b*) More complicated logarithmic amplifiers are given in Sec. 16-11. What purpose is served by these circuits?

16-22 In schematic form indicate how to multiply two analog voltages with log-antilog amplifiers.

16-23 Explain how to multiply two analog voltages using a DIFF AMP.

16-24 (*a*) Draw a schematic diagram of a D/A converter. Use resistance values whose ratios are multiples of 2.
    (*b*) Explain the operation of the converter.

16-25 Indicate two possible implementations for the digitally controlled switch of a D/A converter.

16-26 Repeat Rev. 16-24 for a ladder network whose resistances have one of two values, R or $2R$.

16-27 Explain how a DAC may function as a programmable attenuator for an analog signal.

16-28 (*a*) Draw the block diagram for a counting A/D converter.
    (*b*) Explain the operation of this system.

16-29 Repeat Rev. 16-28 for a servo ADC.

16-30 Repeat Rev. 16-28 for a 2-bit parallel-comparator A/D converter.

SEVENTEEN

WAVESHAPING AND WAVEFORM GENERATORS

The comparator as a basic analog building block is introduced. Waveshaping applications are discussed. The regenerative comparator (Schmitt trigger) is used to generate the following signals: square, triangular, pulse, sweep, and staircase waveforms. Pulse amplitude, pulse width, and frequency modulation of a square wave are considered. Sinusoidal oscillator theory and some configurations which generate sinusoids are included.

17-1 COMPARATORS[1, 2]

An analog *comparator* or *detector* has two inputs (one is usually a constant reference voltage V_R and the other is a time-varying signal v_i) and one output v_o. The ideal comparator, whose characteristic is shown in Fig. 17-1a, has an output which is constant with $v_o = V(0)$ if $v_i < V_R$, and it has a different constant value $v_o = V(1)$ if $v_i > V_R$. Clearly the input is *compared* with the reference and the output is digitized into one of two states: a 0 level of voltage $V(0)$ and a 1 level of voltage $V(1)$. In other words, the comparator behaves as a 1-bit analog-to-digital converter. Voltages $V(0)$ and $V(1)$ compatible with TTL, ECL, or MOS logic may be obtained. Other limiting voltages, for example, ± 10 V, are also available.

Note that the comparator performs highly nonlinear waveshaping because the output bears no resemblance to the input waveform. It is often used to transform a signal v_i which varies slowly with time to another v_o, which exhibits an abrupt change when v_i reaches a specific amplitude V_R.

The DIFF AMP input-output curve of Fig. 15-9 approximates the ideal-comparator characteristic. Note that the total input swing between the two extreme

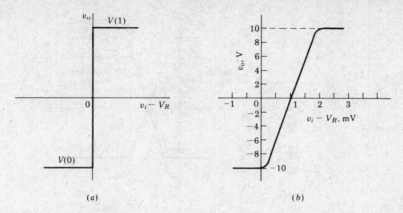

Figure 17-1 The transfer characteristic of (*a*) an ideal comparator and (*b*) a commercial comparator.

output levels is $\sim 8V_T = 200$ mV. This range may be reduced drastically by cascading the DIFF AMP with other high-gain stages. Since this configuration corresponds to the OP AMP topology of Fig. 15-10, an OP AMP may be used (open-loop) as a comparator. A typical OP AMP transfer characteristic is given by the solid curve in Fig. 17-1*b*. It is now observed that the change in the output state takes place with an increment in input Δv_i of only 2 mV. Note that the input offset voltage contributes an error in the point of comparison between v_i and V_R of the order of 1 mV. For some applications this offset may be too large and it will be necessary to balance it out, as indicated in Fig. 15-25.

A number of operational amplifiers have been designed specifically for detector applications and are designated on the manufacturer's specification sheets as *voltage comparator/buffer* instead of OP AMP. Since a comparator is not intended to be used with negative feedback, frequency-compensation components may be omitted, so that greater bandwidth (higher speed) is attainable than with OP AMPS. The designation "buffer" denotes that the comparator will not load down the signal source because of the very high input resistance of the detector. Among the many comparator chips available are the Fairchild μA710 and 760, the National LM111 and 160 and the Harris HA 2111. The uncertainty region Δv_i may be as small as 15 μV and the *response time* (the interval necessary for the comparator to change state) ranges from about 20 to 200 ns. Packages with two or four independent comparators are also available. Some chips are designed with a digital-signal strobe input so that the comparator may be disabled during input transients.

To obtain limiting output voltages, which are independent of the power-supply voltages, a resistor R and two back-to-back Zener diodes are added to clamp the output of the comparator, as indicated in Fig. 17-2*a*. The resistance value is chosen so that the avalanche diodes operate at the recommended Zener

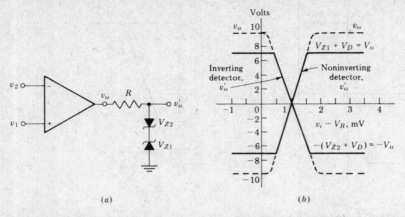

Figure 17-2 (*a*) A comparator cascaded with a resistor-Zener-diode combination. (*b*) The transfer curves at v_o and v'_o. The noninverting characteristics are obtained if $v_1 = v_i$ and $v_2 = V_R$, and the inverting curves result if $v_2 = v_i$ and $v_1 = V_R$.

current. The solid curves give the output v'_o across the diodes, whereas the dashed curves represent the output v_o from the comparator. If the input signal v_i is applied at the noninverting terminal and the reference V_R is tied to the inverting terminal, the noninverting detector is obtained. If the positions of v_i and v_R are interchanged, the inverting-comparator characteristic results. The limiting voltages of v'_o are $V_{Z1} + V_D \equiv V_o$ and $-(V_{Z2} + V_D) \equiv -V_o$, where V_D (~ 0.7 V) is a *p-n* diode forward voltage. A second advantage of adding the Zener diodes is that the limiting may be much sharper for v'_o than for v_o. A disadvantage is the poor transient response of the avalanche diodes.

17-2 APPLICATIONS OF COMPARATORS

If V_R is set equal to zero, the output will change from one state to the other very rapidly (limited by the slew rate) every time that the input passes through zero. Such a configuration is called a *zero-crossing detector*. Among the many applications of the zero-crossing detector are the following.

Square Waves from a Sine Wave

If the input to a comparator is a sine wave, the output is a square wave. If a zero-crossing detector is used (Fig. 17-3*a*), a symmetrical square wave results, as shown in Fig. 17-3*c*. This idealized waveform is shown with vertical sides which, in reality, should extend over a range of a fraction of a millivolt of input voltage v_i.

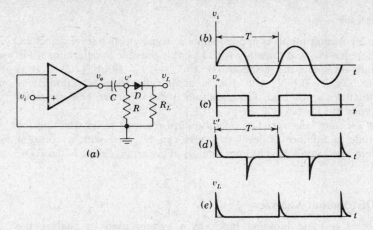

Figure 17-3 A zero-crossing detector converts a sinusoid v_i into a square wave v_o. The pulse waveforms v' and v_L result if v_o is fed into a short time-constant RC circuit in cascade with a diode clipper.

Timing-Markers Generator from a Sine Wave

The square-wave output v_o of the preceding application is applied to the input of an RC series circuit (Fig. 17-3a). If the time constant RC is very small compared with the period T of the sine-wave input, the voltage v' across R is a series of positive and negative pulses, as indicated in Fig. 17-3d. If v' is applied to a clipper with an ideal diode (Fig. 17-3a), the load voltage v_L contains only positive pulses (Fig. 17-3e). Thus the sinusoid has been converted into a train of positive pulses whose spacing is T. These may be used for timing markers (on the sweep voltage of a cathode-ray tube, for example).

Note that the waveshaping performed by the configuration in Fig. 17-3a is very drastic—a sinusoid having been converted into either a square wave or a pulse train.

Phasemeter

The phase angle between two voltages can be measured by a method based on the circuit of Fig. 17-3. Both voltages are converted into pulses, and the time interval between the pulse of one wave and that obtained from the second sine wave is measured. This time interval is proportional to the phase difference. Such a phasemeter can measure angles from 0 to 360°.

If $V_R \neq 0$, the comparator is referred to as a *level detector*. Such a level detector is used in the analog-to-digital converter (ADC) of Fig. 16-42. Several other applications are outlined in the following.

Window Detector

If a pulse (or a constant input) is applied simultaneously to two level detectors, one with a reference V_{R1} and the other with V_{R2}, it is possible to determine if the input amplitude lies within the "window" $W \equiv V_{R2} - V_{R1}$ (Prob. 17-2). If W is held constant (say, at 0.1 V) and V_{R1} is made adjustable, a rudimentary pulse-height analyzer is obtained. The amplitude of the input can be determined to within W volts.

If greater accuracy and speed of operation are required in determining the input magnitude, a number of level detectors can be used with a constant window or "bin" between adjacent comparators. This arrangement is the ADC of Fig. 16-44.

Amplitude-Distribution Analyzer

A comparator is a basic building block in a system used to analyze the amplitude distribution of the noise generated in an active device or the voltage spectrum of the pulses developed by a nuclear-radiation detector, etc. To be more specific, suppose that the output of the comparator is 10 V if $v_i > V_R$ and 0 V if $v_i < V_R$. Let the input to the comparator be noise. A dc meter is used to measure the average value of the output square wave. For example, if V_R is set at zero, the meter will read 10 V, which is interpreted to mean that the probability that the amplitude is greater than zero is 100 percent. If V_R is set at some value V_R' and the meter reads 7 V, this is interpreted to mean that the probability that the amplitude of the noise is greater than V_R' is 70 percent, etc. In this way the cumulative amplitude probability distribution of the noise is obtained by recording meter readings as a function of V_R.

Many other applications of comparators—particularly for the generation of specific types of periodic waveforms—are outlined in this chapter.

Spurious positive- and negative-voltage spikes, called *noise*, superimposed upon the input signal in the neighborhood of the amplitude V_R, may cause the output to "chatter" (change from one binary voltage to the other) several times before settling down to the correct level. This difficulty can be avoided, and also reduced values of transition time can be obtained if *positive feedback* or *regeneration* is added to a comparator, as discussed in the following section.

17-3 REGENERATIVE COMPARATOR (SCHMITT TRIGGER)[1, 3]

The transfer characteristic in Fig. 17-1b makes the change in output from -7 to $+7$ V for a swing in input of about 1.0 mV. Hence, the voltage gain is 14,000. By employing positive feedback the gain may be increased greatly. Consequently the total output excursion takes place in a time interval during which the input is changing by much less than 1 mV. Theoretically, if the loop gain $-\beta A_V$ is adjusted to be unity, then the gain with feedback A_{Vf} becomes infinite [Eq.

(12-4)]. Such an idealized situation results in an abrupt (zero rise time) transition between the extreme values of output voltage. If a loop gain in excess of unity is chosen, the output waveform continues to be virtually discontinuous at the comparison voltage. However, the circuit now exhibits a phenomenon called *hysteresis*, or *backlash*, which is explained in the following.

The regenerative comparator of Fig. 17-4a is commonly referred to as a *Schmitt trigger* (after the inventor of a vacuum-tube version of this circuit). The input voltage is applied to the inverting terminal 2 and the feedback voltage to the noninverting terminal 1. The feedback factor is $\beta = R_2/(R_1 + R_2)$. For $R_2 = 100 \ \Omega$, $R_1 = 10 \ \text{k}\Omega$, and $A_V = -14{,}000$, the loop gain is

$$- \beta A_V = \frac{0.1 \times 14{,}000}{10.1} = 139 \gg 1$$

It is easily verified that the feedback is regenerative. If the output *increases* by Δv_o, then $\beta \, \Delta v_o$ is fed back to v_1, the noninverting input terminal. Hence v_o will *increase* further by $\beta A_V \Delta v_o$, indicating positive feedback.

Let $V_o \equiv V_Z + V_D$ and assume that $v_i < v_1$ so that $v_o = + V_o$. From Fig. 17-4 we find that the voltage at the noninverting terminal is given by

$$v_1 = V_R + \frac{R_2}{R_1 + R_2}\,(V_o - V_R) \equiv V_1 \qquad (17\text{-}1)$$

If v_i is now increased, then v_o remains constant at V_o, and $v_1 = V_1 = $ constant

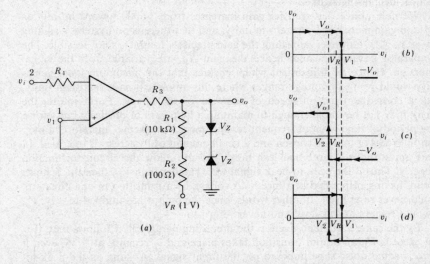

Figure 17-4 (*a*) An inverting Schmitt trigger. The transfer characteristics for (*b*) increasing v_i and (*c*) decreasing v_i. (*d*) The composite input-output curve.

until $v_i = V_1$. At this *threshold*, *critical*, or *triggering voltage*, the output regenera-tively switches to $v_o = - V_o$ and remains at this value as long as $v_i > V_1$. This transfer characteristic is indicated in Fig. 17-4*b*.

The voltage at the noninverting terminal for $v_i > V_1$ is

$$v_1 = V_R - \frac{R_2}{R_1 + R_2} (V_o + V_R) \equiv V_2 \qquad (17\text{-}2)$$

For the parameter values given in Fig. 17-4 and with $V_o = 7$ V,

$$V_1 = 1 + \frac{0.1 \times 6}{10.1} = 1 + 0.059 = 1.059 \text{ V}$$

$$V_2 = 1 - \frac{0.1 \times 8}{10.1} = 1 - 0.079 = 0.921 \text{ V}$$

Note that $V_2 < V_1$, and the difference between these two values is called the *hysteresis* V_H.

$$V_H = V_1 - V_2 = \frac{2 R_2 V_o}{R_1 + R_2} = 0.138 \text{ V} \qquad (17\text{-}3)$$

If we now decrease v_i, then the output remains at $- V_o$ until v_i equals the voltage at terminal 1 or until $v_i = V_2$. At this voltage a regenerative transition takes place and, as indicated in Fig. 17-4*c*, the output returns to $+ V_o$ almost instantaneously. The complete transfer function is indicated in Fig. 17-4*d*, where the portions without arrows may be traversed in either direction, but the other segments can only be obtained if v_i varies as indicated by the arrows. Note that because of the hysteresis, the circuit triggers at a higher voltage for increasing than for decreasing signals.

We note above that transfer gain increases from 14,000 toward infinity as the loop gain increases from zero to unity, and that there is no hysteresis as long as $- \beta A_V \leqslant 1$. However, adjusting the gain precisely to unity is not feasible. The comparator parameters and, hence the gain A_V, are variable over the signal excursion. Hence an adjustment which ensures that the maximum loop gain is unity would result in voltage ranges where this amplification is less than unity, with a consequent loss in speed of response of the circuit. Furthermore, the circuit may not be stable enough to maintain a loop gain of precisely unity for a long period of time without frequent readjustment. In practice, therefore, a loop gain in excess of unity is chosen and a small amount of hysteresis is tolerated. In some applications a large backlash range will not allow the circuit to function properly. Thus if the peak-to-peak signal were smaller than V_H, then the Schmitt circuit, having responded at a threshold voltage by a transition in one direction, would never reset itself. In other words, once the output has jumped to, say, V_o, it would remain at this level and never return to $- V_o$.

By the same argument given in the preceding paragraph it follows that, if v_i just exceeds V_1, an output transition takes place and v_o remains at $- V_o$, even if there is some noise superimposed on the input signal. As long as the peak-to-peak noise voltage does not exceed the hysteresis V_H, v_i cannot fall below V_2 and, therefore, a change of state back to $+ V_o$ is avoided. In other words, the noise chattering mentioned in Sec. 17-2 is eliminated.

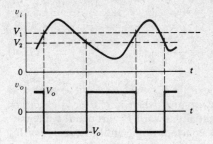

Figure 17-5 Response of the inverting Schmitt trigger to an arbitrary input signal.

The resistor R_4 in Fig. 17-4a is chosen equal to $R_B \equiv R_1 \| R_2$. Consequently the offset due to input current is $I_{io}R_B$, where I_{io} is the input offset current rather than $I_B R_B$ (if R_4 were missing), where $I_B \gg I_{io}$ = the bias current.

The most important use made of the Schmitt trigger is to convert a very slowly varying input voltage into an output having an abrupt (almost discontinuous) waveform, occurring at a precise value of input voltage. This regenerative comparator may be used in all the applications listed in Sec. 17-2. For example, the use of the Schmitt trigger as a squaring circuit is illustrated in Fig. 17-5. The input signal is arbitrary except that it has a large enough excursion to carry the input beyond the limits of the hysteresis range V_H. The output is a square wave as shown, the amplitude of which is independent of the peak-to-peak value of the input waveform. The output has much faster leading and trailing edges than does the input.

From Eqs. (17-1) and (17-2) it follows that, if $V_R = 0$, $V_2 = -V_1 = -R_2 V_o/(R_1 + R_2)$. If an input sinusoid of frequency $f = 1/T$ is applied to such a comparator, the output will be a symmetrical square wave of half period $T/2$. The vertical edge of the output waveform will not occur at the time the sine wave passes through zero, but will be shifted in phase by θ, where $\sin \theta = V_1/V_m$ and V_m is the peak sinusoidal voltage.

A noninverting Schmitt trigger is obtained if v_i and V_R are interchanged in Fig. 17-4 (Prob. 17-6).

Special-purpose Schmitt triggers are commercially available. For example, the TI-13, TI-14, and TI-132 chips behave as positive NAND gates with totem-pole outputs and with a hysteresis of 0.8 V. The positive-going threshold is $V_1 = 1.7$ V and the negative-going threshold is $V_2 = 0.9$ V. The TI-132 package consists of four two-input NAND Schmitt triggers.

17-4 SQUARE-WAVE AND TRIANGULAR-WAVE GENERATORS[4]

Integrating the output of the Schmitt comparator of Fig. 17-4 by means of an RC low-pass combination and applying the capacitor voltage to the inverting terminal in place of the external signal result in a free-running generator of square waves. The network is shown in Fig. 17-6, where a fraction $\beta = R_2/(R_1 + R_2)$ of the output is fed back to the noninverting input terminal. The differential input voltage v_i is given by

$$v_i = v_c - \beta v_o = v_c - \frac{R_2}{R_1 + R_2} v_o \qquad (17\text{-}4)$$

From the ideal-comparator characteristic $v_o = V_Z + V_D = V_o$ if $v_i < 0$ and $v_o = -V_o$ if $v_i > 0$. Consider an instant of time when $v_i < 0$ or $v_c < \beta v_o = \beta V_o$. The capacitor C now charges exponentially toward V_o through the integrating RC combination. The output remains constant at V_o until v_c equals $+\beta V_o$, at which time the comparator output reverses to $-V_o$. Now v_c charges exponentially toward $-V_o$. The output voltage v_o and capacitor voltage v_c waveforms are shown in Fig. 17-6b. If we let $t = 0$ when $v_c = -\beta V_o$ for the first half cycle, we have (since v_c approaches V_o exponentially with a time constant RC),

$$v_c(t) = V_o\big[1 - (1 + \beta)\epsilon^{-t/RC}\big] \qquad (17\text{-}5)$$

Since at $t = T/2$, $v_c(t) = +\beta V_o$, we find T, solving Eq (17-5), to be given by

$$T = 2RC \ln \frac{1 + \beta}{1 - \beta} = 2RC \ln \left(1 + \frac{2R_1}{R_2}\right) \qquad (17\text{-}6)$$

Note that T is independent of V_o.

This square-wave generator is particularly useful in the frequency range of 10 Hz to 10 kHz. At higher frequencies the slew rate of the operational amplifier limits the slope of the output square wave. Waveform symmetry depends upon the matching of the two Zener diodes (Prob. 17-8).

The circuit of Fig. 17-6 is called an *astable multivibrator* because it has two quasistable states. The output remains in one of these states for a time T_1 and

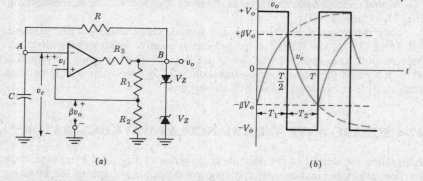

Figure 17-6 (a) A square-wave generator. (b) Output and capacitor voltage waveforms. Note that $\beta \equiv R_2/(R_1 + R_2)$ and $V_o \equiv V_Z + V_D$.

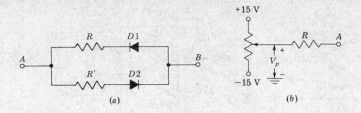

Figure 17-7 (a) In order to obtain an unsymmetrical square wave, the network shown may be used between nodes A and B to replace R in Fig. 17-6. (b) Alternatively, the configuration indicated may be tied to node A in Fig. 17-6 so that $T_1 \neq T_2$.

then abruptly changes to the second state for a time T_2, and the cycle of period $T = T_1 + T_2$ repeats.

If it is desired that $V_{o1} = V_{o2} = V_o$ but that $T_1 \neq T_2$ in Fig. 17-6, the resistor R between nodes A and B is replaced by the network shown in Fig. 17-7a. During the interval when the output is positive, $D1$ conducts but $D2$ is OFF. Hence, the circuit reduces to that in Fig. 17-6 except that V_o is reduced by the diode drop. Since the period is independent of V_o, then T_1 is given by $T/2$ in Eq. (17-6). During the interval when the output is negative, $D1$ is OFF and $D2$ conducts. Hence, the discharge-time constant is now $R'C$, and, therefore, T_2 is given by $T/2$ in Eq. (17-6) with R replaced by R'. If $R' = 2R$, clearly, $T_2 = 2T$.

An alternative method for obtaining an unsymmetrical square wave is to connect the network in Fig. 17-7b to node A of Fig. 17-6a. Assume that the potentiometer resistance is small compared with R and that the voltage from the potentiometer arm to ground is V_p. Then the capacitor charges with a time constant $RC/2$ toward $(V_p + V_o)/2$, but C discharges toward $(V_p - V_o)/2$ (with the same time constant). Hence, $T_1 \neq T_2$.

Triangle-Wave Generator

We observe from Fig. 17-6b that v_c has a triangular waveshape but that the sides of the triangles are exponentials rather than straight lines. To linearize the triangles, it is required that C be charged with a constant current rather than the exponential current supplied through R in Fig. 17-6. A JFET operating in the constant-current (saturation) region may be used to approximate such a linear ramp (Prob. 17-10). An improved system is that shown in Fig. 17-8, where an OP AMP integrator is used to supply constant current to C so that the output is linear. Because of the inversion through the integrator, this voltage is fed back to the noninverting terminal of the comparator in this circuit rather than to the inverting terminal, as shown in Fig. 17-6. In other words, the comparator behaves as a noninverting Schmitt trigger.

To find the maximum value of the triangular waveform assume that the square-wave voltage v_o is at its negative value, $-(V_Z + V_D) \equiv -V_o$. With a

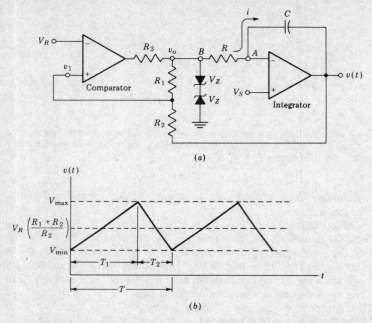

(a)

(b)

Figure 17-8 *(a)* Triangular waveform generator. *(b)* Output waveform. *Note:* $T_1 = T_2$, if $V_S = 0$. Also, $V_{max} = V_o R_2 / R_1 = - V_{min}$, if $V_R = 0$. The square-wave output is $- V_o$ for the interval T_1 and $+ V_o$ during T_2.

negative input, the output $v(t)$ of the integrator is an *increasing* ramp. The voltage at the noninverting comparator input v_1 is given by (using superposition)

$$v_1 = - \frac{V_o R_2}{R_1 + R_2} + \frac{v R_1}{R_1 + R_2} \qquad (17\text{-}7)$$

When v_1 rises to V_R, the comparator changes state, $v_o = + V_o$, and $v(t)$ starts *decreasing* linearly. Hence, the peak V_{max} of the triangular waveform occurs for $v_1 = V_R$. From Eq. (17-7),

$$V_{max} = V_R \frac{R_1 + R_2}{R_1} + V_o \frac{R_2}{R_1} \qquad (17\text{-}8)$$

By a similar argument it is found that

$$V_{min} = V_R \frac{R_1 + R_2}{R_1} - V_o \frac{R_2}{R_1} \qquad (17\text{-}9)$$

The peak-to-peak swing is

$$V_{max} - V_{min} = 2 V_o \frac{R_2}{R_1} \qquad (17\text{-}10)$$

The triangular waveform is indicated in Fig. 17-8b. From Eqs. (17-8) and (17-9) it should be clear that the average value is $V_R(R_1 + R_2)/R_1$. Note that, if $V_R = 0$, the waveform extends between $-V_o R_2/R_1$ and $+V_o R_2/R_1$. Its displacement in voltage is controlled by an adjustment of V_R and the peak-to-peak-swing is varied by changing the ratio R_2/R_1.

We now calculate the sweep times T_1 and T_2 for $V_S = 0$. The capacitor-charging current is

$$i = C \frac{dv_c}{dt} = -C \frac{dv}{dt} \tag{17-11}$$

where $v_c = -v$ is the capacitor voltage. For $v_o = -V_o$, $i = -V_o/R$, and the positive-sweep speed is $dv/dt = V_o/RC$. Hence,

$$T_1 = \frac{V_{max} - V_{min}}{V_o/RC} = \frac{2R_2 RC}{R_1} \tag{17-12}$$

where use was made of Eq. (17-10). Since the negative-sweep speed has the same magnitude as that calculated above, $T_2 = T_1 = T/2 = 1/2f$, where the frequency f is given by

$$f = \frac{R_1}{4R_2 RC} \tag{17-13}$$

Note that the frequency is independent of V_o. The maximum frequency is limited by either the slew rate of the integrator or its maximum output current, which determines the charging rate of C. The slowest sweep is limited by the bias current of the OP AMP. Decade changes in frequency are obtained by switching capacitance values by factors of 10, and increments of frequency within a decade result from continuous variations of resistance R.

Duty-Cycle Modulation

If unequal sweep intervals $T_1 \neq T_2$ are desired, then R in Fig. 17-8a may be replaced by the network in Fig. 17-7a. An alternative method is to apply an adjustable voltage $V_S \neq 0$ to the noninverting terminal of the integrator, as indicated in Fig. 17-8a. The positive-sweep speed is now $(V_o + V_S)/RC$ and the negative-ramp slope is $(V_o - V_S)/RC$. (Why?) The peak-to-peak triangular amplitude is unaffected by the symmetry control voltage V_S. Hence,

$$\frac{T_1}{T_2} = \frac{V_o - V_S}{V_o + V_S} \tag{17-14}$$

The oscillation frequency can be shown (Prob. 17-9) to be given by Eq. (17-13) multiplied by $[1 - (V_S/V_o)^2]$. The frequency is lowered for $V_S \neq 0$.

The *duty cycle* δ of a square- or triangular-wave oscillator is defined as T_1/T, where $T = T_1 + T_2$. From Eq. (17-14) it follows that

$$\delta \equiv \frac{T_1}{T} = \frac{1}{2}\left(1 - \frac{V_S}{V_o}\right) \tag{17-15}$$

The system in Fig. 17-8 with the addition of V_S is a *duty-cycle modulator*. The duty cycle varies linearly with V_S and extends from 0 for $V_S = V_o$, to $\frac{1}{2}$ for $V_S = 0$, and to 1 for $V_S = -V_o$.

Voltage-controlled Oscillator (VCO)

Note that V_S in Fig. 17-8 not only modifies the duty cycle but also affects the period $T = 1/f$. This is an example of a VCO or of a *voltage-to-frequency converter*. However, f is a nonlinear function of V_S, since the frequency depends upon $1 - (V_S/V_o)^2$.

A system for obtaining a square- or triangular-waveform generator whose frequency depends *linearly* upon a modulation voltage v_m is indicated in Fig. 17-9. Note that a CMOS digitally controlled switch (Fig. 16-40b) has been added between the comparator and the integrator. This system differs functionally from that of Fig. 17-8 in that now the sweep speed is determined by v_m but

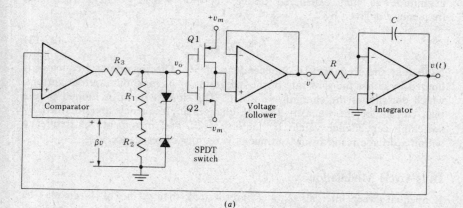

(a)

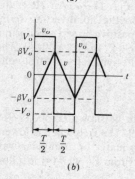

(b)

Figure 17-9 (a) A voltage-controlled oscillator whose frequency varies linearly with the modulating voltage v_m. (b) The square wave v_o and the triangular wave v. Note that it is possible to choose $R_1 = 0$ and $R_2 = \infty$ so that $\beta = 1$.

the waveform amplitude continues to be fixed by the comparator parameters, as in Fig. 17-6, namely, $\pm \beta V_o$. The negative voltage $-v_m$ is obtained from an OP AMP unity-gain inverter.

Assume that the Schmitt comparator output is $v_o = V_o$, where V_o exceeds the maximum value of v_m. Then, in the CMOS digitally controlled SPDT, switch $Q1$ is OFF and $Q2$ is ON. The input v' to the integrator (the output of the voltage follower) is $-v_m$. Hence, $v(t)$ increases linearly with a sweep speed of v_m/RC V/s until v reaches the comparator threshold level $\beta V_o = V_o R_2/(R_1 + R_2)$. Then the Schmitt output changes state to $v_o = -V_o$, as depicted in Fig. 17-9b. Now $Q1$ is ON, $Q2$ is OFF, and the CMOS switch output becomes $+v_m$, resulting in a linear negative ramp $v = -v_m t/RC$ until the negative threshold $-\beta V_o$ is reached. Clearly, the two half cycles are identical and

$$\frac{v_m}{RC} \frac{T}{2} = \beta V_o - (-\beta V_o) = 2 \frac{R_2}{R_1 + R_2} V_o \qquad (17\text{-}16)$$

The frequency of the oscillator is given by $f = 1/T$, or

$$f = \frac{R_1 + R_2}{4RCR_2} \frac{v_m}{V_o} \qquad (17\text{-}17)$$

clearly indicating that this VCO frequency varies linearly with the modulation voltage v_m. Experimentally, it is found[6] that this linearity extends over more than three decades (from below 2 mV to above 2 V). The system of Fig. 17-9 is that of a *frequency-modulated* square or triangular waveform.

A Positive-Negative Controlled-Gain Amplifier

The SPDT digitally controlled switch of Fig. 17-9 may be implemented in a number of ways. For example, the positive-negative gain amplifier[6] (also referred to as a *sign-reversing* or *biphase amplifier*), indicated in Fig. 17-10, may

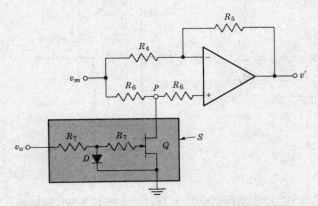

Figure 17-10 A positive-negative controlled-gain amplifier. $A = \pm 1$ if $R_5 = R_4$. The resistances R_6 are large compared with $r_{DS(ON)}$ of the JFET. The resistances R_7 and the diode D limit the gate current of Q.

replace the network between v_o and v' in Fig. 17-9a. The block S is a single-pole switch which connects point P to ground if v_o is positive, and leaves P floating if v_o is negative. If $v_o = -V_o$ so that the JFET is OFF, then S is open and the amplifier gain is $A = +1$ by the following argument: The inverting gain is $-R_5/R_4$ and the noninverting gain is $+(1 + R_5/R_4)$, so that, for $R_5 = R_4$, $A = -1 + 2 = +1$ and $v' = +v_m$. However, if $v_o = +V_o$, then Q is driven ON (the diode D prevents the FET gate from drawing excessive current) and S acts as a very-low-resistance closed switch, effectively grounding P and the non-inverting terminal. Under these circumstances $A = -R_5/R_4 = -1$ and $v' = -v_m$. These considerations indicate that the sign-reversing amplifier is operationally equivalent to the SPDT-controlled switch in Fig. 17-9a.

17-5 PULSE GENERATORS

A monostable multivibrator (abbreviated *monostable multi*) has one stable state and one quasistable state. The circuit remains in its stable state until a triggering signal causes a transition to the quasistable state. Then, after a time T, the circuit returns to its stable state. Hence a single pulse has been generated, and the circuit is referred to as a *one-shot* or *single-shot*.

The square-wave generator of Fig. 17-6 is modified in Fig. 17-11 to operate as a monostable multi by adding a diode ($D1$) clamp across C and by introducing a narrow negative triggering pulse through $D2$ to the noninverting

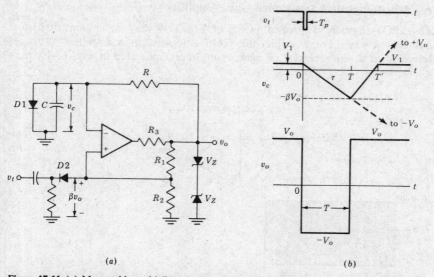

(a)   (b)

Figure 17-11 (a) Monostable multivibrator, or pulse generator. (b) The negative narrow triggering pulse v_t, the capacitor waveform v_c, and the generated negative output pulse v_o (with $T \gg T_p$).

terminal. To see how the circuit operates, assume that it is in its stable state with the output at $v_o = + V_o$ and with the capacitor clamped at the diode $D1$ ON voltage $V_1 \approx 0.7$ V (with $\beta V_o > V_1$). If the trigger amplitude is greater than $\beta V_o - V_1$, it will cause the comparator to switch to an output $v_o = - V_o$. As indicated in Fig. 17-11b, the capacitor will now charge exponentially with a time constant $\tau = RC$ through R toward $- V_o$ because $D1$ becomes reverse-biased. When v_c becomes more negative than $- \beta V_o$, the comparator output swings back to $+ V_o$. The capacitor now starts charging toward $+ V_o$ through R until v_c reaches V_1 and C becomes clampled again at $v_c = V_1$. In Prob. 17-12 we find that the pulse width T is given by

$$T = RC \ln \frac{1 + V_1/V_o}{1 - \beta} \qquad (17\text{-}18)$$

If $V_o \gg V_1$ and $R_2 = R_1$ so that $\beta = \frac{1}{2}$, then $T = 0.69RC$.

The triggering pulse width T_p must be much smaller than the duration T of the generated pulse. The diode $D2$ is not essential but it serves to avoid malfunctioning if any positive noise spikes are present in the triggering line.

Since the one-shot generates a rectangular waveform which starts at a definite instant of time and, hence, can be used to gate other parts of a system, it is called a *gating circuit*. Furthermore, since it generates a fast transition at a predetermined time T after the input trigger, it is also referred to as a *time-delay circuit*.

Note that the capacitor voltage v_c in Fig. 17-11b does not reach its quiescent value $v_c = V_1$ until time $T' > T$. Hence, there is a *recovery time* $T' - T$ during which the circuit may not be triggered again. In other words, the next synchronizing trigger must be delayed from the previous input pulse by at least T' seconds. An alternative monostable circuit[2] with a faster recovery time is given in Prob. 17-13.

The Signetics NE/SE 555 *timer* (operating in a monostable mode) produces accurate time delays over the range from microseconds through hours, by using an external resistor and capacitor. It also functions in the astable mode. The free-running frequency and duty cycle can be controlled with external components.

The Intersil 8038 *waveform generator and voltage-controlled oscillator* can produce pulse, square, triangular, sawtooth, and sinusoidal waveforms of high accuracy. The frequency can be selected externally over a tremendous range $(10^{-3}$ to 10^{+6} Hz) and is highly stable with changes in temperature and supply voltage.

A Retriggerable Monostable Multivibrator[4]

Consider the configuration in Fig. 17-12a. In the quiescent state (before a trigger is applied) the JFET is cut off by the reverse-biased gate-to-source voltage $- V'$ of Q. The capacitor is charged to the supply voltage V so that the voltage at the inverting terminal of the comparator is $v_c = V$. The noninverting input voltage

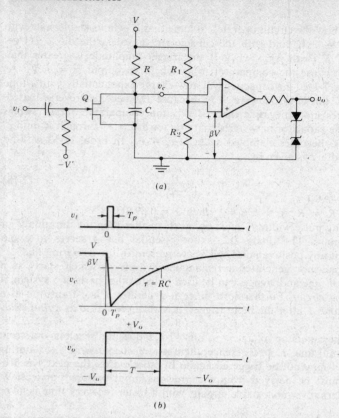

(a)

(b)

Figure 17-12 (a) A retriggerable monostable multivibrator. (b) The positive narrow triggering pulse v_t, the capacitor waveform v_c, and the generated positive output gate v_o (with $T \gg T_p$).

is constant and equals $\beta V = VR_2/(R_1 + R_2)$. Since $v_c > \beta V$, the comparator output is at its low level, $v_o = -V_o$.

Assume that at $t = 0$, a narrow, positive triggering signal v_t is applied, with the pulse amplitude approximately equal to V'. The JFET conducts with a large constant current (Fig. 8-3) and rapidly discharges C linearly toward ground. For small voltages, v_c no longer falls linearly, but it approaches zero exponentially with a time constant $r_{DS(ON)}C$ (Sec. 8-2). The waveforms for v_c and v_o are plotted in Fig. 17-12b. As soon as v_c falls below βV, the comparator output changes to its high level, $v_o = +V_o$.

We assume that the pulse width T_p is large enough so that $v_c \approx 0$ at the end of the input signal. Then, at $t = T_p$, the capacitor charges exponentially with a time constant RC toward V. When $v_c = \beta V$, the comparator switches again and, for $v_c > \beta V$, v_o remains at $-V_o$, thereby generating the positive gating waveform of width T shown in Fig. 17-12b. It can be shown that

$$T = RC \ln\left(1 + \frac{R_2}{R_1}\right) \tag{17-19}$$

In deriving this equation it is assumed that $T \gg T_p$. A better approximation is to add T_p to the right-hand side of Eq. (17-19).

Note that, unlike most monostable configurations (for example, that in Fig. 17-11), no recovery time is required before the system in Fig. 17-12 can be triggered again. If a second positive input pulse appears at any time t' (less than or greater than T), the JFET reduces the voltage on C to zero and the waveforms indicated in Fig. 17-12b are generated at $t = t'$ instead of $t = 0$. Therefore a new gating interval T is initiated at $t = t'$. Such a circuit is called a *retriggerable monostable multi*.

17-6 VOLTAGE TIME-BASE GENERATORS[7]

A linear time-base generator is one that provides an output waveform, a portion of which exhibits a linear variation of voltage or current with time. A very important application of such a waveform is in connection with a cathode-ray oscilloscope ("scope"). The display on the screen of a scope of the variation with respect to time of an arbitrary waveform requires that there be applied to one set of deflecting plates a voltage which varies linearly with time. Since this waveform is used to *sweep* the electron beam horizontally across the screen, it is called a *sweep voltage*. There are, in addition, many other applications for time-base circuits, such as in radar and television indicators, in precise time measurements, and in time modulation.

The typical form of a time-base voltage is shown in Fig. 17-13a. The voltage, starting from some initial value, increases linearly with time to a maximum amplitude V_s, after which it drops to its initial value. The time T_r required for the return to the initial value is called the *restoration time*, the *return time*, or the *flyback time*. Very frequently the shape of the waveform during the restoration time and the interval T_r are unimportant.

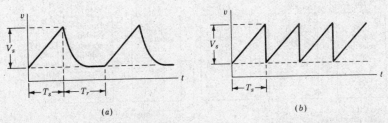

(a)   (b)

Figure 17-13 (a) A general sweep voltage. The sweep time is T_s and the return time is T_r. The sweep amplitude is V_s. (b) A sawtooth voltage waveform of period T_s.

In some cases, however, a restoration time is desired which is very short in comparison with the time occupied by the linear portion of the waveform. If it should happen that the restoration time is extremely short and that a new linear voltage is initiated at the instant the previous one is terminated, then the waveform will appear as in Fig. 17-13b. This figure suggests the designation *sawtooth generator* or *ramp generator*. It is customary to refer to waveforms of the type indicated in Fig. 17-13 as *sweep* waveforms even in applications not involving the deflection of an electron beam.

Clearly, the triangular voltage of Fig. 17-8b is a sweep waveform with a sweep time T_1 and a return time T_2. A sawtooth waveshape is obtained by making $T_2 \ll T_1$ by one of the methods discussed in Sec. 17-4. The flyback time cannot be reduced to zero because of the limitations set by the slew rate of the integrator or by its maximum output current I (since the sweep speed is $dv/dt = I/C$).

The Triggered Sweep

A waveform may not be periodic but may occur rather at irregular intervals. In such a case it is desirable that the sweep circuit, instead of running continuously, should remain quiescent and wait to be initiated by the waveform itself. Even if it should happen that the waveform does recur regularly, it may be that the interesting part of the waveform is short in time duration in comparison with the period of the waveform. For example, the waveform might consist of 1-μs pulses with a time interval of 100 μs between pulses. In this case the fastest recurrent sweep which will provide a synchronized pattern will have a period of 100 μs. If, typically, the time base is spread out over 10 cm, the pulse will occupy 1 mm and none of the detail of form of the pulse will be apparent. If, on the other hand, a sweep of period 1 μs or somewhat larger could be used, the pulse would be spread across the entire screen. Therefore, what is required here is a sweep set for, say, a 1.5-μs interval which remains quiescent until it is initiated by the pulse. Such a monostable circuit is known as a *driven* sweep or a *triggered* sweep.

A block diagram for a time-base system for a cathode-ray tube (CRT) is indicated in Fig. 17-14. The waveform v_s to be observed is applied through a

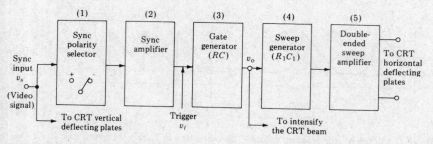

Figure 17-14 A block diagram of a time-base system for a cathode-ray tube (CRT).

high-quality video amplifier (not indicated in Fig. 17-14) to the vertical-deflecting plates of the CRT. This signal is simultaneously applied to the sweep system as the synchronizing input. In block (1) the sync polarity selection is made by taking the output across either a collector or an emitter resistor. The sync amplifier (2) need not operate linearly, since all that is required is that the output v_t be large and fast enough to be able to trigger the (monostable) gate generator. In some scopes a Schmitt trigger is used to obtain a sharp pulse on either the rising or falling portion of the signal, as desired. Since this trigger is used to start the sweep, a selected portion of the input signal appears on the scope face.

The third block in Fig. 17-14 is a monostable multivibrator (Sec. 17-5) whose gate width is determined by the time constant RC (Fig. 17-11a). A negative gating waveform (v_o in Fig. 17-11b) is applied to the sweep generator (4), whose sweep speed depends upon a resistor R_1 and a capacitor C_1 (Fig. 17-15). The sweep generator output is amplified linearly (5) and applied to the horizontal-deflecting plates of the CRT.

In a case in which the sweep time is short in comparison with the time between sweeps the CRT beam will remain in one place most of the time. If the intensity is reduced to prevent screen burns, the fast trace will be very faint. To intensify the trace during the sweep, a positive gate which is derived from the output of the multi is applied to the CRT grid. As a matter of fact, in the presence of this "unblanking" or "intensifier signal" the beam brightness may be adjusted so that the spot is initially invisible but the trace will become visible as soon as the sweep starts.

Sweep Generators

The simplest sweep is obtained by charging a capacitor C_1 through a resistor R_1 from a supply voltage V, as indicated in Fig. 17-15a. At $t = 0$ the switch S is opened and the sweep $v(t)$ is given by

$$v = V(1 - \epsilon^{-t/R_1 C_1}) \tag{17-20}$$

For the present discussion the physical form of the switch S is unimportant. After an interval T_s, when the sweep amplitude reaches V_s, the switch again closes. The resulting sweep waveform is indicated in Fig. 17-15b (assuming zero switch resistance).

Note that the sweep voltage is exponential and not linear. In the case of a cathode-ray oscilloscope, an important requirement of the sweep is that the sweep speed be constant. Hence, a reasonable definition of the deviation from linearity is given by the *slope* or *sweep-speed error e_s*.

$$e_s \equiv \frac{\text{Difference in slope at beginning and end of sweep}}{\text{Initial value of slope}} \tag{17-21}$$

If this definition is applied to Eq. (17-20), we find (Prob. 17-15) that, *independent*

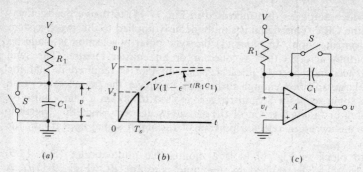

(a)   (b)   (c)

Figure 17-15 (a) Charging a capacitor through a resistor from a fixed voltage; (b) the resultant exponential waveform; (c) a Miller integrator sweep generator.

of the time constant, for a fixed sweep amplitude V_s and power supply voltage V,

$$e_s = \frac{V_s}{V} \qquad (17\text{-}22)$$

The linearity improves as the ratio V_s/V decreases. Hence, the simple circuit of Fig. 17-15a is useful only in applications requiring sweep voltages of the order of volts or tens of volts. For example, a 20-V sweep can be obtained with a sweep-speed error of less than 10 percent by using a supply voltage of at least 200 V. Time-base voltages of hundreds of volts require power supplies of thousands of volts, which are inconveniently large.

A tremendous improvement in linearity is obtained by using the OP AMP (Miller) integrator [7] of Fig. 17-15c instead of the simple circuit in Fig. 17-15a. If the magnitude of the amplifier voltage gain is A, if the input resistance is $R_i = \infty$, and if the output resistance is $R_o = 0$, then $v = Av_i$. The input v_i is V_s/A when the sweep amplitude at the amplifier output is V_s. Hence, from Eq. (17-22), $e_s = V_s/AV$ which indicates that

$$e_s(\text{Fig. } 17\text{-}15c) = \frac{1}{A}\, e_s(\text{Fig. } 17\text{-}15a) \qquad (17\text{-}23)$$

Since $A \approx 100,000$, the Miller circuit generates extremely linear ramp voltages.

An approximately linear sweep may also be obtained by replacing the resistor R_1 in Fig. 17-15a by a JFET operating as a constant-current source (Prob. 17-18) or by using the *bootstrap*[7] configuration of Prob. 17-19.

The switch S may be a JFET driven by a gate generator, as indicated in Fig. 17-16a (corresponding to Fig. 17-15a). The video signal v_s to be observed is indicated in Fig. 17-16b. As shown in Fig. 17-14, this signal is amplified to form a trigger v_t for the monostable multi, whose output v_o is pictured in Fig. 17-16c. There is a small delay (not shown) between the beginning of the pulse in Fig. 17-16b and the beginning of the gate in Fig. 17-16c. In the quiescent state Q is ON because $v_o = +V_o$, and the capacitor voltage is held close to zero since $r_{DS(ON)} \ll R_1$. During the gating interval T_s the FET is cut off by the gate voltage

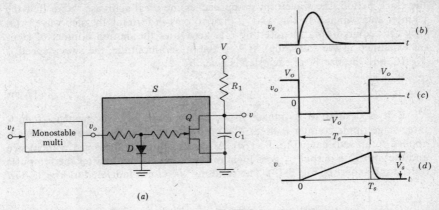

Figure 17-16 (a) A triggered sweep generator. The switch S in Fig. 17-15a or c is implemented by the block S in Fig. 17-10 and is driven from a gate generator. (b) A video pulse-type waveform v_s to be observed on the scope. (c) The gate output v_o of the monostable multi. (d) The generated sweep which is synchronized with the input signal.

$- V_o$ and the capacitor charges, thus generating the sweep v in Fig. 17-16d. At the end of the interval T_s, v_o returns to $+V_o$, which drives the FET ON, discharging C rapidly for a short retrace time, as explained in connection with the waveform v_c in Fig. 17-12b. (The diode D prevents the gate of Q from drawing appreciable current.)

We have already noted that the sweep speed is determined by $R_1 C_1$ in the sweep generator, whereas the gate width is determined by RC in the gate generator. If the sweep amplitude is to remain nominally constant, the gate controls R and C must be adjusted whenever the sweep speed controls R_1 and C_1 are varied. Capacitors C_1 and C are switched simultaneously to change the range of sweep speed, and resistor R_1, which is used for continuous variation of sweep speed, is ganged to R. No attempt is made to maintain constant amplitude with any precision. The sweep amplitude is deliberately made so large that the end of the sweep occurs at a point well off the CRT screen, so that variations of amplitude are not observed.

17-7 STEP (STAIRCASE) GENERATORS

An example of a staircase waveform is that of Fig. 16-42b, generated in the A/D converter system of Fig. 16-42a. A simpler implementation (not requiring the D/A block in Fig. 16-42a) is indicated in Fig. 17-17a.

A train of negative clock pulses v_p (Fig. 17-17b) is applied to an OP AMP integrator. The output v of the integrator rises linearly for the short duration T_p of each pulse and remains constant in between pulses (Fig. 17-17c). If $T_p \ll T =$

the clock period, the waveform v approaches the ideal staircase. Note that the counter and switch S in Fig. 17-17a play no part in forming the step waveform; they are required only for resetting v to zero after the desired number of steps (as explained in the following). If V is the pulse amplitude, the sweep speed is V/RC, and the size V' of each step is given by

$$V' = \frac{VT_p}{RC} \tag{17-24}$$

If it is desired to terminate the staircase after, say, seven steps, then a three-stage ripple counter is used. The output of each of the three FLIP-FLOPS is applied to an AND gate (Fig. 17-17a). After the 7th pulse there is a coincidence and the AND gate output v_A goes high and remains high until after the 8th pulse (refer to the chart of Fig. 7-14). The resulting waveform indicated in Fig. 17-17d

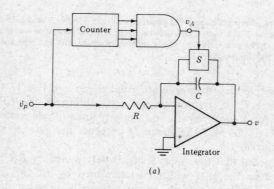

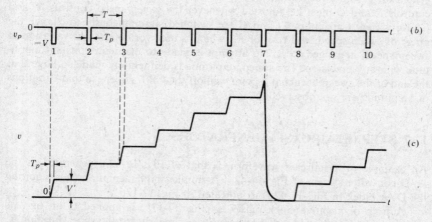

Figure 17-17 (a) A staircase waveform v is obtained by applying a train of narrow negative pulses v_p to a Miller integrator. The counter, AND gate, and controlled switch S perform the resetting operation. The waveforms v_p and v are drawn in Fig. 17-17 b and c, respectively.

is used to control the switch S of Fig. 17-16a, which discharges C rapidly to zero, as shown in Fig. 17-17c. Resetting at any desired step can be accomplished by modifying a ripple counter with an appropriate feedback gate, as explained in connection with Fig. 7-16.

A Storage Counter

The change from one step to the next in Fig. 17-17c takes place in the time T_p (one pulse width). A much more abrupt rise may be obtained with the *storage-counter* configuration of Fig. 17-18. To understand the operation, assume that capacitor C_1 is uncharged and C_2 is charged to a voltage v. An input pulse will cause the capacitor C_1 to charge through the diode $D1$. The time constant with which C_1 charges is the product of C_1 times the sum of the diode and the voltage-follower resistances. This time constant can be very small in comparison with the duration of the pulse, and C_1 will charge fully to the value $v_1 = V$, with the polarity indicated. During the charging time of C_1, the diode $D2$ does not conduct and the voltage across C_2 remains at v. At the termination of the input pulse, the capacitor C_1 is left with the voltage $v_1 = V$, which now appears across $D1$. The polarity of this voltage is such that $D1$ will not conduct. The capacitor C_1 will, however, discharge through $D2$ and the amplifier output resistance into C_2. The virtual ground at the input terminals of the operational amplifier takes no current. Hence, all the charge $C_1 V$ which leaves C_1 must transfer to C_2. The increase in voltage across C_2 is, therefore,

$$V' = \frac{C_1 V}{C_2} \tag{17-25}$$

and the voltage across C_1 is reduced to zero. By the foregoing argument, the next pulse again charges C_1 to the voltage V during T_p and abruptly transfers the charge $C_1 V$ to C_2 at the end of the pulse, so that v decreases by another step of the same size V' given by Eq. (17-25).

Applications

The staircase waveshape is frequently useful to vary some voltage in a step fashion. A (very-high-frequency) sampling scope uses such a step generator.[9]

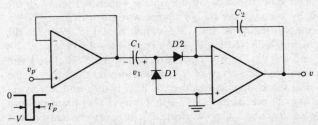

Figure 17-18 A storage-counter staircase generator. The resetting circuitry is identical with that used in Fig. 17-17.

The staircase waveform may also be used to trace out a family of BJT or FET volt-ampere characteristics on a CRT. In this application each step of the staircase corresponds to a particular constant value of base current or of gate voltage.

17-8 MODULATION OF A SQUARE WAVE[4]

The variation of a high-frequency-*carrier* characteristic proportional to a lower-frequency signal is called *modulation*. The parameter being modulated may be frequency, amplitude, or pulse width. The VCO system of Fig. 17-9 is an example of a frequency-modulated (FM) square waveform. Equation (17-17) shows that the frequency f is proportional to the modulating-signal magnitude v_m.

Amplitude Modulation (AM)

By multiplying any carrier waveform by a modulating signal v_m an amplitude-modulated signal is obtained because the instantaneous value of the carrier is proportional to v_m. An analog multiplier, such as the transconductance multiplier of Fig. 16-38, is used in this application for a sinusoidal carrier.

If the carrier v_o is a square wave, the multiplication can be performed very simply with a positive-negative controlled-gain amplifier. The configuration of Fig. 17-10 (with the frequency of the modulating voltage v_m much smaller than that of v_o) is a very simple amplitude-modulated system. In Fig. 17-19a the modulating signal v_m is shown as a piecewise linear signal (for ease of drawing), and the carrier is the square wave v_o in Fig. 17-19b. The amplitude-modulated wave is sketched in Fig. 17-19c. Note that, when $-v_o$ is positive, $v' = v_m$ and when $-v_o$ is negative, $v' = -v_m$. In other words, the square wave is multiplied by the modulating signal. This system is often referred to as a *pulse-height modulator* or a *pulse-amplitude modulator* (PAM).

A Chopper Modulator[10]

A very simple amplitude modulator is obtained by "chopping" the signal with a switch which is controlled synchronously by the square wave. In Fig. 17-20 the switch S_1 is controlled by the negative of the square waveform in Fig. 17-19b. An excellent implementation for S_1 is the JFET switch S of Fig. 17-16 or the CMOS analog switch of Fig. 8-28. During T_2 when v_o (in Figs. 17-19 and 17-20) is negative, S_1 is open and $v = v_m$. During T_1 when v_o is positive, S_1 is closed, and $v = 0$, assuming that the closed resistance of S_1 is much smaller than R. For the modulating signal v_m and the chopping signal v_o in Fig. 17-19a and b, respectively, the waveform v is as indicated in Fig. 17-20b. Observe that the waveform v is a *chopped* or *sampled* version of the waveform v_m. It is for this reason that the circuit of Fig. 17-20a is called a *chopper*.

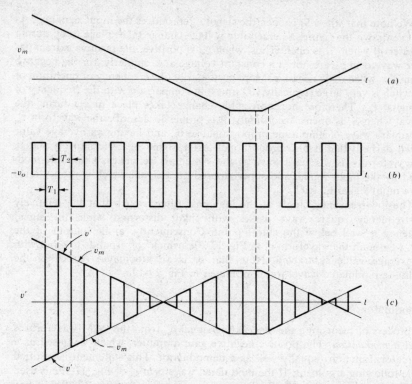

Figure 17-19 (a) A modulating signal. (b) A constant-frequency square-wave carrier. (c) The amplitude modulated waveform.

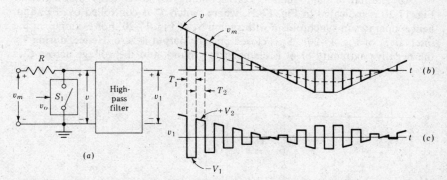

Figure 17-20 (a) A chopper modulator. (b) A chopped reproduction of the modulating signal of Fig. 17–19a. (c) The amplitude-modulated waveform. *Note*: See Fig. 17-19b for v_o.

We note that when S_1 is open the signal v reproduces the input signal v_m. As we have drawn the figure, a perceptible voltage change takes place in v_m during any interval when S_1 is open. Thus, when v_m is positive, the positive extremities of the waveform v_m are not at a constant voltage and, similarly, for the negative extremities when v_m is negative. More customarily, the frequency of operation of the switch is very large (typically 100 times) in comparison with the frequency of the signal v_m. Therefore no appreciable change takes place in v_m during the interval when S_1 is open. Accordingly, it is proper to describe the waveform v_m as a square wave of amplitude proportional to v_m and having an average value (shown dashed) that is also proportional to the signal v_m. Alternatively stated, the waveform v is a square wave at the switching frequency, amplitude-modulated by the input signal, and superimposed on a signal which is proportional to the input signal v_m itself.

The low-frequency cutoff of the high-pass filter is such that the relatively high-frequency square wave passes with small distortion while the signal frequency is well below the cutoff point. Consequently, at the output of the filter, we obtain the waveform v_1 in Fig. 17-20c, which corresponds to v but with the average value subtracted. Note that v_1 is an attenuated replica of the amplitude-modulated waveform v' obtained in Fig. 17-19c.

Demodulators

The process of recovering the modulating signal v_m from the PAM waveform is called *demodulation*. The positive-negative gain amplifier, which was used as a modulator, functions equally well as a demodulator. This statement is justified by the following argument: If the modulated waveform v' of Fig. 17-19c is used as the input v_m to Fig. 17-10, then, in the interval T_1 (Fig. 17-19b), when $v' = -v_m$, the gain A is -1, and in the next half period T_2, when $v' = v_m$, $A = +1$. Hence, the output v' (in Fig. 17-10) in any interval is v_m (in Fig. 17-19). Clearly, we have reconstituted the original signal v_m.

An alternative demodulator, corresponding to the chopper modulator of Fig. 17-20 is indicated in Fig. 17-21, where switch S_2 is controlled by $+v_o$ and, hence, operates in synchronism with switch S_1 of Fig. 17-20. For example, in the interval T_1 of Fig. 17-20c, S_2 is closed and the output is zero. Hence, during T_1, the negative extremity of v_1 is clamped to ground, and the voltage across C is

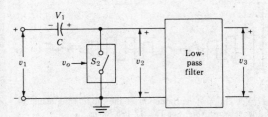

Figure 17-21 A synchronous demodulator.

$- V_1$, as indicated in Fig. 17-21. In the next half cycle T_2 of the square wave, S_2 is open, $v_1 = + V_2$, and $v_2 = V_2 + V_1$, which is the amplitude of v (Fig. 17-20b) during T_2. As a consequence of the clamping action of C and the controlled switch S_2, the waveform v_1 is reconverted into the chopped modulated signal v of Fig. 17-20b. If this waveform v is passed through the low-pass filter of Fig. 17-21, which rejects the high-frequency square wave and transmits the low-frequency signal, the resulting waveform v_3 is the modulating waveform v_m of the Fig. 17-19a. The combination of the capacitor C, the switch S_2, and the low-pass filter constitutes a *synchronous demodulator*.

A Chopper-stabilized Amplifier

A modulator-demodulator system which has a particularly interesting application will now be discussed. Let us assume that it is required to amplify a small signal $v_m(t)$ (say, of the order of millivolts) and that dv_m/dt is extremely small. For example, if the signal is periodic, it may be that the period is minutes or even hours in duration. An ac amplifier with the customary coupling between stages would not be feasible, since these blocking capacitances would be impractically large. Instead it would be necessary to use direct coupling between stages. With such a dc amplifier we would not be able to distinguish between a change in output voltage as a result of a variation in input voltage or as a consequence of a drift in some active device or some component, perhaps because of a temperature change. If the amplifier has high gain, even a tiny shift in the operating point of the first stage, amplified by the following stages, might cause a large variation in output. In summary: For this application an extremely stable (that is, drift-free) dc amplifier is required.

A method of circumventing the above difficulty is to use an ac amplifier, but to precede it by a modulator and to follow it by a demodulator. This system is indicated in Fig. 17-22. Since the slowly varying input signal v_m is chopped, it is easily handled by a conventional ac amplifier (which is a high-pass system). The amplified waveform is then demodulated to reconstitute an enlarged replica of the input v_m. This system is called a chopper-stabilized amplifier. Note, however, that the amplifier is not stabilized by the choppers but rather that the synchronous modulator-demodulator combination eliminates the necessity for an extremely drift-free direct-coupled amplifier.

The frequency response of a chopper-stabilized amplifier is very low. However, high-frequency-stabilized amplifiers are available from several manufacturers, because they augment[11] the chopper with a high-frequency ac-coupled OP

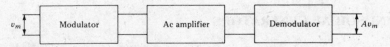

Figure 17-22 A chopper-stabilized amplifier.

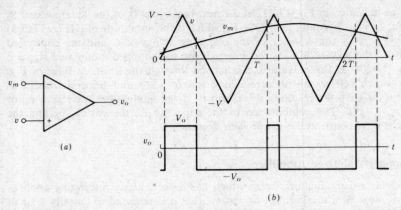

Figure 17-23 (a) A comparator used as a pulse-width modulator. (b) A triangular waveform v is used for the reference, v_m is the modulating signal, and v_o is the output pulse train.

AMP so that the overall response extends down to zero frequency. For example, the Harris Semiconductor HA-2900 or the Burr-Brown Research Corporation 3292 have the following excellent characteristics: Offset voltage drift of ± 0.3 μV/°C, offset current drift of ± 1 pA/°C, a unity-gain bandwidth of 3 MHz, and a minimum open-loop gain of 140 dB.

Pulse-Width Modulation

If a triangular waveform $v(t)$ is applied to a comparator whose reference voltage V_R is not constant but rather is an audio signal $v_m(t)$, a succession of pulses is obtained. The width of these pulses reflects the audio information. Such a *pulse-width modulation* system is indicated in Fig. 17-23a.

If $v > v_m$ then the comparator output $v_o = V_o$, and if $v < v_m$, then $v_o = -V_o$, as indicated in Fig. 17-23b. When $v_m = 0$ the pulse width is $T/2$, where T is the period of the triangular wave. As v_m increases, it linearly decreases the width of the output pulses v_o. The pulse train has an average value which is proportional to the modulating signal. Hence, an average-value detector may be used as a demodulator. Note that equal negative and positive switching delays in the comparator cancel out and do not affect the pulse width.

The system just described is also a *linear duty-cycle modulator* (Sec. 17-4). The duty cycle is given by $\delta = 0.5 (1 - v_m/V)$, where V is the peak value of the triangular wave.

17-9 SINUSOIDAL GENERATORS

The configurations considered in the preceding sections of this chapter involve extremely high-gain amplifiers so that the loop gain is much larger than unity. Under these circumstances the OP AMP (comparator) acts as a voltage-controlled

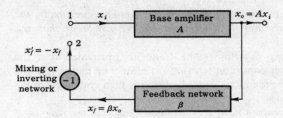

switch which changes the output between binary levels. The operation is highly nonlinear, and the previous discussions were based on transient analyses—first with the comparator in one state and then in the other. We shall assume in the remainder of this chapter that each system considered operates essentially linearly. If the input to a linear network is sinusoidal, the output is also sinusoidal of the *same* frequency, although the magnitude and phase may be altered. Hence, if a *linear* feedback amplifier (without input-signal excitation) oscillates, the output waveform is sinusoidal. We now investigate the basic principles governing these sinusoidal generators.

Figure 17-24 shows an amplifier, a feedback network, and an input mixing circuit not yet connected to form a closed loop. The amplifier provides an output signal x_o as a consequence of the signal x_i applied directly to the amplifier input terminal. The output of the feedback network is $x_f = \beta x_o = A\beta x_i$, and the output of the mixing circuit (which is now simply an inverter) is

$$x_f' = -x_f = -A\beta x_i$$

From Fig. 17-24 the loop gain is

$$\text{Loop gain} = \frac{x_f'}{x_i} = \frac{-x_f}{x_i} = -\beta A \tag{17-26}$$

Suppose it should happen that matters are adjusted in such a way that the signal x_f' is *identically* equal to the externally applied input signal x_i. Since the amplifier has no means of distinguishing the source of the input signal applied to it, it would appear that, if the external source were removed and if terminal 2 were connected to terminal 1, the amplifier would continue to provide the same output signal x_o as before. Note, of course, that the statement $x_f' = x_i$ means that the instantaneous values of x_f' and x_i are exactly equal at all times. The condition $x_f' = x_i$ is equivalent to $-A\beta = 1$; *the loop gain must equal unity*.

The Barkhausen Criterion

For a sinusoidal waveform the equation $x_f' = x_i$ is equivalent to the condition that the *amplitude*, *phase*, and *frequency* of x_i and x_f' must be identical. Hence, we have the following important principle:

The frequency at which a sinusoidal oscillator will operate is the frequency for which the total shift introduced, as a signal proceeds from the input terminals, through the amplifier and feedback network, and back again to the input, is *precisely zero (or, of course, an integral multiple of 2π). Stated more simply, the frequency of a sinusoidal oscillator is determined by the condition that the loop-gain phase shift is zero.*

The principle given above determines the frequency, provided that the circuit oscillates at all. Another condition which must clearly be met is that the magnitude of x_i and x_f' must be identical. This condition is then embodied in the following principle:

Oscillations will not be sustained if, at the oscillator frequency, the magnitude of the product of the transfer gain of the amplifier and the magnitude of the feedback factor of the feedback network (the magnitude of the loop gain) is less than unity.

The condition of *unity loop gain* $-A\beta = 1$ is called the *Barkhausen criterion.* This condition implies, of course, both that $|A\beta| = 1$ and that the phase of $-A\beta$ is zero. The above principles are consistent with the feedback formula $A_f = A/(1 + \beta A)$. For if $-\beta A = 1$, then $A_f \to \infty$, which may be interpreted to mean that there exists an output voltage even in the absence of an externally applied signal voltage.

In the discussion of the stability (Secs. 14-9 and 14-10) and compensation of feedback amplifiers (Sec. 15-10), the Barkhausen criterion $(-A\beta = 1)$ was applied to *prevent* the amplifier from breaking out into oscillation. The present emphasis is quite different; no excitation is applied to a feedback system and the Barkhausen criterion is used to determine how to *cause* the system to oscillate and to find the frequency of the sinusoid generated.

Practical Considerations

Referring to Fig. 13-8, it appears that if $|\beta A|$ at the oscillator frequency is precisely unity, then, with the feedback signal connected to the input terminals, the removal of the external generator will make no difference. If $|\beta A|$ is less than unity, the removal of the external generator will result in a cessation of oscillations. But now suppose that $|\beta A|$ is greater than unity. Then, for example, a 1-V signal appearing initially at the input terminals will, after a trip around the loop and back to the input terminals, appear there with an amplitude larger than 1 V. This larger voltage will then reappear as a still larger voltage, and so on. It seems, then, that if $|\beta A|$ is larger than unity, the amplitude of the oscillations will continue to increase without limit. But of course, such an increase in the amplitude can continue only as long as it is not limited by the onset of nonlinearity of operation in the active devices associated with the amplifier. Such a nonlinearity becomes more marked as the amplitude of oscillation increases. This onset of nonlinearity to limit the amplitude of oscillation is an essential feature of the operation of all practical oscillators, as the following considerations will show: The condition $|\beta A| = 1$ does not give a range of acceptable values of $|\beta A|$, but rather a single and precise value. Now suppose that initially it were even possible to satisfy this condition. Then, because circuit components and, more importantly, transistors change characteristics (drift) with age, temperature, voltage, etc., it is clear that if the entire oscillator is left to itself, in a very short time $|\beta A|$ will become either less or larger than unity. In the former case the oscillation simply stops, and in the latter case we are back to the point of requiring nonlinearity to limit the amplitude. An oscillator in which the loop gain is exactly unity is an abstraction completely unrealizable in practice. It is accordingly necessary, in the adjustment of a practical oscillator, always to arrange to have $|\beta A|$ somewhat larger (say 5 percent) than unity in order to ensure that, with incidental variations in transistor and circuit parameters, $|\beta A|$ shall not fall below unity. While the first two principles stated above must be satisfied on purely theoretical grounds, we may add a third general principle dictated by practical considerations, that is:

In every practical oscillator the loop gain is slightly larger than unity, and the amplitude of the oscillations is limited by the onset of nonlinearity.

17-10 THE PHASE-SHIFT OSCILLATOR

We select the so-called *phase-shift oscillator* (Fig. 17-25) as a first example because it exemplifies very simply the principles set forth above. Here a JFET amplifier of conventional design is followed by three cascaded arrangements of a capacitor C and a resistor R, the output of the last RC combination being returned to the gate. If the loading of the phase-shift network on the amplifier can be neglected, the amplifier shifts by $180°$ the phase of any voltage which appears on the gate, and the network of resistors and capacitors shifts the phase by an additional amount. At some frequency the phase shift introduced by the RC network will be precisely $180°$, and at this frequency the total phase shift from the gate around the circuit and back to the gate will be exactly zero. This particular frequency will be the one at which the circuit will oscillate, provided that the magnitude of the amplification is sufficiently large.

From the mesh equations of the feedback network of Fig. 17-25b we find (Prob. 17-22) that the transfer function of the RC network, which is also the (negative of the) feedback factor, is

$$- \beta = \frac{V_f'}{V_o} = \frac{1}{1 - 5\alpha^2 - j(6\alpha - \alpha^3)} \tag{17-27}$$

where $\alpha \equiv 1/\omega RC$. The phase shift of V_f'/V_o is $180°$ for $\alpha^2 = 6$, or

$$f = 1/(2\pi RC \sqrt{6}) \tag{17-28}$$

At this frequency of oscillation, $\beta = +\frac{1}{29}$. In order that $|\beta A|$ shall not be less than unity, it is required that $|A|$ be at least 29. Hence a FET with $\mu < 29$ cannot be made to oscillate in such a circuit [see Eq. (11-87)].

The FET in Fig. 17-25a may be replaced by an OP AMP as indicated in Fig. 17-25c. Because of the virtual ground, the resistance from the input node P to ground is $R_2 = R$ and, hence, the phase-shift network in Fig. 17-25c is identical with that in Fig. 17-25a. Therefore, the oscillation frequency is given by Eq.

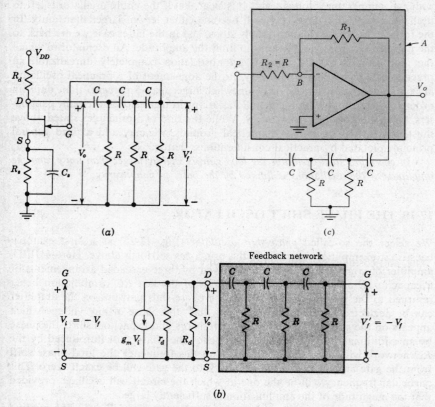

Figure 17-25 (*a*) A FET phase-shift oscillator. (*b*) The small-signal equivalent circuit. The feedback network is indicated shaded and $\beta = V_f'/V_o = -V_f/V_o$. (*c*) An OP AMP phase-shift oscillator.

(17-28). Since the OP AMP gain $A = -(1 + R_1/R)$ and $|A|$ must be at least 29, then R_1/R must be greater than 28 (by about 5 percent).

Note that there are two feedback loops in Fig. 17-25*c*: voltage-series feedback through the phase-shift network to point P of the inverting OP AMP A and voltage-shunt feedback through R_1 within the shaded block to point B.

It is possible to replace the OP AMP in Fig. 17-25*c* by a single transistor stage with $R_1 = \infty$ and $R_2 = R - h_{ie}$. Such an arrangement has only one feedback path, corresponding to voltage-shunt feedback through the phase-shift network to the base B of the transistor. This oscillator is analyzed in Prob. 17-24.

It should be pointed out that it is not always necessary to make use of an amplifier with transfer gain $|A| > 1$ to satisfy the Barkhausen criterion. It is only necessary that $|\beta A| > 1$. Passive network structures exist for which the transfer function $|\beta|$ is greater than unity at some particular frequency. In Prob. 17-26 we show an oscillator circuit consisting of a source follower and the RC circuit of Fig. 17-25 appropriately connected.

Variable-Frequency Operation

The phase-shift oscillator is particularly suited to the range of frequencies from several hertz to several hundred kilohertz, and so includes the range of audio frequencies. The frequency of oscillation may be varied by changing any of the impedance elements in the phase-shifting network. For variations of frequency over a large range, the three capacitors are usually varied simultaneously. Such a variation keeps the input impedance to the phase-shifting network constant (Prob. 17-23) and also keeps constant the magnitude of β and $A\beta$. Hence the amplitude of oscillation will not be affected as the frequency is adjusted. The phase-shift oscillator is operated in class A in order to keep distortion to a minimum.

Two active phase shifters may be used in place of the passive feedback network of Fig. 17-25c to obtain a sinusoidal oscillator with quadrature outputs, sine, and cosine waveforms (Prob. 17-30).

17-11 A GENERAL FORM OF OSCILLATOR CONFIGURATION

Many oscillator circuits fall into the general form shown in Fig. 17-26a. In the analysis that follows we assume an active device with extremely high input resistance such as a FET, or an operational amplifier. Figure 17-26b shows the

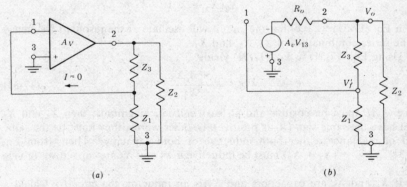

(a)   (b)

Figure 17-26 (a) The basic configuration for many resonant-circuit oscillators. (b) The linear equivalent circuit using an operational amplifier, having very high input resistance and an open-circuit voltage gain $-A_v$.

linear equivalent circuit of Fig. 17-26a, using an amplifier with an open-circuit negative gain $-A_v$ and output resistance R_o. Clearly the topology of Fig. 17-26 is that of voltage-series feedback.

The Loop Gain

The value of $-A\beta$ will be obtained by considering the circuit of Fig. 17-26a to be a feedback amplifier with output taken from terminals 2 and 3 and with input terminals 1 and 3. The load impedance Z_L consists of Z_2 in parallel with the series combination of Z_1 and Z_3. The gain without feedback is $A = -A_v Z_L/(Z_L + R_o)$. The feedback factor is $\beta = -V_f'/V_o = -Z_1/(Z_1 + Z_3)$. The loop gain is found to be

$$-A\beta = \frac{-A_v Z_1 Z_2}{R_o(Z_1 + Z_2 + Z_3) + Z_2(Z_1 + Z_3)} \tag{17-29}$$

Reactive Elements Z_1, Z_2, and Z_3

If the impedances are pure reactances (either inductive or capacitive), then $Z_1 = jX_1$, $Z_2 = jX_2$, and $Z_3 = jX_3$. For an inductor, $X = \omega L$, and for a capacitor, $X = -1/\omega C$. Then

$$-A\beta = \frac{+A_v X_1 X_2}{jR_o(X_1 + X_2 + X_3) - X_2(X_1 + X_3)} \tag{17-30}$$

For the loop gain to be real (zero phase shift),

$$X_1 + X_2 + X_3 = 0 \tag{17-31}$$

and

$$-A\beta = \frac{A_v X_1 X_2}{-X_2(X_1 + X_3)} = \frac{-A_v X_1}{X_1 + X_3} \tag{17-32}$$

From Eq. (17-31) we see that the circuit will oscillate at the resonant frequency of the series combination of X_1, X_2, and X_3.

Using Eq. (17-31) in Eq. (17-32) yields

$$-A\beta = \frac{+A_v X_1}{X_2} \tag{17-33}$$

Since $-A\beta$ must be positive and at least unity in magnitude, then X_1 and X_2 must have the same sign (A_v is positive). In other words, they must be the same kind of reactance, either both inductive or both capacitive. Then, from Eq. (17-31), $X_3 = -(X_1 + X_2)$ must be inductive if X_1 and X_2 are capacitive, or vice versa.

If X_1 and X_2 are capacitors and X_3 is an inductor, the circuit is called a *Colpitts oscillator*. If X_1 and X_2 are inductors and X_3 is a capacitor, the circuit is called a *Hartley oscillator*. In this latter case, there may be mutual coupling between X_1 and X_2 (and the above equations will then not apply).

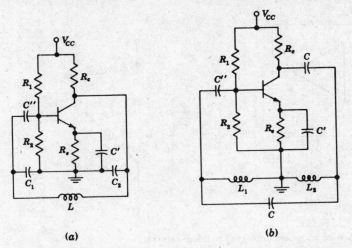

Figure 17-27 (*a*) A transistor Colpitts oscillator. (*b*) A transistor Hartley oscillator.

Transistor versions of the above types of LC oscillators are possible. As an example, a transistor Colpitts oscillator is indicated in Fig. 17-27*a*. Qualitatively, this circuit operates in the manner described above. However, the detailed analysis of a transistor oscillator circuit is more difficult, for two fundamental reasons. First, the low input impedance of the transistor shunts Z_1 in Fig. 17-26*a*, and hence complicates the expressions for the loop gain given above. Second, if the oscillation frequency is beyond the audio range, the simple low-frequency h-parameter model is no longer valid. Under these circumstances the more complicated high-frequency hybrid-π model of Fig. 13-10 must be used. A transistor Hartley oscillator is shown in Fig. 17-27*b*.

17-12 THE WIEN BRIDGE OSCILLATOR

An oscillator circuit in which a balanced bridge is used as the feedback network is the Wien bridge oscillator shown in Fig. 17-28*a*. The "bridge" is clearly indicated in Fig. 17-28*b*. The four arms of the bridge are Z_1, Z_2, R_1, and R_2. The input to the bridge is the output V_o of the OP AMP, and the output of the bridge between nodes 1 and 2 supplies the differential input to the OP AMP.

There are two feedback paths in Fig. 17-28*a*: *positive* feedback through Z_1 and Z_2, whose components determine the frequency of oscillation, and *negative* feedback through R_1 and R_2, whose elements affect the amplitude of oscillation. The loop gain is given by $-\beta A$ where

$$\beta = -\frac{V_f'}{V_o} = -\frac{Z_2}{Z_1 + Z_2} \quad \text{and} \quad A = 1 + \frac{R_1}{R_2}$$

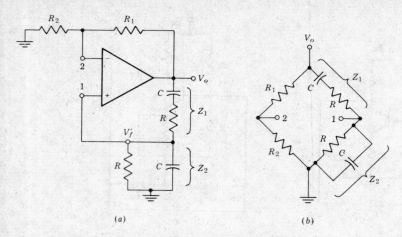

Figure 17-28 (a) A Wien bridge oscillator. (b) The bridge network.

In Prob. 17-28, it is found that, with $\alpha \equiv \omega RC$,

$$- A\beta = \frac{\alpha}{3\alpha - j(1 - \alpha^2)} \left(1 + \frac{R_1}{R_2}\right) \tag{17-34}$$

The Barkhausen condition that $-A\beta = 1$ requires that $\alpha = 1$ and $\frac{1}{3}(1 + R_1/R_2) = 1$. Hence,

$$f = \frac{1}{2\pi RC} \quad \text{and} \quad R_1 = 2R_2 \tag{17-35}$$

The maximum frequency of oscillation is limited by the slew rate of the amplifier. Continuous variation of frequency is accomplished by varying simultaneously the two capacitors (ganged variable-air capacitors). Changes in frequency range are accomplished by switching in different values for the two identical resistors R.

Amplitude Stabilization

We consider modification of the circuit of Fig. 17-28, which serves to stabilize the amplitude against variations due to fluctuations occasioned by the aging of transistors, components, etc. One modification consists simply in replacing the resistor R_2 by a sensistor (Sec. 1-10) which has a positive thermal coefficient.

The amplitude of oscillation is determined by the extent to which the loop gain $-\beta A$ is greater than unity. If the output V_o increases (for any reason), the current in R_2 increases and A decreases. The regulation mechanism introduced by the sensistor operates by automatically changing A so as to keep the loop gain more constant. The temperature of R_2 is determined by the root-mean-square (rms) value of the current which passes through it. If the rms value of the

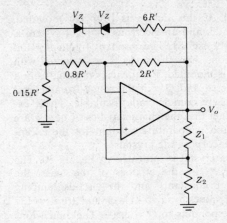

Figure 17-29 Avalanche diodes are used for automatically controlling the gain of the oscillator and, hence, stabilizing the amplitude of the sinusoid.

current changes, then, because of the thermal lag of the sensistor, the temperature will be determined by the average value over a large number of cycles of the rms value of the current. An important fact to keep in mind about the mechanism just described is that, because of the thermal lag of the sensistor, the resistance of the sensistor during the course of a single cycle is very nearly absolutely constant. Therefore, at any fixed amplitude of oscillation, the sensistor behaves entirely like an ordinary linear resistor.

A thermistor which has a negative temperature coefficient can also be used but it must replace R_1 rather than R_2.

Another method[12] of stabilizing the amplitude is indicated in Fig. 17-29. Initially, both Zener diodes are nonconducting, and the loop gain is

$$\frac{1}{3}\left(1 + \frac{R_1}{R_2}\right) = \frac{1}{3}\left(1 + \frac{2R'}{0.15R' + 0.8R'}\right) = 1.04 > 1$$

and, hence, oscillations start. Because the loop gain exceeds unity, the amplitudes of these oscillations grows until the peaks exceed the diode breakdown voltage V_Z. When this happens, the shunting action of the resistor $6R'$ reduces the gain and limits the amplitude to approximately V_Z. Distortion can be reduced to approximately 0.5 percent with this circuit.

The two methods of amplitude stabilization described in the foregoing are examples of automatic gain control (AGC). An active AGC loop may also be used with a FET as a voltage-controlled resistor, somewhat as indicated in Fig. 11-32.

17-13 CRYSTAL OSCILLATORS

If a piezoelectric crystal, usually quartz, has electrodes plated on opposite faces and if a potential is applied between these electrodes, forces will be exerted on the bound charges within the crystal. If this device is properly mounted, deformations take place within the crystal, and an electromechanical system is formed which will vibrate when properly excited. The resonant frequency and the Q depend upon the crystal dimensions, how the surfaces are oriented with respect to its axes, and how the device is mounted. Frequencies ranging from a few kilohertz to a few hundred megahertz, and Q's in the range from several thousand to several hundred thousand, are commercially available. These extraordinarily high values of Q and the fact that the characteristics of quartz are extremely stable with respect to time and temperature account for the exceptional frequency stability of oscillators incorporating crystals.

The electrical equivalent circuit of a crystal is indicated in Fig. 17-30. The inductor L, capacitor C, and resistor R are the analogs of the mass, the compliance (the reciprocal of the spring constant), and the viscous-damping factor of the mechanical system. Typical values for a 90-kHz crystal are $L = 137$ H, $C = 0.0235$ pF, and $R = 15$ kΩ, corresponding to $Q = 5,500$. The dimensions of such a crystal are 30 by 4 by 1.5 mm. Since C' represents the electrostatic capacitance between electrodes with the crystal as a dielectric, its magnitude (~ 3.5 pF) is very much larger than C.

If we neglect the resistance R, the impedance of the crystal is a reactance jX whose dependence upon frequency is given by

$$jX = -\frac{j}{\omega C'} \frac{\omega^2 - \omega_s^2}{\omega^2 - \omega_p^2} \tag{17-36}$$

where $\omega_s^2 = 1/LC$ is the series resonant frequency (the zero impedance frequency), and $\omega_p^2 = (1/L)(1/C + 1/C')$ is the parallel resonant frequency (the infinite impedance frequency). Since $C' \gg C$, then $\omega_p \approx \omega_s$. For the crystal whose parameters are specified above, the parallel frequency is only three-tenths of 1 percent higher than the series frequency. For $\omega_s < \omega < \omega_p$, the reactance is

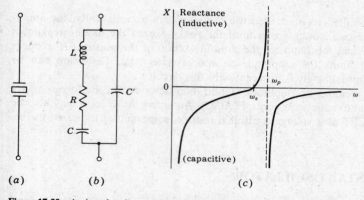

Figure 17-30 A piezoelectric crystal. (a) Symbol; (b) electrical model; (c) the reactance function (if $R = 0$).

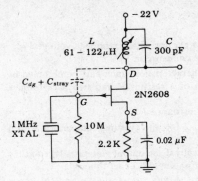

Figure 17-31 A 1-MHz FET crystal oscillator. (*Courtesy of Siliconix Co.*)

inductive, and outside this range it is capacitive, as indicated in Fig. 17-30.

A variety of crystal-oscillator circuits is possible. If in the basic configuration of Fig. 17-26a a crystal is used for Z_1, a tuned LC combination for Z_2, and the capacitance C_{dg} between drain and gate for Z_3, the resulting circuit is as indicated in Fig. 17-31. From the theory given in the preceding section, the crystal reactance, as well as that of the LC network, must be inductive. For the loop gain to be greater than unity, we see from Eq. (17-33) that X_1 cannot be too small. Hence the circuit will oscillate at a frequency which lies between ω_s and ω_p but close to the parallel-resonance value. Since $\omega_p \approx \omega_s$, the oscillator frequency is essentially determined by the crystal, and not by the rest of the circuit.

REFERENCES

1. Kaufman, M., and D. F. Stout: "Handbook of Operational Amplifier Circuit Design," chap. 5, McGraw-Hill Book Company, New York, 1976.
2. Connelly, J. A. (ed.):"Analog Integrated Circuits," chap. 4, John Wiley and Sons, New York, 1975.
3. Sifferlen, T. P., and V. Vartanian: "Digital Electronics with Engineering Applications," Prentice Hall Inc., Englewood, New Jersey, 1970.
4. Graeme, J. G.: "Applications of Operational Amplifiers—Third Generation Techniques," chap. 5, McGraw-Hill Book Company, New York, 1973.
5. Roberge, J. K.: "Operational Amplifiers: Theory and Practice," sec. 12.2, John Wiley and Sons, New York, 1975.
6. Norris, B. (ed.): "Digital Integrated Circuits and Operational-Amplifier and Optoelectronic Circuit Design," p. 144, McGraw-Hill Book Company, New York, 1976.
7. Millman, J. and H. Taub: "Pulse Digital and Switching Waveforms," chap. 14, McGraw-Hill Book Company, New York, 1965.
8. Ref. 1, sec. 27.3.
9. Ref. 7, sec. 17-22.
10. Ref. 7, secs. 17-13 and 17-20.
11. Ref. 5, p. 522–525.
    Ref. 2, sec. 6.2.
12 Ref. 4, sec. 5.1.1.

REVIEW QUESTIONS

17-1 (*a*) Sketch the characteristic of an ideal comparator with a reference voltage V_R.
  (*b*) Repeat (*a*) for a commercially available comparator.

17-2 (*a*) List two improvements in comparator characteristics which may be obtained by cascading the OP AMP with a series combination of a resistor R and two back-to-back avalanche diodes.
  (*b*) What determines the magnitude of the resistance R?

17-3 (*a*) Sketch the system indicated in Rev. 17-2 for an inverting comparator with a reference V_R.
  (*b*) Draw the transfer characteristics realistically if the output voltage is taken at the OP AMP output terminal and also if it is taken across the two Zener diodes.

17-4 Sketch the circuit for converting a sinusoid (*a*) into a square wave and (*b*) into a series of positive pulses, one per cycle.

17-5 Explain how to measure the phase difference between two sinusoids.

17-6 If noise spikes are present on the input signal to a comparator in the neighborhood of the amplitude V_R, explain why the output may "chatter."

17-7 (*a*) Sketch a regenerative comparator system and explain its operation.
  (*b*) What parameters determine the loop gain?
  (*c*) What parameters determine the hysteresis?
  (*d*) Sketch the transfer characteristic and indicate the hysteresis.

17-8 (*a*) Draw the system of a square-wave generator using one comparator.
  (*b*) Explain its operation by drawing the capacitor and output voltage waveforms.
  (*c*) Indicate one method for obtaining a nonsymmetrical square wave ($T_1 \neq T_2$.)

17-9 (*a*) Using a comparator and an integrator, draw the system of a triangular-waveform generator with $T_1 = T_2$.
  (*b*) Explain its operation by drawing the capacitor voltage waveform.

17-10 (*a*) Draw the configuration for a positive-negative controlled-gain amplifier.
  (*b*) Explain its operation.

17-11 (*a*) In a VCO, what oscillator characteristic is controlled by the externally applied voltage?
  (*b*) What is meant by *duty-cycle modulation*?

17-12 (*a*) Draw the configuration of a pulse generator (a one-shot) using a comparator.
  (*b*) Explain its operation by referring to the capacitor and output waveforms.

17-13 (*a*) A capacitor C is charged through a resistor R from a supply V. An n-channel JFET is used as a switch across C and is biased so that the transistor Q is OFF. The capacitor voltage v_c is applied to the inverting terminal of a comparator having a reference voltage $V_R < V$. At $t = 0$ a triggering pulse v_t turns Q ON. Sketch the waveforms v_t, v_c, and v_o (the comparator output).
  (*b*) Explain the operation and show that this configuration functions as a retriggerable monostable multi.

17-14 Draw a block diagram of a time-base system for a CRT.

17-15 Sketch the configuration for a triggered-sweep generator with an output waveform which is (*a*) exponential and (*b*) linear.
  (*c*) Indicate one form for the reset switch.

17-16 (*a*) Sketch the system using an integrator for generating a staircase waveform v, starting with a pulse train v_p.

(*b*) Sketch v_p and v and explain the operation.

(*c*) Explain how to reset the system after N pulses.

17-17 Repeat Rev. 17-16 for a storage-counter step generator.

17-18 Explain how to amplitude-modulate a sinusoidal carrier v_c with a lower-frequency waveform v_m.

17-19 (*a*) Explain how to modulate the amplitude of a square-wave carrier v_c with a lower-frequency waveform v_m, using a positive-negative controlled-gain amplifier A.

(*b*) Explain why A may also be used as an amplitude demodulator.

17-20 Sketch the system of a chopper modulator and explain its operation.

17-21 What is a chopper-stabilized amplifier? Explain carefully.

17-22 (*a*) Explain how a comparator is used as a pulse-width modulator.

(*b*) Draw the modulating waveform v_m and the corresponding output waveform v_o.

17-23 Give the Barkhausen conditions required in order for sinusoidal oscillations to be sustained.

17-24 Sketch the phase-shift oscillator using (*a*) an OP AMP and (*b*) a JFET.

17-25 (*a*) Sketch the topology for a generalized resonant-circuit oscillator, using impedances Z_1, Z_2, Z_3.

(*b*) At what frequency will the circuit oscillate?

(*c*) Under what conditions does the configuration reduce to a Colpitts oscillator? A Hartley oscillator?

17-26 (*a*) Sketch the circuit of a Wien bridge oscillator.

(*b*) Which components determine the frequency of oscillation?

(*c*) Which elements determine the amplitude of oscillation?

17-27 (*a*) Draw the electrical model of a piezoelectric crystal.

(*b*) Sketch the reactance versus frequency function.

(*c*) Over what portion of the reactance curve do we desire oscillations to take place when the crystal is used as part of a sinusoidal oscillator? Explain.

17-28 Sketch a circuit of a crystal-controlled oscillator.

EIGHTEEN

POWER CIRCUITS AND SYSTEMS

An amplifying system usually consists of several stages in cascade. The input and intermediate stages operate in a small-signal class A mode. Their function is to amplify the small-input excitation to a value large enough to drive the final device. This output stage feeds a transducer such as a cathode-ray tube, a loudspeaker, a servo-motor, etc., and hence must be capable of delivering a large voltage or current swing or an appreciable amount of power. In this chapter we study such large-signal amplifiers. Thermal considerations are very important with power amplifiers and are discussed here. Both bipolar and FET power transistors are introduced.

Almost all electronic circuits require a dc source of power. For portable low-power systems, batteries may be used. More frequently, however, electronic equipment is energized by a power supply, a circuit which converts the ac waveform of the power lines to direct voltage of constant amplitude. The process of ac-to-dc conversion is examined in Chap. 10. In this chapter we consider the regulator circuits used to control the amplitude of a dc supply voltage. These circuits are a special class of feedback amplifiers. We also introduce dc-to-dc conversion (switching regulators).

18-1 LARGE-SIGNAL AMPLIFIERS

A simple transistor amplifier that supplies power to a pure resistance load R_L is indicated in Fig. 18-1. Using the notation of Table 11-3, i_C represents the total instantaneous collector current, i_c designates the instantaneous variation from

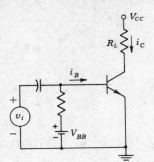

Figure 18-1 The circuit of a simple transistor amplifier.

the quiescent value I_c of the collector current. Similarly, i_B, i_b, and I_B represent corresponding base currents. The total instantaneous collector-to-emitter voltage is given by v_C, and the instantaneous variation from the quiescent value V_C is represented by v_c.

Let us assume that the static output characteristcs are equidistant for equal increments of input base current i_b, as indicated in Fig. 18-2. Then, if the input signal i_b is a sinusoid, the output current and voltage are also sinusoidal, as shown. Under these circumstances the nonlinear distortion is negligible, and the power output may be found as follows:

$$P = V_c I_c = I_c^2 R_L \tag{18-1}$$

where V_c and I_c are the rms values of the output voltage v_c and current i_c, respectively, and R_L is the load resistance. The numerical values of V_c and I_c can be determined graphically in terms of the maximum and minimum voltage and current swings, as indicated in Fig. 18-2. If $I_m(V_m)$ represents the peak

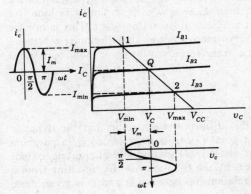

Figure 18-2 The output characteristics and the current and voltage waveforms for the amplifier of Fig. 18-1. The excitation is a sinusoidal base current.

sinusoidal current (voltage) swing, it is seen that

$$I_c = \frac{I_m}{\sqrt{2}} = \frac{I_{max} - I_{min}}{2\sqrt{2}} \tag{18-2}$$

and

$$V_c = \frac{V_m}{\sqrt{2}} = \frac{V_{max} - V_{min}}{2\sqrt{2}} \tag{18-3}$$

so that the power becomes

$$P = \frac{V_m I_m}{2} = \frac{I_m^2 R_L}{2} = \frac{V_m^2}{2R_L} \tag{18-4}$$

which may also be written in the form

$$P = \frac{(V_{max} - V_{min})(I_{max} - I_{min})}{8} \tag{18-5}$$

This equation allows the output power to be calculated very simply. All that is necessary is to plot the load line on the volt-ampere characteristics of the device and to read off the values of V_{max}, V_{min}, I_{max}, and I_{min}.

18-2 HARMONIC DISTORTION

In the preceding section the active device is idealized as a perfectly linear device. In general, however, the dynamic transfer characteristic (i_c versus i_b) is not a straight line. This nonlinearity arises because the static output characteristics are not equidistant straight lines for constant increments of input excitation. Referring to Fig. 11-5 we see that the waveform of the output voltage differs from that of the input signal. Distortion of this type is called *nonlinear*, or *amplitude, distortion*.

The reader may wonder why has the question of distortion not been addressed in the earlier chapters on amplification. The answer is signal magnitude. The underlying precept of Chap. 11 is that any device, independent of the transfer characteristic, can be treated analytically in a linear fashion for sufficiently small excursions about a quiescent operating point. This is not the case with power amplifiers. By its very nature a power amplifier must generate a large output signal, and the entire transfer curve, linear or nonlinear, must therefore be examined.

Second Harmonic Distortion

To investigate the magnitude of this distortion we assume that the dynamic curve with respect to the quiescent point Q can be represented by a parabola rather than a straight line. Thus, instead of relating the alternating output current i_c with the input excitation i_b by the equation $i_c = Gi_b$ resulting from a linear circuit, we assume that the relationship between i_c and i_b is given more accurately by the expression

$$i_c = G_1 i_b + G_2 i_b^2 \tag{18-6}$$

where the G's are constants. Actually, these two terms are the beginning of a power-series expansion of i_c as a function of i_b.

If the input waveform is sinusoidal and of the form

$$i_b = I_{bm} \cos \omega t \tag{18-7}$$

the substitution of this expression in Eq. (18-6) leads to

$$i_c = G_1 I_{bm} \cos \omega t + G_2 I_{bm}^2 \cos^2 \omega t$$

Since $\cos^2 \omega t = \frac{1}{2} + \frac{1}{2} \cos 2\omega t$, the expression for the instantaneous total current i_C reduces to the form

$$i_C = I_C + i_c = I_C + B_0 + B_1 \cos \omega t + B_2 \cos 2\omega t \tag{18-8}$$

where the B's are constants which may be evaluated in terms of the G's. The physical meaning of this equation is evident. It shows that the application of a sinusoidal signal on a parabolic dynamic characteristic results in an output current which contains, in addition to a term of the same frequency as the input, a second-harmonic term, and also a constant current. This constant term B_0 adds to the original dc value I_C to yield a total dc component of current $I_C + B_0$. *Parabolic nonlinear distortion introduces into the output a component whose frequency is twice that of the sinusoidal input excitation. Also, since a sinusoidal input signal changes the average value of the output current rectification takes place.*

The amplitudes B_0, B_1, and B_2 for a given load resistor are readily determined from the static characteristics. We observe from Fig. 18-2 that

$$\text{When } \omega t = 0: \quad i_C = I_{max}$$

$$\text{When } \omega t = \frac{\pi}{2}: \quad i_C = I_C \tag{18-9}$$

$$\text{When } \omega t = \pi: \quad i_C = I_{min}$$

By substituting these values in Eq. (18-8), there results

$$I_{max} = I_C + B_0 + B_1 + B_2$$
$$I_C = I_C + B_0 - B_2 \tag{18-10}$$
$$I_{min} = I_C + B_0 - B_1 + B_2$$

This set of three equations determines the three unknowns B_0, B_1, and B_2. It follows from the second of this group that

$$B_0 = B_2 \tag{18-11}$$

By subtracting the third equation from the first, there results

$$B_1 = \frac{I_{max} - I_{min}}{2} \tag{18-12}$$

With this value of B_1, the value for B_2 may be evaluated from either the first or the last of Eqs. (18-10) as

$$B_2 = B_0 = \frac{I_{\max} + I_{\min} - 2I_C}{4} \qquad (18\text{-}13)$$

The second-harmonic distortion D_2 is defined as

$$D_2 \equiv \frac{|B_2|}{|B_1|} \qquad (18\text{-}14)$$

(To find the percent second-harmonic distortion, D_2 is multiplied by 100.) The quantities $I_{\max}$, $I_{\min}$, and I_C appearing in these equations are obtained directly from the characteristic curves of the transistor and from the load line.

If the dynamic characteristic is given by the parabolic form (18-6) and if the input contains two frequencies ω_1 and ω_2, then the output will consist of a dc term and sinusoidal components of frequencies ω_1, ω_2, $2\omega_1$, $2\omega_2$, $\omega_1 + \omega_2$, and $\omega_1 - \omega_2$ (Prob. 18-1). The sum and difference frequencies are called *intermodulation*, or *combination*, frequencies.

Higher-Order Harmonic Generation

The preceding analysis assumes a parabolic dynamic characteristic. This approximation is usually valid for amplifiers where the swing is small. For a power amplifier with a large input swing, however, it is necessary to express the dynamic transfer curve with respect to the Q point by a power series of the form

$$i_c = G_1 i_b + G_2 i_b^2 + G_3 i_b^3 + G_4 i_b^4 + \cdots \qquad (18\text{-}15)$$

If we assume that the input wave is a simple cosine function of time, of the form in Eq. (18-7), the output current will be given by

$$i_C = I_C + B_0 + B_1 \cos \omega t + B_2 \cos 2\omega t + B_3 \cos 3\omega t + \cdots \qquad (18\text{-}16)$$

This equation results when Eq. (18-7) is inserted in Eq. (18-15) and the proper trigonometric transformations are made.

Note that now a third harmonic and higher-order harmonics are present. The Fourier coefficients B_0, B_1, B_2, $B_3 \cdots$ may be obtained by an extension of the foregoing procedure[1] used with Eq. (18-16) instead of Eq. (18-8).

The harmonic distortion is defined as

$$D_2 \equiv \frac{|B_2|}{|B_1|} \qquad D_3 \equiv \frac{|B_3|}{|B_1|} \qquad D_4 \equiv \frac{|B_4|}{|B_1|} \qquad (18\text{-}17)$$

where $D_s(s = 2, 3, 4, \ldots)$ represents the distortion of the sth harmonic.

Power Output

If the distortion is not negligible, the power delivered at the fundamental frequency is

$$P_1 = \frac{B_1^2 R_L}{2} \qquad (18\text{-}18)$$

However, the total power output is

$$P = (B_1^2 + B_2^2 + B_3^2 + \cdots)\frac{R_L}{2} = (1 + D_2^2 + D_3^2 + \cdots)P_1$$

or $\qquad P = (1 + D^2)P_1$ $\hspace{4cm}$ (18-19)

Where *the total harmonic distortion* (THD), or *distortion factor*, is defined as

$$D \equiv \sqrt{D_2^2 + D_3^2 + D_4^2 + \cdots} \hspace{2cm} (18\text{-}20)$$

If the total distortion is 10 percent of the fundamental, then

$$P = \left[1 + (0.1)^2\right]P_1 = 1.01P_1$$

The total power output is only 1 percent higher than the fundamental power when the distortion is 10 percent. Hence little error is made in using only the fundamental term P_1 in calculating the power output.

In passing, it should be noted that the total harmonic distortion is not necessarily indicative of the discomfort to someone listening to music. Usually, the same amount of distortion is more irritating, the higher the order of the harmonic frequency.

18-3 AMPLIFIER CLASSIFICATION

It has been tacitly assumed in all previous amplifier design and analysis that the transistor is biased in the middle of its operating range, as indicated in Fig. 18-2 (note the location of the point Q in the $i_C v_C$ plane). This is not always the case with power circuits, and a classification (A, B, AB, and C) has evolved to describe amplifier operation, dependent on the type of biasing employed. The significance of this classification is discussed in the following sections.

Class A

A class A amplifier is one in which the operating point and the input signal are such that the current in the output circuit (in the collector, or drain electrode) flows at all times. A class A amplifier operates essentially over a linear portion of its characteristic.

Class B

A class B amplifier is one in which the operating point is at an extreme end of its characteristic, so that the quiescent power is very small. Hence either the quiescent current or the quiescent voltage is approximately zero. If the signal excitation is sinusoidal, amplification takes place for only one half of a cycle. For example, if the quiescent output circuit current is zero, this current will remain zero for one half of a cycle.

Class AB

A class AB amplifier is one operating between the two extremes defined for class A and class B. Hence the output signal is zero for part but less than one half of an input sinusoidal signal cycle.

Class C

A class C amplifier is one in which the operating point is chosen so that the output current (or voltage) is zero for more than one half of an input sinusoidal signal cycle.

18-4 EFFICIENCY OF A CLASS A AMPLIFIER

If a power amplifier design is constrained either by a limited source of power (as would be the case for a satellite) or by maximum power-dissipation consideration, as in Sec. 18-5, attention must be focused on the issue of power conversion.

Conversion Efficiency

A measure of the ability of an active device to convert the dc power of the supply into the ac (signal) power delivered to the load is called the *conversion efficiency*, or *theoretical efficiency*. This figure of merit, designated η, is also called the *collector-circuit efficiency* for a transistor amplifier. By definition, the percentage efficiency is

$$\eta \equiv \frac{\text{Signal power delivered to load}}{\text{DC power supplied to output circuit}} \times 100 \ \text{percent} \qquad (18\text{-}21)$$

In general,

$$\eta = \frac{\frac{1}{2} B_1^2 R_L}{V_{CC}(I_C + B_0)} \times 100 \ \text{percent} \qquad (18\text{-}22)$$

If the distortion components are negligible, then

$$\eta = \frac{\frac{1}{2} V_m I_m}{V_{CC} I_C} \times 100 = 50 \frac{V_m I_m}{V_{CC} I_C} \ \text{percent} \qquad (18\text{-}23)$$

where $V_m (I_m)$ represents the peak sinusoidal voltage (current) swing. The collector-circuit efficiency differs from the overall efficiency because the power taken by the base is not included in the denominator of Eq. (18-22).

From the definitions in Sec. 18-3, the amplifier of Sec. 18-1 operates in class A. Let us now qualitatively examine its efficiency using two limiting cases.

1. *Small signal.* With a small output signal the output power is correspondingly small. However, the power consumed by the class A biasing remains at $V_{CC} I_C$, which may be substantial, resulting in an extremely small conversion

efficiency. Note also that the load must dissipate a large fraction of the dc power $V_{CC}I_C$ even under zero excitation.

2. *Maximum signal.* With careful selection of the bias point, the transistor may be driven from the edge of saturation to cutoff. It can be shown (Prob. 18-4) that under this condition $I_m = I_C$ and $V_m = \frac{1}{2} V_{CC}$, yielding $\eta = 25$ percent. For every 1 W of output power, 3 W are being consumed internally. Clearly, from an efficiency standpoint, class A operation is a poor choice for power amplification.

18-5 CLASS B PUSH-PULL AMPLIFIERS[2]

If $V_{BB} = 0$ in Fig. 18-1, the quiescent current is $I_C = 0$. From the definitions given in Sec. 18-3, this zero-bias circuit is a class B amplifier. Similarly, the emitter follower in Fig. 18-3a operates in class B. Let us assume that the transistor output characteristics are equally spaced for equal intervals of excitation. For such an idealized transistor the dynamic transfer curve (i_C versus i_B) is a straight line passing through the origin (Fig. 18-3b). The graphical construction from which to determine the collector-current waveshape is indicated. Note that for this class B circuit the load current $i_L \approx i_C$ is sinusoidal during one half of each period and is zero during the second half cycle. In other words, this circuit behaves as a rectifier rather than as a power amplifier.

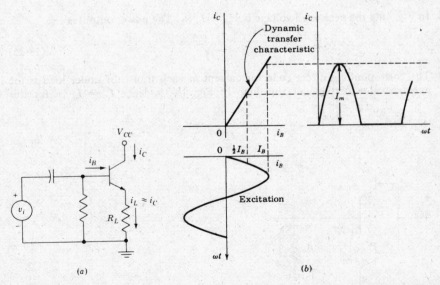

Figure 18-3 (a) An emitter follower with zero bias operating as a class B amplifier. (b) Graphical construction for determining the output current waveform.

The foregoing difficulty is overcome by using the complementary emitter-follower OP AMP output stage of Fig. 15-19, which is repeated in Fig. 18-4 for convenience. This configuration is called a class B *push-pull amplifier*. For positive values of the sinusoidal input v_i, $Q1$ conducts and $Q2$ is OFF ($i_2 = 0$), so that i_1 is the positive half sinewave of Fig. 18-3b. For negative values of v_i, $Q1$ is nonconducting ($i_1 = 0$), and $Q2$ conducts, resulting in a positive half sinusoid for i_2, which is 180° out of phase with that shown in Fig. 18-3b. Since the load current is the difference between the two transistor emitter currents,

$$i_L = i_1 - i_2 \tag{18-24}$$

Consequently, for the idealized transfer characteristic of Fig. 18-3b, the load current is a perfect sinusoid.

The advantages of class B as compared with class A operation are the following: It is possible to obtain greater power output, the efficiency is higher, and there is negligible power loss at no signal. For these reasons, in systems where the power supply is limited, such as those operating from solar cells or a battery, the output power is usually delivered through a push-pull class B transistor circuit. The disadvantages are that the harmonic distortion may be higher and the supply voltages must have good regulation. The power output circuit in most modern IC amplifiers is the complementary emitter-follower push-pull stage.

Efficiency

In Fig. 18-4 the peak load voltage is $V_m = I_m R_L$. The power output is

$$P = \frac{I_m V_m}{2} \tag{18-25}$$

The corresponding direct collector current in each transistor under load is the average value of the half sine loop of Fig. 18-3b. Since $I_{dc} = I_m/\pi$ for this

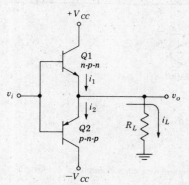

Figure 18-4 A class B push-pull amplifier.

waveform, the dc input power from the supply is

$$P_i = 2 \frac{I_m V_{CC}}{\pi} \qquad (18\text{-}26)$$

The factor 2 in this expression arises because two transistors are used in the push-pull system.

Taking the ratio of Eqs. (18-25) and (18-26), we obtain for the collector-circuit efficiency

$$\eta \equiv \frac{P}{P_i} \times 100 = \frac{\pi}{4} \frac{V_m}{V_{CC}} \times 100 \text{ percent} \qquad (18\text{-}27)$$

If the drop across a transistor is negligible compared with the supply voltage, then $V_m \approx V_{CC}$. Under these conditions, Eq. (18-27) shows that the maximum possible conversion efficiency is $25\pi = 78.5$ percent for a class B system compared with 25 percent for class A operation. This large value of η results from the fact that there is no current in a class B system if there is no excitation, whereas there is a drain from the power supply in a class A system even at zero signal. We also note that in a class B amplifier the dissipation at the collectors is zero in the quiescent state and increases with excitation, whereas the heating of the collectors of a class A system is a maximum at zero input and decreases as the signal increases. Since the direct current increases with signal in a class B amplifier, the power supply must have good regulation.

Dissipation

The dissipation P_C (in both transistors) is the difference between the power input to the collector circuit and the power delivered to the load. Since $I_m = V_m / R_L$,

$$P_C = P_i - P = \frac{2}{\pi} \frac{V_{CC} V_m}{R_L} - \frac{V_m^2}{2R_L} \qquad (18\text{-}28)$$

This equation shows that the collector dissipation is zero at no signal ($V_m = 0$), rises as V_m increases, and passes through a maximum at $V_m = 2V_{CC}/\pi$ (Prob. 18-5). The peak dissipation is found to be

$$P_{C(max)} = \frac{2V_{CC}^2}{\pi^2 R_L} \qquad (18\text{-}29)$$

The maximum power which can be delivered is obtained for $V_m = V_{CC}$ or

$$P_{max} = \frac{V_{CC}^2}{2R_L} \qquad (18\text{-}30)$$

Hence

$$P_{C(max)} = \frac{4}{\pi^2} P_{max} \approx 0.4 P_{max} \qquad (18\text{-}31)$$

If, for example, we wish to deliver 10 W from a class B push-pull amplifier, then

$P_{C(\text{max})} = 4$ W, or we must select transistors which have collector dissipations of approximately 2 W each. In other words, we can obtain a push-pull output of five times the specified power dissipation of a single transistor. On the other hand, if we paralleled two transistors and operated them class A to obtain 10 W out, the collector dissipation of each transistor would have to be at least 20 W (assuming 25 percent efficiency). This statement follows from the fact that $P_i = P/\eta = 10/0.25 = 40$ W. This input power must all be dissipated in the two collectors at no signal, or $P_C = 20$ W per transistor. Hence at no excitation there would be a steady loss of 20 W in each transistor, whereas in class B the standby (no-signal) dissipation is zero. This example clearly indicates the superiority of the push-pull over the parallel configuration.

Distortion

The distortion properties of a push-pull system are rather unique. Consider the operation of Fig. 18-4 when the transfer characteristic is not linear. Either $Q1$ or $Q2$ is conducting, depending upon the polarity of the input signal. If the devices are matched, then the current i_2 is identical with i_1, except shifted in phase by 180°. The current of $Q1$ is given by Eq. (18-16) and is repeated here for convenience:

$$i_1 = I_C + B_0 + B_1 \cos \omega t + B_2 \cos 2\omega t + B_3 \cos 3\omega t + \cdots \quad (18\text{-}32)$$

The output current of transistor $Q2$ is obtained by replacing ωt by $\omega t + \pi$ in the expression for i_1. That is,

$$i_2(\omega t) = i_1(\omega t + \pi) \quad (18\text{-}33)$$

whence

$$i_2 = I_C + B_0 + B_1 \cos (\omega t + \pi) + B_2 \cos \cos 2(\omega t + \pi) + \cdots$$

or $\quad i_2 = I_C + B_0 - B_1 \cos \omega t + B_2 \cos 2\omega t - B_3 \cos 3\omega t + \cdots \quad (18\text{-}34)$

From Eq. (18-24)

$$i_L = i_1 - i_2 = 2(B_1 \cos \omega t + B_3 \cos 3\omega t + \cdots) \quad (18\text{-}35)$$

This expression shows that a push-pull circuit will balance out all even harmonics in the output and will leave the third-harmonic term as the principal source of distortion. This conclusion was reached on the assumption that the two transistors are identical. If their characteristics differ appreciably, the appearance of even harmonics must be expected.

18-6 CLASS AB OPERATION

In addition to the distortion introduced by not using matched transistors and that due to the nonlinearity of the collector characteristics, there is one more source of distortion, that caused by nonlinearity of the input characteristic. As pointed out in Sec. 3-9 and Fig. 3-15, no appreciable base current flows until the

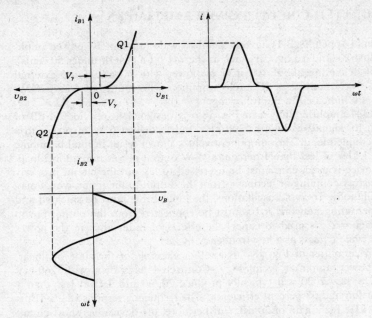

Figure 18-5 Crossover distortion.

emitter junction is forward-biased by the cutin voltage V_γ, which is 0.5 V for silicon. Under these circumstances a sinusoidal base-voltage excitation will not result in a sinusoidal output current. Although already mentioned briefly in Sec. 15-6, the significance of the nonlinear input characteristic merits further discussion.

The distortion caused by this nonlinear curve is indicated in Fig. 18-5. The $i_B - v_B$ curve for each transistor is drawn, and the construction used to obtain the output current (assumed proportional to the base current) is shown. In the region of small currents (for $v_B < V_\gamma$) the output is much smaller than it would be if the response were linear. This effect is called *crossover distortion*. Such distortion would not occur if the driver were a true current generator, in other words, if the base current (rather than the base voltage) were sinusoidal.

To minimize crossover distortion, the transistors must operate in a class AB mode, where a small standby current flows at zero excitation. For example, in the circuit of Fig. 15-20 the difference V between the base voltages of $Q14$ and $Q15$ is adjusted to be approximately equal to $2V_\gamma$. Class AB operation results in less distortion than class B, but the price which must be paid for this improvement is a loss in efficiency and waste of standby power. The calculations of the distortion components in a class AB or class A push-pull amplifier due to the nonlinearity of the collector characteristics is somewhat involved since it requires the construction of composite output curves for the pair of transistors.

18-7 INTEGRATED-CIRCUIT POWER AMPLIFIERS

Manufacturers (Appendix B-1) have available a wide range of IC power amplifiers. An industry standard OP AMP, such as the 741 (at a cost of under 50 cents), is capable of delivering about 100 mW of power with no additional external components. Two examples of IC audio amplifiers with ratings of 4 and 20 W, respectively, are indicated in the following.

The LM384 amplifier shown in Fig. 18-6 is designed to provide 34 dB of amplification for signals as high as 300 kHz and to deliver 5 W of power to a capacitively coupled load. The component values shown result in total harmonic distortion at 1 kHz of less than 1 percent at 5-W output power into an 8-Ω load. When this device is used, care must be exercised to lay out the circuit properly and to avoid stray coupling or feedback from the output to the input, which may result in oscillations. To avoid oscillations, the input cable must be shielded and the lag compensating network $R_1 C_2$ must be connected from the output pin to ground. Capacitor C_3 is used to cancel the effects of inductance in the power supply leads, while C_1 acts as a low-frequency bypass.

The 20-W amplifier of Fig. 18-7 is another example of the state of linear, monolithic, power amplifier technology. Connected as shown with 260-mV input, the SGS TDA2020 will typically produce 20 W into 4 Ω at less than 1 percent distortion and 57 percent efficiency. The frequency response (-3 dB) is 10 Hz to 160 kHz for a gain of 30 dB. Furthermore, the device has short-circuit protection for current overloads and thermal shutdown if the recommended maximum power dissipation limit is exceeded.

Capacitors C_1 through C_4 provide power supply bypass. The $R_3 C_5$ and $R_1 C_6$ networks produce output and input lag compensation, respectively. Further compensation is controlled by C_7. Since the output dc level is set at

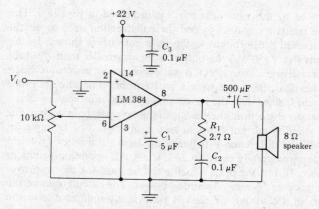

Figure 18-6 A 5-W audio amplifier. (*Courtesy of National Semiconductor.*)

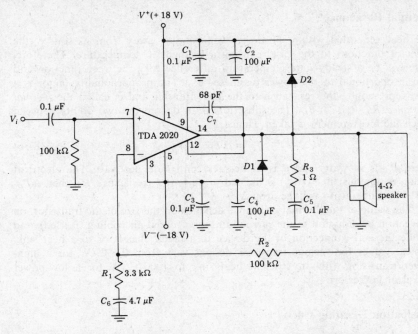

Figure 18-7 A split-supply 20-W audio power amplifier. (*Courtesy of SGS/ATES Corp.*)

$(V^+ + V^-)/2$, split power supply operation yields 0 V dc output, and the load can be direct-coupled, eliminating the need for a very large coupling capacitor. Diodes $D1$ and $D2$ clamp (and thus protect) the output from inductive excursions greater than the supply voltages.

18-8 THERMAL CONSIDERATIONS[2, 3]

The power amplifier of Fig. 18-7 raises a very important question. At 20-W output and 57% efficiency, the input power is $20/0.57 = 35.1$ W. Hence 15.1 W must be dissipated by the transistors. We will now indicate how this heat is removed and what factors must be considered in maintaining proper device operation.

Maximum Junction Temperature

All semiconductor devices have a maximum operating junction temperature $T_{J(\text{max})}$, ranging typically from 125 to 200°C for silicon. Above this temperature catastrophic irreversible failure will occur.

Thermal Resistance

The heat generated within the device will conduct away from its source (the collector junction) to the case, causing a gradient in temperature. Therefore, there will be a steady-state temperature difference ΔT_{JC} between junction and case, proportional to the power dissipated P_D. The proportionality factor is a term representing the resistance to the heat transfer and is called the *thermal resistance* R_{th}, $R_{\theta JC}$, or θ_{JC}. The subscripts on θ denote the two points between which the measurement is taken. Hence

$$T_J - T_C = \Delta T_{JC} = P_D \theta_{JC} \tag{18-36}$$

where P_D is in watts and θ is in degrees centigrade per watt. The electrical analog is obvious: if $P_D(\theta_{JC})$ is likened to current I (resistance R), then ΔT_{JC} must be analogous to voltage drop ΔV.

The value of the thermal resistance depends on the size of the transistor, on convection or radiation to the surroundings, on forced-air cooling (if used), and on the thermal connection of the device to a metal chassis or to a heat sink. Typical values for various transistor designs vary from $0.2°C/W$ for a high-power transistor with an efficient heat sink to $1000°C/W$ for a low-power transistor in free air.

Dissipation Derating Curve

Manufacturers usually present a power-temperature derating curve, such as shown in Fig. 18-8. The maximum junction temperature $T_{J(max)}$ can be deduced by noting that no power (0 W) can be dissipated at $200°C$. Zero power dissipation implies no temperature gradient and, hence, the junction must also be at $200°C$ [Eq. (18-36)].

The specifications for the 2N5671 silicon *n-p-n* transistor are given in Appendix B-8. This transistor has high power (140 W), high current ($I_C = 30$ A, $I_B = 10$ A), and high speed (switching time ~ 1 μs).[†] Since the maximum ordinate in Fig. 18-8 corresponds to $P_{D(max)} = 140$ W, then from Eq. (18-36), $\theta_{JC} = (200 - 25)/140 = 1.25°C/W$. The reciprocal of the thermal resistance is the slope of the line in Fig. 18-8, and is called the *power derating factor* $(1/1.25 = 0.8$ W/$°C)$. The value of the thermal resistance is inversely related to the surface area of the case. The 2N2222A which has a much smaller case than the 2N5671 has $\theta_{JC} = 83°C/W$.

To carry the heat away from the case into the ambient (the surrounding air), a *heat sink* is used with a power transistor. The heat sink is a metallic structure with a relatively large heat-radiating surface to which the transistor case is attached. Figure 18-9 depicts the mounting system for the TDA2020 chip of Fig. 18-7.

[†]The ratings of the 2N5671 power transistor should be compared with those of the 2N2222A small-signal transistor (Appendix B-3).

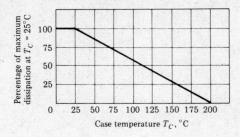

Figure 18-8 Dissipation derating curve for the 2N5671 power transistor. (*Courtesy of RCA Solid State Div.*)

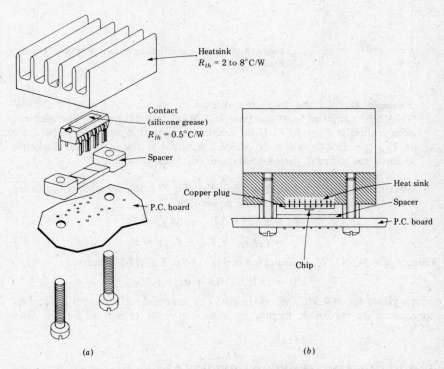

(a)   (b)

Figure 18-9 (*a*) Heat sink mounting of TDA2020. A number of heat sinks are available with thermal resistance values in the range from 2 to 8°C/W. (*b*) Cross section of assembled system. (*Courtesy of SGS/ATES Corp.*)

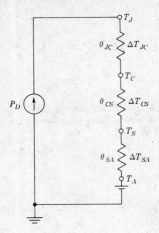

Figure 18-10 The electrical analog of Eq. (18-36) used to solve Example 18-1.

Example 18-1 In the foregoing discussion we observed that the 20-W TDA2020 amplifier must dissipate 15.1 W of internal power. The ambient temperature is $T_A = 30°C$. If the maximum allowable junction temperature is $T_{J(max)} = 150°C$ and if $\theta_{JC} = 3°C/W$, what is the maximum heat-sink thermal resistance θ_{SA} (sink-to-ambient) that can be tolerated?

SOLUTION Using the electrical analog of Eq. (18-36), we obtain the series circuit model of Fig. 18-10 for the power flow.

$$T_J = \Delta T_{JC} + \Delta T_{CS} + \Delta T_{SA} + T_A$$
$$= P_D(\theta_{JC} + \theta_{CS} + \theta_{SA}) + T_A \qquad (18-37)$$

Using $\theta_{CS} = 0.5°C/W$, as indicated in Fig. 18-9a, Eq. (18-37) becomes

$$150 = 15.1(3 + 0.5 + \theta_{SA}) + 30$$

which yields $\theta_{SA} = 4.5°C/W$ maximum. The heat sink in Fig. 18-9 is satisfactory, since its maximum thermal resistance may be chosen to be less than 4.5°C/W.

18-9 REGULATED POWER SUPPLIES[3, 4, 5]

An ideal *regulated power supply* is an electronic circuit designed to provide a predetermined dc voltage V_O which is independent of the current I_L drawn from V_O, of the temperature, and also of any variations in the ac line voltage.

An unregulated power supply consists of a transformer, a rectifier, and a filter, as shown in Figs. 10-14 and 10-18.

There are three reasons why an unregulated power supply is not good enough for many applications. The first is its poor regulation; the output voltage

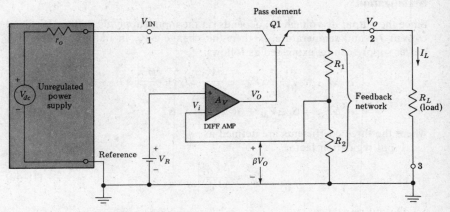

Figure 18-11 A regulated power supply system.

is not constant as the load varies. The second is that the dc output voltage varies with the ac input. In some locations the line voltage (of nominal value 115 V) may vary over as wide a range as 90 to 130 V, and yet it is necessary that the dc voltage remain essentially constant. The third reason is that the dc output voltage varies with the temperature, particularly because semiconductor devices are used. The feedback circuit shown in Fig. 18-11 is used to overcome the above three shortcomings, and also to reduce the ripple voltage. Such a system is called a *regulated power supply*. From Fig. 18-11 we see that the regulated power supply represents a case of voltage-series feedback. If we assume that the voltage gain of the emitter follower $Q1$ ($Q1$ is also called the *pass element*) is approximately unity, then $V_O' \approx V_O$ and

$$V_O' = A_V V_i = A_V(V_R - \beta V_O) \approx V_O \qquad (18\text{-}38)$$

where the feedback factor is

$$\beta \equiv \frac{R_2}{R_1 + R_2} \qquad (18\text{-}39)$$

From Eq. (18-38) it follows that

$$V_O = V_R \frac{A_V}{1 + \beta A_V} \qquad (18\text{-}40)$$

The output voltage V_O can be changed by varying the feedback factor β. The emitter follower $Q1$ is used to provide current gain, because the current delivered by the OP AMP A_V usually is not sufficient. Also, the pass element must absorb the difference between the unregulated input voltage V_{IN} and the regulated output voltage V_O. The dc collector voltage required by the error amplifier A_V is obtained from the unregulated voltage.

Stabilization

Since the output dc voltage V_O depends on the input supply dc voltage V_{dc}, load current I_L, and temperature T, then the change ΔV_O in output voltage of a power supply can be expressed as follows:

$$\Delta V_O = \frac{\partial V_O}{\partial V_{dc}} \Delta V_{dc} + \frac{\partial V_O}{\partial I_L} \Delta I_L + \frac{\partial V_O}{\partial T} \Delta T$$

or

$$\Delta V_O = S_V \Delta V_{dc} + R_o \Delta I_L + S_T \Delta T \qquad (18\text{-}41)$$

where the three coefficients are defined as

Input regulation factor:

$$S_V = \frac{\Delta V_O}{\Delta V_{dc}} \bigg|_{\substack{\Delta I_L = 0 \\ \Delta T = 0}} \qquad (18\text{-}42)$$

Output resistance:

$$R_o = \frac{\Delta V_O}{\Delta I_L} \bigg|_{\substack{\Delta V_{dc} = 0 \\ \Delta T = 0}} \qquad (18\text{-}43)$$

Temperature coefficient:

$$S_T = \frac{\Delta V_O}{\Delta T} \bigg|_{\substack{\Delta V_{dc} = 0 \\ \Delta I_L = 0}} \qquad (18\text{-}44)$$

The smaller the value of the three coefficients, the better the regulation of the power supply. The input-voltage change ΔV_{dc} may be due to a change in ac line voltage or may be ripple because of inadequate filtering.

18-10 MONOLITHIC REGULATORS[3, 4, 5]

It is interesting to note that if we were to construct a discrete component regulator, it would resemble Fig. 18-11 topologically: the amplifier A_V would be an OP AMP (such as the μA741 or LM301A) and the battery V_R would be replaced by a reference diode (an LM103, LM199, or a Zener). With the advent of microelectronics it has become technically and economically feasible to incorporate all components on a monolithic silicon chip. All the benefits of IC's are thus obtained: excellent performance, small size, ease of use, low cost, and high reliability.

An example of a monolithic regulator is the Motorola MC7800C series of three-terminal, positive, fixed-voltage regulators. Figure 18-12 is the standard application, and shows the degree to which user complexity has been all but eliminated. Input capacitor C_i is required to cancel inductive effects associated with long power-distribution leads. Output capacitor C_o improves the transient response. These devices, requiring no adjustment, have an output preset by the

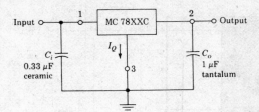

Figure 18-12 A standard three-terminal, positive, fixed-voltage monolithic regulator. The quiescent current is I_Q.

manufacturer to an industry standard voltage of 5, 6, 8, 12, 15, 18, or 24 V. (An MC7824C represents a 24-V regulator.) There must be a minimum of 2 V between input and output. Such regulators are capable of output currents in excess of 1.0 A. They have internal short-circuit protection which limits the maximum current the circuit will pass, thermal shutdown, and output-transistor safe-operating-area protection. Typical values for the stabilization coefficients are:

$$S_V = 3 \times 10^{-3} \qquad R_o = 30 \text{ m}\Omega \qquad S_T = -1 \text{ mV/}^{\circ}\text{C}$$

The level of complexity afforded by monolithic IC techniques can be appreciated by examining Fig. 18-13, the circuit diagram of the MC7800C. To the left of the shaded block is the reference voltage V_R of Fig. 18-11. This is the level shifter of Fig. 15-17a with a Zener diode input to the emitter-follower buffer. The shaded circuit of Fig. 18-13 is the difference amplifier A_V of Fig. 18-11. The design similarity with the 741 OP AMP configuration of Fig. 15-16 should be noted. The resistor divider R_1 and R_2 of Fig. 18-13 corresponds to the same feedback network in Fig. 18-11. The Darlington pair Q' and Q'' of Fig. 18-13 constitutes the pass element $Q1$ of Fig. 18-11.

The protection circuitry is shown in heavy outline and merits explanation. Current limiting is performed by R_3, R_4, and Q_2. Safe-operating protection is accomplished in the following way: If the output is pulled low by an overload, thus increasing the collector-emitter voltage of Q'', Zener $D1$ (which under normal loads is OFF) will conduct. Under these circumstances sufficient base current is supplied to $Q2$ so that it conducts, which, in turn, robs base drive from the $Q'Q''$ Darlington combination. In this manner the volt-ampere product of the pass element is limited to a reasonable power dissipation.

Consider next the thermal overload protection. A fraction of the reference voltage appearing across R_5 is applied to the base-emitter junction of $Q3$. For a fixed value of V_{BE3}, the collector current I_3 increases rapidly with increasing temperature. Hence, at sufficiently elevated temperatures (caused by either power dissipation or a high ambient), transistor $Q3$ will conduct heavily and once again starve the pass transistors $Q'Q''$ of base drive, thereby providing thermal shutdown.

Using monolithic regulators, it is possible to distribute unregulated voltage through electronic equipment and provide regulation locally, for example, on

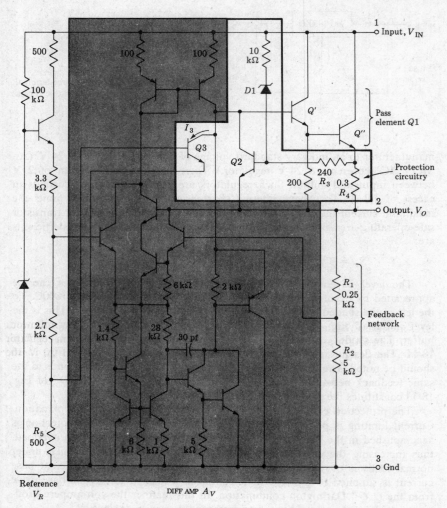

Figure 18-13 Schematic diagram for MC7800C series monolithic regulator. (*Courtesy of Motorola Semiconductor, Inc.*)

individual printed circuit (PC) boards. Among the advantages of this approach are greater flexibility in voltage levels, regulation for individual stages, and improved isolation and decoupling of these stages.

Monolithic regulators are available in a multitude of performance levels: fixed or variable, positive or negative output voltage, high output current (> 1 A), high output voltage (> 24 V), and single or dual (±) outputs. The engineer can also use the standard three-terminal regulator (Fig. 18-12) as a basic building block to tailor its performance to specific needs. Such techniques are considered in Probs. 18-14 and 18-15.

18-11 A SWITCHING REGULATOR[6-9]

The pass regulators of the previous section, despite their usefulness, have three drawbacks.

1. In a power supply that includes ac-to-dc conversion (transformer, bridge rectifier, and filter) the polarity and magnitude of the raw (unregulated) dc voltage can be a design parameter, and thus no inherent problem exists. If, however, in a system with one dc supply voltage (such as $+5$ V for TTL gates) there exists a need for ± 15 V for OP AMP operation, it may be economically (or physically) impracticable to add the facility for additional raw dc voltages.
2. A system operated from a battery, such as a communication system in a field environment or on a satellite in deep space, has no ac source available and hence must generate all voltages (positive or negative) from the single dc voltage source. Such a system is a *dc-to-dc converter*.
3. The input-voltage magnitude must be greater than the output magnitude, and series pass regulators are inherently inefficient. The greater the input-output differential for a given current, the greater the losses. A TTL system regulator operating from 10 V is at best 50 percent efficient, and from 20 V the efficiency drops to 25 percent.

Basic Switching Regulator Topology

All three difficulties can be avoided with the use of a *switching regulator*. The basic regulating control loop is shown in Fig. 18-14. The unregulated input voltage is V_{IN} and the regulated output voltage is V_O. The output current delivered to the load R_L is to be large (say, several amperes). The shaded block contains low-power circuits, fabricated on a single IC chip. The *reference regulator* is the series pass regulator described in Sec. 18-10 whose output is the regulated reference voltage V_{REF} which serves as the power supply voltage for all circuits on the chip. Since the current drawn from V_{REF} is small (say, 10 mA), the small power loss in the pass regulator does not affect appreciably the overall efficiency of the system.

The topology in Fig. 18-14 is that of a feedback system (voltage-series feedback, Sec. 12-2), and the comparison of the fixed input V_{REF} with a fraction $R_1/(R_1 + R_2)$ of the output V_O is made with the DIFF AMP (error amplifier). A triangular waveform generator of period T (circuit not indicated in Fig. 18-14) is also on the chip, and its output v is applied to the noninverting terminal of a comparator which functions as a *pulse-width modulator* (PWM). The error

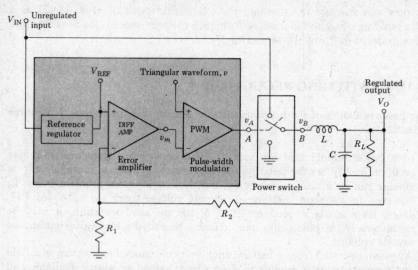

Figure 18-14 Basic switching regulator topology. The circuits indicated in the shaded block are fabricated on a single IC chip. All other components are discrete elements added external to the chip.

amplifier output voltage v_m is applied to the inverting terminal of the PWM, as shown in Fig. 18-14. This modulator operates as described in Sec. 17-8, producing a square wave v_A of period T, whose duty cycle δ varies linearly with v_m. The output v_A of the PWM drives a power switch (indicated by the SPDT block of Fig. 18-14), creating a square wave (of period T and duty cycle δ), whose minimum value is 0 and whose maximum is V_{IN}. This square wave is filtered by the LC combination, which acts as a low-pass filter. If the reactance of C is much smaller than that of L at the fundamental frequency, then all Fourier components in the square wave are greatly attenuated. In other words, if $T/2\pi C \ll 2\pi L/T$ or if $\sqrt{LC} \gg T/2\pi$, then V_O will be a constant, equal to the average value of the square wave.

The Regulated Output Voltage

Since there is a virtual short circuit between the input terminals of the error amplifier, $V_{REF} = R_1 V_O/(R_1 + R_2)$ and the output is given by

$$V_O = V_{REF}(1 + R_2/R_1) \tag{18-45}$$

Note that this regulated voltage is independent of variations in the raw input voltage V_{IN} and of changes in load current. It depends only on the constancy of the regulated voltage V_{REF} and the ratio R_2/R_1. If, for example, the reference voltage is the supply for TTL logic gates, so that $V_{REF} = 5$ V, and if an output voltage $V_O = 15$ V is desired, it is only necessary to select $R_2 = 2R_1$. As noted above, V_O is the dc value of the power-switch square-wave output voltage v_B, whose peak value is V_{IN}. Hence this configuration can be used only if $V_{IN} > V_O$. This control system operates in such a manner that an error voltage v_m is generated automatically, so that the PWM has the correct duty cycle δ to cause v_B to have a dc value V_O, given by Eq. (18-45).

Efficiency

An inspection of Fig. 18-14 reveals that the output current passes from V_{IN} through the power switch and the inductor through the load. Hence, using a switch with low losses (a transistor switch with small $V_{CE(\text{sat})}$ and high switching speed) and a filter with high Q (an inductor with low resistance), the conversion efficiency can exceed 90 percent.

The Power Switch

The action of the SPDT switch in Fig. 18-14 may be obtained by the combination in Fig. 18-15a of a diode and a SPST switch (to be replaced by a transistor in Fig. 18-15b). The LC filter, the load R_L, and the PWM block driving the switch are also included, but R_2, R_1, and the IC block are omitted for simplicity.

The circuit operates as follows: When the switch is closed the diode is reverse-biased by V_{IN} and the load current I_L is supplied from $v_B = V_{IN}$ through L. In the second portion of the cycle when the switch is opened, the inductor current cannot decrease instantaneously. (If it did, then the inductor voltage $L\, di/dt$ would be negative infinity.) Hence, at the instant that the switch opens, i_L remains constant and the current path must be from ground through the diode and the inductor into the load. Neglecting the drop across the diode, $v_B = 0$. Hence, v_B is a square wave of period T and duty cycle δ, with a minimum value of 0 and a maximum value of V_{IN}. This waveform is identical with that for v_B in Fig. 18-14. Therefore the circuit in Fig. 18-15a operates exactly as that in Fig. 18-14. Since, when the switch opens, v_B flies back from V_{IN} to zero, the configuration in Fig. 18-15 gives rise to the name *flyback converter*.

The SPST switch in Fig. 18-15a may be simulated with a p-n-p power transistor $Q1$, as indicated in Fig. 18-15b. If the load current is 1 A, then the collector current of $Q1$ is 1 A and, for $\beta = 100$, the base current is 10 mA. The transistor $Q2$ is used to supply this large base current. Note that $Q1$ and $Q2$ form a Darlington pair (Sec. 11-16). To drive these transistors with the proper voltage polarity the PWM output voltage v_A must be inverted and, hence, the transistor $Q3$ is needed to complete the switch in Fig. 18-15b.

For v_A positive, $Q3$ conducts and its collector current (through the switch resistors) biases $Q1$ and $Q2$ ON, so that $v_B \approx V_{IN}$. On the other hand, for v_A negative or 0, $Q3$ is nonconducting and there is no current in the biasing

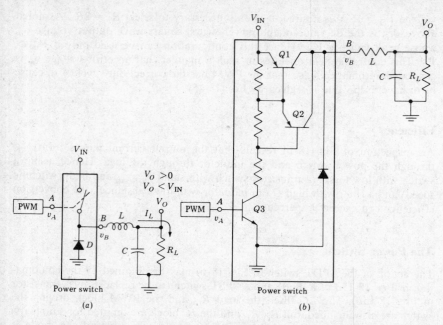

Figure 18-15 (*a*) The SPDT switch of Fig. 18-14 is replaced by a SPST switch and a flyback diode. (*b*) A practical implementation of the power switch using transistors. The output is positive and smaller than the input.

resistors. Hence $Q1$ and $Q2$ are OFF and the switch is open. By the action described in the preceding paragraph, the flyback diode goes ON and $v_B = 0$. This behavior indicates that the power switch in Fig. 18-15*b* is the practical implementation of the idealized switch in Fig. 18-15*a*. Incidentally, the low-power transistor $Q3$ is fabricated as part of the IC chip shown in the shaded block of Fig. 18-14.

18-12 ADDITIONAL SWITCHING REGULATOR TOPOLOGIES

For the configuration in Fig. 18-15 the output voltage is positive and less than the input voltage ($V_O < V_{IN}$), as verified in the preceding section. This restriction is removed by using the configuration in Fig. 18-16, as will now be demonstrated. Consider the interval T_1 when the switch is closed. The diode is reverse-biased by the positive voltage V_O, the feedback loop is open, and C discharges through R_L. By choosing $CR_L \gg T_1$, the drop in V_O (the ripple voltage) is small. During this interval the input voltage is across L and the inductor i_L *increases* by $di_L = V_{IN} \, dt/L = V_{IN}T_1/L$.

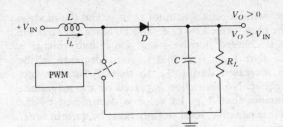

Figure 18-16 For this arrangement of the power components in a switching regulator the output is positive and greater than the input.

Consider now the interval T_2 during which the switch is open. Since the current in an inductor can not change instantaneously, $i_L(T_1 -) = i_L(T_1 +)$ and, hence, the diode goes ON and i_L passes through the diode and into C. In the steady state the voltage across C must be the same at the end of the period $T = T_1 + T_2$ as it was at the beginning, $t = 0$. Similarly, the current must *decrease* ($di_L/dt < 0$) during T_2 by the amount $V_{IN}T_1/L$ by which it increased during T_1. Neglecting the diode voltage we obtain from Fig. 18-16 that v_o (the instantaneous output voltage) is given by

$$v_o = V_{IN} - L\, di/dt > V_{IN}$$

because di_L/dt is negative. This argument verifies that the output V_O exceeds the input for this configuration. Incidentally, the switch action is obtained by using the Darlington pair $Q1Q2$ driven by $Q3$ in a manner similar to that indicated in Fig. 18-15b.

Negative Output Voltages

To obtain a negative supply from a positive raw dc voltage, the configuration of Fig. 18-17 is used for the power components. We assume that $V_O < 0$ and then justify this assumption. The argument is similar to that used in the preceding paragraph. During the interval T_1 when the switch is closed, the diode is OFF because the cathode voltage is $+V_{IN}$ and the anode voltage is negative. The capacitor discharges slightly through the load and the inductor current increases

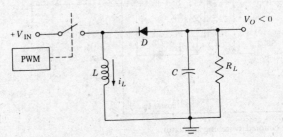

Figure 18-17 This topology of the power components in a switching regulator results in a negative output voltage.

by $V_{IN}T_1/L$. At the instant the switch opens, i_L cannot change, and the diode is forced ON so that i_L flows in the loop formed by L, C, and D. Since i_L enters the bottom plate of C, this plate is charged positively and the output voltage is negative. An alternative proof that $V_O < 0$ is that i_L must decrease in the interval T_2 by the amount it increased during T_1, so that $di_L/dt < 0$ and, therefore, $v_o \approx L \, di_L/dt$ is negative. No restriction is placed on the magnitude of V_O; it may be larger or smaller than V_{IN}. Its value is determined by the control loop of Fig. 18-14. If V_O is negative, level shifting must be used in order for the effective feedback voltage to be positive. This configuration is indicated in Prob. 18-16.

Transformer-Coupled Push-Pull DC-to-DC Converter

This switching regulator configuration has the most flexibility because the output V_O may be greater than or less than the raw dc input V_{IN}, and the sign of V_O may be the same as, or opposite to, that of V_{IN}. The topology of the power components is indicated in Fig. 18-18, which uses an iron-core transformer with a center-tapped primary ($v_{P1} = v_{P2}$) as well as a center-tapped secondary ($v_{S1} = v_{S2}$). The number of turns in the secondary is n times that in the primary, so that $v_{S1} = nv_{P1}$ and $v_{S2} = nv_{P2}$. If $n > 1$, then it is possible to obtain $V_O > V_{IN}$, whereas for $n \leqslant 1$, $V_O < V_{IN}$.

The two switches $SW1$ and $SW2$ are controlled by waveforms v_{A1} and v_{A2}, which are obtained from the PWM output v_A (as explained in Fig. 18-20). The waveforms v_A, v_{A1}, and v_{A2} are sketched in Fig. 18-19a, b, and c, respectively. The waveform v_A is obtained from the PWM in the shaded block of Fig. 18-14. Note that $SW1$ and $SW2$ are closed for the same duty cycle, but each is

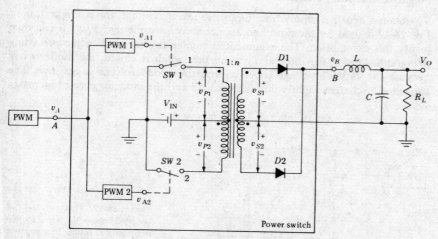

Figure 18-18 Push-pull transformer-coupled switching regulator.

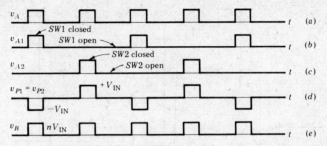

Figure 18-19 The waveforms in Fig. 18-18. (For simplicity all amplitudes are drawn equal.)

operated only once for every other period of the PWM waveform v_A. In other words, each switch operates at one half the frequency of the single-ended converter of Fig. 18-14. Switch $SW1$ $(SW2)$ is a transistor whose base waveform is $v_{A1}(v_{A2})$.

From Fig. 18-18 it follows that the primary voltages are given by

$$v_{P1} = v_{P2} = \begin{cases} -V_{\text{IN}} & \text{if } SW1 \text{ is closed and } SW2 \text{ is open} \\ +V_{\text{IN}} & \text{if } SW1 \text{ is open and } SW2 \text{ is closed} \\ 0 & \text{if } SW1 \text{ is open and } SW2 \text{ is open} \end{cases} \quad (18\text{-}46)$$

This waveform is indicated in Fig. 18-19d. The secondary voltages $v_{S1} = v_{S2}$ have this same waveshape but are n times as large.. During the intervals when $v_{S1} = v_{S2}$ is positive $D1$ conducts, $D2$ is OFF, and $v_B = nV_{\text{IN}}$. If $v_{S1} = v_{S2}$ is negative $D2$ conducts, $D1$ is OFF, and $v_B = nV_{\text{IN}}$ again. When $v_{S1} = v_{S2} = 0$ the two diodes are connected in parallel from point B to ground and, hence, act as a flyback diode, as shown in Fig. 18-15, so that $v_B = 0$ during this interval. Consequently, the waveform v_B is as indicated in Fig. 18-19e. Note that v_B is proportional to v_A. Because of the LC filter, the dc output voltage V_O equals the average value of the v_B waveform and may be greater (less) than V_{IN}, depending upon whether n exceeds (is smaller than) unity. If the diodes are reversed, the sign of V_O is negative. The block between A and B in Fig. 18-18 replaces the power-switch block in the feedback loop of Fig. 18-15b. The regulated output is given by Eq. (18-45).

Generating the Switching Waveforms

We now indicate how to obtain the two switching waveforms v_{A1} and v_{A2} from the PWM waveform v_A. The block diagram is shown in Fig. 18-20 and the waveforms in Fig. 18-21. The square-wave oscillator waveform v_{OSC} in Fig. 18-21a is used to generate the triangular voltage needed for the pulse-width modulator. This PWM waveshape v_A is given in Fig. 18-21b. The duty cycle δ of v_A is $T_1/(T_1 + T_2)$. The FLIP-FLOP is used as a divide-by-2 circuit, whose input is

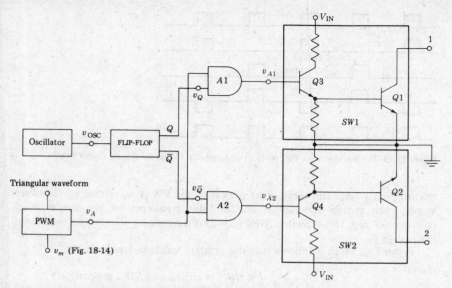

Figure 18-20 The block diagram of the system for obtaining the waveforms v_{A1} and v_{A2} of Fig. 18-19 from the pulse-width modulator output v_A. Also shown are the switch power transistors $Q1$ and $Q2$ and the base-drive transistors $Q3$ and $Q4$.

v_{OSC} and the two complementary FLIP-FLOP outputs v_Q and $v_{\bar{Q}}$ are shown in Fig. 18-21c and d, respectively. The inputs to AND gate $A1$ ($A2$) are $v_Q(v_{\bar{Q}})$ and v_A, and the outputs v_{A1} and v_{A2} are drawn in Fig. 18-21e and f respectively. These are the waveforms used in Fig. 18-19b and c.

The switch $SW1$ ($SW2$) of Fig. 18-18 is replaced by the power transistor $Q1$ ($Q2$) in Fig. 18-20. The base currents for $Q1$ and $Q2$ are supplied by transistors $Q3$ and $Q4$, which are driven by waveforms v_{A1} and v_{A2}, respectively. The

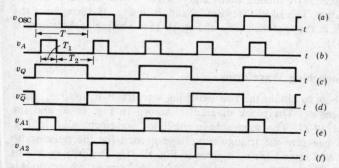

Figure 18-21 The waveforms in Fig. 18-20. (For simplicity all amplitudes are drawn equal.)

complexity of the switching regulator system would preclude its usefulness were it not for the increased level of sophistication attainable in modern microelectronics. The Silicon General SG1524 package is such an example. Contained on this chip are all of the following circuits: reference regulator, pulse-width modulator (consisting of the saw-tooth oscillator and comparator), error amplifier, two uncommitted transistors (for $Q3$ and $Q4$ in Fig. 18-20 or $Q3$ in Fig. 18-15), steering FLIP-FLOP and two AND gates (Fig. 18-20), and provisions for current-limiting and shutdown.

The SG1524 is placed in the feedback loop of Fig. 18-14 to form a switching regulator by adding the feedback resistors R_1 and R_2 and the discrete power-switch components of Fig. 18-15b or Fig. 18-18. For $V_{IN} = 28$ V, it is possible to obtain a regulated output voltage V_O of 5 V at 1 A for the single-ended system and 5 V at 5 A for the push-pull system. For $L \approx 1$ mH and $C \approx 1,000$ μF as filter components with the SG1524, line and load regulation of 0.2 percent with less than 1 percent maximum variation is achieved. The control circuitry operates at a reference voltage of 5 V, draws less than 10 mA, and is capable of operation beyond 100 kHz (external resistors R_T and C_T set the frequency). The output transistors on the chip are rated at 100 mA and are short-circuit protected. The feedback loop is stabilized by adding an RC lag network.

18-13 POWER FIELD-EFFECT TRANSISTOR (VMOS)[10]

In 1976 Siliconix Inc. introduced a new type of FET power transistor which overcomes many of the limitations of the bipolar power transistor. This new device is an n-channel enhancement MOSFET, but it is fabricated so that the current flows vertically. Hence this transistor is designated VMOS. This construction distinguishes the VMOS from the low-power MOSFETs described in Chapter 8, where the carriers flow horizontally from source to drain.

The fabrication of the power FET starts with a silicon n^+ substrate onto which is grown an n^- epitaxial layer. Two successive diffusions then take place, the first with p-type and the second with n-type impurities, as indicated in Fig. 18-22. The structure obtained at this stage of the construction is identical with

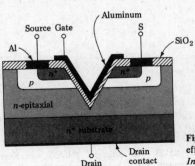

Figure 18-22 Cross-sectional view of a vertical field-effect power transistor, VMOS. (*Courtesy of Siliconix, Inc.*)

the discrete bipolar transistor shown in Fig. 4-6b. In the BJT the top (bottom) n^+ region is the emitter (collector), whereas in Fig. 18-22 the top (bottom) n^+ section becomes the source (drain). In Fig. 4-6b the p region is the base, but in the VMOS the p section is the n channel. In order to be able to place a control gate over the channel, extending from source to drain, a novel fabrication step is introduced; an isosceles V-shaped groove is anisotropically etched into the silicon, as indicated in Fig. 18-22. Continuing with the standard fabrication processes described in Chap. 4, a thin silicon dioxide layer is grown and then metallization is used to form the gate electrode and the source contact.

Note that the drain area (at the bottom of Fig. 18-22) is large and can be placed in contact with a heat sink for optimum removal of the power dissipated within the device. The channel length L (the vertical extent of the p region) is determined by the difference in the depths of the p and n^+ (source) diffusions. Hence, L can be made reliably quite small; for example, $L \approx 1.5$ μm. It should be recalled that in the standard (horizontal) MOSFET the channel length is determined by masking, etching, and the *lateral* diffusion of the source and drain, so that L is much longer (Table 8-1) than for the VMOS. The V-shaped gate controls two vertical MOSFETs, one on each side of the notch. Hence, by paralleling the two S terminals in Fig. 18-22 the current capacity is doubled.

The low-power MOSFETs of Chapter 8 are symmetrical devices, between source and drain. Clearly, from Fig. 18-22, the VMOS is built unsymmetrically so that S and D may not be interchanged.

VMOS Characteristics

The volt-ampere curves of a vertical FET are indicated in Fig. 18-23 and should be compared with the low-power n-channel enhancement MOSFET of Fig. 8-16. The peak VMOS current is 2 A (contrasted with 50 mA for the horizontal MOSFET). Also, note that the characteristics in the saturation region of Fig. 18-23 are much flatter than in Fig. 8-16 (I_D = constant and hence the output conductance is very small). For the 2N6657 family shown the spacing of the characteristics (above $I_D = 0.4$ A) are constant for equal increments of gate voltage. Therefore, the transconductance g_m is constant (≈ 0.25 A/V) for

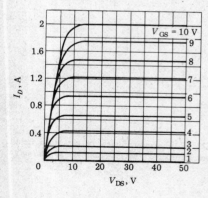

Figure 18-23. The Siliconix 2N6657 n-channel enhancement VMOS output characteristics. $P_D = 25$ W. (80 μs duty-cycle pulse test.)

$I_D \geqslant 0.4$ A. On the other hand, for a low-power MOSFET, g_m varies as the square root of the drain current [Eq. (11-81)], rather than remaining constant.

The VMOS has many advantageous properties, including the following:

1. The transfer characteristic I_D versus V_{GS} is linear (g_m = constant) for $I_D \geqslant 0.4$ A.
2. Switching is very fast because there is no minority carrier storage. For example, 2 A can be turned ON or OFF in less than 10 ns.
3. Thermal runaway (Sec. 11-2) is not possible because the drain-source resistance has a positive temperature coefficient and the current becomes limited as the device heats up. (No hot spots develop and secondary breakdown does not occur.)
4. There is no "current hogging" (Sec. 5-13) when VMOS devices are operated in parallel to increase the current capacity. If one transistor tries to take more than its share of the current the positive drain-to-source temperature coefficient increases V_{DS}, thereby limiting I_D.
5. Because of its very high input resistance the VMOS requires extremely small input power and may be driven from CMOS logic gates. The power gain is extremely high.
6. The ON resistance is very low. From the slope of the curves at the origin in Fig. 18-22, we see that $r_{DS(ON)} \approx 3 \ \Omega$.
7. Power FETs have extremely low noise figures.
8. The threshold voltage V_T ranges from 0.8 to 2 V, so that VMOS devices are compatible with TTL logic.
9. From Fig. 18-22 it is seen that the overlap of the gate and drain (and therefore the capacitance between these electrodes) is quite small. Hence, the capacitive feedback between output and input is minimized and VMOS devices may be used for high-frequency (broadband) circuits ($f_T \approx 600$ MHz).
10. The VMOS breakdown voltage between drain and source is high. This feature results from the fact that the epitaxial layer absorbs the depletion region from the reverse-biased body-drain p-n diode.

Applications

The VMOS may be used as the output stage of an audio or RF power amplifier or of a switching regulator power supply. However, push-pull operation is not possible since complementary VMOS transistors are not available (1978). Other applications include industrial controls such as process control, motor control, solenoid or relay driver, plasma display driver, ultrasonic transducer driver, etc. VMOS will also find wide use in telecommunication switching systems.

REFERENCES

1. Millman, J., and C. C. Halkias: "Integrated Electronics: Analog and Digital Circuits and Systems," pp. 682–683, McGraw-Hill Book Company, New York, 1972.

2. Bohn, D. (Ed.): "Audio Handbook," National Semiconductor Corporation, 1976.
3. Spencer, J. D., and D. E. Pippenger: "The Voltage Regulator Handbook," Texas Instruments, Inc., 1977.
4. Sevastopoulos, N., et al: "Voltage Regulator Handbook," National Semiconductor Corporation, 1975.
5. Wurzburg, H.: "Voltage Regulator Handbook," Motorola, Inc., 1976.
6. Mammano, B.: *Simplifying Converter Design with a New Integrated Regulating Pulse-Width Modulator*, Silicon General Application Note, June, 1976.
7. RCA "Power Transistor Data Book, SSD-204C," 1975.
8. Texas Instruments, Inc.: "Power Semiconductor Data Book for Design Engineers, CC-404," 1976.
9. Unitrode: "Switching Regulator Guide Publication U-68A," 1976.
10. Shaeffer, L., and D. Hoffman: *VMOS-A Breakthrough in Power MOSFET Technology*, Siliconix Application Note AN76-3, 1977.

REVIEW QUESTIONS

18-1 Derive an expression for the output power of a class A large-signal amplifier in terms of V_{max}, V_{min}, I_{max}, and I_{min}.

18-2 Discuss how rectification may take place in a power amplifier.

18-3 Define *intermodulation distortion*.

18-4 Define *total harmonic distortion*.

18-5 Define a (*a*) class A, (*b*) class B, and (*c*) class AB amplifier.

18-6 (*a*) Define the *conversion efficiency* η of a power stage.
(*b*) Derive a simple expression for η for a class A amplifier.
(*c*) What is the theoretical maximum efficiency for a class A amplifier?

18-7 (*a*) Draw the circuit of a class B power stage.
(*b*) For a sinusoidal input, what is the output waveform?

18-8 (*a*) Draw the circuit of a class B push-pull power amplifier.
(*b*) State three advantages of class B over class A.

18-9 Derive a simple expression for the output power of an idealized class B push-pull power amplifier.

18-10 Show that the maximum conversion efficiency of the idealized class B push-pull circuit is 78.5 percent.

18-11 Obtain the expression for the collector dissipation of a class B push-pull stage in terms of V_m and R_L.

18-12 Demonstrate that even harmonics are eliminated in a balanced push-pull amplifier.

18-13 (*a*) Explain the origin of crossover distortion.
(*b*) Describe a method to minimize this distortion.

18-14 (*a*) Define *thermal resistance* θ.
(*b*) Sketch a dissipation derating curve for a power amplifier.
(*c*) How is θ related to the curve in (*b*)?

18-15 (*a*) What is a *heat sink*?
(*b*) Explain why a heat sink must be used with a power amplifier.

18-16 List three reasons why an unregulated supply is not good enough for some applications.

18-17 Define *input regulation factor*, *output resistance*, and *temperature coefficient* for a voltage regulator.

18-18 (*a*) Draw a simplified circuit diagram of a regulated power supply.

(*b*) What type of feedback is employed by this regulator?

18-19 List three disadvantages of pass regulators which may be overcome with a switching regulator.

18-20 (*a*) Draw the basic switching-regulator topology.

(*b*) Explain how the dc output voltage is determined by this feedback system.

18-21 Explain why a switching regulator is capable of very high conversion efficiency.

18-22 (*a*) Draw the power switch of a switching regulator as a SPDT switch. How is the switch controlled and what is the switch output waveform?

(*b*) Verify that the switch in part (*a*) is equivalent to a SPST switch in series with a diode to ground. Why is the diode referred to as a *flyback diode*?

18-23 (*a*) Indicate the SPDT power switch of the basic regulator as a combination of three transistors and a diode.

(*b*) Explain the function of each transistor and the diode.

18-24 (*a*) Draw the power components in a switching regulator for which V_O is positive and greater than V_{IN}.

(*b*) Verify that for this configuration $V_O > V_{IN}$.

(*c*) What determines the numerical value of V_O?

18-25 (*a*) Repeat Rev. 18-24*a* for a regulator for which V_O is negative.

(*b*) Give an argument to justify that $V_O < 0$.

18-26 (*a*) Draw the power switch for a push-pull transformer-coupled switching regulator.

(*b*) Indicate the pulse-width-modulator waveform v_A, and also the waveforms v_{A1} and v_{A2} controlling the two SPST switches in series with the transformer primaries.

(*c*) Sketch the transformer secondary waveforms.

(*d*) Draw the waveform from the output switch (the input voltage to the filter).

18-27 (*a*) Draw in block-diagram form the system for obtaining the waveforms v_{A1} and v_{A2} in Rev. 18-26*b*.

(*b*) Explain the operation of the system with the aid of a waveform chart.

(*c*) Show the switches controlled by v_{A1} and v_{A2} simulated by transistors.

18-28 List all the low-power control circuits which are fabricated on a single IC chip and used with a switching regulator.

18-29 (*a*) Sketch the cross-sectional view of a power FET.

(*b*) Explain briefly how this device is fabricated.

18-30 Give two important differences in the output characteristics of a VMOS and a low-power MOSFET.

18-31 List six advantages of a VMOS.

CONSTANTS AND CONVERSION FACTORS

A-1 PROBABLE VALUES OF GENERAL PHYSICAL CONSTANTS[†]

Constant	Symbol	Value
Electronic charge	q	1.602×10^{-19} C
Electronic mass	m	9.109×10^{-31} kg
Ratio of charge to mass of an electron	q/m	1.759×10^{11} C/kg
Mass of atom of unit atomic weight (hypothetical)	$\ldots$	1.660×10^{-27} kg
Mass of proton	m_p	1.673×10^{-27} kg
Ratio of proton to electron mass	m_p/m	1.837×10^{3}
Planck's constant	h	6.626×10^{-34} J · s
Boltzmann constant	$\bar{k}$	1.381×10^{-23} J/K
	k	8.620×10^{-5} eV/K
Stefan-Boltzmann constant	σ	5.670×10^{-8} W/(m²)(K⁴)
Avogadro's number	N_A	6.023×10^{23} molecules/mol
Gas constant	R	8.314 J/(deg)(mol)
Velocity of light	c	2.998×10^{8} m/s
Faraday's constant	F	9.649×10^{3} C/mol
Volume per mole	V_o	2.241×10^{-2} m³
Acceleration of gravity	g	9.807 m/s²
Permeability of free space	μ_o	1.257×10^{-6} H/m
Permittivity of free space	ϵ_o	8.849×10^{-12} F/m

[†] E. A. Mechtly, "The International System of Units: Physical Constants and Conversion Factors," National Aeronautics and Space Administration, NASA SP-7012, Washington, D.C., 1964.

A-2 CONVERSION FACTORS AND PREFIXES

1 ampere (A)	$= 1$ C/s		1 lumen per square foot	$= 1$ footcandle (fc)
1 angstrom unit (Å)	$= 10^{-10}$ m		mega (M)	$= \times 10^6$
	$= 10^{-4}\ \mu$m		1 meter (m)	$= 39.37$ in
1 atmosphere pressure	$= 760$ mm Hg		micro (μ)	$= \times 10^{-6}$
1 coulomb (C)	$= 1$ A · s		1 micron	$= 10^{-6}$ m
1 electron volt (eV)	$= 1.60 \times 10^{-19}$ J			$= 1\ \mu$m
1 farad (F)	$= 1$ C/V		1 mil	$= 10^{-3}$ in
1 foot (ft)	$= 0.305$ m			$= 25\ \mu$m
1 gram-calorie	$= 4.185$ J		1 mile	$= 5,280$ ft
giga (G)	$= \times 10^9$			$= 1.609$ km
1 henry (H)	$= 1$ V · s/A		milli (m)	$= \times 10^{-3}$
1 hertz (Hz)	$= 1$ cycle/s		nano (n)	$= \times 10^{-9}$
1 inch (in)	$= 2.54$ cm		1 newton (N)	$= 1$ kg · m/s^2
1 joule (J)	$= 10^7$ ergs		pico (p)	$= \times 10^{-12}$
	$= 1$ W · s		1 pound (lb)	$= 453.6$ g
	$= 6.25 \times 10^{18}$ eV		1 tesla (T)	$= 1$ Wb/m^2
	$= 1$ N · m		1 ton	$= 2,000$ lb
	$= 1$ C · V		1 volt (V)	$= 1$ W/A
kilo (k)	$= \times 10^3$		1 watt (W)	$= 1$ J/s
1 kilogram (kg)	$= 2.205$ lb		1 weber (Wb)	$= 1$ V · s
1 kilometer (km)	$= 0.622$ mile		1 weber per square	
1 lumen	$= 0.0016$ W		meter (Wb/m^2)	$= 10^4$ gauss
	(at 0.55 μm)			

B

SEMICONDUCTOR MANUFACTURERS AND DEVICE SPECIFICATIONS

B-1 ELECTRONIC DEVICE MANUFACTURERS

Databooks and applications information may be obtained from the following semiconductor companies:

Advanced Micro Devices   901 Thompson Pl., Sunnyvale, CA 94086
American Microsystems Inc.   3800 Homestead Road, Santa Clara, CA. 95051
Burr-Brown Research Corp.   6730 S. Tucson Blvd., Tucson, Arizona 85734
Fairchild Semiconductor   464 Ellis St., Mt. View, CA 94042
Ferranti Electric   E. Bethpage Rd., Plainview, N. Y. 11803
General Electric Co.   Schenectady, N. Y. 13201
General Instrument Corp.   600 West John St., Hicksville, N. Y. 11802
Harris Semiconductor   Box 833, Melbourne, FL 32901
Hitachi America, Ltd.   111 E. Wacker Dr., Chicago, IL 60601
Imsai   14860 Wicks Blvd., San Leandro, CA 94577
Intel Corp.   3065 Bowers Ave., Santa Clara, CA 95051
Intersil Inc.   10900 N. Tantau Ave., Cupertino, CA 95014
ITT Semiconductors   74 Commerce Way, Woburn, MA 01801
Monolithic Memories, Inc.   1165 E. Argues Ave., Sunnyvale, CA 94086
Mostek Corp.   1215 W. Crosby Rd., Carollton, Texas 75006
Motorola Semiconductor Products   Box 20912, Phoenix, Ariz. 85036
National Semiconductor, Inc.   2900 Semiconductor Dr., Santa Clara, CA 95051
Plessey Semiconductors   1674 McGraw Ave., Santa Ana, CA 92705
Raytheon Semiconductor   350 Ellis St., Mt. View, CA 94042
RCA Solid State Division   Box 3200, Somerville, N. J. 08876
SGS/ATES Semiconductor Corp.   796 Massasoit Street, Waltham, MA 03254
Signetics Corp.   811 E. Argues Ave., Sunnyvale, CA 94086
Silicon General   73826 Bolsoo Ave., Westminster, CA 92683
Siliconix, Inc.   2201 Laurelwood Road, Santa Clara, CA 95054
Stewart-Warner Microcircuits   730 E. Evelyn Ave., Sunnyvale, CA 94086
Teledyne Semiconductor   1300 Terra Bella Ave., Mt. View, CA 94043
Texas Instruments Semiconductor Group   Box 5012, Dallas, Texas 75222
Toshiba America   280 Park Ave., New York, N. Y. 10017
TRW Microelectronics Center   One Space Park, Redondo Beach, CA 90278
Unitrode Corporation   580 Pleasant St., Watertown, MA 02172

B-2 SPECIFICATIONS FOR 1N4153 SILICON DIODE (Courtesy of Texas Instruments, Inc.)

High-speed switching diodes for computer and general-purpose applications.

Table B2-1 Absolute maximum ratings (25 °C)

		1N4151	1N4152	1N4153	1N4154	Unit
V_{RM}	Peak reverse voltage	75	40	75		V
$V_{RM(wkg)}$	Working peak reverse voltage	50	30	50	25	V
P	Continuous power dissipation at (or below) 25°C free-air temperature*			500		mW
T_{stg}	Storage temperature range			−65 to 200		°C
T_L	Lead temperature $\frac{1}{16}$ in from case for 10 s			300		°C

* Derate linearly to 200 °C at the rate of 2.85 mW/°C.

Table B2-2 Electrical characteristics (25°C unless otherwise noted)

Parameter		Test conditions	1N4153		Unit
			Min	Max	
$V_{(BR)}$	Reverse breakdown voltage	$I_R = 5\ \mu A$	75		V
I_R	Static reverse current	$V_R = $ rated $V_{RM(wkg)}$		0.05	μA
		$V_R = $ rated $V_{RM(wkg)}$ $T_A = 150°C$		50	μA
V_F	Static forward voltage	$I_F = 0.1$ mA	0.49	0.55	V
		$I_F = 0.25$ mA	0.53	0.59	V
		$I_F = 1$ mA	0.59	0.67	V
		$I_F = 2$ mA	0.62	0.70	V
		$I_F = 10$ mA	0.70	0.81	V
		$I_F = 20$ mA	0.74	0.88	V
C_T	Total capacitance	$V_R = 0$ $f = 1$ MHz		2	pF
t_{rr}	Reverse recovery time	$I_F = 10$ mA, $I_{RM} = 10$ mA $R_L = 100\ \Omega$		4	ns
		$I_F = 10$ mA, $V_R = 6$ V $R_L = 100\ \Omega$		2	ns

B-3 SPECIFICATIONS FOR 2N2222A N-P-N SILICON BIPOLAR JUNCTION TRANSISTOR (Courtesy of Motorola, Inc.)

Widely used "Industry Standard" transistor for applications as medium-speed switches and as amplifiers from audio to VHF frequencies. Complements to p-n-p transistor 2N2907A.

Table B3-1 Absolute maximum ratings*

Characteristic	Symbol	Rating	Unit
Collector-emitter voltage	V_{CEO}	40	V
Collector-base voltage	V_{CB}	75	V
Emitter-base voltage	V_{EB}	6.0	V
Collector current—continuous	I_C	800	mA
Total device dissipation @ $T_A = 25°C$	P_D	0.5	W
Derate above 25°C		3.33	mW/°C
Total device dissipation @ $T_C = 25°C$	P_D	1.8	W
Derate above 25°C		12	mW/°C
Operating and storage junction Temperature range	T_J, T_{stg}	65 to +200	°C

* T_A = ambient, T_C = case, and T_J = junction temperature.

Table B3-2 Electrical characteristics ($T_A = 25\,°C$ unless otherwise noted)

OFF characteristics	Symbol	Min	Max	Unit
Collector-emitter breakdown voltage $(I_C = 10$ mA, $I_B = 0)$	BV_{CEO}	40		V
Collector-base breakdown voltage $(I_C = 10$ μA, $I_E = 0)$	BV_{CBO}	75		V
Emitter-base breakdown voltage $(I_E = 10$ μA, $I_C = 0)$	BV_{EBO}	60		V
Collector cutoff current $(V_{CE} = 60$ V, $V_{EB(OFF)} = 3.0$ V$)$	I_{CEX}		10	nA
Collector cutoff current $(V_{CB} = 60$ V, $I_E = 0, T_A = 150°C)$	I_{CBO}		10	μA
Emitter cutoff current $(V_{EB} = 3.0$ V, $I_C = 0)$	I_{EBO}		10	nA
Base cutoff current $(V_{CE} = 60$ V, $V_{EB(OFF)} = 3.0$ V$)$	I_{BL}		20	nA

ON characteristics

See Fig. 3-13 for $V_{CE(sat)}$ and $V_{BE(sat)}$. See Fig. 3-14 for h_{FE}.

Table B3-3 Electrical characteristics (continued)

Small-signal characteristics	Symbol	Min	Max	Unit
Current-gain–bandwidth product	f_T	300		MHz
($I_C = 20$ mA, $V_{CE} = 20$ V, $f = 100$ MHz)				
Output capacitance	C_{ob}		8.0	pF
($V_{CB} = 10$ V, $I_E = 0$, $f = 100$ kHz)				
Input capacitance	C_{ib}		25	pF
($V_{EB} = 0.5$ V, $I_C = 0$, $f = 100$ kHz)				
Input impedance	h_{ie}			kΩ
($I_C = 1.0$ mA, $V_{CE} = 10$ V, $f = 1.0$ kHz)		2.0	8.0	
($I_C = 10$ mA, $V_{CE} = 10$ V, $f = 1.0$ kHz)		0.25	1.25	
Voltage feedback ratio	h_{re}			$\times 10^{-4}$
($I_C = 1.0$ mA, $V_{CE} = 10$ V, $f = 1.0$ kHz)			8.0	
($I_C = 10$ mA, $V_{CE} = 10$ V, $f = 1.0$ kHz)			4.0	
Small-signal current gain	h_{fe}			
($I_C = 1.0$ mA, $V_{CE} = 10$ V, $f = 1.0$ kHz)		50	300	
($I_C = 10$ mA, $V_{CE} = 10$ V, $f = 1.0$ kHz)		75	375	
Output admittance	h_{oe}			μmhos
($I_C = 1.0$ mA, $V_{CE} = 10$ V, $f = 1.0$ kHz)		5.0	35	
($I_C = 10$ mA, $V_{CE} = 10$ V, $f = 1.0$ kHz)		25	200	
Collector-base time constant	$r_b' C_c$		150	ps
($I_E = 20$ mA, $V_{CB} = 20$ V, $f = 31.8$ MHz)				
Noise figure	NF		4.0	dB
($I_C = 100$ μA, $V_{CE} = 10$ V,				
$R_S = 1.0$ kΩ, $f = 1.0$ kHz)				

Switching characteristics	Symbol	Min	Max	Unit
Delay time ($V_{CC} = 30$ V,	t_d		10	ns
$V_{BE(\text{OFF})} = 0.5$ V				
$I_C = 150$ mA,				
Rise time $I_{B1} = 15$ mA)	t_r		25	ns
Storage time ($V_{CC} = 30$ V,	t_x		225	ns
$I_C = 150$ mA,				
$I_{B1} = 15$ mA				
Fall time $I_{B2} = 15$ mA)	t_f		60	ns
Active-region time constant ($I_C = 150$ mA,	T_A		2.5	ns
$V_{CE} = 30$ V)				

B-4 SPECIFICATIONS FOR 2N4869 DEPLETION-MODE N-CHANNEL SILICON JUNCTION FIELD-EFFECT TRANSISTOR
(Courtesy of Siliconix, Inc.)

Specifically designed for audio or subaudio frequency applications where noise must be at an absolute minimum.

Table B4-1 Absolute maximum ratings (25°C)

Gate-drain or gate-source voltage*	-40 V
Gate current or drain current	50 mA
Total device dissipation	
(derate 1.7 mW/°C)	300 mW
Storage temperature range	-65 to $+200$°C

* Due to symmetrical geometry, these units may be operated with source and drain leads interchanged.

Table B4-2 Electrical characteristics (25°C unless otherwise noted)

		Characteristic	Min	Max	Unit	Test conditions	
S	I_{GSS}	Gate reverse current		-0.25	nA	$V_{GS} = -30$ V, $V_{DS} = 0$	
T				-0.25	μA		150°C
A	BV_{GSS}	Gate-source breakdown voltage	-40		V	$I_G = -1$ μA, $V_{DS} = 0$	
T	$V_{GS(\text{OFF})}$	Gate-source cutoff voltage	-1.8	-5	V	$V_{DS} = 20$ V, $I_D = 1$ μA	
I	I_{DSS}	Saturation drain current*	2.5	7.5	mA	$V_{DS} = 20$ V, $V_{GS} = 0$	
C							
D	g_{fs}	Common-source forward transconductance*	1,300	4,000	μ℧		$f = 1$ kHz
Y	g_{os}	Common-source output conductance		10	μ℧	$V_{DS} = 20$ V, $V_{GS} = 0$	$f = 1$ kHz
N							
A	C_{rss}	Common-source reverse transfer capacitance		5	pF		$f = 1$ MHz
M							
I	C_{iss}	Common-source input capacitance		25	pF		$f = 1$ MHz
C							

* Pulse test duration = 2 ms.

B-5 SPECIFICATIONS FOR 3N163 P-CHANNEL ENHANCEMENT-TYPE SILICON MOS FIELD-EFFECT TRANSISTOR (Courtesy of Siliconix, Inc.)

Normally OFF MOSFET for analog and digital switching general-purpose amplifiers.

Table B5-1 Absolute maximum ratings (25°C)

Drain-source or gate-source voltage	-40 V
Transient gate-source voltage	± 150 V
Drain current	-50 mA
Storage temperature	-65 to $+200°$C
Operating junction temperature	-55 to $+150°$C
Total device dissipation (derate 3.0 mW/°C to 150°C)	375 mW
Lead temperature $\frac{1}{16}$ in from case for 10 s	265°C

Table B5-2 Electrical characteristics (25°C and $V_{BS} = 0$ unless otherwise noted)

		Characteristic	Min	Max	Unit	Test conditions
S T A T I C	I_{GSS}	Gate-body leakage current		-10 -25	pA pA	$V_{GS} = -40$ V, $V_{DS} = 0$, 125°C
	BV_{DSS}	Drain-source breakdown voltage	-40		V	$I_D = -10\,\mu$A, $V_{GS} = 0$
	BV_{SDS}	Source-drain breakdown voltage	-40		V	$I_S = -10\,\mu$A, $V_{GD} = V_{BD} = 0$
	V_{GS}	Gate-source voltage	-3	-6.5	V	$V_{DS} = -15$ V, $I_D = -0.5$ mA
	$V_{GS(th)}$	Gate-source threshold voltage	-2	-5	V	$V_{DS} = V_{GS}, I_D = -10\,\mu$A
	I_{DSS}	Drain cutoff current		-200	pA	$V_{DS} = -15$ V, $V_{GS} = 0$
	I_{SDS}	Source cutoff current		-400	pA	$V_{SD} = -20$ V, $V_{GD} = 0$, $V_{DB} = 0$
	$I_{D(ON)}$	ON drain current	-5	-30	mA	$V_{DS} = -15$ V, $V_{GS} = -10$ V
	$r_{DS(ON)}$	Drain-source ON resistance		250	Ω	$V_{GS} = -20$ V, $I_D = -100\,\mu$A
D Y N A M I C	g_{fs}	Common-source forward transconductance	2,000	4,000	$\mu\mho$	$V_{DS} = -15$ V, $I_D = -10$ mA, $f = 1$ kHz
	g_{oss}	Common-source output conductance		250	$\mu\mho$	
	C_{iss}	Common-source input capacitance		2.5	pF	
	C_{rss}	Common-source reverse transfer capacitance		0.7	pF	$V_{DS} = -15$ V, $I_D = -10$ mA,
	C_{oss}	Common-source output capacitance		3	pF	$f = 1$ MHz
S W	t_d	Turn-ON delay time		12	ns	$V_{DD} = -15$ V
	t_r	Rise time		24	ns	$I_{D(ON)} = -10$ mA
	t_{OFF}	Turn-OFF time		50	ns	$R_G = R_L = 1.5$ kΩ

B-6 SPECIFICATIONS FOR SCHOTTKY LOW-POWER TTL POSITIVE NAND GATE (LS5410 OR LS7410) WITH TOTEM-POLE OUTPUT (Courtesy of Texas Instruments, Inc.)

The other TTL families have very similar characteristics for NAND gates or inverters.

Table B6-1 Recommended operating conditions

Parameter	Family	Min	Nom	Max	Unit
Supply voltage, V_{CC}	54	4.5	5	5.5	V
	74	4.75	5	5.25	
High-level output current, I_{OH}	54			-400	μA
	74			-400	
Low-level output current, I_{OL}	54			4	mA
	74			8	
Operating free-air temperature, T_A	54	-55		125	°C
	74	0		70	

Table B6-2 Electrical characteristics over recommended operating free-air temperature range (unless otherwise noted)

Parameter		Test conditions*	Family	Min	Typ†	Max	Unit
V_{IH}	High-level input voltage			2			V
V_{IL}	Low-level input voltage		54			0.7	V
			74			0.8	
V_I	Input clamp voltage	$V_{CC} = $ min, $I_I = -18$ mA				-1.5	V
V_{OH}	High-level output voltage	$V_{CC} = $ min, $V_{IL} = V_{IL}$ max	54	2.5	3.4		V
		$I_{OH} = $ max	74	2.7	3.4		
V_{OL}	Low-level output voltage	$V_{CC} = $ min, $V_{IH} = 2$ V	54		0.25	0.4	V
		$I_{OL} = $ max	74		0.35	0.5	
I_I	Input current at maximum input voltage	$V_{CC} = $ max, $V_I = 5.5$ V				0.1	mA
I_{IH}	High-level input current	$V_{CC} = $ max, $V_{HH} = 2.7$ V			20		μA
I_{IL}	Low-level input current	$V_C = $ max, $V_{IL} = 0.4$ V				-0.36	mA
I_{OS}	Short-circuit output current ‡	$V_{CC} = $ max	54	-6		-40	mA
			74	-5		-42	

* For conditions shown as min or max, use the appropriate value specified under recommended operating conditions.
† All typical values are at $V_{CC} = 5$ V, $T_A = 25$°C.
‡ Not more than one output should be shorted at a time.

B-7 SPECIFICATIONS FOR LM741 OPERATIONAL AMPLIFIER
(Courtesy of National Semiconductor, Inc.)

A high-performance monolithic OP AMP intended for a wide range of analog applications. It is short-circuit protected and requires no external components for frequency compensation. The LM741C is identical to the LM741 except that the LM741C has its performance guaranteed over a 0 to 70°C temperature range, instead of −55 to 125°C.

Table B7-1 Absolute maximum ratings

Supply voltage LM741	± 22 V
LM741C	± 18 V
Power dissipation*	500 mW
Differential input voltage	± 30 V
Input voltage†	± 15 V
Output short-circuit duration	Indefinite
Storage temperature range	$-65°C$ to $150°C$
Lead temperature (soldering, 10 s)	$300°C$

* The maximum junction temperature of the LM741 is 150 °C, while that of the LM741C is 100 °C. For operating at elevated temperatures, devices in the TO-5 package must be derated based on a thermal resistance of 150°C/W, junction to case.

† For supply voltages less than ± 15 V, the absolute maximum input voltage is equal to the supply voltage.

Table B7-2 Electrical characteristics*

Parameter	Conditions	LM741 Min	Typ	Max	LM741C Min	Typ	Max	Units
Input offset voltage	$T_A = 25°C$, $R_s < 10$ kΩ		1.0	5.0		1.0	6.0	mV
Input offset current	$T_A = 25°C$		30	200		30	200	nA
Input bias current	$T_A = 25°C$		200	500		200	500	nA
Input resistance	$T_A = 25°C$	0.3	1.0		0.3	1.0		MΩ
Supply current	$T_A = 25°C$, $V_S = \pm 15$ V		1.7	2.8		1.7	2.8	mA
Large-signal voltage gain	$T_A = 25°C$, $V_S = \pm 15$ V							
	$V_{OUT} = \pm 10$ V, $R_L > 2$ kΩ	50	160		25	160		V/mV
Input offset voltage	$R_s < 10$ kΩ			6.0			7.5	mV
Input offset current				500			300	nA
Input bias current				1.5			0.8	μA
Large-signal voltage gain	$V_S = \pm 15$ V, $V_{OUT} = \pm 10$ V							
	$R_L > 2$ kΩ	25			15			V/mV
Output voltage swing	$V_S = \pm 15$ V, $R_L = 10$ kΩ	± 12	± 14		± 12	± 14		V
	$R_L = 2$ kΩ	± 10	± 13		± 10	± 13		V
Input voltage range	$V_S = \pm 15$ V	± 12			± 12			V
Common-mode rejection ratio	$R_s < 10$ kΩ	70	90		70	90		dB
Supply voltage rejection ratio	$R_s < 10$ kΩ	77	96		77	96		dB

* These specifications apply for $V_S = \pm 15$ V and $-55°C < T_A < 125°C$, unless otherwise specified. With the LM741C, however, all specifications are limited to $0°C < T_A < 70°C$ and $V_S = \pm 15$ V.

B-8 SPECIFICATIONS FOR 2N5671 N-P-N SILICON POWER TRANSISTOR (Courtesy of RCA Solid State Division.)

This transistor has high-current- and high-power-handling capability and fast-switching speed. It is especially suitable for switching-control amplifiers, power gates, switching regulators, power-switching circuits, converters, inverters, control circuits, dc-rf amplifiers, and power oscillators.

Table B8-1 Absolute maximum ratings

Characteristic	Symbol	Rating	Unit
Collector-to-base voltage	V_{CBO}	120	V
Collector-to-emitter sustaining voltage:			
With base open	$V_{CEO(sus)}$	90	V
With external base-to-emitter resistance (R_{BE}) $\leqslant$ 50 Ω	$V_{CER(sus)}$	110	V
With external base-to-emitter resistance $\leqslant$ 50 Ω and $V_{BE} = -1.5$ V	$V_{CEX(sus)}$	120	V
Emitter-to-base voltage	V_{EBO}	7	V
Collector current	I_C	30	A
Base current	I_B	10	A
Transistor dissipation at case temperatures up to 25°C and V_{CE} up to 24 V	P_D	140	W
Temperature range	—	up to 200	°C

Table B8-2   Electrical characteristics, case temperature (T_C) = 25°C

Characteristic	Symbol	DC collector voltage (V) V_{CB}	V_{CE}	DC emitter or base voltage (V) V_{EB}	V_{BE}	DC current (A) I_C	I_E	I_B	2N5671 Min	Max	Units
	I_{CEO}		80					0		10	mA
Collector cutoff current	I_{CEV}		110		−1.5					12	mA
Emitter cutoff current	I_{EBO}			7		0				10	mA
Collector-to-emitter sustaining voltage:											
With base open	$V_{CEO\,(sus)}$					0.2		0	90*		V
With external base-to-emitter resistance (R_{BE}) ≤ 50 Ω	$V_{CER\,(sus)}$					0.2		0	110*		V
With base-emitter junction reverse biased and R_{BE} ≤ 50 Ω	$V_{CEX\,(sus)}$				−1.5	0.2			120*		V
Base-to-emitter saturation voltage	$V_{BE\,(sat)}$					15		1.2		1.5	V
Base-to-emitter voltage	V_{BE}		5			15				1.6	V
Collector-to-emitter saturation voltage	$V_{CE\,(sat)}$					15		1.2		0.75	V
DC forward-current transfer ratio	h_{FE}		2			15			20	100	
			5			20			20		
Gain-bandwidth product	f_T		10			2			50		MHz
Output capacitance (at 1 MHz)	C_{ob}	10					0			900	pF
Saturated switching turn-on time (delay time + rise time)	t_{ON}	$V_{CC}=$ 30 V				15		$I_{B_1} =$ $I_{B_2} =$ 1.2		0.5	μs
Saturated switching storage time	t_s	$V_{CC}=$ 30 V				15		$I_{B_1} =$ $I_{B_2} =$ 1.2		1.5	μs
Saturated switching fall time	t_f	$V_{CC}=$ 30 V				15		$I_{B_1} =$ $I_{B_2} =$ 1.2		0.5	μs
Thermal resistance (junction-to-case)	$\theta_{J\text{-}C}$		40			0.5				1.25	°C/W

* *Caution:* The sustaining voltages $V_{CEO\,(sus)}$, $V_{CER\,(sus)}$, and $V_{CEX\,(sus)}$ must not be measured on a curve tracer.

SUMMARY OF NETWORK THEORY

In this book we use linear passive elements such as resistors, capacitors, and inductors in combination with voltage and/or current sources and solid-state devices to form networks. The theorems discussed in this appendix are used frequently in the analysis of these electronic circuits.

C-1 RESISTIVE NETWORKS

Voltage and Current Sources

In this section we review some basic concepts and theorems in connection with resistive networks containing voltage and current sources. The circuit symbols and reference directions of independent voltage and current sources are shown in Fig. C-1. An ideal voltage source is defined as a voltage generator whose output voltage $v = v_s$ is independent of the current delivered by the generator. The output voltage is usually specified as a function of time such as, for example, $v_s = V_m \cos \omega t$ or a constant dc voltage. Similarly, an ideal current source delivers an arbitrary current $i = i_s$ independent of the voltage between the two terminals of the current source. The reference polarity for the voltage source v_s means that 1 C of positive charge moving from the negative to the positive terminal through the voltage source acquires v_s joules of energy. In the same way the arrow reference for the current source i_s indicates that i_s coulombs

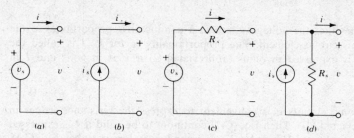

Figure C-1 (*a*), (*b*) ideal and (*c*), (*d*) practical voltage and current sources. A circle with a + and − sign is the symbol for the ideal voltage generator. An arrow inside a circle is the symbol for an ideal current generator. The source resistance is designated by R_s, drawn either in series with a voltage source v_s or in parallel with a current source i_s.

of positive charge per second pass through the source in the direction of the arrow. In a practical voltage or current source there is always some energy converted to heat in an irreversible energy-conversion process. This energy loss can be represented by the loss in a series or parallel source resistance R_s, as shown in Fig. C-1*c* and *d*.

A *controlled* or *dependent* source is one whose voltage or current is a function of the voltage or current elsewhere in the circuit. For example, Fig. C-2*a* represents a small-signal circuit model of a transistor at low frequencies. At the input there is a dependent voltage generator $h_r v_o$ whose *voltage* is proportional to the output *voltage* v_o and the proportionality factor is h_r. At the output there is a dependent current generator $h_f i_1$ whose *current* is proportional to the input *current* i_1 and the proportionality factor is h_f.

Another active device studied in this book is the MOSFET, and its equivalent low-frequency small-signal model is indicated in Fig. C-2*b*. Note that at the output there is a dependent *current* source $g v_s$ controlled by the input *voltage* v_s and the proportionality factor is g.

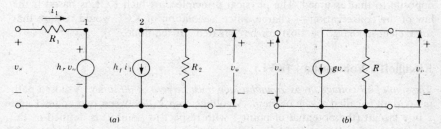

Figure C-2 (*a*) A transistor circuit model contains a voltage-controlled voltage generator $h_r v_o$ and a current-controlled current generator $h_f i_1$ (the proportionality factors h_r and h_f are dimensionless). (*b*) A MOSFET circuit model contains a voltage-controlled current source $g v_s$ (the proportional factor g has the dimensions of A/V).

Resistance

Ohm's law states that the voltage V across a conductor is proportional to the current I in this circuit element. The proportionality factor V/I is called the *resistance* and is expressed in ohms (abbreviated Ω) if V is in volts and I in amperes.

$$V = IR \qquad (C-1)$$

In most electronic circuits it is convenient to express the resistance values in kilohms (abbreviated $k\Omega$). Then Eq. (C-1) continues to be valid if I is expressed in milliamperes (mA) and V in volts (V). If a conductor does not obey Eq. (C-1), it is said to be a nonlinear (or nonohmic) resistor.

To find the resistance R seen between two points in a network, an external voltage source V is considered to be applied between these two points and the current I drawn from the source V is determined. The effective resistance is $R = V/I$, provided that in the above procedure each *independent* source in the network is replaced by its internal source resistance R_s; an ideal voltage source by a short circuit, and an ideal current source by an open circuit (Fig. C-1). *All dependent sources, however, must be retained in the network.*

The two basic laws which allow us to analyze electric networks (linear or nonlinear) are known as *Kirchhoff's current law* (KCL) and *Kirchhoff's voltage law* (KVL).

Kirchhoff's Current Law (KCL)

The sum of all currents toward a node must be zero at any instant of time. A node is a point where two or more circuit components meet such as points 1 or 2 in Fig. C-2a. When we apply this law, currents directed toward a node are usually taken as positive and those directed away are taken as negative. The opposite convention can also be used as long as we are consistent for all nodes of the network. The positive reference direction of the current through any resistor of the network can be assigned arbitrarily with the understanding that, if the computed current is determined to be negative, the actual current direction is opposite to that assumed. The physical principle on which KCL is based is the law of the conservation of charge, since a violation of KCL would require that some electric charge be "lost" or be "created" at the node.

Kirchhoff's Voltage Law (KVL)

The sum of all voltage drops around a loop must be zero at all times. A closed path in a circuit is called a *loop* or *mesh*. A voltage *drop* V_{12} between two nodes 1 and 2 in a circuit (the potential of point 1 with respect to point 2) is defined as the energy in joules (J) removed from the circuit when a positive charge q of 1 C moves from point 1 to point 2. For example, a voltage drop of $+5$ V across the terminal nodes 1 and 2 of a resistor means that 5 J of energy are removed from the circuit and dissipated as heat in the resistor when a positive charge of 1 C moves *from point 1 to point 2*. If the voltage is -5 V, then point 2 is at a higher voltage than point 1 ($V_{12} = -5$ V represents a voltage *rise*), and a positive charge of 1 C moving from point 1 to point 2 gains 5 J of energy. This, of course, is impossible when a resistor is connected between the two nodes 1 and 2, and is possible only if the negative terminal of a battery is connected to node 1 and the positive terminal to node 2.

It should be clear that KVL is a consequence of the law of conservation of energy. In writing the KVL equations, we go completely around a loop, add all voltage drops, and set the sum equal to zero. Remember these two rules:

(1) *There is a positive drop in the direction of the current in a resistor.* (2) *There is a positive drop through a battery (or dc source) in the direction from the + to the − terminal, independent of the direction of the current.*

The two fundamental laws (KCL and KVL) will be illustrated in the following examples. Consider first the situation where a resistor R_L is placed directly across the output terminals of a real (nonidealized) voltage source (Fig. C-1c). This added component is called the *load resistor* or, simply, the *load*. The result is that a single mesh is formed as indicated in Fig. C-3. We wish to find the voltage v across R_L.

The current i around the loop flows through R_s and R_L. Traversing this mesh in the assumed direction of current starting at node 2, adding all voltage drops, and setting the sum to zero (as required by KVL) yields

$$-v_s + iR_s + iR_L = 0$$

or $$i = \frac{v_s}{R_s + R_L} \quad \text{and} \quad v = iR_L = \frac{R_L v_s}{R_s + R_L} \tag{C-2}$$

Note that under *open-circuit* conditions (defined as $R_L \to \infty$), $v = v_s$. This result is obviously correct since no current can flow in an open circuit so that $i = 0$, $iR_s = 0$, and $v = v_s = $ open-circuit voltage. Also note that under *short-circuit* conditions (defined as $R_L = 0$; an ideal zero-resistance wire), the output voltage

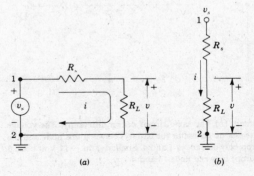

(a)   (b)

Figure C-3 (*a*) A load R_L is placed across a voltage source whose internal resistance is R_s. (*b*) The *same* circuit is redrawn in a different way. The tiny circle at node 1 is used to indicate that a power supply v_s exists between this node and node 2 which has been designated as the reference node. Since one terminal of a generator is usually connected to the metal chassis on which it is built, this terminal is called *ground*. The standard symbol for a ground is shown at node 2.

drops to zero, $v = 0$. Now the current is a maximum (with respect to variations in the value of R_L) and $i = v_s/R_s = $ *short-circuit current*. The voltage v_s may be a function of time, and then v will also be a function of time.

An equivalent alternative way to draw the circuit of Fig. C-3a is shown in Fig. C-3b. The caption explains the meaning of the symbols at nodes 1 and 2. This configuration is referred to as a *voltage divider* or an *attenuator*. Note that v is less than v_s (for any finite R_L), and that the fraction of v_s appearing across R_L is the *attenuation*.

$$\frac{v}{v_s} = \frac{R_L}{R_s + R_L} \tag{C-3}$$

Example C-1 Find the current flowing in the loop shown in Fig. C-4.

SOLUTION We arbitrarily, assign (guess) the positive current direction as shown in Fig. C-4. As a consequence of the assumed positive current flow, the resistor voltage drops are positive in the directions indicated in the figure. Let us go around the loop in the direction of the assumed positive current flow (counter clockwise) and sum the voltage drops, starting at ground (node 4).

$$V_{43} + V_{32} + V_{21} + V_{14} = 0 \tag{C-4}$$

Expressing I in milliamperes, the individual voltage drops in volts are: $V_{43} = +14$, $V_{32} = IR_2 = 9I$, $V_{21} = IR_1 = I$, and $V_{14} = +6$. Substituting the above in Eq. (C-4) we find $14 + 9I + I + 6 = 0$ or $I = -20/10 = -2$ mA. Thus, the magnitude of the current is 2 mA and the minus sign indicates that the current flows clockwise around the loop, opposite to the assumed direction of positive current.

In the next example we must make use of both KCL and KVL equations because the network has more than one mesh.

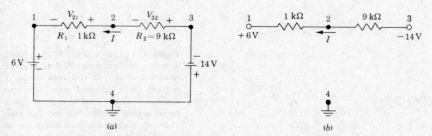

(a)   (b)

Figure C-4 (*a*) A single-loop resistive circuit. (*b*) The same circuit drawn differently. The voltage +6 V shown at node 1 indicates that there is a dc generator (or battery) between this node and the reference (ground) node 4, giving a 6-V drop between nodes 1 and 4. Similarly, the −14 V at node 3 means that there is a 14-V rise (negative drop) between nodes 3 and 4.

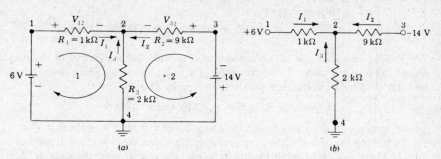

Figure C-5 (*a*) Two-loop resistive network. (*b*) The same network with the voltages (with respect to ground) at nodes 1 and 3 indicated, but with the battery symbols omitted.

Example C-2 (*a*) Find the currents I_1, I_2, and I_3 in the circuit shown in Fig. C-5. (*b*) Find the voltage drop V_{24}.

SOLUTION (*a*) We assign the arbitrary reference directions of positive currents as shown in the figure. We must sum the voltage drops in each loop by going around each mesh in the arbitrary direction shown by the loop arrows, called the *mesh currents*. Note that the current in R_1 is the mesh current I_1 and that in R_2 is the mesh current I_2. However, the current in R_3 is the sum of I_1 and I_2. Applying KVL we obtain the following equations:

$$\text{Loop 1} \qquad V_{12} + V_{24} + V_{41} = 0 \qquad \text{(C-5)}$$
$$\text{Loop 2} \qquad V_{32} + V_{24} + V_{43} = 0 \qquad \text{(C-6)}$$

where the individual voltage drops are given below:

$$V_{12} = I_1 R_1 = I_1 \qquad V_{24} = -I_3 R_3 = -2I_3 \qquad V_{41} = -6$$
$$V_{32} = I_2 R_2 = 9I_2 \qquad V_{43} = 14$$

Substituting these values into Eqs. (C-5) and (C-6) gives

$$I_1 - 2I_3 - 6 = 0$$
$$9I_2 - 2I_3 + 14 = 0$$

Since we have only two equations for our three unknowns, we must use the KCL equation at node 2 to obtain the additional equation

$$I_1 + I_2 + I_3 = 0 \qquad \text{or} \qquad I_3 = -(I_1 + I_2)$$

Substituting this value of I_3 into the equations for I_1 and I_2 gives

$$3I_1 + 2I_2 = 6$$
$$2I_1 + 11I_2 = -14$$

Solving these simultaneous algebraic equations, we find

$$I_1 = 3.242 \qquad I_2 = -1.862 \qquad \text{and} \qquad I_3 = -1.379 \text{ mA}$$

(b) The voltage drop V_{24} is

$$V_{24} = -I_3R_3 = 1.379 \times 2 = 2.758 \text{ V}$$

The voltage drop between any two nodes in a network is independent of the path chosen between the nodes. For example, V_{24} may be found by going from 2 to 1 to 4 and adding all voltage drops along this path. Thus,

$$V_{24} = -I_1R_1 + 6 = -3.242 + 6 = 2.758 \text{ V}$$

which agrees with the value found by going directly from 2 to 4 through R_3.

In solving the above illustrative problem, we chose the two internal meshes 1 and 2. There is a third mesh in this network; the one around the outside loop 4-1-2-3-4. However, this outside mesh is not independent of the other two meshes. *An independent loop is one whose KVL equation includes at least one voltage not included in the other equations.* The number of independent KVL equations is equal to the number of independent loops.

A *junction* is defined as a point where three or more circuit elements meet. Of the four nodes in Fig. C-5, nodes 2 and 4 are junctions. *The number of independent KCL equations is equal to one less than the number of junctions.* Hence, in solving the above problem only one KCL equation is required.

Series and Parallel Combinations of Resistors

The circuit of Fig. C-6a consists of three resistors in *series*, which means that the same current flows in each resistor. From KVL,

$$-V + IR_1 + IR_2 + IR_3 = 0$$

The equivalent resistance R between nodes 1 and 2 is, by definition, given by

$$R \equiv \frac{V}{I} = R_1 + R_2 + R_3 \tag{C-7}$$

To find the total resistance in a series circuit, add the individual values of resistance.

Resistors are in parallel when the same voltage appears across each resistor. Hence, Fig. C-6b shows three resistors in parallel:

$$I_1 = \frac{V}{R_1} = G_1V \qquad I_2 = \frac{V}{R_2} = G_2V \qquad I_3 = \frac{V}{R_3} = G_3V$$

where $G \equiv 1/R$ is called the *conductance*. Its dimensions are A/V or reciprocal ohms, called *mhos* (℧). Applying KCL to Fig. C-6b yields

$$I = I_1 + I_2 + I_3 = (G_1 + G_2 + G_3)V$$

The equivalent conductance between nodes 1 and 2 is, by definition,

$$G \equiv \frac{I}{V} = G_1 + G_2 + G_3 \tag{C-8}$$

To find the total conductance in a parallel circuit, add the individual values of

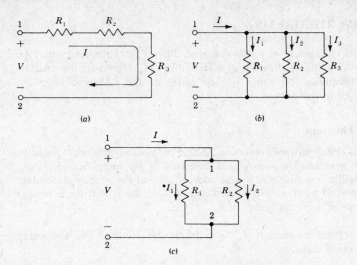

Figure C-6 (*a*) Resistors in series. (*b*) Resistors in parallel. (*c*) A current divider.

conductance. Equation (C-8) is equivalent to

$$\frac{1}{R} = \frac{1}{R_1} + \frac{1}{R_2} + \frac{1}{R_3} \tag{C-9}$$

Of course, the number of resistors in the series or parallel circuits of Fig. C-6 is not limited to three; it can be any number, two or more. For the special case of two resistors, Eq. (C-9) is equivalent to

$$R = R_1 \| R_2 = \frac{R_1 R_2}{R_1 + R_2} \tag{C-10}$$

where the symbol $\|$ is to be read "in parallel with." It follows from this equation that two resistors in parallel have an effective resistance which is *smaller* than either resistor.

Just as a series circuit gives voltage attenuation [Fig. C-3*b* and Eq. (C-3)] so a parallel circuit gives current attenuation. In Fig. C-6*c* the current I_1 in R_1 (or I_2 in R_2) is less than the current I entering node 1. Thus, using Eq. (C-10), we have

$$V = IR = \frac{IR_1 R_2}{R_1 + R_2} = I_1 R_1$$

or

$$I_1 = \frac{R_2 I}{R_1 + R_2} \tag{C-11}$$

Note that if $R_1 = 0$, $I_1 = I$. This result is intuitively correct since all the current should flow in the short circuit. On the other hand if $R_1 \to \infty$, then $I_1 = 0$, which is certainly true because no current can flow in an open circuit.

C-2 NETWORK THEOREMS

The currents and voltages in any network regardless of its complexity may be obtained by a systematic application of KCL and KVL. However, the analysis may often be simplified by using one or more of the additional network theorems discussed in this section.

Superposition Theorem

The response of a linear network containing several independent sources is found by considering each generator separately and then adding the individual responses. When evaluating the response due to one source, each of the other independent generators is replaced by its internal resistance, that is, set $v_s = 0$ for a voltage source and $i_s = 0$ for a current generator.

Example C-3 Find the currents I_1, I_2, and I_3 in the circuit of Fig. C-5, using the superposition theorem.

SOLUTION First consider the currents I_1', I_2', and I_3' due to the 6-V supply. Then node 3 must be short-circuited to node 4, so as to eliminate the response due to the -14-V source. This connection puts R_2 and R_3 in parallel, as indicated in Fig. C-7a. From Eq. (C-10) this parallel combination has a resistance of

$$\frac{R_2 R_3}{R_2 + R_3} = \frac{9 \times 2}{9 + 2} = 1.636 \ \text{k}\Omega$$

The resistance seen by the 6-V supply is R_1 plus the above value, hence

$$I_1' = \frac{6}{1 + 1.636} = 2.276 \ \text{mA}$$

From the current attenuation formula, Eq. (C-11),

$$I_2' = \frac{-I_1' R_3}{R_2 + R_3} = \frac{-2.276 \times 2}{9 + 2} = -0.414 \ \text{mA}$$

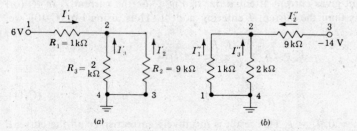

(a)   (b)

Figure C-7 Superposition is applied to the network of Fig. C-5. The circuit from which to calculate the response due to (a) the 6-V supply and (b) the -14-V supply.

Similarly,

$$I_3' = \frac{-I_1'R_2}{R_2 + R_3} = \frac{-2.276 \times 9}{2 + 9} = -1.862 \text{ mA}$$

We now find the currents I_1'', I_2'', and I_3'' due to the -14-V supply. To eliminate the effect of the 6-V source, nodes 1 and 4 are connected together as shown in Fig. C-7b. Proceeding as above, we find

$$I_2'' = \frac{-14}{9 + (1 \times 2)/3} = -1.448 \text{ mA}$$

$$I_1'' = +1.448 \times \tfrac{2}{3} = 0.9655 \text{ mA}$$

$$I_3'' = +1.448 \times \tfrac{1}{3} = 0.4826 \text{ mA}$$

The net current is the algebraic sum of the currents due to each excitation. Thus

$$I_1 = I_1' + I_1'' = 2.276 + 0.966 = 3.242 \text{ mA}$$

$$I_2 = I_2' + I_2'' = -0.414 - 1.448 = -1.862 \text{ mA}$$

$$I_3 = I_3' + I_3'' = -1.862 + 0.483 = -1.379 \text{ mA}$$

These are the same values obtained in the previous section. Note that for this particular network the analysis in Sec. C-1 using KVL and KCL is simpler than that given here using superposition.

Thévenin's Theorem

Any linear network may, with respect to a pair of terminals, be replaced by a voltage generator V_{Th} (equal to the open-circuit voltage) in series with the resistance R_{Th} seen between these terminals. To find R_{Th} all *independent* voltage sources are short-circuited and all *independent* current sources are open-circuited. This theorem is often used to reduce the number of meshes in a network. For example, the two-mesh circuit of Fig. C-5 may be reduced to a single loop by replacing the components to the left of terminals 2 and 4 (including R_3) by the Thévenin equivalent. For convenience, the circuit of Fig. C-5 is redrawn in Fig. C-8a. The components in the shaded box are those to the right of nodes 2 and 4, and they are redrawn unaltered in Fig. C-8b. The other circuit elements do not appear in Fig. C-8b but are replaced by V_{Th} and R_{Th}. Thévenin's theorem states that I_2 and V_{24} calculated from this reduced circuit are identical to the corresponding values in Fig. C-5.

The open-circuit voltage V_{Th} is obtained by disconnecting the components in the box from Fig. C-8a. From the voltage attenuator formula Eq. (C-2),

$$V_{Th} = \frac{6 \times 2}{1 + 2} = 4 \text{ V}$$

To find the resistance seen to the left of nodes 2 and 4, the 6-V supply is imagined reduced to zero, which is equivalent to connecting the top of the 1-kΩ

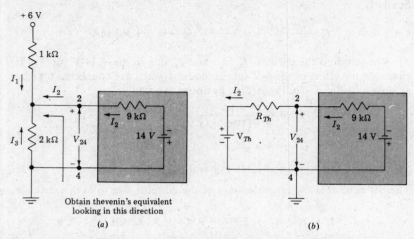

Figure C-8 The circuit of Fig. C-5 redrawn. (b) Thévenin's theorem applied to the circuit in (a) looking to the left of nodes 2 and 4.

resistor to ground. Hence, this resistor is now placed in parallel with the 2-kΩ resistor, and

$$R_{Th} = \frac{1 \times 2}{1 + 2} = 0.667 \text{ k}\Omega$$

From the equivalent circuit of Fig. C-8b we obtain

$$I_2 = \frac{-(14 + V_{Th})}{9 + R_{Th}} = \frac{-18}{9.667} = -1.862 \text{ mA}$$

and
$$V_{24} = -9I_2 - 14 = 9 \times 1.862 - 14 = 2.758 \text{ V}$$

These two values agree with the numerical values found in Sec. C-1. The currents I_3 and I_1 do not appear in Fig. C-8b and must be found from Fig. C-8a. Thus,

$$I_3 = \frac{-V_{24}}{2} = \frac{-2.758}{2} = -1.379 \text{ mA}$$

and
$$I_1 = \frac{6 - V_{24}}{1} = 6 - 1.758 = 3.242 \text{ mA}$$

which are the same currents found previously.

Norton's Theorem

Any linear network may, with respect to a pair of terminals, be replaced by a current generator (equal to the short-circuit current) in parallel with the resistance seen between the two terminals.

From Thévenin's and Norton's theorems it follows that a voltage source V in series with a resistance R is equivalent to a current source I in parallel with R, provided that $I = V/R$. These equivalent circuits are indicated in Figs. C-1c and d with $v_s = V$, $R_s = R$, and $i_s = I = V/R_s$.

Open-Circuit-Voltage—Short-Circuit-Current Theorem

As corollaries to Thévenin's and Norton's theorems we have the following relationships. If V represents the *open-circuit voltage*, I the *short-circuit current*, and $R(G)$ the resistance (conductance) between two terminals in a network, then

$$V = IR = \frac{I}{G} \qquad I = \frac{V}{R} = GV \qquad R = \frac{V}{I} \tag{C-12}$$

In spite of their disarming simplicity, these equations (reminiscent of Ohm's law) should not be overlooked because they are most useful in analysis. For example, the first equation, which states "open-circuit voltage equals short-circuit current divided by conductance," is often the simplest way to find the voltage between two points in a network, as the following problem illustrates.

Example C-4 Find the voltage between nodes 2 and 4 in the circuit of Fig. C-5, using the *open-circuit-voltage—short-circuit-current theorem*.

SOLUTION If a zero-resistance wire connects node 2 to 4, then the current in this short circuit due to the 6-V battery is $\frac{6}{1} = 6$ mA, and due to the other battery, it is $-\frac{14}{9} = -1.556$ mA. Hence, the total short-circuit current is $I = 6 - 1.556 = 4.444$ mA.

The conductance between 2 and 4 with nodes 1 and 3 grounded, corresponds to R_1, R_2, and R_3 all in parallel. Hence,

$$G = \tfrac{1}{1} + \tfrac{1}{2} + \tfrac{1}{9} = 1.611 \text{ mA/V}$$

Therefore,

$$V_{24} = \frac{I}{G} = \frac{4.444}{1.611} = 2.759 \text{ V} \tag{C-13}$$

which agrees with the value found previously. The currents I_1 and I_2 can now be found by applying KVL to each loop in Fig. C-5. For example,

$$I_2 = \frac{-(14 + V_{24})}{9} = \frac{-(14 + 2.759)}{9} = -1.862 \text{ mA}$$

which is the same value obtained previously. Note that this method, or the use of Thévenin's theorem, results in a simpler solution than the straightforward analysis, using KVL and KCL.

Nodal Method of Analysis

When the number of junction voltages (with respect to the reference, or ground, node) is less than the number of independent meshes, then the choice of nodal voltages as the unknowns leads to a simpler solution than considering the mesh currents as the unknowns. For example, the circuit of Fig. C-5 has two independent meshes but only one independent junction voltage. In terms of the one unknown independent voltage V_{24}, the currents are

$$I_1 = \frac{6 - V_{24}}{1} \qquad I_2 = \frac{-14 - V_{24}}{9} \qquad I_3 = \frac{-V_{24}}{2}$$

By KCL the sum of these three currents (which enter node 2) must equal zero. Hence,

$$\frac{6}{1} - \frac{V_{24}}{1} - \frac{14}{9} - \frac{V_{24}}{9} - \frac{V_{24}}{2} = 0$$

or

$$V_{24}\left(\tfrac{1}{1} + \tfrac{1}{9} + \tfrac{1}{2} \right) = \tfrac{6}{1} - \tfrac{14}{9} = 4.444 \text{ mA}$$

which is equivalent to Eq. (C-13) namely, $V_{24}G = I$, where I is the short-circuit current. Hence, the nodal method in this simple case is equivalent to the analysis using the open-circuit-voltage—short-circuit-current theorem.

C-3 THE SINUSOIDAL STEADY STATE

If a sinusoidal excitation (a voltage or current) is applied to a linear network, then the response (the voltage between any two nodes or the current in any branch of the network) will also be sinusoidal. (It is assumed that all transients have died down so that a steady state is reached.) Let us verify this general statement for the simple parallel combination of resistor R and capacitor C in Fig. C-9 to which has been applied the sinusoidal source voltage

$$v = V_m \cos \omega t = V_m \cos 2\pi ft \tag{C-14}$$

where f is the *frequency* of the source in hertz (Hz), $\omega = 2\pi f$ is called the *angular frequency*, and V_m is the *maximum* or *peak* value of voltage. We shall now prove that the generator current i is also a sinusoidal waveform.

A capacitor C is a component (say, two metals separated by a dielectric) which stores charge q (coulombs) proportional to the applied voltage v (volts) so

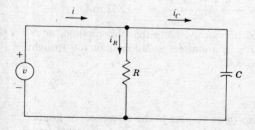

Figure C-9 A parallel RC combination excited by a sinusoidal voltage.

that

$$q = Cv \tag{C-15}$$

where the proportionality factor C is called the *capacitance*. The dimensions of C are coulombs per volt, which is abbreviated as *farads* (F). The capacitor current i_C is therefore

$$i_C = \frac{dq}{dt} = C\frac{dv}{dt} \tag{C-16}$$

or, using Eq. (C-14),

$$i_C = -\omega C V_m \sin \omega t \tag{C-17}$$

From Ohm's law, the resistor current i_R is given by

$$i_R = \frac{v}{R} = \frac{V_m}{R}\cos \omega t \tag{C-18}$$

From Kirchhoff's current law (KCL), $i = i_R + i_C$, or

$$i = \frac{V_m}{R}\cos \omega t - \omega C V_m \sin \omega t \tag{C-19}$$

which has the form

$$i = I_m \cos \theta \cos \omega t - I_m \sin \theta \sin \omega t \tag{C-20}$$

where $\quad I_m \cos \theta \equiv \dfrac{V_m}{R} \quad$ and $\quad I_m \sin \theta \equiv \omega C V_m \tag{C-21}$

From the trigonometric identity,

$$\cos(\theta + \alpha) = \cos \theta \cos \alpha - \sin \theta \sin \alpha \tag{C-22}$$

then Eq. (C-20), with $\alpha \equiv \omega t$, is equivalent to

$$i = I_m \cos(\omega t + \theta) \tag{C-23}$$

We have thus verified that the generator current is indeed a sinusoid. The peak current is I_m, and i is shifted in phase by the angle θ with respect to the source voltage $V_m \cos \omega t$. We say that "the generator current *leads* its voltage by the phase angle θ."

The peak current I_m and the phase θ are obtained from Eqs. (C-21). If the two equations are each squared and then added, we obtain

$$I_m^2 \cos^2 \theta + I_m^2 \sin^2 \theta = \frac{V_m^2}{R^2} + \omega^2 C^2 V_m^2 \tag{C-24}$$

Since $\cos^2 \theta + \sin^2 \theta = 1$, then

$$I_m = V_m \sqrt{\frac{1}{R^2} + \omega^2 C^2} \tag{C-25}$$

Dividing the second equation in Eq. (C-21) by the first yields

$$\frac{I_m \sin \theta}{I_m \cos \theta} = \frac{\omega C V_m}{V_m / R}$$

or $$\tan \theta = \omega CR \qquad\qquad (C-26)$$

For a more complicated network than the one in Fig. C-9, the analysis would involve a prohibitive amount of trigonometric manipulation. Hence, we now present a simpler alternative general method for solving sinusoidal networks in the steady state. Some important concepts (such as phasors, complex plane, and impedance) are first introduced.

Phasors

Each current (or voltage) in a network is a sinusoid, which has a peak value and a phase angle. Hence it can be represented by a vector, which is a directed line segment, having a length and direction. For a sinusoid this vector is called a *phasor*. Its magnitude represents the effective or rms value and is given by the peak value divided by $\sqrt{2}$ [Eq. (10-20)]. The direction of the phasor is the phase θ in the sinusoidal waveform $I_m \cos(\omega t + \theta)$, and the angle θ is measured counterclockwise with respect to the horizontal axis. In this section we use boldface $\mathbf{I}$ $(\mathbf{V})$ to denote a phasor current (voltage). In phasor notation, the current in Eq. (C-23) is written

$$\mathbf{I} = I \angle \theta \qquad\qquad (C-27)$$

where $I = I_m/\sqrt{2}$. This phasor is indicated in Fig. C-10a.

The applied voltage phasor is, from Eq. (C-14), $\mathbf{V} = V \angle 0$, where $V = V_m/\sqrt{2}$, and the current in the resistor is, from Eq. (C-18) $\mathbf{I}_R = V/R \angle 0$. These phasors are indicated in Fig. C-10b. Note that *the current in a resistor is in phase with the voltage across the resistor.*

Since Eq. (C-17) may be written $i_C = \omega C V_m \cos(\omega t + 90°)$, the phasor representing the capacitor current is

$$\mathbf{I}_C = \omega C V \angle 90° \qquad\qquad (C-28)$$

where $V = V_m/\sqrt{2}$ is the rms voltage. Note that *the current in a capacitor leads*

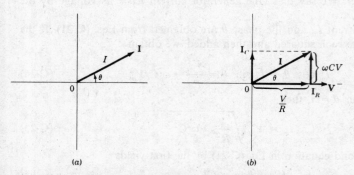

(a)   (b)

Figure C-10 (a) The current represented as a phasor of magnitude I and phase θ. (b) Phasor addition, representing $\mathbf{I} = \mathbf{I}_R + \mathbf{I}_C$.

the voltage across the capacitor by 90°. The phasor $\mathbf{I}_C$ is plotted in Fig. C-10*b*. The generator current is the sum of the resistor current and the capacitor current, or in phasor notation

$$\mathbf{I} = \mathbf{I}_R + \mathbf{I}_C = \frac{V}{R} \angle 0 + \omega C V \angle 90° \qquad (C\text{-}29)$$

This vector sum is indicated in Fig. C-10*b*, where it is found that

$$|I|^2 = \frac{V^2}{R^2} + \omega^2 C^2 V^2 \quad \text{and} \quad \tan \theta = \omega C R$$

in agreement with Eqs. (C-25) and (C-26). Note how simple the phasor method of analysis is compared with the above solution, with the use of instantaneous values of current and voltage and manipulating the equations with the aid of trigonometric identities. By introducing the concept of the complex plane, the analysis may be further simplified. An essentially algebraic, rather than a trigonometric, solution is now obtained.

The *j* Operator

A useful convention is to take the symbol *j* to represent a *phase lead* of 90°. In place of Eq. (C-28) we now write $\mathbf{I}_C = j\omega C V$, and for the total current in Eq. (C-29) we have

$$\mathbf{I} = \frac{V}{R} + j\omega C V \qquad (C\text{-}30)$$

This equation is interpreted to mean that **I** is a phasor formed by combining the phasor V/R horizontally (at zero phase) with $\omega C V$ plotted vertically (at a phase of 90°). Hence, the vertical axis is also called the *j* axis. The current **I**, shown in Fig. C-10*b*, is identical with that found above.

From the definition of *j* it follows that $j\mathbf{I}$ is a phasor whose magnitude is that of **I** but whose phase is 90° greater than the phase of **I**. In other words, *j*

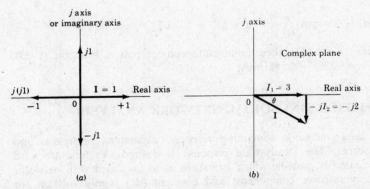

Figure C-11 (*a*) Concerning the operator *j*. (*b*) A phasor current plotted in the complex plane.

"multiplying" a phasor I is an operator which rotates I in the counterclockwise direction by 90°. Consider $I = 1$, a phasor of magnitude 1 and phase 0. Then $jI = j1$ has a magnitude 1 and phase 90°, as indicated in Fig. C-11. Then $j(j1)$ represents a rotation of $j1$ by 90°, which results in a phasor of unit magnitude pointing along the negative horizontal axis, as indicated in Fig. C-11. In a purely formal manner we may write

$$j(j1) = j^2 1 = -1 \quad \text{or} \quad j = \sqrt{-1} \quad \text{(C-31)}$$

Because of this formalism the vertical axis is called the j or *imaginary axis* and the horizontal axis is designated as the *real axis*. The plane of Fig. C-11 is now called the *complex plane*.

Note that higher powers of j are easily found. Thus,

$$j^3 = j(j^2) = j(-1) = -j \quad \text{(C-32)}$$

which represents a phasor of magnitude 1 and phase $-90°$. The reciprocal of j is $-j$, as is easily verified. Thus

$$\frac{1}{j} = \frac{1}{j}\frac{j}{j} = \frac{j}{j^2} = -j \quad \text{(C-33)}$$

because $j^2 = -1$ from Eq. (C-31). A point in the complex plane is called a *complex number*, and it is evident that a phasor is a complex number. Hence, the analysis of sinusoidal circuits is carried out most simply by treating currents and voltages as complex numbers representing phasors.

Assume that a complicated circuit is analyzed (by the general method outlined in Sec. C-4) and that the following complex current is obtained

$$I = I_1 - jI_2 = 3 - j2 \quad \text{mA} \quad \text{(C-34)}$$

This phasor is indicated in the complex plane in Fig. C-11b. From this diagram it follows that the rms current $|I|$ and the phase angle θ are given by

$$|I| = \sqrt{I_1^2 + I_2^2} = \sqrt{13} = 3.61 \text{ mA}$$

and

$$\theta = -\arctan\frac{I_2}{I_1} = -\arctan\frac{2}{3} = -33.7° = -0.588 \text{ rad}$$

If the frequency f is 1 kHz, then the instantaneous current is, from Eq. (C-23), $i = 3.61\sqrt{2} \cos(6,280t - 0.588)$ mA.

C-4 SIMPLIFIED SINUSOIDAL NETWORK ANALYSIS

Consider a linear network containing resistors, capacitors, inductors, and sinusoidal sources. The steady-state response is desired. A straightforward method of solution is possible which is analogous to that used with networks containing only resistive components and constant (dc) supply voltages (or currents). The analysis consists of writing the KVL and KCL equations for the network and then solving for the complex (phasor) currents and voltages. In order to carry out such an analysis, it is first necessary to introduce the concept of *complex resistance* or *reactance*. After defining reactance, a number of specific circuits are solved by using this simple method of analysis.

Reactance

The ratio of the voltage[†] V across a passive circuit component to the current I through the element for each of the three basic components is as follows:

$$\text{Resistance:} \quad \frac{V}{I} = R$$

$$\text{Capacitance:} \quad \frac{V}{I} = \frac{1}{j\omega C} = \frac{-j}{\omega C} \qquad (C\text{-}35)$$

$$\text{Inductance:} \quad \frac{V}{I} = j\omega L$$

The first equation of Eqs. (C-35) is Ohm's law. The second equation follows from Eq. (C-28). An inductor is a component (say, a coil of wire) whose terminal voltage v is proportional to the rate of change of current. The proportionality factor L (henrys, H) is called the *inductance*. From $v = L\,di/dt$, the third equation of Eq. (C-35) can be obtained in a manner analogous to that used in the preceding section to obtain Eq. (C-28).

From Eqs. (C-35) it follows that a capacitor behaves as a "complex resistance" $-j/\omega C$ and an inductor acts like a "complex resistance" $j\omega L$. A more commonly used phrase for complex resistance is *reactance*, denoted by the real positive symbol X.

$$\text{Capacitive reactance} = -jX_C \text{ where } X_C \equiv 1/\omega C$$

and $$\text{Inductive reactance} = +jX_L \text{ where } X_L \equiv \omega L$$

In applying KVL to a circuit containing reactive elements, it must be remembered that the drop across a capacitor is $-jX_C I = -jI/\omega C$ and the drop across an inductor is $jX_L I = j\omega L I$. It follows from the above considerations that KVL applied to the series circuit of Fig. C-12 yields

$$V = RI + j\omega L I - \frac{j}{\omega C} I \qquad (C\text{-}36)$$

or $$I = \frac{V}{R + j(\omega L - 1/\omega C)} = \frac{V}{R + jX} \qquad (C\text{-}37)$$

where the total series reactance is $X \equiv \omega L - 1/\omega C$. The current may be

---

[†] In this section, and throughout the text, bold face **V** (or **I**) used to designate a complex (phasor) voltage (or current) is replaced by an italic symbol V (or I).

expressed in standard complex number form $I = I_1 + jI_2$ by multiplying both the numerator and the denominator by the complex conjugate (change j to $-j$) of the denominator. Thus,

$$I = \frac{V}{R + jX} \frac{R - jX}{R - jX} = \frac{V}{R^2 + X^2}(R - jX) \qquad \text{(C-38)}$$

From this equation we find that the magnitude and phase of I are given by

$$|I| = \frac{V}{\sqrt{R^2 + X^2}} \qquad \text{and} \qquad \tan\theta = -\frac{X}{R} \qquad \text{(C-39)}$$

Impedance

The ratio of the voltage between any two points A and B of a network to the current in this portion of the circuit is called the *impedance Z* between A and B. For the circuit of Fig. C-12

$$Z \equiv \frac{V}{I} = R + j\left(\omega L - \frac{1}{\omega C}\right) \qquad \text{(C-40)}$$

from Eq. (C-36). Since the generator V is placed directly between A and B, then Z is the impedance "seen" by the source V. Note that for a series circuit the impedance equals the sum of the resistances plus reactances in the loop. This statement is analogous to the law for a dc series circuit, which states that the total resistance is the sum of the resistances in series. It should be emphasized that, whereas Z is a complex quantity, it is not a phasor, since it does *not* represent a current or a voltage varying sinusoidally with time.

Two impedances Z_1 and Z_2 in parallel represent an equivalent impedance Z given by

$$Z = \frac{Z_1 Z_2}{Z_1 + Z_2} \qquad \text{(C-41)}$$

corresponding to Eq. (C-10) for two resistors in parallel. For the parallel combination of a resistor R and a capacitor C as in Fig. C-9, $Z_1 = R$, $Z_2 = -j/\omega C = 1/j\omega C$, and from Eq. (C-41),

$$Z = \frac{R(1/j\omega C)}{R + 1/j\omega C} = \frac{R}{1 + j\omega CR} \qquad \text{(C-42)}$$

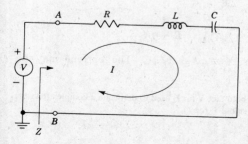

Figure C-12 An *RLC* series circuit.

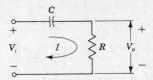

Figure C-13 A high-pass RC circuit.

This same result is obtained by applying KCL to Fig. C-9. Using phasor notation, we have

$$I = I_R + I_C = \frac{V}{R} + \frac{V}{1/j\omega C} = \frac{V}{R} + j\omega CV \tag{C-43}$$

and $Z = V/I$ gives the result in Eq. (C-42).

As another example of a one-mesh circuit, consider the high-pass RC configuration of Fig. 13-1, which is repeated in Fig. C-13 for easy reference. The output voltage is

$$V_o = IR = \frac{V_i}{Z} R = \frac{R}{Z} V_i = \frac{RV_i}{R + 1/j\omega C} \tag{C-44}$$

If $j\omega$ is replaced by s, then

$$V_o = \frac{V_i}{1 + 1/sRC} = \frac{sV_i}{s + 1/RC} \tag{C-45}$$

which is Eq. (11-1) with $R = R_1$ and $C = C_1$.

Loop or Mesh Equations

This general method of analysis described in Sec. C-1 in connection with Fig. C-5 is to apply KVL around each independent loop. We illustrate this method with reference to the two-mesh circuit of Fig. C-14. For the assumed directions of the loop currents it is clear that the current in R_1 is $I_1 - I_2$ so that KCL is satisfied at node A. The sum of the voltage *drops* around loop 1 *in the direction of I_1* is

$$- V_1 + I_1\left(\frac{-j}{\omega C_1} + R_1\right) - I_2 R_1 = 0 \tag{C-46}$$

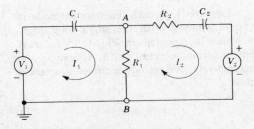

Figure C-14 A two-mesh sinusoidal network.

and KVL around mesh 2 *in the direction of* I_2 is

$$+ V_2 - I_1R_1 + I_2\left(R_1 + R_2 - \frac{j}{\omega C_2}\right) = 0 \qquad \text{(C-47)}$$

Equations (C-46) and (C-47) are solved simultaneously for the two currents I_1 and I_2. It is assumed that the two voltage sources have the same frequency. If this is not true, then the principle of superposition (Sec. C-2) must be used. In Eqs. (C-46) and (C-47) set $V_2 = 0$, $\omega = \omega_1$, and solve the resulting simultaneous equations for the currents. The result will be a sinusoid of frequency f_1 for I_1 (and also for I_2). Then set $V_1 = 0$ and $\omega = \omega_2$ and again solve Eqs. (C-46) and (C-47) for the currents, which will now be sinusoids of frequency f_2. The total current i_1 will be obtained by *adding the two instantaneous values of time,* and will be of the form

$$i_1 = A_1 \cos(\omega_1 t + \theta_1) + A_2 \cos(\omega_2 t + \theta_2)$$

Note that *it is meaningless to add phasors of different frequencies.*

Admittance

The reciprocal of the impedance is called the admittance and is designated by Y, so that

$$Y \equiv \frac{1}{Z} = G + jB \qquad \text{(C-48)}$$

The real part of Y is the *conductance G* and the imaginary part is the *susceptance B.* If a resistor is under consideration then $Z = R$, $G = 1/R$, and $B = 0$. On the other hand if the circuit element is a capacitor $Z = 1/j\omega C$ and $Y = j\omega C$ so that $B = \omega C$ and $G = 0$.

Since $I = V/Z$, then $I = YV$. For a resistor $I_R = GV_R$ and for a capacitor $I_C = j\omega C V_C$. For the circuit of Fig. C-9, with R and C in parallel, $V_R = V_C = V$, and the total current is

$$I = I_R + I_C = (G + j\omega C)V$$

The admittance of this combination is $Y = I/V = G + j\omega C$, which agrees with Eq. (C-43) if $G = 1/R$.

Nodal Equations

This general method of solution is to apply KCL at each independent node in the network. We illustrate this type of analysis with the single-stage high-frequency transistor of Fig. 13-16 which is repeated in Fig. C-15 for easy reference. There are two independent junction voltages which, *with respect to ground,* are designated V_1 and V_2 respectively. We introduce the conductances $G = 1/R$ and $g = 1/r$ for convenience.

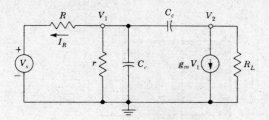

Figure C-15 A CE high-frequency transistor stage.

The currents away from node 1 are gV_1 in r, $j\omega C_e V_1$ in C_e, $G(V_1 - V_s)$ in R, and $j\omega C_c(V_1 - V_2)$ in C_c. Hence, KCL at this node yields

$$0 = (g + j\omega C_e)V_1 + G(V_1 - V_s) + j\omega C_c(V_1 - V_2) \qquad (C\text{-}49)$$

This equation agrees with Eq. (13-36), with the following change of notation: $V_1 = V_{b'e}$, $V_2 = V_o$, $G = G_s'$, $g = g_{b'e}$, and $s = j\omega$.

The currents away from node 2 are the dependent current $g_m V_1$, $j\omega C_c(V_2 - V_1)$ in C_c and V_2/R_L in R_L. At this node KCL is, with $s = j\omega$,

$$0 = (g_m - sC_c)V_1 + \left(sC_c + \frac{1}{R_L}\right)V_2 \qquad (C\text{-}50)$$

which is equivalent to Eq. (13-37).

In passing, note that the nodal equations implicitly satisfy Kirchhoff's voltage laws. For example, KVL around the left-hand loop in Fig. C-15 is

$$-V_s - I_R R + V_1 = 0$$

or
$$I_R = \frac{V_1 - V_s}{R} = G(V_1 - V_s)$$

which is precisely the expression used, in applying KCL to node V_1, for the current in R.

Network Theorems

The network theorems discussed in Sec. C-2 in connection with resistive networks apply equally well to networks containing capacitors and inductors as well as resistors in the sinusoidal steady state. These include *superposition*, *Thévenin's* theorem, *Norton's* theorem, and the *open-circuit-voltage—short-circuit-current theorem*. However, in applying these theorems, the impedance Z must be used in place of the resistance R. For example, Eq. (C-12) is now generalized to

$$V = IZ = \frac{I}{Y} \qquad I = \frac{V}{Z} = VY \qquad Z = \frac{V}{I} \qquad (C\text{-}51)$$

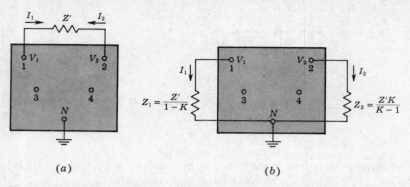

$$(a) \qquad\qquad\qquad (b)$$

Figure C-16 Pertaining to Miller's theorem. By definition, $K \equiv V_2/V_1$. The networks in (a) and (b) have identical node voltages. Note that $I_1 = -I_2$.

Miller's Theorem

This theorem is particularly useful in connection with transistor high-frequency amplifiers. Consider an arbitrary circuit configuration with N distinct nodes 1, 2, 3, ..., N, as indicated in Fig. C-16. Let the node voltages be V_1, V_2, V_3, ..., V_N, where $V_N = 0$, since N is the reference, or ground, node. Nodes 1 and 2 (referred to as N_1 and N_2) are interconnected with an impedance Z'. We postulate that we know the ratio V_2/V_1. Designate this ratio V_2/V_1 by K, which in the sinusoidal steady state will be a complex number. We shall now show that the current I_1 drawn from N_1 through Z' can be obtained by disconnecting Z' from terminal 1 and by bridging an impedance $Z'/(1 - K)$ from N_1 to ground, as indicated in Fig. C-16b.

The current I_1 is given by

$$I_1 = \frac{V_1 - V_2}{Z'} = \frac{V_1(1 - K)}{Z'} = \frac{V_1}{Z'/(1 - K)} = \frac{V_1}{Z_1} \qquad (C\text{-}52)$$

Therefore, if $Z_1 \equiv Z'/(1 - K)$ were shunted across terminals $N_1\text{-}N$, the current I_1 drawn from N_1 would be the same as that from the original circuit. Hence the same expression is obtained for I_1 in terms of the node voltages for the two configurations (Fig. C-16a and b).

In a similar way, it may be established that the correct current I_2 drawn from N_2 may be calculated by removing Z' and by connecting between N_2 and ground an impedance Z_2, given by

$$Z_2 \equiv \frac{Z'}{1 - 1/K} = \frac{Z'K}{K - 1} \qquad (C\text{-}53)$$

Since identical nodal equations (KCL) are obtained from the configurations of Fig. C-16a and b, these two networks are equivalent. It must be emphasized that this theorem will be useful in making calculations only if it is possible to find the value of K by some independent means.

The first application of Miller's theorem in this text is made in Sec. 13-9. Miller's theorem is also valid for resistive networks with Z' replaced by a resistor R'.

C-5 STEP RESPONSE OF AN RC CIRCUIT

The most common transient problem encountered in electronic circuits is that resulting from a step change in dc excitation applied to a series combination of a resistor and capacitor. Consider the high-pass RC circuit of Fig. C-17 to which is applied a step of voltage v_i. The output voltage v_o is taken across the resistor.

The High-Pass RC Circuit

A *step voltage* is one which maintains the value zero for all times $t < 0$ and maintains the value V for all times $t > 0$. The transition between the two voltage levels takes place at $t = 0$ and is accomplished in an arbitrarily short time interval. Thus in Fig. C-18, $v_i = 0$ immediately before $t = 0$ (to be referred to as time $t = 0 -$), and $v_i = V$ immediately after $t = 0$ (to be referred to as time $t = 0 +$).

From elementary considerations, the response of the network is exponential, with a time constant $RC \equiv \tau$, and the output voltage is of the form

$$v_o = B_1 + B_2 \epsilon^{-t/\tau} \tag{C-54}$$

The constant B_1 is equal to the steady-state value of the output voltage because as $t \to \infty$, $v_o \to B_1$. If this final value of output voltage is called V_f, then $B_1 = V_f$. The constant B_2 is determined by the initial output voltage, say V_i, because at $t = 0$, $v_o = V_i = B_1 + B_2$ or $B_2 = V_i - V_f$. Hence the general solution for a single-time-constant circuit having initial and final values V_i and V_f, respectively, is

$$v_o = V_f + (V_i - V_f) \epsilon^{-t/\tau} \tag{C-55}$$

This basic equation is used many times throughout this text.

The constants V_f and V_i must now be determined for the circuit of Fig. C-17. The input is a constant ($v_i = V$) for $t > 0$. Since $i = C(dv_C/dt)$, then in the steady state $i = 0$, and the final output voltage iR is zero, or $V_f = 0$.

The above result may also be obtained by the following argument: We have already emphasized that a capacitor C behaves as an open circuit at zero

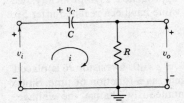

Figure C-17 The high-pass RC circuit.

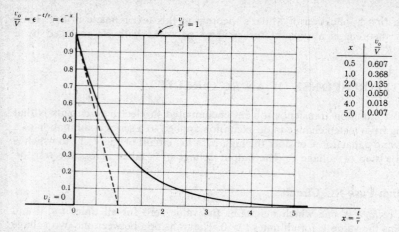

x	$\frac{v_o}{V}$
0.5	0.607
1.0	0.368
2.0	0.135
3.0	0.050
4.0	0.018
5.0	0.007

Figure C-18 Step-voltage response of the high-pass RC circuit. The dashed line is tangent to the exponential at $t = 0 +$.

frequency (because the reactance of C varies inversely with f). Hence, any constant (dc) input voltage is "blocked" and cannot reach the output. Hence, $V_f = 0$.

The value of V_i is determined from the following basic considerations. If the instantaneous current through a capacitor is i, then the change in voltage across the capacitor in time t_1 is $(1/C)\int_0^{t_1} i\, dt$. Since the current is always of finite magnitude, the above integral approaches zero as $t_1 \to 0$. Hence, it follows that *the voltage across a capacitor cannot change instantaneously*.

Applying the above principle to the network of Fig. C-17, we must conclude that since, at $t = 0$, the input voltage changes discontinuously by an amount V, the output must also change abruptly by this same amount. If we assume that the capacitor is initially uncharged, then the output at $t = 0 +$ must jump to V. Hence, $V_i = V$ and since $V_f = 0$, Eq. (C-55) becomes

$$v_o = V\epsilon^{-t/\tau} \tag{C-56}$$

Input and output are shown in Fig. C-18. Note that the output is 0.61 of its initial value at 0.5τ, 0.37 at 1τ, and 0.14 at 2τ. The output has completed more than 95 percent of its total change after 3τ and more than 99 percent of its swing if $t > 5\tau$. Hence, although the steady state is approached asymptotically, we may assume for most applications that the final value has been reached after 5τ.

Discharge of a Capacitor Through a Resistor

Consider a capacitor C charged to a voltage V. At $t = 0$, a resistor R is placed across C. We wish to obtain the capacitor voltage v_o as a function of time. Since the action of shunting C by R cannot instantaneously change the voltage,

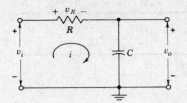

Figure C-19 The low-pass RC circuit.

$v_o = V$ at $t = 0 +$. Hence, $V_i = V$. Clearly at $t = \infty$, the capacitor will be completely discharged by the resistor and, therefore, $V_f = 0$. Substituting these values of V_i and V_f into Eq. (C-55), we obtain Eq. (C-56), and the capacitor discharge is indicated in Fig. C-18.

The Low-Pass RC Circuit

The response of the circuit of Fig. C-19 to a step input is exponential with a time constant RC. Since the capacitor voltage cannot change instantaneously, the output starts from zero and rises toward the steady-state value V, as shown in Fig. C-20. The output is given by Eq. (C-55), or

$$v_o = V(1 - \epsilon^{-t/RC}) \qquad\qquad \text{(C-57)}$$

Note that the circuits of Figs. C-17 and C-19 are identical except that the output $v_o = v_R$ is taken across R in Fig. C-17, whereas the output in Fig. C-19 is $v_o = v_C$. From Fig. C-19,

$$v_C = v_i - v_R = V - V\epsilon^{-t/RC}$$

where v_R is given by Eq. (C-56). This result for v_C agrees with Eq. (C-57).

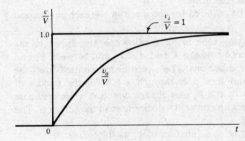

Figure C-20 Step-voltage response of the low-pass RC circuit.

PROBLEMS

Chapter 1

1-1 (*a*) An electron is emitted from an electrode with a negligible initial velocity and is *accelerated* by a potential V. Find the value of V if the final velocity of the particle is 1.88×10^7 m/s.

(*b*) A deuterium ion (heavy hydrogen ion, atomic weight 2.01) is introduced into an accelerating electric field with the same magnitude as in part (*a*) with an initial velocity of 10^5 m/s. Calculate the final velocity of the particle.

1-2 An electron having an initial kinetic energy of 10^{-17} J at the surface of one of two parallel-plane electrodes and moving normal to the surface is slowed down by the retarding field caused by a 65-V potential applied between the electrodes.

(*a*) Will the electron reach the second electrode?

(*b*) What retarding potential would be required for the electron to reach the second electrode with zero velocity?

1-3 The essential features of the displaying tube of an oscilloscope are shown in the accompanying figure. The voltage difference between K and A is V_a and between P_1 and P_2 is V_p. Neither electric field affects the other one. The electrons are emitted from the electrode K with initial zero velocity, and they pass through a hole in the middle of electrode A. Because of the field between P_1 and P_2 they change direction while they pass through these plates and, after that, move with constant velocity toward the screen S. The distance between plates is d.

(*a*) Find the velocity v_x of the electrons as a function of V_a as they cross A.

(*b*) Find the Y component of velocity v_y of the electrons as they come out of the field of plates P_1 and P_2 as a function of V_p, l_d, d, and v_x.

(*c*) Find the distance from the middle of the screen (d_s), when the electrons reach the screen, as a function of tube distances and applied voltages.

(*d*) For $V_a = 1.0$ kV, and $V_p = 100$ V, $l_d = 1.27$ cm, $d = 0.5$ cm, and $l_s = 20$ cm, find the numerical values of v_x, v_y, and d_s.

(*e*) If we want to have a deflection of $d_s = 5$ cm of the electron beam, what must be the value of V_a?

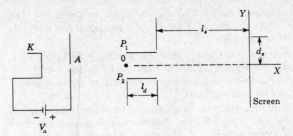

Prob. 1–3

1-4 A diode consists of a plane emitter and a plane-parallel anode separated by a distance of 0.5 cm. The anode is maintained at a potential of 10 V negative with respect to the cathode.

(a) If an electron leaves the emitter with a speed of 10^6 m/s, and is directed toward the anode, at what distance from the cathode will it intersect the potential-energy barrier?

(b) With what speed must the electron leave the emitter in order to be able to reach the anode?

1-5 A particle when displaced from its equilibrium position is subject to a linear restoring force $f = -kx$, where x is the displacement measured from the equilibrium position. Show by the energy method that the particle will execute periodic vibrations with a maximum displacement which is proportional to the square root of the total energy of the particle.

1-6 A particle of mass m is projected vertically upward in the earth's gravitational field with a speed v_o.

(a) Show by the energy method that this particle will reverse its direction at the height of $v_o^2/2g$, where g is the acceleration of gravity.

(b) Show that the point of reversal corresponds to a "collision" with the potential-energy barrier.

1-7 Prove that the concentration n of free electrons per cubic meter of a metal is given by

$$n = \frac{dv}{AM} = \frac{A_o dv \times 10^3}{A}$$

where d = density, kg/m^3
  v = valence, free electrons per atom
  A = atomic weight
  M = weight of atom of unit atomic weight, kg (Appendix A)
  A_o = Avogadro's number, molecules/mol

1-8 An aluminum wire has a resistivity of 3.44×10^{-8} Ω-m, a diameter of 2 mm, and a total length of 0.5 m. What is the voltage drop across the wire for a current of 30 mA?

1-9 The resistance of a copper wire of diameter 1.03 mm is 2.14×10^{-4} Ω/cm. The concentration of free electrons in copper is 8.40×10^{28} electrons/m^3. If the current density is 2×10^6 A/m^2 find the (a) current, (b) drift velocity, (c) mobility, and (d) conductivity.

1-10 (a) Calculate the electric field required to give an electron in silicon an average energy of 1 eV.

(b) Is it practical to generate electron-hole pairs by applying a voltage across a bar of silicon? Explain.

1-11 The specific density of tungsten is 18.8 g/cm^3 and its atomic weight is 184. The concentration of free electrons is 1.23×10^{23}/cm^3. Calculate the number of free electrons per atom. Refer to Example 1-2 in Sec. 1-8.

1-12 (a) Compute the conductivity of monovalent copper for which $\mu = 34.8$ cm^2/V · s, the atomic weight is 63.54, and $d = 8.9$ g/cm^3. Refer to Example 1-2 in Sec. 1-8.

(b) If an electric field is applied across such a copper bar with an intensity of 500 V/m, find the average velocity of the free electrons.

1-13 Compute the mobility of the free electrons in aluminum for which the density is 2.70 g/cm^3 and the resistivity is 3.44×10^{-6} Ω · cm. Assume that aluminum has three valence electrons per atom. The atomic weight is 26.98. Refer to Example 1-2 in Sec. 1-8.

1-14 (a) Determine the concentration of free electrons and holes in a sample of germanium at 300 K which has a concentration of donor atoms equal to 2×10^{14} atoms/cm^3 and a concentration of acceptor atoms equal to 3×10^{14} atoms/cm^3. Is this p- or n-type germanium? In other words, is the conductivity due primarily to holes or to electrons?

(b) Repeat part a for equal donor and acceptor concentrations of 10^{15} atoms/cm^3. Is this p- or n-type germanium?

(c) Repeat part a for a donor concentration of 10^{16} atoms/cm^3 and an acceptor concentration 10^{14} atoms/cm^3.

1-15 (a) Find the concentration of holes and of electrons in p-type germanium at 300 K if the resistivity is 0.02 Ω · cm.

(b) Repeat part a for n-type silicon if the resistivity is 20 Ω · cm.

1-16 (a) Show that the resistivity of intrinsic germanium at 300 K is 44.64 Ω · cm.

(b) If a donor-type impurity is added to the extent of 1 atom per 10^8 germanium atoms, prove that the resistivity drops to 3.73 Ω · cm.

1-17 (a) Find the resistivity of intrinsic silicon at 300 K.

(b) A donor-type impurity is added and the resistivity decreases to 9.6 Ω · cm. Compute the ratio of donor atoms to Si atoms per unit volume.

1-18 Calculate the resistance of a bar of intrinsic silicon at 300 K. The length of the bar is 5 cm and its cross section is 2 mm by 4 mm.

1-19 Consider intrinsic silicon at room temperature (300 K). By what percent does the conductivity increase per degree rise in temperature? Assume that μ is independent of T.

1-20 Repeat Prob. 1-19 for intrinsic germanium.

1-21 Determine the concentration of free electrons and holes in a sample of silicon at 500 K which has a concentration of donors equal to $N_D = 1.874 \times 10^{13}$ atoms/cm^3 and of acceptor atoms equal to $N_A = 3.748 \times 10^{13}$ atoms/cm^3. Show that the sample is essentially intrinsic. Explain why this should be expected physically.

1-22 A sample of germanium is doped to the extent of 10^{14} donor atoms/cm^3 and 7×10^{13} acceptor atoms/cm^3. At the temperature of the sample the resistivity of pure (intrinsic) germanium is 60 Ω · cm. Assume that the value of the mobility of holes and electrons is (approximately) the same as at 300 K. If the total conduction current density is 52.3 mA/cm^2, find the applied electric field.

1-23 (a) Verify the numerical value given in Table 1-1 for the concentration of atoms in germanium.

(b) If a donor-type impurity is added to the extent of 1 part in 10^8 germanium atoms, find the resistivity.

(c) If germanium were a monovalent metal, find the ratio of its conductivity to that of the n-type semiconductor in part (b).

1-24 If silicon were a monovalent metal, find the ratio of its conductivity to that of intrinsic silicon at 300 K.

1-25 (a) Find the magnitude of the Hall voltage V_H in an n-type silicon bar used in Fig. 1-10, which has a majority-carrier concentration $N_D = 10^{13}/cm^3$. Assume $B_z = 0.2$ Wb/m², $d = 5$ mm, and $\mathcal{E}_x = 5$ V/cm.

(b) What happens to V_H if an identical p-type silicon bar having $N_A = 10^{12}/cm^3$ is used in part (a)?

1-26 The Hall effect is used to determine the mobility of holes in a p-type silicon bar used in Fig. 1-10. Assume the bar resistivity is 200,000 $\Omega \cdot cm$, the magnetic field $B_z = 0.1$ Wb/m², and $d = w = 2$ mm. The measured values of the current and Hall voltage are 5 μA and 30 mV, respectively. Find μ_p.

1-27 An n-type silicon bar whose resistivity is 1,000 $\Omega \cdot m$ is used in Fig. 1-10 with $w = 1$ cm. If the current is 10 μA and the Hall voltage is 40 mV, what is the intensity B of the applied magnetic field?

1-28 The hole concentration in a semiconductor specimen is shown.

(a) Find an expression for and sketch the hole current density $J_p(x)$ for the case in which there is no externally applied electric field.

(b) Find an expression for and sketch the built-in electric field that must exist if there is to be no net hole current associated with the distribution shown.

(c) Find the value of the potential between the points $x = 0$ and $x = W$ if $p(0)/p_o = 10^3$.

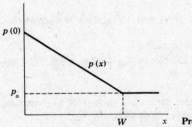

Prob. 1–28

1-29 (a) Consider an open-circuited graded semiconductor as in Fig. 1-12a. Verify the Boltzmann equation for electrons [Eq. (1-42)].

(b) For the step-graded semiconductor of Fig. 1-12b verify the expression for the contact potential V_o given in Eq. (1-45), starting with $J_n = 0$.

1-30 (a) Consider the step-graded germanium semiconductor of Fig. 1-12b with $N_D = 10^2 N_A$ and with N_A corresponding to 1 acceptor atom per 10^8 germanium atoms. Calculate the contact difference of potential V_o at room temperature.

(b) Repeat part a for a silicon p-n junction.

1-31 Consider an open-circuited p-n junction diode at 300 K. If N_D is changed by a factor of 10,000 and N_A remains unchanged, determine the change in the contact difference of potential.

1-32 (a) The resistivities of the two sides of a step-graded germanium diode are 2 Ω-cm (p side) and 1 $\Omega \cdot cm$ (n side). Calculate the height V_o of the potential barrier.

(b) Repeat part (a) for a silicon p-n junction.

Chapter 2

2-1 Sketch logarithmic plots of carrier concentration versus distance for an abrupt silicon junction if $N_D = 5 \times 10^{14}$ atoms/cm^3 and $N_A = 5 \times 10^{16}$ atoms/cm^3. Give numerical values for ordinates. Label the n, p, and depletion regions.

2-2 Repeat Prob. 2-1 for an abrupt germanium junction.

2-3 The resistivities of the two sides of an abrupt silicon junction are 9.6 $\Omega \cdot$ cm (p side) and 100 $\Omega \cdot$ cm (n side). Sketch the logarithmic plots of carrier concentration versus distance. Give numerical values for the ordinates. Label the n, p, and depletion regions.

2-4 (a) For what voltage will the reverse current in a p-n junction silicon diode reach 95 percent of its saturation value at room temperature?

(b) What is the ratio of the current for a forward bias of 0.1 V to the current for the same magnitude of reverse bias?

(c) If the reverse saturation current is 10 nA, calculate the forward currents for voltages of 0.5, 0.6, and 0.7 V, respectively.

2-5 (a) If the reverse saturation current in a p-n junction silicon diode is 1 nA, find the applied voltage for a forward current of 0.5 μA.

(b) What current would result if the voltage found in part (a) were applied across a germanium diode in the forward direction? (Assume a reverse saturation current of 20 μA.)

2-6 (a) A silicon diode at room temperature (300 K) conducts 5 mA at 0.7 V. If the voltage increases to 0.8 V, calculate the diode current. Assume $\eta = 2$.

(b) Calculate the reverse saturation current.

2-7 (a) What is the change in voltage at 300 K for a tenfold increase in current for a silicon diode operating in the conducting region?

(c) Repeat part (a) for a one-hundredfold increase.

2-8 (a) Evaluate η in Eq. (2-3) from the slope of the plot in Fig. 2-6 for $T = 25°C$. Draw the best-fit line over the current range 0.01 to 10 mA.

(b) Repeat for $T = -55$ and 150°C.

2-9 (a) Calculate the factor by which the reverse saturation current of a germanium diode is multiplied when the temperature is increased from 25 to 100°C.

(b) Repeat part a for a silicon diode over the range 25 to 200°C.

2-10 (a) What increase in temperature would result in a reverse saturation current which is 50 times its value at room temperature?

(b) What decrease in temperature would result in a reverse saturation current which is one tenth its value at room temperature?

2-11 It is predicted that, for germanium, the reverse saturation current should increase by 0.11°C^{-1}. It is found experimentally in a particular diode that at a reverse voltage of 10 V, the reverse current is 5 μA and the temperature dependence is only 0.07°C^{-1}. What is the leakage resistance shunting the diode?

2-12 The reverse saturation current of a germanium diode is 10 μA. The diode is shunted by a resistance whose value is 1.25 kΩ. What is the applied voltage if the observed current is 40 μA? *Hint*: Use a graphical method of solution.

2-13 A diode is mounted on a chassis in such a manner that, for each degree of temperature rise above ambient, 0.1 mW is thermally transferred from the diode to its surroundings. (The "thermal resistance" of the mechanical contact between the diode and its surroundings is 0.1 mW/°C.) The ambient temperature is 25°C. The diode temperature is not to be allowed to increase by more than 10°C above ambient. If the reverse saturation current is 5.0 μA at 25°C and increases at the rate $0.07°C^{-1}$, what is the maximum reverse-biasing voltage which may be maintained across the diode?

2-14 A silicon diode operates at a forward voltage of 0.7 V. Calculate the factor by which the current will be multiplied when the temperature is decreased from 25 to $-55°C$.

2-15 An ideal silicon p-n junction diode has at a temperature of 125°C a reverse saturation current of 0.1 μA. At a temperature of 105°C find the dynamic resistance for 0.8 V bias in (a) the forward direction and (b) the reverse direction.

2-16 An ideal germanium diode at room temperature has a static resistance of 4.57 Ω at a point where $I = 43.8$ mA. Find the dynamic resistance for a forward bias of 0.1 V.

2-17 Prove that for an alloy p-n junction (with $N_A \ll N_D$), the width W of the depletion layer is given by

$$W = \left(\frac{2\epsilon\mu_p V_j}{\sigma_p} \right)^{1/2}$$

where V_i is the junction potential under the condition of an applied diode voltage V_d.

2-18 (a) Prove that for an alloy silicon p-n junction (with $N_A \ll N_D$), the depletion-layer capacitance in picofarads per square centimeter is given by

$$C_T = 2.913 \times 10^{-4} \left(\frac{N_A}{V_j} \right)^{1/2}$$

(b) If the resistivity of the p material is 4 $\Omega \cdot$ cm, the barrier height V_o is 0.3 V, the applied reverse voltage is 4 V, and the cross-sectional area is circular of 50 mils in diameter, find C_T.

2-19 Find the resistivity of the p-type material in a p-n silicon junction whose cross-sectional area is circular of 40 mils in diameter and C_T is 61 pF. The barrier height V_o is 0.35 V and the applied reverse voltage is 5 V. *Hint*: Use the result of Prob. 2-18a.

2-20 (a) For the junction of Fig. 2-8, find the expression for the $\mathcal{E}$ and V as a function of x in the n-type side for the case where N_A and N_D are of comparable magnitude. *Hint*: Shift the origin of x so that $x = 0$ at the junction.

(b) Show that the total barrier voltage is given by Eq. (2-15) multiplied by $N_A/(N_A + N_D)$ and with $W = W_p + W_n$.

(c) Prove that $C_T = [qN_A N_D \epsilon / 2(N_A + N_D)]^{1/2} V^{-1/2}$.

(d) Prove that $C_T = \epsilon A / (W_p + W_n)$.

2-21 For a silicon p-n diode $N_D = 10^{15}$ cm$^{-3} \ll N_A$ and $V_o = 0.5$ V. If the applied reverse-biased voltage is 10 V, find (a) the width W of the space-charge region, (b) the value of the electric field at the junctions, and (c) the capacitance per square mil.

2-22 Reverse-biased diodes are frequently employed as electrically controllable variable capacitors. The transition capacitance of an abrupt junction diode is 10 pF at 4 V. Compute the decrease in capacitance for a 0.5-V increase in bias.

2-23 Calculate the barrier capacitance of a germanium p-n junction whose area is 0.5 mm by 0.5 mm and whose space-charge thickness is 3×10^{-4} cm. The dielectric constant of germanium (relative to free space) is 16.

2-24 The zero-voltage barrier height at an alloy-silicon p-n junction is 0.6 V. The concentration N_A of acceptor atoms in the p side is much smaller than the concentration of donor atoms in the n material, and $N_A = 5 \times 10^{16}$ cm^{-3}. Calculate the width of the depletion layer for an applied reverse voltage of (a) 5.6 V, (b) 0.2 V, and (c) for a forward bias of 0.5 V.

(d) If the cross-sectional area of the diode is 1 mm^2, evaluate the space-charge capacitance corresponding to the values of applied voltage in parts (a) and (b).

2-25 (a) Consider a grown junction for which the uncovered charge density ρ varies linearly with distance. If $\rho = ax$, prove that the barrier voltage V_j is given by

$$V_j = \frac{aW^3}{12\epsilon}$$

(b) Verify that the barrier capacitance C_T is given by Eq. (2-17)

2-26 Find the current of a forward-biased silicon diode if the diffusion capacitance is $C_D = 1$ nF. The diffusion length is 2.6×10^{-6} m. Assume that the doping of the p side is much greater than that of the n side.

2-27 The derivation of Eq. (2-26) for the diffusion capacitance assumes that the p side is much more heavily doped than the n side, so that the current at the junction is entirely due to holes. Derive an expression for the total diffusion capacitance when this approximation is not made.

2-28 (a) wProve that the magnitude of the maximum electric field $\mathcal{E}_m$ at a step-graded junction with $N_A \gg N_D$ is given by

$$\mathcal{E}_m = \frac{2V_j}{W}$$

(b) It is found that Zener breakdown occurs when $\mathcal{E}_m = 2 \times 10^7$ V/m $\equiv \mathcal{E}_Z$. Prove that Zener voltage V_Z is given by

$$V_Z = \frac{\epsilon \mathcal{E}_Z^2}{2qN_D}$$

Note that the Zener breakdown voltage can be controlled by controlling the concentration of donor ions.

2-29 (a) Zener breakdown occurs in germanium at a field intensity of 2×10^7 V/m. Prove that the breakdown voltage is $V_Z = 50.93/\sigma_p$, where σ_p is the conductivity of the p material in $(\Omega \cdot \text{cm})^{-1}$. Assume that $N_A \ll N_D$.

(b) If the p material is essentially intrinsic, calculate V_Z.

(c) For a doping of 1 part in 10^8 of p-type material, the resistivity drops to 3.7 $\Omega \cdot$ cm. Calculate V_Z.

(d) For what resistivity of the p-type material will $V_Z = 10$ V?

2-30 (a) A silicon diode is in series with a 2-kΩ resistor and a 10-V power supply. Approximately what is the current in the circuit, if the diode is forward-biased?

(b) If the measured diode drop is 0.6 V at 1 mA, obtain a more accurate value for the current in the circuit.

(c) If the battery is reversed and if the diode breakdown voltage is 7 V, find the current in the circuit.

(d) A second identical diode is added in series opposing (the two anodes are connected together). Approximately what is the current in the circuit?

(e) If the supply voltage in part (d) is reduced to 4 V, what is the current?

2-31 (a) Two p-n silicon diodes are connected in series opposing. The reverse saturation current is 10 nA. A 6-V battery is connected to this series arrangement. If the Zener (avalanche) voltage is 10 V, find the current I in the circuit and the voltage across each diode.

(b) Now assume that $V_Z = 5$ V. Repeat part (a), neglecting the Zener ohmic resistance.

2-32 A series combination of a 15-V avalanche diode and a forward-biased silicon diode is to be used to construct a zero-temperature-coefficient voltage reference. The temperature coefficient of the silicon diode is -1.7 mV/°C. Express in percent per degree centigrade the required temperature coefficient of the Zener diode.

2-33 The saturation currents of the two diodes are 1 and 2 μA. The breakdown voltages of the diodes are the same and are equal to 100 V. Assume $\eta = 2$.

  (a) Calculate the current and voltage for each diode if $V = 80$ V and $V = 120$ V.

  (b) Repeat part (a) if each diode is shunted by a 8-MΩ resistor.

Prob. 2–33

2-34 (a) A silicon p-n junction diode is connected in series with a resistor R of 10 MΩ. A 1-V battery is connected across this series combination so as to reverse-bias the diode. Find the voltage across the diode by assuming that the reverse saturation current is 30 nA.

  (b) Repeat part (a) with the battery connected in the opposite direction. *Hint*: Use the graphical method of solution, indicated in Fig. 2-14.

2-35 (a) In the circuit of Prob. 2-31, the Zener breakdown voltage is 4.0 V. If the ohmic resistance could be neglected, what would be the current?

  (b) If the ohmic resistance is 200 Ω, what is the current? *Note*: Answer part (b) by plotting Eq. (2-3) and drawing a load line. Verify your answer analytically by a method of successive approximations.

2-36 A p-n germanium junction diode at room temperature has a reverse saturation current of 10 μA, negligible ohmic resistance, and a Zener breakdown voltage of 100 V. A 1-kΩ resistor is in series with this diode, and a 30-V battery is impressed across this combination. Find the current (a) if the diode is forward-biased and (b) if the battery is inserted into the circuit with the reverse polarity.

  (c) Repeat parts (a) and (b) if the Zener breakdown voltage is 10 V.

2-37 A 5-V battery is connected across the series combination of a 10-Ω resistor and a device whose V-I characteristics are given by (with I in amperes and V in volts)

$$I = +0.2\sqrt{V - 1} \quad \text{for} \quad V \geq 1$$
$$I = 0 \quad \text{for} \quad V < 1$$

  (a) Find the current in the circuit by using a graphical method.

  (b) Verify the answer analytically.

2-38 Each diode is described by a linearized volt-ampere characteristic, with incremental resistance r and offset voltage V_γ. Diode $D1$ is germanium with $V_\gamma = 0.2$ V and $r = 20$ Ω, whereas $D2$ is silicon with $V_\gamma = 0.6$ V and $r = 15$ Ω. Find the diode currents if (a) $R = 10$ kΩ, (b) $R = 1$ kΩ.

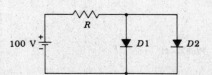

Prob. 2–38

2-39 For the circuit shown the cutin voltage of a diode is 0.6 V and the drop across a conducting diode is $V' = 0.7$ V. Calculate v_o for the following input voltages and indicate the state of each diode (ON or OFF). Justify your assumptions about the state of each diode.

    (a) $v_1 = 10$ V, $v_2 = 0$ V.
    (b) $v_1 = 5$ V, $v_2 = 0$ V.
    (c) $v_1 = 10$ V, $v_2 = 5$ V.
    (d) $v_1 = 5$ V, $v_2 = 5$ V.

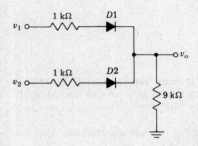

Prob. 2–39

2-40 Assume that the diodes of the circuit are ideal, that is, $R_f = 0$, $V_\gamma = 0$, $R_r = \infty$. Find v_o for the following cases:

    (a) $v_1 = v_2 = 5$ V.
    (b) $v_1 = 5$ V, $v_2 = 0$ V.
    (c) $v_1 = v_2 = 0$ V.

Justify your assumptions about the state of each diode in each part.

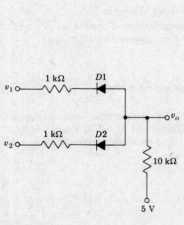

Prob. 2–40

Prob. 2–41

2-41 The diodes of the circuit shown are the same as those of Prob. 2-39. Calculate I_1, I_2, I_3, and v_o under the following conditions:

  (a) $v_1 = 0$ V, $v_2 = 25$ V.
  (b) $v_1 = v_2 = 25$ V.

Chapter 3

3-1 The transistor of Fig. 3-3a has the characteristics given in Figs. 3-6 and 3-7. Let $V_{CC} = 1.2$ V, $R_L = 40$ Ω, and $I_E = 5$ mA.

  (a) Find I_C and V_{CB}.
  (b) Find V_{EB} and V_L.
  (c) If I_E changes by $\Delta I_E = 10$ mA symmetrically around the point of part (a) and with constant V_{CC}, find the corresponding change in I_C.

3-2 The CB transistor used in the circuit of Fig. 3-3a has the characteristics given in Figs. 3-6 and 3-7. Let $I_C = -15$ mA, $V_{CB} = -3$ V, and $R_L = 100$ Ω.

  (a) Find V_{CC} and I_E.
  (b) If the supply voltage V_{CC} decreases from its value in part a by 1 V while I_E retains its previous value, find the new values of I_C and V_{CB}.

3-3 If $\alpha = 0.98$ and $V_{BE} = 0.7$ V, find R_1 in the circuit shown for an emitter current $I_E = -2$ mA. Neglect the reverse saturation current.

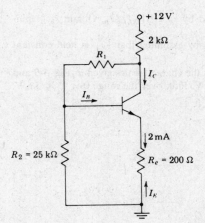

Prob. 3–3

3-4 Draw the idealized output and input characteristics for a CE n-p-n transistor with the following parameters: $h_{FE} = 100$ independent of current; $V_{CE(sat)} = 0.2$ V, $V_{BE(sat)} = 0.8$ V; I_C in the range 0 to 20 mA.

3-5 For the circuit shown, $\alpha_1 = 0.99$, $\alpha_2 = 0.98$, $V_{CC} = 20$ V, $R_c = 100$ Ω, and $I_E = -120$ mA. Neglecting the reverse saturation currents and assuming both transistors in the active region, determine (a) the currents I_{C1}, I_{B1}, I_{E1}, I_{C2}, I_{B2}, and I_C; (b) V_{CE}; (c) I_C/I_B, I_C/I_E.

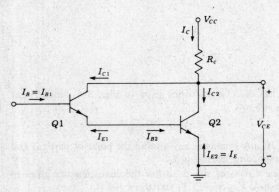

Prob. 3-5

3-6 The CE transistor used in the circuit of Fig. 3-11a has the characteristics given in Figs. 3-9 and 3-10. Let $R_b = 30$ kΩ.

(a) Find V_{BB} if $V_{CC} = 12$ V, $V_{CE} = 6$ V, and $R_c = 240$ Ω.

(b) If $V_{CC} = 6$ V, find R_c so that $I_C = 16$ mA and $V_{CE} = 2$ V. Find V_{BB}.

3-7 The CE transistor whose characteristics are shown in Figs. 3-9 and 3-10 is used in the circuit of Fig. 3-11 with $R_b = 10$ kΩ, $V_{CC} = 8$ V, and $R_c = 200$ Ω.

(a) Calculate V_{BB}.

(b) The input dynamic resistance is defined by $r_b \equiv \Delta V_{BE}/\Delta I_B$. Obtain r_b graphically.

(c) Find the voltage gain $A \equiv \Delta V_{CE}/\Delta V_{BB}$ by assuming that V_{CC} is held constant but V_{BB} is allowed to vary.

3-8 The transistor used in the circuit shown has the characteristics given in Fig. 3-9 and 3-10. Let $R_b = 6$ kΩ, $R_c = 200$ Ω, and $V_{CC} = 8$ V. Plot, over the range $0 \leqslant V_i \leqslant 3.6$ V, (a) I_B versus V_i; (b) I_C versus V_i; (c) V_o versus V_i.

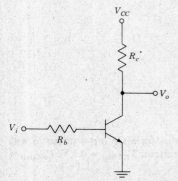

Prob. 3-8

3-9 (*a*)   Find R_c and R_b in the circuit of Fig. 3-11*a* if $V_{CC} = 12$ V and $V_{BB} = 6$ V, so that $I_C = 12$ mA and $V_{CE} = 6$ V. A silicon transistor with $\beta = 100$, $V_{BE} = 0.7$ V, and negligible reverse saturation current is under consideration.

   (*b*)   Repeat part (*a*) if a 200-Ω emitter resistor is added to the circuit.

3-10 In the circuit shown, $V_{CC} = 20$ V, $R_c = 5$ kΩ, and $R_e = 100$ Ω. If a silicon transistor is used with $\beta = 100$ and if $V_{CE} = 4$ V, find R. Neglect the reverse saturation current.

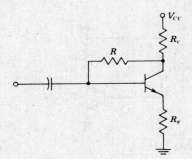

Prob. 3-10

3-11 Calculate the collector and base currents in the transistor shown. Assume $h_{FE} = 50$. Verify any assumptions you make. *Hint:* Apply Thévenin's theorem (Sec. C-2) looking to the left of the base terminal.

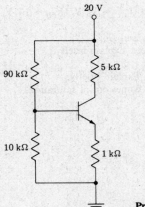

Prob. 3-11

3-12 For the circuit shown, transistors $Q1$ and $Q2$ operate in the active region with $V_{BE1} = V_{BE2} = 0.7$ V, $\beta_1 = 100$, and $\beta_2 = 50$. The reverse saturation currents may be neglected.

   (*a*)   Find the currents I_{B2}, I_1, I_2, I_{C2}, I_{B1}, I_{C1}, and I_{E1}.

(b) Find the voltages V_{o1} and V_{o2}.

Hint: Apply Thévenin's theorem (Sec. C-2) looking to the left of the 100 kΩ resistor.

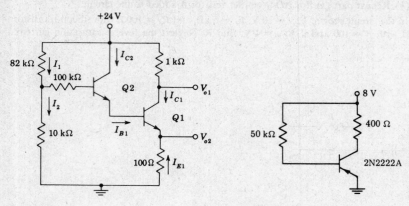

Prob. 3-12   Prob. 3-13

3-13 From the characteristic curves of the 2N2222A type transistor given in Figs. 3-9, 3-10, and 3-12, find the voltages V_{BE}, V_{CE}, and V_{BC} for the circuit shown. *Hint:* As a first approximation assume $V_{BE} = 0.7$ V.

3-14 A silicon transistor with $\beta = h_{FE} = 100$ is used in the circuit of Prob. 3-8. Find the maximum value of R_b for which the transistor remains at saturation. Assume $V_i = 5$ V, $V_{CC} = 10$ V, and $R_c = 4.66$ kΩ.

3-15 The silicon transistor shown has an $h_{FE} = 50$. Let $V_{CC} = 25$ V, $V_{BB} = 10$ V, $R_b = 40$ kΩ, $R_c = 15$ kΩ, and $R_e = 5$ kΩ.

    (a) Assume that Q is in the active region and find I_B and I_C.

    (b) Verify that the assumption in part (a) is *not* correct. Explain briefly.

    (c) Assume that Q is in saturation and find I_B and I_C.

    (d) Verify that the assumption in part (c) is justified. Explain briefly.

    (e) Find the value of R_e for which the transistor just comes out of saturation.

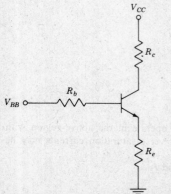

Prob. 3-15

3-16 In the circuit of Prob. 3-15 let $V_{BB} = 10$ V, $V_{CC} = 25$ V, $R_c = 3$ kΩ, $R_e = 2$ kΩ, and $R_b = 50$ kΩ. The silicon transistor has an $h_{FE} = 100$. Use the parameters of Table 3-1.
    (a) Is the transistor in the cutoff, active, or saturation region?
    (b) Find the smallest value of R_b so that the transistor is in its active region.

3-17 In the circuit shown find V_o for (a) $V = 15$ V, (b) $V = 30$ V. A silicon transistor where $\beta = 40$ is used. *Hint:* Apply Thévenin's theorem (Sec. C-2) to both the base and collector circuits.

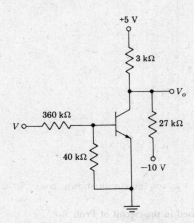

Prob. 3-17

3-18 For the circuit of Prob. 3-8, assume $h_{FE} = 100$. Use the values of Table 3-1 for a silicon transistor, and let $V_{CC} = 5$ V, $R_c = 2$ kΩ, $R_b = 12$ kΩ. Over the range $0 \leqslant V_i \leqslant 2$, plot V_o, I_B, and I_C versus V_i. Indicate the regions in each plot where the transistor is operating at cutoff, in the active mode, or in saturation.

3-19 (a) Is the transistor in the active mode or in saturation? Assume $\beta = 100$, and neglect junction voltages.
    (b) Calculate V_o for the circuit.
    (c) What is the minimum β of a transistor that will saturate in the above circuit?

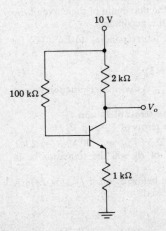

Prob. 3-19

3-20 Silicon transistors with $h_{FE} = 100$ are used in the circuit shown. Neglect the reverse saturation current.

    (a) Find V_o when $V_i = 0$ V. Assume $Q1$ is OFF and justify the assumption.

    (b) Find V_o when $V_i = 6$ V. Assume $Q2$ is OFF and justify this assumption.

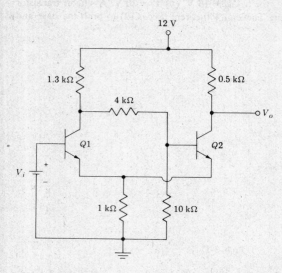

Prob. 3-20

3-21 From Fig. 3-9 find h_{fe} for the 2N2222A transistor at $V_{CE} = 6$ V and $I_B = 120$ μA.

3-22 (a) Show that the relationship between $\beta = h_{FE}$ and the small-signal CE current gain $\beta' = h_{fe}$ is

$$h_{fe} = \frac{h_{FE}}{1 - (I_{CO} + I_B)(\partial h_{FE}/\partial I_C)}$$

Hint: Differentiate Eq. (3-12) with respect to I_C.

    (b) Show that $h_{fe} > h_{FE}$ for currents to the left of the maximum of the curve of h_{FE} versus I_C (Fig. 3-14), and that $h_{fe} < h_{FE}$ to the right of the maximum.

3-23 For the inverted mode of operation, we may write [corresponding to Eq. (3-5) for the normal mode],

$$I_E = -\alpha_I I_C - I_{EO}(\epsilon^{V_E/V_T} - 1) \tag{1}$$

where α_I is the inverted common-base current gain and I_{EO} is the emitter-junction reverse saturation current.

    (a) Derive the explicit expressions for V_E and V_C in terms of I_C and I_E.

    (b) Derive the explicit expressions for I_C and I_E in terms of V_C and V_E.

3-24 (a) Consider the circuit of Fig. 3-11a with $V_{CC} = 12$ V, $V_{BB} = -20$ V, $R_c = 2$ kΩ, and a silicon transistor with $h_{FE} = 30$. For what value of R_b will the transistor be in saturation?

    (b) For what values of R_b will the transistor remain below cutoff if $I_{CBO} = 100$ μA and if the polarity of the base supply voltage V_{BB} is reversed?

3-25 (*a*) The reverse saturation current of the silicon in Fig. 3-18 is 10 nA at room temperature (25°C) and increases by a factor of 2 for each temperature increase of 10°C. The bias $V_{BB} = 8$ V. Find the maximum allowable value for R_b if the transistor is to remain cut off at a temperature of 185°C.

(*b*) If $V_{BB} = 2.0$ V and $R_b = 20$ kΩ, how high may the temperature increase before the transistor comes out of cutoff?

3-26 If the silicon transistor used in the circuit shown has a minimum value of $\beta = h_{FE}$ of 30 and if $I_{CBO} = 10$ nA at 25°C:

(*a*) Find V_o for $V_i = 12$ V and show that Q is in saturation.

(*b*) Find the minimum value of R_1 for which the transistor in part (*a*) is in the active region.

(*c*) If $R_1 = 15$ kΩ and $V_i = 1$ V, find V_o and show that Q is at cutoff.

(*d*) Find the maximum temperature at which the transistor in part (*c*) remains at cutoff.

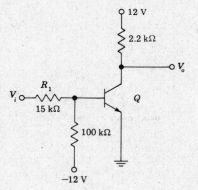

Prob. 3-26

3-27 (*a*) Show that for an *n-p-n* silicon transistor of the alloy type in which the resistivity ρ_B of the base is much larger than that of the collector, the punch-through voltage V is given by $V = 9.42 \times 10^8 W_B^2/\rho_B$, where V is in volts, ρ_B in ohm-centimeters, and W in centimeters.

(*b*) Calculate the punch-through voltage if $W = 2$ μm and $\rho_B = 1$ $\Omega \cdot$ cm.

Chapter 4

4-1 List in order the steps required in fabricating a monolithic silicon integrated transistor by the epitaxial-diffused method. Sketch the cross section after each oxide growth. Label materials clearly. No buried layer is required.

4-2 Sketch *to scale* the cross section of a monolithic transistor fabricated on a 5-mil-thick silicon substrate. *Hint*: Refer to Sec. 4-2 and Figs. 4-7 and 4-8 for typical dimensions.

4-3 From Fig. 4-7, determine the (*a*) epitaxial-collector concentration, (*b*) the distance from the surface at which the collector and emitter junctions are found, (*c*) base thickness, (*d*) concentration of boron at the surface, (*e*) concentration of phosphorus at the surface.

4-4 (*a*) Consider an IC *n-p-n* transistor *Q*1 built upon a *p*-type substrate *S*. Show that between the four terminals *E*, *B*, *C*, and *S* there exists a *p-n-p* transistor *Q*2 in addition to *Q*1.

   (*b*) If *Q*1 is in its active region, in what mode is *Q*2 operating? Explain.

   (*c*) Repeat part (*b*) if *Q*1 is in saturation.

   (*d*) Repeat part (*b*) if *Q*1 is in cutoff.

4-5 Sketch the five basic diode connections (in circuit form) for the monolithic integrated circuits. Which will have the lowest forward-voltage drop? Highest breakdown voltage?

4-6 The Schottky barrier diode prevents *Q* from saturating. Assume a drop of 0.4 V across this diode when it conducts. If $h_{FE} = 100$, calculate the base and collector currents as well as the currents in each resistor.

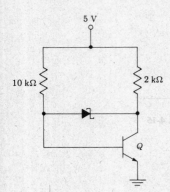

Prob. 4-6

4-7 If the base sheet resistance can be held to within ± 10 percent and resistor line widths can be held to ± 0.1 mil, plot the maximum tolerance (percent change in resistance) of a diffused resistor as a function of line width *w* in mils over the range $0.5 \leqslant w \leqslant 5.0$. (Neglect contact-area and contact-placement errors.)

4-8 A 1-mil-thick silicon wafer has been doped uniformly with phosphorus to a concentration of 10^{17} cm^{-3}, plus boron to a concentration of 5×10^{16} cm^{-3}. Find its sheet resistance.

4-9 (*a*) Calculate the resistance of a diffused crossover 5 mils long, 2 mils wide, and 1 μm thick, given that its sheet resistance is 2.4 Ω/square.

   (*b*) Repeat part (*a*) for an aluminum metalizing layer 0.4 μm thick of resistivity 2.8×10^{-6} Ω · cm. Note the advantage of avoiding diffused crossovers.

4-10 (*a*) What is the total length required to fabricate a 20-kΩ resistor whose width is 1 mil if $R_S = 200$ Ω/square?

   (*b*) What is the width required to fabricate a 5-kΩ resistor whose length is 1 mil?

4-11 An integrated junction capacitor has an area of 2,000 mils^2 and is operated at a reverse barrier potential of 1.5 V. The acceptor concentration of 10^{16} atoms/cm^3 is much smaller than the donor concentration. Calculate the capacitance (the relative dielectric constant $\epsilon_r = 12$).

4-12 A thin-film capacitor has a capacitance of 0.4 pF/mil^2. The thickness of the SiO_2 layer is 500 Å. Calculate the relative dielectric constant ϵ_r of silicon dioxide.

4-13 An MOS capacitor is fabricated with an oxide thickness of 500 Å. How much chip area is required to obtain a capacitance of 200 pF? The relative dielectric constant ϵ_r of silicon dioxide is 3.5.

4-14 The n-type epitaxial isolation region shown is 10 mils long, 5 mils wide, and 1 mil thick and has a resistivity of 0.2 $\Omega \cdot$ cm. The resistivity of the p-type substrate is 20 $\Omega \cdot$ cm. Find the parasitic capacitance between the isolation region and the substrate under 5-V reverse bias. Assume that the sidewalls contribute 0.1 pF/mil^2.

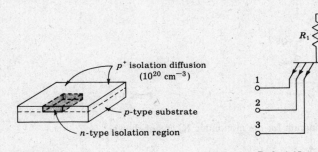

Prob. 4-14   **Prob. 4-15**

4-15 For the circuit shown, find (*a*) the *minimum* number, (*b*) the *maximum* number, of isolation regions.

4-16 (*a*) What is the minimum number of isolation regions required to realize in monolithic form the logic gate shown?

(*b*) Draw a monolithic layout of the gate in the fashion of Fig. 4-23b.

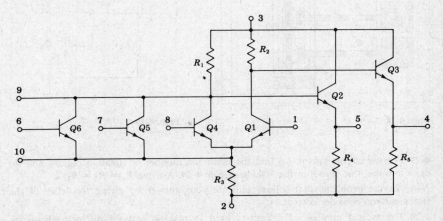

Prob. 4-16

4-17 Repeat Prob. 4-16 for the difference amplifier shown.

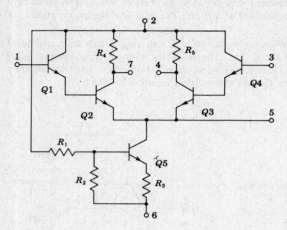

Prob. 4-17

4-18 Draw the circuit diagram for the monolithic IC layout shown.

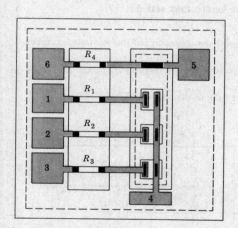

Prob. 4-18   **Prob. 4-19**

4-19 For the circuit shown, (a) find the minimum number of isolation regions and (b) draw a monolithic layout in the fashion of Fig. 4-24. Assume $R_1 = R_2 = R_3/2$.

Note: In the problems that follow, indicate your answer by giving the letter of the statement you consider correct.

4-20 The typical number of diffusions used in making epitaxial-diffused silicon integrated circuits is (a) 6, (b) 3, (c) 4, (d) 5, and (e) 2.

4-21 The "buried layer" in an integrated transistor on a p-type substrate in an integrated transistor is (a) used to reduce the parasitic capacitance; (b) p^+ doped; (c) located in the emitter region; (d) n^+ doped.

4-22 Epitaxial growth is used in integrated circuits (IC's):

(a) Because it produces low parasitic capacitance.

(b) Because it yields back-to-back isolating p-n junctions.

(c) To grow single-crystal n-doped silicon on a single-crystal p-type substrate.

(d) To grow selectively single-crystal p-doped silicon of one resistivity on a p-type substrate of a different resistivity.

4-23 Silicon dioxide (SiO_2) is used in IC's:

(a) To control the location of diffusion and to protect and insulate the silicon surface.

(b) Because it facilitates the penetration of diffusants.

(c) To control the concentration of diffusants.

(d) Because of its high heat conduction.

4-24 When a hole is opened in the SiO_2 and impurities are introduced, they will diffuse vertically (a) a greater distance than laterally; (b) the same distance as laterally; (c) a shorter distance than laterally; (d) twice the lateral distance.

4-25 The p-type substrate in a monolithic circuit should be connected to (a) any dc ground point; (b) nowhere, i.e., be left floating; (c) the most positive voltage available in the circuit; (d) the most negative voltage available in the circuit.

4-26 Monolithic integrated circuit systems offer greater reliability than discrete-component systems because (a) silicon is used; (b) there are fewer interconnections; (c) reduction in size is achieved; (d) they are hermetically sealed.

4-27 The collector-substrate junction in the epitaxial collector structure is, approximately, (a) an exponential junction; (b) a linearly graded junction; (c) a step-graded junction; (d) none of the foregoing.

4-28 The sheet resistance of a semiconductor is (a) a parameter whose value is important in a thin-film resistance; (b) a characteristic whose value determines the required area for a given value of integrated capacitance; (c) an important characteristic of a diffused region, especially when used to form diffused resistors; (d) an undesirable parasitic element.

4-29 Isolation in IC's is required to (a) minimize electrical interaction between circuit components; (b) simplify interconnections between devices; (c) protect the components from mechanical damage; (d) protect the transistor from possible "thermal runaway."

4-30 Almost all resistors are made in a monolithic IC (a) during metalization; (b) during the emitter diffusion; (c) while growing the epitaxial layer; (d) during the base diffusion.

4-31 Increasing the yield of an integrated circuit (a) results in a lower number of good chips per wafer; (b) means that more transistors can be fabricated on the same size wafer; (c) reduces individual circuit cost; (d) increases the cost of each good circuit.

4-32 In a monolithic-type IC (a) each transistor is diffused into a separate isolation region; (b) resistors and capacitors of any value may be made; (c) all isolation problems are eliminated; (d) all components are fabricated into a single crystal of silicon.

4-33 The main purpose of the metalization process is to (a) supply a bonding surface for mounting the chip; (b) act as a ground plane; (c) protect the chip from oxidation; (d) interconnect the various circuit elements.

4-34 Two types of metal semiconductor junctions are (a) ohmic and nonrectifying; (b) rectifying and capacitive; (c) ohmic and capacitive; (d) rectifying and ohmic.

4-35 A Schottky transistor (*a*) has multiple emitters; (*b*) enables faster circuit operation; (*c*) requires a smaller fabrication area; (*d*) is a low-frequency, high-gain device.

Chapter 5

5-1 Convert the following decimal numbers to binary form: (*a*) 753, (*b*) 432, (*c*) 258.

5-2 The parameters in the diode OR circuit of Fig. 5-3 are $V(0) = +12$ V, $V(1) = -2$ V, $R_s = 600 \ \Omega$, $R = 10 \ \text{k}\Omega$, $R_f = 0$, $R_r = \infty$, and $V_\gamma = 0.6$ V. Calculate the output levels if one input is excited and if (*a*) $V_R = +12$ V, (*b*) $V_R = +10$ V, (*c*) $V_R = +14$ V, and (*d*) $V_R = 0$ V.

　　(*e*) Repeat part (*a*) if three inputs are excited. For which of the above values of V_R is the OR function satisfied (except possibly for a shift in level between input and output)?

5-3 For the circuit shown, the digital voltage levels are 2 and 4 V. Assume identical diodes, each with a conducting voltage of 0.7 V and negligible resistance. Verify the OR truth table of Fig. 5-2. For each line in the table indicate which diodes are ON and OFF and justify. Find the current in each diode and in each resistor. What is the purpose of D3?

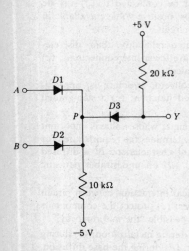

Prob. 5-3

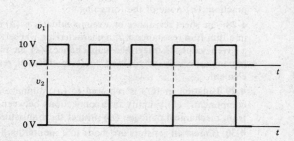

Prob. 5-4

5-4 Consider a two-input positive-logic diode OR gate (Fig. 5-3 with the diodes reversed) and with $V_R = 0$. The inputs are the square waves v_1 and v_2 indicated. Sketch the output waveform if the ratio of the amplitude of v_2 to v_1 is (*a*) 2 and (*b*) $\frac{1}{2}$. Assume ideal diodes ($R_f = 0$, $R_r = \infty$, and $V_\gamma = 0$) and $R_s = 0$.

5-5 Consider two signals, a 1-kHz sine wave and a 6-kHz square wave of zero average value, applied to the OR circuit of Fig. 5-3 with $V_R = 0$. Draw the output waveform if the sine-wave amplitude (*a*) exceeds the square-wave amplitude, (*b*) is less than the square-wave amplitude.

5-6 Consider a two-input positive-logic diode AND circuit (Fig. 5-5b) with $V_R = 10$ V, $R = 10$ kΩ, and $R_s = 0$. Assume ideal diodes and neglect capacitances. The input waveforms are v_1 and v_2 sketched in Prob. 5-4. Sketch the output waveform if the ratio of the amplitude of v_2 to v_1 is (a) 2, (b) 1, and (c) $\frac{1}{2}$. Repeat part (b) if $R_s = 1$ kΩ.

5-7 The binary input levels for the AND circuit shown are $V(0) = 0$ V and $V(1) = 25$ V. Assume ideal diodes. If $v_1 = V(0)$ and $v_2 = V(1)$, then v_o is to be at 2 V. However, if $v_1 = v_2 = V(1)$, then v_o is to rise above 10 V.

(a) What is the range of values of V_R which may be used?

(b) If $V_R = 15$ V, what is v_o at a coincidence $[v_1 = v_2 = V(1)]$? What are the diode currents?

(c) Repeat part (b) if $V_R = 40$ V.

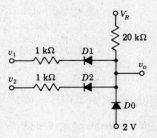

Prob. 5-7

5-8 The two-input-diode AND circuit shown uses diodes with $R_f = 500$ Ω, $R_r = \infty$, and $V_\gamma = 0$. The quiescent current in $D0$ is 6 mA, and $R' = 36.25$ kΩ.

(a) Calculate the quiescent output voltage v_o, the value of R, and the currents through $D1$ and $D2$.

(b) Calculate the output voltage when one input diode is cut off. Calculate this result approximately by assuming that the currents through R and the remaining input diode do not change. Also, calculate the result exactly.

(c) Assume that diode $D0$ is omitted, that the currents in $D1$ and $D2$ remain the same as in part (a), and that the output v_o is the same as that found in part (a). Find R and R'.

(d) If the conditions are as indicated in part (c) but one of the diodes is cut off, find the output voltage v_o. Compare with the result in part (b).

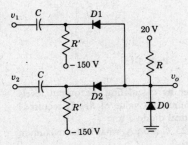

Prob. 5-8

5-9 Find v_o and v' if (a) there are no pulses at either A or B, (b) there is a 30-V positive pulse at A or B, and (c) there are positive pulses at both A and B.

(d) What is the minimum pulse amplitude which must be applied in order that the circuit operate properly? Assume ideal diodes.

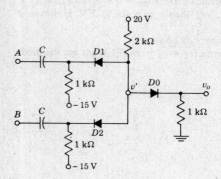

Prob. 5-9

5-10 The binary input levels for the circuit shown are $V(0) = 0$ V and $V(1) = 10$ V. Assume ideal diodes.

(a) Analyze the circuit and find a logical expression for Y in terms of A, B, and C.

(b) What is the minimum value of R_2 (in terms of R_1) for proper operation?

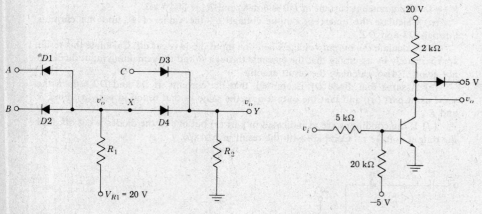

Prob. 5-10

Prob. 5-11

5-11 (a) Verify that the circuit shown is an inverter by calculating the output levels corresponding to input levels of 0 and 5 V. What minimum value of h_{FE} is required? Neglect junction saturation voltages and assume an ideal diode.

(b) If the reverse collector saturation current at 25°C is 5 μA, what is the maximum temperature at which this inverter will operate properly?

5-12 Consider the silicon inverter shown. The output Y feeds eight inverters, identical to the one shown, (i.e., fan-out is 8).

(a) If the input is $V_i = V(0) = 0.2$ V, calculate I, assuming that the eight driven transistors are in saturation.

(b) Find the base current of each transistor and find $h_{FE(min)}$ so that the output transistors are indeed in saturation.

(c) Calculate the output voltage v_o.

(d) If $h_{FE} = 50$, what is the maximum fan-out?

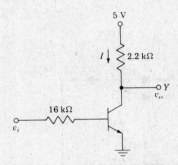

Prob. 5-12

5-13 The inverter of Fig. 5-7 has the following parameters: $R_1 = R$ kΩ, $R_2 = 50$ kΩ, $R_c = 2$ kΩ, $V_{CC} = 6$ V, and $V_R = -6$ V. The circuit is to operate properly in the temperature range of -50 to $+145°$C. At the lower temperature $h_{FE(min)} = 50$ and at the upper temperature $h_{FE(min)} = 150$. The logic level $V(1)$ may be anywhere in the range of 5 to 7 V and $V(0)$ in the range of 0 to 0.5 V. The value of I_{CBO} is 10 nA $= 10^{-5}$ mA at 25°C and doubles for every 10°C increase in temperature. Each junction voltage decreases 2.5 mV for every degree centigrade increase in temperature. Assume that at any temperature $V_{BE} \leqslant 0$ for the transistor to be OFF.

(a) Using the worst possible conditions at the lowest temperature, calculate the minimum and maximum values of R for proper inverter operation.

(b) Repeat part (a) at the highest temperature.

(c) What is the range of values of R for proper operation for $-50°C \leqslant T \leqslant 145°$C?

5-14 The inverter of Fig. 5-7 feeds one terminal of a three-input AND gate of Fig. 5-5b with $V_R = 12$ V and $R = 15$ kΩ. Verify that this circuit satisfies the truth table of Fig. 5-8b for the inhibitor operation.

5-15 A half adder is a combination of OR, NOT, and AND gates. It has two input and two outputs and the following truth table:

Input 1	Input 2	Output 1	Output 2
0	0	0	0
0	1	1	0
1	0	1	0
1	1	0	1

Draw the logic block diagram for a half adder.

5-16 The four inputs v_1, v_2, v_3, and v_4 are voltages from zero-impedance sources whose values are either $V(0) = 10$ V or $V(1) = 20$ V. The diodes are ideal. $V_R = 25$ V, $R_1 = 5$ kΩ, and $R_2 = 10$ kΩ.

    (a) If $v_1 = v_2 = 10$ V and $v_3 = v_4 = 20$ V, find v_o and the currents in each diode.

    (b) If $v_1 = v_3 = 10$ V and $v_2 = v_4 = 20$ V, find v_o and the currents in each diode.

    (c) Sketch in block-diagram form the logic performed by this circuit.

    (d) Verify that in order for the circuit to operate properly the following inequality must be satisfied:

$$R_2 > \frac{V_R - V(0)}{V(0)} R_1$$

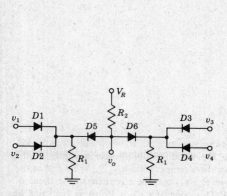

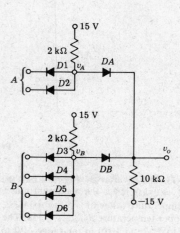

Prob. 5-16

Prob. 5-17

5-17 (a) In block-diagram form indicate the logic performed by the diode system shown. The input levels are $V(0) = -5$ V and $V(1) = 0$ V. Neglect source resistance and assume that the diodes are ideal. Justify your answer by calculating the voltages v_A, v_B, and v_o (and indicating which diodes are conducting) under the following circumstances: (i) all inputs are at $V(0)$; (ii) some but not all inputs in A are at $V(1)$ and all inputs in B are at $V(0)$; (iii) all inputs in A are at $V(1)$ and some inputs in B are at $V(1)$; and (iv) all inputs are at $V(1)$.

    (b) If the 2-kΩ resistance were increased, at what maximum value would the circuit no longer operate in the manner described above?

5-18 (a) Verify De Morgan's law [Eq. (5-26)] in a manner analogous to that given in the text in connection with the proof of Eq. (5-25).

    (b) Prove Eq. (5-26) by constructing a truth table for each side and verifying that these two tables have the same outputs.

5-19 Verify the auxiliary Boolean identities in Table 5-2.

5-20 Using Boolean algebra, verify:

    (a) $\overline{(\overline{A} + B)} + \overline{(A + \overline{B})} + \overline{(\overline{A}B)(\overline{AB})} = 1$.

    (b) $AB + AC + B\overline{C} = AC + B\overline{C}$

Hint: Multiply the first term on the left-hand side by $C + \overline{C} = 1$.

    (c) $AB + A\overline{B} + \overline{A}B = A + B$.

5-21 Using Boolean algebra, verify:

    (a) $(A + B)(B + C)(C + A) = AB + BC + CA$.

    (b) $(A + B)(\bar{A} + C) = AC + \bar{A}B$.

    (c) $AB + \bar{B}C + AC = AB + \bar{B}C$.

Hint: A term may be multiplied by $B + \bar{B} = 1$.

5-22 Given two N-bit characters which are available in parallel form. Indicate in block-diagram form a system whose output is 1 if and only if *all* corresponding bits are equal, that is, only if the two characters are equal.

5-23 A, B, and C represent the presence of pulses. The logic statement "A or B and C" can have two interpretations. Which are they? In block-diagram form draw the circuit to perform each of the two logic operations. How can this ambiguity be avoided?

5-24 In block-diagram form draw a circuit to perform the following logic: If pulses A_1, A_2, and A_3 occur simultaneously or if pulses B_1 and B_2 occur simultaneously, an output pulse is delivered, provided that pulse C does not occur at the same time. No output is to be obtained if A_1, A_2, A_3, B_1, and B_2 occur simultaneously.

5-25 In block-diagram form draw a circuit which satisfies simultaneously the conditions a, b, and c as follows:

    (a) The output is excited if any pair of inputs A_1, A_2, and A_3 is excited, provided that B is also excited.

    (b) The output is 1 if any one (and only one) of the inputs A_1, A_2, or A_3 is 1, provided that $B = 0$.

    (c) No output is excited if A_1, A_2, and A_3 are simultaneously excited.

5-26 For the illustrative NAND gate of Fig. 5-14:

    (a) What is the maximum noise voltage (superimposed upon the logic level) which will still permit the circuit to operate properly? Consider the following two cases: (i) a complete coincidence and (ii) all inputs but one in the 1 state.

    (b) What is the maximum value of the source resistance which will still permit proper circuit operation? Assume a 0.7-V drop across a conducting diode.

5-27 The circuit shown uses silicon diodes and a silicon transistor. The input A or B is obtained from the output Y of a similar gate.

    (a) What are the logic levels? Take junction voltages into account.

    (b) Find $h_{FE(\text{min})}$ in order for the circuit to satisfy the NAND operation.

    (c) What is the maximum allowable value of I_{CBO}?

    (d) Now neglect junction voltages and I_{CBO} and verify that the circuit satisfies the NOR operation.

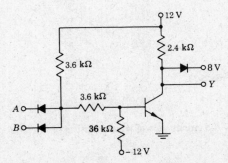

Prob. 5-27

5-28 Use only NAND gates and INVERTERS, and with the two inputs A and B generate the following functions: AB, $\overline{AB}$, $A\overline{B}$, $\overline{A}B$, $\overline{A}B + A\overline{B}$, $A + B$. What is the minimum number of NAND and NOT gates required?

5-29 Verify that the NOR-NOR topology is equivalent to an OR-AND system.

5-30 Verify that the circuit shown, consisting of NOR gates only, is an EXCLUSIVE OR.

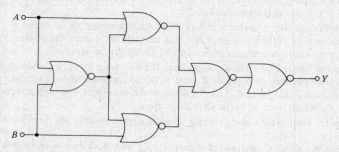

Prob. 5-30

5-31 Verify that the logic operations OR, AND, and NOT may be implemented by using only (a) NOR gates, (b) NAND gates.

5-32 Consider a three-input NAND gate which is used in a system with only two input variables A and B. Verify that the gate operates properly if (a) one of the terminals is tied to $V_{CC} = V(1)$ and (b) two of the input terminals are tied together and treated as a single terminal. What is the advantage of (a) over (b)?

(c) Consider the analogous problem of converting a three-input to a two-input NOR gate.

5-33 Verify that for the block diagram shown,

$$Y = AB + CD + \overline{E}$$

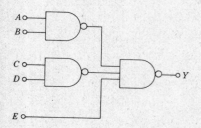

Prob. 5-33

5-34 What logic operation is performed by the circuit shown, which consists of interconnected NAND gates?

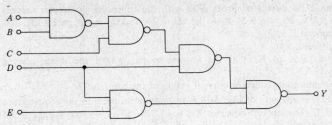

Prob. 5-34

5-35 (*a*) Implement the EXCLUSIVE-OR gate using NAND gates.
(*b*) Repeat part (*a*) for the half adder of Prob. 5-15.

5-36 (*a*) The discrete-components circuit of a DTL gate shown uses a silicon transistor with worst-case values of $V_{BE(\text{sat})} = 1.0$ V and $V_{CE(\text{sat})} = 0.5$ V. The voltage across any silicon diode (when conducting) is 0.7 V. Assume that $D1$ consists of two diodes in series.

The inputs to this switch are obtained from the outputs of similar gates. Verify that the circuit functions as a positive NAND. In particular, for proper operation, calculate the minimum value of the clamping voltage V' and h_{FE}.

(*b*) Will the circuit operate properly if $D1$ is (i) a single diode or (ii) three diodes in series?

(*c*) What is the maximum allowed fan-in, assuming that the diodes are ideal? What is a practical limitation on fan-in?

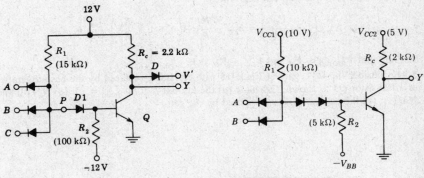

Prob. 5-36　　　　　　　　　　　　　　　　**Prob. 5-37**

5-37 The DTL gate shown uses silicon devices with $V_{BE(\text{sat})} = 0.8$ V, $V_{CE(\text{sat})} = 0.2$ V, $V_{\gamma} = 0.5$ V, and the drop across a conducting diode $= 0.7$ V. The inputs to this switch are obtained from the outputs of similar gates $V_{BB} = 0$ V.

(*a*) Verify that the circuit functions as a positive NAND and calculate $h_{FE(\text{min})}$. Assume that the transistor is essentially cut off if the base-to-emitter voltage is at least 0.1 V smaller than the cutin voltage V_{γ}.

(*b*) Assume that the diode reverse saturation current is equal to the transistor reverse saturation collector current. Find $I_{CBO(\text{max})}$.

(*c*) Calculate $NM(0)$ and $NM(1)$.

5-38 (*a*) Analyze the DTL circuit of Prob. 5-37 with the following parameter values: $V_{BB} = 3$ V, $V_{CC1} = 5.2$ V, $V_{CC2} = 3$ V, $R_1 = 3.3$ kΩ, $R_2 = 15$ kΩ, and $R_c = 1.1$ kΩ.

(*b*) Find the fan-out if $h_{FE(min)} = 25$.

(*c*) Find $NM(0)$ and $NM(1)$.

(*d*) What is the value of $V(1)$ at the fan-out in part (*b*) if the diode reverse saturation current is $1\mu A$?

(*e*) What is the minimum collector-current rating of the transistor?

5-39 The positive DTL NAND gate of Fig. 5-18 is to operate properly in the temperature range of -50 to $160°C$. The silicon transistor has $h_{FE(min)} = 50$ at $-50°C$, $h_{FE(min)} = 65$ at $25°C$, and $h_{FE(min)} = 100$ at $160°C$. The reverse saturation collector current of the transistor at $25°C$ is $I_{CBO} = 0.5$ nA, and is equal to the reverse saturation current of the silicon diode. The maximum current rating of the transistor is 50 mA. The gate will be used in a system with power-supply voltage of 10 V, and the allowed variation in $V(1)$ is ± 0.5 V. The desired absolute value of the noise margin is 1.5 V and the desired fan-out is 10. The transistor is considered OFF if $V_{BE} \leqslant 0.4$ V. [Take Eqs. (2-5) and (2-6) into account.]

(*a*) Calculate the minimum and also the maximum required number of diodes between P and the base of the transistor.

(*b*) Calculate the maximum value of R_2.

(*c*) If $R_c = R_2 = 5$ kΩ, find the minimum and also the maximum values of R_1 for proper operation.

5-40 For the integrated positive DTL NAND gate of Fig. 5-18, prove that:

(*a*) The maximum number of diodes that can be used in series between P and B is given by $n_{max} = (V_{CC} - V_{BE(sat)})/V_D$, where V_D is the voltage drop across a diode.

(*b*) The maximum fan-out is given by

$$N_{max}\left(1 - \frac{V_D}{V_{CC}}\right) = h_{FE} - \frac{R_1}{R_c} - \left(n + 1 + \frac{R_1}{R_2}\right)\frac{V_D}{V_{CC}}h_{FE}$$

Assume that $V_{BE(sat)} \approx V_D$ and $V_{CC} - V_D \gg V_{CE(sat)}$.

5-41 Consider the DTL gate of Fig. 5-18 with $D2$ being replaced by the base-to-emitter of a transistor $Q1$ as shown. Calculate (*a*) the fan-out N if $h_{FE(min)} = 25$; (*b*) $NM(0)$ and $NM(1)$; (*c*) the average power dissipated by the gate.

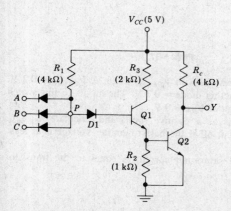

Prob. 5-41

5-42 In the DTL gate of Fig. 5-19, the parameters are: $R_1 = 1.6$ kΩ, $R_2 = 2.15$ kΩ, $R_3 = 5$ kΩ, $R_c = 6$ kΩ, and $V_{CC} = 5$ V.

(a) Calculate $h_{FE(min)}$ for a fan-out of $N = 20$.

(b) Calculate $NM(0)$ and $NM(1)$.

(c) Calculate the power dissipation of the gate in the two possible states.

(d) Calculate $V(1)$ at 175°C, assuming a diode reverse saturation current of 1 nA at 25°C.

5-43 For the integrated positive DTL gate shown, the inputs are obtained from the outputs of similar gates, and its output drives similar gates.

(a) Verify its function as a NAND gate and specify the state of each transistor when at least one input is low and also at a coincidence.

(b) For $h_{FE(min)} = 30$, calculate the fan-out of this gate.

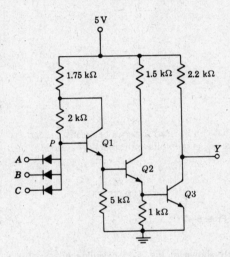

Prob. 5-43

5-44 In the integrated positive DTL NAND gate of Fig. 5-19 a Schottky diode is fabricated between base and collector to prevent the transistor $Q2$ from saturating. The anode of the Schottky diode is at the base and the drop across the diode when conducting is 0.4 V.

(a) Explain why the transistor $Q2$ does not go into saturation.

(b) Verify the operation of the gate as a NAND gate and calculate noise margins.

(c) Find the Schottky diode current for zero fan-out.

(d) Find the logic levels and maximum fan-out if the inputs of this gate are obtained from similar gates and the output drives similar gates and $h_{FE} = 30$.

5-45 Verify that this wired circuit performs the EXCLUSIVE OR function.

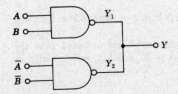

Prob. 5-45

5-46 (*a*) For the high-threshold logic NAND gate of Fig. 5-20, if V_2 of the diode is 6.9 V, verify that this circuit functions as a positive NAND and calculate $h_{FE(min)}$. The inputs of this gate are obtained from the output of similar gates.

(*b*) Calculate noise margins.

(*c*) Calculate the fan-out of this gate if $h_{FE(min)} = 40$.

(*d*) Calculate the average power dissipation of the gate for a fan-out of 10.

5-47 If the output in Fig. 5-20 is capacitively loaded (by C), then the rise time as Y goes from its low to its high state will be long because of the high load resistance (15 kΩ) of $Q2$. To reduce this time constant the active pull-up circuit indicated in the dashed block is added across the 15-kΩ resistor.

(*a*) Explain how the circuit works.

(*b*) Why not simply replace 15 kΩ by 1.5 kΩ?

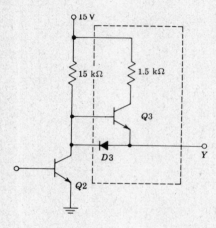

Prob. 5-47

5-48 (*a*) For the IC positive TTL NAND gate shown in Fig. 5-21 calculate $h_{FE(min)}$ for proper operation. For $Q1$, $h_{FEI} = 0.5$. Assume that $Q2$ and $Q3$ both saturate if all inputs are high.

(*b*) Repeat part (*a*) assuming that $Q2$ remains in the active region but $Q3$ saturates with all inputs at $V(1)$.

(*c*) Calculate $NM(0)$ and $NM(1)$.

(*d*) Calculate the fan-out if $h_{FE(min)} = 20$. Is $Q2$ in the active region or in saturation?

(*e*) Calculate the average power dissipation of the gate.

5-49 The output of the TTL gate with the totem-pole driver of Fig. 5-23 is accidentally short-circuited to ground. Calculate the short-circuit current if (*a*) all inputs are high and (*b*) at least one input is low. Assume $h_{FE} = 20$.

5-50 Solve Prob. 5-48 for the TTL gate with the totem-pole output stage of Fig. 5-23 with $h_{FEI} = 1$, $R_{b1} = 5$ kΩ, and $R_{c2} = 2$ kΩ.

5-51 For the IC positive NAND TTL gate shown, if the inputs are obtained from the outputs of similar gates and $h_{FE(min)}$ of the transistors is 30 and $h_{FEI} = 0.5$, verify its operation as a NAND gate when the fan-out is 10.

(a) At coincidence, find the state of each transistor and all currents and voltages of the circuit. Assume $Q5$ operates in the active region and justify this assumption.

(b) Repeat part (a) if at least one input is low. Assume $Q5$ operates in saturation and justify this assumption.

(c) Find the logic levels.

(d) Calculate the peak current drawn from the supply during the transient.

(e) Calculate maximum fan-out for proper operation of the gate.

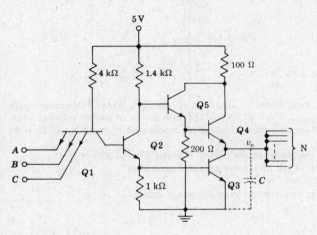

Prob. 5-51

5-52 For an RTL IC positive NOR gate prove that the maximum fan-out can be approximated by the formula

$$N_{max} = h_{FE(min)} - h_{FE(min)} \frac{0.6}{V_{CC}} - \frac{R_b}{R_c}$$

5-53 The inputs of the RTL IC positive NOR gate shown in Fig. 5-25 are obtained from the outputs of similar gates and the outputs drive similar gates. If the supply voltage of the system is 5 V and the temperature range for proper operation of the gate is -50 to $150°C$, calculate the maximum and minimum permissible values of the resistances R_b and R_c, respectively. Assume $h_{FE} = 30$ at $-50°C$, $I_{CBO} = 10$ nA at $25°C$, and the desired fan-out is 10.

5-54 A low-power RTL gate uses 1.5-kΩ base resistors, 3.6-kΩ collector resistors, and a 5-V supply. For a fan-out of 3, calculate (a) the output voltage levels V_o; (b) $NM(0)$ and $NM(1)$; (c) $h_{FE(min)}$; (d) the power dissipation of the gate when all the input transistors are OFF, and also when all input transistors are ON.

5-55 (a) Verify that the DCTL circuit shown with the fan-in transistors in series satisfies the NAND operation. Assume that for the silicon transistors, $V_{CE(sat)} = 0.2$ V and $V_{BE(sat)} = 0.8$ V. Calculate the collector currents in each transistor when all inputs are high. The input to each base is taken from the output of a similar gate. Also calculate $h_{FE(min)}$.

(b) If $v_1 = 0.6$ V, $v_2 = v_3 = 1.5$ V, find the base and emitter currents and the collector-to-emitter voltages of each transistor. Refer to Fig. 3-12.

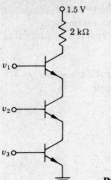

Prob. 5-55

5-56 (*a*) Calculate the logic levels at output Y of the ECL Texas Instruments gate shown. Assume that $V_{BE(active)} = 0.7$ V. To find the drop across an emitter follower when it behaves as a diode assume a piecewise-linear diode model with $V_\gamma = 0.6$ V and $R_f = 20$ Ω.

    (*b*) Find the noise margin when the output Y is at $V(0)$ and also at $V(1)$.

    (*c*) Verify that none of the transistors goes into saturation.

    (*d*) Calculate R so that $Y' = \bar{Y}$.

    (*e*) Find the average power taken from the power source.

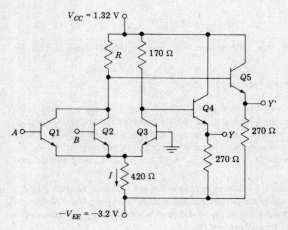

Prob. 5-56

5-57 Verify that, if the outputs of two (or more) ECL gates are tied together as in Fig. 5-30, the OR function is satisfied.

5-58 (*a*) For the system in Fig. 5-30 obtain an expression for Y which contains three terms.

    (*b*) If in Fig. 5-30 $\bar{Y}_1$ and $\bar{Y}_2$ are tied together, verify that the output is $Y = \bar{A}\bar{B} + \overline{CD}$.

Chapter 6

6-1 (*a*) How many leads does a two-wide, four-input AOI package have?

   (*b*) Repeat for a four-wide, 4-2-3-2-input AOI.

6-2 (*a*) Indicate how to implement S_n of Eq. (6-1) with AND, OR, and NOT gates.

   (*b*) Verify that the sum S_n in Eq. (6-1) for a full adder can be put in the form

$$S_n = A_n \oplus B_n \oplus C_{n-1}$$

6-3 (*a*) For convenience, let $A_n = A$, $B_n = B$, $C_{n-1} = C$, and $C_n = C'$. Using Eq. (6-4) for C', verify Eq. (6-5) with the aid of the Boolean identities in Table 5-2; in other words, prove that

$$\overline{C}' = \overline{B}\,\overline{C} + \overline{C}\,\overline{A} + \overline{A}\,\overline{B}$$

   (*b*) Evaluate $D \equiv (A + B + C)\overline{C}'$ and prove that S_n in Eq. (6-1) is given by

$$S_n = D + ABC$$

6-4 Consider a digital system for majority logic. There are three inputs A, B, and C. The output Y is to equal 1 if two or three inputs are 1.

   (*a*) Write the truth table.

   (*b*) From the truth table obtain the Boolean expression for Y.

   (*c*) Minimize Y and show the logic block diagram.

6-5 The time to add two numbers in parallel is limited by the time it takes to propagate the carry through the word. At the expense of more logic this carry propagation time can be avoided by generating a carry-look-ahead signal. Prove that if two 4-bit words ($A_3A_2A_1A_0$ and $B_3B_2B_1B_0$ with A_3 as the MSB) are added, then the carry-out C_3 is given by

$$\overline{C}_3 = \overline{C}_{-1}(\overline{B_0A_0})(\overline{B_1A_1})(\overline{B_2A_2})(\overline{B_3A_3}) + (\overline{A_0 + B_0})(\overline{B_1A_1})(\overline{B_2A_2})(\overline{B_3A_3})$$
$$+ (\overline{A_1 + B_1})(\overline{B_2A_2})(\overline{B_3A_3}) + (\overline{A_2 + B_2})(\overline{B_3A_3}) + (\overline{A_3 + B_3})$$

where C_{-1} is the input carry (Fig. 6-4). Note that the carry out is given as a function of only input variables and does not involve intermediate carries. *Hint:* Apply Eq. (6-5) recursively four times (for $n = 0$, 1, 2, and 3). Start with Eq. (6-5) in the form $\overline{C}_n = \overline{C}_{n-1}(\overline{B_nA_n}) + (\overline{A_n + B_n})$.

6-6 The system shown is called a true/complement-zero/one element. Verify the truth table.

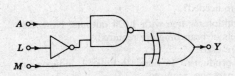

Control inputs		Output
L	M	Y
0	0	$\overline{A}$
0	1	A
1	0	1
1	1	0

Prob. 6-6

6-7 (*a*) Verify that an EXCLUSIVE-OR gate is a true/complement unit.

   (*b*) One input is A, the other (control) input is C, and the output is Y. Is $Y = A$ for $C = 1$ or $C = 0$?

6-8 (*a*) Make a truth table for a binary half subtractor A minus B (corresponding to the half adder of Fig. 6-3). Instead of a carry C, introduce a *borrow* P.

(b) Verify that the digit D is satisfied by an EXCLUSIVE-OR gate and that P follows the logic "B but not A."

6-9 Consider an 8-bit comparator. Justify the connections $C' = C_L$, $D' = D_L$, and $E' = E_L$ for the chip handling the more significant bits. *Hint:* Add 4 to each subscript in Fig. 6-12. Extend Eq. (6-12) for E and Eq. (6-13) for C to take all 8 bits into account.

6-10 Consider a comparator which has as input two n-bit words, and has as outputs E, C, and D as in Fig. 6-12, but input leads E', C', and D' are not available. What additional logic is required to compare two $2n$-bit numbers using two n-bit comparators?

6-11 Consider two 5-bit words $S_A A_3 A_2 A_1 A_0$ and $S_B B_3 B_2 B_1 B_0$, where S_A and S_B are the sign bits, while the remaining bits indicate the magnitude of the word. S_A $(S_B) = 0$ means the corresponding word is positive and S_A $(S_B) = 1$ means the corresponding word is negative. Design a system to compare the two words, using a 4-bit comparator to compare the magnitudes and a 1-bit comparator to compare the sign bits.

6-12 (a) By means of a truth table verify the Boolean identity

$$Y = (A \oplus B) \oplus C = A \oplus (B \oplus C)$$

(b) Verify that $Y = 1$ (0) if an odd (even) number of variables equals 1. This result is *not* limited to three inputs, but is true for any number of inputs. It is used in Sec. 6-5 to construct a parity checker.

6-13 Construct the truth table for the EXCLUSIVE-OR tree of Fig. 6-13 for all possible inputs A, B, C, and D. Include $A \oplus B$ and $C \oplus D$ as well as the output Z. Verify that $Z = 1$ (0) for odd (even) parity.

6-14 (a) Draw the logic circuit diagram for an 8-bit parity check/generator system.

(b) Verify that the output is 0 (1) for odd (even) parity.

6-15 (a) Indicate an 8-bit parity checker as a block having 8 input bits (collectively designated A_1), an output P_1, and an input control P_1'. Consider a second 8-bit unit with inputs A_2, output P_2, and control P_2'. Show how to cascade the two packages in order to check for odd parity of a 16-bit word. Verify that the system operates properly if $P_1' = 1$. Consider the four possible parity combinations of A_1 and A_2.

(b) Show how to cascade three units to obtain the parity of a 24-bit word. Should $P_1' = 0$ or 1 for odd parity?

(c) Show how to cascade units to obtain the parity of a 10-bit word.

6-16 (a) Draw a 4-to-10-line decoder. (In other words, complete Fig. 6-16.)

(b) Show how to convert this to a 3-to-8-line decoder.

6-17 (a) Draw a block diagram of a demultiplexer tree with 32 outputs using $N_1 = 8$ and $N_2 = 4$. Explain the operation with reference to line 25.

(b) How many equivalent packages are needed?

6-18 (a) Draw a block diagram of a demultiplexer tree with 1,024 outputs. Note that $1,024 = 16$ times 8 times 8. Hence, two levels of branching are required.

(b) How many equivalent packages are used?

(c) If 1,024 were broken up into the product of 16 times 16 times 4, indicate the system and state the number of equivalent packages required.

6-19 (a) How many NAND-gate inputs does a 1-to-16 demultiplexer have?

(b) How many gate inputs does a 1-to-16-tree demultiplexer (using only 1-to-4 demultiplexers) have?

6-20 (a) Draw a logic diagram for a 6-to-1-line multiplexer.

(b) How is the network in part (a) augmented to become an 8-to-1-line multiplexer?

6-21 Design a system to convert two 1-out-of-16 data-selector chips to form a 1-out-of-32 data selector. Explain the operation of this sytem. *Hint:* The enable input S_2 to the higher-order chip is the complement of S_1 to the lower-order chip. Also, the outputs Y_1 and Y_2 from the two chips are the inputs to an OR gate whose output Y is the output of the system.

6-22 (*a*) Draw a block diagram of a 32-to-1-line selector as in Fig. 6-21, but with $N_2 = 4$ and $N_1 = 8$. Explain the operation with respect to the input X_{25}.

(*b*) How many equivalent packages are needed?

6-23 Repeat Prob. 6-22 for a 64-to-1 multiplexer, using identical chips.

6-24 (*a*) Draw a block diagram of a multiplexer using 512 inputs, noting the fact that 512 = 16 times 8 times 4.

(*b*) How many packages are needed?

(*c*) Repeat part (*b*) noting that 512 = 16 times 16 times 2.

6-25 (*a*) Generate Eq. (6-1) for the sum S_n of a full adder using a multiplexer. Find the X's in terms of C, $\bar{C}$, 0, and 1. *Note:* for simplicity drop the subscripts on A, B, and C and let $Y \equiv S_n$.

(*b*) Generate Eq. (6-2) for the carry C_n using a multiplexer. *Note:* Let $C_{n-1} \equiv C$ and $C_n \equiv Y$.

(*c*) Can the same multiplexer be used for S_n and C_n? Explain.

6-26 Use a multiplexer to generate the combinational-logic equation

$$Y = \bar{D}\bar{C}\bar{B}\bar{A} + D\bar{C}\bar{B}A + D\bar{C}B\bar{A} + \bar{D}\bar{C}B\bar{A} + DC\bar{B}\bar{A} + \bar{D}C\bar{B}A + DCB\bar{A} + \bar{D}CB\bar{A}$$

How many data inputs are needed? Find the values of the data inputs X.

6-27 Consider a digital system for majority logic. There are four inputs A, B, C, and D. The output Y is equal to 1, if three or four inputs are 1.

(*a*) Write the Boolean expression for Y.

(*b*) Use a multiplexer/selector to satisfy this majority logic. What are the values of the data inputs X?

6-28 Design an encoder satisfying the following truth table, using a diode matrix.

Inputs				Outputs			
W_3	W_2	W_1	W_0	Y_3	Y_2	Y_1	Y_0
0	0	0	1	0	1	1	1
0	0	1	0	1	1	0	0
0	1	0	0	1	1	0	1
1	0	0	0	0	0	1	0

6-29 (*a*) Design an encoder, using multiple-emitter transistors, to satisfy the following truth table.

(*b*) How many transistors are needed and how many emitters are there is each transistor?

Inputs			Outputs				
W_2	W_1	W_0	Y_4	Y_3	Y_2	Y_1	Y_0
0	0	1	1	0	1	1	0
0	1	0	1	1	1	0	0
1	0	0	0	1	0	1	1

6-30 A block diagram of a three-input (A, B, and C) and eight-output (Y_0 to Y_7) decoder matrix is indicated. The bit Y_5 is to be 1 (5 V) if the input code is 101 corresponding to decimal 5.

(a) Indicate how diodes are to be connected to line Y_5.

(b) Repeat for Y_2, Y_3, and Y_4.

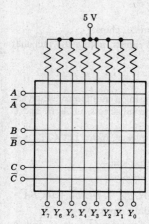

Prob. 6-30

6-31 For the priority encoder in Table 6-3 verify that:

(a) $Y_3 = W_9 + W_8$.

(b) $Y_2 = (\overline{W_9} + \overline{W_8})(W_7 + W_6 + W_5 + W_4)$.

6-32 For the 10-line-decimal–to–4-line BCD priority encoder verify that

$$Y_0 = W_9 + \overline{W_8}\left(W_7 + \overline{W_6}W_5 + \overline{W_6}\,\overline{W_4}W_3 + \overline{W_6}\,\overline{W_4}\,\overline{W_2}W_1\right)$$

6-33 (a) Fill in the truth table for an 8-data-line–to–3-line binary (octal) priority encoder, using X to indicate a don't care state.

(b) Obtain the expression for Y_0.

6-34 Repeat Prob. 6-33 for Y_1.

6-35 (a) Implement the code conversion indicated below using a Read-Only Memory (ROM). Indicate *all* connections between the X inputs and the Y outputs. Use the standard symbols for INVERTERS, for AND gates, and for OR gates.

Inputs		Outputs			
X_1	X_0	Y_3	Y_2	Y_1	Y_0
0	0	1	0	1	1
0	1	0	1	0	1
1	0	0	1	1	1
1	1	1	1	0	0

(b) Draw the OR gates as multiple-emitter transistors.

6-36 (a) Draw a block diagram of a 512 × 4-bit ROM using two-dimensional addressing with a 5-to-32-line decoder.

(b) How many NAND gates are required?

(c) How many transistors must be used in the memory matrix and how many emitters must each transistor have?

6-37 Consider a 256 × 8-bit ROM using two-dimensional addressing with 8-to-1 selectors.

(a) How many bits are needed to address the ROM?

(b) How many bits are needed for the X-address?

(c) How many NAND gates are required?

(d) Specify the number of transistors in the memory matrix and the number of emitters in each transistor.

6-38 (a) Write the expressions for Y_0 and Y_2 in the binary-to-Gray-code converter.

(b) Indicate how to implement the relationship for Y_0 with diodes.

6-39 (a) Give the relationships between the output and input bits for the Gray-to-binary-code translator for Y_3 and Y_2.

(b) Indicate how to implement the equation for Y_3 with transistors.

6-40 (a) Write the sum-of-products canonical form for Y_5 of Table 6-5 for the seven-segment indicator code.

(b) Verify that this expression can be minimized to $Y_5 = \overline{DCA} + \overline{CBA} + BA$.

6-41 Minimize the number of terms in Eq. (6-32) to obtain Eq. (6-33).

Chapter 7

7-1 (a) Verify that it is not possible for both outputs in Fig. 7-1 to be in the same state.

(b) Verify that $B_1 = B_2 = 0$ in Fig. 7-1b is not allowed.

7-2 Consider the chatterless switch of Fig. 7-2. At time $t_1' > t_6$ the key is depressed so that the pole is moved from 1 to 2. It reaches 2 at time t_2' and then bounces three times. Indicate the waveforms B_2, B_1, and Q and explain your reasoning.

7-3 (a) Verify that the AOI topology shown gives the same logic as the latch of Fig. 7-3.

(b) Transform the block diagram so that it becomes equivalent to that of Fig. 7-3.

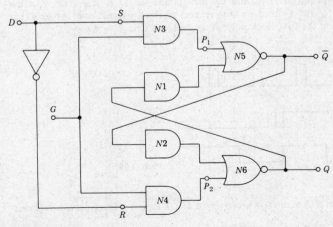

Prob. 7-3

7-4 Show how to build the bistable latch of Fig. 7-3 by using the AOI configuration.

7-5 The excitation table for a J-K FLIP-FLOP is shown. An X in the table is to be interpreted to mean that it does not matter whether this entry is a 1 or a 0. It is referred to as a "don't care" condition. Thus the second row indicates that if the output is to change from 0 to 1, the J input must be 1, whereas K can be either 1 or 0. Verify this excitation table by referring to the truth table of Fig. 7-6.

Q_n	Q_{n+1}	J_n	K_n
0	0	0	X
0	1	1	X
1	0	X	1
1	1	X	0

7-6 Verify that the J-K FLIP-FLOP truth table is satisfied by the difference equation $Q_{n+1} = J_n \bar{Q}_n + \bar{K}_n Q_n$.

7-7 (a) Show that the J-K FLIP-FLOP of Fig. 7-7 will preset correctly ($Pr = 0$, $Cr = 1$) only if $\bar{K} + \overline{Ck} = 1$.

    (b) Show that the J-K FLIP-FLOP will clear correctly ($Pr = 1$, $Cr = 0$) only if $\bar{J} + \overline{Ck} = 1$.

    (c) Verify that $Cr = Pr = Ck = 0$ leads to an indeterminate state.

    (d) Show that $Pr = 1$, $Cr = 1$ will enable the FLIP-FLOP.

7-8 (a) Verify that there is no race-around difficulty in the J-K circuit of Fig. 7-7 for any data input combination except $J = K = 1$.

    (b) Explain why the race-around condition does not exist (even for $J = K = 1$) provided that $t_p < \Delta t < T$.

7-9 (a) For the master-slave J-K FLIP-FLOP of Fig. 7-8 assume $Q = 0$, $\bar{Q} = 1$, $Ck = 1$, $J = 0$, and K arbitrary. What is Q_M?

    (b) If J changes to 1, what is Q_M?

    (c) If J returns to 0, what is Q_M? Note that Q_M does not return to its initial value. Hence J (and K) must not vary during the pulse.

7-10 The indicated waveforms J, K, and Ck are applied to a J-K FLIP-FLOP. Plot the output waveform for Q and $\bar{Q}$ lined up with respect to clock pulses. *Note:* Assume that the output $Q = 0$ when the first clock pulse is applied and that $Pr = Cr = 1$.

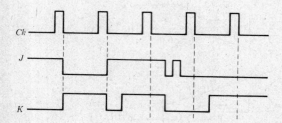

Prob. 7-10

7-11 (a) Verify that an S-R FLIP-FLOP is converted to a T type if S is connected to $\bar{Q}$ and R to Q.

    (b) Verify that a D-type FLIP-FLOP becomes a T type if D is tied to $\bar{Q}$.

7-12 The truth-table for an A-B FLIP-FLOP is as shown. Show how to build this FLIP-FLOP using a J-K FLIP-FLOP and any additional logic required.

A_n	B_n	Q_{n+1}
0	0	$\overline{Q}_n$
1	0	Q_n
0	1	1
1	1	0

7-13 A 4-bit cascadable priority register (TI 278) consisting of D-type latches is shown.

(a) Set $P_0 = 0$, $D_0 = D_1 = D_3 = 0$, and $D_2 = 1$. Verify that $Y_2 = 1$ and all other outputs are 0.

(b) Set $P_0 = 0$, $D_0 = D_1 = 0$, and $D_2 = D_3 = 1$. Verify that only $Y_2 = 1$.

(c) Generalize the above results to show that the lowest-order D_n among those in the high (1) state is transferred to make the corresponding Y_n high.

(d) Cascade two such 4-bit packages. Put $P_0 = 0$ for the lower-order chip. For the higher-order chip tie P_0 to the complement of the P_1 output of the lower-order package. Demonstrate that this cascaded system functions as an 8-bit priority register.

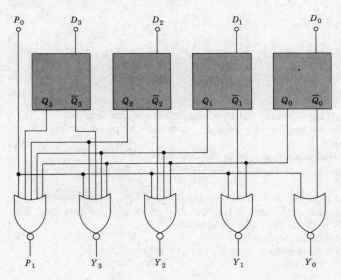

Prob. 7-13

7-14 For the bidirectional shift register of Fig. 7-12 verify the mode of operation indicated in Table 7-4 for (a) the second row, (b) the third row, and (c) the fourth row.

7-15 Augment the shift register of Fig. 7-11 with a four-input NOR gate whose output is connected to the *serial input* terminal. The NOR-gate inputs are Q_4, Q_3, Q_2, and Q_1.

(a) Verify that regardless of the initial state of each FLIP-FLOP, when power is applied, the register will assume correct operation as a ring counter after P clock pulses, where $P \leqslant 4$.

(b) If initially $Q_4 = 0$, $Q_3 = 1$, $Q_2 = 1$, $Q_1 = 0$, and $Q_0 = 1$, sketch the waveform at Q_0 for the first 16 pulses.

(c) Repeat part (b) if $Q_4 = 1$, $Q_3 = 1$, $Q_2 = 0$, $Q_1 = 1$, and $Q_0 = 0$.

7-16 (a) Draw a waveform chart for the twisted-ring counter; i.e., indicate the waveforms Q_4, Q_3, Q_2, Q_1, and Q_0 for, say, 12 pulses. Assume that initially $Q_0 = Q_1 = Q_2 = Q_3 = Q_4 = 0$.

(b) Write the truth table after each pulse.

(c) By inspection of the table show that two-input AND gates can be used for decoding. For example, pulse 1 is decoded by $Q_4\bar{Q}_3$. Why?

7-17 (a) For the modified ring counter shown, assume that initially $Q_0 = 0$, $Q_1 = 0$, and $Q_2 = 1$. Make a table of the readings Q_0, Q_1, Q_2, J_2, and K_2 after each clock pulse. How many pulses are required before the system begins to operate as a divide-by-N counter? What is N?

(b) Repeat part (a) if initially $Q_0 = 0$, $Q_1 = 1$, and $Q_2 = 0$.

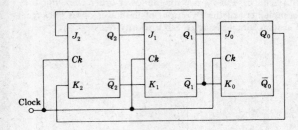

Clock

Prob. 7-17

7-18 A 25 : 1 ripple counter is desired. (a) How many FLIP-FLOPS are required?

(b) If 4-bit FLIP-FLOPS are available on a chip, how many chips are needed? How are these interconnected?

(c) Indicate the feedback connections to the clear terminals.

7-19 (a) Indicate a divide-by-20 ripple-counter block diagram. Include a latch in the clear input.

(b) What are the inputs to the feedback NAND gate for a 125 : 1 ripple counter?

7-20 Consider the operation of the latch in Fig. 7-16. Make a table of the quantities Ck, Q_1, Q_3, P_1, $\overline{Ck}$, and $P_2 = Cr$ for the following conditions:

(a) Immediately after the tenth pulse.

(b) After the tenth pulse and assuming Q_1 has reset before Q_3.

(c) During the eleventh pulse.

(d) After the eleventh pulse.

This table should demonstrate that

(a) The tenth pulse sets the latch to clear the counter.

(b) The latch remains set until all FLIP-FLOPS are cleared.

(c) The positive edge of the eleventh pulse resets the latch so that $Cr = 1$.

(d) The negative edge of the eleventh pulse initiates the new counting cycle.

7-21 (a) Indicate a divide-by-11 ripple counter in the block-diagram form. Indicate the connections to J, K, and Ck for each FLIP-FLOP and the inputs to the feedback gate to the clear inputs. (Your may omit the latch.) The preset inputs are held at the 1 level.

(b) There is a second method of obtaining an 11-to-1 ripple counter. The clear inputs are now held at the 1 level and the feedback gate excites the preset inputs. Draw the block diagram for such a programmable ripple counter. Indicate all connections carefully.

7-22 (a) For the logic diagram of the decade counter (TI90A) shown write the truth table for Q_0, Q_1, Q_2, and Q_3 (starting with 0000) after each pulse. If no connection is shown to a J or K input, then this terminal is understood to be high (a 1). Verify that this system is a 10 : 1 counter.

(b) How can this system be used as a 5 : 1 counter?

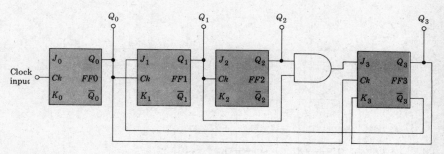

Prob. 7-22

7-23 Modify the logic diagram of Prob 7-22 as follows. Remove the clock from the input to $FF0$ and apply the Q_3 output to this input. Remove Q_0 from Ck of $FF1$. Apply the clock input to Ck of $FF1$. Change no other connections. Write the truth table for Q_0, Q_1, Q_2, and Q_3 (starting with 0000) after each pulse. Verify that this system is a 10 : 1 counter. This is called a *biquinary* counter because a symmetrical square wave is obtained at Q_0. Your truth table should show that this statement is true.

7-24 (a) For the block diagram shown (TI92A) write the truth table for Q_0, Q_1, Q_2, and Q_3 (starting with 0000) after each pulse. Verify that this is a 12 : 1 counter.

(b) How can this system be used as a 6 : 1 counter?

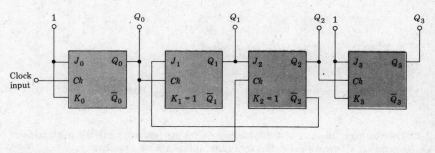

Prob. 7-24

7-25 (*a*) The circuit shown is a *programmable* ripple counter. Initially $Ck = 0$ and the counter is cleared by momentarily setting $Cr = 0$. Thereafter it is understood that $J = K = Cr = 1$, and that the latch in Fig. 7-16 exists between P_1 and P_2. If $Pr_0 = Pr_1 = 0$ and $Pr_2 = Pr_3 = 1$, and *if a pulse from an external source* (not shown) is applied to the preset input, to what state is each FLIP-FLOP set? If a clock-pulse train is now applied to the counter input, what is the count N? Explain the operation of the system carefully.

(*b*) Why is the latch required?

(*c*) Generalize the result of part (*a*) as follows. The counter has n stages and is to divide by N, where $2^n > N > 2^{n-1}$. How must the preset inputs be programmed?

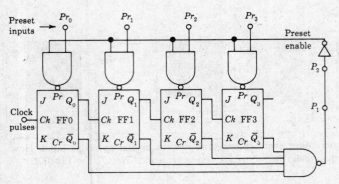

Prob. 7-25

7-26 Draw the logic diagram of a 5-bit UP-DOWN synchronous counter with series carry.

7-27 Verify that the system shown is a 3 : 1 synchronous counter. Start with $Q_0 = Q_1 = 0$ and show the state of Q_0 and Q_1 after each pulse.

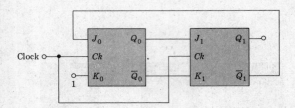

Prob. 7-27

7-28 For the logic diagram of the synchronous counter shown, write the truth table of Q_0, Q_1, and Q_2 after each pulse and verify that this is a 5 : 1 counter.

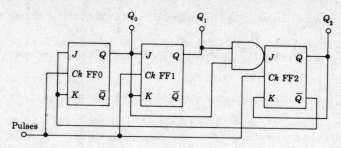

Prob. 7-28

7-29 Consider a two-stage synchronous counter (both stages receive the pulses at the Ck input). In each counter $K = 1$. If $J_0 = \bar{Q}_1$ and $J_1 = Q_0$, draw the circuit. From a truth table of Q_0 and Q_1 after each pulse, demonstrate that this is a 3 : 1 counter.

7-30 Draw the waveform chart for a 6 : 1 divider from Fig. 7-14 and deduce the connections for a synchronous counter. Draw the logic block diagram.

7-31 Solve Prob. 7-30 for 5 : 1 divider.

7-32 Assume you have a crystal-oscillator circuit which provides a series of clock pulses at a frequency of 131.0 kHz. Construct a system whose output is a light-emitting diode flickering approximately once every second, and use the crystal pulses as input. How many seconds does the system "miss" over a one-hour period? *Hint:* $2^{17} = 131,072$.

Chapter 8

8-1 Consider an n-channel device with donor concentration of N_D atoms/cm^3 and a heavily doped gate with acceptor concentration of N_A atoms/cm^3, such that $N_A \gg N_D$ and an abrupt channel-gate junction. Assume that $V_{DS} = 0$ and that the junction contact potential is much smaller than $|V_P|$. Prove that, for the geometry of Fig. 8-1,

$$|V_P| = \frac{qN_D}{2\epsilon} a^2$$

where ϵ = dielectric constant of the channel material and q = magnitude of the electronic charge.

(*a*) Find V_P for a silicon n-channel FET with $a = 2$ μm and $N_D = 7 \times 10^{14}$ atoms/cm^3.

(*b*) Find the resistivity ρ of the p-channel of a germanium FET with $a = 2$ μm and $V_P = 3.94$ V.

8-2 The resistance R_d in series with the drain of an n-channel FET with the source grounded is 5 kΩ. The FET is operating at a quiescent point $V_{DS} = 17.5$ V, and $I_{DS} = 2.5$ mA, and its characteristics are given in Fig. 8-3.

(a) To what value must the gate voltage be changed if the drain current is to change to 3 mA?

· (b) To what value must the voltage V_{DD} be changed if the drain current is to be brought back to its previous value? The gate voltage is maintained constant at the value found in part (a).

8-3 (a) Derive Eq. (8-2).

(b) For the FET whose characteristics are plotted in Fig. 8-3, determine $r_{DS(\text{ON})}$ for $V_{GS} = 0$ V.

(c) Consider an n-channel silicon FET whose structure is indicated in Fig. 8-1 with $a = 4$ μm, $w = 120$ μm, and channel length $L = 6$ μm. Find $r_{DS(\text{ON})}$ for $V_{GS} = 0$ V if the channel resistivity is $\rho = 10\ \Omega \cdot$ cm.

8-4 The silicon FET of Fig. 8-1 has channel length $L = 20$ μm, $a = 4$ μm, $w = 16$ μm, and an abrupt channel-gate junction. If the pinch-off voltage V_P is -6 V, find $r_{DS(\text{ON})}$ for $V_{GS} = 0$ V. *Hint:* Use the formula for V_P given in Prob. 8-1a.

8-5 The MOSFET in the circuit shown has the characteristics of Fig. 8-8a.

(a) Find I_D and V_{DS}.

(b) If it is desired that $V_{DS} = -25$ V, what should be the value of V_{GG}?

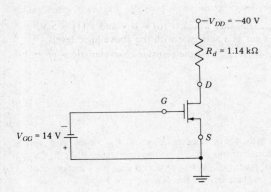

Prob. 8-5

8-6 (a) Draw the transfer curve for $V_{DS} = -20$ V of the MOSFET whose drain characteristics are shown in Fig. 8-8a.

(b) On Fig. 8-8a plot the locus of points for which $V_{GS} - V_{DS} = V_T$; observe that this is the quadratic curve, defined by Eq. (8-4), which separates the triode from the saturation regions.

8-7 (a) Graphically obtain the transfer characteristic for the inverter of Fig. 8-14b, using the output curves supplied for this problem (change the NMOS to PMOS devices and V_{DD} to $-V_{DD}$). Proceed as in Fig. 8-16. Assume $V_{DD} = 20$ V.

(b) Repeat part (a) by assuming that $Q1$ and $Q2$ do not have the same geometry and that the resistance of $Q2$ is much higher than that in part (a).

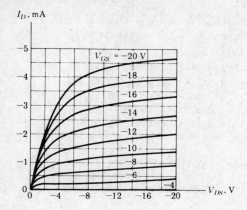

Prob. 8-7

Prob. 8-8

8-8 Repeat Prob. 8-7 using the drain characteristics supplied for this problem. Assume $V_{DD} = 10$ V.

8-9 (a) Draw the circuit of an inverter (using p-channel enhancement MOSFETs) which is compatible with positive TTL logic. In other words, $V(0) = 0$ V and $V(1) = 5$ V.

(b) Verify that your circuit does act as a NOT gate with the above logic levels.

8-10 (a) For the inverter of Fig. 8-14b verify that the transfer characteristic is given by

$$V_o = -\frac{k_D}{k_L}(V_i - V_T) + V_{DD} - V_T$$

for $V_i \geqslant V_T$, as long as both the driver and the load remain in saturation. Use Eq. (8-4) with $k \equiv \mu C_o w/2L$, where the subscripts D and L of k refer to the driver and load.

(b) What is the slope of the transfer characteristic? Does Fig. 8-17 agree with this result?

8-11 For the inverter of Fig. 8-14b verify that the transfer characteristic is given by

$$V_o^2(1 + \beta_R) - 2V_o[V_{DD} - V_T + \beta_R(V_i - V_T)] + (V_{DD} - V_T)^2 = 0$$

provided that the load is in saturation but that V_i is large enough so the driver is operating in the ohmic region. Use Eqs. (8-3) and (8-4) and let $\beta_R \equiv k_D/k_L$, where $k \equiv \mu C_o w/2L$ and the subscripts D and L of k refer to the driver and load.

8-12 (a) Graphically obtain the transfer characteristic for the inverter of Fig. 8-14c by using the output curves supplied in Prob. 8-7. (Change NMOS to PMOS devices, V_{DD} to $-V_{DD}$, and V_{GG} to $-V_{GG}$.) Proceed as in Fig. 8-18. Assume $V_{DD} = 20$ V and $V_{GG} = 28$ V.

(b) Repeat part (a) by assuming that $Q1$ and $Q2$ do not have the same geometry and that the resistance of $Q2$ is much higher than that in part (a).

8-13 Repeat Prob. 8-12 using the drain characteristics supplied in Prob. 8-8. Assume $V_{DD} = 10$ V, $V_{GG} = 17$ V.

8-14 (*a*) For the inverter of Fig. 8-14*c*, the load operates in the ohmic region and the driver in saturation (Fig. 8-18). Use Eqs. (8-3) and (8-4) and show that the transfer characteristic is given by

$$k_L\left[2(V_{DD} - V_o + V' - V_T)(V_{DD} - V_o) - (V_{DD} - V_o)^2\right] = k_D(V_i - V_T)^2$$

where $V_i \geqslant V_T$ and $k \equiv \mu C_o w/2L$, and the subscript D and L on k refer to the driver and load, and $V' \equiv V_{GG} - V_{DD}$.

(*b*) Verify that $V_o = V_{DD}$ for $V_i = V_T$ independent of k_L, k_D, V_T, or V'.

(*c*) For $V_{DD} = 10$ V, $V_{GG} = 16$ V, $V_T = 2$ V and $k_L = k_D$, calculate V_o for $V_i = 6$ and also $V_i = 10$ V. Compare the calculated values with those obtained graphically in curve A of Fig. 8-19.

8-15 (*a*) Graphically obtain the transfer characteristic for the inverter of Fig. 8-20*a* by using PMOS instead of NMOS devices (change V_{DD} to $-V_{DD} = -20$ V). The driver output characteristics are indicated in Prob. 8-7. The depletion-load characteristic for $Q2$ is marked A in the graph supplied with this problem.

(*b*) Repeat part (*a*) by using the depletion curve B.

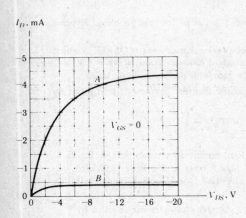

Prob. 8-15   **Prob. 8-16**

8-16 Repeat Prob. 8-15 using the enhancement characteristics of $Q1$ supplied in Prob. 8-8 and the depletion characteristics of $Q2$ supplied with this problem.

8-17 (*a*) Verify that the circuit of Fig. 8-23 satisfies the AND-OR-INVERT logic by considering all possible combinations of the four inputs V_1, V_2, V_3, and V_4 and the corresponding output V_o.

(*b*) Apply the wired-AND concept to Fig. 8-23 and thereby show that the circuit represents an AND-OR-INVERTER gate.

8-18 Draw a MOSFET circuit satisfying the logic equation $Y = \overline{ABC + D}$, where Y is the output corresponding to the four inputs A, B, C, and D.

8-19 Repeat Prob. 8-18 for $Y = \overline{A + B} + \overline{C + D}$.

8-20 Draw the circuit of a MOSFET positive AND gate and explain its operation.

8-21 Draw the circuit of a two-input EXCLUSIVE-OR circuit using MOSFETs. *Hint*: Use Eq. (5-22).

8-22 Consider the 1-bit memory of Fig. 8-24.

   (*a*) Assume $S = R = 0$. Verify that the circuit has two possible states $Q = 1$, $\overline{Q} = 0$, and $Q = 0$, $\overline{Q} = 1$.

   (*b*) Assume that $S = 1$ and $R = 0$. Verify that $Q = 1$ and $\overline{Q} = 0$.

8-23 (*a*) Augment the latch of Fig. 8-24 with two FETs $Q7$ and $Q8$ in the manner indicated in Sec. 8-8 to convert it to a clocked *S-R* FLIP-FLOP.

   (*b*) Verify that the FLIP-FLOP does not change state between pulses where $Ck = 0$.

   (*c*) Show that the *S-R* truth table (Fig. 7-5*b*) is satisfied. Consider each row in turn and indicate the state of each MOSFET (ON or OFF).

8-24 Construct an *S-R* clocked FLIP-FLOP in the manner indicated in Sec. 8-8. Add an inverter whose input is *S* and whose output goes to *R*. Show that the resulting circuit is a *D*-type FLIP-FLOP with $D = S$; that is, verify the truth table

D_n	Q_{n+1}
1	1
0	0

8-25 Construct an *S-R* clocked FLIP-FLOP in the manner indicated in Sec. 8-8. Augment this circuit so that it becomes a *J-K* master-slave FLIP-FLOP equivalent to that shown in Fig. 7-8. Indicate each MOSFET in the circuit.

8-26 Draw a CMOS inverter using negative logic.

8-27 (*a*) Draw a CMOS inverter with the *p*-well and the *n* substrate connected to the proper voltages. Verify that no isolation islands are required; that is, show that all *p-n* junctions are reverse-biased.

   (*b*) Repeat part (*a*) for the CMOS NAND gate of Fig. 8-27*b*.

8-28 Show that a CMOS gate in which the drivers and the loads are all placed in series will *not* function properly as a NAND gate (or as a NOR gate).

8-29 Draw the circuit of a CMOS 2-input NAND gate using negative logic.

8-30 Draw the circuit of CMOS 2-input NOR gate by using positive logic and explain its operation.

8-31 Using positive logic with CMOS devices draw a circuit to perform AOI logic $Y = \overline{AB + CD}$.

8-32 Repeat Prob. 8-31 for $Y = \overline{A(B + C)}$

8-33 (*a*) Consider two isolated NMOS INVERTERS. The input to one is *A* and to the other is *B*. The two outputs are now wired together and the common output is *Y*. What is the logic relationship between *Y*, *A*, and *B*?

   (*b*) Since the two load FETs are in parallel, omit one load. Draw the circuit and show that it performs the correct logic of part (*a*).

   (*c*) If the INVERTERS are built from CMOS devices, is it possible to wire-AND the outputs? Explain.

8-34 (*a*) Draw the circuit of a bistable *R-S* latch using CMOS devices, corresponding to the topology in Fig. 8-24.

   (*b*) Verify that, if $S = 1$ and $R = 0$, then $Q = 1$.

8-35 Consider the analog transmission gate of Fig. 8-28 with the control voltages $V(0) = -5$ V and $V(1) = +5$ V, and a 5-V peak-input sinusoid. Assume that the threshold voltage V_T is zero.

   (*a*) Verify that the entire sinusoid appears at the output if the binary control *C* corresponds to the voltage $V(1)$.

(*b*) Show that transmission is inhibited if the control voltage is $V(0) = -5$ V.

(*c*) Repeat parts (*a*) and (*b*) if $V_T = 2$ V. Indicate the range of input voltage over which both $Q1$ and $Q2$ conduct.

(*d*) Assume a peak sinusoidal input voltage of 8 V. If the control voltage is $V(1) = +5$ V, sketch the output voltage.

(*e*) Repeat part (*d*) if the control voltage is $V(0) = -5$ V and $V_T = 2$ V.

Chapter 9

9-1 (*a*) Modify the dynamic MOS inverter of Fig. 9-1 by adding another FET $Q4$ in series with $Q1$. Designate the input to $Q4$ ($Q1$) by V_4 (V_1). Verify that this circuit performs the function of a dynamic NAND gate. The input levels of V_1 and V_4 are 0 and 10 V.

(*b*) Show that this circuit dissipates less power than the corresponding static NAND gate of Fig. 8-22.

9-2 Modify the circuit of Fig. 9-1 by adding another FET $Q4$ in parallel with $Q1$. Repeat Prob. 9-1 (with the word NAND replaced by NOR and Fig. 8-22 replaced by Fig. 8-21).

9-3 (*a*) Consider the shift-register stage of Fig. 9-2 but with unclocked loads; that is, the gates of $Q2$ and $Q5$ are tied to V_{DD} instead of being excited by the clock waveforms. Explain the operation of the circuit.

(*b*) Show that more power is dissipated in this cell than in the clocked-load version of Fig. 9-2.

9-4 (*a*) An NMOS dynamic shift-register stage is shown. The two-phase waveforms ϕ_1 and ϕ_2 are sketched in Fig. 9-2*b* or Fig. 9-5*b*. Carefully explain·the operation of this circuit. Assume $C_1 \gg C_2$.

(*b*) Are the inverters ratioed or ratioless? Explain.

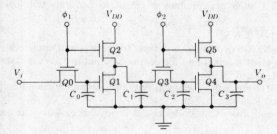

Prob. 9-4

9-5 Verify Eq. (9-1). *Hint:* When the transmission gate $Q3$ closes, the same charge which leaves C_1 must be added to C_2.

9-6 (*a*) Consider the two-phase NMOS inverter shown which uses the clock waveforms pictured in Fig. 9-5*b*. Explain the circuit operation by considering first the interval t_1–t_2, then t_2–t_3, etc.

(b) Is this a ratioed or ratioless inverter? Explain.

(c) Using two such inverters and two bidirectional gates, sketch one stage of a shift register. *Hint:* Interchange ϕ_1 and ϕ_2 in the second inverter and read the output during ϕ_1.

(d) Explain the operation of this shift-register cell.

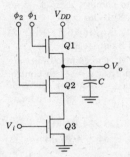

Prob. 9-6

9-7 Consider the four-phase dynamic NMOS shift-register cell shown. Note that the four clock pulses are nonoverlapping so that, if one of the phases is high, then the other three are low. Explain the operation and verify that V_o equals the value which V_i had one period earlier.

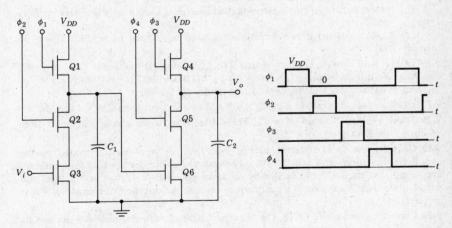

Prob. 9-7

9-8 Consider a 4-kb ROM with four output bits. If the encoder matrix is square, how many bits are needed for (a) the X address? (b) the Y address?

(c) Sketch the block diagram of the system.

9-9 Consider an 8-kb ROM with eight output bits. If the memory matrix has 128 rows, how many bits are needed for (a) the X address? (b) The Y address?

(c) Repeat parts (a) and (b) if there are 64 rows in the encoder.

(d) How many words does this ROM have and how many bits are needed to decode these words? Check your answer against the sum of the bits in the X and Y addresses for each of the two ROMs considered in this problem.

9-10 Two 16-kb ($2,048 \times 8$) ROMs are available. Show how to connect these so as to obtain (a) a 32-kb ($2,048 \times 16$) ROM and (b) a 32-kb ($4,096 \times 8$) ROM.

9-11 Indicate in block diagram form how to assemble thirty-two 16-kb ($2,048 \times 8$) ROMs to obtain an equivalent ROM with 16 address lines and 8 output lines.

9-12 (a) A 32×8 ROM is to be converted into a 64×4 ROM. The eight outputs are $0_0 \cdots 0_7$ and the addresses are $A_0 \cdots A_4$. Add one more address $X = A_5$ to control AND-OR gates (as in the up-down counter of Fig. 7-18) so that, with $X = 1$, the four outputs $0_0 \cdots 0_3$ are used and for $X = 0$ the four outputs $0_4 \cdots 0_7$ are used. Indicate this 64×4 ROM system.

(b) Indicate how to convert two 32×8 ROM chips into a 128×4 ROM.

9-13 (a) Show the block diagram of a system for converting a 64×8 ROM into a 512×1 ROM, using a selector/multiplexer.

(b) Repeat part (a) for converting a 64×8 ROM into a 256×2 ROM.

9-14 A 128-bit RAM consists of 32 words of 4 bits each. If linear selection is used, show a block diagram of the system organization. *Note:* Use one rectangle to represent the 1-bit read/write cell of Fig. 9-15, with three terminals: X for the *address input*, W for the *write input*, and R for the *read output*.

9-15 (a) How many NAND gates and how many inputs to each gate are there in the decoder (or decoders) for a 1024×1 RAM, if linear selection is used?

(b) Repeat part (a) if two-dimensional addressing is used to give a square memory array.

(c) Repeat part (a) if two-dimensional addressing is used to produce a 64×16 memory array.

9-16 In Fig. 9-19 chip (0) contains words 0 to 1,023, chip (1) has words 1,024 to 2,047, etc. What word is decoded by (a) $A_{11} \cdots A_0 = 011100101011$? (b) 111000010110?

(c) What address must be applied to obtain word 2,600?

9-17 Draw a block diagram of a $2,048 \times 4$ RAM system built from 256×1 RAMs.

9-18 Draw a block diagram of a $131,072 \times 4$-bit read/write system assembled from 16-kb $\times$ 1-bit RAMs.

9-19 Consider the CCD structure in Fig. 9-26 (but use n channel) operated by the two-phase waveforms of Fig. 9-30. All odd-numbered electrodes are tied to ϕ_1, and all even-numbered electrodes are excited by ϕ_2. Draw potential profiles as in Fig. 9-26, and demonstrate that this system is unsatisfactory because the direction of charge transfer is indeterminate.

9-20 Consider a two-phase CCD. The effective length of each electrode is 8 μm and its width is 8 μm. The separation between rows of electrodes is also 8 μm.

(a) Calculate the area in square mils occupied by a memory cell.

(b) Mnemonics, Inc. has built a 65-kb (65,536 bits) memory using the cell described in part (a). The chip size is 218×235 mils. What fraction of the chip area is occupied by the auxiliary circuits (input, output, clocks, etc.)?

9-21 Consider the two-phase CCD structure of Fig. 9-29a excited by the positive clock pulses of Fig. 9-32a. Assume that $V_2 = V$ and $V_1 = \frac{1}{2} V$. Draw the potential-energy profiles under the first four electrodes for the five times $t_1 \cdots t_5$ indicated in Fig. 9-32. Start with charge under E_1 at $t = t_1$ and demonstrate that it is shifted to E_2 at $t = t_5$. Use quadrilled paper.

9 -22 Consider the two-phase CCD structure of Fig. 9-29a excited by the negative clock pulses of Fig. 9-32b. Assume that $V_2 = V$ and $V_1 = 0$. Draw the potential-energy profiles under the first four electrodes for the five times $t_1 \cdots t_5$ indicated in Fig. 9-32. Start with a bit stored under E_1 at $t = t_1$ and show that the information is transferred to the well under E_2 at $t = t_5$. Use quadrilled paper.

9-23 (a) Consider a single-phase CCD structure (Fig. 9-29a or 9-31a). The odd elec- trodes are biased to a constant voltage $\frac{1}{2} V$. The even electrodes are excited by the positive-pulse waveform ϕ_2 shown in Fig. 9-32a with $V_1 = 0$ and $V_2 = V$. Draw the potential-energy profiles under the first four electrodes for the times t_2, t_3, and t_4 of Fig. 9-32a. Start with electrons stored under E_1 at time t_2 and show that the charge is retained in the well under E_2 at $t = t_4$. Use quadrilled paper.

(b) Draw the potential-energy profile for a time t'_4 (where $\phi_2 = \frac{1}{4} V$), for t_5 or t_6 (where $\phi_2 = 0$), and for t_7 (where $\phi_2 = \frac{1}{2} V$). Demonstrate that the information under E_1 has been transferred to E_3 in one clock period.

9-24 Draw the potential-energy profiles for a four-phase CCD under electrodes $E_3 \cdots E_7$ during the interval $t_8 - t_7$ of Fig. 9-33. Start with the electrons stored under E_4, as in Fig. 9-34g. Use quadrilled paper.

9-25 (a) Consider the four-phase CCD of Fig. 9-34a excited by symmetrical square waves. The first clock is ϕ_1 of Fig. 9-30. The second clock is this same waveform delayed by $T/4$. The third clock is ϕ_1 delayed by $T/2$, which is the waveform labeled ϕ_2 in Fig. 9-30. The fourth clock is ϕ_1 delayed by $3T/4$. Draw these clock waveforms carefully on quadrilled paper.

(b) Draw the potential-energy profiles. Start at $t = t_1$, where $\phi_1 = \phi_2 = V$ and $\phi_3 = \phi_4 = 0$. Assume that charge is stored under E_2 at t_1. Increase t in intervals in order to show how the charge transfers to E_4 without being stored at E_3. Indicate the shift interval and the I/O interval in one half cycle. What is E/B?

9-26 Consider three logic variables A, B, and C at the collectors of three I^{2}L inverters. Connect these three outputs together. Show by a physical argument that at the common node the logic variable is $Y = ABC$. In other words, justify the wired-AND operation for injection logic.

9-27 Given the four external variables A, B, C, and D, draw an I^{2}L connection diagram for the AOI output $Y = AB + CD$.

9-28 The three inputs to a decoder are A, B, and C. Draw an I^{2}L connection diagram to obtain the eight outputs.

9-29 Consider a 2-to-1-line multiplexer without a strobe (Fig. 6-20). Draw an I^{2}L connection diagram for this data selector.

9-30 The carry in a full adder is of the form $C' = AB + BC + CA$. Draw an I^{2}L connection diagram for C'.

9-31 Draw an I^{2}L connection diagram for the J-K clocked FLIP-FLOP of Fig. 7-7a.

Chapter 10

10-1 (a) For the application in Sec. 10-1, plot the voltage across the diode for one cycle of the input voltage v_i. Let $V_m = 3.0$ V, $V_\gamma = 0.6$ V, $R_f = 20$ Ω, and $R_L = 200$ Ω.

(b) By direct integration find the average value of the diode voltage and the load voltage in terms of the symbols V_m, V_γ, ϕ, etc. Note that these two answers are identical and explain why.

10-2 A symmetrical 10-kHz square wave whose output varies between $+20$ and -20 V is impressed upon the clipping circuit of Fig. 10-5d. Assume $R_f = 0$, $R_r = 2$ MΩ, $V_R = 5$ V, $R = 1$ MΩ, and $V_\gamma = 0$. Sketch the steady-state output waveform, indicating the maximum and minimum values it attains.

10-3 For the diode clipping circuit of Fig. 10-5c assume that $V_R = 10$ V, $v_i = 20 \sin \omega t$, and that the diode forward resistance is $R_f = 10$ Ω while $R_r = \infty$ and $V_\gamma = 0$. Neglect all capacitances. Draw to scale the input and output waveforms and label the maximum and minimum values if (a) $R = 50$ Ω, (b) $R = 500$ Ω, and (c) $R = 2$ kΩ.

10-4 Repeat Prob. 10-3 for the case where the reverse resistance is $R_r = 20$ kΩ.

10-5 In the diode clipping circuit of Fig. 10-5a and b, $v_i = 20 \sin \omega t$, $R = 1$ kΩ, and $V_R = 10$ V. The reference voltage is obtained from a tap on a 10-kΩ divider connected to a 50-V source. Neglect all capacitances. The diode forward resistance is 10 Ω, $R_r = \infty$, and $V_\gamma = 0$. In both cases draw the input and output waveforms to scale. Which circuit is the better clipper? *Hint:* Apply Thévenin's theorem to the reference-voltage-divider network.

10-6 For the clipping circuits shown in Fig. 10-5b and d derive the transfer characteristic v_o versus v_i, taking into account R_f and V_γ and considering $R_r = \infty$.

10-7 (a) Assume an ideal diode in the circuit of Fig. 10-5c, so that $V_\gamma = 0$, $R_f = 0$, $R_r = \infty$. Plot the transfer characteristic v_o versus v_i for $-25 \leqslant v_i \leqslant +25$ V, if $R = 2$ kΩ and $V_R = -20$ V.

(b) Assume that the diode is no longer ideal but that $V_\gamma = 0.6$ V, $R_f = 10$ Ω, and $R_r = 10$ MΩ. If $v_i = -25$ V, find v_o.

10-8 (a) In the circuit of Fig. 10-5d assume that v_i increases linearly with time at the rate 10 V/μs, and has the value -20 V at $t = 0$. Sketch v_i and v_o versus time for $0 \leqslant t \leqslant 4$ μs, assuming and ideal diode and $V_R = 0$ V. *Note:* The above circuit is called a comparator. It is used to mark the instant when an arbitrary waveform attains some reference level.

(b) The diode-resistor comparator of part (a) is connected to a device which responds when the comparator output attains a level of 0.1 V. The silicon diode has a reverse saturation current of 1 nA. Initially, $R = 10$ kΩ and the 0.1-V output level is attained at a time $t = t_1$. If we now set $R = 100$ kΩ, what will be the corresponding change in t_1? $V_R = 0$. *Hint:* Use Eq. (2-3).

10-9 The clipping circuit shown employs temperature compensation. The dc voltage source V_γ represents the diode offset voltage; otherwise the diodes are assumed to be ideal with $R_f = 0$ and $R_r = \infty$.

(a) Sketch the transfer curve v_o versus v_i.

(b) Show that the maximum value of the input voltage v_i so that the current in $D2$ is always in the forward direction is

$$v_{i(\max)} = V_R + \frac{R}{R'}(V_R - V_\gamma)$$

(c) What is the temperature dependence of the point on the input waveform at which clipping occurs?

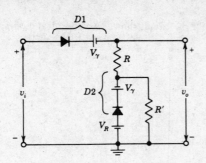

Prob. 10-9

Prob. 10-10

10-10 (*a*) In the clipping circuit shown, $D2$ compensates for temperature variations. Assume that the diodes have infinite back resistance, a forward resistance of 20 Ω, and a break point at the origin ($V_\gamma = 0$). Calculate and plot the transfer characteristic v_o against v_i. Show that the circuit has an extended break point, that is, two break points close together.

(*b*) Find the transfer characteristic that would result if $D2$ were removed and the resistor R were moved to replace $D2$.

(*c*) Show that the double break of part (*a*) would vanish and only the single break of part (*b*) would appear if the diode forward resistances were made vanishingly small in comparison with R.

10-11 (*a*) In the peak clipping circuit shown, add another diode $D2$ and a resistor R' in a manner that will compensate for drift with temperature.

(*b*) Show that the break point of the transmission curve occurs at V_R. Assume $R_r \gg R \gg R_f$.

(*c*) Show that if $D2$ is always to remain in conduction it is necessary that

$$v_i < v_{i(\text{max})} = V_R + \frac{R}{R'}(V_R - V_\gamma)$$

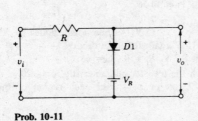

Prob. 10-11

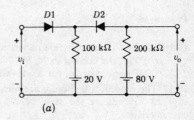

(*a*)

Prob. 10-12

10-12 The input voltage v_i to the two-level clipper shown in the figure varies linearly from 0 to 100 V. Sketch the output voltage v_o to the same time scale as the input voltage. Assume ideal diodes.

10-13 Sketch the transfer-characteristic curve for the circuit shown. Assume ideal diodes.

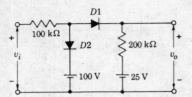

Prob. 10-13

10-14 The circuit of Fig. 10-6a is used to "square" a 10-kHz input sine wave whose peak value is 60 V. It is desired that the output voltage waveform be flat for 95 percent of the time. Diodes are used having a forward resistance of 100 Ω and a backward resistance of 500 kΩ.
   (a) Find the values of V_{R1} and V_{R2}.
   (b) What is a reasonable value to use for R?

10-15 (a) Repeat Prob. 10-13 for the circuit shown for $-5\text{ V} < v_i < 5\text{ V}$.
   (b) Repeat for the case where the diodes have an offset voltage $V_\gamma = 1$ V.

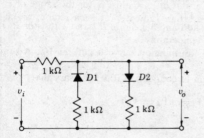

Prob. 10-15   **Prob. 10-16**

10-16 (a) Plot the transfer characteristic for $0 < v_i < 15$ V. Assume ideal diodes.
   (b) Plot the current in R_1 for $0 < v_i < 15$ V again for ideal diodes.
   (c) Repeat part (a) for diodes with $V_\gamma = 0.7$ V, $R_f = 0$, $R_r = \infty$.

10-17 The diodes in the figure are ideal. Sketch the transfer characteristic for $-20 \leqslant v_i \leqslant +20$ V. Indicate the state of $D1$ and $D2$ (ON or OFF) over each region of the characteristic.

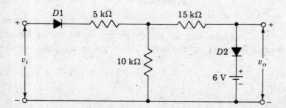

Prob. 10-17

10-18 The diodes in the figure are ideal. Plot the transfer characteristic over the range $0 \leqslant v_i \leqslant 50$ V. Proceed as follows:

(a) Find v_o for $v_i = 0$. What are the states (ON or OFF) of the diodes?

(b) Find the equation for v_o in terms of v_i if $D1$ is ON and $D2$ OFF. Over what range of v_i are these states valid?

(c) Find v_o for $v_i = 50$ V (use superposition).

(d) Now plot v_o versus v_i and indicate in each region which diodes are conducting.

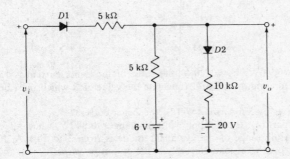

Prob. 10-18

10-19 Assume that the diodes are ideal. Make a plot of v_o against v_i for the range of v_i from 0 to 50 V. Indicate all slopes and voltage levels. Indicate, for each region, which diodes are conducting.

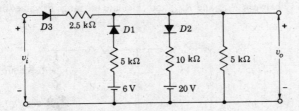

Prob. 10-19

10-20 (a) Construct a circuit that exhibits the terminal characteristic shown.

(b) Modify the circuit so that the two slopes in the figure are not the same.

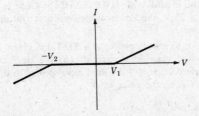

Prob. 10-20

10-21 Construct a circuit whose transfer characteristic (v_o versus v_i) has the form shown in the figure. Use ideal diodes and give the numerical values of all elements in your circuit.

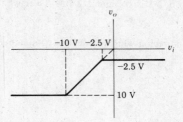

Prob. 10-21

10-22 (a) In Fig. 10-10, $V = 300$ V, $V_Z = 220$ V, the value of the Zener current is 15 mA, and the value of the load current is 25 mA. Calculate the value of R which must be used.

(b) If the load decreases by 5 mA, what will be the Zener current?

(c) The load is as in part (a). If the supply voltage changes to 340 V, what is I_Z?

(d) The normal operating range of the avalanche diode is from 3 to 50 mA. If $R = 1.5$ kΩ and $V = 340$ V, over what load current can the output be varied?

10-23 (a) The avalanche diode regulates at 40 V over a range of diode currents from 10 to 50 mA. The supply voltage is 200 V. Calculate R to allow voltage regulation from a load R_L from ∞ to $R_{L(min)}$, the minimum possible value of R_L.

(b) What is the maximum possible load current and $R_{L(min)}$?

(c) If V can have any value between 160 and 300 V when $R_L = 2$ kΩ, calculate the maximum R_{max} and minimum R_{min} allowed values for R.

(d) Choose R halfway between R_{min} and R_{max} and calculate the range of values of the Zener current.

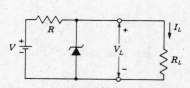

Prob. 10-23

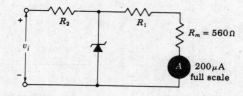

Prob. 10-24

10-24 The Zener diode can be used to prevent overloading of sensitive meter movements without affecting meter linearity. The circuit shown represents a dc voltmeter which reads 25 V full scale. The meter resistance is 560 Ω, and the full scale is 0.2 mA. If the diode is a 20-V Zener, find R_1 and R_2 so that, when $V_i > 25$ V, the Zener diode conducts and the overload current is shunted away from the meter.

10-25 A diode whose internal resistance is 10 Ω is to supply power to a 500-Ω load from a 100-V (rms) source of supply. Calculate (a) the peak load current; (b) the dc load current; (c) the ac load current; (d) the dc diode voltage; (e) the total input power to the circuit; (f) the percentage regulation from no load to the given load.

10-26 Show that the maximum dc output power $P_{dc} \equiv V_{dc}I_{dc}$ in a half-wave single-phase circuit occurs when the load resistance equals the diode resistance R_f.

10-27 The efficiency of rectification η_r is defined as the ratio of the dc output power $P_{dc} \equiv V_{dc}I_{dc}$ to the input power $P_i = \dfrac{1}{2\pi} \displaystyle\int_0^{2\pi} v_i i \, d\alpha$.

 (a) Show that, for the half-wave-rectifier circuit,

$$\eta_r = \frac{40.5}{1 + R_f/R_L} \text{ percent}$$

 (b) Show that, for the full-wave rectifier, η_r has twice the value given in part (a).

10-28 Prove that the regulation of both the half-wave and the full-wave rectifier is given by

$$\text{Percent regulation} = \frac{R_f}{R_L} \times 100 \text{ percent}$$

10-29 (a) Prove Eqs. (10-24) and (10-25) for a full-wave-rectifier circuit.

 (b) Find the dc voltage across a diode by direct integration.

10-30 A full-wave single-phase rectifier consists of a double-diode vacuum tube, the internal resistance of each element of which may be considered to be constant and equal to 300 Ω. These feed into a pure resistance load of 1 kΩ. The secondary transformer voltage to center tap is 200 V. Calculate (a) the dc load current; (b) the direct current in each tube; (c) the ac voltage across each diode; (d) the dc output power.

10-31 In the full-wave single-phase bridge, can the transformer and the load be interchanged? Explain carefully.

10-32 The bridge-rectifier system shown in Fig. 10-16 is used to construct an ac voltmeter. The forward resistance of the diodes is 100 Ω, the resistance R is 50 Ω, and the ammeter resistance is negligible. The signal voltage is given by $v_s = 100 \sin \omega t$.

 (a) Sketch the waveform of the current i_L through the ammeter. Calculate maximum instantaneous value on your sketch.

 (b) Write down an integral whose value will give the reading of the dc ammeter. Evaluate this expression and find I_{dc}.

 (c) Sketch realistically the voltage waveform across diode $D1$. Indicate maximum instantaneous values on your sketch. Evaluate the average diode voltage.

 (d) Write an integral whose value will give the reading of an rms voltmeter placed across $D1$. (This meter does not have a series blocking capacitor.) Find the value of this rms diode voltage.

10-33 A 10-mA dc meter whose resistance is 20 Ω is calibrated to read rms volts when used in a bridge circuit with semiconductor diodes. The effective resistance of each element may be considered to be zero in the forward direction and infinite in the inverse direction. The sinusoidal input voltage is applied in series with a 10 kΩ resistance. What is the full-scale reading of this meter?

10-34 (a) Consider the bridge voltage-doubler circuit of Fig. 10-17 with $R_L = \infty$. Show that in the steady state each capacitor charges up to peak transformer voltage V_m and, hence, that $v_o = 2V_m$. Assume ideal diodes.

 (b) What is the peak inverse voltage across each diode?

10-35 The circuit shown is a half-wave voltage doubler. Analyze the operation of this circuit by sketching as a function of time the waveforms v_i, v_{c1}, v_{D1}, v_{D2}, and v_o. Assume that the capacitors are uncharged at $t = 0$. Calculate (a) the maximum possible voltage across each capacitor and (b) the peak inverse voltage of each diode. Compare this circuit with the bridge voltage doubler of Fig. 10-17. In this circuit the output voltage is negative with respect to ground. Show that if the connections to the cathode and anode of each diode are interchanged, the output voltage will be positive with respect to ground.

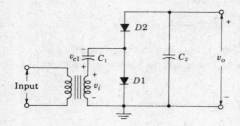

Prob. 10-35

10-36 The circuit of Prob. 10-35 can be extended from a doubler to a quadrupler by adding two diodes and two capacitors as shown. In the figure, parts (a) and (b) are alternative ways of drawing the same circuit.

(a) Analyze the operation of this circuit.

(b) Answer the same questions as asked in parts (a) and (b) of Prob. 10-35.

(c) Generalize the circuit of this and of Prob. 10-35 so as to obtain n-fold multiplication when n is any even number. In particular, sketch the circuit for sixfold multiplication.

(d) Show that n-fold multiplication, with n odd, can also be obtained provided that the output is properly chosen.

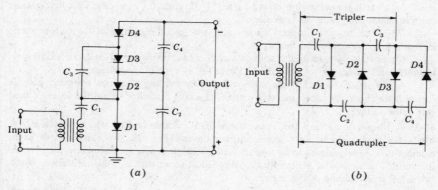

Prob. 10-36

10-37 (a) Consider the capacitor filter of Fig. 10-18. Show that, during the interval when the diode conducts, the diode current is given by $i = I_m \sin(\omega t + \psi)$, where

$$I_m \equiv V_m \sqrt{\frac{1}{R_L^2} + \omega^2 C^2} \qquad \text{and} \qquad \psi \equiv \arctan \omega C R_L$$

(b) Find the cutout angle ωt_1 in Fig. 10-20.

10-38 A single-phase full-wave rectifier uses a semiconductor diode. The transformer voltage is 40 V rms to center tap. The load consists of a 50-μF capacitance in parallel with a 300-Ω resistor. The diode and the transformer resistances and leakage reactance may be neglected. The frequency of operation is 60 Hz.

(*a*) Calculate the cutout angle.

(*b*) Plot to scale the output voltage and the diode current as in Fig. 10-20. Determine the cutin point graphically from this plot, and find the peak diode current corresponding to this point.

(*c*) Repeat (*a*) and (*b*), using a 150-μF instead of a 50-μF capacitance.

10-39 For the circuit of Fig. 10-23a $v_i = V_m \sin \omega t$. Assume $V_m > V_R$ and an ideal diode. Let $R = \infty$ and sketch the v_c and v_o waveforms for two cycles, where v_c is the capacitor voltage. Assume that at $t = 0$, $v_c = 0$.

Chapter 11

11-1 (*a*) Determine the quiescent currents and the collector-to-emitter voltage for a silicon transistor with $\beta = 100$ in the self-biasing arrangement of Fig. 11-4. The circuit component values are $V_{CC} = 12$ V, $R_c = 1.5$ kΩ, $R_e = 0.1$ kΩ, $R_1 = 90$ kΩ, and $R_2 = 10$ kΩ.

(*b*) Repeat part (*a*) for a germanium transistor.

11-2 A *p-n-p* silicon transistor is used in the self-biasing arrangement of Fig. 11-4. The circuit component values are $V_{CC} = 18$ V, $R_c = 2$ kΩ, $R_e = 200$ Ω, $R_2 = 3$ kΩ, and $R_1 = 27$ kΩ. If $\beta = 50$:

(*a*) Find the quiescent point.

(*b*) Recalculate these values if the base-spreading resistance of 500 Ω is taken into account.

11-3 An *n-p-n* silicon transistor with $\beta = 50$ is used in the circuit of Fig. 11-4. The circuit component values are $V_{CC} = 20$ V, $R_c = 2$ kΩ, and $R_e = 100$ Ω. If it is desired that the quiescent value of I_C be 1.26 mA, find the values of R_1 and R_2. Assume that the transistor is in the active region. For ac purposes it is required that the parallel combination of R_1 and R_2 be 4.76 kΩ.

11-4 For the circuit shown:

(*a*) Calculate I_B, I_C, and V_{CE} if a silicon transistor is used with $\beta = 100$.

(*b*) Specify a value for R_b so that $V_{CE} = 6.5$ V.

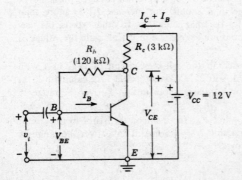

Prob. 11-4

11-5 (*a*) Verify Eq. (11-10).

(*b*) Verify Eq. (11-12).

11-6 For the self-bias circuit of Fig. 11-4, β may have any value between 40 and 120 at room temperature. Neglect I_{CO}. The nominal bias point is $I_C = 1.5$ mA, $V_{CE} = 6.0$ V, and $V_{BE} = 0.7$ V, obtained with $V_{CC} = 15$ V and $R_c = 2$ kΩ. Find R_e, R_1, and R_2 if I_C is to be in the range 1.35 to 1.65 mA as β varies from 40 to 120. Neglect I_B compared with I_C in the calculation of R_e.

11-7 In the two-stage circuit shown, assume $\beta = 100$ for each transistor.

(*a*) Determine R so that the quiescent conditions are $V_{CE1} = -5$ V and $V_{CE2} = -6$ V. Assume $V_{BE} = -0.6$ V.

(*b*) Explain how quiescent-point stabilization is obtained.

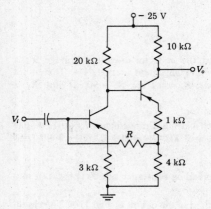

Prob. 11-7

11-8 Assume that a silicon transistor with $\beta = 60$, $V_{BE(\text{active})} = 0.7$ V, $V_{CC} = 20$ V, and $R_c = 5$ kΩ is used in Fig. 11-4a. It is desired to establish a Q point at $V_{CE} = 10$ V, $I_C = 1.3$ mA, such that, if V_{BE} and β remain constant, $\Delta I_C \leq 4\Delta I_{CO}$. Find R_e, R_1, and R_2.

11-9 A transistor with $\beta = 50$ and $V_{BE} = 0.80$ V is used in the self-bias circuit with $V_{CC} = 20$ V. The quiescent point is $I_C = 2$ mA and $V_{CE} = 14$ V. The transistor is replaced by another with $\beta = 200$ and $V_{BE} = 0.60$ V. (I_{CO} does not change appreciably.) It is desired that the effect of the change in β should not increase I_C by more than 0.1 mA and that the same should be true for the change in V_{BE}. In other words, the new value of I_C with transistor replacement should not exceed 2.2 mA. Calculate the values of the four resistors R_e, R_c, R_1, and R_2.

11-10 Consider a germanium transistor used in the self-bias circuit of Fig. 11-4 with $R_e = 4.7$ kΩ, $R_b = 7.75$ kΩ, and $I_C = 1.5$ mA at 25°C. Using Table 11-2 calculate I_C at -65 and at $+75$°C. Is the effect of V_{BE} or I_{CO} more important on I_C over the range 25 to 75°C?

11-11 Calculate the change ΔI_C due to I_{CO} and V_{BE} for a silicon transistor in the circuit of Fig. 11-4 with $R_e = 4.7$ kΩ, $R_b = 7.75$ kΩ, and $I_C = 1.5$ mA at 25°C over the range 25 to 145°C. Which of the two affect I_C the most?

11-12 In the circuit of Fig. 11-4, let $R_c = 6$ kΩ, $R_e = 1$ kΩ, $R_1 = 100$ kΩ, $R_2 = 20$ kΩ, $I_C = 2$ mA at 25°C. Using the transistor of Table 11-1, find I_C at +175 and −65°C.

11-13 Repeat Prob. 11-12 for the transistor of Table 11-2 at +75 and −65°C.

11-14 In an emitter-follower circuit, $R_e = 1$ kΩ and V_{CC} and V_{EE} are adjusted to give $I_C = 2$ mA at 25°C. Using the transistor of Table 11-1, find I_C at +175 and −65°C.

11-15 Repeat Prob. 11-14 for the transistor of Table 11-2 at +75 and −65°C.

11-16 Find R_e for an emitter-follower by using a silicon transistor to meet the specifications of Example 11-4 of Sec. 11-4.

11-17 A silicon n-p-n transistor has a collector current of 1.5 mA at 25°C, and $r_{b'e} = 2$ kΩ. Find a good approximation for h_{fe}.

11-18 A CE stage has a load resistor R_L between collector and the supply voltage V_{CC}. If $r_{bb'} \ll r_{b'e}$ show that the voltage gain is

$$A_V = -g_m R_L = -\frac{|V_{CC} - V_o|}{V_T}$$

where V_o is the quiescent output voltage.

11-19 The amplifier shown uses a transistor whose h parameters are $h_{fe} = 200$ and $h_{ie} = 4$ kΩ. Apply the rules given in Sec. 11-8 and use the approximate model to calculate (a) $A_I \equiv I_o/I_i$; (b) $R_i = V_i/I_i$; (c) $A_V = V_o/V_i$; (d) $A_{Vs} = V_o/V_s$; (e) R_o; (f) R_o'. The capacitance of C may be assumed to be arbitrarily large, so that it acts as a short circuit at the signal frequency.

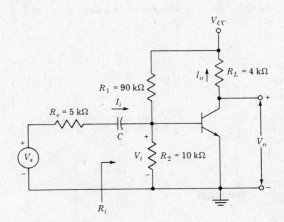

Prob. 11-19

11-20 Apply the rules given in Sec. 11-8 to the circuit shown to draw its approximate h-parameter model. Assuming $h_{fe} = 50$ and $h_{ie} = 1.1$ kΩ, calculate (a) $A_I \equiv -I_2/I_b$; (b) $R_i \equiv V_b/I_b$; (c) $R_i' \equiv V_b/I_2$; (d) $A_I' \equiv -I_2/I_1$; (e) $A_V \equiv V_o/V_b$; (f) $A_{Vs} \equiv V_o/V_s$; (g) R_o; (h) R_o'.

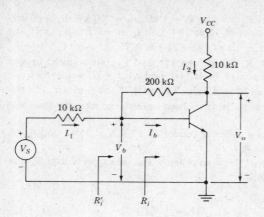

Prob. 11-20

11-21 (*a*) Draw the equivalent circuits for the CE and CC configurations using the approximate model of Fig. 11-6*a* subject to the restriction that $R_L = 0$. Show that the input resistances of the two circuits are identical.

(*b*) Draw the circuits for the CE and CC configurations subject to the restriction that the input is open-circuited. Show that the output resistances of the two circuits are identical. Prove that $R_o' = R_L$.

11-22 The circuit shown is an amplifier using a *p-n-p* and an *n-p-n* transistor in parallel. The two transistors have identical characteristics. Find the expressions for the voltage gain and the input resistance of the amplifier, using the simplified hybrid model of Fig. 11-6*a*.

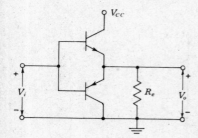

Prob. 11-22

11-23 For a CB connection derive the simplified expressions given in Table 11-4.

11-24 Verify the numerical values in Table 11-5 for the following configurations: (*a*) CB, (*b*) CC, and (*c*) CE.

(*d*) Find R_o' and A_{V_s} for each of the three configurations.

11-25 Apply the rules of Sec. 11-8 to the circuit shown to draw its approximate h-parameter model. Find expressions for the voltage gain V_o/V_s and input resistance R_i.

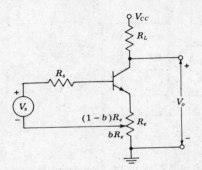

Prob. 11-25

11-26 Modify the circuit of Prob. 11-19 by placing a resistor R_e of 1 kΩ between the emitter and ground. For the resulting circuit, compute $A_I = I_o/I_i$, $R_i = V_i/I_i$, A_V, A_{Vs}, R_o, and R_o'.

11-27 Consider the circuit of Prob. 11-26, but take the output voltage V_o' across R_e. Find $A_{Vs}' = V_o'/V_s$.

11-28 (a) For the two-stage cascade shown, compute the input and output resistances and the individual and overall voltage and current gains, using the approximate formulas in Table 11-4 with $h_{fe} = 100$, and $h_{ie} = 3$ kΩ. *Note:* Assume each capacitor is arbitrarily large so that it may be considered a short circuit at the signal frequency.

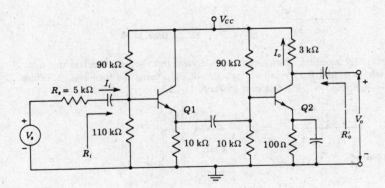

Prob. 11-28

11-29 Compute A_I, A_V, A_{Vs}, R_i, and R_o' for the two-stage cascade shown, using the approximate formulas in Table 11-4 with $h_{fe} = 200$ and $h_{ie} = 4$ kΩ. See note in Prob. 11-28.

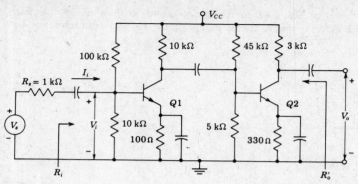

Prob. 11-29

11-30 For the circuit shown, compute A_I, A_V, A_{Vs}, R_i, and R_o' using the formulas of Table 11-4 with $h_{fe} = 100$ and $h_{ie} = 4$ kΩ.

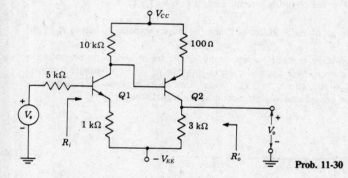

Prob. 11-30

11-31 The three-stage amplifier shown contains identical transistors. Calculate the voltage gain of each stage and the overall voltage gain V_o/V_s, using the formulas of Table 11-4 with $h_{fe} = 50$ and $h_{ie} = 2$ kΩ. See note in Prob. 11-28.

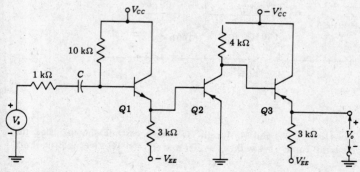

Prob. 11-31

11-32 From the definitions of h_{fe} and h_{oe} in Sec. 11-15, obtain numerical values for these parameters at the Q point $I_C = 25$ mA, $I_B = 120$ μA, and $V_{CE} = 5$ V for the 2N2222A transistor whose characteristics are given in Fig. 3-9.

11-33 Using the four-parameter model of Fig. 11-17, draw the equivalent circuit for the CE amplifier of Fig. 11-8a. Show that

(a) $A_I \equiv \dfrac{I_o}{I_b} = -\dfrac{h_{fe}}{1 + h_{oe}R_L}$.

(b) $R_i \equiv \dfrac{V_i}{I_b} = h_{ie} + h_{re}A_I R_L$.

(c) $A_V \equiv \dfrac{V_o}{V_i} = \dfrac{A_I R_L}{R_i}$.

(d) $Y_o = \dfrac{1}{R_o} = h_{oe} - \dfrac{h_{fe}h_{re}}{h_{ie} + R_s}$.

11-34 The voltage gain A_V of a CE stage is proportional to the load R_L if the approximate low-frequency small-signal model is used. However, if R_L is so large that the inequality Eq. (11-55) is not satisfied, then the four-parameter model must be used.

(a) If an ideal constant-current source is used as a load (so that $R_L = \infty$), show that

$$A_v = \frac{-h_{fe}}{h_{ie}h_{oe}}\frac{1}{\gamma} \qquad \text{where} \qquad \gamma \equiv 1 - \frac{h_{re}h_{fe}}{h_{ie}h_{oe}}$$

(b) Verify that $R_i = h_{ie}\gamma$.

(c) Evaluate A_v and R_i for the transistor whose parameter values are given in Eq. (11-54).

11-35 Repeat Prob. 11-34, using the hybrid-π model of Fig. 11-18. Make use of the inequalities $r_{b'c} \gg r_{ce} \gg r_{b'e} \gg r_{bb'}$, and $g_m \gg 1/r_{b'c}$ to verify that:

(a) $A_v = \dfrac{-g_m r_{b'e}}{1 + g_m r_{ce}r_{bb'}/r_{b'c}} \approx -g_m r_{ce}$.

(b) $R_i = \dfrac{r_{b'e}}{1 + g_m r_{ce}r_{b'e}/r_{b'c}}$.

(c) Evaluate A_v and R_i, using the parameter values in Eq. (11-56).

11-36 (a) In the circuit of Fig. 11-13a, find the input resistance R_i in terms of the four CE h parameters, R_L and R_e. *Hint:* Follow the rules given in Sec. 11-8.

(b) If $R_L = R_e = 2$ kΩ, calculate R_i with the parameter values in Eq. (11-54).

11-37 For a CE configuration, what is the maximum value of R_L for which R_i differs by no more than 10 percent of its value at $R_L = 0$? Use the transistor parameters given in Eq. (11-54) and the results of Prob. 11-33.

11-38 (a) Consider the hybrid-π circuit. If the load resistance is $R_L = 1/g_L$, prove that

$$K \equiv \frac{V_{ce}}{V_{b'e}} = \frac{-g_m + g_{b'c}}{g_{b'c} + g_{ce} + g_L}$$

Hint: Use the theorem that the voltage between C and E equals the short-circuit current times the resistance seen between C and E, with the input voltage $V_{b'e}$ short-circuited.

11-39 Draw the equivalent circuit for the CE amplifier of Fig. 11-8a, using the hybrid-π model with $r_{b'c} = \infty$ (open-circuited). Obtain an expression for $A_I \equiv I_o/I_b$ and show that it is equivalent to that of Prob. 11-33a.

11-40 The h-parameters for a 2N2222A transistor operating at 13 mA at room temperature are $h_{ie} = 700 \ \Omega$, $h_{re} = 10^{-4}$, $h_{fe} = 250$, and $h_{oe} = 10^{-4} \ \Omega$. Find the parameters of the hybrid-π model.

11-41 For the emitter follower with $R_s = 1 \ \text{k}\Omega$ and $R_L = 3 \ \text{k}\Omega$, calculate A_I, R_i, A_V, A_{Vs} and R_o. Assume $h_{fe} = 100$, $h_{ie} = 2 \ \text{k}\Omega$, $h_{oe} = 10 \ \mu\text{A/V}$.

11-42 (a) Calculate R_L for an emitter follower having $R_i = 600 \ \text{k}\Omega$ and $R_o = 25 \ \Omega$. Assume $h_{fe} = 80$, $h_{ie} = 2 \ \text{k}\Omega$, $h_{oe} = 20 \ \mu\text{A/V}$.

(b) Find A_I and A_V for the emitter follower of part (a).

(c) Find R_i and R_L so that $A_V = 0.999$.

11-43 For the amplifier shown, using a transistor whose four parameters are given in Eq. (11-54), compute $A_I = I_o/I_i$, A_V, A_{Vs}, and R_i. *Hint:* Follow the rules given in Sec. 11-8. (Refer to note in Prob. 11-28.)

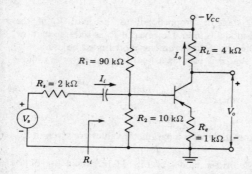

Prob. 11-43

11-44 (a) Prove that the output conductance $G_o = 1/R_o$ for an emitter-follower, taking all four h parameters into account, is given by Eq. (11-73). Note that $h_{re} \ll 1$.

(b) For the Darlington composite emitter follower of Sec. 11-16 calculate the output resistance if $R_s = 3 \ \text{k}\Omega$. Compare this value with R_{o1} for a single CC stage.

(c) Repeat part (b) if $R_s = 0$.

11-45 Assume $h_{fe1} = h_{fe2} = h_{fe}$ and $h_{ie1} = (1 + h_{fe})h_{ie2}$ for the Darlington circuit of Fig. 11-20b. Take $h_{fe} = 100$, $h_{ie2} = 2.1 \ \text{k}\Omega$, $h_{oe1} = h_{oe2} = 10^{-5} \ \Omega$, and a source resistance $R_s = 10 \ \text{k}\Omega$. Calculate (a) R_i, (b) A_V, and (c) R_o; use Eq. (11-73).

(d) Repeat part (c) for $R_s = 0$.

(e) Calculate R_o for a single stage with $h_{ie} = 2.1 \ \text{k}\Omega$ for $R_s = 10 \ \text{k}\Omega$ and also for $R_s = 0$. Note that in part (c) R_o is less than that of an emitter follower and in part (d) R_o is larger than that of an emitter follower.

11-46 Assume $h_{fe1} = h_{fe2} = h_{fe} \gg 1$, $h_{ie1} = (1 + h_{fe})h_{ie2}$, and $h_{oe} = 0$ for the Darlington circuit of Fig. 11-20b.

(a) Use Eq. (11-73) to prove that $R_o \approx 2h_{ie2}/h_{fe}$.

(b) Prove that the output resistance is less than that of a single CC stage (with $h_{ie} = h_{ie2}$) if $R_s > h_{ie2}$, but may be greater than that of a single CC stage if $R_s < h_{ie2}$.

11-47 In the boot-strapped circuit of Fig. 11-21b assume that C is very large so that it acts as a short circuit at the signal frequency. Verify that the effective emitter resistor consists of the following four resistors in parallel: R_e, R_1, R_2, and $A_V R_3/(A_V - 1)$. Show that one of these resistors is negative but that the parallel combination is positive.

11-48 In the circuit of Fig. 11-21b let $R_1 = R_2 = 20$ kΩ, $R_3 = 10$ kΩ, and $R_e = 2$ kΩ. Use the concept of the effective resistance [Eq. (11-74)] and the result of Prob. 11-47 to find (*a*) R_i, (*b*) A_V, (*c*) R_i', and (*d*) $A_I = I_o/I_i$.

    (*e*) If C' were missing in Fig. 11-21b, calculate R_i'. Compare with part (*c*). Use a transistor with $h_{fe} = 100$, $h_{ie} = 2$ kΩ, $h_{oe} \approx 0$, and $h_{re} \approx 0$. *Hint:* Initially assume $A_V \approx 1$ to account for the effect of R_3 on the input and output circuits.

11-49 Consider an emitter follower using two power supplies as described in Sec. 11-16. The input signal is a sinusoid with respect to ground. It is desired to operate at a quiescent current of 2 mA at 25°C, and that this current should vary by no more than 10 percent from this value at the extreme temperatures for silicon in Table 11-1.

    (*a*) Find the minimum value that can be used for R_e.

    (*b*) Calculate the corresponding V_{EE}.

11-50 The CS amplifier stage shown in Fig. 11-22 has the following parameters: $R_d = 10$ kΩ, $R_g = 0.9$ MΩ, $R_s = 500$ Ω, $V_{DD} = 25$ V, C_s is arbitrarily large, $I_{DSS} = 2.5$ mA, and $V_P = -2.2$ V. Determine (*a*) the gate-to-source bias voltage V_{GS}; (*b*) the drain current I_D; (*c*) the quiescent voltage V_{DS}.

11-51 FET 2N3684, whose transfer curves are shown in Fig. 11-25, is used in the circuit of Fig. 11-24b. It is desired to bias the circuit so that $I_{D(min)} = 0.4$ mA and $I_{D(max)} = 0.9$ mA for $V_{DD} = 20$ V. Find (*a*) R_s, and (*b*) the values of R_1 and R_2 for which $R_g = 0.5$ MΩ. *Note:* Use ($V_{GS} = -1$ V, $I_D = 0.4$ mA) and $(-4, 0.8)$ as two points on the bias line in Fig. 11-25.

11-52 The FET used in the circuit of Fig. 11-24b has $I_{DSS} = 3$ mA and $V_P = -3$ V. Let $R_1 = 1.5$ MΩ, $R_2 = 300$ kΩ, $R_d = 20$ kΩ, $R_s = 5$ kΩ, and $V_{DD} = 65$ V. Calculate the quiescent values I_D, V_{GS}, and V_{DS}.

11-53 The FET shown has the following parameters: $I_{DSS} = 5$ mA and $V_P = -4.5$ V.

    (*a*) If $v_i = 0$, find v_o.

    (*b*) If $v_i = 12$ V, find v_o.

    (*c*) If $v_o = 0$, find v_i.

    Note: v_i and v_o are constant voltages (and not small-signal voltages).

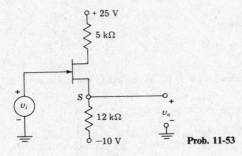

Prob. 11-53

11-54 The drain current in milliamperes of the enhancement-type MOSFET shown is given by

$$I_D = 0.3(V_{GS} - V_P)^2$$

in the region $V_{DS} \geqslant V_{GS} - V_P$. If $V_P = +4$ V, calculate the quiescent values I_D, V_{GS}, and V_{DS}.

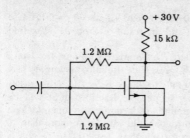

Prob. 11-54

11-55 The drain current in milliamperes of the depletion-type FET shown is given by

$$I_D = 16\left(1 + \frac{V_{GS}}{4}\right)^2$$

Calculate the quiescent current I_D and the quiescent value of the transconductance g_m.

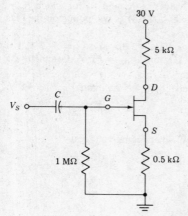

Prob. 11-55

11-56 (a) Show that the transconductance g_m of a JFET is related to the drain current I_{DS} by

$$g_m = \frac{2}{|V_P|} \sqrt{I_{DSS}I_{DS}}$$

(b) If $V_P = -3$ V and $I_{DSS} = 3$ mA, plot g_m versus I_{DS}.

(c) Show that for small values of V_{GS} compared with V_P, the drain current is given approximately by $I_D \approx I_{DSS} + g_{mo}V_{GS}$.

11-57 (a) Verify Eq. (11-79).

(b) Starting with the definitions of g_m and r_d, show that, if two FETs are connected in parallel, r_d and μ are given by

$$\frac{1}{r_d} = \frac{1}{r_{d1}} + \frac{1}{r_{d2}} \qquad \text{and} \qquad \mu = \frac{\mu_1 r_{d2} + \mu_2 r_{d1}}{r_{d1} + r_{d2}}$$

11-58 (a) For a source follower verify that the output conductance g_o is equal to the transconductance g_m, if $\mu \gg 1$.

(b) If a FET is used as a load resistance by connecting gate and drain (Fig. 8-14b), show that the dynamic conductance equals g_m.

11-59 If an input signal V_i is impressed between gate and ground, find the amplification $A_V = V_o/V_i$. The FET parameters are $\mu = 40$ and $r_d = 10$ kΩ. The input V_i is impressed from the source S to ground.

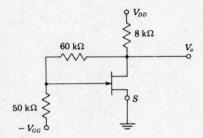

Prob. 11-59

11-60 If in Prob. 11-59 the signal V_i is impressed in series with the 50-kΩ resistor (instead of from gate to ground), find $A_V = V_o/V_i$. *Hint:* Apply the rules of Sec. 11-8, and use the model in Fig. 11-27.

11-61 The circuit shown is called the common-gate amplifier. For this circuit find (a) the voltage gain, (b) the input impedance, (c) the output impedance. Power supplies are omitted for simplicity. *Hint:* Apply the rules of Sec. 11-8 and use the model in Fig. 11-27.

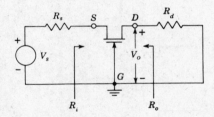

Prob. 11-61

11-62 (a) Calculate the voltage gain $A_V = V_o/V_i$ at 1 kHz for the circuit shown. The FET parameters are $g_m = 1$ mA/V and $r_d = 15$ kΩ. Neglect capacitances.

(b) Repeat part (a) if the capacitance 0.002 μF is taken under consideration.

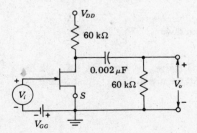

Prob. 11-62

11-63 (a) If in the amplifier stage shown the positive supply voltage V_{DD} changes by $\Delta V_{DD} = v_a$, how much does the drain-to-ground voltage change? *Hint:* Consider v_a as an externally applied signal.

(b) How much does the source-to-ground voltage change under the conditions in part (a)?

(c) Repeat parts (a) and (b) if V_{DD} is constant but V_{SS} changes by $\Delta V_{SS} = v_b$.

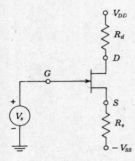

Prob. 11-63

11-64 Find an expression for the signal voltage across R_L. The two FETs are identical, with parameters μ, r_d, and g_m. *Hint:* Use the equivalent circuits in Fig. 11-30 at S_2 and D_1.

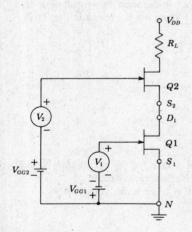

Prob. 11-64

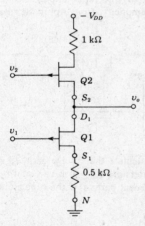

Prob. 11-65

11-65 FET $Q1$ in the circuit shown has $r_{d1} = 10$ kΩ and $g_{m1} = 3$ mA/V. FET $Q2$ has $r_{d2} = 15$ kΩ and $g_{m2} = 2$ mA/V. Using the equivalent circuits in Fig. 11-30 at S_2 and D_1, find the gain (a) v_o/v_1 if $v_2 = 0$, (b) v_o/v_2 if $v_1 = 0$.

11-66 (a) If the two FETs are identical, verify that the voltage gain is given by

$$A_V = \frac{-\mu[r_d + (\mu + 1)R_1]}{2r_d + (\mu + 1)(R_1 + R_2)}$$

(b) Show that the output conductance is given by

$$G_o = \frac{1}{R_o} = \frac{1}{r_d + (\mu + 1)R_1} + \frac{1}{r_d + (\mu + 1)R_2}$$

(c) If $R_1 = R_2 = R$, find A_V and R_o.

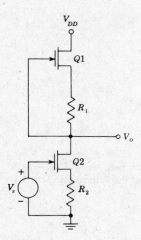

Prob. 11-66

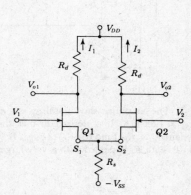

Prob. 11-67

11-67 (a) In the circuit shown $V_2 = 0$. Solve for the current I_2 by drawing the equivalent circuit, looking to the left into the source of $Q1$ and looking to the right into the source S_2 (Fig. 11-30). The source resistance may be taken as arbitrarily large. Show that $V_{o1} = -V_{o2}$ so that this circuit is a phase inverter.

  (b) Obtain the same answer by proceeding as follows: First draw the equivalent circuit looking into the source of $Q1$. Then replace $Q2$ by the equivalent circuit looking into its drain.

11-68 In the circuit of Prob. 11-67, assume that $V_2 = 0$, $R_d = r_d = 15$ kΩ, $R_s = 2$ kΩ, and $\mu = 24$. If the output is taken from the drain of $Q2$, find (a) the voltage gain and (b) the output resistance. *Hint:* Use the equivalent circuits in Fig. 11-30.

11-69 (a) In the circuit of Prob. 11-67, $V_2 \neq V_1$ and $R_s \neq \infty$. Draw the equivalent circuit as suggested in Prob. 11-67a. Find the voltage gains A_1 and A_2 defined by $V_{o2} = A_1V_1 + A_2V_2$.

  (b) If $R_s = \infty$, show that $A_2 = -A_1$ so that $V_{o2} = A_1(V_1 - V_2)$, and hence the circuit behaves as a *difference amplifier*.

11-70 If $h_{ie} \ll R_d$, $h_{ie} \ll r_d$, $h_{fe} \gg 1$, and $\mu \gg 1$ for the circuit, show that

$$(a) \ A_{V1} = \frac{v_{o1}}{v_i} \approx \frac{g_m h_{fe} R_s}{1 + g_m h_{fe} R_s} \qquad (b) \ A_{V2} = \frac{v_{o2}}{v_i} \approx \frac{g_m h_{fe} (R_s + R_c)}{1 + g_m h_{fe} R_s}$$

where g_m is the FET transconductance. Use the approximate model for the bipolar transistor.

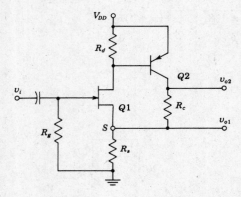

Prob. 11-70

11-71 The amplifier shown utilizes an n-channel FET for which $V_P = -2.0$ V and $I_{DSS} = 1.65$ mA. It is desired to bias the circuit at $I_D = 0.8$ mA, using $V_{DD} = 24$ V. Assume $r_d \gg R_d$. Find (a) V_{GS}, (b) g_m, (c) R_s, (d) R_d, such that the voltage gain is at least 20 dB, with R_s bypassed with a very large capacitance C_s.

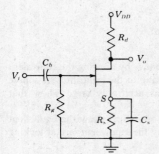

Prob. 11-71

Chapter 12

12-1 For the circuit shown, with $R_c = 3$ kΩ, $R_L = 3$ kΩ, $R_b = 30$ kΩ, and $R_s = 2$ kΩ, the transistor parameters are $h_{ie} = 2.1$ kΩ and $h_{fe} = 100$. Find: (a) The current gain $I_L/I_s = A_I$; (b) the voltage gain V_o/V_s, where $V_s \equiv I_s R_s$; (c) the transconductance $I_L/V_s = G_M$.

(d) the transresistance $V_o/I_s = R_M$; (e) the input resistance seen by the source; (f) the output resistance seen by the load.

Make reasonable approximations. Neglect all capacitive effects.

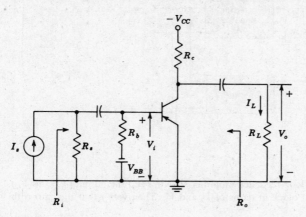

Prob. 12-1

12-2 Repeat Prob. 12-1 for the circuit shown, with $g_m = 3$ mA/V and $r_d = 80$ kΩ. Note that $V_s \equiv I_s R_s$.

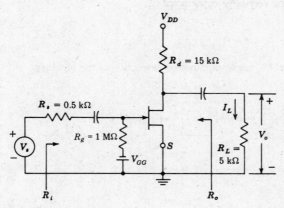

Prob. 12-2

12-3 (a) For the circuit shown, find the ac voltage V_i as a function of V_s and V_f. Assume that the inverting-amplifier input resistance is infinite, that $A = A_V = -2,000$, $\beta = V_f/V_o = \frac{1}{150}$, $R_s = R_e = 1$ kΩ, $R_c = 3$ kΩ, $h_{ie} = 2$ kΩ, $h_{fe} = 200$, and $h_{re} = h_{oe} = 0$.
(b) Find $A_{Vf} = V_o/V_s$.

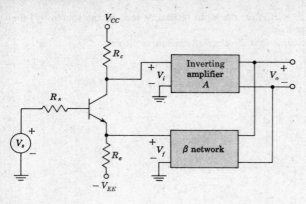

Prob. 12-3

12-4 (*a*) Consider the two-loop feedback amplifier shown. Find the output (V_{i1} and V_{i2}) of each mixer (comparator) block in terms of V_i and V_o. Then verify that the gain with feedback is

$$A_{Vf} = \frac{V_o}{V_i} = \frac{A_1 A_2}{1 + A_2 B_2 + A_1 A_2 \beta_1}$$

(*b*) from the result of part (*a*) it is clear that

$$A_{Vf} = \frac{A_1 A_{2f}}{1 + A_1 A_{2f} \beta_1}$$

where

$$A_{2f} \equiv \frac{A_2}{1 + A_2 \beta_2}$$

Explain how this expression for A_{Vf} can be obtained by inspection of the topology.

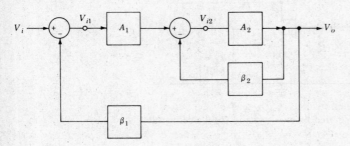

Prob. 12-4

12-5 In the block diagram shown, V_1 is a disturbance (say, signal noise) introduced at the input, V_2 is a disturbance introduced within the basic amplifier (perhaps due to power supply hum), and V_3 is a disturbance introduced at the amplifier output. The amplifier open-loop gain is $A = A_1 A_2$. Verify that

$$V_o = \frac{A[(V_s + V_1) + V_2/A_1 + V_3/A]}{1 + \beta A}$$

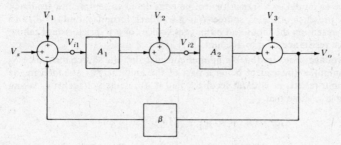

Prob. 12-5

12-6 (*a*) Prove Eq. (12-6).

(*b*) An amplifier consists of three identical stages connected in cascade. The output voltage is sampled and returned to the input in series opposing. If it is specified that the relative change dA_f/A_f in the closed-loop voltage gain A_f must not exceed Ψ_f, show that the minimum value of the open-loop gain A of the amplifier is given by

$$A = 3A_f \frac{|\Psi_1|}{|\Psi_f|}$$

where $\Psi_1 \equiv dA_1/A_1$ is the relative change in the voltage gain of each stage of the amplifier.

12-7 An amplifier with open-loop voltage gain $A_V = 2{,}000 \pm 150$ is available. It is necessary to have an amplifier whose voltage gain varies by no more than ± 0.2 percent.

(*a*) Find the reverse transmission factor β of the feedback network used.

(*b*) Find the gain with feedback.

12-8 An amplifier without feedback gives a fundamental output of 30 V with 10 percent second-harmonic distortion when the input is 0.025 V.

(*a*) If 1.5 percent of the output is fed back into the input in a negative voltage-series feedback circuit, what is the output voltage?

(*b*) If the fundamental output is maintained at 30 V but the second-harmonic distortion is reduced to 1 percent, what is the input voltage?

12-9 An amplifier with an open-loop voltage gain of 1,500 delivers 15 W of output power at 5 percent second-harmonic distortion when the input signal is 12 mV. If 45-dB negative voltage-series feedback is applied and the output power is to remain at 15 W, determine (*a*) the required input signal and (*b*) the percent harmonic distortion.

12-10 (*a*) Verify Eq. (12-16) for the input resistance of the current-series feedback amplifier.

(*b*) Repeat part (*a*) for Eq. (12-25) for the voltage-shunt amplifier.

(*c*) Verify Eq. (12-32) for the output resistance of the voltage-shunt feedback amplifier. Reduce the input signal to zero, replace the load by a voltage source V, and find $R_{of} = V/I$, where I is the current drawn from V.

(*d*) Repeat part (*c*) for Eq. (12-38) for the current-series feedback amplifier.

12-11 The output resistance may be calculated as the ratio of the open-circuit voltage to the short-circuit current. Using this method, evaluate R_{of} and R'_{of} for (a) voltage-series feedback, (b) current-series feedback, (c) current-shunt feedback, and (d) voltage-shunt feedback.

12-12 The h-parameter model of a transistor can be considered to represent a feedback amplifier due to the presence of the h_{re} source. Using feedback formulas, find (a) R_{if} and (b) $Y_{of} = 1/R_{of}$, representing the input and output resistances of a transistor stage taking h_{re}, h_{oe}, and a source resistance R_s into account.

12-13 (a) For the voltage-series feedback amplifier consider that the source resistance R_s is external to the amplifier (instead of being a part of the amplifier, as assumed in the text). If R'_i is the input resistance without feedback and if A_{Vs} is the voltage gain, taking R_L and R_s into account, show that

$$R_{if} = (R_s + R'_i)(1 + \beta A_{Vs})$$

(b) If A_V is the voltage amplification without feedback, taking R_L but not R_s into account, verify from the above equation that

$$R_{if} = R_s + R'_i(1 + \beta A_V)$$

Interpret this equation physically.

12-14 Assume that the parameters of the circuit are $r_d = 8$ kΩ, $R_g = 1$ MΩ, $R_1 = 50$ Ω, $R_d = 40$ kΩ, and $g_m = 5$ mA/V. Neglect the reactances of all capacitors. Find the voltage gain and output impedance of the circuit at the terminals (a) AN and (b) BN.

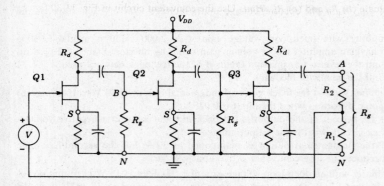

Prob. 12-14

12-15 The transistors in the feedback amplifier shown are identical and $h_{fe} = 150$, $h_{ie} = 2.5$ kΩ. All resistance values are given in kilohms. Neglect the reactances of all capacitors and the shunting effect of all resistors whose numerical values are not indicated.

(a) Identify the topology and calculate A_{Vf}.
Evaluate (b) R_{if}, (c) R_{of}, and (d) R'_{of}.

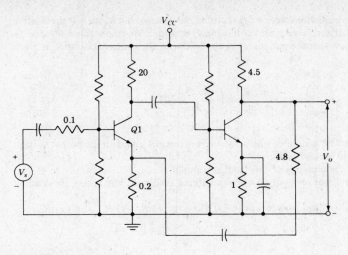

Prob. 12-15

12-16 The FETs in the feedback amplifier shown are identical and $\mu = 30$, $r_d = 10$ kΩ. Assume $R_d = 50$ kΩ, $R_s = 0.3$ kΩ, $R = 10$ kΩ, and $R_g = 1$ MΩ. Neglect the reactances of all capacitors.

    (a) Identify the topology and calculate A_{Vf}.

    Evaluate (b) R_{if} and (c) R'_{of}. *Hint:* Use the equivalent circuit in Fig. 11-30 for each FET stage.

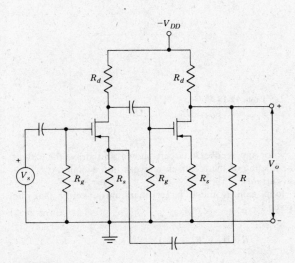

Prob. 12-16

12-17 Consider the transistor stage of Fig. 12-16a, and assume that $h_{fe} \gg 1$. If the relative change dA_f/A_f of the voltage gain A_f must not exceed a specified value Ψ_f due to variations of h_{fe}, show that the minimum required value of the emitter resistor R_e is given by

$$R_e = \frac{R_s + h_{ie}}{h_{fe}} \left(\frac{dh_{fe}/h_{fe}}{\Psi_f} - 1 \right)$$

12-18 Consider the FET amplifier with a source resistance R and with the output voltage V_o from the drain to ground as shown.

(a) Identify the topology of this feedback amplifier.

(b) Find the input and output circuits without feedback, but taking the loading into account.

Find (c) G_{Mf}, (d) A_{Vf} [compare with Eq. (11-86)], (e) R_{if}, (f) R_{of}, and (g) R'_{of}.

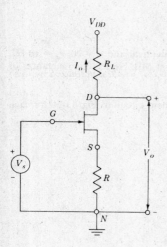

Prob. 12-18

12-19 (a) Identify the topology for the feedback circuit shown and draw the circuit without feedback but taking the loading of the feedback network into account.

(b) Find β. Neglect base currents.

(c) Assuming that the loop gain is much larger than unity, verify that the voltage-gain with feedback is given by

$$A_{Vf} \approx - \frac{R_3(R_4 + R_5 + R')}{R_4 R_5}$$

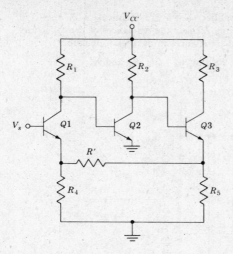

Prob. 12-19

12-20 In the feedback amplifier shown all resistances are in kilohms. Do *not* worry about quiescent conditions. The transistors are identical with the following parameters: $h_{fe} = 100$, $h_{ie} = 2$ kΩ.

(*a*) Identify the topology. Calculate the transfer gain A_f which is stabilized by this amplifier.

Find (*b*) A_{Vf}, (*c*) R_{if} and R'_{if}, and (*d*) R_{of} and R'_{of}.

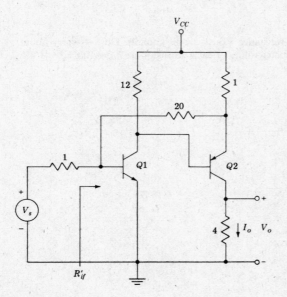

Prob. 12-20

12-21 The transistors in the feedback amplifier shown are identical, with $h_{fe} = 150$, $h_{ie} = 3$ kΩ. All resistance values are in kilohms. Neglect the reactances of the capacitors. Note that the load consists of the two 4.7-kΩ resistances in parallel. Calculate

(a) $A_{If} \equiv \dfrac{I_o}{I_s}$; (b) $A_{Vf} \equiv \dfrac{V_o}{V_s}$; (c) R_{if} and R'_{if}; (d) R_{of} and R'_{of}.

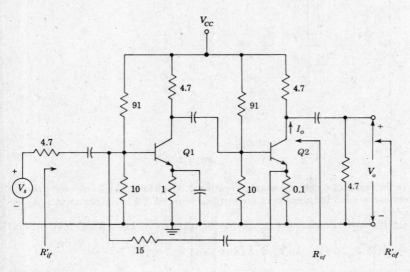

Prob. 12-21

12-22 For the circuit shown all resistance values are in kilohms. Do *not* worry about quiescent conditions. The *h*-parameter values of each transistor are $h_{fe} = 100$, $h_{ie} = 1$ kΩ, $h_{re} = 0$, $h_{oe} = 0$.

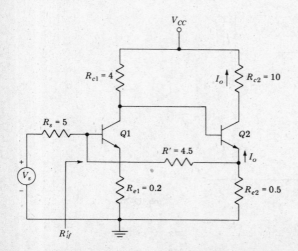

Prob. 12-22

(a) Identify the topology if the output is taken from the collector of $Q2$. Calculate the transfer gain A_f which is stabilized by this feedback amplifier.

Find (b) A_{Vf}, (c) R_{if} and the resistance seen by V_s, and (d) R_{of} and R'_{of}.

12-23 For the transistor feedback-amplifier stage shown, $h_{fe} = 150$, $h_{ie} = 2$ kΩ, while h_{re} and h_{oe} are negligible. All resistance values are in kilohms. Determine with $R_e = 0$:

    (a) $R_{Mf} = \dfrac{V_o}{I_s}$.

    (b) $A_{Vf} = \dfrac{V_o}{V_s}$, where $I_s = \dfrac{V_s}{R_s}$.

    (c) R_{if}.

    (d) R'_{of}.

    (e) Repeat the four preceding calculations if $R_e = 0.5$ kΩ.

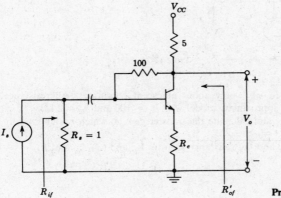

Prob. 12-23

12-24 (a) Identify the topology of the feedback amplifier shown. The capacitor is arbitrarily large. The parameters of the FET are $r_d = 40$ kΩ and $g_m = 2$ mA/V. All resistance values are in kilohms.

Calculate (b) R_{Mf}, (c) A_{Vf}, (d) R_{if}, and (e) R'_{of}.

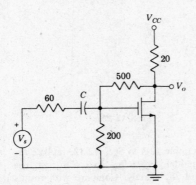

Prob. 12-24

12-25 Repeat Prob. 12-22 if the output is now taken from the emitter of $Q2$.

12-26 For the circuit shown:

(a) Identify the topology and obtain a formula for the transfer gain A_f which is stabilized by the feedback.

(b) What are the conditions for which $A_f \approx 1/\beta$?

(c) Prove that

$$A_{Vf} = \frac{V_o}{V_s} = -\frac{R'}{R} \frac{1}{1 + \dfrac{R'}{R_m}\left(\dfrac{R_i + R'}{R'} + \dfrac{R_i}{R}\right)}$$

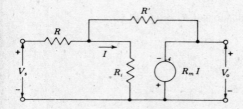

Prob. 12-26

12-27 For the circuit shown do *not* worry about quiescent conditions. All resistance values are in kilohms. Use the approximate model with $h_{fe} = 100$ and $h_{ie} = 2\ \text{k}\Omega$.

(a) Identify the topology and calculate the transfer gain A_f which is stabilized by this amplifier.

Calculate (b) A_{Vf}, (c) R_{if} and the resistance seen by V_s, and (d) R_{of} and R'_{of}.

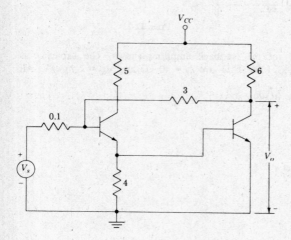

Prob. 12-27

12-28 For the voltage-shunt feedback circuit in Example 12-4 in Sec. 12-12, replace the transistor by the low-frequency approximate model ($h_{oe} = h_{re} = 0$). Do *not* use feedback-analysis methods. Solve for A_{Vf}, R_{if}, and R_{of} exactly. Compare with results obtained in Sec. 12-12.

12-29 (*a*) For the voltage-shunt feedback amplifier of Fig. 12-19, show that

$$\lim_{R_s \to 0} R_M = 0$$

(*b*) Evaluate

$$A_{Vf} = \lim_{R_s \to 0} \frac{R_{Mf}}{R_s}$$

Check your result by obtaining the voltage gain of Fig. 12-19 if $R_s = 0$ (in which case this is no longer a feedback amplifier).

(*c*) Evaluate numerically for the parameter values in Example 12-4 (from Table 12-4)

$$R_m = \lim_{R_c \to \infty} R_M \quad \text{and} \quad R_{of}$$

(*d*) Calculate $R'_{of} = R_{of} \| R_c$.

12-30 (*a*) Show that the first assumption in Sec. 12-3 is not satisfied exactly for the voltage-shunt feedback amplifier of Fig. 12-19. *Hint:* If the amplifier is deactivated by reducing h_{fe} to zero, find the current I_f which passes through the β network (the resistor R') from input to output.

(*b*) In order for the forward transmission through the feedback network to be neglected it is necessary that I_f in part (*a*) be very small compared with I_o (with the amplifier activated). Show that this condition is satisfied if

$$A_{Vf} \gg \frac{R_c}{R_s + R' + R_c}$$

Chapter 13

13-1 (*a*) To show the effect of phase shift on the image seen on a cathode-ray screen, consider the following example: The sinusoidal voltages applied to both sets of plates should be equal in phase and magnitude so that the maximum displacement in either direction on the screen is 1 in. Because of frequency distortion in the horizontal amplifier, the phase of the horizontal voltage is shifted 10° but the magnitude is changed inappreciably. Plot to scale the image that actually appears on the screen, and compare with the image that would be seen if there were no phase shift.

(*b*) If the phase shift in both amplifiers were the same, what would be seen on the cathode-ray screen?

13-2 The input to an amplifier consists of a voltage made up of a fundamental signal and a second-harmonic signal of half the magnitude and in phase with the fundamental. Plot the resultant.

The output consists of the same magnitude of each component, but with the second harmonic shifted 90° (on the fundamental scale). This corresponds to perfect frequency response but bad phase-shift response. Plot the output and compare it with the input waveshape.

13-3 The bandwidth of an amplifier extends from 30 Hz to 15 kHz. Find the frequency range over which the voltage gain is down less than 0.5 dB from its midband value. Assume that the low- and high-frequency response is given by Eqs. (13-4) and (13-6) multiplied by a constant A_{Vo}.

13-4 Prove that over the range of frequencies from $10f_L$ to $0.1f_H$ the voltage amplification is constant to within 0.5 percent and the phase shift to within ± 0.1 rad. Make the same assumption as in Prob. 13-3.

13-5 What is the maximum ratio $f/f_L \equiv a$ of a symmetrical square wave below which the approximation for P given by Eq. (13-13) is within 10 percent of its actual value P^* [obtained from Eq. (13-10)]?

13-6 An ideal 1-μs pulse is fed into an amplifier. Plot the output if the bandpass is (a) 80 MHz, (b) 10 MHz, (c) 1 MHz. Assume $f_L = 0$ and a single-pole amplifier.

13-7 (a) Prove that the response of a two-stage (identical and noninteracting) low-pass amplifier to a unit step is

$$v_o = A_o^2[1 - (1 + x)\epsilon^{-x}]$$

where A_o is the midband voltage gain and $x \equiv t/RC$.

(b) For $t \ll RC$, show that the output varies quadratically with time.

13-8 In Prob. 13-7, let the upper 3-dB frequency of a single stage be f_H and the rise time of the two stages in cascade be t_r. Show that $f_H t_r = 0.53$. Explain why this rise time is larger than that for a single-stage [Eq. (13-9)].

13-9 (a) Prove that the response of a two-stage (identical and noninteracting) high-pass amplifier to a unit step is

$$v_o = A_o^2(1 - x)\epsilon^{-x}$$

where A_o is the midband voltage gain and $x = t/RC$.

(b) For symmetrical square waves whose period T is much smaller than RC show that a good approximation for the percent tilt is

$$P \equiv \frac{T}{RC} \times 100 \text{ percent}$$

13-10 For the amplifier of Fig. 13-8 $h_{fe} = 100$, $h_{ie} = 1$ kΩ, $R_c = 3$ kΩ, $C_b = C_z = 100$ μF.

(a) It is desired that the absolute value of the midband gain A_o be more than 160, and that f_L be at most 90 Hz. Find a range of values for R_s which satisfies these two requirements.

(b) Is there a value of R_s for which $|A_o| > 165$ and $f_L < 85$ Hz?

13-11 Consider the circuit of Fig. 13-8 with C_b arbitrarily large, so as to find the effect of only the bypass capacitor C_z on the low-frequency response. Make the same assumptions as in Sec. 13-3 except that the inequality Eq. (13-16) is no longer true.

(a) Verify that

$$\frac{A_V}{A_o} = \frac{1}{1 + R'/R} \frac{1 + jf/f_o}{1 + jf/f_p}$$

where

$$A_o = -\frac{h_{fe}R_c}{R} \qquad R \equiv R_s + h_{ie} \qquad R' \equiv (1 + h_{fe})R_e$$

$$f_o \equiv \frac{1}{2\pi C_z R_e} \qquad f_p \equiv Bf_o \qquad B \equiv 1 + \frac{R'}{R}$$

(b) Prove that the low 3-dB frequency is given by

$$f_L = \frac{\sqrt{B^2 - 2}}{2\pi C_z R_e}$$

What is the physical meaning of the condition $B < \sqrt{2}$?

(c) If $B^2 \gg 2$, show that $f_L \approx f_p$.

(d) Make a sketch of $y = 20 \log|A_V/A_o|$ versus f. Indicate the values of y as $f \to 0$ and as $f \to \infty$. Also indicate f_o and f_p on your sketch, assuming that $f_p \gg f_o$.

13-12 (*a*) Use the approximate model for the transistor in the circuit shown to obtain the lower 3-dB frequency f_L.

(*b*) Calculate the percentage tilt in the output if the input current I is a 200-Hz square wave.

(*c*) What is the lowest-frequency square wave which will suffer less than 2 percent tilt?

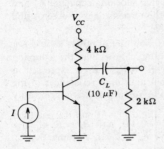

Prob. 13-12

13-13 There is a single coupling capacitor C_b between the two stages of an amplifier which uses bipolar junction transistors. Assume that R_1, R_2, and C_z (Fig. 13-9) are arbitrarily large so that the emitters are effectively grounded. The parameters are $h_{fe} = 100$, $h_{ie} = 2$ kΩ, $R_c = 3$ kΩ, and $C_b = 5$ μF.

(*a*) Obtain the 3-dB frequency f_L.

(*b*) At what frequency does the voltage gain of the second stage drop 12 dB below its midband value?

13-14 Show that the low-frequency voltage gain of the FET stage shown with $r_d \gg R_L + R_s$ is given by

(*a*)
$$\frac{A_V}{A_o} = \frac{1}{1 + g_m R_s} \frac{1 + jf/f_o}{1 + jf/f_p}$$

where

$$A_o \equiv -g_m R_L \qquad f_o \equiv \frac{1}{2\pi C_s R_s} \qquad f_p = \frac{1 + g_m R_s}{2\pi C_s R_s}$$

(*b*) If $g_m R_s \gg 1$ and $g_m = 3$ mA/V, find C_s so that a 60-Hz square-wave input will suffer no more than 10 percent tilt.

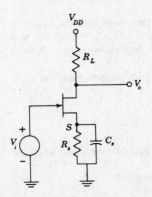

Prob. 13-14

13-15 (*a*) In Prob. 13-14 assume also that the reactance of C_s is much smaller than R_s at all frequencies of interest. Show (without using the results of Prob. 13-14) that the response has one pole and one zero. Verify that

$$A_o = -g_m R_L \quad \text{and} \quad f_L = \frac{g_m}{2\pi C_s}$$

(*b*) If the excitation V_s is applied to the gate through a blocking capacitor C_b and a resistor R_g (with R_g from gate to ground), prove that the response has two zeros and two poles. What are the values of the zeros and the poles?

13-16 It is desired that the voltage gain of an *RC*-coupled amplifier at 60 Hz should not decrease by more than 5 percent from its midband value. Show that the coupling capacitance C_b must be at least equal to $8.07/R'$, where $R' = R'_o + R'_i$ is expressed in kilohms, and C_b in microfarads. (R'_o and R'_i are the output and input resistances of the previous and next stage, respectively, as "seen" by C_b.) Assume that C_z is arbitrarily large.

13-17 Consider the circuit of Fig. 13-9 with $R_1 = R_2 = 60$ kΩ, and $R_c = R_s = R_e = 1.5$ kΩ. The transistors are identical with $h_{fe} = 100$ and $h_{ie} = 2$ kΩ. Assume that C_z represents a short circuit in the frequency range of interest.

(*a*) Find the midband gain of $Q2$.

(*b*) Find the value of C_b necessary to give a lower 3-dB frequency of 10 Hz for $Q2$.

(*c*) Find the value (or values) of C_b necessary to ensure less than 5 percent tilt for a 200-Hz square-wave input.

13-18 (*a*) Draw the circuit of a cascade of *RC*-coupled amplifiers using FETs in the common-source mode with the source resistance R_s bypassed by a capacitance C_z.

(*b*) Assume that C_z is arbitrarily large so that the source is effectively grounded. Show that for every stage

$$f_L = \frac{1}{2\pi(R_g + R)C_b}$$

where R_g is the resistance from gate to ground and $R = R_d \| r_d$ for every stage except the first. R is the resistance of the input signal source for the first stage. Usually $R_g \gg R$.

(*c*) A two-stage FET *RC*-coupled amplifier has the following parameters: $g_m = 5$ mA/V, $r_d = 12$ kΩ, $R_d = 12$ kΩ, and $R_g = 0.5$ MΩ for each stage. What must be the value of C_b in order that the frequency characteristic of each stage be flat within 0.5 dB down to 20 Hz?

(*d*) Repeat part (*c*) if the overall gain of both stages is to be down 0.5 dB at 20 Hz.

(*e*) What is the overall midband voltage gain?

13-19 A three-stage *RC*-coupled amplifier uses field-effect transistors with the following parameters: $g_m = 3$ mA/V, $r_d = 8$ kΩ, $R_d = 10$ kΩ, $R_g = 0.2$ MΩ, $C_b = 0.005$ μF, and $C_s = \infty$. Evaluate (*a*) the overall midband voltage gain in decibels and (*b*) f_L of each individual stage.

13-20 The following low-frequency parameters are known for a given transistor at $I_C = 5$ mA, $V_{CE} = 8$ V, and at room temperature.

$$h_{ie} = 1 \text{ k}\Omega \qquad h_{oe} = 4 \times 10^{-5} \text{ A/V}$$

$$h_{fe} = 100 \qquad h_{re} = 10^{-4}$$

At the same operating point, $f_T = 10$ MHz and $C_{ob} = 2$ pF, compute the values of all the hybrid-π parameters.

13-21 A silicon p-n-p transistor has an $f_T = 300$ MHz. What is the base thickness? Use Table 1-1.

13-22 Assume a germanium p-n-p transistor whose base width is 200 μm. At room temperature and for a dc emitter current of 1.5 mA, find (a) the emitter diffusion capacitance and (b) f_T. Use Table 1-1.

13-23 Given the following transistor measurements made at $I_C = 3$ mA, $V_{CE} = 12$ V, and at room temperature:

$$h_{fe} = 100 \qquad\qquad h_{ie} = 1.1 \text{ k}\Omega$$

$$[A_{ie}] = 20 \text{ at 5 MHz} \qquad\qquad C_c = 3 \text{ pF}$$

Find f_β, f_T, C_e, $r_{b'e}$, and $r_{bb'}$.

13-24 (a) Redraw the CE hybrid-π equivalent circuit of Fig. 13-12 with the base as the common terminal and the output terminals, collector, and base short-circuited. Taking account of typical values of the transistor parameters, show that C_c may be neglected.

  (b) Using the circuit in part (a), prove that the CB short-circuit current gain is

$$A_{ib} = \frac{g_m}{g_{b'e} + g_m + j\omega C_e} = \frac{\alpha_o}{1 + jf/f_\alpha}$$

where

$$\alpha_o = \frac{h_{fe}}{1 + h_{fe}} \qquad \text{and} \qquad f_\alpha = \frac{g_m}{2\pi C_e \alpha_o} \approx \frac{f_\beta}{1 - \alpha_o}$$

13-25 Consider a single-stage CE amplifier with a load R_L and driven by an ideal current source I_i. Apply Miller's theorem (Appendix C-4) to simplify the hybrid-π model, proceeding as in Sec. 13-7.

  (a) Verify that the low-frequency (midband) gain is $A_{Io} = -h_{fe}$.

  (b) Verify that the high frequency f_H at which the current gain decreases by 3-dB is given by $f_H = g_{b'e}/C$, where $C = C_e + C_c(1 + g_m R_L)$.

  (c) Verify that

$$|A_{Io} f_H| = \frac{f_T}{1 + 2\pi f_T C_c R_L}$$

13-26 Repeat Prob. 13-25 if the current source I_i is not ideal but has a resistance R_s (in parallel with the generator I_i). Verify that:

(a) $A_{Iso} = \dfrac{-h_{fe} R_s}{R_s + h_{ie}}$.

(b) $f_H = \dfrac{1}{2\pi RC}$, where $C = C_e + C_c(1 + g_m R_L)$, $R = R_s' \| r_{b'e}$, and $R_s' = R_s + r_{bb'}$.

(c) $|f_H A_{Iso}| = \dfrac{f_T}{1 + 2\pi f_T R_L C_c} \dfrac{R_s}{R_s + r_{bb'}}$.

13-27 Find by inspection the number of poles and the number of zeros in the transfer function of each circuit shown. Explain carefully.

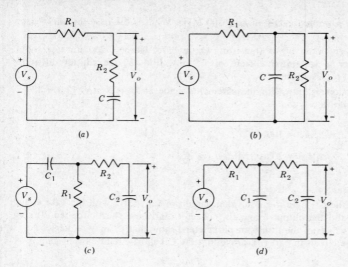

(a)

(b)

(c)

(d)

Prob. 13-27

13-28 Consider a transfer function with no zeros and two poles at frequencies f_{p1} and f_{p2}.

(a) Plot the magnitude $|A|$ (dB) versus $\log f$ curve for $f_{p1} = 5$ kHz and $f_{p2} = 20$ kHz. For convenience, assume a midband gain of 10. If the 3-dB frequency is $f_H = Ff_{p1}$, what is F?

(b) Verify analytically that, if $f_{p2} = 4f_{p1}$, the 3-dB frequency is 5.4 percent smaller than f_{p1}.

13-29 (a) Obtain the equivalent circuit of a CE amplifier driven by an ideal voltage source V_s and a resistive load at high frequencies employing the hybrid-π model of Fig. 13-10.

(b) Find by inspection the number of poles and the number of zeros in the transfer function $A_V \equiv V_o/V_s$.

(c) Prove by inspection that the zero s_o is given by

$$s_o = \frac{g_m - g_{b'c}}{C}$$

13-30 (a) Verify the nodal equations for the single-stage CE amplifier of Sec. 13-9.

(b) Obtain Eq. (11-38) for the voltage gain V_o/V_s.

13-31 (a) Verify the values of K_1, s_o, s_1, and s_2 given in Sec. 13-9 for the CE stage of Fig. 13-16.

(b) Evaluate the gain at zero frequency.

(c) Evaluate the magnitude of the gain at 1 MHz and check with Fig. 13-17.

(d) Evaluate the phase of the gain at 1 MHz and check with Fig. 13-17.

13-32 Consider a single-stage CE amplifier with $R_L = 2$ kΩ and $R_s = 1$ kΩ. The hybrid-π parameters of the transistor are given in Eq. (13-29). Using Miller's theorem and the approximate analysis, compute (a) the upper 3-dB frequency of the current gain $A_I = I_L/I_i$, where I_i is the input base currnet, and (b) the magnitude of the voltage gain $A_{Vs} = V_o/V_s$ at the frequency of part (a).

13-33 For a single-stage CE transistor amplifier whose hybrid-π parameters have the average values given in Eq. (13-29), what value of source resistance R_s will give a 3-dB frequency f_H which is (a) half the value for $R_s = 0$, (b) twice the value for $R_s = \infty$? Do these values of R_s depend upon the magnitude of the load R_L? Use Miller's theorem and the approximate analysis.

13-34 A single-stage CE amplifier is measured to have a voltage-gain bandwidth f_H of 4 MHz with $R_L = 600\ \Omega$. Assume $h_{fe} = 100$, $g_m = 50$ mA/V, $r_{bb'} = 100\ \Omega$, $C_c = 2$ pF, and $f_T = 300$ MHz.

    (a) Find the value of the source resistance that will give the required bandwidth.

    (b) With the value of R_s found in part a, find the midband voltage gain V_o/V_s.

Hint: Use the approximate analysis.

13-35 Consider a single-stage CE transistor amplifier with the load resistor R_L shunted by a capacitance C_L. Use the model of Fig. 13-12 and apply Miller's theorem.

    (a) Prove that the internal voltage gain $K = V_{ce}/V_{b'e}$ is

$$K \approx \frac{-g_m R_L}{1 + j\omega(C_c + C_L)R_L}$$

    (b) Prove that the 3-dB frequency is given by

$$f_H \approx \frac{1}{2\pi(C_c + C_L)R_L}$$

provided that the following condition is satisfied:

$$g_{b'e}R_L(C_c + C_L) \gg C_e + C_c(1 + g_m R_L)$$

13-36 The hybrid-π parameters of the transistor used in the circuit shown are given in Eq. (13-29). The input to the amplifier is an abrupt current step 0.3 mA in magnitude. Find the output voltage as a function of time (a) if $C_L = 0$ (neglect the output time constant); (b) if $C_L = 0.2\ \mu$F (neglect the input time constant).

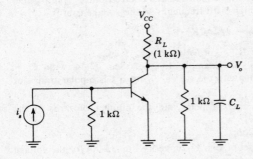

Prob. 13-36

13-37 (a) Verify the nodal equations in Sec. 13-11 for the emitter follower.

    (b) Find the gain V_e/V_s as a function of s.

13-38 Delete all capacitors from the emitter-follower equivalent circuit of Fig. 13-20b.

    (a) Find the input resistance.

    (b) Find the output resistance.

    (c) Show that these results are consistent with the low-frequency formulas in Table 11-4.

13-39 (*a*) For the emitter follower of Fig. 13-20 at high frequencies, obtain $K = V_e / V_i'$ and (with $g \equiv g_m + g_{b'e}$) verify that

$$K = \frac{gR_L}{1 + gR_L} \; \frac{1 + j\omega(C_e/g)}{1 + j\omega\left(\dfrac{C_L + C_e}{1 + gR_L}\right)R_L}$$

(*b*) If $gR_L \gg 1$ and $C_L \gg C_e$, show that

$$K \approx \frac{1}{1 + jf/f_H} \qquad \text{where} \qquad f_H = \frac{g_m + g_{b'e}}{2\pi C_L}$$

13-40 Verify that the transfer function of the two-stage interacting amplifier of Fig. 13-24 is given by Eq. (13-70).

13-41 Consider a three-stage CE amplifier having the same parameter values as the two-stage amplifier of Fig. 13-22.

(*a*) Draw the equivalent circuit by replacing each transistor by its hybrid-π model of Fig. 13-12.

(*b*) Use the Miller approximation to obtain a simplified network.

(*c*) Verify the voltage-gain expression in Eq. (13-72).

13-42 The complex voltage gain of a FET is $A_V = A_1 + jA_2$. The input impedance may be represented as a resistance R_i in parallel with a capacitance C_i. Find R_i and C_i in terms of A_1, A_2, and the interelectrode capacitances.

13-43 A MOSFET connected in the CS configuration works into a 90-kΩ resistive load. Calculate the complex voltage gain, the input capacitance, and the input resistance of the system for frequencies of 100 and 100,000 Hz. Take the interelectrode capacitances into consideration. The MOSFET parameters are $\mu = 50$, $r_d = 20$ kΩ, $C_{gs} = 4.0$ pF, $C_{ds} = 1$ pF, and $C_{gd} = 2.5$ pF. Compare these results with those obtained when the interelectrode capacitances are neglected.

13-44 (*a*) Consider a CS FET amplifier whose load consists of a resistance R_d in parallel with a capacitance C_d. Prove that the high 3-dB frequency of the voltage gain is

$$f_H = 1/2\pi C R_d'$$

where $C = C_d + C_{ds} + C_{gd}$, and $R_d' = R_d \| r_d$, provided that $g_m \gg \omega C_{gd}$.

(*b*) Calculate f_H for the FET of Example 13-2 of Sec. 13-14 with $C_d = 100$ pF and $R_d = 50$ kΩ. Compare this with the typical f_H for a CE bipolar transistor amplifier (Sec. 13-9).

13-45 (*a*) Find by inspection the number of poles and the number of zeros in the transfer function of Fig. 13-25*b*.

(*b*) Repeat part (*a*) for Fig. 13-26*b*.

13-46 (*a*) Starting with the circuit model of Fig. 13-26, verify Eq. (13-80) for the voltage gain of the source follower, taking interelectrode capacitances into account.

(*b*) Verify Eq. (13-83) for the input admittance.

(*c*) Verify Eq. (13-84) for the output admittance.

Hint: For part (*c*) set $V_i = 0$ and impress an external voltage V_o from S to N; the current drawn from V_o divided by V_o is Y_o.

13-47 Starting with the circuit model for a FET at high frequencies, show that, for the CG amplifier stage with $R_s = 0$ and $C_{ds} = 0$,

(*a*) $A_V = \dfrac{(g_m + g_d)R_d}{1 + R_d(g_d + j\omega C_{gd})}$.

(b) $Y_i = g_m + g_d(1 - A_V) + j\omega C_{sg}$.

(c) Repeat part (a), taking the source resistance R_s into account.

(d) Repeat part (b), taking the source resistance R_s into account.

13-48 (a) For the source follower with $g_m = 3$ mA/V, $R_s = 50$ kΩ, $r_d = 30$ kΩ, and with each internode capacitance 2 pF, find the frequency at which the reactive component of the output admittance equals the resistive component.

(b) At the frequency found in part (a) calculate the gain and compare it with the low-frequency value.

13-49 Verify Eq. (13-88).

13-50 Consider a transfer function with n poles and k zeros. Assume that all the zeros occur at much higher frequencies than the poles. Verify that the 3-dB frequency is given by Eq. (13-86).

13-51 The transfer function V_o/V_s of an amplifier has n poles and no zeros. The pole frequencies are $f_1 > f_2 > \cdots > f_n$. If $n > 1$, then $f_H < f_1$. Why? Show that

(a) An approximate expression for the high 3-dB frequency f_H is given by

$$\frac{1}{f_H} \approx \sqrt{\frac{1}{f_1^2} + \frac{1}{f_2^2} + \cdots + \frac{1}{f_n^2}}$$

(b) An expression which is more accurate is Eq. (13-92). Compare the results obtained with Eq. (13-92) and Eq. (13-87) for the case of (i) two identical poles $f_1 = f_2$; (ii) three identical poles $f_1 = f_2 = f_3$. Calculate the error.

(c) Repeat the calculations in part (b) for the formula in part (a).

13-52 Consider a transfer function with poles at 2 MHz and 4 MHz. Assume all other poles and zeros are much larger than 4 MHz. Calculate the high 3-dB frequency. Compare your result with the approximate value obtained from Eq. (13-92).

13-53 If two cascaded single-pole stages have very unequal bandpasses, show that the combined bandwidth is essentially that of the smaller. Assume noninteracting stages.

13-54 Obtain the high 3-dB frequency for each of the two stages of Fig. 13-24. Calculate f_H for the amplifier by using Eq. (13-92). Compare the result with the exact value of f_H obtained in Fig. 13-23.

13-55 Three identical cascaded stages have an overall upper 3-dB frequency of 25 kHz and a lower 3-dB frequency of 10 Hz. What are f_L and f_H of each stage? Assume noninteracting stages.

Chapter 14

14-1 A single-stage RC-coupled amplifier with a midband voltage gain of 1,000 is made into a feedback amplifier by feeding 5 percent of its output voltage in series with the input opposing. Assume that the amplifier gain without feedback may be approximated at low frequencies by Eq. (13-2) and at high frequencies by Eq. (14-2).

(a) As the frequency is varied, to what value does the voltage gain of the amplifier without feedback fall before gain of the amplifier with feedback falls 3 dB?

(b) What is the ratio of the half-power frequencies with feedback to those without feedback?

(c) If $f_L = 10$ Hz and $f_H = 30$ kHz for the amplifier without feedback, what are the corresponding values after feedback has been added?

14-2 Design an amplifier with the following requirements: Overall midband gain of at least 3,000, and 3-dB frequencies $f_{Lf} = 20$ Hz (or lower), and $f_{Hf} = 30$ kHz (or higher). The building blocks you have at your disposal are identical amplifiers with a midband gain of 100. Their behavior can be approximated by Eqs. (13-2) and (14-2) at low and high frequencies, respectively, with $f_L = 40$ Hz and $f_H = 20$ kHz. Assume noninteracting stages. Find the range of values of the feedback factor β which will satisfy the above specifications.

14-3 (a) Verify Eqs. (14-9) and (14-10) for A_f for the two-pole transfer gain.

(b) Verify that for $Q = Q_{min}$, the roots are ω_1 and ω_2.

14-4 Verify Eqs. (14-14) and (14-16) for the transfer function of the circuit of Fig. 14-4.

14-5 Verify that the transfer function of the circuit shown can be put in the form of Eq. (14-16). How are ω_o and Q defined in this case?

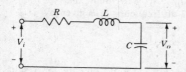

Prob. 14-5

14-6 (a) Show that the two-pole closed-loop magnitude of the gain A_f is given by Eq. (14-18) as a function of frequency.

(b) Verify that the peak on the frequency response occurs at $\omega/\omega_o = \sqrt{1 - 2k^2}$ and has a value given by Eq. (14-20).

14-7 Plot the phase response (versus ω/ω_o) of a double-pole transfer function for $Q = 0.5, 1.5, 4.5$.

14-8 Derive Eqs. (14-23), (14-24), (14-25), and (14-26), for the step response of the two-pole feedback amplifier. *Hint:* For the overdamped case, assume $k^2 \gg 1$ and expand $(1 - 1/k^2)^{1/2}$ in Taylor series.

14-9 Verify Eq. (14-27) for the positions x_m and magnitude y_m of the oscillatory response maxima and minima.

14-10 Define the normalized settling time x_s to be the time at which the first peak (or dip) in Fig. 14-6 that is within the error band of $\pm P$ percent occurs. Show that the value of m corresponding to x_s is given by the smallest value of m that satisfies

$$100\epsilon^{-\pi k m/(1 - k^2)^{1/2}} \leqslant P$$

14-11 (a) Given a two-pole amplifier with corner frequencies at $\omega_1 = 2$ Mrad/s and $\omega_2 = 0.5$ Mrad/s. What are the maximum decibels of feedback which will give the fastest rise time without overshoot?

(b) What is the rise-time improvement for the condition in part (a)? In other words, find the ratio of the rise time for $k = 1$ to the rise time for zero feedback.

14-12 (a) If -31.84 dB of feedback are applied to the amplifier of Prob. 14-11a, find the rise time.

(b) What is the percent overshoot in this case?

14-13 Given a two-pole amplifier with corner frequencies at $\omega_1 = 12$ MHz and $\omega_2 = 2$ MHz.

(a) Find the maximum value of the loop gain for which the step-response over-shoot will be 5 percent.

(b) At what time will the peak occur?

(c) Calculate the magnitude of the first minimum of the step response and the time it occurs.

(d) Verify that for $k = 0.6$ the maximum overshoot is approximately 9.5 percent.

14-14 An amplifier has two poles on the negative real axis: $s_1 = -4 \ \mu s^{-1}$, $s_2 = -10 \ \mu s^{-1}$.

(a) Plot the root locus of the amplifier with negative feedback.

(b) Find the value of βA_o for which the maximum overshoot of the amplifier step response with feedback is 9.5 percent.

14-15 (a) If $k = 0.707$, calculate the percent maximum overshoot in the step response for a two-pole feedback amplifier.

(b) If there is a 5 percent overshoot in the frequency response, what is the percent overshoot in the step response?

14-16 The roots of a closed-loop two-pole amplifier are $s_1 = -\sigma + j\omega$, $s_2 = -\sigma - j\omega$. Find the relationship between Q and $|\omega/\sigma|$.

14-17 For the three-pole feedback amplifier, verify Eq. (14-29) and show that

$$\omega_o^3 = \omega_1\omega_2\omega_3(1 + \beta A_o)$$

$$a_2 = \frac{\omega_1 + \omega_2 + \omega_3}{\omega_o}$$

$$a_1 = \frac{\omega_1\omega_2 + \omega_2\omega_3 + \omega_1\omega_3}{\omega_o^2}$$

14-18 (a) Consider a three-pole open-loop transfer function with all three poles at $s = -\omega_1$. Find an expression for the closed-loop gain.

(b) Show that as negative feedback is added, one pole s_{3f} moves along the negative real axis while the other two poles become complex conjugates and move toward the right-hand complex plane as indicated in Fig. 14-8.

(c) Verify that the system is unstable for $|\beta A_o| > 8$; and that for $|\beta A_o| = 8$, the poles s_{1f} and s_{2f} are $\pm j\omega_1\sqrt{3}$ and $s_{3f} = -3\omega_1$.

14-19 Consider the amplifier with a transfer function $A(s) = A_1/[s(s+2)^2]$.

(a) Find the value of βA_1 corresponding to the breakaway point (the point in the complex plane where the real poles become complex).

(b) Find the value of βA_1 for which the amplifier with negative feedback becomes unstable.

(c) Plot approximately the root locus.

14-20 Consider a transfer function with three poles s_1, s_2, s_3. Find s_1, s_2, s_3 knowing that

$$|s_1| = |s_2| = |s_3| = 2$$

and the Q of the complex pole pair (s_2, s_3) is 2.

14-21 An amplifier has the following transfer function:

$$A(s) = \frac{A_o \times 2 \times 10^{-5}}{(s + 0.01)(s + 0.02)(s + 0.1)}$$

Feedback is applied to this amplifier. Find the poles and βA_o if the Q of the complex pole pair is 1.

14-22 (a) Verify that a two-pole feedback amplifier *cannot* have a closed-loop dominant pole if $Q > 0.4$.

(b) Find an expression for the maximum value of βA_o for which a closed-loop dominant pole exists.

(c) Calculate βA_o for $n = 4$. What is the physical interpretation of this result?

14-23 Verify Eq. (14-33), using Eqs. (14-32) and (14-31).

14-24 A two-pole amplifier has a value $f_2/f_1 = 10$ and a dominant pole at 1 MHz (in the absence of feedback). Feedback is applied such that $\beta A_o = 0.8$.

(a) Find the two new poles.

(b) Find the upper 3-dB frequency.

(c) Compare the gain-bandwidth product with and without feedback.

(d) Obtain a better approximation for the high 3-dB frequency and repeat part (c).

14-25 Verify the expression for R_M in Eq. (14-34).

14-26 Repeat the analysis of Sec. 14-5 with $R' = 30$ kΩ to find the upper 3-dB frequency of R_{Mf}.

14-27 Verify the expression for G_M in Eq. (14-40).

14-28 Repeat the analysis of Sec. 14-6 with $R_e = 200$ Ω to find the upper 3-dB frequency of G_{Mf}.

14-29 For the voltage-series feedback pair in Sec. 14-8 verify that a dominant pole exists and that A_{Vf} can be approximated by Eq. (14-47).

14-30 (a) Show that the Bode magnitude plot for an n-pole transfer function with no zeros is equal to the sum of the magnitude plots of each pole considered separately.

(b) Repeat part (a) for the Bode phase plot.

(c) Generalize the results of parts (a) and (b) for a transfer function with m zeros and n poles.

14-31 Consider a transfer characteristic with two poles such that $f_{p2} = 4f_{p1}$.

(a) Plot the idealized and true Bode magnitude curves. Obtain the actual 3-dB frequency graphically.

(b) Plot the idealized and true Bode phase curves.

14-32 Repeat Prob. 14-31 for poles at $f_{p2} = 2f_{p1}$.

14-33 Obtain the Bode magnitude and phase plots for the single-zero transfer function

$$A(jf) = A_o\left(1 + \frac{jf}{f_z}\right)$$

Sketch both the actual and the piecewise linear plots.

14-34 Consider the transfer function given in Eq. (13-1) which has one pole and a zero at $f = 0$. Draw the piecewise linear Bode plots for the pole, the zero, and the resultant for (a) amplitude and (b) phase.

14-35 Sketch the idealized Bode amplitude and phase plots for a transfer function with one zero f_z and one pole f_p if (a) $f_p < f_z$ and (b) $f_p > f_z$. Take the two frequencies at 1 and 10 kHz.

14-36 Assume that $A_o = -10$ in the amplifier of Fig. 14-19.

(a) Find the phase and gain margins for $\beta = 0.5$. *Hint:* Use $\log A/A_o = \log A\beta - \log A_o\beta$.

(b) Find the value of β which will produce a phase margin of 45°.

14-37 A three-pole feedback amplifier has a dc gain without feedback of -10^3. All three open-loop poles are at $f = 1$ MHz.

(a) What is the maximum value of β for which the amplifier is stable?

(b) Assume that one of the poles is shifted to $f_1 = 200$ kHz. Using the value of β found in part a, what is the gain margin of the modified circuit?

14-38 Consider a voltage-series feedback amplifier whose loop gain is $A_V\beta$.

(a) The poles of A_V are located at 1.0, 5.0, and 30 MHz and the midband gain is -10^4. The amplifier has no zeros. Write the expression for A_V as a function of frequency.

(b) Draw the idealized Bode magnitude plot for the amplifier of part (a) on the top portion of semilog paper. Label the axes properly. The horizontal axis extends from 0.1 to 1,000 MHz. The vertical axis is gain with each major division equal to 10 dB. *Label the slopes of each portion of the graph.*

(c) Draw the idealized Bode phase plot on the bottom portion of the semilog paper. Each major division vertically should equal 45°. Indicate with a tiny circle on the phase plot every frequency where the plot changes slope. *Give the values of these frequencies in your solution.*

(d) At what frequency will this feedback amplifier oscillate (if it oscillates at all)? To prevent oscillation, what restriction must you place on the magnitude of the midband loop gain? Explain.

(e) In order to have a phase margin of at least 45°, what is the amount of feedback (in decibels at low frequencies) which can be added? Is this the maximum or minimum feedback which can be used? Explain.

14-39 Repeat Prob. 14-38 if the poles of A_V are 1, 3, and 20 MHz.

Chapter 15

15-1 Consider the inverting OP AMP of Fig. 15-3.

(a) Write the equation at the output node. Evaluate the ratio V_o/V_i and verify Eq. (15-3).

(b) Write the equation at the input node. Evaluate V_o/V_s and verify Eq. (15-2).

(c) Evaluate the effective input impedance Z'_{if} in parallel with R_i and show that it equals $(Z' + R_o)/(1 - A_v)$. Note that Z'_{if} is very small because A_v is a large negative number.

(d) Show that the effective output impedance Z_{of} is given by

$$Z_{of} = R_o \frac{1 + Z'Y}{A_v + R_o Y}$$

You may neglect the large input resistance R_i in your circuit. *Hint:* Z_{of} equals the open-circuit voltage ($\approx -Z'/Z$) divided by the short-circuit current. Note that Z_{of} is very small because A_v is very large.

15-2 Replace Z' in Fig. 15-3 by its Miller's impedance Z_1 from the input terminal to ground and Z_2 from the output terminal to ground (Sec. C-4). Verify Eqs. (15-2) and (15-3).

15-3 Find an expression for A_{Vf} in Fig. 15-3 by using feedback concepts. Show that this expression agrees with Eq. (15-2) if $1 \gg R_o Y'$.

15-4 The amplifier shown uses an OP AMP with input resistance R_i, voltage gain $A_v < 0$, and zero output resistance. Assume also that the OP AMP is unilateral from input to output.

    (*a*) Show that the amplifier satisfies the three fundamental assumptions of Sec. 12-3

    (*b*) Show that the transresistance of the amplifier without feedback is

$$\frac{A_v R_i R R'}{R R' + (R_i + R_1)(R + R')}$$

    (*c*) Show that

$$A_{Vf} = \frac{V_o}{V_s} = \frac{A_v R_i R'}{R R' + (R_i + R_1)(R + R') - A_v R_i R}$$

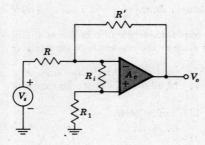

Prob. 15-4

15-5 Consider the noninverting OP AMP of Fig. 15-4. For the circuit model of the amplifier use that indicated in Fig. 15-1*b*; that is, an input resistance R_i, an output resistance R_o, and an open-circuit gain A_v. Write the equations at the input and output nodes and obtain two equations involving V_s, V_i, and V_o. Evaluate $A_{Vf} \equiv V_o/V_s$. Show that this equation reduces to Eq. (15-4) if $A_v \to \infty$.

15-6 Without using feedback formulas, verify that the input impedance for the noninverting OP AMP of Fig. 15-4 is

$$R_{if} = R_i \left(1 + \frac{Z}{Z + Z'} A_V \right)$$

where R_i is large and $R_o = 0$. A_V is the difference gain *without feedback*.

15-7 Consider the noninverting OP AMP of Fig. 15-4 with the following parameters: $Z = R = 1$ kΩ, $R_i = 10$ kΩ, $Z' = R' = 9$ kΩ, $R_o = 5$ kΩ, and $A_v = 10{,}000$. Using the feedback analysis in Chap. 12 for a voltage-series configuration, calculate (*a*) A_{Vf}, (*b*) R_{if}, and (*c*) R_{of}.

15-8 (*a*) Repeat Prob. 15-4 for the noninverting amplifier shown.

    (*b*) Show that the voltage gain of the amplifier without feedback is

$$A_V = \frac{-A_v R_i (R + R')}{(R + R')(R_1 + R_i) + R R'}$$

(c) Show that

$$A_{Vf} = \frac{-A_v R_i (R + R')}{RR' + (R_i + R_1)(R + R') - A_v RR_i} = \frac{V_o}{V_s}$$

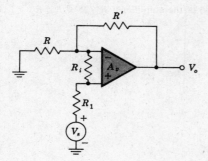

Prob. 15-8

15-9 For the circuit of this problem with $R_i = \infty$, show that $Y_{of} = 1/R_{of}$ is given by

$$Y_{of} = \frac{1}{R_o}\left(1 - A_v \frac{R}{R + R'}\right) + \frac{1}{R + R'}$$

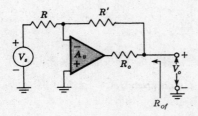

Prob. 15-9

15-10 (a) Apply the 4 h-parameter model of Fig. 11-17 to Fig. 15-7b and solve for the common-mode gain A_c for DIFF AMP. Subject to the following reasonable (why?) approximations,

$$R_e \gg R_c \quad \text{and} \quad 1 \gg h_{re}$$

verify that

$$A_c = \frac{V_o}{V_s} = \frac{(2h_{oe}R_e - h_{fe})R_c}{2R_e(1 + h_{fe}) + (R_s + h_{ie})(2h_{oe}R_e + 1)}$$

(b) Calculate the short-circuit ($R_c = 0$) load current (I_{sc}) if $R_3 \equiv 2R_e$ and $R_3 \gg h_{re}f_{fe}/h_{oe}$. Calculate the open-circuit load voltage V_{oc} (with $R_c = \infty$). Then verify that R_o for $Q3$ in Fig. 15-8 is given by

$$\frac{V_{oc}}{I_{sc}} \equiv R_o = \frac{(R_s + h_{ie})(1 + h_{oe}R_3) + (1 + h_{fe})R_3}{h_{oe}(R_s + h_{ie} + R_3 - (h_{fe}h_{re}/h_{oe}))}$$

(c) Evaluate R_o if $R_3 = R_s = 1$ kΩ and use the parameter values in Eq. (11-54). The value of R_o is the effective emitter resistance R_e in Fig. 15-6.

15-11 The circuit shown is a differential amplifier using an ideal OP AMP.

(a) Find the output voltage v_o.

(b) Show that the output corresponding to the common-mode voltage $v_c = \frac{1}{2}(v_1 + v_2)$ is equal to zero if $R'/R = R_1/R_2$. Find v_o in this case.

(c) Find the common-mode rejection ratio of the amplifier if $R'/R \neq R_1/R_2$.

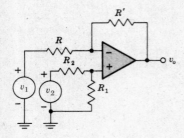

Prob. 15-11

15-12 (a) Verify Eqs. (15-13) and (15-14) for the difference amplifier.

(b) Assuming $R_s \ll h_{ie}$ and $r_{bb'} \ll r_{b'e}$, verify that

$$A_d = \tfrac{1}{2} g_m R_c \quad \text{and} \quad g_{md} = \frac{I_O}{4V_T}$$

15-13 For the circuit of Fig. 15-6 assume that $R_s = 0$, $h_{oe}(R_c + 2R_e) \ll 1$, $h_{fe} \gg 1$, and $h_{ie} \ll 2R_e h_{fe}$.

(a) Verify that the common-mode rejection ratio is given by

$$\rho = \frac{h_{fe} R_e}{h_{ie}}$$

(b) If $r_{bb'} \ll r_{b'e}$ verify that $\rho = g_m R_e \approx V/2V_T$, where V is the quiescent voltage across R_e.

15-14 Draw the h-parameter model for the common-mode gain in a DIFF AMP. *Without* solving for A_c show from the circuit that A_c must be zero if $h_{fe}/h_{oe} = 2R_e$.

15-15 (a) Verify Eq. (15-22) for the transfer characteristic of the DIFF AMP.

(b) Verify Eq. (15-23) for g_{md}.

15-16 (a) From Eq. (15-22), for the transfer characteristic of the DIFF AMP find the range $\Delta V = \Delta(V_{B1} - V_{B2})$ over which each collector current increases from 0.1 to 0.9 its peak value.

(b) Repeat part (a) for a collector-current variation from 50 to 99 percent of I_O.

(c) Compare your result with Fig. 15-9.

15-17 The differential amplifier of Fig. 15-8 is modified by putting two resistors R_e in series with the emitter lead of $Q1$ and $Q2$.

(a) Express $V_{B1} - V_{B2}$ as a function of $V_{BE1} - V_{BE2}$ and I_{C1}.

(b) Find the transfer characteristic I_{C1}/I_O versus $(V_{B1} - V_{B2})/V_T$ if $R_e = 50\ \Omega$ and $I_O = 2$ mA. Solve *graphically* by using Fig. 15-9 and part (a).

(c) Find the transconductance

$$g'_{md} = \frac{dI_{C1}}{d(V_{B1} - V_{B2})}$$

evaluated at $V_{B1} = V_{B2}$.

(d) Express g'_{md} in terms of g_{md} given in Eq. (15-23).

15-18 (a) In the current repeater of Fig. 15-13 there are two transistors $Q1$ and $Q2$ in addition to the diode-connected transistor Q. If all transistors are identical except that the collector area of $Q1$ ($Q2$) is K_1 (K_2) times the collector area of Q, prove that the current in $Q1$ is

$$I_{C1} = \frac{K_1 \beta I_1}{1 + \beta + K_1 + K_2}$$

(b) Under what condition is $I_{C1} = K_1 I_1$?

(c) If $\beta = 50$, $K_1 = 2$, and $K_2 = 3$, evaluate I_{C1}/I_1.

15-19 (a) For the current repeater of Fig. 15-14a apply KCL to the base node of $Q3$ and verify Eq. (15-29).

(b) Assuming p-n-p transistors with $\beta = 5$, evaluate I_2/I_1 for the present circuit and compare it with this ratio for the current mirror of Fig. 15-12.

15-20 For the current repeater shown, using identical transistors, note that the collector current I_C of $Q1$ is identical to that of $Q2$. Why? In terms of I_C and β find (a) I_2, (b) I_1, and (c) calculate I_2/I_1.

(d) Show that this ratio is slightly closer to unity than is the same ratio for the current mirror of Fig. 15-14a. Calculate the ratio for $\beta = 5$ for each circuit.

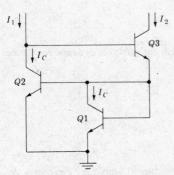

Prob. 15-20

15-21 (a) It is desired to obtain $I_{C2} = 10 \ \mu A$ in the circuit of Fig. 15-12. If $V_{CC} = 15$ V, find R. Is this a reasonable value?

(b) Add an emitter resistor R_2 to $Q2$. Verify Eq. (15-30).

(c) If $R = 10$ kΩ, find I_{C1}. Also evaluate R_2 so that $I_{C2} = 10 \ \mu A$. *Hint:* Neglect base currents compared with collector currents and assume the exponential relationship between I_C and V_{BE} given by Eq. (15-21).

15-22 (a) For the circuit of Fig. 15-14b verify that

$$\frac{I_2}{I_1} = \frac{R_1}{R_2}\left(1 - \frac{V_T \ln I_2/I_1}{R_1 I_1}\right)$$

Hint: Use Eq. (15-21) and neglect base currents.

(b) Choose $R_1 I_1 = 1$ V. Over the range $0.1 < I_2/I_1 < 10$, what is the maximum percentage error made in assuming that $I_2/I_1 = R_1/R_2$?

15-23 For the level-shift network shown, calculate $V_2 - V_1$. Assume identical silicon transistors with very large values of β.

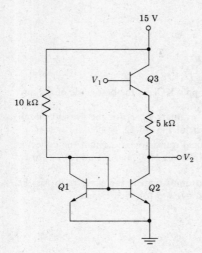

Prob. 15-23

15-24 (a) For the voltage source of Fig. 15-18 show that the effective dynamic resistance is given by

$$R_o = \frac{R_3 + R_4'}{1 + g_m R_4'}$$

where $R_4' \equiv R_4 \| r_{b'e}$ and $r_{bb'} \ll r_{b'e}$.

(b) If $g_m \gg 1/R_4$ and $g_m \gg g_{b'e}$, prove that

$$R_o = \frac{R_3}{h_{fe}} + \frac{R_3 + R_4}{g_m R_4}$$

15-25 The circuit shown in the Motorola MC1530 operational amplifier. Perform a dc analysis of this configuration. Assume that all base currents can be neglected and that all diode forward voltages and base-to-emitter voltages are 0.7 V. Start with the current source $Q1$ and calculate (a) the base-to-ground voltage, V_{BN1}; (b) I_o and $I_{C2} = I_{C3}$; (c) V_{BN4}; (d) V_{EN4}; (e) $I_{C4} = I_{C5}$; (f) V_3; (g) V_4; (h) I_{C7}; (i) V_{BN8}; (j) I_9; (k) I_{10}; (l) V_O.

(m) Assume $h_{fe} = 100$ and use Eq. (15-25) to find h_{ie}. Calculate the gain of the first stage $A_{V1} = V_2/V_1$. Remember that the load resistance is 7.75 kΩ in parallel with the input resistance of $Q4$.

(n) Calculate the complete amplifier gain $A_V = V_o/V_1$.

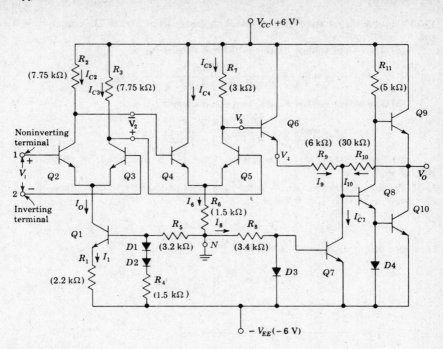

Prob. 15-25

15-26 (a) For the 741 network of Fig. 15-16 calculate I_4 under quiescent conditions. Assume a conducting diode voltage of 0.7 V.

(b) Obtain I_3 from Eq. (15-30) using a graphical solution.

(c) Prove that $I_1 = I_2$ is given by

$$I_1 = \frac{1}{2}\left(\frac{\beta^2 + 2\beta}{\beta^2 + 2\beta + 2}\right)I_3$$

where β is the CE current gain of the *p-n-p* transistors. Evaluate I_1 for $\beta = 4$.

15-27 (a) For Example 15-2 in Sec. 15-7 solve part (c) by writing $I_{B1} = I_{B2} - I_{io}$. Apply the superposition theorem to the two current sources I_{B2} and I_{io} and solve for V_o.

(b) Do not use superposition, but find V_o in terms of I_{B1} and I_{B2}.

15-28 (a) Consider an OP AMP whose slew rate is given in Table 15-1. What is the maximum frequency of an output sinusoid of peak value 5 V, at which distortion sets in due to the slew-rate limitation?

(b) If a sinusoid of 10-V peak output is specified, what is the full-power bandwidth?

15-29 Consider the OP AMP of Fig. 15-24b with the model of Fig. 15-23. Assume $R_o = 0$ and $I_{B1} = I_{B2} = 0$. Owing to the input offset voltage V_{io} show that V_o is given by

$$V_o = \frac{-A_v R_i (R + R') V_{io}}{RR' + (R_i + R_1)(R + R') - RR_i A_v}$$

15-30 Consider the OP AMP of Fig. 15-24b with the model of Fig. 15-23. Assume $R_o = 0$ and $V_{io} = 0$. Prove that:

(a) The output voltage V_{o2} due to the bias current I_{B2} is

$$V_{o2} = \frac{-R'RR_iA_v}{(R_i + R_1)(R' + R) + RR' - A_vRR_i} I_{B2}$$

(b) The output voltage V_{o1} due to the bias current I_{B1} is

$$V_{o1} = \frac{R_iR_1(R + R')A_v}{(R + R')(R_1 + R_i) - A_vRR_i + RR'} I_{B1}$$

(c) Show that if $I_{B2}/I_{B1} \approx 1$, then $V_{o1} + V_{o2}$ is minimized by taking $R_1 = RR'/(R + R')$.

15-31 For the amplifier shown, V_1 and V_2 represent undesirable voltages. Show that, if $R_i = \infty$, $R_o = 0$, and $A_{v1} < 0$ and $A_{v2} < 0$,

$$V_o = A_{v2}[A_{v1}(V' - V_1) - V_2] \quad \text{where} \quad V' = V_o \frac{R}{R + R'}$$

Show also that, if $A_{v2}A_{v1}R/(R + R') \gg 1$,

$$V_o = \left(1 + \frac{R'}{R}\right)\left(V_1 + \frac{V_2}{A_{v1}}\right)$$

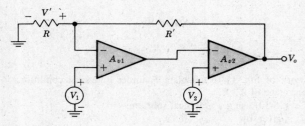

Prob. 15-31

15-32 (a) Use the parameter values in Table 15-1. If $R'/R = 1,000$ for the circuit of Fig. 15-24b, find the output voltage due to V_{io} at 175°C.

(b) At what temperature is the output voltage equal to 0.55 V if $R'/R = 100$?

15-33 (a) For the inverting OP AMP with $R_i = \infty$, draw the model of the amplifier without feedback but taking the loading of R' into account (Sec. 12-7). Refer to Fig. 15-2 with $Z = R$ and $Z' = R'$.

(b) Verify that $\beta = -1/R'$ and $R_M = A_V R_{11}$, where $R_{11} = R \| R'$.

(c) Verify that

$$A_{Vf} = \frac{R_{Mf}}{R} = \frac{R_M/R}{1 + \beta R_M}$$

and show that this expression reduces to Eq. (15-48).

15-34 Verify Eq. (15-48) for the inverting amplifier without using feedback concepts.

Hint: Find V_i by applying the superposition theorem to V_s and V_o and then set $V_i = V_o/A_V$.

15-35 Verify Eq. (15-52) for the noninverting amplifier of Fig. 15-4 without using feedback concepts. *Hint:* Find V_i by applying KVL to the input loop and then set $V_i = V_o/A_V$.

15-36 For the Motorola MC1530 the first three open-loop poles are at 1, 6, and 22 MHz and the open-loop gain is 10^4 (80 dB). The amplifier has no zeros. It is connected as an inverting OP AMP.

(*a*) Draw the idealized Bode open-loop gain and phase-shift characteristics on semilog paper. The horizontal axis extends from 0.1 to 1,000 MHz. Use 10 dB and 45° divisions vertically.

(*b*) At what frequency will the circuit oscillate if too much feedback is added?

(*c*) What is the minimum closed-loop gain A_{Vf} at which the circuit oscillates?

(*d*) Is the circuit stable as a unity-gain follower? Explain.

(*e*) Calculate the minimum closed-loop gain for which the circuit is stable, with a 45° phase margin.

(*f*) A dominant pole f_d is added such that $|A_V|$ passes through $f = 1$ MHz and 0 dB. Find f_d.

15-37 The first two open-loop poles of an OP AMP are 10 Hz and 1 MHz and all other poles are at very high frequencies. The open-loop gain is 100,000 (100 dB).

(*a*) What is the unity-gain frequency of the dominant pole?

(*b*) Plot the idealized Bode gain and phase characteristics.

(*c*) Show that this amplifier is stable as a unity-gain system by finding the phase margin.

15-38 Consider an amplifier with the transfer function given in Eq. (14-55). A dominant pole is added at a frequency of 1 kHz.

(*a*) Carefully construct Bode amplitude and phase plots of the new transfer gain. Use semilog paper and cover the frequency range from 1 kHz to 5 MHz.

(*b*) What is the phase margin?

(*c*) What is the maximum midband loop gain for this phase margin?

(*d*) Can this OP AMP be used as a voltage follower with this dominant-pole compensation? Explain.

15-39 Verify Eqs. (15-55) and (15-56) for the transfer function of the pole-zero-compensation network.

15-40 (*a*) Find the bandwidth and the phase margin for 25 dB of feedback at midband applied to the uncompensated amplifier, whose Bode characteristics are given in Fig. 14-20, and to the two compensated (pole-zero and dominant-pole) amplifiers whose curves are shown in Fig. 15-32. Which system has the best stability? Explain.

(*b*) Pole-zero compensation is used in the amplifier of Fig. 14-20. If the added pole f_p is located so that $|A'_V|$ passes through $f_2 = 10$ MHz and 0 dB, find f_p.

(*c*) Repeat part (*a*) for the OP AMP of part (*b*).

(*d*) Find the phase margin for a voltage follower using the amplifier of part (*b*).

15-41 (*a*) Verify Eq. (15-59). *Hint:* Draw the network without feedback but taking the loading of R' into account.

(*b*) Draw on the same figure the following Bode plots:

(i) Open-circuit voltage gain of the amplifier without compensation. Assume that the amplifier has three poles.

(ii) Open-circuit voltage gain of the amplifier if pole-zero cancellation is achieved, using Eq. (15-59).

15-42 In Prob. 15-41 the transfer function of the amplifier without compensation has three poles at 1, 4, and 40 MHz and a low-frequency open-loop gain of 72 dB.

    (*a*) Find R_c and C_c as a function of R_1, R, and R' if the gain of the compensated amplifier is zero dB at a frequency of 4 MHz.

    (*b*) Find R_c and C_c if $R_1 = RR'/(R + R')$, $R = 1$ kΩ, and $R' \gg R$. Find also the bandwidth of the compensated amplifier without feedback.

15-43 Derive Eqs. (15-63) through (15-66) in Sec. 15-12 which relate to pole-zero cancellation using the Miller-effect technique.

15-44 (*a*) Apply feedback concepts to the circuit of Fig. 15-36. Find β and R_M and verify Eq. (15-67).

    (*b*) Prove Eq. (15-68).

15-45 In Fig. 15-36 a capacitor C is shunted across R (instead of C' across R'). Apply feedback concepts to this circuit. Find β and R_M and βR_M. Show that this configuration represents lag compensation.

15-46 The OP AMP shown has internal feedback because of R_1 and R_1' in addition to the overall voltage-shunt feedback due to R and R'. The compensating capacitor C is across R_1. Apply feedback concepts to find β, R_M, and βR_M. Show that this configuration corresponds to lead compensation. Neglect R_o and assume A_2 to be extremely large.

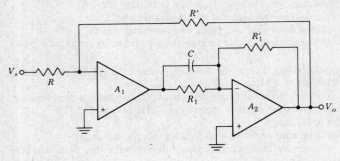

Prob. 15-46

Chapter 16

16-1 (*a*) Find the transfer gain for the circuit shown.

    (*b*) If the input is a constant V, show that the output $v_o(t)$ is determined by the

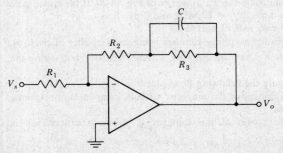

Prob. 16-1

differential equation

$$C \frac{dv_o}{dt} + \frac{v_o}{R_3} + \frac{V}{R_1}\left(1 + \frac{R_2}{R_3}\right) = 0$$

16-2 Given the operational amplifier circuit of Fig. 16-1, consisting of R and L in series for Z, and C for Z'. If the input voltage is a step of magnitude V, find the output v_o as a function of time.

16-3 Consider the operational amplifier circuit of Fig. 16-1 with Z consisting of a resistor R in parallel with a capacitor C, and Z' consisting of a resistor R'. The input is a sweep voltage $v = \alpha t$. Show that the output voltage v_o is a sweep voltage that starts with an initial step. Thus prove that

$$v_o = -\alpha R'C - \alpha \frac{R'}{R} t$$

16-4 Find the transfer functions for the OP AMP configurations shown.

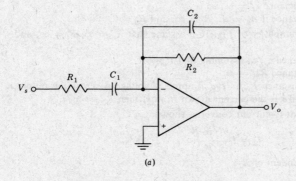

(a)

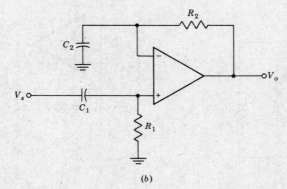

(b)

Prob. 16-4

16-5 Find v_o in terms of the input voltages for the adder-subtracter shown.

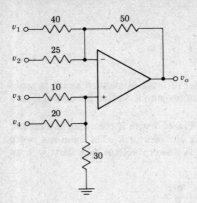

Prob. 16-5

16-6 In Fig. 16-4b show that i_L is equal to $-v_s/R_2$ if $R_3/R_2 = R'/R_1$.

16-7 For the transconductance amplifier of Fig. 16-4a (with Z_L replaced by R_L) assume $A_v \neq \infty$, $R_i = \infty$, and $R_o = 0$.

    (a) Find the transconductance i/v_s.

    (b) Find the input resistance if $A_v \neq \infty$, $R_i \neq \infty$, and $R_o \neq 0$.

16-8 For the transresistance amplifier of Fig. 16-5 assume that $A_v \neq \infty$, $R_i \neq \infty$, and $R_o \neq 0$. Let $R_1 \equiv R_s \| R_i$.

    (a) Find the transresistance v_o/i_s, assuming that $R_o \ll R'$.

    (b) Find the input resistance R_{in}.

    (c) Find the output resistance R_{out}. *Hint:* Impress a voltage v across the output, remove the excitation i_s, and if i is the current drawn from v, then $R_{\text{out}} = v/i$.

16-9 Prove that for the current-to-current converter shown,

$$I_o = I_i \frac{R_f + R}{R}$$

Note that this result is independent of R_L.

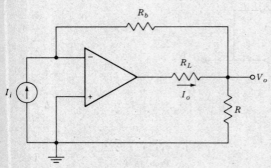

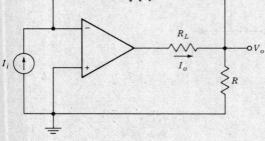

Prob. 16-9

Prob. 16-10

16-10 (a) Consider the current amplifier shown. Verify that the load current I_L is independent of the load resistance and that the current gain is

$$A_{If} = \frac{I_L}{I_i} = -\left(1 + \frac{R_1}{R_2}\right)$$

(b) If $A_V \gg 1$ but $A_V \neq \infty$, $R_i = \infty$, and $R_o = 0$, show that the answer in part (a) is divided by $1 - R_L/R_2A_V$.

16-11 To obtain high gain with low-value resistances the feedback resistance R' in an inverting OP AMP is replaced by the T network shown. Verify that the transfer gain is given by

$$\frac{V_o}{V_i} = \frac{R_2 + R_3 + R_2R_3/R_4}{R_1}$$

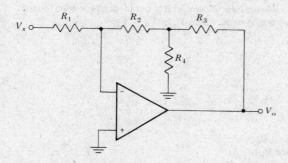

Prob. 16-11

16-12 An amplifier whose gain can be adjusted to be ± 1 by means of a switch is shown. Demonstrate that, if S is closed (open), the transfer gain $A = v_o/v_i = -1 \ (+1)$.

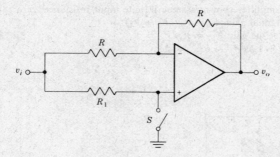

Prob. 16-12

16-13 (a) For the phase-shifter circuit shown verify that the transfer gain is

$$A = \frac{V_o}{V_i} = \frac{1 - j\omega RC}{1 + j\omega RC}$$

What is $|A|$? Over what range does the phase ϕ vary as R varies from 0 to ∞?

(b) If R and C are interchanged, show that the gain is the negative of the equation in part (a). Over what range does the phase vary as R varies from 0 to ∞?

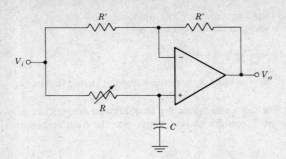

Prob. 16-13

16-14 Consider the DIFF AMP configuration shown which has a double-ended output. Verify that Eq. (16-9) is valid for this system, where $V_o = V_4 - V_3$.

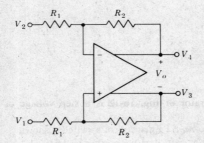

Prob. 16-14

16-15 For the differential-input amplifier shown, assume infinite input resistance, zero output resistance, and finite differential gain $A_V = V_o/(V_2 - V_1)$.

    (a) Obtain an expression for the gain $A_{Vf} = V_o/V_s$.

    (b) Show that $\lim A_{Vf} = n + 1$ as $A_V \to \infty$.

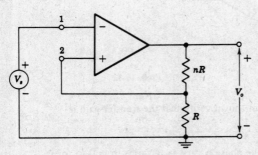

Prob. 16-15

16-16 Verify Eq. (16-10) for the three-OP AMP instrumentation amplifier.

16-17 For the instrumentation amplifier shown verify that

$$V_o = \left(1 + \frac{R_2}{R_1} + \frac{2R_2}{R}\right)(V_2 - V_1)$$

Note that the gain may be adjusted by varying R.

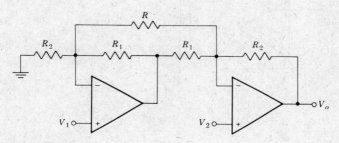

Prob. 16-17

16-18 Given the signals v_1 and v_2. Use two inverting amplifiers to obtain an output $v_o = k(v_2 - v_1)$, where k is a positive number.

16-19 Verify Eq. (16-11) for the bridge amplifier.

16-20 (*a*) The input to the operational integrator of Fig. 16-12 is a step voltage of magnitude V. Show that the output is

$$v_o = A_V V (1 - \epsilon^{-t/RC(1-A_V)})$$

(*b*) Compare this result with the output obtained if the step voltage is impressed upon a simple RC integrating network (without the use of an operational amplifier). Show that for large values of RC, both solutions represent a voltage which varies approximately linearly with time. Verify that if $-A_V \gg 1$, the slope of the ramp output is approximately the same for both circuits. Also prove that the deviation from linearity for the amplifier circuit is $1/(1 - A_V)$ times that of the simple RC circuit.

16-21 Derive Eq. (16-18).

16-22 Given an operational amplifier with Z consisting of R in series with C, and Z' consisting of R' in parallel with C'. The input is a step voltage of magnitude V.

(*a*) Show by qualitative argument that the output voltage must start at zero, reach a maximum, and then again fall to zero.

(*b*) Show that if $R'C' \neq RC$, the output is given by

$$v_o = \frac{R'CV}{R'C' - RC}(\epsilon^{-t/RC} - \epsilon^{-t/R'C'})$$

16-23 An op amp is used to integrate three voltages $v_1(t)$, $v_2(t)$, and $v_3(t)$ and to obtain the sum of the integrations. Two supply voltages $+10$ and -10 V are available to allow for an initial output voltage which may be anywhere between -5 and $+5$ V. Indicate the system using two single-pole triple-throw ganged switches so that in position 1 the initial condition is set, in position 2 the integration takes place, and in position 3 the output is held constant at the end of the integration time.

16-24 Obtain the transfer function of the network shown. Verify that $v_o = (1/RC)\int v_s\,dt$ so that noninverting integration is performed.

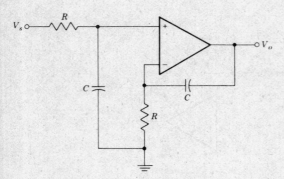

Prob. 16-24

16-25 Prove that the network shown is a noninverting integrator with $v_o = (2/RC)\int v_s(t)\,dt$.

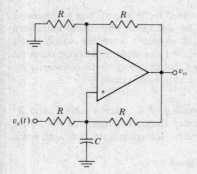

Prob. 16-25

16-26 Verify that the system shown, which uses only one OP AMP, is a double integrator. In other words, prove that the transfer gain is

$$\frac{V_o}{V_s} = -\frac{1}{(RCs)^2}$$

Hint: Evaluate I_1 and I_2 independently and set $I_1 = -I_2$. Why?

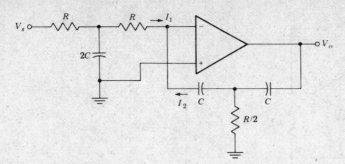

Prob. 16-26

16-27 Sketch in block-diagram form an analog computer, using operational amplifiers, to solve the differential equation

$$\frac{d^2v}{dt^2} + 2v - 5 \sin \omega t = 0$$

where

$$v(0) = -1$$

and

$$\left.\frac{dv}{dt}\right|_{t=0} = 0$$

An oscillator is available which will provide a signal $\sin \omega t$. Use only resistors and capacitors.

16-28 Set up a computer in block-diagram form, using operational amplifiers, to solve the following differential equation:

$$\frac{d^3y}{dt^3} - 5\frac{d^2y}{dt^2} + 4\frac{dy}{dt} + 3y = x(t)$$

where

$$y(0) = 2 \qquad \left.\frac{dy}{dt}\right|_{t=0} = 0$$

and

$$\left.\frac{d^2y}{dt^2}\right|_{t=0} = 3$$

Assume that a generator is available which will provide the signal $x(t)$.

16-29 Verify that the input impedance $V_i/I_i = -Z$. This circuit is called a *negative-impedance converter.*

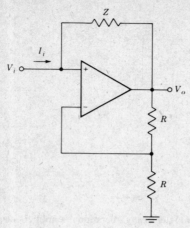

Prob. 16-29

·16-30 In the system shown OP AMP (2) supplies current to the input of OP AMP (1) and, hence, increases the input resistance R_i. Find (a) V_1/V_i, (b) V_2/V_i; (c) Verify that

$$R_i = \frac{V_i}{I_i} = \frac{R_1 R_3}{R_3 - R_1}$$

Note that $R_i = \infty$ if $R_1 = R_3$ and R_i is negative if $R_1 > R_3$.

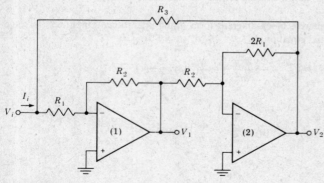

Prob. 16-30

16-31 (a) Verify that the gyrator shown has an input impedance $V_i/I_i = R_1 R_2/Z$. *Hint:* Find the voltages at P_1 and P_2 and apply KCL to the input node.

(b) If Z is a capacitor C, show that this system behaves as an inductor.

(c) Find the value of C in order to obtain a 1-H inductance, if $R_1 = R_2 = 1$ kΩ.

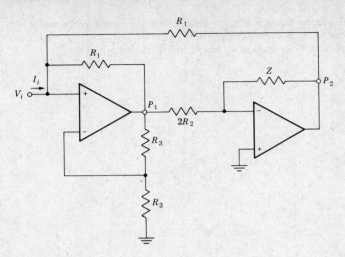

Prob. 16-31

16-32 Using the value of $B_n(s)$ from Table 16-1 verify that (with $s = j\omega$),

$$[B_n(s)]^2 = B_n(j\omega)B_n(-j\omega) = 1 + \omega^{2n}$$

for (a) $n = 1$ and (b) $n = 2$.

16-33 Repeat Prob. 16-32 for $n = 3$.

16-34 Repeat Prob. 16-32 for $n = 4$.

16-35 Proceeding as outlined in the text, verify Eq. (16-28) for the low-pass active filter.

16-36 Design an active fifth-order Butterworth low-pass active filter with a cutoff frequency of 1 kHz. Choose $C = 0.01$ μF. Specify the value of all resistances. (If some resistance values may be chosen arbitrarily, state which these are.)

16-37 Repeat Prob. 16-36 for a sixth-order high-pass Butterworth filter with a cutoff frequency of 5 kHz.

16-38 Repeat Prob. 16-36 for a third-order bandpass Butterworth filter with the desired passband between 2 and 4 kHz.

16-39 (a) Obtain the transfer function V_o/V_s for the system shown and verify that it is of the form of a second-order Butterworth low-pass filter.

(b) Evaluate C_1 and C_2 in terms of ω_o and k, with R chosen arbitrarily.

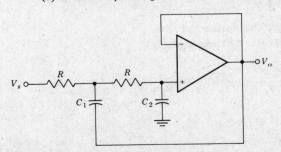

Prob. 16-39

16-40 (*a*) Obtain the transfer function V_o/V_s for the system shown and verify that it is of the form of a second-order Butterworth high-pass filter.

(*b*) Evaluate R_1 and R_2 in terms of ω_o and k, with C chosen arbitrarily.

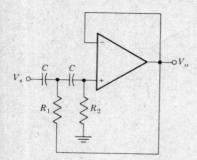

Prob. 16-40

16-41 A third-order low-pass filter is indicated. Verify that $B_3(s) = V_o/V_s$ is given by

$$B_3(s) = s^3R^3C_1C^2 + 2s^2R^2C(C + C_1) + sR(3C + C_1) + 1$$

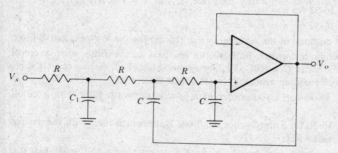

Prob. 16-41

16-42 Verify Eq. (16-47) for the RC bandpass active filter.

16-43 For the circuit shown verify that the transfer gain is

$$\frac{V_o}{V_s} = \frac{-Y_1Y_3}{Y_3Y_4 + Y_5(Y_1 + Y_2 + Y_3 + Y_4)}$$

in terms of the five admittances Y, where $Y = 1/Z$.

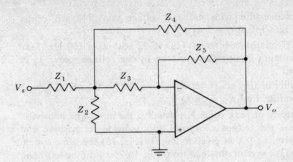

Prob. 16-43

16-44 (*a*) In the bandpass filter of Fig. 16-24 interchange R_3 with the vertical C and interchange R_2 with the horizontal C. Verify that the transfer function of the modified system is

$$\frac{V_o}{V_s} = \frac{-R_3/R_1}{s^2C^2R_2R_3 + sC(R_2 + R_3 + R_2R_3/R_1) + 1}$$

Note that this is the transfer function of a Butterworth low-pass filter (with a low-frequency gain of $-R_3/R_1$) and *not that of a bandpass filter*.

    (*b*) For a gain of -1, a frequency ω_o, and a damping factor k, find implicit expressions for R_1, R_2, R_3, and C. How many of these parameters may be fixed arbitrarily?

16-45 (*a*) For the system shown verify that (for $A_{VO} \equiv A = 1 + R_3/R_2$) the transfer function is

$$\frac{V_o}{V_s} = \frac{(A/CR)s}{s^2 + \{[2R + R_1(2 - A)]/CRR_1\}s + 2/C^2RR_1}$$

Note that this represents a resonant bandpass active filter.

    (*b*) Obtain implicit expressions for R, R_1, R_2, R_3, and C in terms of A_o, ω_o, and Q. How many parameters may be fixed arbitrarily?

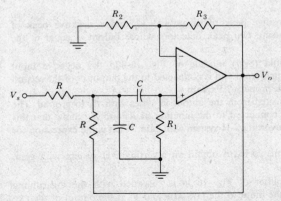

Prob. 16-45

16-46 Design a bandpass RC active filter with midband voltage gain of 40 dB, center frequency of 200 Hz, and $Q = 12$. Choose $C = 0.01\ \mu F$.

16-47 Design the resonant RLC bandpass filter of Fig. 16-22 with $f_o = 160$ Hz, 3-dB bandwidth $B = 16$ Hz, and minimum input resistance seen by the voltage source V_s of 1,000 Ω. Is this a practical circuit?

16-48 In the bandpass active filter of Fig. 16-24, let $R_2 = \infty$ and $C = 0.1\ \mu F$. If $Q = 2$, $\omega_o = 500$ Hz, and A_o is not specified, find R_1, R_3, and A_o.

16-49 Sketch a sinusoidal waveform of peak value V_m, which is the input to a half-wave rectifier. Directly below it draw the output waveform and indicate its positive and negative peak values, if the system is (a) that given in Fig. 16-26a; (b) the same system with the two diodes reversed; (c) the system obtained from Fig. 16-26a with the left-hand side of R grounded and v_i impressed on the noninverting terminal; (d) the system in part (c) with the diodes reversed.

16-50 (a) Verify that the circuit shown gives full-wave rectification provided that $R_2 = KR_1$. Find K.

   (b) What is the peak value of the rectified output?

   (c) Draw carefully the waveforms $v_i = 10 \sin \omega t$, v_p, and v_o if $R_3 = R_2$.

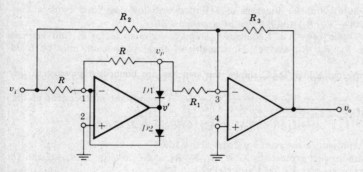

Prob. 16-50

16-51 If a waveform has a positive peak of magnitude V_1 and a negative peak of magnitude V_2, draw a circuit using two peak detectors whose output is equal to the peak-to-peak value $V_1 - V_2$.

16-52 (a) In the sample-and-hold (S-H) module of Fig. 16-30b the negative input terminal of $A1$ is removed from its output and is connected to the output v_o of the second OP AMP. Will the system function properly? Explain.

   (b) A resistor R_2 is connected from the output in series with R_1 to ground. The inverting terminal of $A1$ is now connected to the junction of R_2 and R_1. Show that this configuration operates as a noninverting S-H system with gain. What is the expression for the gain?

   (c) Modify the connections so as to obtain an inverting S-H system with gain. Evaluate the gain.

16-53 (a) The exponential amplifier of Fig. 16-36 is cascaded with the logarithmic amplifier of Fig. 16-35. If V_s is the input to the LOG AMP and V_o' is the output of the EXP AMP, prove that $V_o' = V_s$.

(b) Assume that the resistors R_1, R_2, R_3, and R_4 in Fig. 16-35 are not identical with the corresponding resistors in Fig. 16-36. Designate the constants in Eq. (16-68) by $K_1' \neq K_1$ and $K_2' \neq K_2$. For the cascaded arrangement in part (a), prove that V_o' is proportional to a power n of V_s, where $n = K_1/K_1'$.

(c) Assume that R_3 in the EXP AMP is adjustable, but that all other resistance values are as indicated in Figs. 16-35 and 16-36. Calculate R_3 so that $n = 3$. Repeat for $n = \frac{1}{3}$.

16-54 For the feedback circuit shown, the nonlinear feedback network β gives an output proportional to the product of the two inputs to this network, or $V_f = \beta V_2 V_o$. Prove that if $A = \infty$, then $V_o = KV_1/V_2$, where K is a constant.

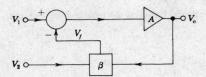

Prob. 16-54

16-55 (a) For the D/A converter of Fig. 16-41 the third most significant bit $N - 3$ is 1 and all other bits are zero. Find the voltages at nodes $N - 3$, $N - 2$, $N - 1$, and at the output V_o in terms of V_R and the resistors.

(b) For a 5-bit DAC with the least significant bit equal to 1 and all other bits equal to 0, find the voltage at all nodes 0, 1, 2, . . . , and at the output.

16-56 In the inverted-ladder DAC shown the switches are connected directly to the OP AMP input.

(a) Show that the current I drawn from V_R is a constant independent of the

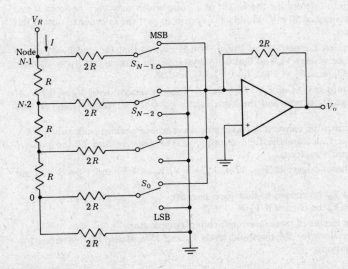

Prob. 16-56

digital word. Explain why propagation-delay-time transients are eliminated with this system.

(b) What is the switch current and V_o if the MSB is 1 and all other bits are zero?

(c) Repeat part (b) if the next MSB is 1 and all other bits are zero.

(d) Calculate V_o for the LSB in the 4-bit D/A converter with all other bits zero.

Chapter 17

17-1 (a) For the comparator circuit shown plot the transfer characteristic if the OP AMP gain is infinite and $V_{Z1} = V_{Z2} = 10$ V. Explain.

(b) Repeat part (a) if the large-signal gain is 50,000.

(c) Repeat part (a) if a voltage of 4 V is applied between the *negative terminal and ground*.

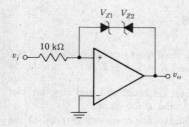

Prob. 17-1

17-2 (a) Using two comparators and an AND gate, draw a system whose output is logic 1 if and only if the input lies in the window between V_{R1} and V_{R2}. Explain the operation.

(b) It is desired to determine the height of a pulse which may vary between 0 and 10 V with an uncertainty of 50 mV. Modify the system in part (a) in order to obtain this pulse-height analyzer.

17-3 In the regenerative comparator of Fig. 17-4 it is desired that the threshold voltage V_1 equal the reference voltage V_R and that 0.1-V hysteresis be obtained. If $A_V = 100,000$, the loop gain is 1,000, and $R_2 = 1$ kΩ, find V_R, V_Z, and R_1.

17-4 (a) The Schmitt trigger of Fig 17-4 uses 6-V Zener diodes, with $V_D = 0.7$ V. If the threshold voltage V_1 is zero and the hysteresis is $V_H = 0.2$ V, calculate R_1/R_2 and V_R.

(b) This comparator converts a 1-kHz sine wave whose peak-to-peak value is 4 V into a square wave. Calculate the time duration of the negative and of the positive portions of the output waveform.

17-5 (a) In the Schmitt trigger of Fig. 17-4, $V_o = 8$ V, $V_1 = 4$ V, and $V_2 = 3$ V. Find R_1/R_2 and V_R.

(b) How must V_R be chosen so that V_2 is negative?

(c) How must V_R be chosen if $V_1 = -V_2$?

17-6 (a) Sketch the circuit of a *noninverting* Schmitt comparator.

(b) Find expressions for the threshold levels V_1 and V_2 and also the hysteresis V_H. Explain your calculation.

(c) Show the transfer characteristic corresponding to Fig. 17-4.

(d) Draw the output waveform lined up in time with a sinusoidal input signal.

17-7 (a) For the comparator shown find expressions for the threshold voltages V_1 and V_2 in terms of R_1, R_2, and the OP AMP limited value (of magnitude V_o). Explain your calculations.

(b) Show the transfer characteristic corresponding to Fig. 17-4d.

(c) The peak output is 10 V and $R_2 = 5 R_1$. If a 4-V peak sinusoid is applied, draw the output lined up in time with the input.

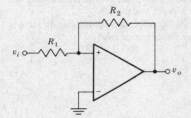

Prob. 17-7

17-8 (a) Consider the square-wave generator of Fig. 17-6, where nonidentical avalanche diodes V_{Z1} and V_{Z2} are used. If the output is either $+ V_{o1}$ or $- V_{o2}$, where $V_{o1} \equiv V_{Z1} + V_D$ and $V_{o2} \equiv V_{Z2} + V_D$, verify that the duration of the positive section is given by

$$T_1 = RC \ln \frac{1 + \beta V_{o2}/V_{o1}}{1 - \beta}$$

(b) Verify that T_2 (the duration of the negative section) is given by the same equation with V_{o1} and V_{o2} interchanged.

(c) If $V_{o1} > V_{o2}$, is T_1 greater or less than T_2? Explain.

17-9 The triangular waveform generator of Fig. 17-8 has a symmetry control voltage V_S added to the noninverting terminal of the integrator.

(a) Verify that the sweep speed for the positive ramp is $(V_o + V_S)/RC$.

(b) Find T_1, T_2, and f.

(c) Verify that the duty cycle is given by Eq. (17-15).

17-10 (a) Draw Fig. 17-6 with R replaced by the two JFETs indicated. Explain the system operation qualitatively. *Hint:* During the positive-going ramp, $Q1$ behaves as a constant-current source and $Q2$ as a forward-biased diode.

Use the 2N4869 JFET characteristics of Fig. 8-3. It is desired that the charging current be 3 mA, the discharging current 1 mA, and the period 1 ms. If $V_o = 20$ V and $R_1 = R_2$, calculate (b) R_{S1}, (c) R_{S2} and (d) C.

(e) Indicate the waveform across $Q1$, as well as v_o and v_c.

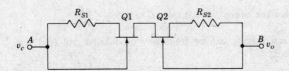

Prob. 17-10

17-11 Show a system using a zero-crossing detector and a positive-negative controlled-gain amplifier to obtain an output which is the absolute value of the input. Explain its operation.

17-12 Verify Eq. (17-18) for the pulse width T of the monostable multi of Fig. 17-11.

17-13 (a) Consider the pulse generator shown. In the quiescent state (before a trigger is applied) find v_2, v_o, and v_1.

(b) At $t = 0$ a narrow, positive, triggering pulse v_t whose magnitude exceeds V_R is applied. At $t = 0 +$ find v_o and v_1. (Remember that the voltage across a capacitor cannot change instantaneously.)

Now plot the waveforms v_o and v_1 as a function of time. Demonstrate that the circuit behaves as a monostable multivibrator with a pulse width T.

(c) Find v_o and v_1 at $t = T +$ and continue the waveforms until the steady state is reached. What is the recovery-time constant? (Is the diode ON or OFF?)

(d) Verify that T is given by

$$T = RC \ln \frac{2V_o}{V_R}$$

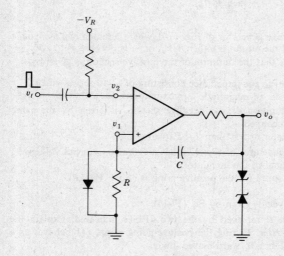

Prob. 17-13

17-14 (a) A free-running sweep is obtained from the system shown, where $-V$ is a constant negative voltage and where $R' \ll R$. Explain the operation and sketch the waveform $v(t)$.

(b) Find V_R and V_s so that the sweep extends from 0 to V_s.

(c) Find the sweep time T_s.

(d) Explain how the sweep voltage can be frequency modulated and find f in terms of the modulating voltage v_m.

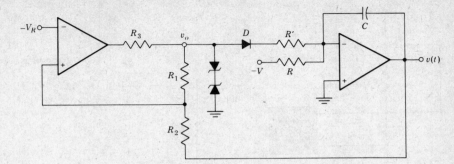

Prob. 17-14

17-15 (a) Verify Eq. (17-22) for the sweep-speed error e_s of an exponential sweep.

(b) In the circuit of Fig. 17-15a a resistor R_2 is shunted across C_1. Show that e_s is now multiplied by $(R_1 + R_2)/R_2$.

17-16 For the OP AMP sweep generator of Fig. 17-15c assume finite R_i, finite A, and nonzero R_o.

(a) Draw the amplifier model with R_i at the input and Av_i in series with R_o across the output.

(b) Apply Miller's theorem to the impedance consisting of C_1 in series with R_o.

(c) Making reasonable order-of-magnitude approximations, show that the expression for the slope error is

$$e_x = \frac{V_s}{AV} \frac{R_1 + R_i}{R_i}$$

17-17 In the Miller sweep of Fig. 17-15c, $V = 45$ V, $R_1 = 1$ MΩ, and the OP AMP gain is $A = 50,000$. The output sweep amplitude is 25 V. The longest sweep is 5 s and the shortest is 5 μs.

(a) For $T_s = 5$ s, find C_1.

(b) Calculate the sweep-speed error e_s.

(c) Repeat parts (a) and (b) for $T_s = 5$ μs.

(d) How can a 5-μs sweep be obtained with a more reasonable value of C_1 than that found in part (c), say, 100 times as large.

17-18 (a) In the triggered sweep of Fig. 17-16 $R_1 = 100$ kΩ, $V = 30$ V, and the monostable pulse width is 100 μs. Find C_1 if the sweep amplitude is to be 10 V.

(b) The resistor R_1 is replaced by the JFET shown, which acts as a constant-current source I. If Q is a 2N4869 depletion-mode JFET whose characteristics are given in Fig. 8-3, find R_s so that $I = 1$ mA.

(c) By adding a sample-and-hold circuit across C_1, this configuration became a *pulse-width decoder*. Verify that the voltage across C_1 is proportional to the pulse width T_s. Find V_s for $T_s = 100$ μs, if $C_1 = 0.005$ μF.

(d) What is the maximum value of T_s which may be measured with this system?

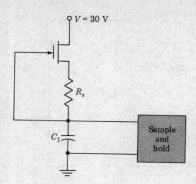

$V = 30$ V

R_s

C_1

Sample and hold

Prob. 17-18

17-19 (a) For the *bootstrap sweep* shown in part (a) of the figure, the capacitor C_1 may be taken as arbitrarily large. The drop across the ideal diode D may be neglected during conduction, and it may be assumed that any negative voltage turns D OFF. The OP AMP is ideal ($R_i = \infty$, $R_o = 0$, and $A = \infty$). With the switch S closed, what is the voltage across C_1 and R? With S open and C charged to v_c, what is the voltage across C_1 and R? Show that a precisely linear sweep is obtained and that $v_c = Vt/RC$.

(b) A linear sweep with a pair of symmetrical outputs ($v_{o1} = -v_{o2}$) is to be obtained from the system shown in part (b) of the figure. Find the values of R'/R and R''/R.

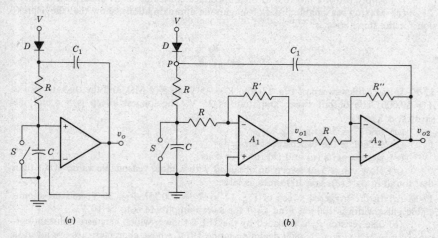

(a)

(b)

Prob. 17-19

17-20 (a) Consider the chopper modulator of Fig. 17-20a with S_1 controlled by $+v_o$ instead of $-v_o$. For the modulating waveform v_m in Fig. 17-20b, sketch the first five pulses. Call this waveform v_+. Lined up in time with v_+ draw v_-, the output of the chopper, when S_1 is controlled by $-v_o$ ($v_- = v$ in Fig. 17-20b).

(b) Indicate how to combine v_+ and v_- with OP AMPS in order to obtain the AM waveform v' of Fig. 17-19c.

17-21 The switch S is controlled by a square wave $-v_o$. The modulating waveform v_m is a sinusoid of peak value V_m. The period of v_m is eight times the period of v_o, and v_m passes through zero and is increasing when v_o passes through zero and is increasing. Sketch the following waveforms lined up in time. (a) One cycle of v_m, (b) v_o, (c) v_1, (d) v_2.

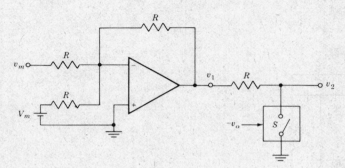

Prob. 17-21

17-22 Verify Eq. (17-27) for the feedback factor of the phase-shift network of Fig. 17-25, assuming that this network does not load the amplifier. Prove that the phase shift of V_f'/V_o is 180° for $\alpha^2 = 6$ and that at this frequency $\beta = \frac{1}{29}$.

17-23 (a) For the network of Prob. 17-22, show that the input impedance is given by

$$Z_i = R \frac{1 - 5\alpha^2 - j(6\alpha - \alpha^3)}{3 - \alpha^2 - j4\alpha}$$

(b) Show that the input impedance at the frequency of the oscillator, $\alpha = \sqrt{6}$, is $(0.83 - j2.70)R$.

Note that if the frequency is varied by varying C, the input impedance remains constant. However, if the frequency is varied by varying R, the impedance is varied in proportion to R.

17-24 (a) Draw the equivalent circuit of the transistor phase-shift oscillator shown. Assume that $h_{oe}R_c < 0.1$ so that the approximate hybrid model may be used to characterize the low-frequency behavior of the transistor. Let $R_3 = R - h_{ie}$. Neglect the effect of the biasing resistors R_1, R_3, and R_e. Imagine the loop broken at the base between B_1 and B_2, but in order not to change the loading on the feedback network, place $R_i = h_{ie}$ between B_2 and ground.

(b) Find the current loop gain $-x_f/x_i = I_3/I_b$ by writing Kirchhoff's voltage equations for the meshes of the phase-shift network. Verify that the Barkhausen condition that the phase of I_3/I_b must equal zero leads to the following expression for the frequency of oscillation:

$$f = \frac{1}{2\pi RC} \frac{1}{\sqrt{6 + 4k}}$$

where $k \equiv R_c/R$. The requirement that the magnitude of I_3/I_b must exceed unity in order for oscillations to start leads to the inequality

$$h_{fe} > 4k + 23 + \frac{29}{k}$$

(c) Find the value of k which gives a minimum for h_{fe}. Show that $h_{fe(min)} = 44.5$. A transistor with $h_{fe} < 44.5$ cannot be used in a phase-shift-oscillator configuration.

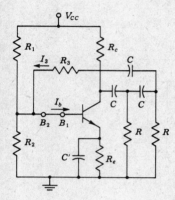

Prob. 17-24

17-25 Design a phase-shift oscillator to operate at a frequency of 5 kHz. Use a MOSFET with $\mu = 55$ and $r_d = 5.5$ kΩ. The phase-shift network is not to load down the amplifier.

(a) Find the minimum value of the drain-circuit resistance R_d for which the circuit will oscillate.

(b) Find the product RC.

(c) Choose a reasonable value for R, and find C.

17-26 For the FET oscillator shown, find (a) V_f'/V_o, (b) the frequency of oscillation, (c) the minimum gain of the source follower required for oscillations.

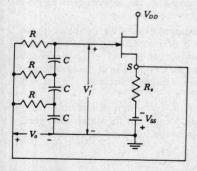

Prob. 17-26

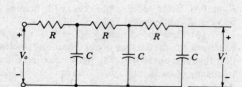

Prob. 17-27

17-27 (a) For the feedback network shown, find the transfer function V_f'/V_o.

(b) This network is used with an OP AMP to form an oscillator. Find the frequency of oscillation and the minimum amplifier voltage gain.

(c) Draw the network connected to the OP AMP to form the oscillator.

17-28 Verify Eq. (17-34) for the loop gain of a Wien-bridge oscillator.

17-29 (a) For the network shown prove that

$$\frac{V_f'}{V_o} = \frac{1}{3 + j(\omega RC - 1/\omega RC)}$$

(b) This network is used with an OP AMP to form an oscillator. Show that the frequency of oscillation is $f = 1/2\pi RC$ and that the gain must exceed 3.

(c) Draw the network connected to the OP AMP.

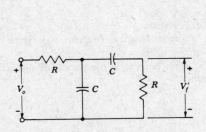

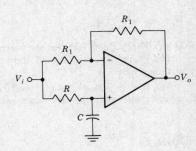

Prob. 17-29   **Prob. 17-30**

17-30 (a) Verify that the transmission characteristic of the OP AMP configuration shown is given by

$$\frac{V_o}{V_i} = \frac{1 - j\omega RC}{1 + j\omega RC}$$

Note the magnitude $|V_o/V_i| = 1$ for all frequencies and that the output leads the input by a phase $\phi = -2\arctan \omega RS$. If R is varied as C remains constant, this configuration acts as a constant-amplitude phase shifter. The phase is 0 for $R = 0$ and $-180°$ for $R \rightarrow \infty$.

(b) Cascade two identical phase shifters of the type given in part (a). Complete the loop with an inverting OP AMP. Show that this system will oscillate at the frequency $f = 1/2\pi RC$ provided that the amplifier gain exceeds unity (in magnitude).

(c) Demonstrate that two quadrature sinusoids (sine waves differing in phase by 90°) are obtained.

17-31 In the Wien-bridge topology of Fig. 17-28, Z_1 consists of R, C, and L in series, and Z_2 is a resistor R_3. Find the frequency of oscillation and the minimum ratio R_1/R_2.

17-32 For the oscillator shown, find the frequency of oscillation and the minimum value of R.

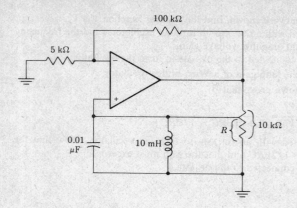

Prob. 17-32

17-33 (*a*) Verify Eq. (17-36) for the reactance of a crystal.

(*b*) Prove that the ratio of the parallel- to series-resonant frequencies is given approximately by $1 + \frac{1}{2}C/C'$.

(*c*) If $C = 0.04$ pF and $C' = 2.0$ pF, by what percent is the parallel-resonant frequency greater than the series-resonant frequency?

Chapter 18

18-1 (*a*) Nonlinear distortion results in the generation of frequencies in the output that are not present in the input. If the dynamic curve can be represented by Eq. (18-6), and if the input signal is given by

$$i_b = I_1 \cos \omega_1 t + I_2 \cos \omega_2 t$$

show that the output will contain a dc term and sinusoidal terms of frequency ω_1, ω_2, $2\omega_1$, $2\omega_2$, $\omega_1 + \omega_2$, and $\omega_1 - \omega_2$.

(*b*) Generalize the results of part (*a*) by showing that if the dynamic curve must be represented by higher-order terms in i_b, the output will contain intermodulation frequencies, given by the sum and difference of integral multiples of ω_1 and ω_2; for example, $2\omega_1 \pm 2\omega_2$, $2\omega_1 \pm \omega_2$, $3\omega_1 \pm \omega_2$, etc.

18-2 A transistor supplies 2 W to a 4-kΩ load. The zero-signal dc collector current is 35 mA, and the dc collector current with signal is 39 mA. Determine the percent second-harmonic distortion.

18-3 The input excitation of an amplifier is $i_b = I_{bm} \sin \omega t$. Prove that the output current can be represented by a Fourier series which contains only odd sine components and even cosine components.

18-4 (*a*) Consider an ideal transistor with no distortion even if the transistor is driven from cutoff to the edge of saturation, where $v_C = V_{min}$. Verify that the conversion efficiency is given by

$$\eta = \frac{25(V_{CC} - V_{min})}{V_{CC}} \times 100 \text{ percent}$$

(b) What is the maximum possible efficiency and under what circumstances is this maximum value obtained?

18-5 For an ideal class B push-pull amplifier show that the collector dissipation P_C is zero at no signal ($V_m = 0$), rises as V_m increases, and passes through a maximum given by Eq. (18-29) at $V_m = 2V_{CC}/\pi$.

18-6 *Mirror symmetry* requires that the bottom portion of a waveform, when shifted 180° along the time axis, will be the mirror image of the top portion. The condition for mirror symmetry is represented mathematically by the equation

$$i(\omega t) = -i(\omega t + \pi)$$

(a) Verify that a class B push-pull system possesses mirror symmetry by using Eq. (18-35).

(b) Without recourse to a Fourier series, prove that a class B push-pull system has mirror symmetry.

18-7 For the ideal class B push-pull amplifier of Fig. 18-4, $V_{CC} = 15$ V and $R_L = 4\ \Omega$. The input is sinusoidal. Determine (a) the maximum output signal power; (b) the collector dissipation in each transistor at this power output; (c) the conversion efficiency.

(d) What is the maximum dissipation of each transistor and what is the efficiency under this condition?

18-8 The ideal class B push-pull amplifier of Fig. 18-4 is operating at the sinusoidal amplitude for which the dissipation is a maximum. Verify that the conversion efficiency is 50%.

18-9 In the circuit shown the base-to-emitter voltage may be assumed to remain constant at the cutin value V_γ for all values of forward bias. The biasing voltage is idealized by two batteries of voltage kV_γ, where $0 < k \leqslant 1$. Assume that $v_i = V_s \sin \omega t$.

(a) If $V_\gamma = 0.6$ V and $V_s = 1$ V, sketch the output v_o as a function of time for $k = 0, 0.5,$ and 1. Calculate the cutin angle for each value of k.

(b) What happens to the distortion as V_s is increased?

(c) What happens if k exceeds unity?

(d) If a resistor R is added between the two emitters, what happens if $k > 1$?

(e) Is the push-pull operation class A, B, AB, or C in parts (a) and (d)?

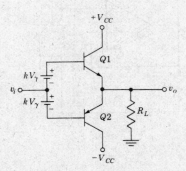

Prob. 18-9

18-10 (*a*) A 2N2905A transistor in a TO5 case has a $\theta_{JA} = 438°C/W$ and a $\theta_{JC} = 97°C/W$. The maximum allowable junction temperature is 200°C. What is the maximum power dissipation at $T_A = 25°C$ if no heat sink is used?

(*b*) An infinite heat sink is attached directly to the case with no electrical isolation. Find $T_{J(max)}$. *Hint*: What must θ_{SA} equal for an infinite mass?

(*c*) If the infinite heat sink is isolated from the case by an insulator with $\theta_{CS} = 4°C/W$, calculate $T_{J(max)}$.

18-11 A complementary push-pull output stage whose transistors are heat-sink-mounted is driven by an asymmetrical waveform causing uneven heating. As a result the dissipation of $Q1$ is 20 W and that of $Q2$ is 10 W. These transistors must be isolated electrically from the heat sink by mica insulators with $\theta_{CS} = 1°C/W$ and $\theta_{JC} = 2°C/W$ for each transistor. Calculate the maximum ambient temperature for $T_{J(max)} = 175°C$, and $\theta_{SA} = 1.5°C/W$? *Hint:* The electrical analog of the thermal circuit has two branches because there are two junction temperatures, T_{J1} for $Q1$ and T_{J2} for $Q2$.

(*b*) Find the temperature of the cooler junction.

18-12 Typical stabilization coefficients for a monolithic regulator are given in Sec. 18-10. The unregulated dc voltage varies by ±0.5 V due to line voltage fluctuations. The load current may change by ±2 A. The peak temperature change from the ambient of 30°C is $\pm50°C$. Calculate the total maximum excursion in output voltage from that at 30°C.

18-13 In Fig. 18-11, $A_V = 10^5$, $R_1 = R_2$, $V_R = 6$ V, and the input offset voltage drift of the OP AMP is 10 $\mu V/°C$.

(*a*) What is the approximate output voltage?

(*b*) What is S_T due to the input offset voltage drift of the OP AMP?

(*c*) What is S_T caused by the base-emitter voltage temperature drift of $Q1$. Assume $S_T = 0$ in part (*b*).

18-14 The output voltage V_{REG} of the monolithic regulator of Fig. 18-12 may be adjusted to a higher value V_O by the circuits shown. Find expressions for V_O in terms of V_{REG} and I_Q, defined in Fig. 18-12. What is the advantage of (*b*) over (*a*)?

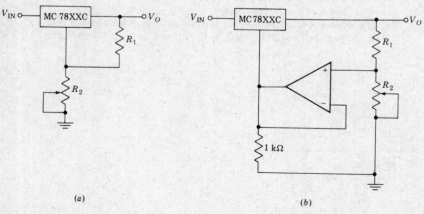

(*a*)   (*b*)

Prob. 18-14

18-15 The three-terminal fixed-voltage regulator is converted into a current regulator by the circuit shown. If the output voltage of the regulator is 5 V, if $R = 5 \, \Omega$, and if $I_Q = 10$ mA, what is the output current I_L? Note that I_L is independent of the load. (b) How can I_L be made independent of I_Q? Hint: See circuit (b) of Prob. 18-14.

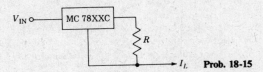

Prob. **18-15**

18-16 If the output voltage V_O of a switching regulator is negative, the level-shifting circuit shown is used at the input of the error amplifier in Fig. 18-14.

    (a) What is the effective feedback voltage?

    (b) Verify that

$$V_O = \frac{1}{2} V_{\text{REF}} \left(1 - \frac{R_2}{R_1} \right)$$

Note that for $V_O < 0$, $R_2 / R_1 > 1$.

    (c) Show that for given values of V_O and V_{REF}, the ratio R_2 / R_1 must be chosen to be

$$\frac{R_2}{R_1} = 1 - \frac{2 V_O}{V_{\text{REF}}}$$

which indicates that for a negative V_O, $R_2 / R_1 > 1$.

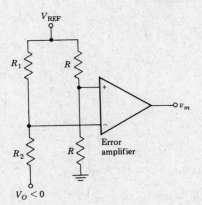

Prob. **18-16**

ANSWERS TO SELECTED PROBLEMS

Chapter 1

1-1 (a) 1,006 V; (b) 3.27×10^5 m/s

1-3 (d) 1.874×10^7 m/s, 2.380×10^6 m/s, 2.54 cm; (e) 508 V

1-4 (a) 0.143 cm; (b) 1.874×10^6 m/s

1-9 (a) 1.666 A; (b) 1.488×10^{-4} m/s; (c) 4.17×10^{-3} m²/V · s; (d) 5.61×10^7 (Ω · cm)$^{-1}$

1-12 (a) 4.697×10^7 (Ω · cm)$^{-1}$; (b) 1.74 m/s

1-14 (a) $p = 1.06 \times 10^{14}$/cm³, $n = 6 \times 10^{12}$/cm³; (b) $n = p = 2.5 \times 10^{13}$/cm³; (c) $n = 10^{16}$/cm³, $p = 6.25 \times 10^{10}$/cm³

1-17 (a) 2.315 Ω · cm; (b) 1 donor atom per 10^8 Si atoms

1-19 8.286% per degree K

1-21 $p = 3.843 \times 10^{14}$/cm³, $n = 3.656 \times 10^{14}$/cm³

1-23 (b) 3.72 Ω · cm; (c) 10^8

1-25 (a) 65 mV; (b) 25 mV

1-27 0.308 Wb/m²

1-30 (a) 0.268 V; (b) 0.659 V

1-32 (a) 0.219 V; (b) 0.666 V

Chapter 2

2-1 (a) $n_n = 5 \times 10^{14}$/cm³, $p_n = 4.5 \times 10^5$/cm³; (b) $p_p = 5 \times 10^{16}$/cm³, $n_p = 4.5 \times 10^3$/cm³

2-3 $p_p = 1.30 \times 10^{15}$/cm³, $n_p = 1.73 \times 10^5$/cm³, $n_n = 4.81 \times 10^{13}$/cm³, $p_n = 4.68 \times 10^6$/cm³

2-6 (a) 34.21 mA; (b) 7.123 nA

2-9 (a) 181.1; (b) 1.854×10^5

2-10 (a) 56.44°C; (b) −33.22°C

2-12 0.027 V

2-14 0.579

2-16 27.77 Ω

2-18 (b) 99.49 pF

2-21 (a) 0.147 mil; (b) -5.624×10^4 V/cm; (c) 0.0184 pF/mil^2

2-24 (a) 4.056×10^{-5} cm; (b) 1.457×10^{-5} cm; (c) 1.629×10^{-7} cm; (d) 2.618×10^{-10} F, 7.288×10^{-10} F

2-30 (a) 4.65 mA; (b) 4.66 mA; (c) 1.5 mA; (d) 1.15 mA; (e) 9.71×10^{-6} mA

2-31 (a) 10 nA, 0.036 V, 5.964 V; (b) 2.25 A, 5 V, 1 V

2-33 (a) 1μA, -79.96 V, -36.04 mV, for $V = 80$ V: 2μA, -100 V, -20 V, for $V = 120$ V; (b) $I_1 = 1$ μA, $I_2 = 2\mu$A, $V_1 = -44$ V, $V_2 = -36$ V, for $V = 80$ V

2-35 (a) 5.054×10^8 A (b) 6.519 mA

2-36 (a) 29.79 mA; (b) -10 μA; (c) 29.79 mA, 20 mA

2-38 (a) $I_1 = 9.96$ mA, $I_2 = 0$ mA; (b) $I_1 = 53.74$ mA, $I_2 = 49.99$ mA

2-40 (a) 5 V; (b) 0.4545 V; (c) 0.2381 V

2-41 (a) $I_1 = 3.6$ mA, $I_2 = 0$ A, $I_3 = 1.815$ mA, $v_o = 4.3$ V; (b) $I_1 = I_2 = 0.349$ mA, $I_3 = 0$ A, $v_o = 26.04$ V

Chapter 3

3-1 (a) -5 mA, -1 V; (b) 0.67 V, 0.2 V; (c) 10 mA

3-5 (a) 2.376, 0.024, -2.4, 117.6, 2.4, 119.98 mA; (b) 8 V; (c) 4,999, 0.9998

3-6 (a) 4.3 V; (b) 250 Ω, 3.08 V

3-9 (a) 0.5, 44.2 kΩ; (b) 0.298, 24 kΩ

3-12 1.94 μA, 0.261, 0.259, 0.097, 0.0989, 9.89, -9.99 mA

3-13 0.71, 0.15, 0.56 V

3-15 (a) 0.0315, 1.575 mA; (c) 0.0686, 1.223 mA; (e) 7.426 kΩ

3-17 (a) 1.10 V; (b) 0.2 V

3-20 (a) 8.5 V; (b) 12 V

3-23 (b) $I_C = \dfrac{\alpha I_{EO}}{1 - \alpha\alpha_I}(\epsilon^{V_E/V_T} - 1) - \dfrac{I_{CO}}{1 - \alpha\alpha_I}(\epsilon^{V_C/V_T} - 1)$

3-25 (a) 12.2 kΩ; (b) 157.9°C

3-26 (a) 0.2 V; (b) 36.96 kΩ; (c) 12 V; (d) 148.9°C

Chapter 4

4-6 0.0275, 2.753, 0.430, 2.350 mA

4-9 (a) 6.0 Ω; (b) 0.175 Ω

4-11 307 pF

4-13 500 mils^2

4-14 4.05 pF

Chapter 5

5-2 (a) -0.641 V; (b) -0.755 V; (c) -0.528 V; (d) -1.320 V; (e) -1.137 V, -1.176 V, -1.098 V, -1.373 V

5-7 (a) 10 V $< V_R < 42$ V; (b) 15 V, all diode currents are zero; (c) $D1$, $D2$ are ON, $D0$ is OFF, 25.37 V, diode currents $= 0.366$ mA

5-8 (a) $v_o = -3$ V, $I_{D1} = I_{D2} = 4$ mA, $R = 11.5$ kΩ; (b) $v_o \approx -1$ V, $v_o = -1.108$ V; (c) $R = 2.875$ kΩ, $R' = 36.25$ kΩ; (d) $v_o = 7.666$ V

5-11 (a) 5 V, 0 V, $h_{FE(\min)} = 13.33$; (b) $81.42°$C

5-16 (a) 10 V, $I_{D1} = I_{D2} = 0.25$ mA, $I_{D3} = I_{D4} = 2$ mA, $I_{D5} = 1.5$ mA, $D6$ is OFF; (b) 20 V, $D1$, $D3$ are OFF, $I_{D2} = I_{D4} = 3.75$ mA, $I_{D5} = I_{D6} = 0.25$ mA

5-26 (a) (i) $NM(0) = -6.20$ V, (ii) $NM(1) = 1.48$ V; (b) 2.853 kΩ

5-27 (a) 0.2 V, 8.7 V; (b) 4.10; (c) 0.236 mA

5-36 (a) 1.1 V, 10.25

5-38 (a) $h_{FE(\min)} = 3.88$; (b) 10; (c) -1.4 V, 0.8 V; (d) 2.989 V, (e) 15.58 mA

5-42 (a) 15.54; (b) -3.4 V, 0.7 V; (c) 5.465 W, 12.22 W; (d) 1.068 V

5-44 (b) $NM(0) = 0.6$ V, $NM(1) = -3.5$ V; (c) 1.289 mA;
(d) $V(0) = 0.3$ V, $V(1) = 5$ V, 37

5-46 (a) 1.457; (b) $NM(1) = 7.0$ V, $NM(0) = -7.2$ V; (c) 77; (d) 99.96 mW

5-48 (a) 2.82; (b) $0.957 < h_{FE} < 2.381$; (c) -3.4 V, 0.7 V; (d) 58, $Q2$ in saturation; (e) 14.39 W

5-50 (a) 23.64; (b) 1.652; (c) -2.0 V, 0.6 V; (d) 55, $Q2$ in saturation; (e) 92.975 W

5-54 (a) $V(0) = 0.2$ V, $V(1) = 1.312$ V; (b) -0.512 V, 0.3 V; (c) 3.91;
(d) 5.12 W, 6.667 W

5-55 (a) $I_{C1} = 0.45$ mA, $I_{C2} = 0.6$ mA, $I_{C3} = 0.85$ mA, $h_{FE(\min)} = 3.0$; (b) $V_{CE3} = V_{CE2}$ $= 0$ V, $V_{CE1} = 1.5$ V, $I_{E1} = 0$, $I_{E2} = -0.4$ mA, $I_{E3} = -0.8$ mA $I_{B1} = 0$, $I_{B2} = I_{B3} =$ 0.4 mA

5-56 (a) -0.39, $+0.45$ V; (b) 190, -250 mV; (d) 144 Ω; (e) 137 mW

Chapter 6

6-1 (a) 11; (b) 14

6-18 (b) 145; (c) 145

6-19 (a) 80; (b) 60

6-23 (b) 9

6-24 (b) $36\frac{1}{2}$; (c) $34\frac{1}{2}$

6-26 8, $X_0 = X_5 = \overline{D}$, $X_1 = X_4 = D$, $X_2 = X_6 = 1$, $X_3 = X_7 = 0$

6-27 (b) $X_7 = 1$, $X_3 = X_5 = X_6 = D$, all other X are zero

6-29 (b) Three transistors; $Q0$ and $Q1$ have two emitters each, $Q2$ has three emitters

6-34 $Y_1 = W_7 + W_6 + W_3\overline{W}_4\overline{W}_5 + W_2\overline{W}_4\overline{W}_5$

6-37 (a) 8; (b) 5; (c) 104; (d) 32 transistors, each with a maximum of 64 emitters

6-38 (a) $Y_0 = W_1 + W_2 + W_5 + W_6 + W_9 + W_{10} + W_{13} + W_{14}$;
(b) $Y_2 = W_4 + W_5 + W_6 + W_7 + W_8 + W_9 + W_{10} + W_{11}$

6-40 (a) $Y_5 = \overline{D}\overline{C}\overline{B}A + \overline{D}\overline{C}B\overline{A} + \overline{D}\overline{C}BA + \overline{D}C\overline{B}A + D\overline{C}B\overline{A} + D\overline{C}BA + DCBA$

Chapter 7

7-17 (*b*) One pulse, $N = 5$
7-18 (*b*) 2
7-25 (*a*) $N = 4$

Chapter 8

8-1 (*a*) 2.11 V; (*b*) $1.99 \times 10^{-2} \, \Omega \cdot m$
8-2 (*a*) -0.75 V; (*b*) 14.8 V
8-3 (*b*) 0.55 kΩ; (*c*) 625 Ω
8-5 (*a*) -20 mA, -16.7 V; (*b*) -10.8 V
8-14 (*c*) 8.35 V, 5.06 V

Chapter 9

9-8 (*a*) 6; (*b*) 4
9-9 (*a*) 7; (*b*) 3; (*c*) 6, 4; (*d*) 1,024
9-16 (*a*) 1,835; (*b*) 3,606; (*c*) 101000101000
9-20 (*a*) 0.397 mil^2; (*b*) 0.492

Chapter 10

10-2 20 V, -3.33 V
10-4 (*a*) 19.98 V, 5V; (*b*) 19.76 V, 9.41 V; (*c*) 19.09 V, 9.85 V
10-6 (*b*) Fig. 10-5*d*: for $v_i \leqslant V_R + V_\gamma$, $v_o = V_R$ for $v_i \geqslant V_R + V_\gamma$,
$v_o = V_R + [R/(R + R_f)][v_i - V_\gamma - V_R]$
10-7 (*a*) For $v_i \leqslant -20$ V, $v_o = -20$ V, otherwise $v_o = v_i$; (*b*) -20.62 V
10-10 (*a*) $v_i \leqslant 19.92$ V, $v_o = 19.92$ V   19.92 V $\leqslant v_i \leqslant 20.08$ V, $v_o = 0.499(v_i + 20)$
$v_i \geqslant 20.08$ V, $v_o = 0.996v_i$; (*b*) $v_i \leqslant 20$ V, $v_o = 20$ V, otherwise, $v_o = 0.996v_i + 0.08$
10-13 $v_i \leqslant 25 \, V$, $v_o = 25$ V $v_i \geqslant 137.5$ V, $v_o = 100$ V
25 V $\leqslant v_i \leqslant 137.5$ V, $v_o = 25 + \frac{2}{3}(v_i - 25)$
10-15 (*a*) $v_o = \frac{1}{2} v_i$ for all v_i; (*b*) -5 V $\leqslant v_i \leqslant -1$ V, $v_o = \frac{1}{2}(v_i - 1)$
-1 V $\leqslant v_i \leqslant 1$ V, $v_o = v_i$ 1 V $\leqslant v_i \leqslant 5$ V, $v_o = \frac{1}{2}(v_i + 1)$
10-16 (*a*) $v_i \leqslant 10$ V, $v_o = 10$ V, otherwise, $v_o = v_i$; (*b*) $v_i \leqslant 10$ V, $i_{R1} = 0$, otherwise,
$i_{R1} = (v_i - 10)$ mA; (*c*) $v_i \leqslant 10$ V, $v_o = 9.3$ V, otherwise, $v_o = v_i - V_\gamma$
10-17 $v_i \leqslant 0$ V: $D1$, $D2$ OFF, $v_o = 0$ V $0 < v_i \leqslant 9$ V: $D1$ ON, $D2$ OFF, $v_o = \frac{2}{3} v_i$
$v_i > 9$ V: $D1$, $D2$ ON, $v_o = 6$ V
10-22 (*a*) 2 kΩ; (*b*) 20 mA; (*c*) 35 mA; (*d*) 30 to 77 mA
10-24 124.4 kΩ, 25 kΩ
10-30 (*a*) 138.5 mA; (*b*) 69.25 mA; (*c*) 252.4 V; (*d*) 19.18 W
10-32 (*a*) 400 mA; (*b*) 254.7 mA; (*c*) 40 V, 0 V; (*d*) 28.28 V
10-33 111.3 V

10-36 (b) The maximum possible voltages across $C1$, $C2$, $C3$, and $C4$ are V_m, $2V_m$, $2V_m$, and $2V_m$, respectively, where $v_i = V_m \sin \omega t$. The peak inverse voltage for each diode is $2V_m$

10-38 (a) 100°; (b) 44°, 904 mA; (c) 93.4°, 62°, 1.67 A

Chapter 11

11-1 (a) 0.026 mA, 2.6 mA, 7.84 V; (b) 0.052 mA, 5.2 mA, 3.67 V

11-4 (a) 0.0267 mA, 2.67 mA, 3.904 V; (b) 321 kΩ

11-6 4 kΩ, 102.65 kΩ, 120.8 kΩ

11-9 1.92 kΩ, 1.04 kΩ, 18.24 kΩ, 5.92 kΩ

11-11 0.0109 mA, 0.0638 mA

11-12 2.982 mA, 1.4 mA

11-15 2.145 mA, 1.755 mA

11-16 1.42 kΩ

11-20 (a) -47.62; (b) 1.1 kΩ; (c) 325 Ω; (d) -14.09; (e) -432.9; (f) -13.63; (g) 4.37 Ω; (h) 4.37 Ω

11-25 $V_o/V_s = -h_{fe}R_L/[R_s + h_{ie} + (h_{fe} + 1)(1 - b)R_e]$
$R_i = R_s + h_{ie} + (h_{fe} + 1)(1 - b)R_e$

11-26 -8.41, 8.62 kΩ, -3.90, -2.47, ∞, 4 kΩ

11-28 $A_{I2} = -100$, $R_{i2} = 3$ kΩ, $A_{V2} = -100$, $R_{o2} = \infty$, $A_{I1} = 101$, $R_{i1} = 188.8$ kΩ, $A_{V1} = 0.984$, $R_{o1} = 74.65$ Ω; overall, $A_V = -98.4$, $R_i = 39.22$, $R_o' = 3$ kΩ, $A_{Vs} = -87.27$, $A_I = -1,287$, $A_{Is} = -145.5$

11-30 $A_V = 118.5$, $R_i = 105$ kΩ, $A_I = 4,149$, $A_{Vs} = 113.1$, $R_o' = 3$ kΩ

11-31 $A_{V3} = 0.987$, $A_{V2} = -97.5$, $A_{V1} = 0.968$, $A_{Vs} = -83.48$

11-35 (c) $-9,524$, 1 kΩ

11-37 26.58 kΩ

11-42 (a) 8.63 kΩ; (b) 69.08, 0.997; (c) 2 MΩ, 48.68 kΩ

11-43 -7.99, -3.87, -3.11, 8.25 kΩ

11-45 (a) 8.65 MΩ; (b) 0.971; (c) 42.1 Ω; (d) 41.2 Ω; (e) 120 Ω, 20.8 Ω

11-48 (a) 171 kΩ; (b) 0.988; (c) 142 kΩ; (d) 83.8; (e) 17.9 kΩ

11-49 (a) 2.49 kΩ; (b) 5.68 V

11-51 (a) 7.5 kΩ; (b) 5 MΩ, 555.6 kΩ

11-52 2.25 mA, -0.404 V, 8.75 V

11-54 1.8 mA, 6 V, 12V

11-55 4 mA, 4 mA/V

11-60 -1.06

11-63 (a) $v_{dn} = \dfrac{r_d + (\mu + 1)R_s}{r_d + R_d + (\mu + 1)R_s} v_a$; (b) $v_{sn} = \dfrac{R_s}{r_d + R_d + (\mu + 1)R_s} v_a$;

(c) $v_{dn} = \dfrac{(\mu + 1)R_d}{r_d + R_d + (\mu + 1)R_s} v_b \qquad v_{sn} = \dfrac{(r_d + R_d)}{r_d + R_d + (\mu + 1)R_s} v_b$

11-65 (a) -0.595; (b) 0.948

11-67 $I_2 = \mu V_1/[2(r_d + R_d)]$

11-68 (a) 4.44; (b) 10.38 kΩ

11-71 (a) -0.61 V; (b) 1.15 mA/V; (c) 763 Ω; (d) 8.70 kΩ

Chapter 12

12-1 (a) -23.61; (b) -35.42; (c) -11.81 mA/V; (d) -70.84 kΩ; (e) 0.99 kΩ; (f) 3 kΩ

12-3 (a) $V_i = -200(V_s - V_f)$; (b) 149.9

12-8 (a) 1.58 V; (b) 0.25 V

12-9 (a) 2.133 V; (b) 0.028%

12-12 (a) $R_{if} = R_s + h_{ie} - h_{re}R_L h_{fe}/(1 + h_{oe}R_L)$; (b) $Y_{of} = -h_{fe}h_{re}/(R_s + h_{ie}) + h_{oe}$

12-14 (a) -1.289×10^4, 2.35 kΩ; (b) -11.77, 2.35 kΩ

12-16 (a) Voltage-series, 29.25; (b) 6.74 MΩ; (c) 0.878 kΩ

12-18 (a) Current-series; (c) $G_{Mf} = -\mu/[R_L + r_d + (\mu + 1)R]$; (d) $A_{Vf} = G_{Mf}R_L$;
    (e) $R_{if} = \infty$; (f) $R_{of} = r_d + (\mu + 1)R$;
    (g) $R'_{of} = \{R_L[r_d + (\mu + 1)R]\}/[r_d + (\mu + 1)R + R_L]$

12-20 (a) Current-shunt, 19.83; (b) 79.32; (c) 36.36 Ω, 37.73 Ω; (d) ∞, 4 kΩ

12-21 (a) 137.4; (b) 68.7; (c) 125 Ω, 128 Ω; (d) ∞, 2.35 kΩ

12-23 (a) -70.28 kΩ; (b) -70.28; (c) 184.9 Ω; (d) 1.415 kΩ; (e) -8.266 kΩ, -8.266, 896.8 Ω, 4.369 kΩ

1225 (a) Voltage-shunt, -4.006 kΩ; (b) -0.801; (c) 235.8 Ω, 5.247 kΩ;
    (d) 5.034 Ω, 4.98 Ω

12-27 (a) Voltage-shunt, -2.282 kΩ; (b) -22.82; (c) 23.15 Ω, 130.1 Ω; (d) 520 Ω, 479 Ω

12-29 (c) $-1,760$ kΩ, 890 Ω; (d) 728 Ω

Chapter 13

13-3 $f_1 = 85.9$ Hz, $f_2 = 5.2$ kHz

13-10 (a) 805 Ω $< R_s <$ 875 Ω

13-12 (a) 2.65 Hz; (b) 4.16%; (c) 416.3 Hz

13-13 (a) 6.37 Hz; (b) 1.65 Hz

13-17 (a) -75; (b) 4.72 μF; (c) $C_b \geqslant 14.81$ μF

13-19 (a) 67.47 dB; (b) 155.7 Hz

13-20 $g_m = 192$ mA/V, $r_{b'e} = 521$ Ω, $r_{bb'} = 479$ Ω, $r_{b'c} = 5.21$ MΩ,
$g_{ce} = 2.06 \times 10^{-5}$ A/V, $C_e = 3.06$ nF, $C_c = 2$ pF

13-22 (a) 24.5 pF; (b) 374.8 MHz

13-23 1.02 MHz, 102 MHz, 180 pF, 866.7 Ω, 233.3 Ω

13-27 (a) One zero, one pole; (b) no zero, one pole; (c) one zero, two poles; (d) no zero, two poles

13-31 (b) -87.07; (c) 38.28 dB; (d) -1.11π rad

13-33 (a) 122.2 Ω; (b) 900 Ω

13-34 (a) 478 Ω; (b) -23.27

13-36 (a) $v_o = -7.15(1 - \epsilon^{-t/0.133})$; (b) $v_o = -7.15(1 - \epsilon^{-t/200})$

13-37 (b)

$$\frac{V_e}{V_s} = \frac{G_s'(g + sC_e)}{\begin{aligned}s^2(C_L C_e + C_L C_c + C_e C_c) \\ + s[C_e(g_L + G_s') + C_L(G_s' + g_{b'e}) + C_c(g_L + g)] + g_L(G_s' + g_{b'e}) + G_s'g\end{aligned}}$$

13-43 For $f = 100$ Hz: -40.92, 108.8 pF, ∞; for $f = 0.1$ MHz: $-40.86 + j1.47$, 108.65 pF, 433 kΩ

13-47 (c) $A_V = \dfrac{g_m + g_d}{(1/R_d + g_d + j\omega C_{dg})\left[1 + (g_m + g_d + j\omega C_{sg})R_s\right] - (g_m + g_d)R_s g_d}$;

(d) $Y_i = \dfrac{g_m + g_d(1 - A_V) + j\omega C_{sg}}{1 + R_s\left[g_m + g_d(1 - A_V) + j\omega C_{sg}\right]}$

13-52 1.675 MHz, error = 2.93%

13-54 $f_{H1} = 5.84$ MHz, $f_{H2} = 0.583$ MHz, $f_H = 527$ kHz

Chapter 14

14-1 (a) 19.6 dB; (b) 51 and $\frac{1}{51}$; (c) 0.196 Hz, 1.53 MHz

14-11 (a) 3.86 dB; (b) 0.49

14-13 (a) 3.25; (b) 0.43 μs; (c) 0.998, 0.855 μs

14-15 (a) 4.3%; (b) 10.1%

14-19 (a) 1.185; (b) 16

14-20 $-2, -\frac{1}{2}(1 \pm j\sqrt{15})$

14-21 $\beta A_o = 2.194$, poles: $-0.1054, -0.01231 \pm j0.02132$

14-26 1.74 MHz

14-28 20.55 MHz

14-36 (a) 60°, 41 dB; (b) 1

14-37 (a) 0.008; (b) 1.54 dB

14-38 (c) Corner frequencies: 0.1, 0.5, 3, 10, 50, 300 MHz; (d) 12 MHz, midband loop gain < 30 dB; (e) maximum midband loop gain = 15 dB

Chapter 15

15-7 (a) 9.984; (b) 6.126 MΩ; (c) 4.995 Ω

15-11 (a) $v_o = -(R'/R)v_1 + [(R + R')/R][R_1/(R_1 + R_2)]v_2$; (b) $v_o = -(R_1/R_2)v_d$; (c) $\frac{1}{2}[R'(R_1 + R_2) + R_1(R + R')]/[R'(R_1 + R_2) - R_1(R + R')]$

15-17 (a) $V_{B1} - V_{B2} = (V_{BE1} - V_{BE2}) + R_e(2I_{C1} - I_O)$; (c) $g_{md}' = (I_O/4V_T)/(1 + 2R_e I_O/4V_T)$; (d) $g_{md}' = g_{md}/(1 + 2R_e g_{md})$

15-20 (a) $[(2 + \beta)/(1 + \beta)]I_C$; (b) $\{(\beta^2 + 2\beta + 2)/[\beta(1 + \beta)]\}I_C$; (d) 0.946, 0.938

15-21 (a) 1.43 MΩ; (c) 1.43 mA, 12.9 kΩ

15-23 7.85 V

15-25 (a) -3.13 V; (b) 0.986, 0.493 mA; (c) 2.18 V; (d) 1.48 V; (e) 0.494 mA;
(f) 4.52 V; (g) 3.82 V; (h) 1.56 mA; (i) -4.60 V; (j) 1.40 mA; (k) 0.16 mA; (l)
0.20 V; (m) 59.6; (n) 8, 493

15-26 (a) 0.733 mA; (b) 19 μA; (c) 0.462 I_3

15-28 (a) 31.83 kHz; (b) 15.91 kHz

15-36 (b) 11 MHz; (c) -500; (d) no; (e) 2,112; (f) 100 Hz

15-40 (a)

	Uncompensated Fig. 14-20	Pole-zero Fig. 15-32	Dominant-pole
Bandwidth	13 MHz	3.5 MHz	20 kHz
Phase margin	15°	65°	90°

(b) 10 kHz; (c) 178 kHz, 90°; (d) 45°

15-42 (a) $C_c = 1.628 \times 10^{-4}/[R_1 + RR'/(R + R')]$,
$R_c = 9.77 \times 10^{-4}[R_1 + RR'/(R + R')]$; (b) 195 Ω, 81.5 nF

15-46 $\beta = -1/R'$, $R_M = [RR'A_1/(R + R')](-R_1'/R_1)(1 + j\omega CR_1)$

Chapter 16

16-2 $v_o = -(V/RC)[t + (L/R)(\epsilon^{-Rt/L} - 1)]$

16-4 (b) $[sC_1R_1(1 + sC_2R_2)]/(1 + sC_1R_1)$

16-5 $-1.25v_1 - 2.0v_2 + 2.32v_3 + 1.16v_4$

16-8 (a) $A_vR_1R'/[R' + (1 - A_v)R_1]$; (b) $R_1R'/[R' + (1 - A_v)R_1]$;
(c) $[R_o(R_1 + R')]/[R_o + R' + (1 - A_v)R_1]$

16-36 $R = 15.92$ kΩ, $R_1 = R_2 = R_3 = 10$ kΩ (arbitrary), $R_1' = 13.82$ kΩ, $R_2' = 3.82$ kΩ

16-39 (a) $1/(s^2R^2C_1C_2 + 2sRC_2 + 1)$; (b) $C_1 = 1/\omega_oRk$, $C_2 = k/\omega_oR$

16-44 (b) $1/\omega_o^2 = C^2R_1R_2$, $2k/\omega_o = (R_1 + 2R_2)C$, $R_3 = R_1$

16-46 $R_1 = 9.55$ kΩ, $R_2 = 5.08$ kΩ, $R_3 = 1.91$ MΩ

16-48 5 kΩ, 80 kΩ, -8

16-53 (c) 1.553 kΩ, 165 Ω

16-55 (a) $V_{N-1} = \frac{1}{2}V_{N-2} = \frac{1}{4}V_{N-3} = V_R/12$

16-56 (b) $V_R/2R$, $-V_R$; (c) $V_R/4R$, $-V_R/2$; (d) $-V_R/8$

Chapter 17

17-3 5 V, 4.3 V, 99 kΩ

17-5 (a) 15, 3.73 V; (b) $V_R < R_2V_o/R_1$

17-9 (b) $T_1 = (2R_2RC/R_1)[V_o/(V_o + V_S)]$, $f = (R_1/4R_2RC)[1 - (V_S/V_o)^2]$

17-10 (b) 267 Ω; (c) 2 kΩ; (d) 0.0375 μF

17-14 (b) $V_R = R_2V_o/(R_1 + R_2)$, $V_s = 2V_oR_2/R_1$; (c) $T_s = V_sRC/V$;
(d) $f = v_m/V_sRC$

17-17 (a) 9 μF; (b) 0.0011%; (c) 9 pF, 0.0011%

17-18 (a) 0.00247 μF; (b) 2 kΩ; (c) 20 V; (d) 115 μs

17-19 (*b*) 2, 1
17-25 (*a*) 6.13 kΩ; (*b*) 12.99 μs; (*c*) 30 kΩ, 430 pF
17-26 (*b*) $1/2\pi RC\sqrt{6}$; (*c*) $\frac{29}{30}$
17-31 R/R_3
17-32 15.92 kHz, $R_{\min} = 476$ Ω

Chapter 18

18-2 12.65
18-7 (*a*) 28.13 W; (*b*) 3.84 W; (*c*) 78.55%; (*d*) 5.70 W, 50%
18-9 (*a*) 37°, 17.5°, 0
18-10 (*a*) 0.400 W; (*b*) 1.80 W; (*c*) 1.73 W
18-12 0.112 V
18-13 (*a*) 12 V; (*b*) 20 μV/°C; (*c*) −0.05 μV/°C
18-14 (*a*) $V_O = I_Q R_2 + V_{\text{REG}}(1 + R_2/R_1)$

INDEX